Instructor's Solutions

MW01710584

CALCULUS

Finney/Thomas

Instructor's Solutions Manual Part II

CALCULUS

Finney/Thomas

Michael Schneider
Thomas L. Cochran

Belleville Area College

ADDISON-WESLEY PUBLISHING COMPANY
Reading, Massachusetts • Menlo Park, California • New York
Don Mills, Ontario •Wokingham, England • Amsterdam • Bonn
Sydney • Singapore • Tokyo • Madrid • San Juan

Reproduced by Addison-Wesley from camera-ready copy supplied by the authors.

ISBN 0-201-51394-3
ABCDEFGHIJ-BK-943210

The authors would like to dedicate this book to their wives, Jane and Barbara, the two who have put up with so much for so long.

Michael Schneider
Tom Cochran

The authors have attempted to make this manual as error free as possible but nobody's "perfeck". If you find errors, we would appreciate knowing them. You are more than welcome to write us at:

Belleville Area College
2500 Carlyle Road
Belleville, Illinois 62221

TABLE OF CONTENTS

CHAPTER 9

INFINITE SERIES

9.1 SEQUENCES OF NUMBERS

1. $a_1 = 0, a_2 = -\frac{1}{4}, a_3 = -\frac{2}{9}, a_4 = -\frac{3}{16}$

2. $a_1 = 1, a_2 = \frac{1}{2}, a_3 = \frac{1}{6}, a_4 = \frac{1}{24}$

3. $a_1 = 1, a_2 = -\frac{1}{3}, a_3 = \frac{1}{5}, a_4 = -\frac{1}{7}$

4. $a_1 = 1, a_2 = 3, a_3 = 1, a_4 = 3$

5. $1, \frac{3}{2}, \frac{7}{4}, \frac{15}{8}, \frac{63}{32}$

6. $1, \frac{1}{2}, \frac{1}{6}, \frac{1}{24}, \frac{1}{120}, \frac{1}{720}$

7. $2, 1, \frac{1}{2}, \frac{1}{4}, \frac{1}{8}, \frac{1}{16}$

8. $-2, -1, -\frac{2}{3}, -\frac{1}{2}, -\frac{2}{5}, -\frac{1}{3}$

9. 1, 1, 2, 3, 5, 8

10. 1, 1, 2, 4, 8, 16

11. $\lim_{n\to\infty} a_n = \lim_{n\to\infty} \left(2 + (0.1)^n\right) = 2 \Rightarrow$ converges

12. $\lim_{n\to\infty} a_n = \lim_{n\to\infty} \frac{n + (-1)^n}{n} = \lim_{n\to\infty} \left(1 + \frac{(-1)^n}{n}\right) = 1 \Rightarrow$ converges

13. $\lim_{n\to\infty} a_n = \lim_{n\to\infty} \frac{1 - 2n}{1 + 2n} = -1 \Rightarrow$ converges

14. $\lim_{n\to\infty} a_n = \lim_{n\to\infty} \frac{2n + 1}{1 - 3n} = -\frac{2}{3} \Rightarrow$ converges

15. $\lim_{n\to\infty} a_n = \lim_{n\to\infty} \frac{n}{10} = \infty \Rightarrow$ diverges

16. $\lim_{n\to\infty} a_n = \lim_{n\to\infty} \left(1 + (-1)^n\right)$ does not exist $\Rightarrow$ diverges

17. $\lim_{n\to\infty} a_n = \lim_{n\to\infty} \frac{1 - 5n^4}{n^4 + 8n^3} = -5 \Rightarrow$ converges

18. $\lim_{n\to\infty} a_n = \lim_{n\to\infty} \frac{n + 3}{n^2 + 5n + 6} = 0 \Rightarrow$ converges

19. $\lim_{n\to\infty} a_n = \lim_{n\to\infty} \frac{n^2 - 2n + 1}{n - 1} = \lim_{n\to\infty} n - 1 = \infty \Rightarrow$ diverges

20. $\lim_{n\to\infty} a_n = \lim_{n\to\infty} \frac{1 - n^3}{70 - 4n^2} = \infty \Rightarrow$ diverges

21. $\lim_{n\to\infty} a_n = \lim_{n\to\infty} 5 = 5 \Rightarrow$ converges

22. $\lim_{n\to\infty} a_n = \lim_{n\to\infty} \left(\frac{10^{n+1}}{10^n}\right) = \lim_{n\to\infty} 10 = 10 \Rightarrow$ converges

23. $\lim_{n\to\infty} a_n = \lim_{n\to\infty} 5^n = \infty \Rightarrow$ diverges

24. $\lim_{n\to\infty} a_n = (-1)^n\left(1 - \frac{1}{n}\right) =$ does not exist $\Rightarrow$ diverges

25. $\lim_{n\to\infty} a_n = \lim_{n\to\infty} \left(\frac{n + 1}{2n}\right)\left(1 - \frac{1}{n}\right) = \frac{1}{2} \Rightarrow$ converges

26. $\lim_{n\to\infty} a_n = \lim_{n\to\infty} \left(2 - \frac{1}{2^n}\right)\left(3 + \frac{1}{3^n}\right) = 6$

27. $\lim_{n\to\infty} a_n = \lim_{n\to\infty} \frac{(-1)^{(n+1)}}{2n - 1} = 0$

28. $\lim_{n\to\infty} a_n = \lim_{n\to\infty} \left(-\frac{1}{2}\right)^n = 0$

29. $\lim_{n\to\infty} a_n = \lim_{n\to\infty} \frac{\sin n}{n} = 0$, by the Sandwich Theorem for Sequences

30. $\lim_{n\to\infty} a_n = \lim_{n\to\infty} \frac{\sin^2 n}{2^n} = 0$, because $0 \le \frac{\sin^2 n}{2^n} \le \frac{1}{2^n}$ and the Sandwich Theorem for Sequences

31. $\lim_{n\to\infty} a_n = \lim_{n\to\infty} \sqrt{\frac{2n}{n + 1}} = \sqrt{\lim_{n\to\infty} \frac{2n}{n + 1}} = \sqrt{2}$

32. $\lim_{n\to\infty} a_n = \lim_{n\to\infty} \sin\left(\frac{\pi}{2} + \frac{1}{n}\right) = \sin\left(\lim_{n\to\infty} \frac{\pi}{2} + \frac{1}{n}\right) = \sin\left(\frac{\pi}{2}\right) = 1$

33. $\lim_{n\to\infty} a_n = \lim_{n\to\infty} \tan^{-1} n = \frac{\pi}{2}$

34. $\lim_{n\to\infty} a_n = \lim_{n\to\infty} (\ln(n) - \ln(n + 1)) = \ln\left(\lim_{n\to\infty} \frac{n}{n + 1}\right) = \ln(1) = 0$

35. $\lim_{n\to\infty} a_n = \lim_{n\to\infty} \frac{n}{2^n} = \lim_{n\to\infty} \frac{1}{2^n \ln 2} = 0$

36. $\lim_{n\to\infty} a_n = \lim_{n\to\infty} \frac{3^n}{n^3} = \infty$, due to the growth rates in section 7.6

37. $\lim_{n\to\infty} a_n = \lim_{n\to\infty} \frac{\ln(n+1)}{n} = 0$, due to the growth rates in secion 7.6

38. $\lim_{n\to\infty} a_n = \lim_{n\to\infty} \frac{\ln(n)}{\ln(2n)} = \lim_{n\to\infty} \frac{1/n}{1/n} = 1$

39. $\lim_{n\to\infty} a_n = \lim_{n\to\infty} 8^{1/n} = \lim_{n\to\infty} \exp\left(\frac{\ln 8}{n}\right) = e^0 = 1$

40. $\lim_{n\to\infty} a_n = \lim_{n\to\infty} (0.03)^{1/n} = \lim_{n\to\infty} \exp\left(\frac{\ln 0.03}{n}\right) = e^0 = 1$

41. $\lim_{n\to\infty} a_n = \lim_{n\to\infty} \left(1 + \frac{7}{n}\right)^n = e^7$, due to table 9.1 part 5

42. $\lim_{n\to\infty} a_n = \lim_{n\to\infty} \left(1 - \frac{1}{n}\right)^n = \lim_{n\to\infty} \left(1 + \frac{-1}{n}\right)^n = e^{-1}$, due to table 9.1, part 5

43. $\lim_{n\to\infty} a_n = \lim_{n\to\infty} \frac{1}{(0.9)^n} = \lim_{n\to\infty} \left(\frac{10}{9}\right)^n = \infty \Rightarrow$ diverges

44. $\lim_{n\to\infty} a_n = \lim_{n\to\infty} \frac{(n+1)!}{n!} = \lim_{n\to\infty} (n+1) = \infty \Rightarrow$ diverges

45. $\lim_{n\to\infty} a_n = \lim_{n\to\infty} \sqrt[n]{10\,n} = \lim_{n\to\infty} \exp\left(\frac{\ln(10\,n)}{n}\right) = \exp\left(\lim_{n\to\infty} \frac{\ln 10\,n}{n}\right) = \exp\left(\lim_{n\to\infty} \frac{1}{n}\right) = e^0 = 1$

46. $\lim_{n\to\infty} a_n = \lim_{n\to\infty} \sqrt[n]{n^2} = \left(\lim_{n\to\infty} \sqrt[n]{n}\right)^2 = 1^2 = 1$, due to table 9.1, part 2

47. $\lim_{n\to\infty} a_n = \lim_{n\to\infty} \left(\frac{3}{n}\right)^{1/n} = \frac{\lim_{n\to\infty} \sqrt[n]{3}}{\lim_{n\to\infty} \sqrt[n]{n}} = \frac{1}{1} = 1$, due to table 9.1, parts 2 and 3

48. $\lim_{n\to\infty} a_n = \lim_{n\to\infty} (n+4)^{1/(n+4)} = \lim_{t\to\infty} \sqrt[t]{t} = 1$, due to table 9.1, part 2 and where $t = n + 4$

49. $\lim_{n\to\infty} a_n = \lim_{n\to\infty} \frac{\ln n}{n^{1/n}} = \frac{\lim_{n\to\infty} \ln n}{\lim_{n\to\infty} \sqrt[n]{n}} = \frac{\infty}{1} = \infty$, due to table 9.1, part 2

50. $\lim_{n\to\infty} a_n = \lim_{n\to\infty} \sqrt[n]{4^n n} = 4 \lim_{n\to\infty} \sqrt[n]{n} = 4(1) = 4$, due to table 9.1, part 2

51. $\lim_{n\to\infty} a_n = \lim_{n\to\infty} \left(\frac{1}{3}\right)^n = 0$, due to table 9.1, part 4

52. $\lim_{n\to\infty} a_n = \lim_{n\to\infty} \frac{1}{\sqrt{2^n}} = 0$

53. $\lim_{n\to\infty} a_n = \lim_{n\to\infty} \frac{1}{n!} = 0$, due to table 9.1, part 6

54. $\lim_{n\to\infty} a_n = \lim_{n\to\infty} \frac{(-4)^n}{n!} = 0$, due to table 9.1, part 6

55. $\lim_{n\to\infty} a_n = \lim_{n\to\infty} \left(\frac{1}{n}\right)^{1/\ln(n)} = \lim_{n\to\infty} \exp\left(\frac{\ln(1/n)}{\ln(n)}\right) = \exp\left(\lim_{n\to\infty} \frac{\ln(1/n)}{\ln(n)}\right) = \exp(\lim_{n\to\infty} -1) = e^{-1}$

56. $\lim_{n\to\infty} a_n = \lim_{n\to\infty} \frac{n!}{2^n 3^n} = \lim_{n\to\infty} \frac{n!}{6^n} = \frac{1}{\lim_{n\to\infty} \frac{6^n}{n!}} = \infty \Rightarrow$ diverges

57. $\lim_{n\to\infty} a_n = \lim_{n\to\infty} \frac{n!}{10^{6n}} = \frac{1}{\lim_{n\to\infty} \frac{(10^6)^n}{n!}} = \infty \Rightarrow$ diverges

58. $\lim_{n\to\infty} a_n = \lim_{n\to\infty} \frac{3^n 6^n}{2^{-n} n!} = \frac{36^n}{n!} = 0$, due to table 9.1, part 6

59. $\left|\sqrt[n]{0.5} - 1\right| < 10^{-3} \Rightarrow -\frac{1}{1000} < \left(\frac{1}{2}\right)^{1/n} - 1 < \frac{1}{1000} \Rightarrow n > \frac{\ln(1/2)}{\ln(999/1000)} \Rightarrow n > 692.8 \Rightarrow N = 693$

60. $\left|\sqrt[n]{n} - 1\right| < 10^{-3} \Rightarrow \left(\frac{999}{1000}\right)^n < n < \left(\frac{1001}{1000}\right)^n \Rightarrow n > 9124 \Rightarrow N = 9124$

61. $(0.9)^n < 10^{-3} \Rightarrow n \ln(0.9) < -3 \ln 10 \Rightarrow n > 65.56 \Rightarrow N = 66$

62. $\frac{2^n}{n!} < 10^{-7} \Rightarrow n! > 10^7 2^n \Rightarrow n \geq 15 \Rightarrow N = 15$

63. $x_1 = 1$, $x_2 = 1 + \cos(1) \approx 1.540302305$, $x_3 = 1 + \cos(1) + \cos(1 + \cos(1)) \approx 1.570791601$,
$x_4 = 1 + \cos(1) + \cos(1 + \cos(1)) + \cos(1 + \cos(1) + \cos(1))) \approx 1.570796327$

64. $x_1 = 1$, $x_2 = \cos(1) \approx 0.540302305$, $x^3 \approx x_4 \approx 0.857553215$, $x_5 \approx 0.65428979 \Rightarrow x_{56} \approx 0.73908133$

9.2 INFINITE SERIES

1. $s_n = \frac{a\left(1-r^n\right)}{(1-r)} = \frac{2\left(1-(1/3)^n\right)}{1-1/3} \Rightarrow \lim_{n\to\infty} s_n = \frac{2}{1-1/3} = 3$

2. $s_n = \frac{a\left(1-r^n\right)}{(1-r)} = \frac{(9/100)(1-(1/100)^n)}{(1-1/100)} \Rightarrow \lim_{n\to\infty} s_n = \frac{9/100}{1-1/100} = \frac{1}{11}$

3. $s_n = \frac{a\left(1-r^n\right)}{(1-r)} = \frac{1-(-1/2)^n}{1-(-1/2)} \Rightarrow \lim_{n\to\infty} s_n = \frac{1}{3/2} = \frac{2}{3}$

4. $s_n = \frac{1-(-2)^n}{1-(-2)}$, a geometric series where $|r| > 1 \Rightarrow$ divergence.

5. $\frac{1}{(n+1)(n+2)} = \frac{1}{n+1} - \frac{1}{n+2} \Rightarrow s_n = \left(\frac{1}{2}-\frac{1}{3}\right) + \left(\frac{1}{3}-\frac{1}{4}\right) + \ldots + \left(\frac{1}{n+1}-\frac{1}{n+2}\right) = \frac{1}{2} - \frac{1}{n+2} \Rightarrow$
$\lim_{n\to\infty} s_n = \frac{1}{2}$

6. $\frac{5}{n(n+1)} = \frac{5}{n} - \frac{5}{n+1} \Rightarrow s_n = \left(5-\frac{5}{2}\right) + \left(\frac{5}{2}-\frac{5}{3}\right) + \left(\frac{5}{3}-\frac{5}{4}\right) + \ldots + \left(\frac{5}{n-1}-\frac{5}{n}\right) + \left(\frac{5}{n}-\frac{5}{n+1}\right) =$
$5 - \frac{5}{n+1} \Rightarrow \lim_{n\to\infty} s_n = 5$

7. $1 + \frac{1}{4} + \frac{1}{16} + \ldots$ The sum of this geometric series is $\frac{1}{1-1/4} = \frac{4}{3}$.

8. $\frac{1}{16} + \frac{1}{64} + \frac{1}{256} + \ldots$ The sum of this geometric series is $\frac{1/16}{1-1/4} = \frac{1}{12}$.

9. $\frac{7}{4} + \frac{7}{16} + \frac{7}{64} + \ldots$. The sum of this geometric series is $\frac{7/4}{1-1/4} = \frac{7}{3}$.

10. $5 - \frac{5}{4} + \frac{5}{16} - \frac{5}{64} + \ldots$. The sum of this geometric series is $\frac{5}{1-(-1/4)} = 4$.

11. $(5+1) + \left(\frac{5}{2}+\frac{1}{3}\right) + \left(\frac{5}{8}+\frac{1}{27}\right) + \ldots$ is the sum of two geometric series; the sum
$\frac{5}{1-1/2} + \frac{1}{1-1/3} = 10 + \frac{3}{2} = \frac{23}{2}$.

12. $(5-1) + \left(\frac{5}{2}-\frac{1}{3}\right) + \left(\frac{5}{8}-\frac{1}{27}\right) + \ldots$ is the difference of two geometric series; the sum
$\frac{5}{1-1/2} - \frac{1}{1-1/3} = 10 - \frac{3}{2} = \frac{17}{2}$.

13. $(1+1) + \left(\frac{1}{2}-\frac{1}{5}\right) + \left(\frac{1}{4}+\frac{1}{25}\right) + \left(\frac{1}{8}-\frac{1}{125}\right) + \ldots$ is the sum of two geometric series; the sum
$\frac{1}{1-3/2} + \frac{1}{1+1/5} = 2 + \frac{5}{6} = \frac{17}{6}$.

14. $2 + \frac{4}{5} + \frac{8}{25} + \frac{16}{125} + \ldots = 2\left(1 + \frac{2}{5} + \frac{4}{25} + \frac{8}{125} + \ldots\right)$; the sum of this geometric series
$2\left(\frac{1}{1-2/5}\right) = \frac{10}{3}$.

15. $\frac{4}{(4n-3)(4n+1)} = \frac{1}{4n-3} - \frac{1}{4n+1} \Rightarrow s_n = \left(1-\frac{1}{5}\right) + \left(\frac{1}{5}-\frac{1}{9}\right) + \left(\frac{1}{9}-\frac{1}{13}\right) + \ldots + \left(\frac{1}{4n-7}-\frac{1}{4n-3}\right)$
$+ \left(\frac{1}{4n-3}-\frac{1}{4n+1}\right) = 1 - \frac{1}{4n+1} \Rightarrow \lim_{n\to\infty} s_n = \lim_{n\to\infty} 1 - \frac{1}{4n+1} = 1$

16. $\frac{1}{(4n-3)(4n+1)} = \frac{1/4}{4n-3} - \frac{1/4}{4n+1} \Rightarrow s_n =$
$\frac{1}{4}\left[\left(1-\frac{1}{5}\right)+\left(\frac{1}{5}-\frac{1}{9}\right)+\left(\frac{1}{9}-\frac{1}{13}\right)+\ldots+\left(\frac{1}{4n-7}-\frac{1}{4n-3}\right)+\left(\frac{1}{4n-3}-\frac{1}{4n+1}\right)\right] \Rightarrow$
$\lim_{n\to\infty} s_n = \frac{1}{4} \lim_{n\to\infty} 1 - \frac{1}{4n+1} = \left(\frac{1}{4}\right)(1) = \frac{1}{4}$

17. $\frac{4}{(4n-3)(4n+1)} = \frac{1}{4n-3} - \frac{1}{4n+1} \Rightarrow s_n = \left(\frac{1}{9}-\frac{1}{13}\right)+\left(\frac{1}{13}-\frac{1}{17}\right)+\left(\frac{1}{17}-\frac{1}{21}\right)+\ldots+$
$\left(\frac{1}{4n-7}-\frac{1}{4n-3}\right)+\left(\frac{1}{4n-3}-\frac{1}{4n+1}\right) \Rightarrow \lim_{n\to\infty} s_n = \lim_{n\to\infty} \frac{1}{9} - \frac{1}{4n+1} = \frac{1}{9}$

18. $\frac{2n+1}{n^2(n+1)^2} = \frac{1}{n^2} - \frac{1}{(n+1)^2} \Rightarrow s_n = \left(1-\frac{1}{4}\right)+\left(\frac{1}{4}-\frac{1}{9}\right)+\left(\frac{1}{9}-\frac{1}{16}\right)+\ldots+\left(\frac{1}{(n-1)^2}-\frac{1}{n^2}\right)+$
$\left(\frac{1}{n^2}-\frac{1}{(n+1)^2}\right) \Rightarrow \lim_{n\to\infty} s_n = \lim_{n\to\infty} 1 - \frac{1}{(n+1)^2} = 1$

19. A convergent geomertic series with a sum of $\frac{1}{1-1/\sqrt{2}} = 2+\sqrt{2}$.

20. $\lim_{n\to\infty} a_n = \lim_{n\to\infty} \ln\left(\frac{1}{n}\right) \neq 0 \Rightarrow$ divergence.

21. A convergent geometric series with a sum of $\frac{3/2}{1-(-1/2)} = 1$.

22. A geometric series with $|r| > 1 \Rightarrow$ divergence.

23. $\lim_{n\to\infty} \cos(n\pi) \neq 0 \Rightarrow$ divergence.

24. A convergent geometric series with a sum of $\frac{1}{1-(-1/5)} = \frac{5}{6}$.

25. A convergent geometric series with a sum of $\frac{1}{1-1/e^2} = \frac{e^2}{e^2-1}$.

26. $\lim_{n\to\infty} a_n = \lim_{n\to\infty} \frac{n^2+1}{n} = \infty \neq 0 \Rightarrow$ divergence.

27. $\lim_{n\to\infty} a_n = \lim_{n\to\infty} (-1)^{n+1} n \neq 0 \Rightarrow$ divergence.

28. A convergent geometric series with a sum of $\frac{2/10}{1-1/10} = \frac{2}{9}$.

29. The difference of two convergent geometric series with a sum of $\frac{1}{1-2/3} - \frac{1}{1-1/3} = 3 - \frac{3}{2} = \frac{3}{2}$.

30. $\lim_{n\to\infty} a_n = \lim_{n\to\infty} \left(1+\frac{-1}{n}\right)^n = e^{-1} \neq 0 \Rightarrow$ divergence.

31. $\lim_{n\to\infty} a_n = \lim_{n\to\infty} \frac{n!}{1000^n} = \frac{1}{\lim_{n\to\infty} \frac{1000^n}{n!}} = \infty \neq 0 \Rightarrow$ divergence.

32. A convergent geometric series with a sum of $\frac{1}{1-1/x} = \frac{x}{x-1}$.

33. $\frac{1}{1+x} = \frac{1}{1-(-x)} \Rightarrow a = 1, r = -x$

34. $\frac{1}{1+x^2} = \frac{1}{1-\left(-x^2\right)} \Rightarrow a = 1, r = -x^2$

35. distance $= 4 + 2\left[(4)\left(\frac{3}{4}\right) + (4)\left(\frac{3}{4}\right)^2 + \dots\right] = 4 + 2\left[\frac{3}{1-3/4}\right] = 28$ m

36. time $= \sqrt{\frac{4}{4.9}} + 2\sqrt{\frac{4}{4.9}\left(\frac{3}{4}\right)} + 2\sqrt{\frac{4}{4.9}\left(\frac{3}{4}\right)^2} + 2\sqrt{\frac{4}{4.9}\left(\frac{3}{4}\right)^3} + \dots =$

$$\sqrt{\frac{4}{4.9}} + 2\sqrt{\frac{4}{4.9}}\left[\sqrt{\frac{3}{4}} + \sqrt{\left(\frac{3}{4}\right)^2} + \dots\right] = \frac{2}{\sqrt{4.9}} + \frac{4}{\sqrt{4.9}}\left[\frac{\sqrt{\frac{3}{4}}}{1-\sqrt{\frac{3}{4}}}\right] = \frac{8\sqrt{3}+14}{\sqrt{4.9}} =$$

$$\frac{2\sqrt{10}\left(4\sqrt{3}+7\right)}{7} \approx 12.58 \text{ sec}$$

37. $\frac{234}{10^3} + \frac{234}{10^6} + \frac{234}{10^9} + \dots = \frac{234/10^3}{1-1/10^3} = \frac{26}{111}$

38. $1.24 + \frac{123}{10^5} + \frac{123}{10^8} + \dots = \frac{124}{100} + \frac{123/10^5}{1-1/10^3} = \frac{124}{100} + \frac{123}{99900} = \frac{41333}{33300}$

39. a) $\sum_{n=-2}^{\infty} \frac{1}{(n+4)(n+5)}$ b) $\sum_{n=0}^{\infty} \frac{1}{(n+2)(n+3)}$ c) $\sum_{n=5}^{\infty} \frac{1}{(n-3)(n-2)}$

40. a) $\sum_{n=-1}^{\infty} \frac{5}{(n+2)(n+3)}$ b) $\sum_{n=3}^{\infty} \frac{5}{(n-2)(n-1)}$ c) $\sum_{n=20}^{\infty} \frac{5}{(n-19)(n-18)}$

41. area $= 2^2 + \left(\sqrt{2}\right)^2 + (1)^2 + \left(\frac{1}{\sqrt{2}}\right)^2 + \dots = 4 + 2 + 1 + \frac{1}{2} + \dots = \frac{4}{1-1/2} = 8 \text{ m}^2$

42. area $= 2\left[\frac{\pi(1/2)^2}{2}\right] + 4\left[\frac{\pi(1/4)^2}{2}\right] + 8\left[\frac{\pi(1/8)^2}{2}\right] + \dots = \pi\left[\frac{1}{4} + \frac{1}{8} + \frac{1}{16} + \dots\right] = \pi\left[\frac{1/4}{1-1/2}\right] = \frac{\pi}{2}$

43. $\sum_{n=1}^{\infty} n$ and $\sum_{n=1}^{\infty} (-n)$

44. Both $\sum_{n=1}^{\infty} \left(\frac{1}{4}\right)^n$ and $\sum_{n=1}^{\infty} \left(\frac{1}{8}\right)^n$ converge, but $\sum_{n=1}^{\infty} \frac{(1/4)^n}{(1/8)^n} = \sum_{n=1}^{\infty} 2^n$ diverges.

45. $\sum_{n=0}^{\infty} \left(\frac{1}{5}\right)^n$ converges to $\frac{5}{4}$ and $\sum_{n=0}^{\infty} \left(\frac{1}{2}\right)^n$ converges to 2,

but $\sum_{n=0}^{\infty} \frac{(1/5)^n}{(1/2)^n}$ converges to $\frac{5}{3} \neq \frac{5/4}{2} = \frac{5}{8}$.

46. If $\sum_{n=1}^{\infty} a_n$ converges and $a_n \neq 0$ for all n, then $\lim_{n\to\infty} a_n = 0 \Rightarrow \lim_{n\to\infty} \frac{1}{a_n} \neq 0 \Rightarrow \sum_{n=1}^{\infty} \frac{1}{a_n}$ diverges.

9.3 SERIES WITH NONEGATIVE TERMS: COMPARISON AND INTEGRAL TESTS

1. converges, a geometric series with $r = 1/10$
2. diverges, $\lim_{n \to \infty} a_n = \lim_{n \to \infty} \frac{n}{n+2} = 1 \neq 0$
3. converges, by the Comparison Test for Convergence since, $\frac{\sin^2 n}{2^n} \le \frac{1}{2^n}$
4. diverges, by the Comparison Test for Convergence, since $\frac{1}{n} < \frac{5}{n}$
5. converges, by the Comparison Test for Convergence, since $\frac{1 + \cos n}{n^2} \le \frac{2}{n^2}$
6. converges, a geometric series with $r = 1/8$
7. diverges, by the Comparison Test for Convergence, since $\frac{1}{n} < \frac{\ln n}{n}$ for $n \ge 3$
8. converges, by the Limit Comparison Test when compared with $\sum_{n=1}^{\infty} \frac{1}{n^{3/2}}$
9. converges, a geometric series with $r = 2/3$
10. diverges, factor -2 out of the given series and compare the result with $\sum_{n=1}^{\infty} \frac{1}{n}$ by using the Limit Comparison Test
11. diverges, by the Comparison Test for Convergence, since $\frac{1}{n+1} < \frac{1}{1 + \ln n}$ and $\sum_{n=1}^{\infty} \frac{1}{n+1}$ diverges by the Limit Comparison Test when compared with $\sum_{n=1}^{\infty} \frac{1}{n}$
12. diverges, by the Limit Comparison Test when compared with $\sum_{n=1}^{\infty} \frac{1}{n}$
13. diverges, $\lim_{n \to \infty} a_n = \lim_{n \to \infty} \frac{2^n}{n+1} = \lim_{n \to \infty} \frac{2^n \ln 2}{1} = \infty \neq 0$
14. converges, by the Comparison Test for Convergence, since $\left(\frac{n}{3n+1}\right)^n < \left(\frac{n}{3n}\right)^n < \left(\frac{1}{3}\right)^n$
15. converges, by the Limit Comparison Test when compared with $\sum_{n=1}^{\infty} \frac{1}{n^{3/2}}$
16. diverges, by the Comparison Test for Convergence, since $\frac{1}{n} \le \frac{\sqrt{n}}{n} \le \frac{\sqrt{n}}{\ln n}$ for $n \ge 3$

17. diverges, by the Limit Comparison Test when compared with $\sum_{n=1}^{\infty} \frac{1}{n}$

18. diverges: If $f(x) = \sqrt[x]{x} \Rightarrow f'(x) > 0$ when $1 < x < e$ and $f'(x) < 0$ when $x > e$, then $\sqrt[e]{e} > \sqrt[n]{n}$ for all $n \geq 3$. Also, $3^e > e \Rightarrow 3 > \sqrt[e]{e}$. Consequently, $3n > n\sqrt[n]{n} \Rightarrow \frac{1}{3n} < \frac{1}{n\sqrt[n]{n}} \Rightarrow$

$\sum_{n=1}^{\infty} \frac{1}{n\sqrt[n]{n}}$ diverges, by the Comparison Test for Convergence.

19. diverges, $\lim_{n \to \infty} a_n = \lim_{n \to \infty} \left(1 + \frac{1}{n}\right)^n = e \neq 0$

20. converges, by the Limit Comparison Test when compared with $\sum_{n=1}^{\infty} \frac{1}{n^{3/2}}$

21. converges: $\sum_{n=1}^{\infty} \frac{1-n}{n\,2^n} = \sum_{n=1}^{\infty} \frac{1}{n\,2^n} + \sum_{n=1}^{\infty} \frac{1}{2^n}$, the sum of two convergent series.

$\sum_{n=1}^{\infty} \frac{1}{n\,2^n}$ converges, by the Comparison Test for Convergence, since $\frac{1}{n\,2^n} < \frac{1}{2^n}$.

22. diverges, a geometric series with $r = 1/\ln 2$

23. converges, by the Comparison Test for Convergence, since $\frac{1}{3^{n-1}+1} < \frac{1}{3^{n-1}}$

24. converges, by the Limit Comparison Test when compared with $\sum_{n=1}^{\infty} \frac{1}{n^2}$

25. There are $(13)(365)(24)(60)(60)(10^9)$ seconds in 13 billion years. $s_n \leq 1 + \ln n$ where $n = (13)(365)(24)(60)(60)(10^9)$. $1 + \ln\left((13)(365)(24)(60)(60)(10^9)\right) = 1 + \ln(13) + \ln(365) + \ln(24) + 2\ln(60) + 9\ln(10) \approx 41.55$

26. $\sum_{n=1}^{\infty} \frac{1}{x\,n} = \left(\frac{1}{x}\right) \sum_{n=1}^{\infty} \frac{1}{n}$, but $\sum_{n=1}^{\infty} \frac{1}{n}$ diverges $\Rightarrow \sum_{n=1}^{\infty} \frac{1}{x\,n}$ will diverge for all finite values of x.

27. If $\sum_{n=1}^{\infty} a_n$ converges and $\frac{a_n}{n} \le a_n$, then by the Comparison Test for Convergence,

$\sum_{n=1}^{\infty} \frac{a_n}{n}$ converges.

28. If $\sum_{n=1}^{\infty} b_n$ and $\sum_{n=1}^{\infty} a_n$ converge, then $\lim_{n \to \infty} a_n = 0$. There is a $n_o > 0$ for which $a_n < 1$, hence $a_n b_n < b_n$ for $n > n_o$. $\sum_{n=1}^{\infty} a_n b_n = \sum_{n=1}^{n_o} a_n b_n + \sum_{n=n_o+1}^{\infty} a_n b_n \le \sum_{n=1}^{n_o} a_n b_n + \sum_{n=n_o+1}^{\infty} b_n$ which is convergent $\Rightarrow \sum_{n=1}^{\infty} a_n b_n$ converges, by the Comparison Test for Convergence.

29. If $\{S_n\}$ is nonincreasing with lower bound M, then $\{-S_n\}$ is a nondecreasing sequence with upper bound $-M$. By Theorem 6, $\{-S_n\}$ converges, and hence, $\{S_n\}$ converges. If $\{S_n\}$ has no lower bound, then $\{-S_n\}$ has no upper bound and diverges. Hence, $\{S_n\}$ also diverges.

30. a) $a_{2^n} = \frac{1}{2^n \ln(2^n)} = \frac{1}{2^n n \ln(2)}$. $\sum_{n=2}^{\infty} 2^n a_{2^n} = \sum_{n=2}^{\infty} 2^n \frac{1}{2^n n \ln(2)} = \frac{1}{\ln(2)} \sum_{n=2}^{\infty} \frac{1}{n}$ which diverges. $\therefore \sum_{n=2}^{\infty} \frac{1}{n \ln n}$ is divergent.

b) $a_{2^n} = \frac{1}{2^{np}}$. $\sum_{n=1}^{\infty} 2^n a_{2^n} = \sum_{n=1}^{\infty} 2^n \frac{1}{2^{np}} = \sum_{n=1}^{\infty} \frac{1}{(2^n)^{p-1}}$ which converges when $p > 1$, but diverges for $p \le 1$.

9.4 SERIES WITH NONNEGATIVE TERMS: RATIO AND ROOT TESTS

1. converges, by the Ratio Test for $\lim_{n\to\infty} \left| \frac{(n+1)^2}{2^{n+1}} \frac{2^n}{n^2} \right| = \frac{1}{2} < 1$

2. diverges, by the Ratio Test for $\lim_{n\to\infty} \left| \frac{(n+1)!}{10^{n+1}} \frac{10^n}{n!} \right| = \infty$

3. converges, by the Ratio Test for $\lim_{n\to\infty} \left| \frac{(n+1)^{10}}{10^{n+1}} \frac{10^n}{n^{10}} \right| = \frac{1}{10} < 1$

4. converges, by the Ratio Test for $\lim_{n\to\infty} \left| \frac{(n+1)^2}{e^{n+1}} \frac{e^n}{n^2} \right| = \frac{1}{e} < 1$

5. diverges, for $\lim_{n\to\infty} a_n = \lim_{n\to\infty} \left(\frac{n-2}{n}\right)^n = \lim_{n\to\infty} \left(1 + \frac{-2}{n}\right)^n = e^{-2} \neq 0$

6. converges, by the Comparison Test for Convergence, since $\frac{2+(-1)^n}{(1.25)^n} =$

$\left(\frac{4}{5}\right)^n \left[2 + (-1)^n\right] < \left(\frac{4}{5}\right)^n (3)$

7. diverges, by the Ratio Test for $\lim_{n\to\infty} \left| \frac{(n+1)!\, e^{-(n+1)}}{n!\, e^{-n}} \right| = \infty$

8. converges, a geometric series with $r = -\frac{2}{3}$

9. diverges, $\lim_{n\to\infty} a_n = \lim_{n\to\infty} \left(1 - \frac{3}{n}\right)^n = e^{-3} \neq 0$

10. diverges, $\lim_{n\to\infty} a_n = \lim_{n\to\infty} \left(1 - \frac{1}{n^2}\right)^n = \exp\left[\lim_{n\to\infty} \frac{\ln\left(1 - \frac{1}{n^2}\right)}{\frac{1}{n}}\right] = \exp\left[\lim_{n\to\infty} \frac{\left(\frac{n^2}{n^2-1}\right)\left(\frac{2}{n^3}\right)}{-\frac{1}{n^2}}\right] =$

$\exp\left[\lim_{n\to\infty} \frac{-2n}{n^2-1}\right] = e^0 = 1 \neq 0$

11. converges, by the Comparison Test for Convergence, since $\frac{\ln n}{n^3} < \frac{n}{n^2} = \frac{1}{n^2}$ for $n \geq 2$

12. diverges, $\sum_{n=1}^{\infty} \frac{n-1}{n^2} = \sum_{n=1}^{\infty} \frac{1}{n} - \sum_{n=1}^{\infty} \frac{1}{n^2}$, the difference between a divergent and convegent series diverges

13. diverges, by the Comparison Test for Convergence, since $\frac{\ln n}{n} > \frac{1}{n}$ for $n \geq 3$

14. converges, by the Ratio Test for $\lim_{n\to\infty} \left| \frac{(n+1)\ln(n+1)}{2^{n+1}} \frac{2^n}{n\ln(n)} \right| = \frac{1}{2} < 1$

15. converges, by the Ratio Test for $\lim_{n\to\infty} \left| \frac{(n+2)(n+3)}{(n+1)!} \frac{n!}{(n+1)(n+2)} \right| = 0 < 1$

16. converges, by the Ratio Test for $\lim_{n\to\infty} \left| \frac{(n+1)^3}{e^{n+1}} \frac{e^n}{n^3} \right| = \frac{1}{e} < 1$

17. converges, by the Ratio Test for $\lim_{n\to\infty} \left| \frac{(n+4)!}{3!\,(n+1)!\,3^{n+1}} \frac{3!\,n!\,3^n}{(n+3)!} \right| = \frac{1}{3} < 1$

18. converges, by the Ratio Test for $\lim_{n\to\infty} \left| \frac{(n+1)^2}{2^{n+1}} \frac{2^n}{n^2} \right| = \frac{1}{2} < 1$

19. converges, by the Ratio Test for $\lim_{n\to\infty} \left| \frac{1}{(2n+3)!} \frac{(2n+1)!}{1} \right| = 0 < 1$

20. converges, by the Ratio Test for $\lim_{n\to\infty} \left| \frac{(n+1)!}{(n+1)^{n+1}} \frac{n^n}{n!} \right| = \lim_{n\to\infty} \left(\frac{n}{n+1}\right)^n =$

$$\exp\left[\lim_{n\to\infty} \frac{\ln\left(\frac{n}{n+1}\right)}{\frac{1}{n}}\right] = \exp\left[\lim_{n\to\infty} \frac{-n^2}{n(n+1)}\right] = e^{-1} = \frac{1}{e} < 1$$

21. converges, by the Root Test for $\lim_{n\to\infty} \sqrt[n]{\frac{n}{(\ln n)^n}} = \lim_{n\to\infty} \frac{\sqrt[n]{n}}{\ln n} = \lim_{n\to\infty} \frac{1}{\ln n} = 0 < 1$

22. diverges, by the Comparison Test for Convergence, since $\frac{1}{(\ln n)^2} > \frac{1}{n}$ for $n \geq 2$

23. converges, by the Comparison Test for Convergence, since $\frac{n!}{(n+2)!} = \frac{1}{(n+1)(n+2)} < \frac{1}{n^2}$

24. diverges, by the Ratio Test for $\lim_{n\to\infty} \left| \frac{3^{n+1}}{(n+1)^3\, 2^{n+1}} \frac{n^3\, 2^n}{3^n} \right| = \frac{3}{2} > 1$

25. converges, by the Ratio test for $\lim_{n\to\infty} \left| \frac{(n+1)!}{(2n+3)!} \frac{(2n+1)!}{n!} \right| = \lim_{n\to\infty} \frac{n+1}{(2n+3)(2n+2)} = 0 < 1$

26. converges, by the Comparison Test for Convergence, since $n^n > n! \Rightarrow \left(n^n\right)^{n-1} > (n!)^{n-1} \Rightarrow$

$n! \left(n^n\right)^{n-1} > (n!)^n \Rightarrow \frac{n!}{n^n} > \frac{(n!)^n}{\left(n^n\right)^n} \Rightarrow \frac{n!}{n^n} > \frac{(n!)^n}{n^{n^2}}$ and $\sum_{n=1}^{\infty} \frac{n!}{n^n}$ converges, from exercise 20

27. converges, by the ratio Test for $\lim_{n\to\infty} \frac{\left(\frac{1+\sin n}{n}\right) a_n}{a_n} = 0 < 1$

28. diverges, by the Ratio Test for $\lim_{n\to\infty} \left| \frac{\left(\frac{3n-1}{2n+5}\right) a_n}{a_n} \right| = \frac{3}{2} > 1$

29. diverges, the given series is $\sum_{n=1}^{\infty} \frac{3}{n}$

30. converges, by the Ratio Test for $\lim_{n \to \infty} \left| \frac{\left(\frac{2}{n}\right)(a_n)}{a_n} \right| = 0 < 1$

31. converges, by the Ratio Test for $\lim_{n \to \infty} \left| \frac{a_{n+1}}{a_n} \right| = \lim_{n \to \infty} \left| \frac{(1 + \ln n)\, a_n}{n\, a_n} \right| = \lim_{n \to \infty} \frac{1 + \ln n}{n} = 0 < 1$

32. diverges, $n \geq 22027 \Rightarrow \ln n > 10 \Rightarrow \frac{n + \ln n}{n + 10} > 1 \Rightarrow a_{n+1} > a_n \Rightarrow \lim_{n \to \infty} a_n \neq 0$

33. converges, by the Ratio Test for $\lim_{n \to \infty} \left| \frac{2^{n+1}\, (n + 1)!\, (n + 1)!}{(2n + 2)!} \; \frac{(2n)!}{2^n\, (n!)(n!)} \right| =$

$\lim_{n \to \infty} \frac{2\, (n + 1)^2}{(2n + 2)(2n + 1)} = \frac{1}{2} < 1$

34. diverges, by the Ratio Test for $\lim_{n \to \infty} \left| \frac{(3n + 3)!\, (n + 1)!(n + 2)!\, n!}{(n + 1)!\, (n + 2)!\, (n+3)!\, (3n!)} \right| =$

$\lim_{n \to \infty} \frac{(3n + 1)\, (3n + 2)\, (3n + 1)}{(n + 1)\, (n + 2)\, (n + 3)} = 27 > 1$

35. coverges, by the Comparison Test for Convergence, since $a_1 = 1 = \frac{12}{(1)(3)(2)^2}$, $a_2 = \frac{1 \cdot 2}{3 \cdot 4} =$
$\frac{12}{(2)(4)(3)^2}$, $a_3 = \frac{2 \cdot 3}{4 \cdot 5} \frac{1 \cdot 2}{3 \cdot 4} = \frac{12}{(3)(5)(4)^2}$, $a_4 = \frac{3 \cdot 4}{5 \cdot 6} \frac{2 \cdot 3}{4 \cdot 5} \frac{1 \cdot 2}{3 \cdot 4} = \frac{12}{(4)(6)(5)^2}, \ldots \Rightarrow$

$1 + \sum_{n=1}^{\infty} \frac{12}{(n + 1)(n + 3)(n + 2)^2}$ represents the given series and $\frac{12}{(n + 1)\, (n + 3)\, (n + 2)^2} < \frac{12}{n^4}$

36. The Ratio Test applied to the p–series $\Rightarrow \lim_{n \to \infty} \left| \frac{1}{(n + 1)^p} \frac{n^p}{1} \right| = \left(\lim_{n \to \infty} \frac{n}{n + 1} \right)^p = 1$. The Root Test

applied to the p–series $\Rightarrow \lim_{n \to \infty} \sqrt[n]{\frac{1}{n^p}} = \lim_{n \to \infty} \frac{1}{n^{p/n}} = \exp\left[\lim_{n \to \infty} \frac{\ln n}{n/p} \right] = \exp\left[\lim_{n \to \infty} \frac{1/n}{1/p} \right] = e^0 = 1.$

In either case, the result was 1, which is not conclusive.

9.5 ALTERNATING SERIES AND ABSOLUTE CONVERGENCE

1. converges absolutely, by the Absolute Convergence Theorem, since $\sum_{n=1}^{\infty} |a_n|$ is a convergent p–series

2. converges, by the Alternating Series Theorem, since $f(x) = \frac{1}{\ln x} \Rightarrow f'(x) = -\frac{1}{(\ln x)^2} < 0 \Rightarrow f(x)$ is decreasing, $\lim_{n \to \infty} a_n = \lim_{n \to \infty} \frac{1}{\ln n} = 0$

3. diverges, $\lim_{n \to \infty} a_n \neq 0$

4. diverges, $\lim_{n \to \infty} \frac{10^n}{n^{10}} = \lim_{n \to \infty} \frac{10^n (\ln 10)^{10}}{10!} = \infty$

5. converges, by the Alternating Series Theorem, since $f(x) = \frac{\sqrt{x} + 1}{x + 1} \Rightarrow f'(x) = \frac{1 - 2x - 2\sqrt{x}}{(x + 1)^2} < 0 \Rightarrow$ f(x) is decreasing, $\lim_{n \to \infty} a_n = \lim_{n \to \infty} \frac{\sqrt{n} + 1}{n + 1} = 0$

6. converges, by the Alternating Series Theorem, since $f(x) = \frac{\ln x}{x} \Rightarrow f'(x) = \frac{1 - \ln x}{x^2} < 0$, when $x > e \Rightarrow$ f(x) is decreasing, $\lim_{n \to \infty} a_n = \lim_{n \to \infty} \frac{\ln n}{n} = \lim_{n \to \infty} \frac{1/n}{1} = 0$

7. converges absolutely, by the Absolute Convergence Theorem, since $\sum_{n=1}^{\infty} |a_n|$ is a convergent p–series

8. diverges, $\lim_{n \to \infty} a_n = \lim_{n \to \infty} \frac{\ln n}{\ln n^2} = \frac{1}{2} \neq 0$

9. diverges, $\lim_{n \to \infty} a_n = \lim_{n \to \infty} \frac{3\sqrt{n+1}}{\sqrt{n} + 1} = \lim_{n \to \infty} \frac{3\sqrt{1 + 1/n}}{\sqrt{1} + 1/\sqrt{n}} = 3 \neq 0$

10. converges, by the Alternating Series Theorem, since $f(x) = \ln\left(1 - x^{-1}\right) \Rightarrow f'(x) = -\frac{1}{x(x + 1)} < 0 \Rightarrow$ f(x) is decreasing, $\lim_{n \to \infty} a_n = \lim_{n \to \infty} \ln\left(1 + \frac{1}{n}\right) = \ln\left(\lim_{n \to \infty} 1 + \frac{1}{n}\right) = \ln(1) = 0$

11. converges absolutely, by the Absolute Convergence Theorem, since $\sum_{n=1}^{\infty} |a_n|$ is a convergent geometric series

12. converges conditionally, since $f(x) = x^{-1/2} \Rightarrow f'(x) = \dfrac{-1}{2\sqrt{x^3}} < 0 \Rightarrow f(x)$ is decreasing and

$\lim\limits_{n \to \infty} \dfrac{1}{\sqrt{n}} = 0 \Rightarrow$ the given series converges, by the Alternating Series Test, but $\sum\limits_{n=1}^{\infty} \dfrac{1}{\sqrt{n}}$ is a divergent p–series

13. converges absolutely, by the Absolute Convergence Theorem, since $\sum\limits_{n=1}^{\infty} |a_n|$ converges, by the Limit Comparison Test when compared with $\sum\limits_{n=1}^{\infty} \dfrac{1}{n^2}$

14. diverges, by the Ratio Test for $\lim\limits_{n \to \infty} \left| \dfrac{(n+1)!}{2^{n+1}} \dfrac{2^n}{n!} \right| = \infty$

15. converges conditionally, since $f(x) = \dfrac{1}{x+3} \Rightarrow f'(x) = \dfrac{-1}{(x+3)^2} < 0 \Rightarrow f(x)$ is decreasing and

$\lim\limits_{n \to \infty} \dfrac{1}{n+3} = 0 \Rightarrow$ the given series converges, by the Alternating Series Test, but $\sum\limits_{n=1}^{\infty} \dfrac{1}{n+3}$ diverges, by the Limit Comparison Test when compared with $\sum\limits_{n=1}^{\infty} \dfrac{1}{n}$

16. converges absolutely, by the Absolute Convergence Theorem, since $\left| \dfrac{\sin n}{n^2} \right| < \dfrac{1}{n^2}$

17. diverges, $\lim\limits_{n \to \infty} a_n = \lim\limits_{n \to \infty} \dfrac{3+n}{5+n} = 1 \neq 0$

18. converges conditionally, since $f(x) = \dfrac{1}{\ln x^3} \Rightarrow f'(x) = \dfrac{-3x^2}{\left(\ln x^3\right)^2} < 0 \Rightarrow f(x)$ is decreasing and

$\lim\limits_{n \to \infty} \dfrac{1}{\ln n^3} = 0 \Rightarrow$ the given series converges, by the Alternating Series Test, but $\sum\limits_{n=1}^{\infty} \dfrac{1}{\ln n^3}$ diverges, by the Comparison Test for Convergence, since $\dfrac{1}{\ln n^3} > \dfrac{1}{n}$ for $n \geq 5$

19. converges conditionally, since $f(x) = \frac{1}{x^2} + \frac{1}{x} \Rightarrow f'(x) = -\left(\frac{1}{x^3} + \frac{1}{x}\right) < 0 \Rightarrow f(x)$ is decreasing and

$\lim_{n\to\infty} a_n = \lim_{n\to\infty} \frac{1}{n^2} + \frac{1}{n} = 0 \Rightarrow$ the given series converges, by the Alternating Series Test, but

$\sum_{n=1}^{\infty} \frac{1+n}{n^2} = \sum_{n=1}^{\infty} \frac{1}{n^2} + \sum_{n=1}^{\infty} \frac{1}{n}$, the sum of a convergent and divergent series, diverges

20. converges absolutely, by the Absolute Convergence Theorem, since $\left|\frac{(-2)^{n+1}}{n+5^n}\right| \le \frac{2^{n+1}}{n+5^n} < 2\left(\frac{2}{5}\right)^n$

21. converges absolutely, by the Ratio Test for $\lim_{n\to\infty} \left|\frac{(n+1)^2\left(\frac{2}{3}\right)^{n+1}}{n^2\left(\frac{2}{3}\right)^n}\right| = \frac{2}{3} < 1$

22. diverges, $\lim_{n\to\infty} a_n = \lim_{n\to\infty} 10^{1/n} = 1 \neq 0$

23. converges absolutiely, by the Integral Test, since $\int_1^{\infty} \arctan x \left(\frac{1}{1+x^2}\right) dx =$

$\lim_{t\to\infty} \int_1^t \arctan x \left(\frac{1}{1+x^2}\right) dx = \lim_{t\to\infty} \left[\frac{(\arctan)^2}{2}\right]_1^t = \frac{3\pi^2}{32}$

24. converges conditionally, since $f(x) = \frac{1}{x \ln x} \Rightarrow f'(x) = -\frac{(\ln(x) + 1)}{(x \ln x)^2} < 0 \Rightarrow f(x)$ is decreasing and

$\lim_{n\to\infty} a_n = \lim_{n\to\infty} \frac{1}{n \ln n} = 0 \Rightarrow$ the given series converges, by the Alternating Series Test, but

$\int_2^{\infty} \frac{1}{x \ln x} dx = \lim_{t\to\infty} \int_2^t \left(\frac{1/x}{\ln x}\right) dx = \lim_{t\to\infty} [\ln(\ln x)]_2^t = \lim_{t\to\infty} (\ln(\ln t) - \ln(\ln 2)) = \infty \Rightarrow$

$\sum_{n=1}^{\infty} \frac{(-1)^{n+1}}{n \ln n}$ diverges, by the Integral Test

25. diverges; the given series is $\frac{1}{2}\left(\sum_{n=1}^{\infty} \frac{1}{n}\right)$

26. converges absolutely, by the Absolute Convergence Theorem, since $\left|\frac{(-1)^{n+1}(0.1)^n}{n}\right| =$

$\frac{1}{(10)^n n} < \left(\frac{1}{10}\right)^n$, a convergent geometric series

27. diverges, $\lim_{n\to\infty} a_n = \lim_{n\to\infty} \frac{(-1)^n n}{n+1} = \lim_{n\to\infty} (-1)^n \neq 0$

28. converges conditionally, since $f(x) = \frac{1}{1 + x^{1/2}} \Rightarrow f'(x) = -\frac{1}{2\sqrt{x}\left(1 + \sqrt{x}\right)^2} < 0 \Rightarrow f(x)$ is decreasing and $\lim_{n\to\infty} a_n = \lim_{n\to\infty} \frac{1}{1 + \sqrt{x}} = 0 \Rightarrow$ the given series converges by the Alternating Series Test, but $\sum_{n=1}^{\infty} \frac{1}{1+\sqrt{n}}$ diverges, by The Limit Comparison Test when compared with $\sum_{n=1}^{\infty} \frac{1}{\sqrt{n}}$

29. converges absolutely, by the Absolute Convergence Theorem, since $\left|\frac{-1}{n^2 + 2n + 1}\right| < \frac{1}{n^2}$

30. converges, since $(5)^{-n} = \left(\frac{1}{5}\right)^n$, a convergent geometric series

31. converges absolutely, by the Ratio Test for $\lim_{n\to\infty} \left|\frac{(100)^{n+1}}{(n + 1)!} \frac{n!}{100^n}\right| = \lim_{n\to\infty} \frac{100}{n + 1} = 0 < 1$

32. converges absolutely, by the Absolute Convergence Theorem, since $\left|(-1)^n \left(\frac{\ln n}{\ln n^2}\right)^n\right| = \left(\frac{1}{2}\right)^n$, a convergent geometric series

33. converges absolutely, by the Absolute Convergence Theorem, since $\left|\frac{\cos n\pi}{n\sqrt{n}}\right| = \left|\frac{(-1)^{n+1}}{n^{3/2}}\right| = \frac{1}{n^{3/2}}$, a convergent p–series

34. converges conditionally, since $\frac{\cos n\pi}{n} = (-1)^n \frac{1}{n}$, the nth term of the alternating harmonic series

35. converges conditionally, since $f(x) = \sqrt{x + 1} - \sqrt{x} \Rightarrow f'(x) = \frac{1}{2}\left(\frac{1}{\sqrt{x + 1}} - \frac{1}{\sqrt{x}}\right) < 0 \Rightarrow f(x)$ is decreasing and $\lim_{n\to\infty} a_n = \lim_{n\to\infty} \frac{1}{\sqrt{n} + \sqrt{n + 1}} = 0 \Rightarrow$ the given series converges, by the Alternating Series Test, but $\sum_{n=1}^{\infty} \frac{1}{\sqrt{n} + \sqrt{n + 1}} = \sum_{n=1}^{\infty} \sqrt{n + 1} - \sum_{n=1}^{\infty} \sqrt{n}$, the difference of two diverging series

36. converges absolutely, by the Ratio Test for $\lim_{n\to\infty} \left|\frac{\left((n + 1)!\right)^2}{((2n + 2)!)} \frac{(2n)!}{(n!)^2}\right| =$

$$\lim_{n\to\infty} \frac{(n + 1)^2}{(2n + 2)(2n + 1)} = \frac{1}{4} < 1$$

37. $|\text{error}| < \left|(-1)^6 \frac{1}{5}\right| = 0.2$

38. $|\text{error}| < \left|(-1)^6 \frac{1}{10^5}\right| = 0.00001$

39. $|\text{error}| < \left|(-1)^6 \frac{(0.01)^5}{5}\right| = 2. \times 10^{-11}$

40. $|\text{error}| < \left|(-1)^4 t^4\right| = t^4 < 1$

41. $\frac{1}{(2n)!} < \frac{5}{10^6} \Rightarrow (2n)! > \frac{10^6}{5} = 200000 \Rightarrow 2n = 10 \Rightarrow n = 5 \Rightarrow 1 - \frac{1}{2!} + \frac{1}{4!} - \frac{1}{6!} + \frac{1}{8!} \approx$

$0.540302579 \approx 0.54030$

42. $\frac{1}{n!} < \frac{5}{10^6} \Rightarrow \frac{10^6}{5} < n! \Rightarrow n = 9 \Rightarrow 1 - 1 + \frac{1}{2!} - \frac{1}{3!} + \frac{1}{4!} - \frac{1}{5!} + \frac{1}{6!} - \frac{1}{7!} + \frac{1}{8!} \approx 0.367881944$

43. a) $a_n \ge a_{n+1}$ fails, since $\frac{1}{3} < \frac{1}{2}$

b) $\left(\frac{1}{3} + \frac{1}{9} + \frac{1}{27} + \ldots\right) - \left(\frac{1}{2} + \frac{1}{4} + \frac{1}{8} + \ldots\right) = \frac{1/3}{1 - 1/3} - \frac{1/2}{1 - 1/2} = \frac{1}{2} - 1 = -\frac{1}{2}$

44. $s_{20} = 1 - \frac{1}{2} + \frac{1}{3} - \frac{1}{4} + \ldots + \frac{1}{19} - \frac{1}{20} = 0.668771403,\ s_{20} + \frac{1}{2}\,\frac{1}{21} = 0.692580927$

45. The unused terms are $\sum_{j = n+1}^{\infty} (-1)^{j+1} a_j = (-1)^{n+1}\left(a_{n+1} - a_{n+2}\right) + (-1)^{n+3}\left(a_{n+3} - a_{n+4}\right) + \ldots =$

$(-1)^{n+1}\left[\left(a_{n+1} - a_{n+2}\right) + \left(a_{n+3} - a_{n+4}\right) + \ldots\right]$. Each grouped term is positive, hence the remainder has the same sign as $(-1)^{n+1}$, which is the sign of the first unused term.

46. $s_n = \frac{1}{1 \cdot 2} + \frac{1}{2 \cdot 3} + \frac{1}{3 \cdot 4} + \ldots + \frac{1}{n(n+1)} = \sum_{i=1}^{n} \frac{1}{i(i+1)} = \sum_{i=1}^{n} \left(\frac{1}{i} - \frac{1}{i+1}\right) = 1 - \frac{1}{2} + \frac{1}{2} - \frac{1}{3} + \frac{1}{4} - \frac{1}{5} +$

$\ldots + \frac{1}{n} - \frac{1}{n+1} = s_{2n}$, hence the series are the same. $s_{2n} = \sum_{i=1}^{n} \frac{1}{i} - \frac{1}{i+1} = 1 - \frac{1}{2} + \frac{1}{2} - \frac{1}{3} + \frac{1}{4} - \frac{1}{5} +$

$\ldots + \frac{1}{n-1} - \frac{1}{n} + \frac{1}{n} - \frac{1}{n+1} = 1 - \frac{1}{n+1} \Rightarrow \lim_{n \to \infty} s_{2n} = \lim_{n \to \infty} \left(1 - \frac{1}{n+1}\right) = 1 \Rightarrow$ the series converge

to 1. The sum of the first 2n + 1 terms of $\sum_{i=1}^{\infty} \frac{1}{i} - \frac{1}{i+1}$ is $1 - \frac{1}{2} + \frac{1}{2} - \frac{1}{3} + \frac{1}{4} - \frac{1}{5} + \ldots + \frac{1}{n} -$

$\frac{1}{n+1} + \frac{1}{n+1} = 1.$

9.6 POWER SERIES

1. $\lim_{n \to \infty} \left| \frac{x^{n+1}}{x^n} \right| < 1 \Rightarrow -1 < x < 1$; when $x = -1$ we have $\sum_{n=1}^{\infty} (-1)^n$, a divergent series; when $x = 1$ we have $\sum_{n=1}^{\infty} 1$, a divergent series $\therefore$

   a) $-1 < x < 1$ b) $-1 < x < 1$

2. $\lim_{n \to \infty} \left| \frac{(x+5)^{n+1}}{(x+5)^n} \right| < 1 \Rightarrow -6 < x < -4$; when $x = -6$ we have $\sum_{n=1}^{\infty} (-1)^n$, a divergent series; when $x = -4$ we have $\sum_{n=1}^{\infty} 1$, a divergent series $\therefore$

   a) $-6 < x < -4$ b) $-6 < x < -4$

3. $\lim_{n \to \infty} \left| \frac{(x+1)^{n+1}}{(x+1)^n} \right| < 1 \Rightarrow -2 < x < 0$; when $x = -2$ we have $\sum_{n=1}^{\infty} 1$, a divergent series; when $x = 0$ we have $\sum_{n=1}^{\infty} (-1)^n$, a divergent series $\therefore$

   a) $-2 < x < 0$ b) $-2 < x < 0$

4. $\lim_{n \to \infty} \left| \frac{(x-2)^{n+1}}{n+1} \frac{n}{(x-2)^n} \right| < 1 \Rightarrow 1 < x < 3$; when $x = 1$ we have $\sum_{n=1}^{\infty} \frac{(-1)^n}{n}$, a convergent series; when $x = 3$ we have $\sum_{n=1}^{\infty} \frac{1}{n}$, a divergent series $\therefore$

   a) $1 \le x < 3$ b) $1 < x < 3$

5. $\lim_{n \to \infty} \left| \frac{(x-2)^{n+1}}{10^{n+1}} \frac{10^n}{(x-2)^n} \right| < 1 \Rightarrow -8 < x < 12$; when $x = -8$ we have $\sum_{n=1}^{\infty} (-1)^n$, a divergent series; when $x = 12$ we have $\sum_{n=1}^{\infty} 1$, a divergent series $\therefore$

   a) $-8 < x < 12$ b) $-8 < x < 12$

6. $\lim_{n\to\infty}\left|\frac{(2x)^{n+1}}{(2x)^n}\right| < 1 \Rightarrow -\frac{1}{2} < x < \frac{1}{2}$; when $x = -\frac{1}{2}$ we have $\sum_{n=1}^{\infty}(-1)^n$, a divergent series;

when $x = \frac{1}{2}$ we have $\sum_{n=1}^{\infty} 1$, a divergent series $\therefore$

a) $-\frac{1}{2} < x < \frac{1}{2}$ b) $-\frac{1}{2} < x < \frac{1}{2}$

7. $\lim_{n\to\infty}\left|\frac{(n+1)\,x^{n+1}}{n+3}\,\frac{n+2}{n\,x^n}\right| < 1 \Rightarrow -1 < x < 1$; when $x = -1$ we have $\sum_{n=1}^{\infty}(-1)^n\frac{n}{n+2}$, a divergent

series; when $x = 1$ we have $\sum_{n=1}^{\infty}\frac{n}{n+2}$, a divergent series $\therefore$

a) $-1 < x < 1$ b) $-1 < x < 1$

8. $\lim_{n\to\infty}\left|\frac{(x+2)^{n+1}}{n+1}\,\frac{n}{(x+2)^n}\right| < 1 \Rightarrow -3 < x < -1$; when $x = -3$ we have $\sum_{n=1}^{\infty}\frac{1}{n}$, a divergent series;

when $x = -1$ we have $\sum_{n=1}^{\infty}\frac{(-1)^n}{n}$, a convergent series $\therefore$

a) $-3 \le x < -1$ b) $-3 < x < -1$

9. $\lim_{n\to\infty}\left|\frac{x^{n+1}}{(n+1)\sqrt{n+1}}\,\frac{n\sqrt{n}}{x^n}\right| < 1 \Rightarrow -1 < x < 1$; when $x = -1$ we have $\sum_{n=1}^{\infty}\frac{(-1)^n}{n^{3/2}}$, a convergent

series; when $x = 1$ we have $\sum_{n=1}^{\infty}\frac{1}{n^{3/2}}$, a convergent series $\therefore$

α) $-1 \le x \le 1$ b) $-1 \le x \le 1$

10. $\lim_{n\to\infty}\left|\frac{(x-1)^{n+1}}{\sqrt{n+1}}\,\frac{\sqrt{n}}{(x-1)^n}\right| < 1 \Rightarrow 0 < x < 2$; when $x = 0$ we have $\sum_{n=1}^{\infty}\frac{(-1)^n}{n^{1/2}}$, a convergent series;

when $x = 2$ we have $\sum_{n=1}^{\infty}\frac{1}{n^{1/2}}$, a divergent series $\therefore$

α) $0 \le x < 2$ b) $0 < x < 2$

11. $\lim\limits_{n\to\infty} \left| \dfrac{x^{n+1}}{(n+1)!} \dfrac{n!}{x^n} \right| < 1 \Rightarrow |x| \lim\limits_{n\to\infty} \dfrac{1}{n+1} < 1$ for all $x \Rightarrow$

a) For all x b) For all x

12. $\lim\limits_{n\to\infty} \left| \dfrac{3^{n+1} x^{n+1}}{(n+1)!} \dfrac{n!}{3^n x^n} \right| < 1 \Rightarrow 3|x| \lim\limits_{n\to\infty} \dfrac{1}{n+1} < 1$ for all x

a) For all x b) For all x

13. $\lim\limits_{n\to\infty} \left| \dfrac{x^{2n+3}}{(n+1)!} \dfrac{n!}{x^{2n+1}} \right| < 1 \Rightarrow x^2 \lim\limits_{n\to\infty} \left| \dfrac{1}{n+1} \right| < 1$ for all x

a) For all x b) For all x

14. $\lim\limits_{n\to\infty} \left| \dfrac{(x-3)^{2n+3}}{(n+1)!} \dfrac{n!}{(x-3)^{2n+1}} \right| < 1 \Rightarrow (x-3)^2 \lim\limits_{n\to\infty} \left| \dfrac{1}{n+1} \right| < 1$ for all x

a) For all x b) For all x

15. $\lim\limits_{n\to\infty} \left| \dfrac{x^{n+1}}{\sqrt{n^2+2n+4}} \dfrac{\sqrt{n^2+3}}{x^n} \right| < 1 \Rightarrow |x| \sqrt{\lim\limits_{n\to\infty} \dfrac{n^2+3}{n^2+2n+4}} < 1 \Rightarrow$

$-1 < x < 1$; when $x = -1$ we have $\displaystyle\sum_{n=1}^{\infty} \frac{(-1)^n}{\sqrt{n^2+3}}$, a convergent series;

when $x = 1$ we have $\displaystyle\sum_{n=1}^{\infty} \frac{1}{\sqrt{n^2+3}}$, a divergent series $\therefore$

a) $-1 \le x < 1$ b) $-1 < x < 1$

16. $\lim\limits_{n\to\infty} \left| \dfrac{x^{n+1}}{\sqrt{(n+1)^2+3}} \dfrac{\sqrt{n^2+3}}{x^n} \right| < 1 \Rightarrow |x| \sqrt{\lim\limits_{n\to\infty} \dfrac{n^2+3}{(n+1)^2+3}} < 1 \Rightarrow$

$-1 < x < 1$; when $x = -1$ we have $\displaystyle\sum_{n=1}^{\infty} \frac{1}{\sqrt{n^2+3}}$, a divergent series;

when $x = 1$ we have $\displaystyle\sum_{n=1}^{\infty} \frac{(-1)^n}{\sqrt{n^2+3}}$, a convergent series $\therefore$

a) $-1 < x \le 1$ b) $-1 < x < 1$

17. $\lim_{n\to\infty}\left|\frac{(n+1)x^{n+1}}{n^2+2n+2}\,\frac{n^2+1}{nx^n}\right| < 1 \Rightarrow -1 < x < 1$; when x = – 1 we have $\sum_{n=1}^{\infty}\frac{(-1)^n n}{n^2+1}$, a convergent series; when x = – 1 we have $\sum_{n=1}^{\infty}\frac{n}{n^2+1}$, a divergent series ∴

a) $-1 \le x < 1$ b) $-1 < x < 1$

18. $\lim_{n\to\infty}\left|\frac{(n+1)(x-3)^{n+1}}{n(x-3)^n}\right| < 1 \Rightarrow |x-3| < 1 \Rightarrow 2 < x < 4$; both series: $\sum_{n=1}^{\infty} n(-1)^n$, when x = 2 and $\sum_{n=1}^{\infty} n$, when x = 4 diverge ∴

a) $2 < x < 4$ b) $2 < x < 4$

19. $\lim_{n\to\infty}\left|\frac{\sqrt{n+1}\,x^{n+1}}{3^{n+1}}\,\frac{3^n}{\sqrt{n}\,x^n}\right| < 1 \Rightarrow \frac{|x|}{3}\sqrt{\lim_{n\to\infty}\frac{n+1}{n}} < 1 \Rightarrow -3 < x < 3$; both series: $\sum_{n=1}^{\infty}\sqrt{n}\,(-1)^n$, when x = – 3 and $\sum_{n=1}^{\infty}\sqrt{n}$, when x = 3 diverge ∴

a) $-3 < x < 3$ b) $-3 < x < 3$

20. $\lim_{n\to\infty}\left|\frac{\sqrt[n+1]{n+1}\,(x-1)^{n+1}}{\sqrt[n]{n}\,(x-1)^n}\right| < 1 \Rightarrow |x-1| \lim_{n\to\infty}\frac{\sqrt[n+1]{n+1}}{\sqrt[n]{n}} < 1 \Rightarrow 0 < x < 2$; both series: $\sum_{n=1}^{\infty}\sqrt[n]{n}\,(-1)^n$, when x = 0 and $\sum_{n=1}^{\infty}\sqrt[n]{n}$, when x = 2 diverge ∴

a) $0 < x < 2$ b) $0 < x < 2$

21. $\lim_{t\to\infty}\left|\frac{\left(1+\frac{1}{n+1}\right)^{n+1}x^{n+1}}{\left(1+\frac{1}{n}\right)^n x^n}\right| < 1 \Rightarrow |x| \lim_{t\to\infty}\left(\frac{n(n+2)}{(n+1)^2}\right)^n\left(1+\frac{1}{n+1}\right) < 1 \Rightarrow -1 < x < 1$;

both series: $\sum_{n=1}^{\infty}(-1)^n\left(1+\frac{1}{n}\right)^n$, when x = – 1 and $\sum_{n=1}^{\infty}\left(1+\frac{1}{n}\right)^n$, when x = 1 diverge ∴

a) $-1 < x < 1$ b) $-1 < x < 1$

22. $\lim_{n\to\infty} \left|\frac{\ln(n+1)\, x^{n+1}}{x^n \ln n}\right| < 1 \Rightarrow |x| \lim_{n\to\infty} \left|\frac{1/(n+1)}{1/n}\right| < 1 \Rightarrow -1 < x < 1$; both series: $\sum_{n=1}^{\infty} (-1)^n(\ln n)$, when $x = -1$ and $\sum_{n=1}^{\infty} (\ln n)$, when $x = 1$diverge $\therefore$

a) $-1 < x < 1$ b) $-1 < x < 1$

23. $\lim_{n\to\infty} \left|\frac{(n+1)^{n+1}\, x^{n+1}}{n^n\, x^n}\right| < 1 \Rightarrow |x| \left(\lim_{n\to\infty} \left(1 + \frac{1}{n}\right)^n\right)\left(\lim_{n\to\infty} (n+1)\right) < 1 \Rightarrow$
$e\,|x| \lim_{n\to\infty} (n+1) < 1 \Rightarrow$

a) $x = 0$ b) $x = 0$

24. $\lim_{n\to\infty} \left|\frac{(n+1)!\,(x-4)^{n+1}}{n!\,(x-4)^n}\right| < 1 \Rightarrow |x-4| \lim_{n\to\infty} (n+1) < 1 \Rightarrow$

a) $x = 4$ b) $x = 4$

25. $\lim_{n\to\infty} \left|\frac{(x-3)^{n+1}}{2^{n+1}} \frac{2^n}{(x-3)^n}\right| < 1 \Rightarrow |x-3| < 2 \Rightarrow 1 < x < 5$; both series: $\sum_{n=1}^{\infty} (1)$, when $x = 1$ and $\sum_{n=1}^{\infty} (-1)^n$, when $x = 5$ diverge $\therefore$ the interval of convergence is $1 < x < 5$;

the sum of this convergent geometric series is $\frac{1}{1 + \frac{x-3}{2}} = \frac{2}{x-1}$;

$f(x) = 1 - \frac{1}{2}(x-3) + \frac{1}{4}(x-3)^2 + \ldots + \left(-\frac{1}{2}\right)^n (x-3)^n + \ldots = \frac{2}{x-1} \Rightarrow$

$f'(x) = -\frac{1}{2} + \frac{1}{2}(x-3) + \ldots + \left(-\frac{1}{2}\right)^n n(x-3)^{n-1} + \ldots$ is convergent when $1 < x < 5$, and diverges when $x = 1$ or 5; its sum is $\frac{-2}{(x-1)^2}$, the derivative of $\frac{2}{x-1}$

26. $f(x) = 1 - \frac{1}{2}(x-3) + \frac{1}{4}(x-3)^2 + \ldots + \left(-\frac{1}{2}\right)^n (x-3)^n + \ldots = \frac{2}{x-1} \Rightarrow \int f(x)\,dx =$

$x - \frac{(x-3)^2}{4} + \frac{(x-3)^3}{12} + \ldots + \left(-\frac{1}{2}\right)^n \frac{(x-3)^{n+1}}{n+1} + \ldots$; at $x = 1$ the series $\sum_{n=1}^{\infty} \frac{-2}{n+1}$ diverges,

at $x = 5$ the series $\sum_{n=1}^{\infty} \frac{(-1)^n\, 2}{n+1}$ converges $\therefore$ the interval of convergence is $1 < x \le 5$ and the sum is $2 \ln|x-1| + 3 - \ln 4$, since $\int \frac{2}{x-1}\,dx = 2\ln|x-1| + C$ where $C = 3 - \ln 4$ when $x = 3$

27. a) $\ln|\sec x| + C = \int \tan x\,dx = \int x + \frac{x^3}{3} + \frac{2x^5}{15} + \frac{17x^7}{315} + \ldots dx = \frac{x^2}{2} + \frac{x^4}{12} + \frac{x^6}{45} + \frac{17x^8}{2520} + \ldots + C$, but $x = 0 \Rightarrow C = 0 \Rightarrow \ln|\sec x| = \frac{x^2}{2} + \frac{x^4}{12} + \frac{x^6}{45} + \frac{17x^8}{2520} + \ldots$, when $-\frac{\pi}{2} < x < \frac{\pi}{2}$

b) $\sec^2 x = \frac{d(\tan x)}{dx} = \frac{d\left(x + \frac{x^3}{3} + \frac{2x^5}{15} + \frac{17x^7}{315} + \ldots\right)}{dx} = 1 + x^2 + \frac{2x^4}{3} + \frac{17x^6}{45} + \ldots$, when $-\frac{\pi}{2} < x < \frac{\pi}{2}$

c) $\sec^2 x = (\sec x)(\sec x) = \left(1 + \frac{x^2}{2} + \frac{5x^4}{24} + \frac{61x^6}{720} + \ldots\right)\left(1 + \frac{x^2}{2} + \frac{5x^4}{24} + \frac{61x^6}{720} + \ldots\right) = 1 + \left(\frac{1}{2} + \frac{1}{2}\right)x^2 + \left(\frac{5}{24} + \frac{1}{4} + \frac{5}{24}\right)x^4 + \left(\frac{61}{720} + \frac{5}{48} + \frac{5}{48} + \frac{61}{720}\right)x^6 + \ldots = 1 + x^2 + \frac{2x^4}{3} + \frac{17x^6}{45} + \ldots$

28. a) $\ln|\sec x + \tan x| + C = \int \sec x\,dx = \int 1 + \frac{x^2}{2} + \frac{5x^4}{24} + \frac{61x^6}{720} + \ldots dx = x + \frac{x^3}{6} + \frac{x^5}{24} + \frac{61x^7}{5040} + \ldots + C$, but $x = 0 \Rightarrow C = 0 \Rightarrow \ln|\sec x + \tan x| = x + \frac{x^3}{6} + \frac{x^5}{24} + \frac{61x^7}{5040} + \ldots$, when $-\frac{\pi}{2} < x < \frac{\pi}{2}$

b) $\sec x \tan x = \frac{d(\sec x)}{dx} = \frac{d\left(1 + \frac{x^2}{2} + \frac{5x^4}{24} + \frac{61x^6}{720} + \ldots\right)}{dx} = x + \frac{5x^3}{6} + \frac{61x^5}{120} + \frac{277x^7}{1008} + \ldots$, when $-\frac{\pi}{2} < x < \frac{\pi}{2}$

c) $(\sec x)(\tan x) = \left(1 + \frac{x^2}{2} + \frac{5x^4}{24} + \frac{61x^6}{720} + \ldots\right)\left(x + \frac{x^3}{3} + \frac{2x^5}{15} + \frac{17x^7}{315} + \ldots\right) = x + \left(\frac{1}{3} + \frac{1}{2}\right)x^3 + \left(\frac{2}{15} + \frac{1}{6} + \frac{5}{24}\right)x^5 + \left(\frac{17}{315} + \frac{1}{15} + \frac{5}{72} + \frac{61}{720}\right)x^7 + \ldots = x + \frac{5x^3}{6} + \frac{61x^5}{120} + \frac{277x^7}{1008} + \ldots$

9.7 TAYLOR AND MACLAURIN SERIES

1. $f(x) = \ln x$ $f'(x) = \frac{1}{x}$ $f''(x) = -\frac{1}{x^2}$ $f'''(x) = \frac{2}{x^3}$

$f(1) = \ln 1 = 0$ $f'(1) = 1$ $f''(1) = -1$ $f'''(1) = 2$

$P_0(x) = 0,\ P_1(x) = x - 1,\ P_2(x) = (x-1) - \frac{1}{2}(x-1)^2,\ P_3(x) = (x-1) - \frac{1}{2}(x-1)^2 + \frac{1}{3}(x-1)^3$

2. $f(x) = \ln(1+x)$ $f'(x) = \frac{1}{1+x} = (1+x)^{-1}$ $f''(x) = -(1+x)^{-2}$ $f'''(x) = 2(1+x)^{-3}$

$f(0) = \ln(1) = 0$ $f'(0) = \frac{1}{1} = 1$ $f''(0) = -(1)^{-2} = -1$ $f'''(0) = 2(1)^{-3} = 2$

$P_0(x) = 0,\ P_1(x) = x,\ P_2(x) = x - \frac{x^2}{2},\ P_3(x) = x - \frac{x^2}{2} + \frac{x^3}{3}$

3. $f(x) = \frac{1}{x} + x^{-1}$ $f'(x) = -x^{-2}$ $f''(x) = 2x^{-3}$ $f'''(x) = -6x^{-4}$

$f(2) = \frac{1}{2}$ $f'(2) = -\frac{1}{4}$ $f''(2) = \frac{1}{4}$ $f'''(2) = -\frac{3}{8}$

$P_0(x) = \frac{1}{2},\ P_1(x) = \frac{1}{2} - \frac{1}{4}(x-2),\ P_2(x) = \frac{1}{2} - \frac{1}{4}(x-2) + \frac{1}{8}(x-2)^2,$

$P_3(x) = \frac{1}{2} - \frac{1}{4}(x-2) + \frac{1}{8}(x-2)^2 - \frac{1}{16}(x-2)^3$

4. $f(x) = (x+2)^{-1}$ $f'(x) = -(x+2)^{-2}$ $f''(x) = 2(x+2)^{-3}$ $f'''(x) = -6(x+2)^{-6}$

$f(0) = (2)^{-1} = \frac{1}{2}$ $f'(0) = -(2)^{-2} = -\frac{1}{4}$ $f''(0) = 2(2)^{-3} = \frac{1}{4}$ $f'''(0) = -6(2)^{-6} = -\frac{3}{8}$

$P_0(x) = \frac{1}{2},\ P_1(x) = \frac{1}{2} - \frac{x}{4},\ P_2(x) = \frac{1}{2} - \frac{x}{4} + \frac{x^2}{8},\ P_3(x) = \frac{1}{2} - \frac{x}{4} + \frac{x^2}{8} - \frac{x^3}{16}$

5. $f(x) = \sin x \Rightarrow f'(x) = \cos x \Rightarrow f''(x) = -\sin x \Rightarrow f'''(x) = -\cos x \Rightarrow f(\pi/4) = \sin \pi/4 = \frac{\sqrt{2}}{2},$

$f'(\pi/4) = \cos \pi/4 = \frac{\sqrt{2}}{2},\ f''(\pi/4) = -\sin \pi/4 = -\frac{\sqrt{2}}{2},\ f'''(\pi/4) = -\cos \pi/4 = -\frac{\sqrt{2}}{2} \Rightarrow P_0 = \frac{\sqrt{2}}{2},$

$P_1(x) = \frac{\sqrt{2}}{2} + \frac{\sqrt{2}}{2}\left(x - \frac{\pi}{4}\right),\ P_2(x) = \frac{\sqrt{2}}{2} + \frac{\sqrt{2}}{2}\left(x - \frac{\pi}{4}\right) - \frac{\sqrt{2}}{4}\left(x - \frac{\pi}{4}\right)^2,$

$P_3(x) = \frac{\sqrt{2}}{2} + \frac{\sqrt{2}}{2}\left(x - \frac{\pi}{4}\right) - \frac{\sqrt{2}}{4}\left(x - \frac{\pi}{4}\right)^2 - \frac{\sqrt{2}}{12}\left(x - \frac{\pi}{4}\right)^3$

6. $f(x) = \cos x \Rightarrow f'(x) = -\sin x \Rightarrow f''(x) = -\cos x \Rightarrow f'''(x) = \sin x \Rightarrow f(\pi/4) = \cos \pi/4 = \frac{1}{\sqrt{2}},$

$f'(\pi/4) = -\sin \pi/4 = -\frac{1}{\sqrt{2}},\ f''(\pi/4) = -\cos \pi/4 = -\frac{1}{\sqrt{2}},\ f'''(\pi/4) = \sin \pi/4 = \frac{1}{\sqrt{2}} \Rightarrow P_0(x) = \frac{1}{\sqrt{2}},$

$P_1(x) = \frac{1}{\sqrt{2}} - \frac{1}{\sqrt{2}}\left(x - \frac{\pi}{4}\right),\ P_2(x) = \frac{1}{\sqrt{2}} - \frac{1}{\sqrt{2}}\left(x - \frac{\pi}{4}\right) - \frac{1}{2\sqrt{2}}\left(x - \frac{\pi}{4}\right)^2,$

$P_3(x) = \frac{1}{\sqrt{2}} - \frac{1}{\sqrt{2}}\left(x - \frac{\pi}{4}\right) - \frac{1}{2\sqrt{2}}\left(x - \frac{\pi}{4}\right)^2 + \frac{1}{6\sqrt{2}}\left(x - \frac{\pi}{4}\right)^3$

7. $f(x) = \sqrt{x} = x^{1/2} \Rightarrow f'(x) = (1/2)x^{-1/2} \Rightarrow f''(x) = (-1/4)x^{-3/2} \Rightarrow f'''(x) = (3/8)x^{-5/2} \Rightarrow$

$f(4) = \sqrt{4} = 4^{1/2} = 2,\ f'(4) = (1/2)4^{-1/2} = \frac{1}{4},\ f''(4) = (-1/4)4^{-3/2} = -\frac{1}{32},\ f'''(4) = (3/8)4^{-5/2} = \frac{3}{256} \Rightarrow$

$P_0(x) = 2, P_1(x) = 2 + \frac{1}{4}(x-4),\ P_2(x) = 2 + \frac{1}{4}(x-4) - \frac{1}{64}(x-4)^2,$

$P_3(x) = 2 + \frac{1}{4}(x-4) - \frac{1}{64}(x-4)^2 + \frac{1}{512}(x-4)^3$

8. $f(x) = (x+4)^{1/2} \Rightarrow f'(x) = (1/2)(x+4)^{-1/2} \Rightarrow f''(x) = (-1/4)(x+4)^{-3/2} \Rightarrow f'''(x) = (3/8)(x+4)^{-5/2} \Rightarrow$ $f(0) = (4)^{1/2} = 2,\ f'(0) = (1/2)(4)^{-1/2} = \frac{1}{4},\ f''(0) = (-1/4)(4)^{-3/2} = -\frac{1}{32},\ f'''(0) = (3/8)(4)^{-5/2} = \frac{3}{256} \Rightarrow$ $P_0(x) = 2,\ P_1(x) = 2 + \frac{1}{4}x,\ P_2(x) = 2 + \frac{1}{4}x - \frac{1}{64}x^2,\ P_3(x) = 2 + \frac{1}{4}x - \frac{1}{64}x^2 + \frac{1}{512}x^3$

9. $e^x = \sum_{n=0}^{\infty} \frac{x^n}{n!} \Rightarrow e^{-x} = \sum_{n=0}^{\infty} \frac{(-x)^n}{n!} = 1 - x + \frac{x^2}{2!} - \frac{x^3}{3!} + \frac{x^4}{4!} - \ldots$

10. $e^x = \sum_{n=0}^{\infty} \frac{x^n}{n!} \Rightarrow e^{x/2} = \sum_{n=0}^{\infty} \frac{(x/2)^n}{n!} = 1 + \frac{x}{2} + \frac{x^2}{4 \cdot 2!} + \frac{x^3}{2^3 \cdot 3!} + \ldots$

11. $\sin x = \sum_{n=0}^{\infty} \frac{(-1)^n x^{2n+1}}{(2n+1)!} \Rightarrow \sin 3x = \sum_{n=0}^{\infty} \frac{(-1)^n (3x)^{2n+1}}{(2n+1)!} = 3x - \frac{(3x)^3}{3!} + \frac{(3x)^5}{5!} - \ldots$

12. $\cos x = \sum_{n=0}^{\infty} \frac{(-1)^n x^{2n}}{(2n)!} \Rightarrow 5 \cos \pi x = 5 \sum_{n=0}^{\infty} \frac{(-1)^n (\pi x)^{2n}}{(2n)!} = 5\left(1 - \frac{(\pi x)^2}{2!} + \frac{(\pi x)^4}{4!} - \frac{(\pi x)^6}{6!} + \ldots\right)$

13. $\cos(-x) = \cos(x) = \sum_{n=0}^{\infty} \frac{(-1)^n x^{2n}}{(2n)!} = 1 - \frac{x^2}{2!} + \frac{x^4}{4!} - \frac{x^6}{6!} + \ldots$, since cosine is an even function

14. $x \sin x = x \sum_{n=0}^{\infty} \frac{(-1)^n x^{2n+1}}{(2n+1)!} = \sum_{n=0}^{\infty} \frac{(-1)^n x^{2n+2}}{(2n+1)!}$, since $\sin x = \sum_{n=0}^{\infty} \frac{(-1)^n x^{2n+1}}{(2n+1)!}$

15. $\cosh x = \frac{e^x + e^{-x}}{2} = \frac{1}{2}\left[\left(1 + x + \frac{x^2}{2!} + \frac{x^3}{3!} + \frac{x^4}{4!} + \ldots\right) + \left(1 - x + \frac{x^2}{2!} - \frac{x^3}{3!} + \frac{x^4}{4!} - \ldots\right)\right] =$ $1 + \frac{x^2}{2!} + \frac{x^4}{4!} + \frac{x^6}{6!} + \ldots$

16. $\sinh x + C = \int \cosh x\, dx = \int 1 + \frac{x^2}{2!} + \frac{x^4}{4!} + \frac{x^6}{6!} + \ldots dx = x + \frac{x^3}{3!} + \frac{x^5}{5!} + \ldots + C$, but $x = 0 \Rightarrow$ $C = 0 \Rightarrow \sinh x = x + \frac{x^3}{3!} + \frac{x^5}{5!} + \ldots$

17. $\frac{x^2}{2} - 1 + \cos x = \frac{x^2}{2} - 1 + \sum_{n=0}^{\infty} \frac{(-1)^n x^{2n}}{(2n)!} = \frac{x^2}{2} - 1 + 1 - \frac{x^2}{2!} + \frac{x^4}{4!} - \frac{x^6}{6!} + \ldots = \frac{x^4}{4!} - \frac{x^6}{6!} + \frac{x^8}{8!} - \frac{x^{10}}{10!} + \ldots$

18. $\cos^2 x = \frac{1}{2} + \frac{\cos 2x}{2} = \frac{1}{2} + \frac{1}{2} \sum_{n=0}^{\infty} \frac{(-1)^n (2x)^{2n}}{(2n)!} = \frac{1}{2} + \frac{1}{2}\left[1 - \frac{(2x)^2}{2!} + \frac{(2x)^4}{4!} - \frac{(2x)^6}{6!} + \frac{(2x)^8}{8!} - \ldots\right] =$ $1 - \frac{(2x)^2}{2 \cdot 2!} + \frac{(2x)^4}{2 \cdot 4!} - \frac{(2x)^6}{2 \cdot 6!} + \frac{(2x)^8}{2 \cdot 8!} - \ldots$

19. $f(x) = \frac{1}{x+1} = (1+x)^{-1} \Rightarrow f'(x) = -(1+x)^{-2} \Rightarrow f''(x) = 2(1+x)^{-3} \Rightarrow f'''(x) = -6(1+x)^{-4} \Rightarrow$
$f(0) = 1$, $f'(0) = -(1)^{-2} = -1$, $f''(0) = 2(1)^{-3} = 2 \Rightarrow f'''(c) = -6(1+c)^{-4} = \frac{-6}{(1+c)^4}$, where c
is between 0 and $x \Rightarrow P_2(x) = 1 - x + x^2 + R_2(x)$, where $R_2(x) = \frac{-x^3}{(1+c)^4}$

20. $f(x) = (1+x)^{1/2} \Rightarrow f'(x) = (1/2)(1+x)^{-1/2} \Rightarrow f''(x) = (-1/4)(1+x)^{-3/2} \Rightarrow f'''(x) = (3/8)(1+x)^{-5/2} \Rightarrow$
$f(0) = 1$, $f'(0) = (1/2)(1)^{-1/2} \Rightarrow f''(0) = (-1/4)(1)^{-3/2} \Rightarrow f'''(c) = (3/8)(1+c)^{-5/2} = \frac{3}{8(1+c)^{5/2}}$, where
c is between 0 and $x \Rightarrow P_2(x) = 1 + \frac{1}{2}x - \frac{1}{8}x^2 + R_2(x)$, where $R_2(x) = \frac{3}{8(1+c)^{5/2} \cdot 3!}x^3$

21. $f(x) = \ln(1+x) \Rightarrow f'(x) = (1+x)^{-1} \Rightarrow f''(x) = -(1+x)^{-2} \Rightarrow f'''(x) = 2(1+x)^{-3} \Rightarrow$
$f(0) = \ln(1) = 0$, $f'(0) = 1$, $f''(0) = -1$, $f'''(c) = 2(1+c)^{-3} = \frac{2}{(1+c)^3}$, where c is between
0 and $x \Rightarrow P_2(x) = x - \frac{x^2}{2} + R_2(x)$, where $R_2(x) = \frac{2}{(1+c)^3 \cdot 3!}x^3$

22. $f(x) = (1+x)^k \Rightarrow f'(x) = k(1+x)^{k-1} \Rightarrow f''(x) = k(k-1)(1+x)^{k-2} \Rightarrow$
$f'''(x) = k(k-1)(k-2)(1+x)^{k-3} \Rightarrow f(0) = 1$, $f'(0) = k$, $f''(0) = k(k-1)$,
$f'''(c) = k(k-1)(k-2)(1+c)^{k-3}$, where c is between 0 and $x \Rightarrow P_2(x) = 1 + kx + \frac{k(k-1)}{2!}x^2 + R_2(x)$,
where $R_2(x) = \frac{k(k-1)(k-2)(1+c)^{k-3}}{6!}x^3$

23. $f(x) = \sin x \Rightarrow f'(x) = \cos x \Rightarrow f''(x) = -\sin x \Rightarrow f'''(x) = -\cos x \Rightarrow f(0) = \sin 0 = 0, f'(0) = \cos 0 = 1$,
$f''(0) = -\sin 0 = 0, f'''(c) = -\cos c$, where c is between 0 and $x \Rightarrow P_2(x) = x + R_2(x)$, where
$R_2(x) = \frac{-\cos c}{3!}x^3$

24. $f(x) = \cos x \Rightarrow f'(x) = -\sin x \Rightarrow f''(x) = -\cos x \Rightarrow f'''(x) = \sin x \Rightarrow f(0) = \cos 0 = 1$, $f'(0) = -\sin 0 = 0$,
$f''(0) = -\cos 0 = -1$, $f'''(c) = \sin c$, where c is between 0 and $x \Rightarrow P_2(x) = 1 - \frac{x^2}{2!} + R_2(x)$, where
$R_2(x) = \frac{\sin c}{3!}x^3$

25. If $e^x = \sum_{n=0}^{\infty} \frac{f^{(n)}(a)}{n!}(x-a)^n$ and $f(x) = e^x$, we have $f^{(n)}(a) = e^a$ for all n = 0, 1, 2, 3,...;
$e^x = e^a\left[\frac{(x-a)^0}{0!} + \frac{(x-a)^1}{1!} + \frac{(x-a)^2}{2!} + \ldots\right] = e^a\left[1 + (x-a) + \frac{(x-a)^2}{2!} + \ldots\right]$, at x = a

26. $e^x = e\left[1 + (x-1) + \frac{(x-1)^2}{2!} + \ldots\right]$, exercise 25 with a = 1

27. $|R_3(x)| = \left|\frac{-\cos c}{5!}(x-0)^5\right| < \left|\frac{x^5}{5!}\right| < 5 \times 10^{-4} \Rightarrow -5 \times 10^{-4} < \frac{x^5}{5!} < 5 \times 10^{-4} \Rightarrow$
$-\sqrt[5]{5!(5 \times 10^{-4})} < x < \sqrt[5]{5!(5 \times 10^{-4})} \Rightarrow -0.569679052 < x < 0.569679052$

28. If $\cos x = 1 - \frac{x^2}{2}$ and $|x| < 0.5$, then the $|\text{error}| = |R_3(x)| = \left|\frac{\cos c}{4!}x^4\right| < \left|\frac{(.5)^4}{24}\right| = 0.0026$, where c is between 0 and x; since the next term in the series is positive, the approximation, $1 - \frac{x^2}{2}$, is low, by the Alternating Series Estimation Theorem.

29. $\sin x = x + R_1(x)$, when $|x| < 10^{-3} \Rightarrow |R_1(x)| = \left|\frac{-\cos c}{3!}x^3\right| < \left|\frac{(1)\,x^3}{3!}\right| < \frac{10^{-3}}{3!} = 1.67 \times 10^{-10}$

From exercise 45 in section 9.5, $R_1(x)$ has the same sign as $-\frac{x^3}{3!}$. $x < \sin x \Rightarrow 0 < \sin x - x = R_1(x)$, which has the same sign as $-\frac{x^3}{3!} \Rightarrow x < 0 \Rightarrow -10^{-3} < x < 0$.

30. From exercise 20, we have $\sqrt{1+x} = 1 + \frac{1}{2}x - \frac{1}{8}x^2 + R_2(x)$. If we approximate $\sqrt{1+x}$ with $1 + \frac{1}{2}x$ and $|x| < 0.01$, then the $|\text{error}| < \left|-\frac{1}{8}x^2\right| = \frac{(0.01)^2}{8} = 0.0000125$.

31. $|R_2(x)| = \left|\frac{e^c x^3}{3!}\right| < \frac{3^{(0.1)}(0.1)^3}{3!} = 0.00018602$, where c is between 0 and x.

32. $|\text{error}| < \left|-\frac{x^3}{3!}\right| = \frac{1}{6 \cdot 10^3} = \frac{1}{6000} = 0.000\overline{6}$

33. If we approximate $\sin x$ with $x - \frac{x^3}{3!}$ and $|x| < 0.5$, then the $|\text{error}| < \left|\frac{(\cos c)\, x^5}{5!}\right| < \frac{(0.5)^5}{5!} = \frac{1}{(32)(120)} = 0.000260416$, where c is between 0 and x.

34. If we approximate e^h with $1 + h$ and $0 \le h \le 0.01$, then $|\text{error}| < \left|\frac{e^c h^2}{2}\right| \le \frac{e^{0.01}\, h \cdot h}{2} = \left(\frac{e^{0.01}(0.01)}{2}\right)h = 0.005005h < 0.006h = 0.6\%h$, where c is between 0 and h.

35. $\sin x$, when $x = 0.1$; the sum is $\sin(0.1) \approx 0.99833416$

36. $\cos x$, when $x = \frac{\pi}{4}$; the sum is $\cos\left(\frac{\pi}{4}\right) = \frac{1}{\sqrt{2}} \approx 0.707106781$

37. $\sin x = x - \frac{x^3}{3!} + \frac{x^5}{5!} - \frac{x^7}{7!} + \ldots$; $\frac{d(\sin x)}{dx} = \frac{d\left(x - \frac{x^3}{3!} + \frac{x^5}{5!} - \frac{x^7}{7!} + \ldots\right)}{dx} = 1 - \frac{x^2}{2!} + \frac{x^4}{4!} - \frac{x^6}{6!} + \ldots = \cos x$;

$\frac{d(\cos x)}{dx} = \frac{d\left(1 - \frac{x^2}{2!} + \frac{x^4}{4!} - \frac{x^6}{6!} + \ldots\right)}{dx} = -x + \frac{x^3}{3!} - \frac{x^5}{5!} + \ldots = -\sin x$; $\frac{d\left(e^x\right)}{dx} = \frac{d\left(1 + x + \frac{x^2}{2!} + \frac{x^3}{3!} + \frac{x^4}{4!} + \ldots\right)}{dx} = 1 + x + \frac{x^2}{2!} + \frac{x^3}{3!} + \frac{x^4}{4!} + \ldots$

38. $\int \sin x\, dx = \int x - \frac{x^3}{3!} + \frac{x^5}{5!} - \frac{x^7}{7!} + \ldots dx = \frac{x^2}{2!} - \frac{x^4}{4!} + \frac{x^6}{6!} - \frac{x^8}{8!} + \ldots + C = -\cos x + C$ and $C = -1$

when $x = 0 \Rightarrow \int \sin x\, dx = \int x - \frac{x^3}{3!} + \frac{x^5}{5!} - \frac{x^7}{7!} + \ldots dx = -1 + \frac{x^2}{2!} - \frac{x^4}{4!} + \frac{x^6}{6!} - \frac{x^8}{8!} + \ldots$;

$\int \cos x\, dx = \int 1 - \frac{x^2}{2!} + \frac{x^4}{4!} - \frac{x^6}{6!} + \ldots dx = x - \frac{x^3}{3!} + \frac{x^5}{5!} - \frac{x^7}{7!} + \ldots + C = \sin x + C$ and $C = 0$

when $x = 0 \Rightarrow \int \cos x\, dx = \int 1 - \frac{x^2}{2!} + \frac{x^4}{4!} - \frac{x^6}{6!} + \ldots dx = x - \frac{x^3}{3!} + \frac{x^5}{5!} - \frac{x^7}{7!} + \ldots$;

$\int e^x\, dx = \int 1 + x + \frac{x^2}{2!} + \frac{x^3}{3!} + \frac{x^4}{4!} + \ldots dx = x + \frac{x^2}{2!} + \frac{x^3}{3!} + \frac{x^4}{4!} + \ldots + C$ and $C = 1$ when $x = 0 \Rightarrow$

$\int e^x\, dx = \int 1 + x + \frac{x^2}{2!} + \frac{x^3}{3!} + \frac{x^4}{4!} + \ldots dx = 1 + x + \frac{x^2}{2!} + \frac{x^3}{3!} + \frac{x^4}{4!} + \ldots$

39. $e^x \sin x = \left(1 + x + \frac{x^2}{2!} + \frac{x^3}{3!} + \frac{x^4}{4!} + \ldots\right)\left(x - \frac{x^3}{3!} + \frac{x^5}{5!} - \frac{x^7}{7!} + \ldots\right) = (1)\,x + (1)\,x^2 + \left(-\frac{1}{6} + \frac{1}{2}\right) x^3 + \left(-\frac{1}{6} + \frac{1}{6}\right) x^4 + \left(\frac{1}{120} - \frac{1}{12} + \frac{1}{24}\right) x^5 + \left(\frac{1}{120} - \frac{1}{36}\right) x^6 + \ldots = x + x^2 + \frac{1}{3}x^3 - \frac{1}{30}x^5 - \frac{7}{360}x^6 + \ldots$

40. $e^x \cos x = \left(1 + x + \frac{x^2}{2!} + \frac{x^3}{3!} + \frac{x^4}{4!} + \ldots\right)\left(1 - \frac{x^2}{2!} + \frac{x^4}{4!} - \frac{x^6}{6!} + \ldots\right) = (1) + (1)\,x + \left(-\frac{1}{2} + \frac{1}{2}\right) x^2 + \left(-\frac{1}{2} + \frac{1}{6}\right) x^3 + \left(\frac{1}{24} - \frac{1}{4} + \frac{1}{24}\right) x^4 + \left(\frac{1}{24} - \frac{1}{12}\right) x^5 + \ldots = 1 + x - \frac{x^3}{3} - \frac{x^4}{6} - \frac{5}{24} + \ldots$

41. $\sin x = x - \frac{x^3}{3!} + \frac{x^5}{5!} - \frac{x^7}{7!} + \ldots \Rightarrow \frac{\sin x}{x} = 1 - \frac{x^2}{3!} + \frac{x^4}{5!} - \frac{x^5}{7!} + \ldots$, $s_1 = 1$ and $s_3 = 1 - \frac{x^3}{6}$; If L is the sum of the series representing $\frac{\sin x}{x}$, then $L - s_1 = \frac{\sin x}{x} - 1 < 0$ and $L - s_3 = \frac{\sin x}{x} - \left(1 - \frac{x^3}{6}\right) > 0$, by the Alternating Series Estimation Theorem. $\therefore 1 - \frac{x^3}{6} < \frac{\sin x}{x} < 1$

42. $\cos x = 1 - \frac{x^2}{2!} + \frac{x^4}{4!} - \frac{x^6}{6!} + \ldots \Rightarrow 1 - \cos x = \frac{x^2}{2!} - \frac{x^4}{4!} + \frac{x^6}{6!} - \frac{x^8}{8!} \ldots = \frac{1 - \cos x}{x^2} = \frac{1}{2} - \frac{x^2}{4!} + \frac{x^6}{6!} - \frac{x^8}{8!} + \ldots$

If L is the sum of the series representing $\frac{1 - \cos x}{x^2}$, $L - s_1 = \frac{1 - \cos x}{x^2} - \frac{1}{2} < 0$ and $\frac{1 - \cos x}{x^2} - \left(\frac{1}{2} - \frac{x^4}{4!}\right) > 0$, by the Alternating Series Estimation Theorem. $\therefore \frac{1}{2} - \frac{x^4}{4!} < \frac{1 - \cos x}{x^2} < \frac{1}{2}$

43. a) $e^{-i\pi} = \cos(-\pi) + i\sin(-\pi) = -1 + i(0) = -1$

b) $e^{i\pi/4} = \cos\left(\frac{\pi}{4}\right) + i\sin\left(\frac{\pi}{4}\right) = \frac{1}{\sqrt{2}} + \frac{i}{\sqrt{2}} = \left(\frac{1}{\sqrt{2}}\right)(1 + i)$

c) $e^{-i\pi/2} = \cos\left(-\frac{\pi}{2}\right) + i\sin\left(-\frac{\pi}{2}\right) = 0 + i(-1) = -i$

44. $e^{i\theta} = \cos\theta + i\sin\theta \Rightarrow e^{-i\theta} = e^{i(-\theta)} = \cos(-\theta) + i\sin(-\theta) = \cos\theta - i\sin\theta$; $e^{i\theta} + e^{-i\theta} = \cos\theta + i\sin\theta + \cos\theta - i\sin\theta = 2\cos\theta \Rightarrow \cos\theta = \frac{e^{i\theta} + e^{-i\theta}}{2}$; $e^{i\theta} - e^{-i\theta} = \cos\theta + i\sin\theta - (\cos\theta - i\sin\theta) = 2i\sin\theta \Rightarrow \sin\theta = \frac{e^{i\theta} - e^{-i\theta}}{2i}$

45. $e^x = 1+x+\frac{x^2}{2!}+\frac{x^3}{3!}+\frac{x^4}{4!}+\ldots \Rightarrow e^{i\theta} = 1+i\theta+\frac{(i\theta)^2}{2!}+\frac{(i\theta)^3}{3!}+\frac{(i\theta)^4}{4!}+\ldots$ and $e^{-i\theta} =$

$1-i\theta+\frac{(-i\theta)^2}{2!}+\frac{(-i\theta)^3}{3!}+\frac{(-i\theta)^4}{4!}+\ldots = 1-i\theta+\frac{(i\theta)^2}{2!}-\frac{(i\theta)^3}{3!}+\frac{(i\theta)^4}{4!}-\ldots,$

$$\frac{e^{i\theta}+e^{-i\theta}}{2} = \frac{\left(1+i\theta+\frac{(i\theta)^2}{2!}+\frac{(i\theta)^3}{3!}+\frac{(i\theta)^4}{4!}+\ldots\right)+\left(1-i\theta+\frac{(i\theta)^2}{2!}-\frac{(i\theta)^3}{3!}+\frac{(i\theta)^4}{4!}-\ldots\right)}{2} =$$

$1-\frac{\theta}{2!}+\frac{\theta^4}{4!}-\frac{\theta^6}{6!}+\ldots = \cos\theta;$

$$\frac{e^{i\theta}-e^{-i\theta}}{2} = \frac{\left(1+i\theta+\frac{(i\theta)^2}{2!}+\frac{(i\theta)^3}{3!}+\frac{(i\theta)^4}{4!}+\ldots\right)-\left(1-i\theta+\frac{(i\theta)^2}{2!}-\frac{(i\theta)^3}{3!}+\frac{(i\theta)^4}{4!}-\ldots\right)}{2i} =$$

$\theta-\frac{\theta^3}{3!}+\frac{\theta^5}{5!}-\frac{\theta^7}{7!}+\ldots = \sin\theta$

46. $\frac{de^{(a+ib)x}}{dx} = \frac{de^{ax}[\cos bx + i\sin bx]}{dx} = ae^{ax}[\cos bx + i\sin bx] + be^{ax}[-\sin bx + i\cos bx] =$

$ae^{ax}[\cos bx + i\sin bx] + ibe^{ax}[\cos bx + i\sin bx] = (a+ib)e^{ax}[\cos bx + i\sin bx] = (a+ib)e^{(a+ib)x}$

47. $\frac{a-bi}{a^2+b^2}e^{(a+bi)x} + C_1 + iC_2 = \left(\frac{a-bi}{a^2+b^2}\right)e^{ax}(\cos bx + i\sin bx) + C_1 + iC_2 =$

$\frac{e^{ax}}{a^2+b^2}[a\cos bx + ia\sin bx - ib\cos bx + b\sin bx] + C_1 + iC_2 =$

$\frac{e^{ax}}{a^2+b^2}\left[(a\cos bx + b\sin bx) + (a\sin bx - b\cos bx)i\right] + C_1 + iC_2 =$

$\frac{e^{ax}[a\cos bx + b\sin bx]}{a^2+b^2} + C_1 + \frac{e^{ax}[a\sin bx - b\cos bx]}{a^2+b^2} + iC_2$, $e^{(a+bi)x} = e^{ax}e^{ibx} =$

$e^{ax}[\cos bx + i\sin bx] = e^{ax}\cos bx + ie^{ax}\sin bx$ and given that $\int e^{(a+bi)x}\,dx =$

$\frac{a-bi}{a^2+b^2}e^{(a+bi)x} + C_1 + iC_2$ we may conclude:

$$\int e^{ax}\cos bx\,dx = \frac{e^{ax}[a\cos bx + b\sin bx]}{a^2+b^2} + C_1 \text{ and}$$

$$\int e^{ax}\sin bx\,dx = \frac{e^{ax}[a\sin bx - b\cos bx]}{a^2+b^2} + iC_2$$

9.8 FURTHER EXAMPLES OF TAYLOR SERIES

1. $f(x) = \cos x \Rightarrow f(\pi/3) = 0.5,\ f'(\pi/3) = -\frac{\sqrt{3}}{2},\ f''(\pi/3) = -0.5,\ f'''(\pi/3) = \frac{\sqrt{3}}{2},$
$f^{(4)}(\pi/3) = 0.5;\ \cos x = \frac{1}{2} - \frac{\sqrt{3}}{2}(x - \pi/3) - \frac{1}{4}(x - \pi/3)^2 + \frac{\sqrt{3}}{12}(x - \pi/3)^3 - \ldots$

2. $f(x) = \sin x \Rightarrow f(2\pi) = 0,\ f'(2\pi) = 1,\ f''(2\pi) = 0,\ f'''(2\pi) = -1,\ f^{(4)}(2\pi) = 0,\ f^{(5)}(2\pi) = 1,\ f^{(6)}(2\pi) = 0,$
$f^{(7)}(2\pi) = -1;\ \sin x = (x - 2\pi) - \frac{(x-2\pi)^3}{3!} + \frac{(x-2\pi)^5}{5!} - \frac{(x-2\pi)^7}{7!} + \ldots$

3. $e^x = 1 + x + \frac{x^2}{2!} + \frac{x^3}{3!} + \ldots$

4. $f(x) = \ln x \Rightarrow f(1) = 0,\ f'(1) = 1,\ f''(1) = -1,\ f'''(1) = 2,\ f^{(4)}(1) = -6;\ \ln x = (x-1) - \frac{(x-1)^2}{2} +$
$\frac{(x-1)^3}{3} - \frac{(x-1)^4}{4} + \ldots$

5. $f(x) = \cos x \Rightarrow f(22\pi) = 1,\ f'(22\pi) = 0,\ f''(22\pi) = -1,\ f'''(22\pi) = 0, f^{(4)}(22\pi) = 1,\ f^{(5)}(22\pi) = 0,$
$f^{(6)}(22\pi) = -1;\ \cos x = 1 - \frac{1}{2}(x - 22\pi)^2 + \frac{1}{4!}(x - 22\pi)^4 - \frac{1}{6!}(x - 22\pi)^6 + \ldots$

6. $f(x) = \tan^{-1} x \Rightarrow f(1) = \frac{\pi}{4},\ f'(1) = \frac{1}{2},\ f''(1) = -\frac{1}{2},\ f'''(1) = \frac{1}{2};$
$\tan^{-1} x = \frac{\pi}{4} + \frac{(x-1)}{2} - \frac{(x-1)^2}{4} + \frac{(x-1)^3}{12} + \ldots$

7. $\int_0^{0.1} \frac{\sin x}{x}\,dx = \int_0^{0.1} 1 - \frac{x^2}{3!} + \frac{x^4}{5!} - \frac{x^5}{7!} + \ldots\,dx = \left[x - \frac{x^3}{3\cdot 3!} + \frac{x^5}{5\cdot 5!} - \frac{x^7}{7\cdot 7!} + \ldots\right]_0^{0.1} \approx$
$\left[x - \frac{x^3}{3\cdot 3!} + \frac{x^5}{5\cdot 5!}\right]_0^{0.1} \approx 0.099944461$

8. $\int_0^{0.1} \exp(-x^2)\,dx = \int_0^{0.1} 1 - x^2 + \frac{x^4}{2!} - \frac{x^6}{3!} + \frac{x^8}{4!} - \ldots\,dx = \left[x - \frac{x^3}{3} + \frac{x^5}{10} + \frac{x^7}{42} + \ldots\right]_0^{0.1} \approx$
$\left[x - \frac{x^3}{3} + \frac{x^5}{10} - \frac{x^7}{42}\right]_0^{0.1} \approx 0.099667666$

9. $(1 + x^4)^{1/2} = (1)^{1/2} + \frac{1/2}{1}(1)^{-1/2}(x^4) + \frac{(1/2)(-1/2)}{2!}(1)^{-3/2}(x^4)^2 + \frac{(1/2)(-1/2)(-3/2)}{3!}(1)^{-5/2}(x^4)^3 +$
$\frac{(1/2)(-1/2)(-3/2)(-5/2)}{4!}(1)^{-7/2}(x^4)^4 + \ldots = 1 + \frac{x^4}{2} - \frac{x^8}{8} + \frac{x^{12}}{16} - \frac{5x^{16}}{128} + \ldots;$
$\int_0^{0.1} 1 + \frac{x^4}{2} - \frac{x^8}{8} + \frac{x^{12}}{16} - \frac{5x^{16}}{128} + \ldots\,dx = \left[x + \frac{x^5}{10} - \frac{x^9}{72} + \frac{x^{13}}{208} - \frac{5x^{17}}{2176} + \ldots\right]_0^{0.1} \approx 0.100001$

10. From exercise 42 in section 9.7 we have $\frac{1-\cos x}{x^2} = \frac{1}{2} - \frac{x^2}{4!} + \frac{x^4}{6!} - \frac{x^6}{8!} + \frac{x^8}{10!} - \ldots$.

$\int_0^1 \frac{1-\cos x}{x^2}\,dx = \int_0^1 \frac{1}{2} - \frac{x^2}{4!} + \frac{x^4}{6!} - \frac{x^6}{8!} + \frac{x^8}{10!} - \ldots\,dx =$

$\left[\frac{x}{2} - \frac{x^3}{3 \cdot 4!} + \frac{x^5}{5 \cdot 6!} - \frac{x^7}{7 \cdot 8!} + \frac{x^9}{9 \cdot 10!} - \ldots\right]_0^1 \approx 0.486385345$

11. $\ln\left(\frac{1+x}{1-x}\right) = \ln(1+x) - \ln(1-x) = \left(x - \frac{x^2}{2} + \frac{x^3}{3} - \frac{x^4}{4} + \ldots\right) - \left(-x - \frac{x^2}{2} - \frac{x^3}{3} - \frac{x^4}{4} - \ldots\right) =$
$2\left(x + \frac{x^3}{3} + \frac{x^5}{5} + \ldots\right)$

12. $\ln(1+x) = x - \frac{x^2}{2} + \frac{x^3}{3} - \frac{x^4}{4} + \ldots + \frac{(-1)^{n-1} x^n}{n} + \ldots$ and the $|\text{error}| = \left|\frac{(-1)^{n-1} x^n}{n}\right| = \frac{1}{n\,10^n}$,

when $x = 0.1$; $\frac{1}{n\,10^n} < \frac{1}{10^8} \Rightarrow n\,10^n > 10^8$ when $n \geq 8 \Rightarrow$ 8 terms

13. $\tan^{-1}x = x - \frac{x^3}{3} + \frac{x^5}{5} - \frac{x^7}{7} + \frac{x^9}{9} - \ldots + \frac{(-1)^{n-1} x^{2n-1}}{2n-1} + \ldots$ and the $|\text{error}| =$

$\left|\frac{(-1)^{n-1} x^{2n-1}}{2n-1}\right| = \frac{1}{2n-1}$, when $x = 1$; $\frac{1}{2n-1} < \frac{1}{10^3} \Rightarrow n > \frac{1001}{2} = 500.5 \Rightarrow$ the first term not used is $501^{st} \Rightarrow$ we must use 500 terms

14. $\tan^{-1}x = x - \frac{x^3}{3} + \frac{x^5}{5} - \frac{x^7}{7} + \frac{x^9}{9} - \ldots + \frac{(-1)^{n-1} x^{2n-1}}{2n-1} + \ldots$, $\lim_{n \to \infty} \left|\frac{x^{2n+1}}{2n+1} \frac{2n-1}{x^{2n-1}}\right| =$

$x^2 \lim_{n \to \infty} \left|\frac{2n-1}{2n+1}\right| = x^2 \Rightarrow \tan^{-1}x$ converges for $|x| < 1$; when $x = -1$ we have $\sum_{n=1}^{\infty} \frac{(-1)^n}{2n-1}$, a convergent series; when $x = 1$ we have $\sum_{n=1}^{\infty} \frac{(-1)^{n+1}}{2n-1}$, a convergent series $\Rightarrow$ the series representing $\tan^{-1}x$ diverges for $|x| > 1$

15. $\tan^{-1}x = x - \frac{x^3}{3} + \frac{x^5}{5} - \frac{x^7}{7} + \frac{x^9}{9} - \ldots + \frac{(-1)^{n-1} x^{2n-1}}{2n-1} + \ldots$; when the series representing $48\tan^{-1}\left(\frac{1}{18}\right)$ has an error of magnitude less than 10^{-6}, then the series representing $48\tan^{-1}\left(\frac{1}{18}\right) + 32\tan^{-1}\left(\frac{1}{57}\right) - 20\tan^{-1}\left(\frac{1}{239}\right)$ will also have an error of magnitude less than 10^{-6}; $\frac{\left(\frac{1}{18}\right)^{2n-1}}{2n-1} < \frac{1}{10^6} \Rightarrow n \geq 3 \Rightarrow$ 3 terms

PRACTICE EXERCISES

1. $\lim_{n\to\infty} a_n = \lim_{n\to\infty}\left(1+\frac{(-1)^n}{n}\right) = 1$, converges to 1

2. $\lim_{n\to\infty} a_n = \lim_{n\to\infty}\left(\frac{1-2^n}{2^n}\right) = \lim_{n\to\infty}\left(\frac{1}{2^n}-1\right) = -1$, converges to -1

3. $a_1 = \cos\left(\frac{\pi}{2}\right) = 0,\ a_2 = \cos(\pi) = -1,\ a_3 = \cos\left(\frac{3\pi}{2}\right) = 0,\ a_4 = \cos(2\pi) = 1, \ldots \Rightarrow$ the sequence diverges

4. $\lim_{n\to\infty} a_n = \lim_{n\to\infty}\left(\frac{3}{n}\right)^{1/n} = \frac{\lim_{n\to\infty} 3^{1/n}}{\lim_{n\to\infty} n^{1/n}} = \frac{1}{1} = 1$, converges to 1

5. $\lim_{n\to\infty} a_n = \lim_{n\to\infty}\frac{\ln n^2}{n} = 2\lim_{n\to\infty}\frac{1/n}{1} = 0$, converges to 0

6. $\lim_{n\to\infty} a_n = \lim_{n\to\infty}\left(\frac{n+5}{n}\right)^n = \lim_{n\to\infty}\left(1+\frac{1}{n}\right)^n = e^5$, converges to e^5

7. $\lim_{n\to\infty} a_n = \lim_{n\to\infty}\left(\frac{3^n}{n}\right)^{1/n} = \frac{3}{\lim_{n\to\infty} n^{1/n}} = \frac{3}{1} = 3$, conveges to 3

8. $\lim_{n\to\infty}\frac{1}{3^{2n-1}} = 0$, converges to 0

9. $\lim_{n\to\infty} a_n = \lim_{n\to\infty}\frac{(-4)^n}{n!} = 0$, by table 9.1, converges to 0

10. $\lim_{n\to\infty} a_n = \lim_{n\to\infty}\frac{\ln(2n+1)}{n} = \lim_{n\to\infty}\frac{\frac{2}{n+1}}{1} = 0$, converges to 0

11. $\lim_{n\to\infty} a_n = \lim_{n\to\infty}\frac{(n+1)!}{n!} = \lim_{n\to\infty} n+1 = \infty$, the sequence diverges

12. $\lim_{n\to\infty} a_n = \lim_{n\to\infty}\frac{n^2-n}{2n^2+n} = \frac{1}{2}$, converges to $\frac{1}{2}$

13. $\sum_{n=1}^{\infty}\ln\left(\frac{n}{n+1}\right) = \sum_{n=1}^{\infty}\ln(n)-\ln(n+1) \Rightarrow s_n = \big(\ln(1)-\ln(2)\big)+\big(\ln(2)-\ln(3)\big)+\big(\ln(3)-\ln(4)\big)+$
$\ldots+\big(\ln(n-1)-\ln(n)\big)+\big(\ln(n)-\ln(n+1)\big) = \ln(1)-\ln(n+1) = -\ln(n+1) \Rightarrow$
$\lim_{n\to\infty} s_n = -\infty$, $\Rightarrow$ the series diverges

14. $\sum_{n=2}^{\infty} \frac{-2}{n(n+1)} = \sum_{n=1}^{\infty} \frac{2}{n} - \frac{2}{n+1} \Rightarrow s_n = 2\left(\frac{1}{2} - \frac{1}{3} + \frac{1}{3} - \frac{1}{4} + \frac{1}{4} - \frac{1}{5} + \ldots + \frac{1}{n-1} - \frac{1}{n} + \frac{1}{n} - \frac{1}{n+1}\right) =$
$2\left(\frac{1}{2} - \frac{1}{n+1}\right) \Rightarrow \lim_{n \to \infty} s_n = \lim_{n \to \infty} 2\left(\frac{1}{2} - \frac{1}{n+1}\right) = 1 \Rightarrow$ the series converges to 1

15. $\sum_{n=1}^{\infty} e^{-n} = \sum_{n=1}^{\infty} \left(\frac{1}{e}\right)^n = 1 + \frac{1}{e} + \left(\frac{1}{e}\right)^2 + \left(\frac{1}{e}\right)^3 + \ldots = \frac{1}{1 - \frac{1}{e}} = \frac{e}{e-1}$, a convergent geometric series

16. $\sum_{n=1}^{\infty} (-1)^n \frac{3}{4^n} = 3\left(\sum_{n=1}^{\infty} \left(\frac{-1}{4}\right)^n\right) = 3\left(\frac{-\frac{1}{4}}{1 + \frac{1}{4}}\right) = -\frac{3}{5}$, a convergent geometric series

17. diverges, a p–series where $p = \frac{1}{2}$

18. $\sum_{n=1}^{\infty} \frac{-5}{n} = -5\left(\sum_{n=1}^{\infty} \frac{1}{n}\right)$ diverges, by the corollaries of Theorem 5

19. $f(x) = \frac{1}{x^{1/2}} \Rightarrow f'(x) = -\frac{1}{2x^{1/2}} < 0 \Rightarrow f(x)$ is decreasing, $\lim_{n \to \infty} a_n = \lim_{n \to \infty} \frac{(-1)^n}{\sqrt{n}} = 0$ and $a_{n+1} > a_n \Rightarrow$
$\sum_{n=1}^{\infty} \frac{(-1)^n}{\sqrt{n}}$ converges, by the Alternating Series Theorem; the series $\sum_{n=1}^{\infty} \frac{1}{\sqrt{n}}$ diverges $\Rightarrow$
the given series converges conditionally

20. converges absolutely; compare with $\sum_{n=1}^{\infty} \frac{1}{n^3}$, since $\frac{1}{2n^3} < \frac{1}{n^3}$ for $n \geq 1$

21. the given series does not converge absolutlely, since $\frac{1}{\ln(n+1)} > \frac{1}{n+1}$ and $\sum_{n=1}^{\infty} \frac{1}{n+1}$ diverges;
$f(x) = \frac{1}{\ln(x+1)} \Rightarrow f'(x) = -\frac{1}{(\ln(x+1))^2(x+1)} < 0 \Rightarrow f(x)$ is decreasing, $\lim_{n \to \infty} a_n =$
$\lim_{n \to \infty} \frac{1}{\ln(n+1)} = 0$ and $a_{n+1} > a_n \Rightarrow$ the given series converges conditionally, by
the Alternating Series Test

22. $\int_2^{\infty} \frac{1}{x(\ln x)^2} dx = \lim_{t \to \infty} \int_2^t \frac{1}{x(\ln x)^2} dx = \lim_{t \to \infty} \left[(\ln x)^{-1}\right]_2^t = \lim_{t \to \infty} \left(\frac{1}{\ln t} - \frac{1}{\ln 2}\right) = -\frac{1}{\ln 2} \Rightarrow$
the series converges by the Integral Test

23. $\lim_{n \to \infty} \dfrac{\frac{1}{n\sqrt{n^2+1}}}{\frac{1}{n^2}} = \sqrt{\lim_{n \to \infty} \dfrac{n^2}{n^2+1}} = \sqrt{1} = 1 \Rightarrow$ converges absolutely, by the Limit Comparison Test

24. the given series does not converge absolutely, since $\lim_{n \to \infty} \dfrac{\frac{3n^2}{n^3+1}}{\frac{1}{n}} = \lim_{n \to \infty} \dfrac{3n^3}{n^3+1} = 3$, by the Limit Comparison Test; $f(x) = \dfrac{3x^2}{x^3+1} \Rightarrow f'(x) = \dfrac{3x\left(2 - x^2\right)}{\left(x^2+1\right)^2} < 0$ when $x \geq 2$, $\lim_{n \to \infty} \dfrac{3n^2}{n^3+1} = 0$ and $a_{n+1} > a_n \Rightarrow$ the given series converges conditionally, by the Alternating Series Test

25. converges, by the Ratio Test, since $\lim_{n \to \infty} \left| \dfrac{n+2}{(n+1)!} \, \dfrac{n!}{n+1} \right| = \lim_{n \to \infty} \dfrac{n+2}{(n+1)^2} = 0$

26. diverges, since $\lim_{n \to \infty} a_n = \lim_{n \to \infty} \dfrac{(-1)^n\left(n^2+1\right)}{2n^2+n-1} = \lim_{n \to \infty} \dfrac{n^2+1}{2n^2+n-1} \cdot \lim_{n \to \infty} (-1)^n = \dfrac{1}{2}\left(\lim_{n \to \infty} (-1)^n \right) \neq 0$

27. converges absolutely, by the Ratio Test, since $\lim_{n \to \infty} \left| \dfrac{3^{n+1}}{(n+1)!} \, \dfrac{n!}{3^n} \right| = \lim_{n \to \infty} \left| \dfrac{3}{n+1} \right| = 0$

28. converges absolutely, by the Root Test, since $\lim_{n \to \infty} \sqrt[n]{a_n} = \lim_{n \to \infty} \sqrt[n]{\dfrac{2^n 3^n}{n^n}} = \lim_{n \to \infty} \dfrac{6}{n} = 0$

29. $\lim_{n \to \infty} \left| \dfrac{(x+2)^{n+1}}{3^{n+1}(n+1)} \, \dfrac{3^n n}{(x+2)^n} \right| < 1 \Rightarrow \dfrac{|x+2|}{3} \lim_{n \to \infty} \left| \dfrac{n}{n+1} \right| < 1 \Rightarrow -5 < x < 1$; when $x = -5$ we have $\displaystyle\sum_{n=1}^{\infty} \frac{(-1)^n}{n}$, a convergent series; when $x = 1$ we have $\displaystyle\sum_{n=1}^{\infty} \frac{1}{n}$, a divergent series $\therefore$

a) $-5 \leq x < 1$ b) $-5 < x < 1$

30. $\lim_{n \to \infty} \left| \dfrac{(x-1)^{2n}}{(2n+1)!} \, \dfrac{(2n-1)!}{(x-1)^{2n-2}} \right| < 1 \Rightarrow (x-1)^2 \lim_{n \to \infty} \left| \dfrac{1}{(2n+1)(2n)} \right| < 1 \Rightarrow (x-1)^2(0) < 1 \Rightarrow$

a) For all x b) For all x

31. $\lim_{n\to\infty} \sqrt[n]{a_n} < 1 \Rightarrow \lim_{n\to\infty} \sqrt[n]{\frac{x^n}{n^n}} < 1 \Rightarrow \lim_{n\to\infty} \frac{|x|}{n} < 1 \Rightarrow |x|(0) < 1 \Rightarrow$

a) For all x b) For all x

32. $\lim_{n\to\infty} \left| \frac{n+2}{2n+3} \frac{(x-1)^{n+1}}{2^{n+1}} \frac{2n+1}{n+1} \frac{2^n}{(x-1)^n} \right| < 1 \Rightarrow \frac{|x-1|}{2} \lim_{n\to\infty} \left| \frac{(n+2)(2n+1)}{(2n+3)(n+1)} \right| < 1 \Rightarrow -1 < x < 3;$

when $x = -1$ we have $\sum_{n=1}^{\infty} \frac{(-1)^n(n+1)}{2n+1}$, a divergent series, since $\lim_{n\to\infty} a_n = \lim_{n\to\infty} \frac{(-1)^n(n+1)}{2n+1} =$

$\lim_{n\to\infty} \frac{n+1}{2n+1} \cdot \lim_{n\to\infty} (-1)^n = \frac{1}{2} \lim_{n\to\infty} (-1)^n \neq 0$; when $x = 3$ we have $\sum_{n=1}^{\infty} \frac{n+1}{2n+1}$, a divergent series,

since $\lim_{n\to\infty} a_n = \lim_{n\to\infty} \frac{n+1}{2n+1} = \frac{1}{2} \neq 0 \therefore$

a) $-1 < x < 3$ b) $-1 < x < 3$

33. $\lim_{n\to\infty} \left| \frac{(x-1)^{n+1}}{(n+1)^2} \frac{n^2}{(x-1)^n} \right| < 1 \Rightarrow |x-1| \lim_{n\to\infty} \left| \frac{n^2}{n^2+2n+1} \right| < 1 \Rightarrow 0 < x < 2$; when $x = 0$ we have

$\sum_{n=1}^{\infty} \frac{-1}{n^2}$, a convergent series; when $x = 2$ we have $\sum_{n=1}^{\infty} \frac{(-1)^{n-1}}{n}$, a convergent series $\therefore$

a) $0 \le x \le 2$ b) $0 \le x \le 2$

34. $\lim_{n\to\infty} \left| \frac{x^{n+1}}{\sqrt{n+1}} \frac{\sqrt{n}}{x^n} \right| < 1 \Rightarrow |x| \sqrt{\lim_{n\to\infty} \frac{n}{n+1}} < 1 \Rightarrow -1 < x < 1$; when $x = -1$ we have

$\sum_{n=1}^{\infty} \frac{(-1)^n}{\sqrt{n}}$, a convergent series; when $x = 1$ we have $\sum_{n=1}^{\infty} \frac{1}{\sqrt{n}}$, a divergent series $\therefore$

a) $-1 \le x < 1$ b) $-1 < x < 1$

35. The given series is in the form $1 - x + x^2 - x^3 + \ldots + (-x)^n + \ldots = \frac{1}{1+x}$, where $x = \frac{1}{4}$. The sum is $\frac{1}{1 + 1/4} = \frac{4}{5}$.

36. The given series is in the form $x - \frac{x^2}{2} + \frac{x^3}{3} - \ldots + (-1)^{n-1}\frac{x^n}{n} + \ldots = \ln(1 + x)$, where $x = \frac{2}{3}$.

The sum is $\ln\left(\frac{5}{3}\right) \approx 0.510825623$.

37. The given series is in the form $x - \frac{x^3}{3!} + \frac{x^5}{5!} - \ldots + (-1)^n \frac{x^{2n+1}}{(2n+1)!} + \ldots = \sin x$, where $x = \pi$.

The sum is $\sin\pi = 0$.

38. The given series is in the form $1 - \frac{x^2}{2!} + \frac{x^4}{4!} - \ldots + (-1)^n \frac{x^{2n}}{(2n)!} + \ldots = \cos x$, where $x = \frac{\pi}{3}$.

The sum is $\cos\frac{\pi}{3} = \frac{1}{2}$.

39. The given series is in the form $1 + x + \frac{x^2}{2!} + \frac{x^3}{3!} + \ldots + \frac{x^n}{n!} + \ldots = e^x$, where $x = \ln 2$.

The sum is $e^{\ln(2)} = 2$.

40. The given series is in the form $x - \frac{x^3}{3} + \frac{x^5}{5} - \ldots + (-1)^n \frac{x^{2n-1}}{2n-1} + \ldots = \tan^{-1} x$, where $x = \frac{1}{\sqrt{3}}$.

The sum is $\tan^{-1}\left(\frac{1}{\sqrt{3}}\right) = \frac{\pi}{6}$.

41. $f(x) = \sqrt{3 + x^2} = \left(3 + x^2\right)^{1/2} \Rightarrow f(-1) = 2,\ f'(-1) = -\frac{1}{2},\ f''(-1) = \frac{3}{8},\ f'''(-1) = \frac{9}{32}$;

$$\sqrt{3 + x^2} = 2 - \frac{(x+1)}{2 \cdot 1!} + \frac{3(x+1)^2}{2^3 \cdot 2!} + \frac{9(x+1)^3}{2^5 \cdot 3!} + \ldots$$

42. $f(x) = \frac{1}{1-x} = (1 - x)^{-1} \Rightarrow f(2) = -1,\ f'(2) = 1,\ f''(2) = -2,\ f'''(2) = 6$;

$$\frac{1}{1-x} = -1 + (x-2) - (x-2)^2 + (x-2)^3 - \ldots$$

43. $$\left(x - \frac{x^3}{3!} + \frac{x^5}{5!} - \ldots\right)\left(2 - \frac{2x^2}{2!} + \frac{2x^4}{4!} - \ldots\right) = (2)\,x + (0)\,x^2 + \left(-\frac{2}{2!} - \frac{2}{3!}\right)x^3 + (0)\,x^4 +$$
$$\left(\frac{2}{4!} + \frac{2}{3!\,2!} + \frac{2}{5!}\right)x^5 + (0)\,x^6 + \left(-\frac{2}{6!} - \frac{2}{4!\,3!} - \frac{2}{5!\,2!} - \frac{2}{7!}\right)x^7 + \ldots = 2x - \frac{(2x)^3}{3!} + \frac{(2x)^5}{5!} - \frac{(2x)^7}{7!} + \ldots$$

44. $\cos x = 1 - \frac{x^2}{2!} + \frac{x^4}{4!} - \frac{x^6}{6!} + \ldots \Rightarrow \cos 2x = 1 - \frac{(2x)^2}{2!} + \frac{(2x)^4}{4!} - \frac{(2x)^6}{6!} + \ldots \Rightarrow 1 - \cos 2x =$

$$\frac{(2x)^2}{2!} - \frac{(2x)^4}{4!} + \frac{(2x)^6}{6!} - \frac{(2x)^8}{8!} + \ldots \Rightarrow \frac{1 - \cos 2x}{2} = \frac{2x^2}{2!} - \frac{2^3x^4}{4!} + \frac{2^5x^6}{6!} - \frac{2^7x^8}{8!} + \ldots$$

45. $\displaystyle\int_0^{1/2} \exp\left(-x^3\right)\,dx = \int_0^{1/2} 1 - x^3 + \frac{x^6}{2!} - \frac{x^9}{3!} + \frac{x^{12}}{4!} + \ldots dx =$

$$\left[x - \frac{x^4}{4} + \frac{x^7}{7\cdot 2!} - \frac{x^{10}}{10\cdot 3!} + \frac{x^{13}}{13\cdot 4!} - \ldots\right]_0^{1/2} \approx \frac{1}{2} - \frac{1}{2^4\cdot 4} + \frac{1}{2^7\cdot 7\cdot 2!} - \frac{1}{2^{10}\cdot 10\cdot 3!} \approx 0.484916759$$

46. $\displaystyle\int_0^{1/2} \frac{\tan^{-1}x}{x}\,dx = \int_0^{1/2} 1 - \frac{x^2}{3} + \frac{x^4}{5} - \frac{x^6}{7} + \frac{x^8}{9} - \frac{x^{10}}{11} + \ldots\ dx =$

$$\left[x - \frac{x^3}{9} + \frac{x^5}{25} - \frac{x^7}{49} + \frac{x^9}{81} - \frac{x^{11}}{121} + \ldots\right]_0^{1/2} \approx \frac{1}{2} - \frac{1}{9\cdot 2^3} + \frac{1}{5^2\cdot 2^5} - \frac{1}{7^2\cdot 2^7} + \frac{1}{9^2\cdot 2^9} - \frac{1}{11^2\cdot 2^{11}} +$$

$$\frac{1}{13^2\cdot 2^{13}} - \frac{1}{15^2\cdot 2^{15}} + \frac{1}{17^2\cdot 2^{17}} \approx 0.487222362$$

47. a) the Maclaurin series is 0 + 0 + . . ., since $f^{(n)}(0) = 0$ for all n; this series conveges for all x; f(x) is zero only at x = 0

b) $f(x) = 0 + 0 + \ldots + 0 + R_n(x) \Rightarrow R_n(x) = \exp\left(-\frac{1}{x^2}\right)$

CHAPTER 10

PLANE CURVES AND POLAR COORDINATES

10.1 CONIC SECTIONS AND QUADRATIC EQUATIONS

1. $x = \frac{y^2}{8} \Rightarrow 4p = 8 \Rightarrow p = 2.$ $\therefore$ Focus is (2,0), directrix is $x = -2$.

2. $x = -\frac{y^2}{4} \Rightarrow 4p = 4 \Rightarrow p = 1.$ $\therefore$ Focus is (–1,0), directrix is $x = 1$.

3. $y = -\frac{x^2}{6} \Rightarrow 4p = 6 \Rightarrow p = \frac{3}{2}.$ $\therefore$ Focus is $(0,-\frac{3}{2})$, directrix is $y = \frac{3}{2}$.

4. $y = \frac{x^2}{2} \Rightarrow 4p = 2 \Rightarrow p = \frac{1}{2}.$ $\therefore$ Focus is $(0,\frac{1}{2})$, directrix is $y = -\frac{1}{2}$.

5. $\frac{x^2}{4} - \frac{y^2}{9} = 1 \Rightarrow c = \sqrt{4+9} = \sqrt{13} \Rightarrow$ Foci are $(\pm\sqrt{13},0)$. $e = \frac{c}{a} = \frac{\sqrt{13}}{2} \Rightarrow \frac{a}{e} = \frac{2}{\frac{\sqrt{13}}{2}} = \frac{4}{\sqrt{13}} \Rightarrow$

   Directrices are $x = \pm\frac{4}{\sqrt{13}}$. Asymptotes are $y = \pm\frac{3}{2}x$.

6. $\frac{x^2}{4} + \frac{y^2}{9} = 1 \Rightarrow c = \sqrt{9-4} = \sqrt{5} \Rightarrow$ Foci are $(0,\pm\sqrt{5})$. $e = \frac{c}{a} = \frac{\sqrt{5}}{3} \Rightarrow$ Directrices are $y = \pm\frac{9}{\sqrt{5}}$.

7. $\frac{x^2}{2} + y^2 = 1 \Rightarrow c = \sqrt{2-1} = 1 \Rightarrow$ Foci are $(\pm1,0)$. $e = \frac{c}{a} = \frac{1}{\sqrt{2}} \Rightarrow$ Directrices are $x = \pm\frac{\sqrt{2}}{\frac{1}{\sqrt{2}}} = \pm2$.

8. $\frac{y^2}{4} - x^2 = 1 \Rightarrow c = \sqrt{4+1} = \sqrt{5} \Rightarrow$ Foci are $(0, \pm\sqrt{5})$. $e = \frac{c}{a} = \frac{\sqrt{5}}{2} \Rightarrow$ Directrices are $y = \pm\frac{\sqrt{5}}{\frac{\sqrt{5}}{2}} = \pm 2$.

   Asymptotes are $y = \pm 2x$.

9. $16x^2 + 25y^2 = 400 \Rightarrow \frac{x^2}{25} + \frac{y^2}{16} = 1$

   $\Rightarrow c = \sqrt{a^2 - b^2} = \sqrt{25-16} = 3.$

   $e = \frac{c}{a} = \frac{3}{5}$

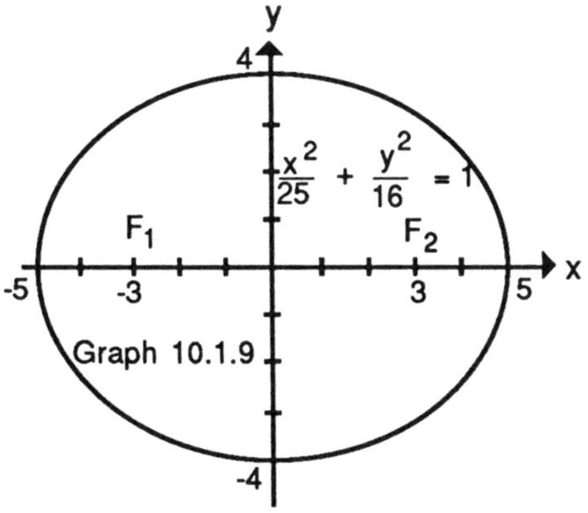

Graph 10.1.9

10. $7x^2 + 16y^2 = 112 \Rightarrow \frac{x^2}{16} + \frac{y^2}{7} = 1$

    $c = \sqrt{a^2 - b^2} = \sqrt{16-7} = 3$

    $e = \frac{c}{a} = \frac{3}{4}$

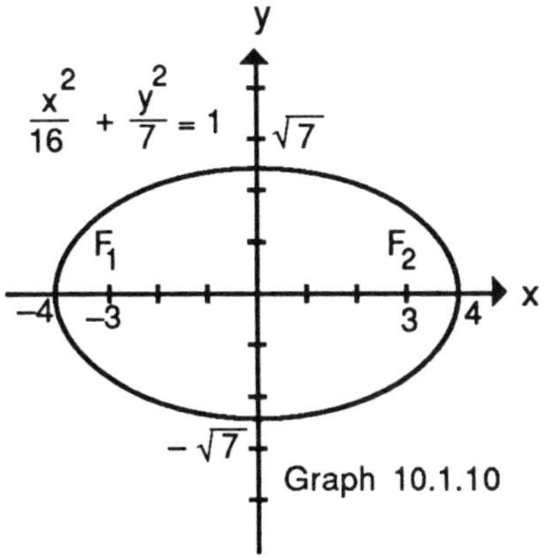

Graph 10.1.10

11. $2x^2 + y^2 = 2 \Rightarrow x^2 + \frac{y^2}{2} = 1$

$\Rightarrow c = \sqrt{a^2 - b^2} = \sqrt{2 - 1} = 1$

$e = \frac{c}{a} = \frac{1}{\sqrt{2}}$

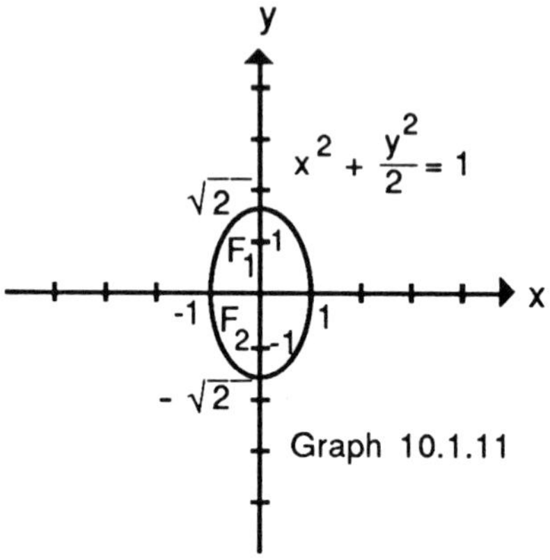

Graph 10.1.11

12. $2x^2 + y^2 = 4 \Rightarrow \frac{x^2}{2} + \frac{y^2}{4} = 1$

$\Rightarrow c = \sqrt{a^2 - b^2} = \sqrt{4 - 2} = \sqrt{2}$

$e = \frac{c}{a} = \frac{\sqrt{2}}{2}$

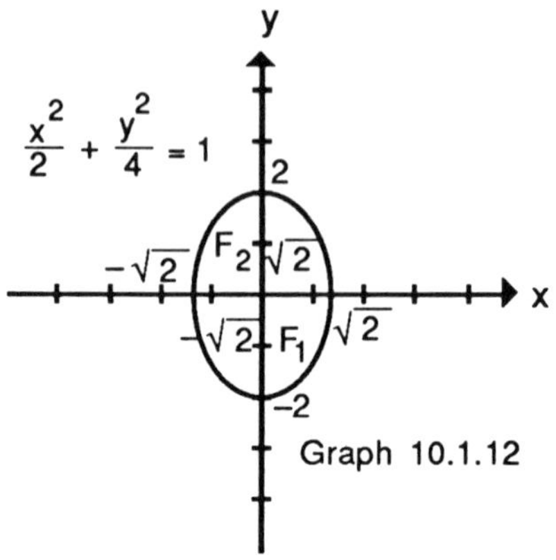

Graph 10.1.12

13. $3x^2 + 2y^2 = 6 \Rightarrow \frac{x^2}{2} + \frac{y^2}{3} = 1$

$\Rightarrow c = \sqrt{a^2 - b^2} = \sqrt{3 - 2} = 1$

$e = \frac{c}{a} = \frac{1}{\sqrt{3}}$

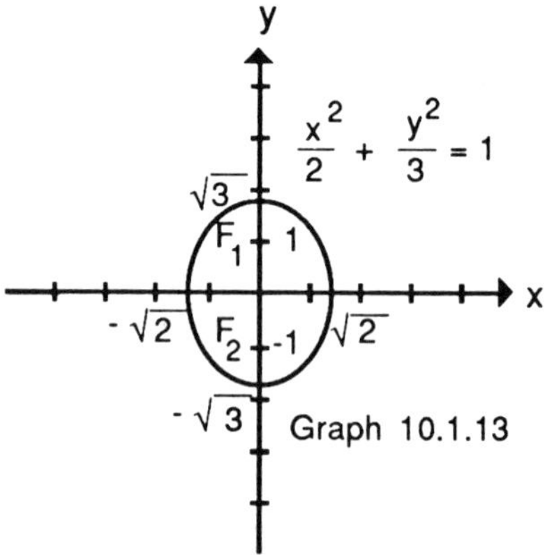

Graph 10.1.13

14. $9x^2 + 10y^2 = 90 \Rightarrow \frac{x^2}{10} + \frac{y^2}{9} = 1$

$c = \sqrt{a^2 - b^2} = \sqrt{10 - 9} = 1$

$e = \frac{c}{a} = \frac{1}{\sqrt{10}}$

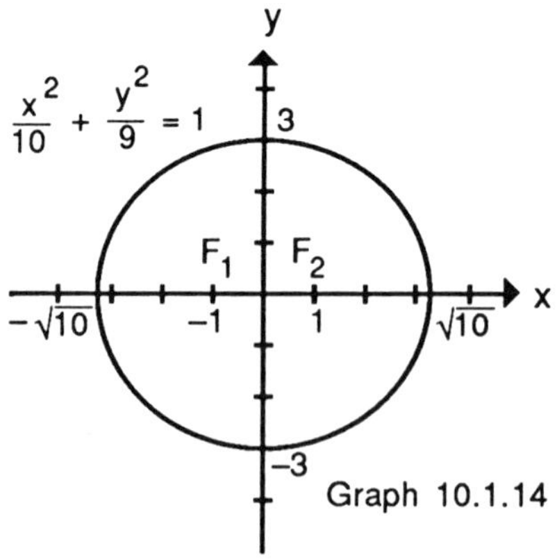

Graph 10.1.14

15. $6x^2 + 9y^2 = 54 \Rightarrow \frac{x^2}{9} + \frac{y^2}{6} = 1$

$c = \sqrt{a^2 - b^2} = \sqrt{9 - 6} = \sqrt{3}$

$e = \frac{c}{a} = \frac{\sqrt{3}}{3}$

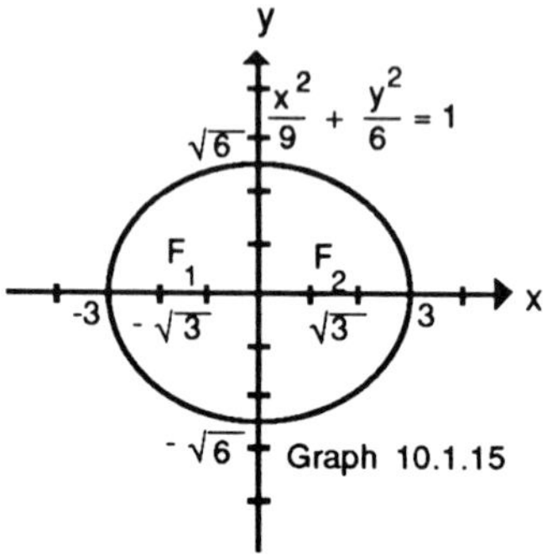

Graph 10.1.15

16. $169x^2 + 25y^2 = 4225 \Rightarrow \frac{x^2}{25} + \frac{y^2}{169} = 1$

$c = \sqrt{a^2 - b^2} = \sqrt{169 - 25} = 12$

$e = \frac{c}{a} = \frac{12}{13}$

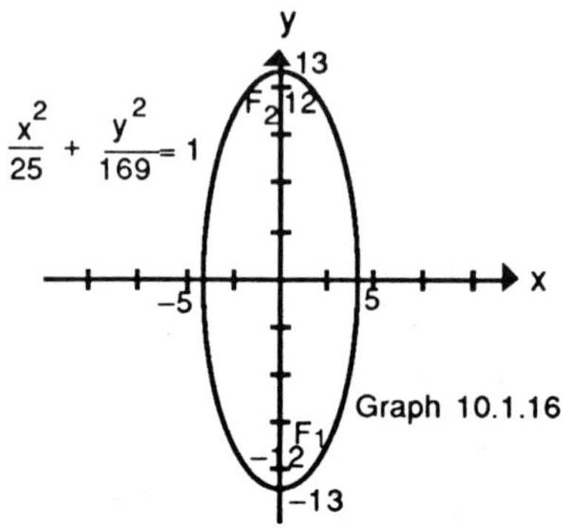

Graph 10.1.16

17. $x^2 - y^2 = 1 \Rightarrow c = \sqrt{a^2 + b^2} = \sqrt{1+1} = \sqrt{2}$

$e = \frac{c}{a} = \frac{\sqrt{2}}{1} = \sqrt{2}$

Asymptotes are $y = \pm x$

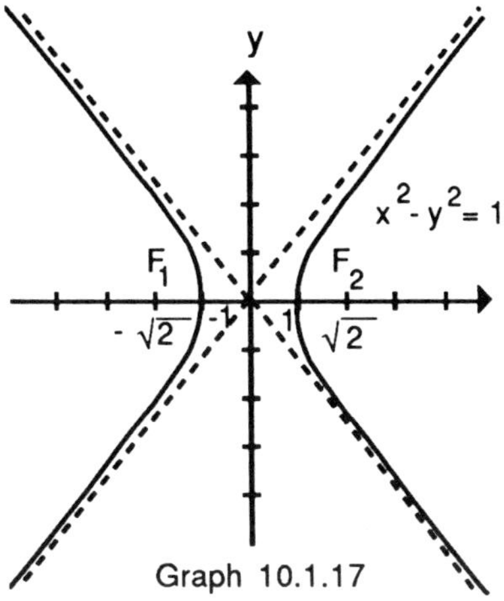

Graph 10.1.17

18. $9x^2 - 16y^2 = 144 \Rightarrow \frac{x^2}{16} - \frac{y^2}{9} = 1$

$\Rightarrow c = \sqrt{a^2 + b^2} = \sqrt{16 + 9} = 5$

$e = \frac{c}{a} = \frac{5}{4}$ Asymptotes are $y = \pm \frac{3}{4} x$

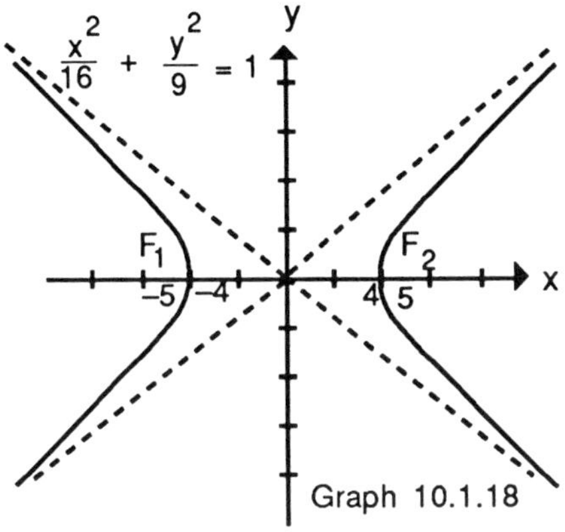

Graph 10.1.18

19. $y^2 - x^2 = 8 \Rightarrow \frac{y^2}{8} - \frac{x^2}{8} = 1$

$\Rightarrow c = \sqrt{a^2 + b^2} = \sqrt{8 + 8} = 4$

$e = \frac{c}{a} = \frac{4}{\sqrt{8}} = \sqrt{2}$ Asymptotes are $y = \pm x$.

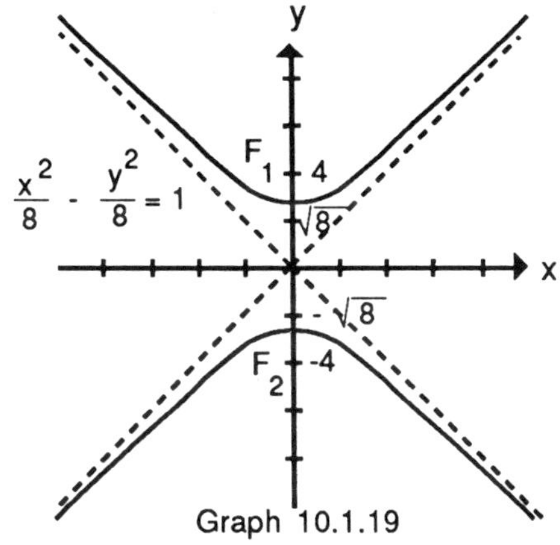

Graph 10.1.19

20. $y^2 - x^2 = 4 \Rightarrow \frac{y^2}{4} - \frac{x^2}{4} = 1$

$\Rightarrow c = \sqrt{a^2 + b^2} = \sqrt{4 + 4} = 2\sqrt{2}$

$e = \frac{c}{a} = \frac{2\sqrt{2}}{2} = \sqrt{2}$ Asymptotes are $y = \pm x$.

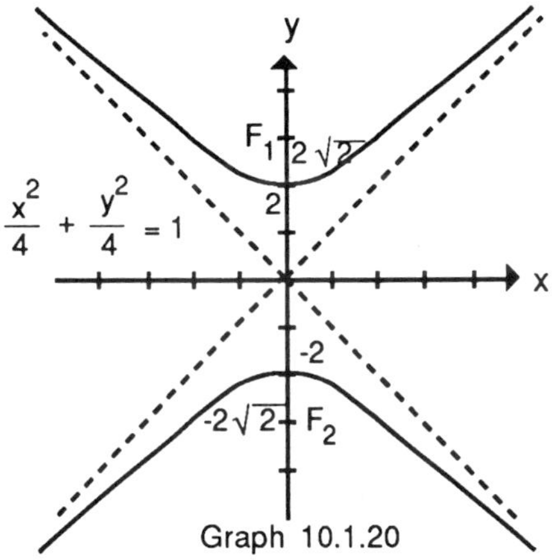

Graph 10.1.20

21. $8x^2 - y^2 = 16 \Rightarrow \frac{x^2}{2} - \frac{y^2}{8} = 1$

$\Rightarrow c = \sqrt{a^2 + b^2} = \sqrt{2 + 8} = \sqrt{10}$

$e = \frac{c}{a} = \frac{\sqrt{10}}{\sqrt{2}} = \sqrt{5}$

Asymptotes are $y = \pm 2x$.

(The graph is on the next page.)

22. $y^2 - 3x^2 = 3 \Rightarrow \frac{y^2}{3} - x^2 = 1$

$\Rightarrow c = \sqrt{a^2 + b^2} = \sqrt{3 + 1} = 2$

$e = \frac{c}{a} = \frac{2}{\sqrt{3}}$

Asymptotes are $y = \pm \sqrt{3}\, x$

(The graph is on the next page.)

21. (Graph)

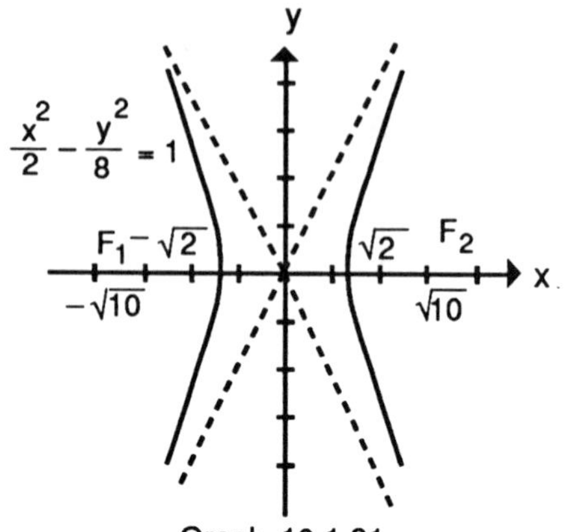

Graph 10.1.21

22. (Graph)

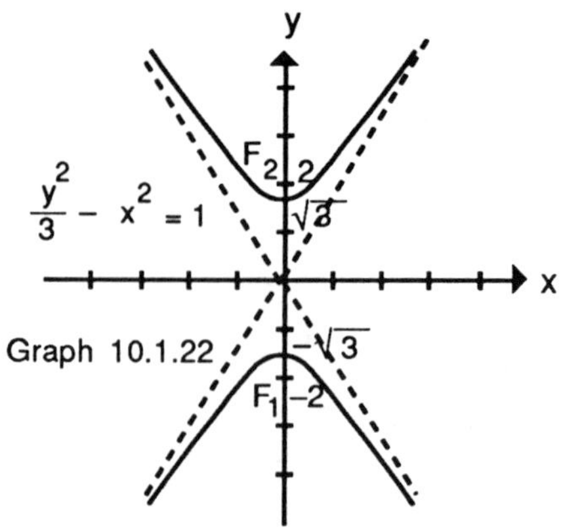

Graph 10.1.22

23. $8y^2 - 2x^2 = 16 \Rightarrow \frac{y^2}{2} - \frac{x^2}{8} = 1$

$\Rightarrow c = \sqrt{a^2 + b^2} = \sqrt{2 + 8} = \sqrt{10}$

$e = \frac{c}{a} = \frac{\sqrt{10}}{\sqrt{2}} = \sqrt{5}$ Asymptotes are $y = \pm \frac{x}{2}$.

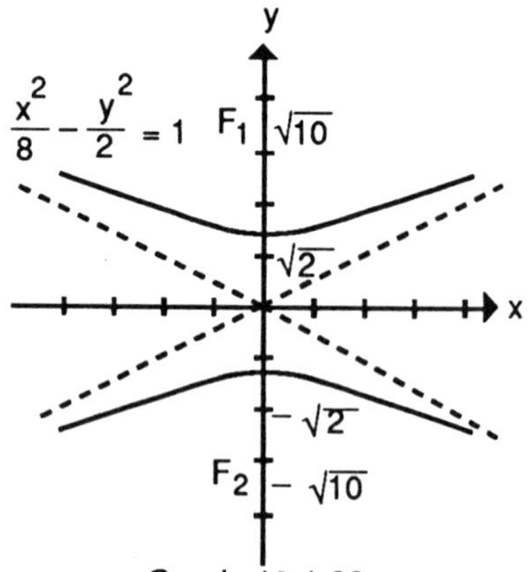

Graph 10.1.23

24. $64x^2 - 36y^2 = 2304 \Rightarrow \frac{x^2}{36} - \frac{y^2}{64} = 1$

$\Rightarrow c = \sqrt{a^2 + b^2} = \sqrt{36 + 64} = 10$

$e = \frac{c}{a} = \frac{10}{6} = \frac{5}{3}$ Asymptotes are $y = \pm \frac{4}{3}x$.

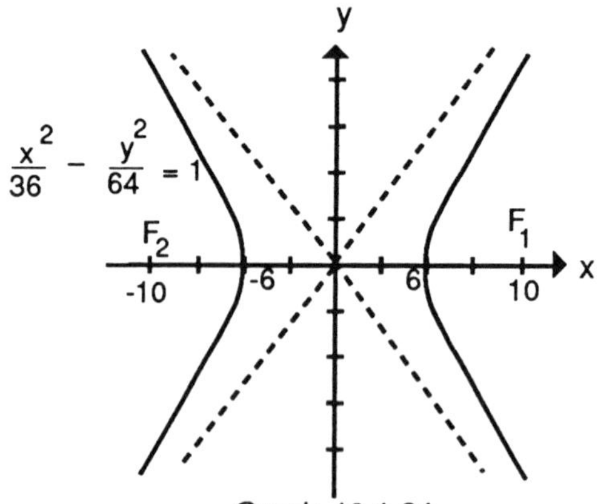

Graph 10.1.24

25. Volume of the Parabolic Solid: $V_1 = \int_0^{b/2} 2\pi x\left(\frac{4h}{b^2}\right)x^2\,dx = \frac{8\pi h}{b^2}\int_0^{b/2} x^3\,dx = \frac{8\pi h}{b^2}\left[\frac{x^4}{4}\right]_0^{b/2} = \frac{\pi h b^2}{8}$

Volume of the Cone: $V_2 = \frac{1}{3}\pi\left(\frac{b}{2}\right)^2 h = \frac{1}{3}\pi\left(\frac{b^2}{4}\right)h = \frac{\pi h b^2}{12}$

$\therefore V_1 = \frac{3}{2}V_2$

26. $y = \int \frac{w}{H}x\,dx = \frac{w}{H}\left(\frac{x^2}{2}\right) + C = \frac{wx^2}{2H} + C$. $y = 0$ when $x = 0 \Rightarrow 0 = \frac{w(0)^2}{2H} + C \Rightarrow C = 0$

$\therefore y = \frac{wx^2}{2H}$ is the equation of the cable's curve.

27. $e = \frac{4}{5} \Rightarrow$ Let c = 4 and a = 5. Then $c^2 = a^2 - b^2 \Rightarrow 16 = 25 - b^2 \Rightarrow$ $b^2 = 9 \Rightarrow b = 3$. $\therefore \frac{x^2}{25} + \frac{y^2}{9} = 1$

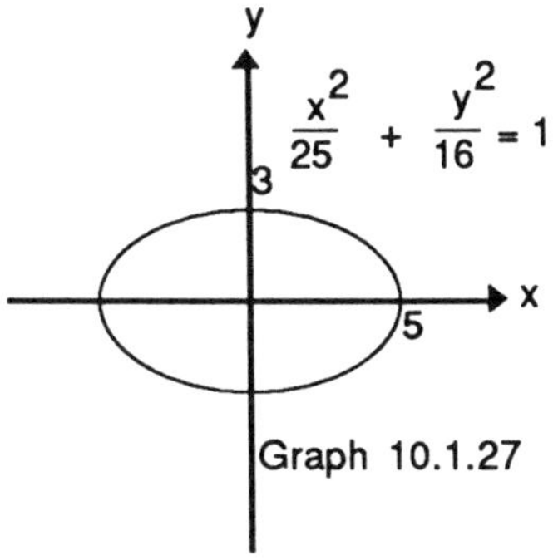

Graph 10.1.27

28. The eccentricity, e, for Pluto is 0.25 (Page 654 in the text) $\Rightarrow e = \frac{c}{a} = 0.25 = \frac{1}{4} \Rightarrow$ Let c = 1 and a = 4. Then $c^2 =$ $c^2 = a^2 - b^2 \Rightarrow 1 = 16 - b^2 \Rightarrow b^2 = 15 \Rightarrow b = \sqrt{15}$.

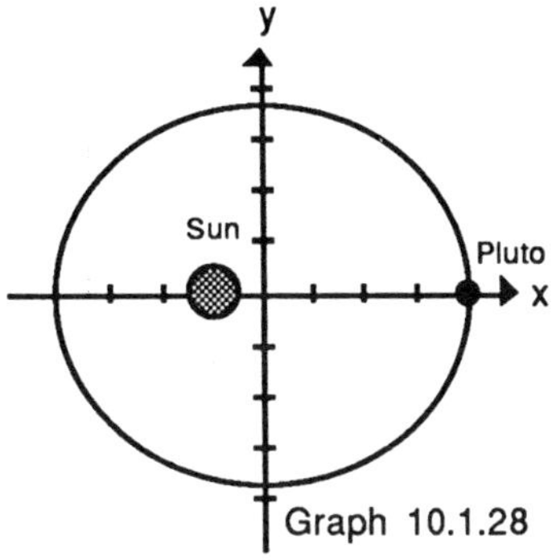

Graph 10.1.28

29. Let $y = \sqrt{1 - \frac{x^2}{4}}$ on the interval $0 \le x \le 2$. Then the area of the inscribed rectangle is given by

$A(x) = 2x\left(2\sqrt{1 - \frac{x^2}{4}}\right) = 4x\sqrt{1 - \frac{x^2}{4}}$ since the length is 2x and the height is 2y.

Then $A'(x) = 4\sqrt{1 - \frac{x^2}{4}} - \frac{x^2}{\sqrt{1 - \frac{x^2}{4}}}$. Let $A'(x) = 0 \Rightarrow 4\sqrt{1 - \frac{x^2}{4}} - \frac{x^2}{\sqrt{1 - \frac{x^2}{4}}} = 0 \Rightarrow 4\left(1 - \frac{x^2}{4}\right) - x^2 = 0$

$\Rightarrow x^2 = 2 \Rightarrow x = \sqrt{2}$ (Only the positive square root is in the interval.) Since $A(0) = 0$, $A(2) = 0$, $A(\sqrt{2}) = 4$ is the maximum area when the length is $2\sqrt{2}$ and the height is $\sqrt{2}$.

30. Let $y = \sqrt{16 - \frac{16}{9}x^2}$ on the interval $-3 \le x \le 3$. Since the plate is symmetric about the y–axis, $\bar{x} = 0$.

$\bar{y} = \frac{M_x}{M}$. $\therefore$ Look at the center of mass of a vertical strip $\Rightarrow (\tilde{x}, \tilde{y}) = \left(x, \frac{\sqrt{16 - \frac{16}{9}x^2}}{2}\right)$.

Length $= \sqrt{16 - \frac{16}{9}x^2}$, Width = dx $\Rightarrow$ Area = dA $= \sqrt{16 - \frac{16}{9}x^2}\ dx \Rightarrow$ Mass = dm = δdA =

$\delta\sqrt{16 - \frac{16}{9}x^2}\ dx$. Distance of center of mass to x–axis $= \tilde{y} = \frac{\sqrt{16 - \frac{16}{9}x^2}}{2}$.

Moment of the strip about the x–axis: $\tilde{y}\ dm = \frac{\sqrt{16 - \frac{16}{9}x^2}}{2}\left(\delta\sqrt{16 - \frac{16}{9}x^2}\right)dx = \delta\left(8 - \frac{8}{9}x^2\right)dx$.

$\therefore$ Moment of the Plate about the x–axis is $M_x = \int \tilde{y}\ dm = \int_{-3}^{3} \delta\left(8 - \frac{8}{9}x^2\right)dx = \delta\left[8x - \frac{8}{27}x^3\right]_{-3}^{3}$

$= 32\delta$.

30. (Continued from the previous page.)

Mass of the Plate = M = $\int_{-3}^{3} \delta\sqrt{16 - \frac{16}{9}x^2}\, dx = \int_{-3}^{3} 4\delta\sqrt{1 - \left(\frac{1}{3x}\right)^2}\, dx =$

$4\delta \int_{-1}^{1} 3\sqrt{1-u^2}\, du = 12\delta \int_{-1}^{1} \sqrt{1-u^2}\, du = 12\delta\left[\frac{1}{2}\left(u\sqrt{1-u^2} + \sin^{-1} u\right)\right]_{-1}^{1} = 6\pi\delta.$

(Let $u = \frac{1}{3}x \Rightarrow 3du = dx$, $x = -3 \Rightarrow u = -1$, $x = 3 \Rightarrow u = 1$)

$\therefore \bar{y} = \frac{M_x}{M} = \frac{32\delta}{6\pi\delta} = \frac{16}{3\pi}$. $\therefore$ The center of mass is $\left(0, \frac{16}{3\pi}\right)$.

31. $x^2 - y^2 = 1 \Rightarrow x = \pm\sqrt{1 + y^2}$ on the interval $-3 \le y \le 3$. $\therefore$ Volume = $\int_{-3}^{3} \pi\left(\sqrt{1+y^2}\right)^2 dy =$

$2\int_{0}^{3} \pi\left(\sqrt{1+y^2}\right)^2 dy = 2\pi \int_{0}^{3} (1 + y^2)dy = 2\pi\left[y + \frac{y^3}{3}\right]_0^3 = 24\pi$

32. $S = \int_{0}^{\sqrt{2}} 2\pi y\sqrt{1 + \left(\frac{dy}{dx}\right)^2}\, dx$. $\frac{dy}{dx} = \frac{1}{2}(x^2 + 1)^{-1/2}(2x) = \frac{x}{\sqrt{x^2+1}} \Rightarrow \left(\frac{dy}{dx}\right)^2 = \frac{x^2}{x^2+1} \Rightarrow$

$\sqrt{1 + \left(\frac{dy}{dx}\right)^2} = \sqrt{1 + \frac{x^2}{x^2+1}} = \sqrt{\frac{2x^2+1}{x^2+1}}.$

$\therefore S = \int_{0}^{\sqrt{2}} 2\pi\sqrt{x^2+1}\left(\sqrt{\frac{2x^2+1}{x^2+1}}\right)dx = \int_{0}^{\sqrt{2}} 2\pi\sqrt{2x^2+1}\, dx = \frac{2\pi}{\sqrt{2}}\int_{0}^{2} \sqrt{u^2+1}\, du$

(Let $u = \sqrt{2}\,x \Rightarrow \frac{1}{\sqrt{2}}du = dx$, $x = 0 \Rightarrow u = 0$, $x = \sqrt{2} \Rightarrow u = 2$)

$= \frac{2\pi}{\sqrt{2}}\left[\frac{1}{2}\left(u\sqrt{u^2+1} + \ln\left(u + \sqrt{u^2+1}\right)\right)\right]_0^2 = \frac{\pi}{\sqrt{2}}\left(2\sqrt{5} + \ln\left(2 + \sqrt{5}\right)\right)$

33. $\frac{dr_A}{dt} = \frac{dr_B}{dt} \Rightarrow \int \frac{dr_A}{dt} = \int \frac{dr_B}{dt} \Rightarrow r_A + C_1 = r_B + C_2 \Rightarrow r_A - r_B = C$, a constant $\Rightarrow$ The points, P(t), lie on a hyperbola with foci A and B.

34. Part 1: $\tan\beta = m_L \Rightarrow \tan\beta = f'(x_0)$ where $f(x) = \sqrt{4px}$. $f'(x) = \frac{1}{2}(4px)^{-1/2}(4p) = \frac{2p}{\sqrt{4px}} \Rightarrow f'(x_0) = \frac{2p}{\sqrt{4px_0}}$

$= \frac{2p}{y_0}$ $\therefore \tan\beta = \frac{2p}{y_0}$. Part 2: $\tan\phi = m_{FP} = \frac{y_0 - 0}{x_0 - p} = \frac{y_0}{x_0 - p}$

Part 3: $\tan\alpha = \frac{\tan\phi - \tan\beta}{1 + \tan\phi\tan\beta} = \frac{\frac{y_0}{x_0 - p} - \frac{2p}{y_0}}{1 + \frac{y_0}{x_0 - p}\left(\frac{2p}{y_0}\right)} = \frac{y_0^2 - 2p(x_0 - p)}{y_0(x_0 - p + 2p)} = \frac{4px_0 - 2px_0 + 2p^2}{y_0(x_0 + p)} = \frac{2p(x_0 + p)}{y_0(x_0 + p)} = \frac{2p}{y_0}$

SECTION 10.2 THE GRAPHS OF QUADRATIC EQUATIONS IN X AND Y

1. $x^2 - y^2 - 1 = 0 \Rightarrow B^2 - 4AC = 0 - 4(1)(-1) = 4 > 0 \Rightarrow$ Hyperbola

2. $25x^2 + 9y^2 - 225 = 0 \Rightarrow B^2 - 4AC = 0^2 - 4(25)(9) = -900 < 0 \Rightarrow$ Ellipse

3. $y^2 - 4x - 4 = 0 \Rightarrow B^2 - 4AC = 0^2 - 4(0)(1) = 0 \Rightarrow$ Parabola

4. $x^2 + y^2 - 10 = 0 \Rightarrow B^2 - 4AC = 0^2 - 4(1)(1) = -4 < 0 \Rightarrow$ Ellipse (Circle)

5. $x^2 + 4y^2 - 4x - 8y + 4 = 0 \Rightarrow B^2 - 4AC = 0^2 - 4(1)(4) = -16 < 0 \Rightarrow$ Ellipse

6. $2x^2 - y^2 + 4xy - 2x + 3y = 6 \Rightarrow B^2 - 4AC = 4^2 - 4(2)(-1) = 24 > 0 \Rightarrow$ Hyperbola

7. $x^2 + 4xy + 4y^2 - 3x = 6 \Rightarrow B^2 - 4AC = 4^2 - 4(1)(4) = 0 \Rightarrow$ Parabola

8. $x^2 + y^2 + 3x - 2y = 10 \Rightarrow B^2 - 4AC = 0^2 - 4(1)(1) = -4 < 0 \Rightarrow$ Ellipse (Circle)

9. $xy + y^2 - 3x = 5 \Rightarrow B^2 - 4AC = 1^2 - 4(0)(1) = 1 > 0 \Rightarrow$ Hyperbola

10. $3x^2 + 6xy + 3y^2 - 4x + 5y = 12 \Rightarrow B^2 - 4AC = 6^2 - 4(3)(3) = 0 \Rightarrow$ Parabola

11. $x^2 - y^2 = 1 \Rightarrow B^2 - 4AC = 0^2 - 4(1)(-1) = 4 > 0 \Rightarrow$ Hyperbola

12. $2x^2 + 3y^2 - 4x = 7 \Rightarrow B^2 - 4AC = 0^2 - 4(2)(3) = -24 < 0 \Rightarrow$ Ellipse

13. $x^2 - 3xy + 3y^2 + 6y = 7 \Rightarrow B^2 - 4AC = (-3)^2 - 4(1)(3) = -3 < 0 \Rightarrow$ Ellipse

14. $25x^2 - 4y^2 - 350x = 0 \Rightarrow B^2 - 4AC = 0^2 - 4(25)(-4) = 400 > 0 \Rightarrow$ Hyperbola

15. $6x^2 + 3xy + 2y^2 + 17y + 2 = 0 \Rightarrow B^2 - 4AC = 3^2 - 4(6)(2) = -39 < 0 \Rightarrow$ Ellipse

16. $3x^2 + 12xy + 12y^2 + 435x - 9y + 72 = 0 \Rightarrow B^2 - 4AC = 12^2 - 4(3)(12) = 0 \Rightarrow$ Parabola

17. $\cot 2\alpha = \frac{A - C}{B} = \frac{0}{1} = 0 \Rightarrow 2\alpha = \frac{\pi}{2} \Rightarrow \alpha = \frac{\pi}{4}$. $\therefore$ $x = x'\cos\alpha - y'\sin\alpha$, $y = x'\sin\alpha + y'\cos\alpha \Rightarrow$
$x = x'\frac{\sqrt{2}}{2} - y'\frac{\sqrt{2}}{2}$, $y = x'\frac{\sqrt{2}}{2} + y'\frac{\sqrt{2}}{2} \Rightarrow \left(\frac{\sqrt{2}}{2}x' - \frac{\sqrt{2}}{2}y'\right)\left(\frac{\sqrt{2}}{2}x' + \frac{\sqrt{2}}{2}y'\right) = 2 \Rightarrow \frac{1}{2}x'^2 - \frac{1}{2}y'^2 = 2 \Rightarrow x'^2 - y'^2 = 4$
$\Rightarrow$ Hyperbola

18. $\cot 2\alpha = \frac{A - C}{B} = \frac{1 - 1}{1} = 0 \Rightarrow 2\alpha = \frac{\pi}{2} \Rightarrow \alpha = \frac{\pi}{4}$. $\therefore$ $x = x'\cos\alpha - y'\sin\alpha$, $y = x'\sin\alpha + y'\cos\alpha \Rightarrow$
$x = x'\frac{\sqrt{2}}{2} - y'\frac{\sqrt{2}}{2}$, $y = x'\frac{\sqrt{2}}{2} + y'\frac{\sqrt{2}}{2} \Rightarrow \left(\frac{\sqrt{2}}{2}x' - \frac{\sqrt{2}}{2}y'\right)^2 + \left(\frac{\sqrt{2}}{2}x' - \frac{\sqrt{2}}{2}y'\right)\left(\frac{\sqrt{2}}{2}x' + \frac{\sqrt{2}}{2}y'\right) + \left(\frac{\sqrt{2}}{2}x' + \frac{\sqrt{2}}{2}y'\right)^2 = 1$
$\Rightarrow \frac{1}{2}x'^2 - x'y' + \frac{1}{2}y'^2 + \frac{1}{2}x'^2 - \frac{1}{2}y'^2 + \frac{1}{2}x'^2 + x'y' + \frac{1}{2}y'^2 = 1 \Rightarrow \frac{3}{2}x'^2 + \frac{1}{2}y'^2 = 1 \Rightarrow 3x'^2 + y'^2 = 2$, Ellipse.

19. $\cot 2\alpha = \frac{A - C}{B} = \frac{3 - 1}{2\sqrt{3}} = \frac{1}{\sqrt{3}} \Rightarrow 2\alpha = \frac{\pi}{3} \Rightarrow \alpha = \frac{\pi}{6}$. $\therefore$ $x = x'\cos\alpha - y'\sin\alpha$, $y = x'\sin\alpha + y'\cos\alpha \Rightarrow$
$x = \frac{\sqrt{3}}{2}x' - \frac{1}{2}y'$, $y = \frac{1}{2}x' + \frac{\sqrt{3}}{2}y' \Rightarrow 3\left(\frac{\sqrt{3}}{2}x' - \frac{1}{2}y'\right)^2 + 2\sqrt{3}\left(\frac{\sqrt{3}}{2}x' - \frac{1}{2}y'\right)\left(\frac{1}{2}x' + \frac{\sqrt{3}}{2}y'\right) + \left(\frac{1}{2}x' + \frac{\sqrt{3}}{2}y'\right)^2 -$
$8\left(\frac{\sqrt{3}}{2}x' - \frac{1}{2}y'\right) + 8\sqrt{3}\left(\frac{1}{2}x' + \frac{\sqrt{3}}{2}y'\right) = 0 \Rightarrow 4x'^2 + 16y' = 0$, Parabola

20. $\cot 2\alpha = \frac{A-C}{B} = \frac{1-2}{-\sqrt{3}} = \frac{1}{\sqrt{3}} \Rightarrow 2\alpha = \frac{\pi}{3} \Rightarrow \alpha = \frac{\pi}{6}$. $\therefore$ $x = x'\cos\alpha - y'\sin\alpha$, $y = x'\sin\alpha + y'\cos\alpha \Rightarrow$

$x = \frac{\sqrt{3}}{2}x' - \frac{1}{2}y'$, $y = \frac{1}{2}x' + \frac{\sqrt{3}}{2}y' \Rightarrow \left(\frac{\sqrt{3}}{2}x' - \frac{1}{2}y'\right)^2 - \sqrt{3}\left(\frac{\sqrt{3}}{2}x' - \frac{1}{2}y'\right)\left(\frac{1}{2}x' + \frac{\sqrt{3}}{2}y'\right) + 2\left(\frac{1}{2}x' + \frac{\sqrt{3}}{2}y'\right)^2 = 1 \Rightarrow$

$\frac{1}{2}x'^2 + \frac{5}{2}y'^2 = 1 \Rightarrow x'^2 + 5y'^2 = 2$, Ellipse

21. $\cot 2\alpha = \frac{A-C}{B} = \frac{1-1}{-2} = 0 \Rightarrow 2\alpha = \frac{\pi}{2} \Rightarrow \alpha = \frac{\pi}{4}$. $\therefore$ $x = x'\cos\alpha - y'\sin\alpha$, $y = x'\sin\alpha + y'\cos\alpha \Rightarrow$

$x = x'\frac{\sqrt{2}}{2} - y'\frac{\sqrt{2}}{2}$, $y = x'\frac{\sqrt{2}}{2} + y'\frac{\sqrt{2}}{2} \Rightarrow \left(\frac{\sqrt{2}}{2}x' - \frac{\sqrt{2}}{2}y'\right)^2 - 2\left(\frac{\sqrt{2}}{2}x' - \frac{\sqrt{2}}{2}y'\right)\left(\frac{\sqrt{2}}{2}x' + \frac{\sqrt{2}}{2}y'\right) + \left(\frac{\sqrt{2}}{2}x' + \frac{\sqrt{2}}{2}y'\right)^2 = 2$

$\Rightarrow y'^2 = 1$, Parallel Horizontal Lines.

22. $\cot 2\alpha = \frac{A-C}{B} = \frac{1-1}{-3} = 0 \Rightarrow 2\alpha = \frac{\pi}{2} \Rightarrow \alpha = \frac{\pi}{4}$. $\therefore$ $x = x'\cos\alpha - y'\sin\alpha$, $y = x'\sin\alpha + y'\cos\alpha \Rightarrow$

$x = x'\frac{\sqrt{2}}{2} - y'\frac{\sqrt{2}}{2}$, $y = x'\frac{\sqrt{2}}{2} + y'\frac{\sqrt{2}}{2} \Rightarrow \left(\frac{\sqrt{2}}{2}x' - \frac{\sqrt{2}}{2}y'\right)^2 - 3\left(\frac{\sqrt{2}}{2}x' - \frac{\sqrt{2}}{2}y'\right)\left(\frac{\sqrt{2}}{2}x' + \frac{\sqrt{2}}{2}y'\right) + \left(\frac{\sqrt{2}}{2}x' + \frac{\sqrt{2}}{2}y'\right)^2 = 5$

$\Rightarrow -\frac{1}{2}x'^2 + \frac{5}{2}y'^2 = 5 \Rightarrow 5y'^2 - x'^2 = 10$, Hyperbola.

23. $\cot 2\alpha = \frac{A-C}{B} = \frac{\sqrt{2}-\sqrt{2}}{2\sqrt{2}} = 0 \Rightarrow 2\alpha = \frac{\pi}{2} \Rightarrow \alpha = \frac{\pi}{4}$. $\therefore$ $x = x'\cos\alpha - y'\sin\alpha$, $y = x'\sin\alpha + y'\cos\alpha \Rightarrow$

$x = x'\frac{\sqrt{2}}{2} - y'\frac{\sqrt{2}}{2}$, $y = x'\frac{\sqrt{2}}{2} + y'\frac{\sqrt{2}}{2} \Rightarrow \sqrt{2}\left(\frac{\sqrt{2}}{2}x' - \frac{\sqrt{2}}{2}y'\right)^2 + 2\sqrt{2}\left(\frac{\sqrt{2}}{2}x' - \frac{\sqrt{2}}{2}y'\right)\left(\frac{\sqrt{2}}{2}x' + \frac{\sqrt{2}}{2}y'\right) +$

$\sqrt{2}\left(\frac{\sqrt{2}}{2}x' + \frac{\sqrt{2}}{2}y'\right)^2 - 8\left(\frac{\sqrt{2}}{2}x' - \frac{\sqrt{2}}{2}y'\right) + 8\left(\frac{\sqrt{2}}{2}x' + \frac{\sqrt{2}}{2}y'\right) = 0 \Rightarrow 2\sqrt{2}\,x'^2 + 8\sqrt{2}\,y' = 0$, Parabola

24. $\cot 2\alpha = \frac{A-C}{B} = \frac{0-0}{1} = 0 \Rightarrow 2\alpha = \frac{\pi}{2} \Rightarrow \alpha = \frac{\pi}{4}$. $\therefore$ $x = x'\cos\alpha - y'\sin\alpha$, $y = x'\sin\alpha + y'\cos\alpha \Rightarrow$

$x = x'\frac{\sqrt{2}}{2} - y'\frac{\sqrt{2}}{2}$, $y = x'\frac{\sqrt{2}}{2} + y'\frac{\sqrt{2}}{2} \Rightarrow \left(\frac{\sqrt{2}}{2}x' - \frac{\sqrt{2}}{2}y'\right)\left(\frac{\sqrt{2}}{2}x' + \frac{\sqrt{2}}{2}y'\right) - \left(\frac{\sqrt{2}}{2}x' - \frac{\sqrt{2}}{2}y'\right) - \left(\frac{\sqrt{2}}{2}x' + \frac{\sqrt{2}}{2}y'\right) + 1 = 0$

$\Rightarrow x'^2 - y'^2 - 2\sqrt{2}\,x' + 2 = 0$, Hyperbola.

25. $\cot 2\alpha = \frac{A-C}{B} = \frac{3-3}{2} = 0 \Rightarrow 2\alpha = \frac{\pi}{2} \Rightarrow \alpha = \frac{\pi}{4}$. $\therefore$ $x = x'\cos\alpha - y'\sin\alpha$, $y = x'\sin\alpha + y'\cos\alpha \Rightarrow$

$x = x'\frac{\sqrt{2}}{2} - y'\frac{\sqrt{2}}{2}$, $y = x'\frac{\sqrt{2}}{2} + y'\frac{\sqrt{2}}{2} \Rightarrow 3\left(\frac{\sqrt{2}}{2}x' - \frac{\sqrt{2}}{2}y'\right)^2 + 2\left(\frac{\sqrt{2}}{2}x' - \frac{\sqrt{2}}{2}y'\right)\left(\frac{\sqrt{2}}{2}x' + \frac{\sqrt{2}}{2}y'\right) + 3\left(\frac{\sqrt{2}}{2}x' + \frac{\sqrt{2}}{2}y'\right)^2$

$= 19 \Rightarrow 4x'^2 + 2y'^2 = 19$, Ellipse.

26. $\cot 2\alpha = \frac{A-C}{B} = \frac{3-(-1)}{4\sqrt{3}} = \frac{1}{\sqrt{3}} \Rightarrow 2\alpha = \frac{\pi}{3} \Rightarrow \alpha = \frac{\pi}{6}$. $\therefore$ $x = x'\cos\alpha - y'\sin\alpha$, $y = x'\sin\alpha + y'\cos\alpha \Rightarrow$

$x = \frac{\sqrt{3}}{2}x' - \frac{1}{2}y'$, $y = \frac{1}{2}x' + \frac{\sqrt{3}}{2}y' \Rightarrow 3\left(\frac{\sqrt{3}}{2}x' - \frac{1}{2}y'\right)^2 + 4\sqrt{3}\left(\frac{\sqrt{3}}{2}x' - \frac{1}{2}y'\right)\left(\frac{1}{2}x' + \frac{\sqrt{3}}{2}y'\right) - \left(\frac{1}{2}x' + \frac{\sqrt{3}}{2}y'\right)^2 = 7 \Rightarrow$

$5x'^2 - 3y'^2 = 7$, Hyperbola.

27\. a) For $xy = 1$, $\alpha = \frac{\pi}{4} \Rightarrow x = x'\frac{\sqrt{2}}{2} - y'\frac{\sqrt{2}}{2}$, $y = x'\frac{\sqrt{2}}{2} + y'\frac{\sqrt{2}}{2} \Rightarrow \left(\frac{\sqrt{2}}{2}x' - \frac{\sqrt{2}}{2}y'\right)\left(\frac{\sqrt{2}}{2}x' + \frac{\sqrt{2}}{2}y'\right) = 1 \Rightarrow$
$x'^2 - y'^2 = 2$.

b) For $xy = a$, $\alpha = \frac{\pi}{4} \Rightarrow x = x'\frac{\sqrt{2}}{2} - y'\frac{\sqrt{2}}{2}$, $y = x'\frac{\sqrt{2}}{2} + y'\frac{\sqrt{2}}{2} \Rightarrow \left(\frac{\sqrt{2}}{2}x' - \frac{\sqrt{2}}{2}y'\right)\left(\frac{\sqrt{2}}{2}x' + \frac{\sqrt{2}}{2}y'\right) = 1 \Rightarrow$
$x'^2 - y'^2 = 2a$.

SECTION 10.3 PARAMETRIC EQUATIONS FOR PLANE CURVES

1\. $x = \cos t$, $y = \sin t$, $0 \le t \le \pi \Rightarrow$
$\cos^2 t + \sin^2 t = 1 \Rightarrow x^2 + y^2 = 1$

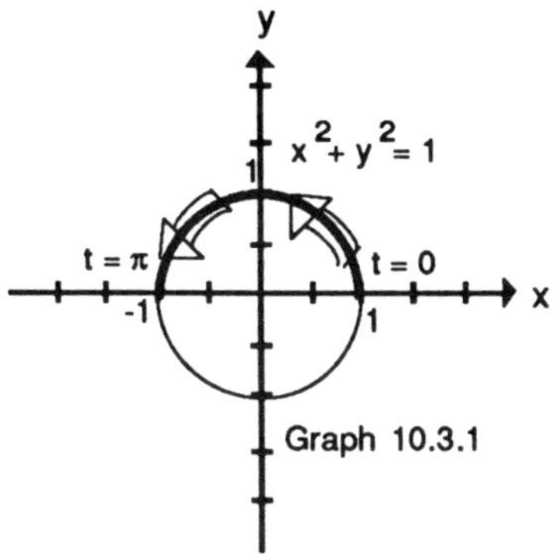

Graph 10.3.1

2\. $x = \cos 2t$, $y = \sin 2t$, $0 \le t \le \pi \Rightarrow$
$\cos^2 2t + \sin^2 2t = 1 \Rightarrow x^2 + y^2 = 1$

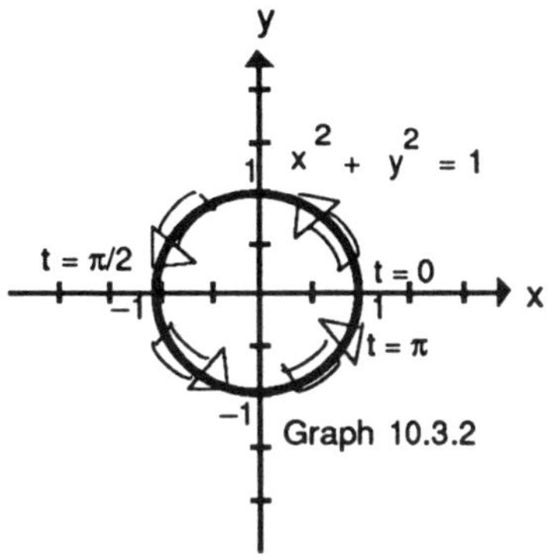

Graph 10.3.2

3\. $x = \sin 2\pi t$, $y = \cos 2\pi t$, $0 \le t \le 1 \Rightarrow$
$\sin^2 2\pi t + \cos^2 2\pi t = 1 \Rightarrow x^2 + y^2 = 1$

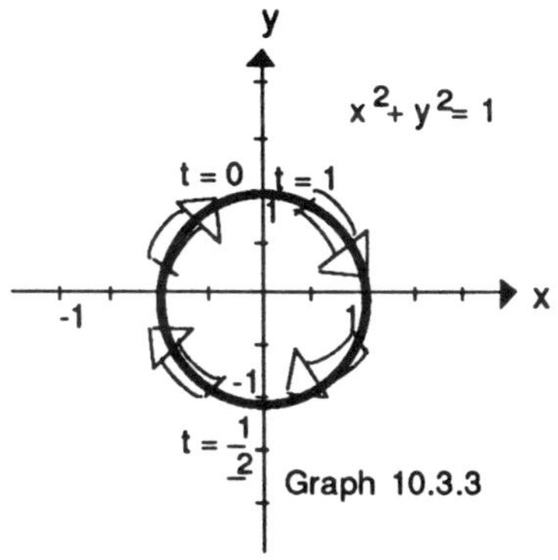

Graph 10.3.3

4\. $x = \cos(\pi - t)$, $y = \sin(\pi - t)$, $0 \le t \le \pi \Rightarrow$
$\cos^2(\pi - t) + \sin^2(\pi - t) = 1 \Rightarrow x^2 + y^2 = 1$

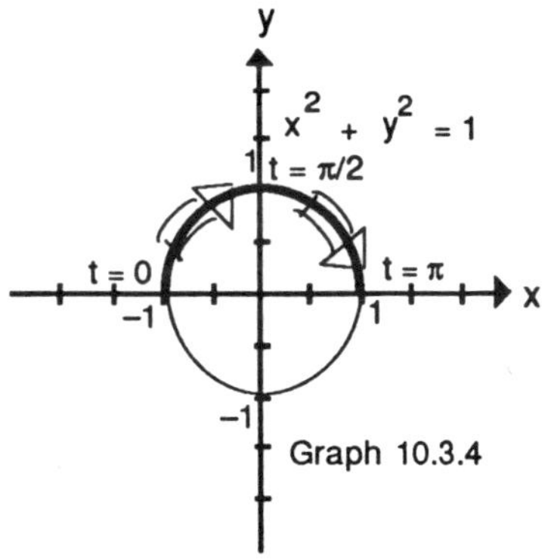

Graph 10.3.4

5\. $x = 4\cos t$, $y = 2\sin t$, $0 \le t \le 2\pi \Rightarrow$
$\frac{16\cos^2 t}{16} + \frac{4\sin^2 t}{4} = 1 \Rightarrow \frac{x^2}{16} + \frac{y^2}{4} = 1$

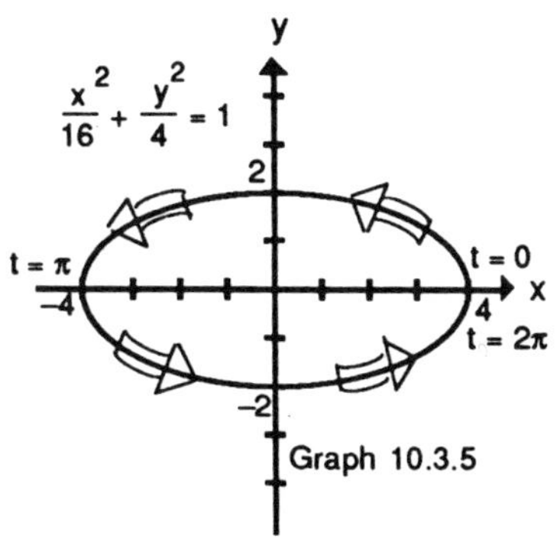

Graph 10.3.5

6\. $x = 4\sin t$, $y = 2\cos t$, $0 \le t \le \pi \Rightarrow$
$\frac{16\sin^2 t}{16} + \frac{4\cos^2 t}{4} = 1 \Rightarrow \frac{x^2}{16} + \frac{y^2}{4} = 1$

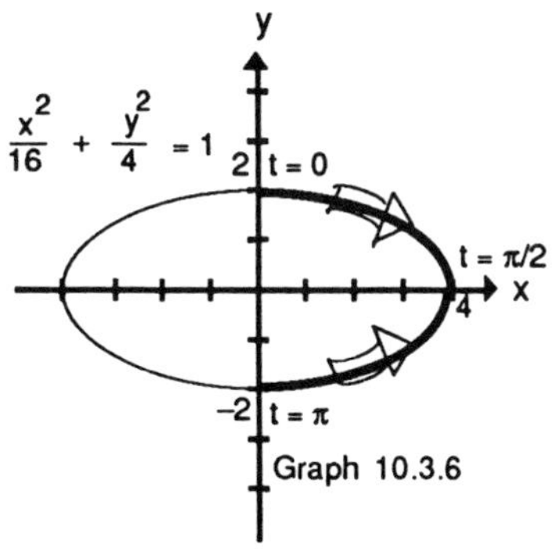

Graph 10.3.6

7. $x = 4\cos t,\ y = 5\sin t,\ 0 \le t \le 2\pi \Rightarrow$
$\frac{16\cos^2 t}{16} + \frac{25\sin^2 t}{25} = 1 \Rightarrow \frac{x^2}{16} + \frac{y^2}{25} = 1$

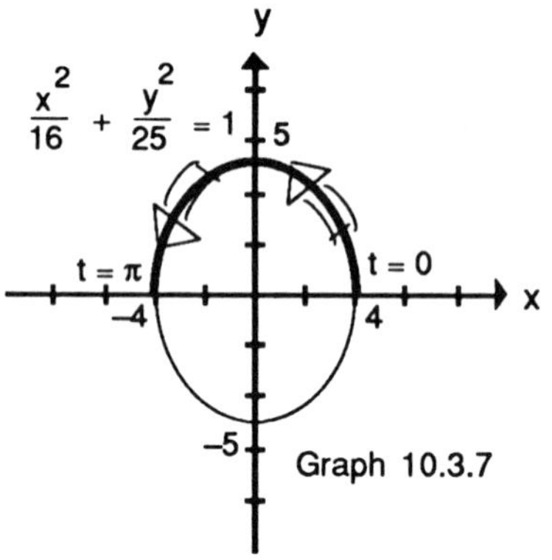

Graph 10.3.7

8. $x = 4\sin t,\ y = 5\cos t,\ 0 \le t \le 2\pi \Rightarrow$
$\frac{16\sin^2 t}{16} + \frac{25\cos^2 t}{25} = 1 \Rightarrow \frac{x^2}{16} + \frac{y^2}{25} = 1$

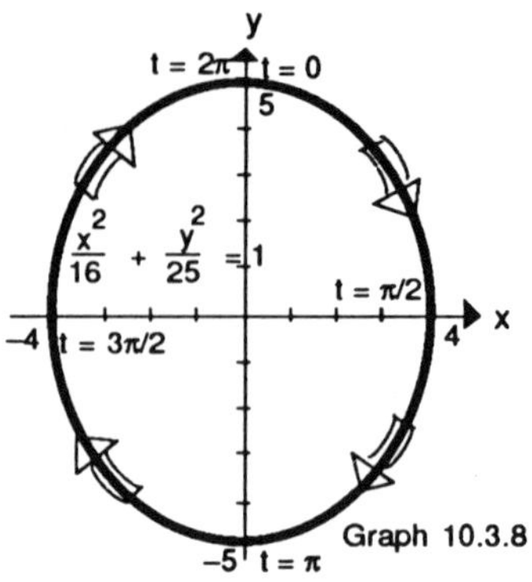

Graph 10.3.8

9. $x = 3t,\ y = 9t^2,\ -\infty < t < \infty \Rightarrow$
$y = x^2$

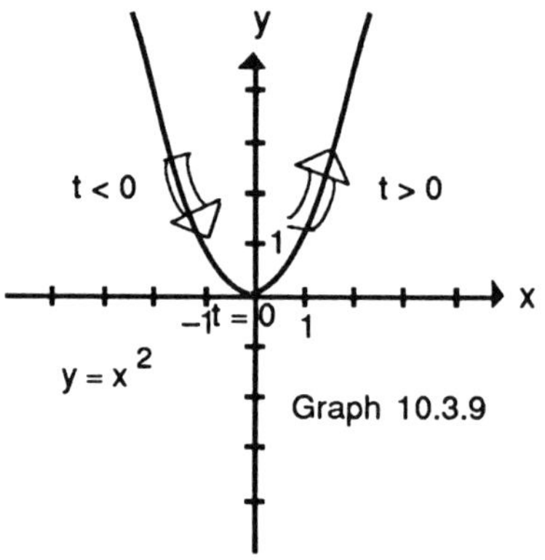

Graph 10.3.9

10. $x = -\sqrt{t},\ y = t,\ t \ge 0 \Rightarrow$
$x = -\sqrt{y}$

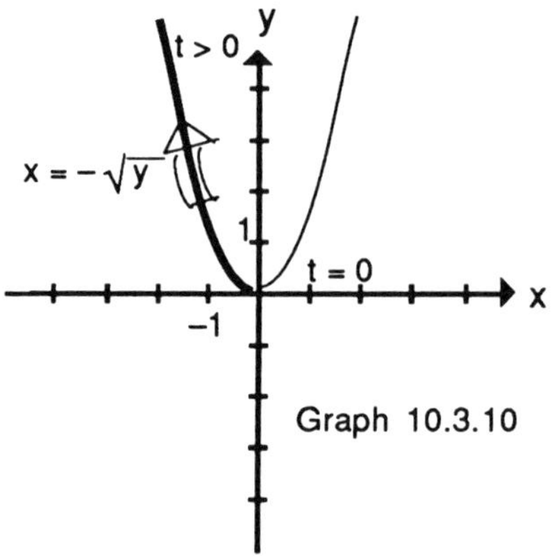

Graph 10.3.10

11. $x = t,\ y = \sqrt{t},\ t \ge 0$
$x = y^2$

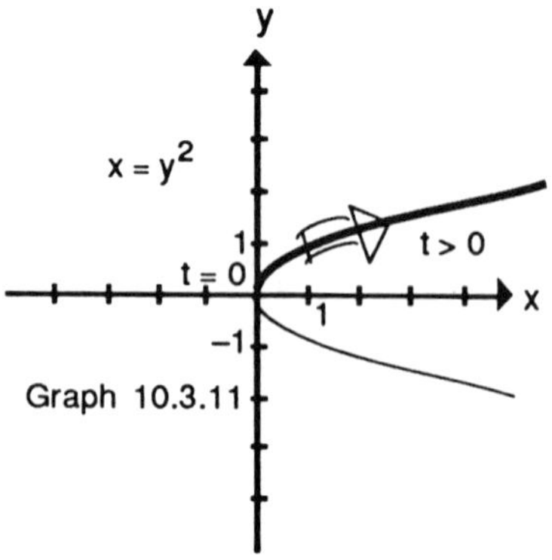

Graph 10.3.11

12. $x = \sec^2 t - 1,\ y = \tan t,\ -\frac{\pi}{2} < t < \frac{\pi}{2} \Rightarrow$
$\sec^2 t - 1 = \tan^2 t \Rightarrow x = y^2$

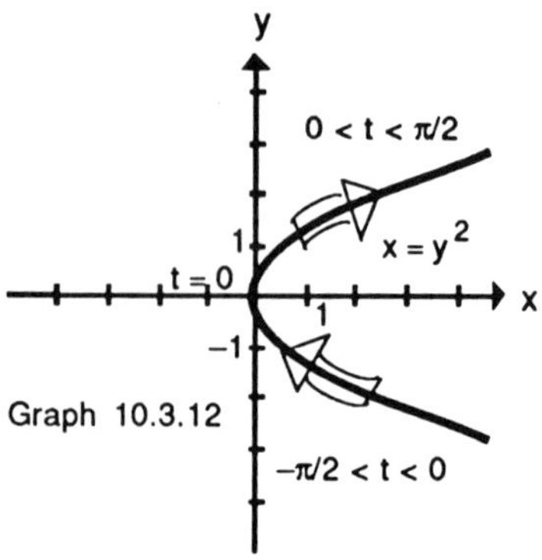

Graph 10.3.12

13. $x = -\sec t,\ y = \tan t,\ -\frac{\pi}{2} < t < \frac{\pi}{2} \Rightarrow$

$\sec^2 t - \tan^2 t = 1 \Rightarrow x^2 - y^2 = 1$

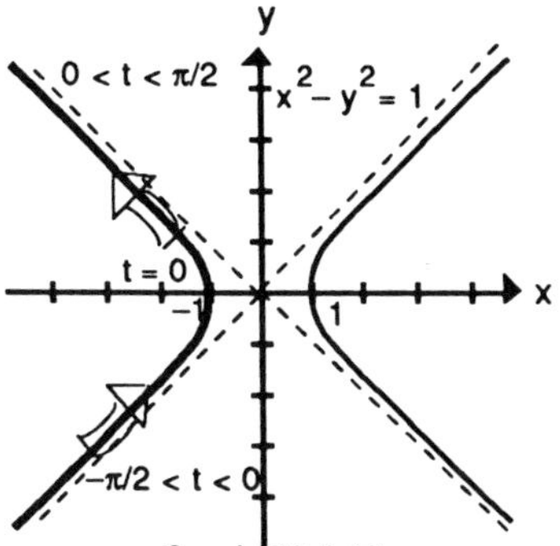

Graph 10.3.13

14. $x = \csc t,\ y = \cot t,\ 0 < t < \pi \Rightarrow$

$1 + \cot^2 t = \csc^2 t \Rightarrow 1 + y^2 = x^2 \Rightarrow x^2 - y^2 = 1$

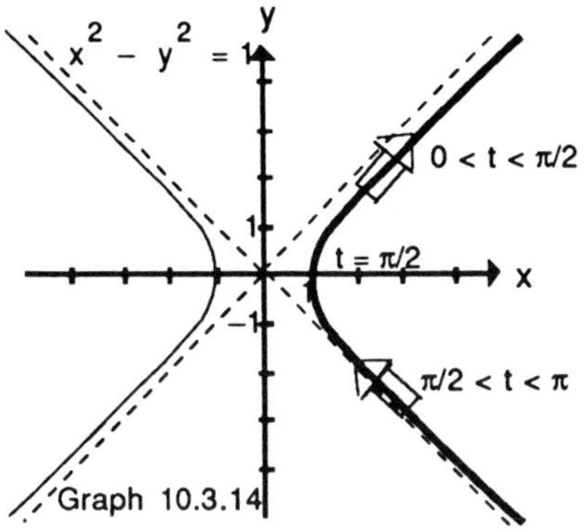

Graph 10.3.14

15. $x = 2t - 5,\ y = 4t - 7,\ -\infty < t < \infty \Rightarrow$

$x + 5 = 2t \Rightarrow 2(x + 5) = 4t \Rightarrow$

$y = 2(x + 5) - 7 \Rightarrow y = 2x + 3$

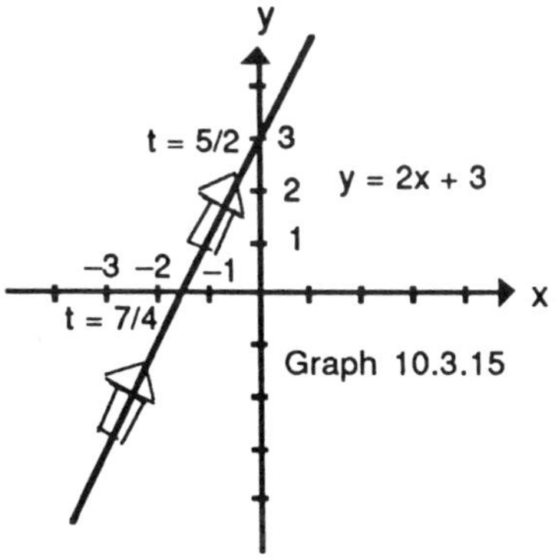

Graph 10.3.15

16. $x = 1 - t,\ y = 1 + t,\ -\infty < t < \infty \Rightarrow$

$1 - x = t \Rightarrow y = 1 + (1 - x) \Rightarrow$

$y = -x + 2$

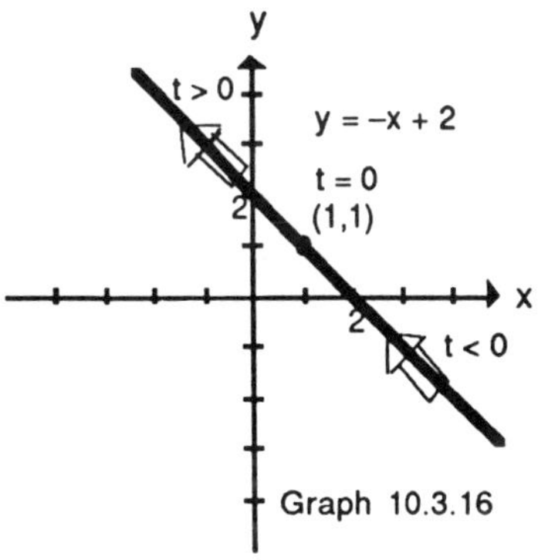

Graph 10.3.16

17. $x = t,\ y = 1 - t,\ 0 \le t \le 1 \Rightarrow$

$y = 1 - x$

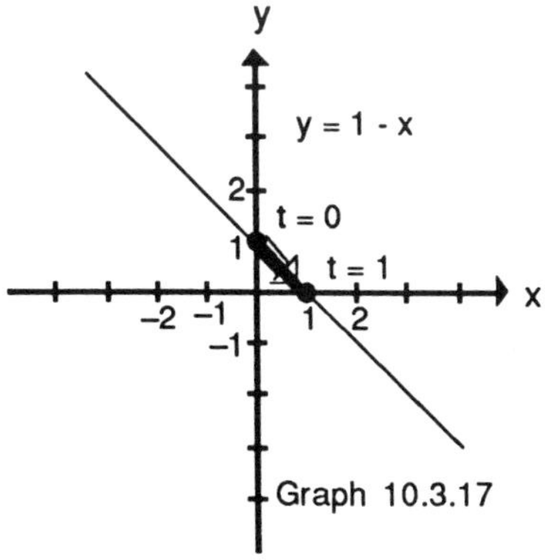

Graph 10.3.17

18. $x = 3t,\ y = 2 - 2t,\ 0 \le t \le 1 \Rightarrow$

$\frac{x}{3} = t \Rightarrow y = 2 - 2\left(\frac{x}{3}\right) \Rightarrow y = 2 - \frac{2}{3}x$

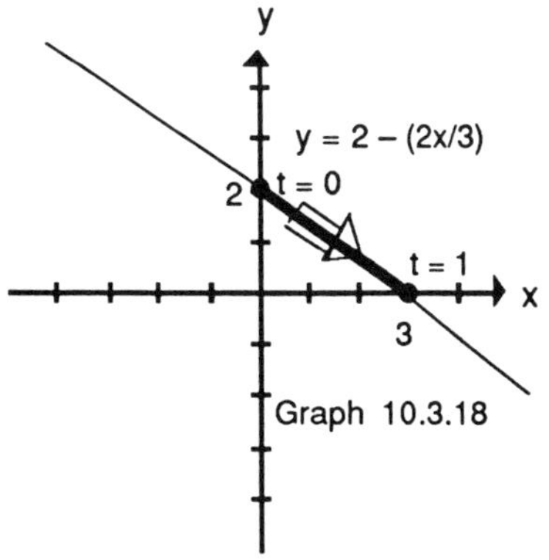

Graph 10.3.18

19. $x = t$, $y = \sqrt{1 - t^2}$, $-1 \le t \le 1 \Rightarrow$
$y = \sqrt{1 - x^2}$

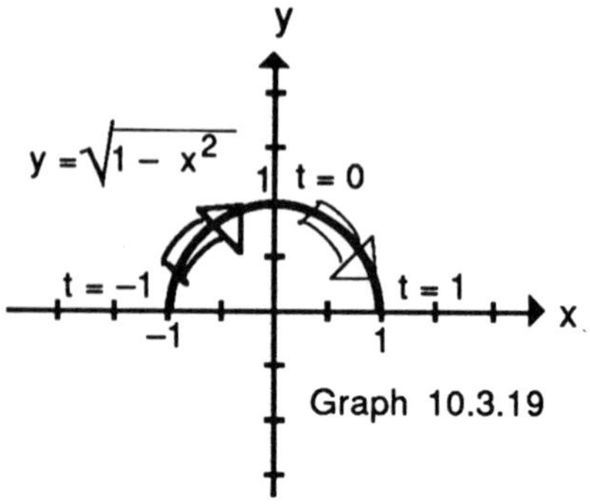

Graph 10.3.19

20. $x = t$, $y = \sqrt{4 - t^2}$, $0 \le t \le 2$
$y = \sqrt{4 - x^2}$

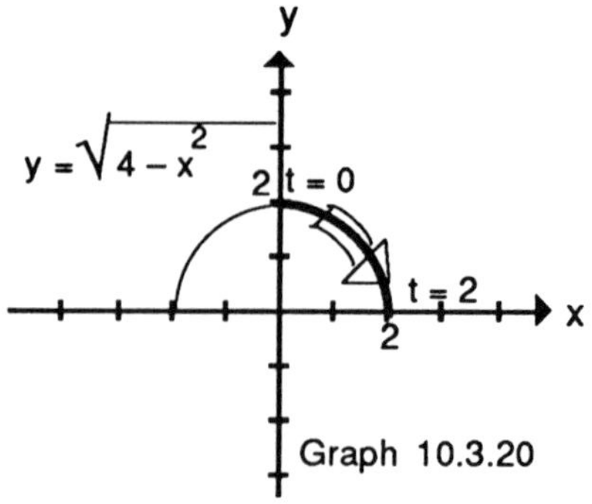

Graph 10.3.20

21. $x = t^2$, $y = \sqrt{t^4 + 1}$, $t \ge 0 \Rightarrow$
$y = \sqrt{x^2 + 1}$, $x \ge 0$

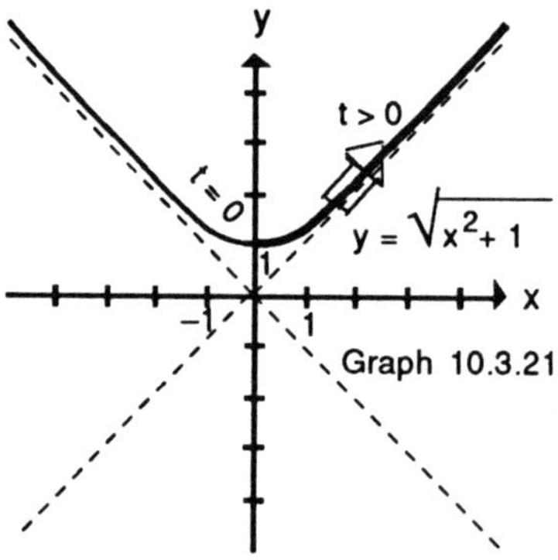

Graph 10.3.21

22. $x = \sqrt{t + 1}$, $y = \sqrt{t}$, $t \ge 0 \Rightarrow$
$y^2 = t \Rightarrow x = \sqrt{y^2 + 1}$, $y \ge 0$

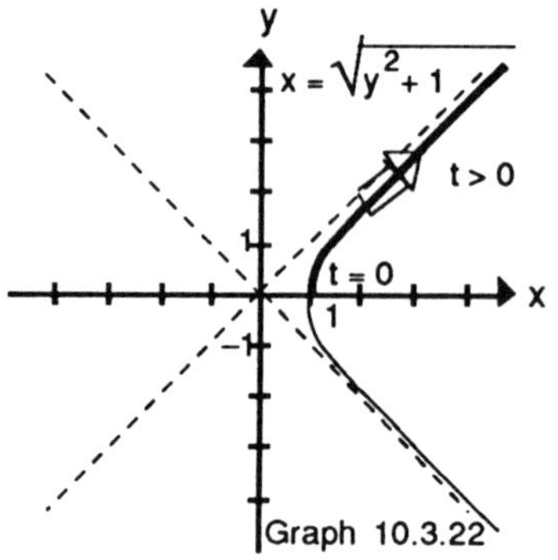

Graph 10.3.22

23. $x = \cosh t$, $y = \sinh t$, $-\infty < t < \infty \Rightarrow$
$\cosh^2 t - \sinh^2 t = 1 \Rightarrow x^2 - y^2 = 1$

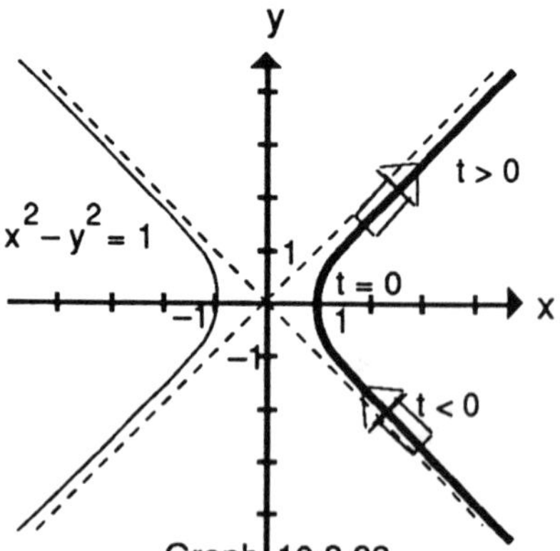

Graph 10.3.23

24. $x = 2\sinh t$, $y = 2\cosh t$, $-\infty < t < \infty \Rightarrow$
$4\cosh^2 t - 4\sinh^2 t = 4 \Rightarrow y^2 - x^2 = 4$

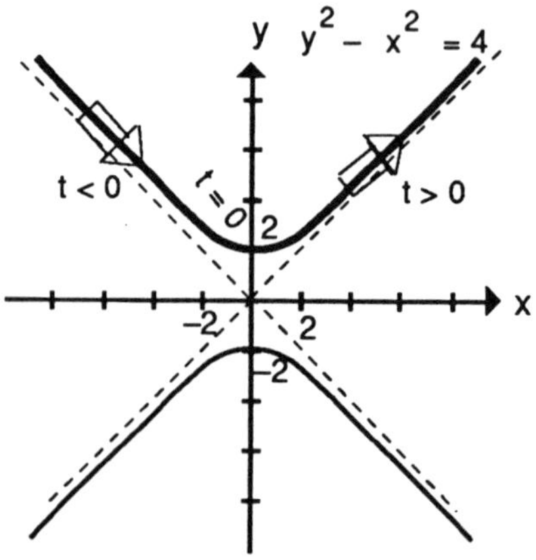

Graph 10.3.24

25. a) $x = a\cos t,\ y = -a\sin t,\ 0 \le t \le 2\pi$ b) $x = a\cos t,\ y = a\sin t,\ 0 \le t \le 2\pi$

c) $x = a\cos t,\ y = -a\sin t,\ 0 \le t \le 4\pi$ d) $x = a\cos t,\ y = a\sin t,\ 0 \le t \le 4\pi$

26. a) $x = a\sin t,\ y = b\cos t,\ \frac{\pi}{2} \le t \le \frac{5\pi}{2}$ b) $x = a\cos t,\ y = b\sin t,\ 0 \le t \le 2\pi$

c) $x = a\sin t,\ y = b\cos t,\ \frac{\pi}{2} \le t \le \frac{9\pi}{2}$ d) $x = a\sin t,\ y = b\cos t,\ 0 \le t \le 4\pi$

27.

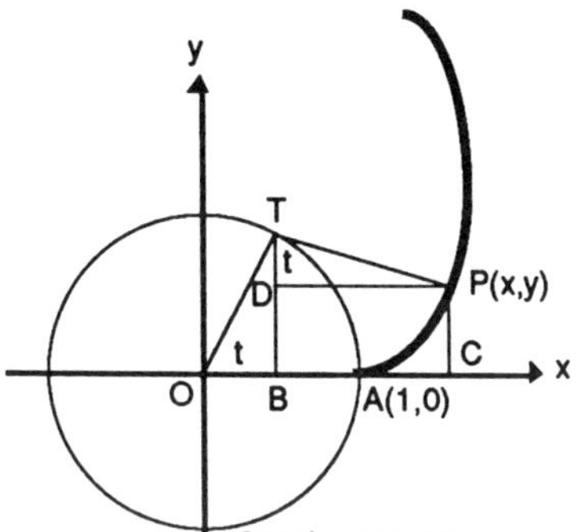

Graph 10.3.27

$\angle PTB = \angle TOB = t$. PT = arc(AT) = t since PT = length of the unwound string = length of arc(AT).

$\therefore\ x = OB + BC = OB + DP = \cos t + t\sin t$.

$y = PC = TB - TD = \sin t - t\cos t$

28.

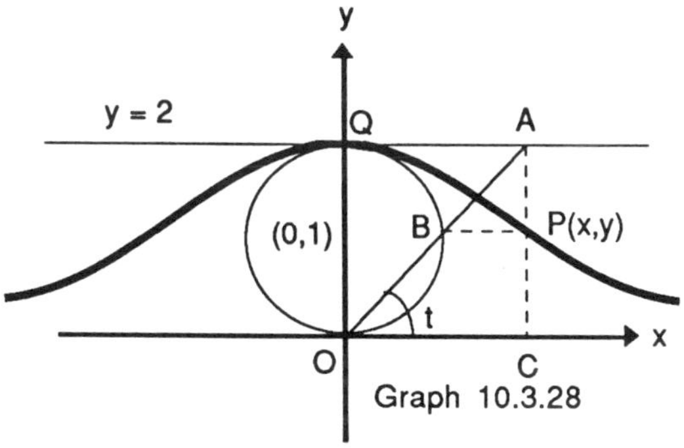

Graph 10.3.28

Extend the vertical line through A to the x–axis, let C be the point of intersection. Then OC = AQ = x and $\tan t = \frac{2}{AC} = \frac{2}{x} \Rightarrow x = \frac{2}{\tan t} = 2\cot t$. $\sin t = \frac{2}{OA} \Rightarrow$ $OA = \frac{2}{\sin t}$. $(AB)(OA) = (AQ)^2 \Rightarrow AB\left(\frac{2}{\sin t}\right) = x^2 \Rightarrow$ $AB\left(\frac{2}{\sin t}\right) = \left(\frac{2}{\tan t}\right)^2 \Rightarrow AB = \frac{2\sin t}{\tan^2 t}$.

$y = 2 - AB\sin t \Rightarrow y = 2 - \left(\frac{2\sin t}{\tan^2 t}\right)\sin t = 2 - \frac{2\sin^2 t}{\tan^2 t} =$

$2 - 2\cos^2 t = 2\sin^2 t$.

$\therefore$ let $x = 2\cot t,\ y = 2\sin^2 t,\ 0 < t < \pi$

29. $d = \sqrt{(x-2)^2 + (y-\frac{1}{2})^2} \Rightarrow d^2 = (x-2)^2 + (y-\frac{1}{2})^2 = (t-2)^2 + (t^2-\frac{1}{2})^2 \Rightarrow d^2 = t^4 - 4t + \frac{17}{4}$

$\frac{d(d^2)}{dt} = 4t^3 - 4$. Let $4t^3 - 4 = 0 \Rightarrow t = 1$ The second derivative is always positve for $t \neq 0 \Rightarrow$ $t = 1$ gives a local minimum which is an absolute minimum since it is the only extremum. $\therefore$ the closest point on the parabola is (1,1).

30. $d = \sqrt{(2\cos t - \frac{3}{4})^2 + (\sin t - 0)^2} \Rightarrow d^2 = (2\cos t - \frac{3}{4})^2 + \sin^2 t$. $\frac{d(d^2)}{dt} = 2(2\cos t - \frac{3}{4})(-2\sin t) +$ $2\sin t\cos t = -2\sin t(3\cos t - \frac{3}{2})$. Let $\frac{d(d^2)}{dt} = 0 \Rightarrow -2\sin t(3\cos t - \frac{3}{2}) = 0 \Rightarrow -2\sin t = 0$ or $3\cos t - \frac{3}{2} = 0 \Rightarrow t = 0, \pi$ or $t = \frac{\pi}{3}, \frac{5\pi}{3}$. Now $\frac{d^2(d^2)}{dt^2} = -6\cos^2 t + 3\cos t + 6\sin^2 t$.

30. (Continued from previous page)
$\frac{d^2(d^2)}{dt^2}(0) = -3 \Rightarrow$ relative maximum. $\frac{d^2(d^2)}{dt^2}(\pi) = -9 \Rightarrow$ relative maximum. $\frac{d^2(d^2)}{dt^2}\left(\frac{\pi}{3}\right) = \frac{9}{2} \Rightarrow$ relative minimum. $\frac{d^2(d^2)}{dt^2}\left(\frac{5\pi}{3}\right) = -\frac{9}{2} \Rightarrow$ relative maximum. $\therefore\ t = \frac{\pi}{3}$ gives the point on the ellipse closest to $\left(\frac{3}{4}, 0\right) \Rightarrow \left(1, \frac{\sqrt{3}}{2}\right)$ is the desired point.

31. a) $x = x_0 + (x_1 - x_0)t,\ y = y_0 + (y_1 - y_0)t \Rightarrow t = \frac{x - x_0}{x_1 - x_0} \Rightarrow y = y_0 + (y_1 - y_0)\left(\frac{x - x_0}{x_1 - x_0}\right) \Rightarrow$

$$y - y_0 = \left(\frac{y_1 - y_0}{x_1 - x_0}\right)(x - x_0)$$ which is the equation of the line through the points.

b) $x = x_1 t,\ y = y_1 t$ (Answer not unique)

c) $x = -1 + t,\ y = t$ or $x = -t,\ y = 1 - t$

32. a)

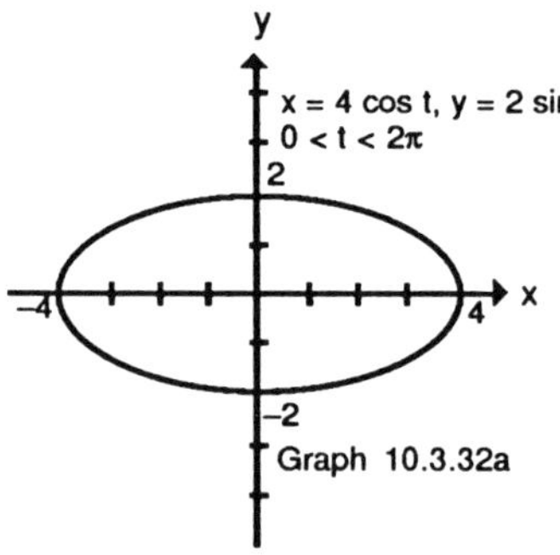

b)

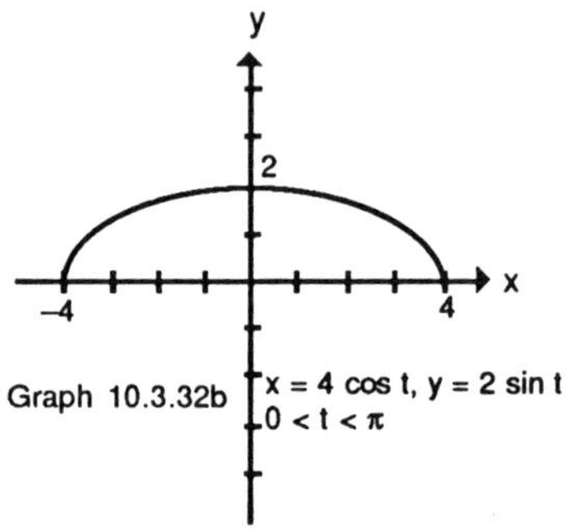

c)

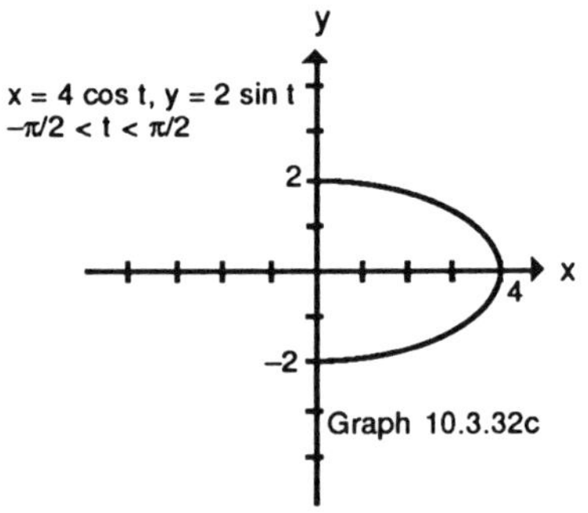

33. a)

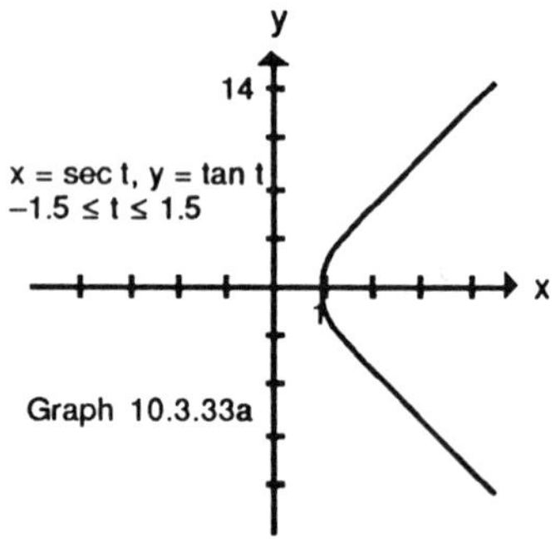

b)

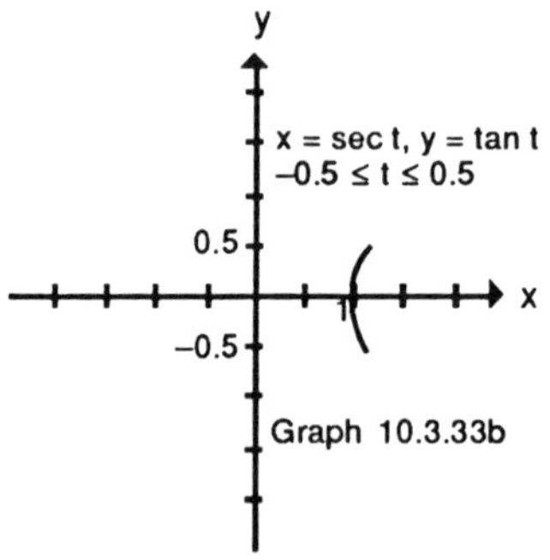

c)

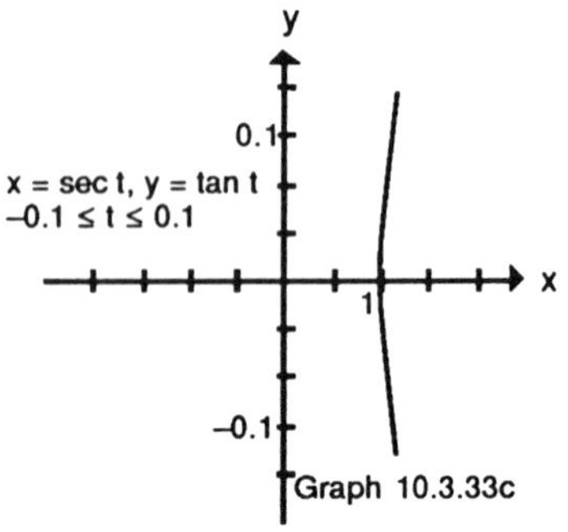

34.

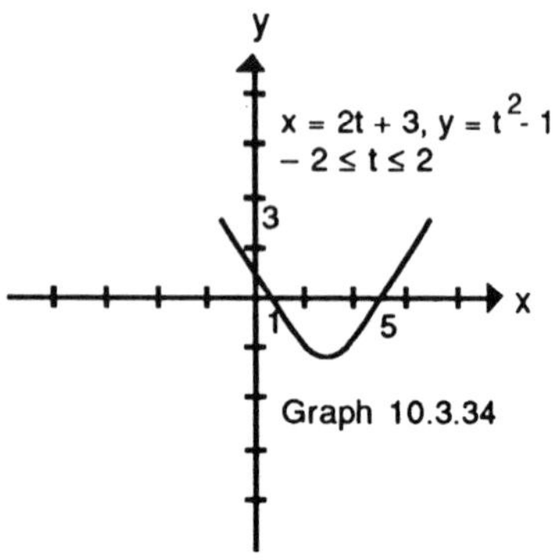

35. a)

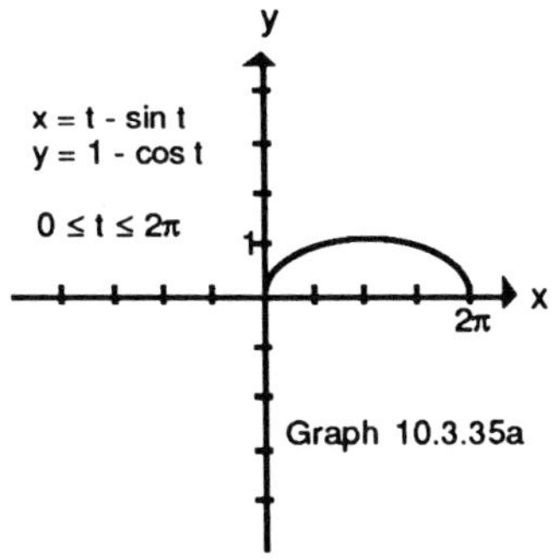

Graph 10.3.35a

b)

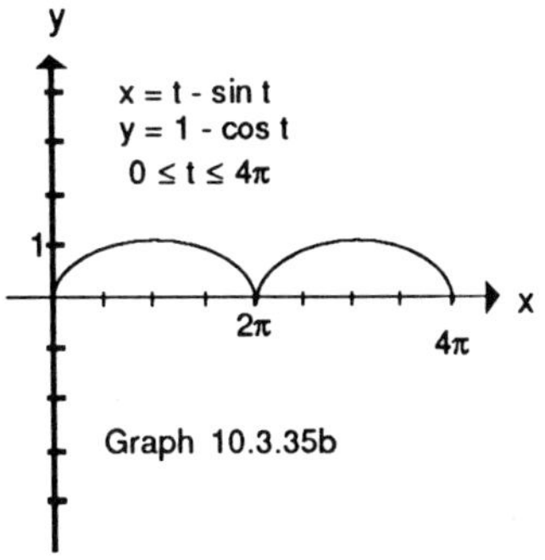

Graph 10.3.35b

c)

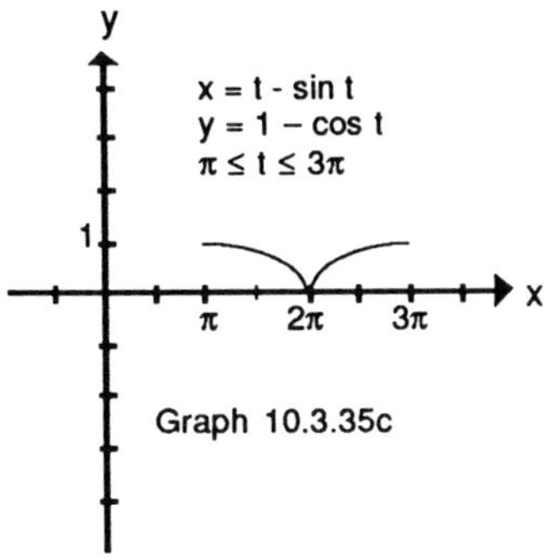

Graph 10.3.35c

36. a)

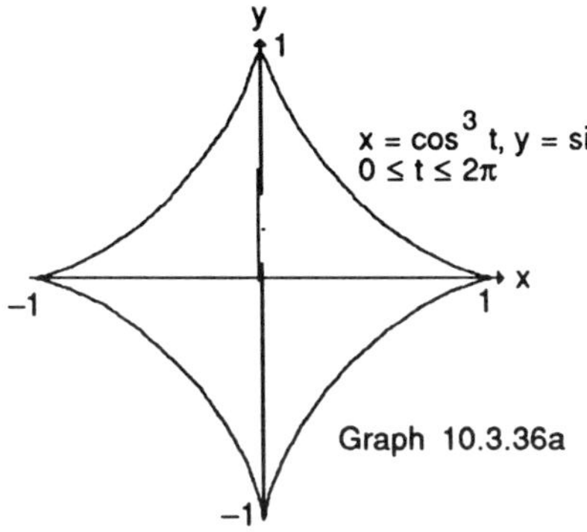

Graph 10.3.36a

b)

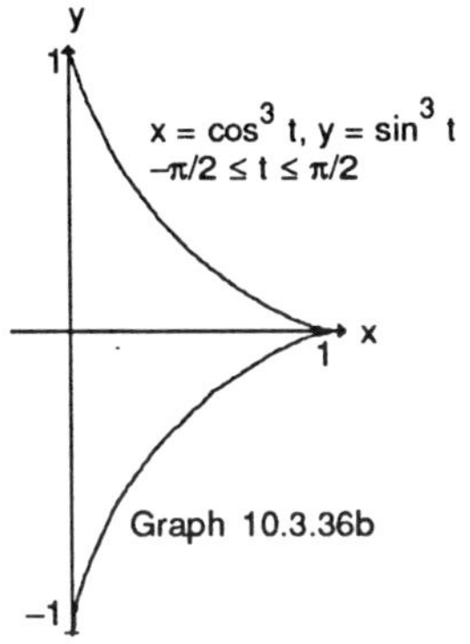

Graph 10.3.36b

37. a)

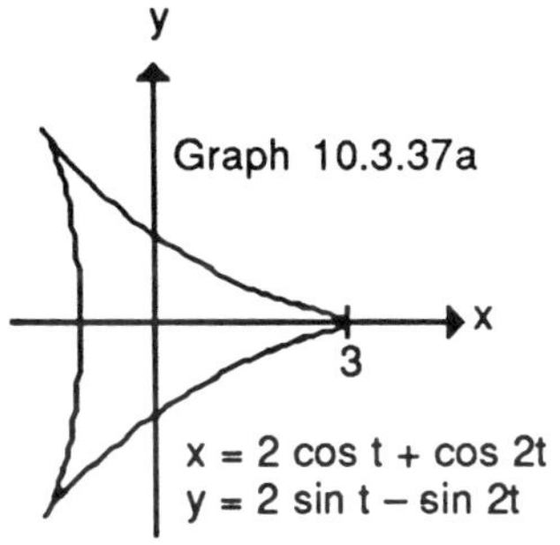

Graph 10.3.37a

b)

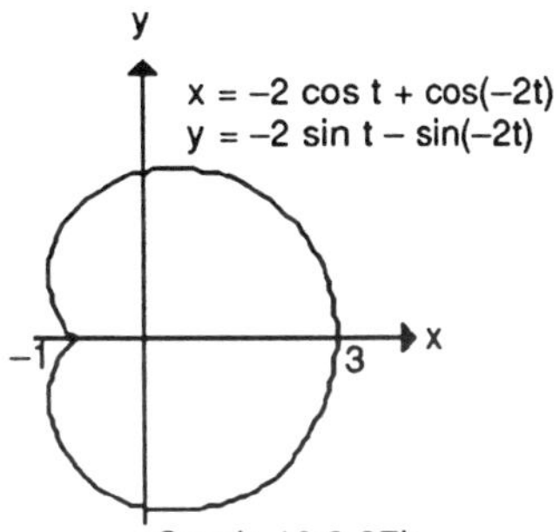

Graph 10.3.37b

38. a)

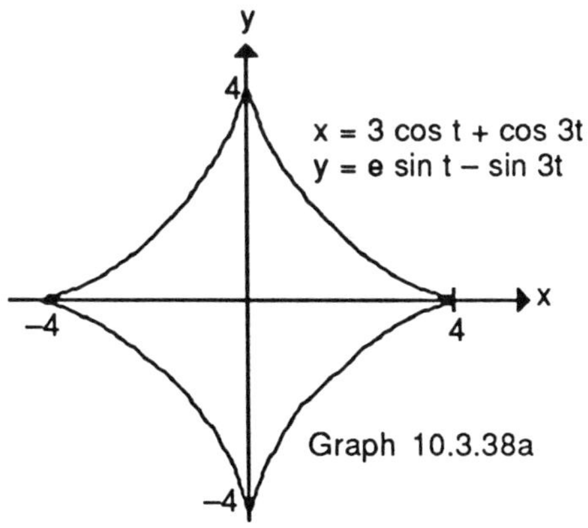

Graph 10.3.38a

b)

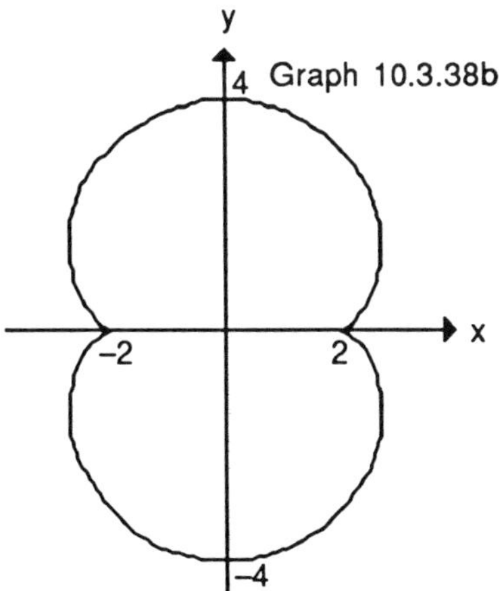

Graph 10.3.38b

x = −3 cos t + cos(−3t)
y = −3 sin t − sin(−3t)

39. a)

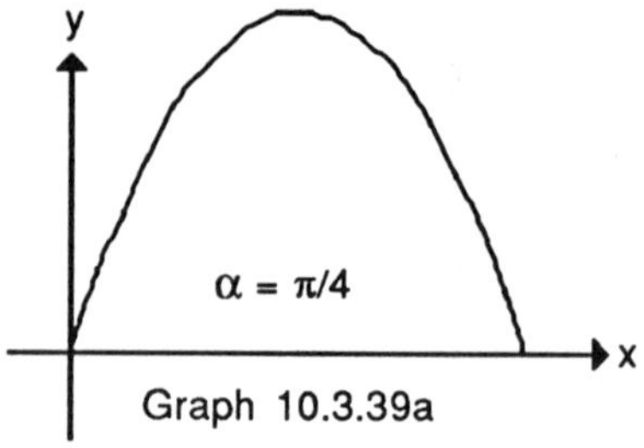

b)

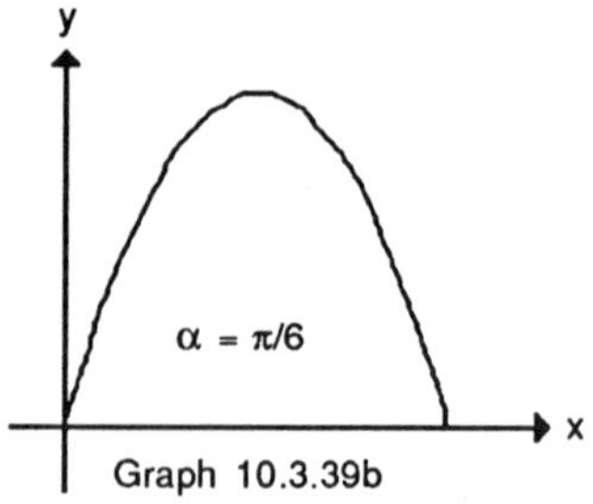

c)

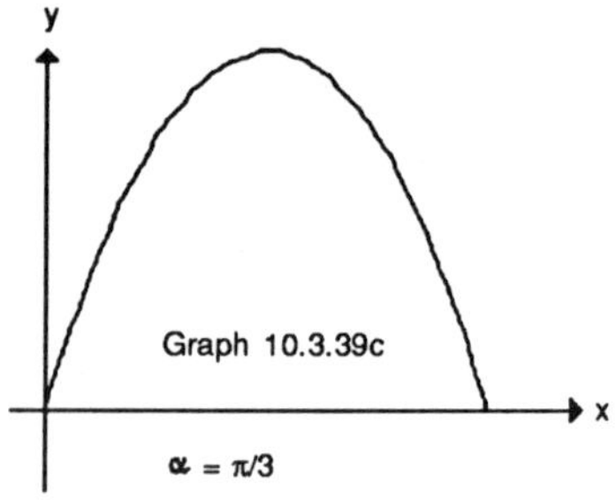

d)

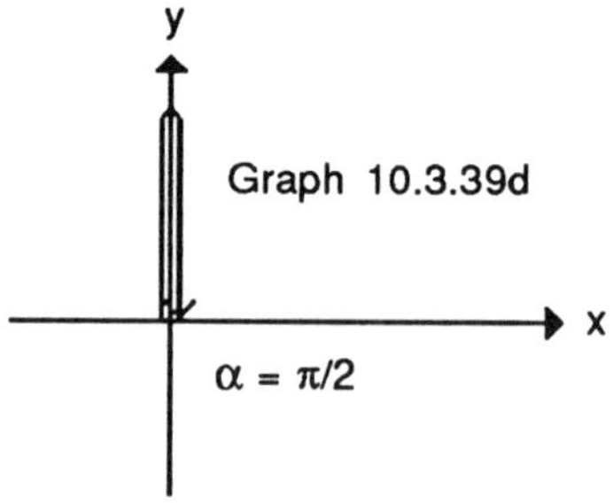

40. a)

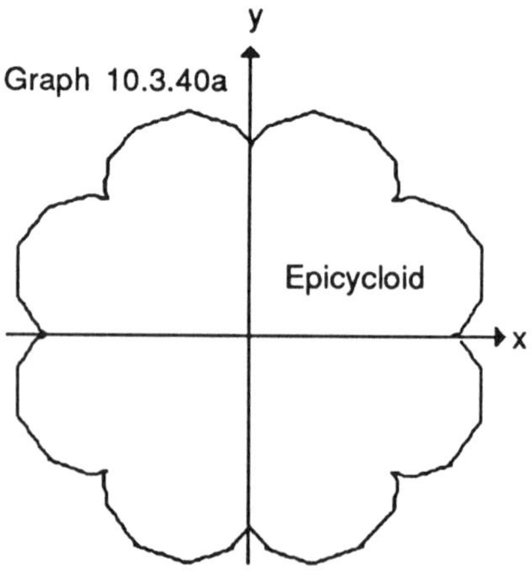

b)

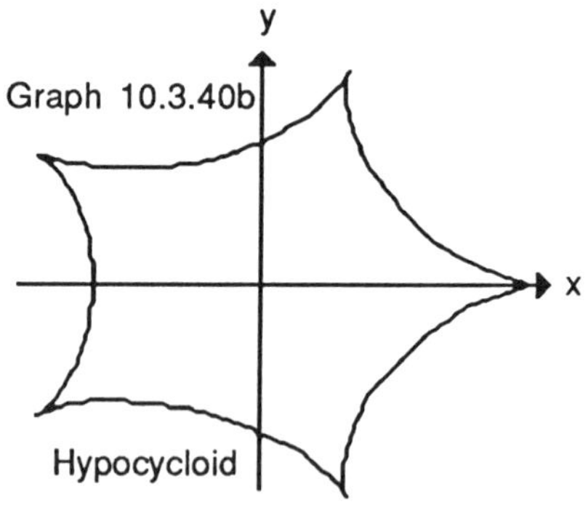

c)

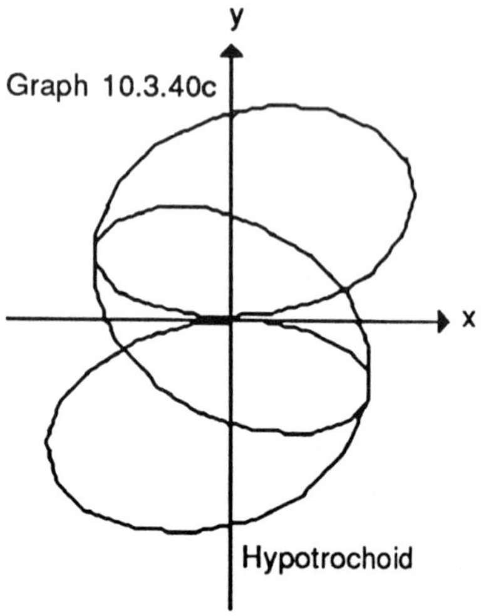

SECTION 10.4 THE CALCULUS OF PARAMETRIC EQUATIONS

1. $t = \frac{\pi}{4} \Rightarrow x = 2\cos\frac{\pi}{4} = \sqrt{2}$, $y = 2\sin\frac{\pi}{4} = \sqrt{2}$. $\frac{dx}{dt} = -2\sin t$, $\frac{dy}{dt} = 2\cos t \Rightarrow \frac{dy}{dx} = \frac{2\cos t}{-2\sin t} = -\cot t$.
$\therefore \frac{dy}{dx}\left(\frac{\pi}{4}\right) = -\cot\frac{\pi}{4} = -1$. $\therefore$ Tangent line is $y - \sqrt{2} = -1(x - \sqrt{2}) \Rightarrow y = -x + 2\sqrt{2}$.
$\frac{dy'}{dt} = \csc^2 t \Rightarrow \frac{d^2y}{dx^2} = \frac{\csc^2 t}{-2\sin t} = -\frac{1}{2\sin^3 t} \Rightarrow \frac{d^2y}{dx^2}\left(\frac{\pi}{4}\right) = -\sqrt{2}$

2. $t = -\frac{1}{6} \Rightarrow x = \sin 2\pi\left(-\frac{1}{6}\right) = \sin\left(-\frac{\pi}{3}\right) = -\frac{\sqrt{3}}{2}$, $y = \cos 2\pi\left(-\frac{1}{6}\right) = \cos\left(-\frac{\pi}{3}\right) = \frac{1}{2}$. $\frac{dx}{dt} = 2\pi\cos 2\pi t$,
$\frac{dy}{dt} = -2\pi\sin 2\pi t \Rightarrow \frac{dy}{dx} = \frac{-2\pi\sin 2\pi t}{2\pi\cos 2\pi t} = -\tan 2\pi t \Rightarrow \frac{dy}{dx}\left(-\frac{1}{6}\right) = -\tan 2\pi\left(-\frac{1}{6}\right) = -\tan\left(-\frac{\pi}{3}\right) = \sqrt{3}$.
$\therefore$ Tangent line is $y - \frac{1}{2} = \sqrt{3}\left(x - \left(-\frac{\sqrt{3}}{2}\right)\right) \Rightarrow y = \sqrt{3}\,x + 2$.
$\frac{dy'}{dt} = -2\pi\sec^2 2\pi t \Rightarrow \frac{d^2y}{dx^2} = \frac{-2\pi\sec^2 2\pi t}{2\pi\cos 2\pi t} = -\frac{1}{\cos^3 2\pi t} \Rightarrow \frac{d^2y}{dx^2}\left(-\frac{1}{6}\right) = 8$

3. $t = \frac{\pi}{4} \Rightarrow x = 4\sin\frac{\pi}{4} = 2\sqrt{2}$, $y = 2\cos\frac{\pi}{4}$. $\frac{dx}{dt} = 4\cos t$, $\frac{dy}{dt} = -2\sin t \Rightarrow \frac{dy}{dx} = \frac{-2\sin t}{4\cos t} = -\frac{1}{2}\tan t \Rightarrow$
$\frac{dy}{dx}\left(\frac{\pi}{4}\right) = -\frac{1}{2}\tan\frac{\pi}{4} = -\frac{1}{2}$. $\therefore$ Tangent line is $y - \sqrt{2} = -\frac{1}{2}(x - 2\sqrt{2}) \Rightarrow y = -\frac{1}{2}x + 2\sqrt{2}$.
$\frac{dy'}{dt} = -\frac{1}{2}\sec^2 t \Rightarrow \frac{d^2y}{dx^2} = \frac{-\frac{1}{2}\sec^2 t}{4\cos t} = -\frac{1}{8\cos^3 t} \Rightarrow \frac{d^2y}{dx^2}\left(\frac{\pi}{4}\right) = -\frac{\sqrt{2}}{4}$

4. $t = \frac{2\pi}{3} \Rightarrow x = \cos\frac{2\pi}{3} = -\frac{1}{2}$, $y = \sqrt{3}\cos\frac{2\pi}{3} = -\frac{\sqrt{3}}{2}$. $\frac{dx}{dt} = -\sin t$, $\frac{dy}{dt} = -\sqrt{3}\sin t \Rightarrow \frac{dy}{dx} = \frac{-\sqrt{3}\sin t}{-\sin t} = \sqrt{3}$
$\Rightarrow \frac{dy}{dx}\left(\frac{2\pi}{3}\right) = \sqrt{3}$. $\therefore$ Tangent line is $y - \left(-\frac{\sqrt{3}}{2}\right) = \sqrt{3}\left(x - \left(-\frac{1}{2}\right)\right) \Rightarrow y = \sqrt{3}\,x$.
$\frac{dy'}{dt} = 0 \Rightarrow \frac{d^2y}{dx^2} = \frac{0}{-\sin t} = 0 \Rightarrow \frac{d^2y}{dx^2}\left(\frac{2\pi}{3}\right) = 0$

5. $t = \frac{1}{4} \Rightarrow x = \frac{1}{4}$, $y = \frac{1}{2}$. $\frac{dx}{dt} = 1$, $\frac{dy}{dt} = \frac{1}{2\sqrt{t}} \Rightarrow \frac{dy}{dx} = \frac{1}{2\sqrt{t}} \Rightarrow \frac{dy}{dx}\left(\frac{1}{4}\right) = \frac{1}{2\sqrt{\frac{1}{4}}} = 1$.
$\therefore$ Tangent line is $y - \frac{1}{2} = 1\left(x - \frac{1}{4}\right) \Rightarrow y = x + \frac{1}{4}$. $\frac{dy'}{dt} = -\frac{1}{4}t^{-3/2} \Rightarrow \frac{d^2y}{dx^2} = -\frac{1}{4}t^{-3/2} \Rightarrow \frac{d^2y}{dx^2}\left(\frac{1}{4}\right) = -2$.

6. $t = -\frac{\pi}{4} \Rightarrow x = \sec^2\left(-\frac{\pi}{4}\right) - 1 = 1$, $y = \tan\left(-\frac{\pi}{4}\right) = -1$. $\frac{dx}{dt} = 2\sec^2 t\tan t$, $\frac{dy}{dt} = \sec^2 t \Rightarrow$
$\frac{dy}{dx} = \frac{\sec^2 t}{2\sec^2 t\tan t} = \frac{1}{2\tan t} = \frac{1}{2}\cot t \Rightarrow \frac{dy}{dx}\left(-\frac{\pi}{4}\right) = \frac{1}{2}\cot\left(-\frac{\pi}{4}\right) = -\frac{1}{2}$. $\therefore$ Tangent line is
$y - (-1) = -\frac{1}{2}(x - 1) \Rightarrow y = -\frac{1}{2}x - \frac{1}{2}$. $\frac{dy'}{dt} = -\frac{1}{2}\csc^2 t \Rightarrow \frac{d^2y}{dx^2} = \frac{-\frac{1}{2}\csc^2 t}{2\sec^2 t\tan t} = -\frac{1}{4}\cot^3 t \Rightarrow \frac{d^2y}{dx^2}\left(-\frac{\pi}{4}\right) =$
$\frac{1}{4}$

7. $t = \frac{\pi}{6} \Rightarrow x = \sec\frac{\pi}{6} = \frac{2}{\sqrt{3}}$, $y = \tan\frac{\pi}{6} = \frac{1}{\sqrt{3}}$. $\frac{dx}{dt} = \sec t \tan t$, $\frac{dy}{dt} = \sec^2 t \Rightarrow \frac{dy}{dx} = \frac{\sec^2 t}{\sec t \tan t} = \csc t \Rightarrow$ $\frac{dy}{dx}\left(\frac{\pi}{6}\right) = \csc\frac{\pi}{6} = 2$. $\therefore$ Tangent line is $y - \frac{1}{\sqrt{3}} = 2\left(x - \frac{2}{\sqrt{3}}\right) \Rightarrow y = 2x - \sqrt{3}$. $\frac{dy'}{dt} = -\csc t \cot t \Rightarrow$ $\frac{d^2y}{dx^2} = \frac{-\csc t \cot t}{\sec t \tan t} = -\cot^3 t \Rightarrow \frac{d^2y}{dx^2}\left(\frac{\pi}{6}\right) = -3\sqrt{3}$

8. $t = 3 \Rightarrow x = -\sqrt{3+1} = -2$, $y = \sqrt{3(3)} = 3$. $\frac{dx}{dt} = -\frac{1}{2}(t+1)^{-1/2}$, $\frac{dy}{dt} = \frac{3}{2}(3t)^{-1/2} \Rightarrow \frac{dy}{dx} = \frac{\frac{3}{2}(3t)^{-1/2}}{-\frac{1}{2}(t+1)^{-1/2}} =$ $-\frac{3\sqrt{t+1}}{\sqrt{3t}} \Rightarrow \frac{dy}{dx}(3) = \frac{-3\sqrt{3+1}}{\sqrt{3(3)}} = -2$. $\therefore$ Tangent line is $y - 3 = -2(x - (-2)) \Rightarrow y = -2x - 1$.

$$\frac{dy'}{dt} = \frac{\sqrt{3t}\left(-\frac{3}{2}(t+1)^{-1/2}\right) + 3\sqrt{t+1}\left(\frac{3}{2}(3t)^{-1/2}\right)}{3t} = \frac{3}{2t\sqrt{3t}\sqrt{t+1}} \Rightarrow \frac{d^2y}{dx^2} = \frac{\frac{3}{2t\sqrt{3t}\sqrt{t+1}}}{\frac{-1}{2\sqrt{t+1}}} = -\frac{3}{t\sqrt{3t}} \Rightarrow$$

$\frac{d^2y}{dx^2}(3) = -\frac{1}{3}$

9. $t = -1 \Rightarrow x = 5$, $y = 1$. $\frac{dx}{dt} = 4t$, $\frac{dy}{dt} = 4t^3 \Rightarrow \frac{dy}{dx} = \frac{4t^3}{4t} = t^2 \Rightarrow \frac{dy}{dx}(-1) = (-1)^2 = 1$. $\therefore$ Tangent line is $y - 1 = 1(x - 5) \Rightarrow y = x - 4$. $\frac{dy'}{dt} = 2t \Rightarrow \frac{d^2y}{dx^2} = \frac{2t}{4t} = \frac{1}{2} \Rightarrow \frac{d^2y}{dx^2}(-1) = \frac{1}{2}$

10. $t = 1 \Rightarrow x = 1$, $y = -2 + \ln 1 = -2$. $\frac{dx}{dt} = -\frac{1}{t^2}$, $\frac{dy}{dt} = \frac{1}{t} \Rightarrow \frac{dy}{dx} = \frac{\frac{1}{t}}{-\frac{1}{t^2}} = -t \Rightarrow \frac{dy}{dx}(1) = -1$. $\therefore$ Tangent line is $y - (-2) = -1(x - 1) \Rightarrow y = -x - 1$. $\frac{dy'}{dt} = -1 \Rightarrow \frac{d^2y}{dx^2} = \frac{-1}{-\frac{1}{t^2}} = t^2 \Rightarrow \frac{d^2y}{dx^2}(1) = 1$

11. $t = \frac{\pi}{3} \Rightarrow x = \frac{\pi}{3} - \sin\frac{\pi}{3} = \frac{\pi}{3} - \frac{\sqrt{3}}{2}$, $y = 1 - \cos\frac{\pi}{3} = 1 - \frac{1}{2} = \frac{1}{2}$. $\frac{dx}{dt} = 1 - \cos t$, $\frac{dy}{dt} = \sin t \Rightarrow \frac{dy}{dx} = \frac{\sin t}{1 - \cos t}$ $\Rightarrow \frac{dy}{dx}\left(\frac{\pi}{3}\right) = \frac{\sin\frac{\pi}{3}}{1 - \cos\frac{\pi}{3}} = \frac{\frac{\sqrt{3}}{2}}{\frac{1}{2}} = \sqrt{3}$. $\therefore$ Tangent line is $y - \frac{1}{2} = \sqrt{3}\left(x - \frac{\pi}{3} + \frac{\sqrt{3}}{2}\right) \Rightarrow y = \sqrt{3}\,x - \frac{\pi\sqrt{3}}{3} + 2$.

$$\frac{dy'}{dt} = \frac{(1 - \cos t)\cos t - \sin t(\sin t)}{(1 - \cos t)^2} = \frac{-1}{1 - \cos t} \Rightarrow \frac{d^2y}{dx^2} = \frac{\frac{-1}{1 - \cos t}}{1 - \cos t} = \frac{-1}{(1 - \cos t)^2} \Rightarrow \frac{d^2y}{dx^2}\left(\frac{\pi}{3}\right) = -4$$

12. $t = \frac{\pi}{2} \Rightarrow x = \cos\frac{\pi}{2} = 0$, $y = 1 + \sin\frac{\pi}{2} = 2$. $\frac{dx}{dt} = -\sin t$, $\frac{dy}{dt} = \cos t \Rightarrow \frac{dy}{dx} = \frac{\cos t}{-\sin t} = -\cot t \Rightarrow \frac{dy}{dx}\left(\frac{\pi}{2}\right) =$ $-\cot\frac{\pi}{2} = 0$. $\therefore$ Tangent line is $y = 2$. $\frac{dy'}{dt} = -\csc^2 t \Rightarrow \frac{d^2y}{dx^2} = \frac{-\csc^2 t}{-\sin t} = \csc^3 t \Rightarrow \frac{d^2y}{dx^2}\left(\frac{\pi}{2}\right) = 1$

13. $\frac{dx}{dt} = -\sin t,\ \frac{dy}{dt} = 1 + \cos t \Rightarrow \sqrt{\left(\frac{dx}{dt}\right)^2 + \left(\frac{dy}{dt}\right)^2} = \sqrt{(-\sin t)^2 + (1 + \cos t)^2} = \sqrt{2 + 2\cos t}$.

$\therefore$ Length $= \int_0^{\pi} \sqrt{2 + 2\cos t}\, dt = \sqrt{2} \int_0^{\pi} \sqrt{\frac{1 - \cos t}{1 - \cos t}(1 + \cos t)}\, dt = \sqrt{2} \int_0^{\pi} \sqrt{\frac{\sin^2 t}{1 - \cos t}}\, dt =$

$\sqrt{2} \int_0^{\pi} \frac{\sin t}{\sqrt{1 - \cos t}} dt$ (since $\sin t \geq 0$ on $[0,\pi]$) $= \sqrt{2} \int_0^{2} u^{-1/2}\, du = \sqrt{2}\left[2u^{1/2}\right]_0^2 = 4.$

(Let $u = 1 - \cos t \Rightarrow du = \sin t\, dt;\ t = 0 \Rightarrow u = 0,\ t = \pi \Rightarrow u = 2$)

14. $\frac{dx}{dt} = 3t^2,\ \frac{dy}{dt} = 3t \Rightarrow \sqrt{\left(\frac{dx}{dt}\right)^2 + \left(\frac{dy}{dt}\right)^2} = \sqrt{(3t^2)^2 + (3t)^2} = \sqrt{9t^4 + 9t^2} = 3t\sqrt{t^2 + 1}$ (since $t \geq 0$ on $[0,\sqrt{3}]$)

$\therefore$ Length $= \int_0^{\sqrt{3}} 3t\sqrt{t^2 + 1}\, dt = \int_1^{4} \frac{3}{2} u^{1/2}\, du = \left[u^{3/2}\right]_1^4 = 7$

(Let $u = t^2 + 1 \Rightarrow \frac{3}{2} du = 3t\, dt;\ t = 0 \Rightarrow u = 1,\ t = \sqrt{3} \Rightarrow u = 4$)

15. $\frac{dx}{dt} = t,\ \frac{dy}{dt} = (2t + 1)^{1/2} \Rightarrow \sqrt{\left(\frac{dx}{dt}\right)^2 + \left(\frac{dy}{dt}\right)^2} = \sqrt{t^2 + \left((2t + 1)^{1/2}\right)^2} = |t + 1| = t + 1$ since $0 \leq t \leq 4$.

$\therefore$ Length $= \int_0^{4} (t + 1)\, dt = \left[\frac{t^2}{2} + t\right]_0^4 = 12.$

16. $\frac{dx}{dt} = (2t + 3)^{1/2},\ \frac{dy}{dt} = 1 + t \Rightarrow \sqrt{\left(\frac{dx}{dt}\right)^2 + \left(\frac{dy}{dt}\right)^2} = \sqrt{\left((2t + 3)^{1/2}\right)^2 + (1 + t)^2} = |t + 2| = t + 2$ since

$0 \leq t \leq 3$. $\therefore$ Length $= \int_0^{3} (t + 2)\, dt = \left[\frac{t^2}{2} + 2t\right]_0^3 = \frac{21}{2}.$

17. $\frac{dx}{dt} = 8t\cos t,\ \frac{dy}{dt} = 8t\sin t \Rightarrow \sqrt{\left(\frac{dx}{dt}\right)^2 + \left(\frac{dy}{dt}\right)^2} = \sqrt{(8t\cos t)^2 + (8t\sin t)^2} = \sqrt{64t^2\cos^2 t + 64t^2\sin^2 t} = |8t|$

$= 8t$ since $0 \leq t \leq \frac{\pi}{2}$. $\therefore$ Length $= \int_0^{\pi/2} 8t\, dt = \left[4t^2\right]_0^{\pi/2} = \pi^2.$

18. $\frac{dx}{dt} = \frac{1}{\sec t + \tan t}(\sec t \tan t + \sec^2 t) - \cos t = \sec t - \cos t,\ \frac{dy}{dt} = -\sin t \Rightarrow \sqrt{\left(\frac{dx}{dt}\right)^2 + \left(\frac{dy}{dt}\right)^2} =$

$\sqrt{(\sec t - \cos t)^2 + (-\sin t)^2} = \sqrt{\sec^2 t - 1} = \sqrt{\tan^2 t} = |\tan t| = \tan t$ since $0 \leq t \leq \frac{\pi}{3}$.

$\therefore$ Length $= \int_0^{\pi/3} \tan t\, dt = \int_0^{\pi/3} \frac{\sin t}{\cos t} dt = \left[-\ln(\cos t)\right]_0^{\pi/3} = -\ln\frac{1}{2} = \ln 2.$

19. $\frac{dx}{dt} = -\sin t$, $\frac{dy}{dt} = \cos t \Rightarrow \sqrt{\left(\frac{dx}{dt}\right)^2 + \left(\frac{dy}{dt}\right)^2} = \sqrt{(-\sin t)^2 + (\cos t)^2} = 1.$

$\therefore$ Area $= \int_0^{2\pi} 2\pi(2 + \sin t)1\, dt = 2\pi[2t - \cos t]_0^{2\pi} = 8\pi^2$

20. $\frac{dx}{dt} = t^{1/2}$, $\frac{dy}{dt} = t^{-1/2} \Rightarrow \sqrt{\left(\frac{dx}{dt}\right)^2 + \left(\frac{dy}{dt}\right)^2} = \sqrt{(t^{1/2})^2 + (t^{-1/2})^2} = \sqrt{t + t^{-1}} = \sqrt{\frac{t^2+1}{t}}.$

$\therefore$ Area $= \int_0^{\sqrt{3}} 2\pi\left(\frac{2}{3}t^{3/2}\right)\sqrt{\frac{t^2+1}{t}}\, dt = \int_0^{\sqrt{3}} \frac{4\pi}{3} t\sqrt{t^2+1}\, dt = \int_1^4 \frac{2\pi}{3}\sqrt{u}\, du = \left[\frac{4\pi}{9}u^{3/2}\right]_1^4 = \frac{28\pi}{9}.$

(Let $u = t^2 + 1 \Rightarrow du = 2tdt$; $t = 0 \Rightarrow u = 1$, $t = \sqrt{3} \Rightarrow u = 4$)

Note: $\int_0^{\sqrt{3}} 2\pi\left(\frac{2}{3}t^{3/2}\right)\sqrt{\frac{t^2+1}{t}}\, dt$ is an improper integral but $\lim_{t \to 0^+} f(t)$ exists and is equal to 0.

21. $\frac{dx}{dt} = 1$, $\frac{dy}{dt} = t + \sqrt{2} \Rightarrow \sqrt{\left(\frac{dx}{dt}\right)^2 + \left(\frac{dy}{dt}\right)^2} = \sqrt{1^2 + (t + \sqrt{2})^2} = \sqrt{t^2 + 2\sqrt{2}\,t + 3}.$

$\therefore$ Area $= \int_{-\sqrt{2}}^{\sqrt{2}} 2\pi(t + \sqrt{2})\sqrt{t^2 + 2\sqrt{2}\,t + 3}\, dt = \int_1^9 \pi\sqrt{u}\, du = \left[\frac{2}{3}\pi u^{3/2}\right]_1^9 = \frac{52}{3}\pi$

(Let $u = t^2 + 2\sqrt{2}\,t + 3 \Rightarrow du = 2t + 2\sqrt{2}$; $t = -\sqrt{2} \Rightarrow u = 1$, $t = \sqrt{2} \Rightarrow u = 9$)

22. $\sqrt{\left(\frac{dx}{dt}\right)^2 + \left(\frac{dy}{dt}\right)^2} = \tan t$ (See Exercise 18). $\therefore$ Area $= \int_0^{\pi/3} 2\pi\cos t \tan t\, dt = 2\pi \int_0^{\pi/3} \sin t\, dt =$

$2\pi[-\cos t]_0^{\pi/3} = \pi.$

23. $\frac{dx}{dt} = 2$, $\frac{dy}{dt} = 1 \Rightarrow \sqrt{\left(\frac{dx}{dt}\right)^2 + \left(\frac{dy}{dt}\right)^2} = \sqrt{2^2 + 1^2} = \sqrt{5}.$ $\therefore$ Area $= \int_0^1 2\pi(t + 1)\sqrt{5}\, dt =$

$2\pi\sqrt{5}\left[\frac{t^2}{2} + t\right]_0^1 = 3\pi\sqrt{5}.$ The slant height is $\sqrt{5} \Rightarrow$ Area $= \pi(1 + 2)\sqrt{5} = 3\pi\sqrt{5}.$

24. $\frac{dx}{dt} = h$, $\frac{dy}{dt} = r \Rightarrow \sqrt{\left(\frac{dx}{dt}\right)^2 + \left(\frac{dy}{dt}\right)^2} = \sqrt{h^2 + r^2}.$ $\therefore$ Area $= \int_0^1 2\pi rt\sqrt{h^2 + r^2}\, dt =$

$2\pi r\sqrt{h^2 + r^2} \int_0^1 t\, dt = 2\pi r\sqrt{h^2 + r^2}\left[\frac{t^2}{2}\right]_0^1 = \pi r\sqrt{h^2 + r^2}.$ The slant height is $\sqrt{h^2 + r^2} \Rightarrow$

Area $= \pi r\sqrt{h^2 + r^2}$

25. a) $\frac{dx}{dt} = -2\sin 2t$, $\frac{dy}{dt} = 2\cos 2t \Rightarrow \sqrt{\left(\frac{dx}{dt}\right)^2 + \left(\frac{dy}{dt}\right)^2} = \sqrt{(-2\sin 2t)^2 + (2\cos 2t)^2} = 2.$

$\therefore$ Length = $\int_0^{\pi/2} 2\,dt = [2t]_0^{\pi/2} = \pi.$

b) $\frac{dx}{dt} = \pi\cos \pi t$, $\frac{dy}{dt} = -\pi\sin \pi t \Rightarrow \sqrt{\left(\frac{dx}{dt}\right)^2 + \left(\frac{dy}{dt}\right)^2} = \sqrt{(\pi\cos \pi t)^2 + (-\pi\sin \pi t)^2} = \pi.$

$\therefore$ Length = $\int_{-1/2}^{1/2} \pi\,dt = [\pi t]_{-1/2}^{1/2} = \pi.$

26. a) $a = 1,\ e = \frac{1}{2} \Rightarrow$ Length $= 4\int_0^{\pi/2} \sqrt{1 - \frac{1}{4}\cos^2 t}\,dt = 2\int_0^{\pi/2} \sqrt{4 - \cos^2 t}\,dt.$

Use Trapezoid Rule with $n = 10 \Rightarrow h = \frac{b-a}{n} = \frac{\frac{\pi}{2} - 0}{10} = \frac{\pi}{20}.$

x_i	x_i	$f(x_i)$	m	$mf(x_i)$
x_0	0	1.732050808	1	1.732050808
x_1	$\frac{\pi}{20}$	1.739100843	2	3.478201686
x_2	$\frac{\pi}{10}$	1.759400893	2	3.518801786
x_3	$\frac{3\pi}{20}$	1.790560631	2	3.581121262
x_4	$\frac{\pi}{5}$	1.82906848	2	3.658136959
x_5	$\frac{\pi}{4}$	1.870828693	2	3.741657387
x_6	$\frac{3\pi}{10}$	1.911676881	2	3.823353762
x_7	$\frac{7\pi}{20}$	1.947791731	2	3.895583461
x_8	$\frac{2\pi}{5}$	1.975982919	2	3.951965839
x_9	$\frac{9\pi}{20}$	1.993872679	2	3.987745357
x_{10}	$\frac{\pi}{2}$	2	1	2

$\sum_{n=0}^{10} m(f(x_n)) = 37.3686183 \Rightarrow$

$T = \frac{h}{2}(37.3686183) =$

$\frac{\pi}{40}(37.3686183\ ) = 2.934924418$

$\approx \int_0^{\pi/2} \sqrt{4 - \cos^2 t}\,dt \Rightarrow$

Length $\approx 2(2.934924418) =$

5.869848837.

b) $|f''(t)| < 1 \Rightarrow M = 1.\ \therefore\ |E_T| \le$

$\frac{b-a}{12} h^2 M \le \frac{\frac{\pi}{2} - 0}{12}\left(\frac{\pi}{20}\right)^2 1 \le$

0.00032.

27. $\frac{dx}{dt} = -\cos t$, $\frac{dy}{dt} = -2\cos 2t \Rightarrow \frac{dy}{dx} = \frac{-2\cos 2t}{-\cos t} = \frac{2(2\cos^2 t - 1)}{\cos t}$. Let $\frac{2(2\cos^2 t - 1)}{\cos t} = 0 \Rightarrow 2\cos^2 t - 1 = 0$

$\Rightarrow \cos t = \pm\frac{1}{\sqrt{2}} \Rightarrow t = \frac{\pi}{4}, \frac{3\pi}{4}, \frac{5\pi}{4}, \frac{7\pi}{4}$. In the 1st quadrant, $t = \frac{\pi}{4} \Rightarrow x = \sin\frac{\pi}{4} = \frac{\sqrt{2}}{2}$, $y = \sin 2\left(\frac{\pi}{4}\right) = 1$

$\Rightarrow \left(\frac{\sqrt{2}}{2}, 1\right)$ is the point in the 1st quadrant where the tangent line is horizontal.

$x = 0, y = 0 \Rightarrow \sin t = 0 \Rightarrow t = 0$ or $t = \pi$ and $\sin 2t = 0 \Rightarrow t = 0, \frac{\pi}{2}, \pi, \frac{3\pi}{2}$. $\therefore$ $t = 0$ and $t = \pi$ give the tangent lines at the origin. $\frac{dy}{dx}(0) = 2 \Rightarrow y = 2x$ and $\frac{dy}{dx}(\pi) = -2 \Rightarrow y = -2x.$

28. $\frac{dx}{dt} = 2\cos 2t$, $\frac{dy}{dt} = 3\cos 3t \Rightarrow \frac{dy}{dx} = \frac{3\cos 3t}{2\cos 2t} = \frac{3(\cos 2t \cos t - \sin 2t \sin t)}{2(2\cos^2 t - 1)} =$

$\frac{3((2\cos^2 t - 1)\cos t - 2\sin t \cos t \sin t)}{2(2\cos^2 t - 1)} = \frac{3\cos t(2\cos^2 t - 1 - 2\sin^2 t)}{2(2\cos^2 t - 1)} = \frac{3\cos t(4\cos^2 t - 3)}{2(2\cos^2 t - 1)}$.

Let $\frac{dy}{dx} = 0 \Rightarrow \frac{3\cos t(4\cos^2 t - 3)}{2(2\cos^2 t - 1)} = 0 \Rightarrow 3\cos t = 0$ or $4\cos^2 t - 3 = 0$. $3\cos t = 0 \Rightarrow t = \frac{\pi}{2}, \frac{3\pi}{2}$.

$4\cos^2 t - 3 = 0 \Rightarrow \cos t = \pm\frac{\sqrt{3}}{2} \Rightarrow t = \frac{\pi}{6}, \frac{5\pi}{6}, \frac{7\pi}{6}, \frac{11\pi}{6}$. In the 1st quadrant, $t = \frac{\pi}{6}$ gives the point where the graph has a horizontal tangent line. $t = \frac{\pi}{6} \Rightarrow x = \sin 2\left(\frac{\pi}{6}\right) = \frac{\sqrt{3}}{2}$ and $y = \sin 3\left(\frac{\pi}{6}\right) = 1 \Rightarrow$ The point is $\left(\frac{\sqrt{3}}{2}, 1\right)$.

$x = 0$, $y = 0 \Rightarrow \sin 2t = 0$ and $\sin 3t = 0$. $\sin 2t = 0 \Rightarrow t = 0, \frac{\pi}{2}, \pi, \frac{3\pi}{2}, 2\pi$.

$\sin 3t = 0 \Rightarrow t = 0, \frac{\pi}{3}, \frac{2\pi}{3}, \pi, \frac{4\pi}{3}, \frac{5\pi}{3}, 2\pi$ $\therefore$ $t = 0$ and $t = \pi$ give the tangent lines at the origin.

$\frac{dy}{dx}(0) = \frac{3\cos 3(0)}{2\cos 2(0)} = \frac{3}{2} \Rightarrow y = \frac{3}{2}x$. $\frac{dy}{dx}(\pi) = \frac{3\cos 3\pi}{2\cos 2\pi} = -\frac{3}{2} \Rightarrow y = -\frac{3}{2}x$.

29.

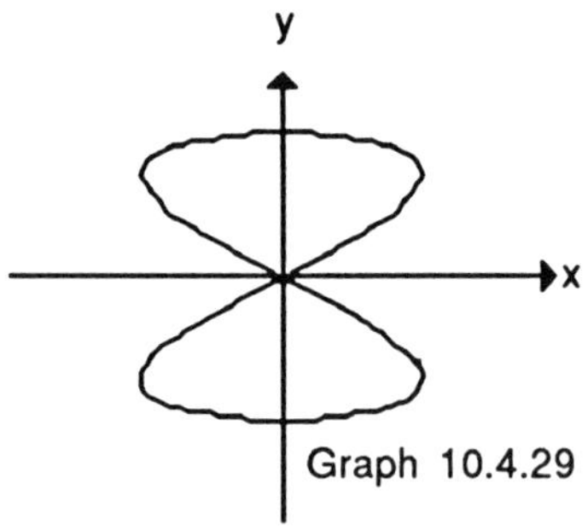

Graph 10.4.29

30.

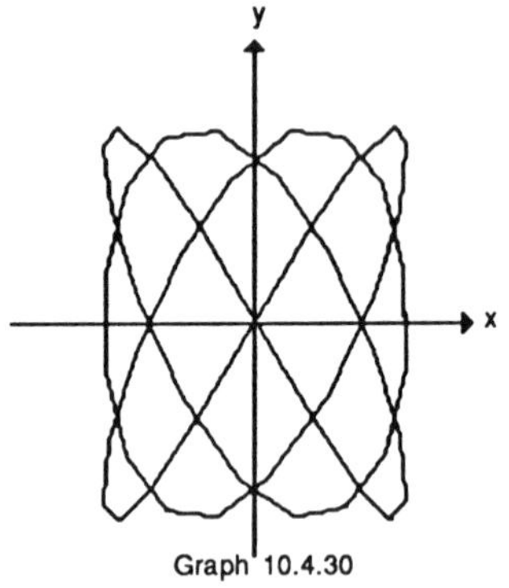

Graph 10.4.30

31.

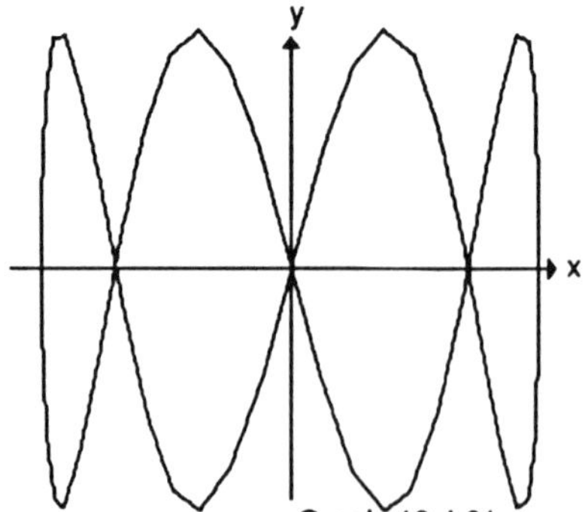

Graph 10.4.31

32.

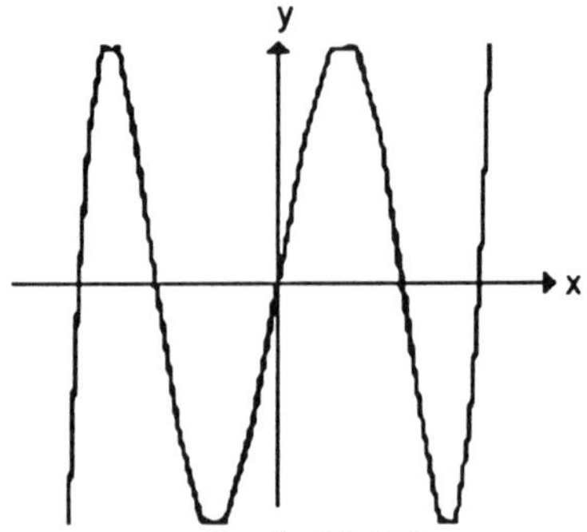

Graph 10.4.32

33.

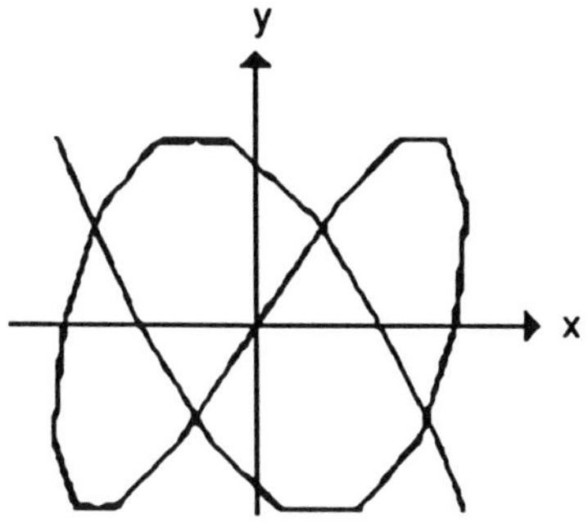

Graph 10.4.33

34.

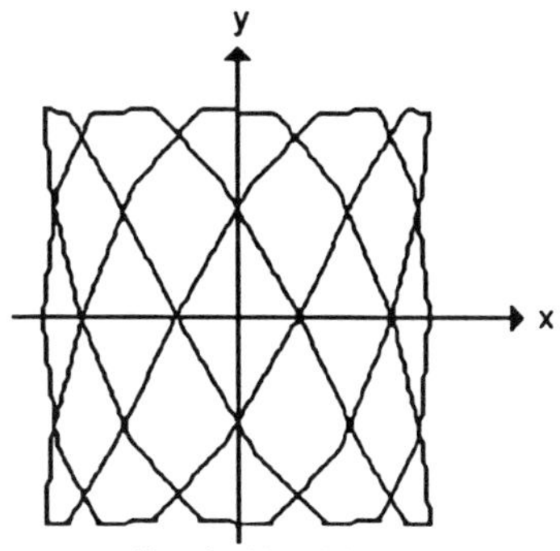

Graph 10.4.34

35.

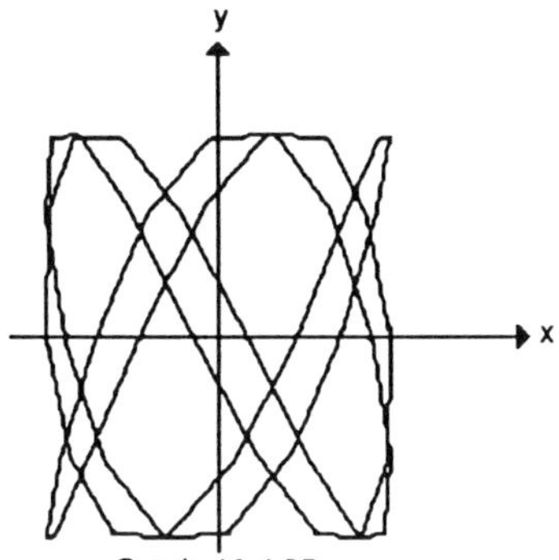

Graph 10.4.35

SECTION 10.5 POLAR COORDINATES

1. a, c; b, d; e, k; g, j; h, f; i, l; m, o; n, p

2. a) (3,0) b) (−3,0) c) (3,0) d) (−3,0) e) $(-1,\sqrt{3})$ f) $(-1,-\sqrt{3})$

g) $(1,\sqrt{3})$ h) $(-1,-\sqrt{3})$ i) $(1,-\sqrt{3})$ j) $(1,\sqrt{3})$ k) $(-1,\sqrt{3})$ l) $(1,-\sqrt{3})$

3. a) $\left(2,\frac{\pi}{2}+2n\pi\right)$ and $\left(-2,\frac{\pi}{2}+(2n+1)\pi\right)$, n an integer

b) $(2, 2n\pi)$ and $(-2,(2n+1)\pi)$, n an integer

c) $\left(2,\frac{3\pi}{2}+2n\pi\right)$ and $\left(-2,\frac{3\pi}{2}+(2n+1)\pi\right)$, n an integer

d) $(2,(2n+1)\pi)$ and $(-2,2n\pi)$, n an integer

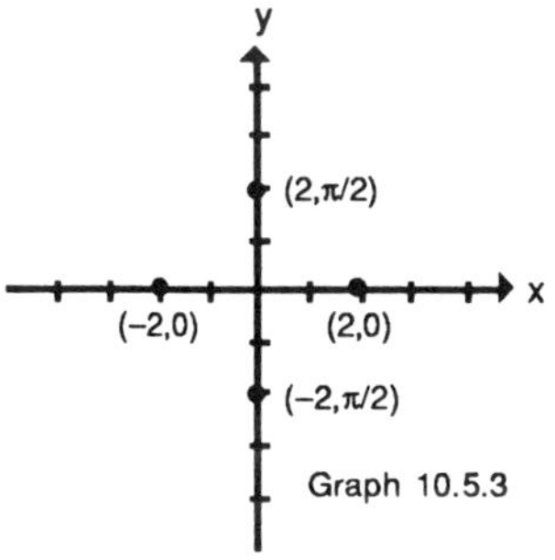

Graph 10.5.3

4. a) $\left(3,\frac{\pi}{4}+2n\pi\right)$ and $\left(-3,\frac{5\pi}{4}+2n\pi\right)$, n an integer

b) $\left(-3,\frac{\pi}{4}+2n\pi\right)$ and $\left(3,\frac{5\pi}{4}+2n\pi\right)$, n an integer

c) $\left(3,-\frac{\pi}{4}+2n\pi\right)$ and $\left(-3,\frac{3\pi}{4}+2n\pi\right)$, n an integer

d) $\left(-3,-\frac{\pi}{4}+2n\pi\right)$ and $\left(3,\frac{3\pi}{4}+2n\pi\right)$, n an integer

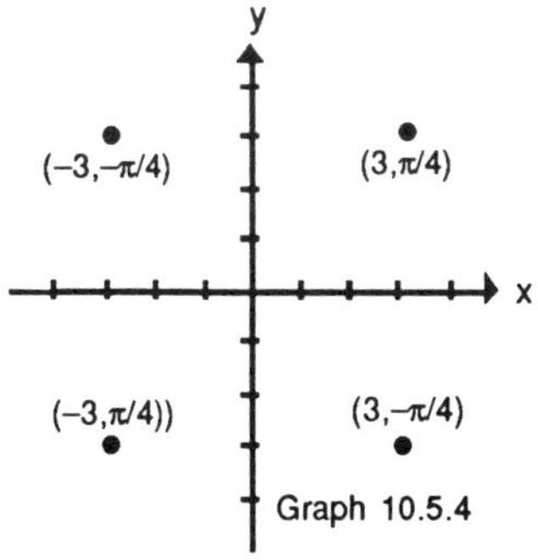

Graph 10.5.4

5. a) $x = \sqrt{2}\cos\frac{\pi}{4} = 1,\ y = \sqrt{2}\sin\frac{\pi}{4} = 1 \Rightarrow (1,1)$

b) $x = 1\cos 0 = 1,\ y = 1\sin 0 = 1 \Rightarrow (1,0)$

c) $x = 0\cos\frac{\pi}{2} = 0,\ y = 0\sin\frac{\pi}{2} = 0 \Rightarrow (0,0)$

d) $x = -\sqrt{2}\cos\frac{\pi}{4} = -1,\ y = -\sqrt{2}\sin\frac{\pi}{2} = -1 \Rightarrow (-1,-1)$

e) $x = -3\cos\frac{5\pi}{6} = \frac{3\sqrt{3}}{2},\ y = -3\sin\frac{5\pi}{6} = -\frac{3}{2}$
$\Rightarrow \left(\frac{3\sqrt{3}}{2},-\frac{3}{2}\right)$

f) $x = 5\cos(\tan^{-1}\frac{4}{3}) = 3,\ y = 5\sin(\tan^{-1}\frac{4}{3}) = 4 \Rightarrow$
(3,4)

5. (Continued)

g) $x = -1\cos 7\pi = 1,\ y = -1\sin 7\pi = 0 \Rightarrow (-1,0)$ h) $x = 2\sqrt{3}\cos\frac{2\pi}{3} = -\sqrt{3},\ y = 2\sqrt{3}\sin\frac{2\pi}{3} = 3$
$\Rightarrow (-\sqrt{3}, 3)$

6. $(0,\theta)$ where θ is any angle.

7.

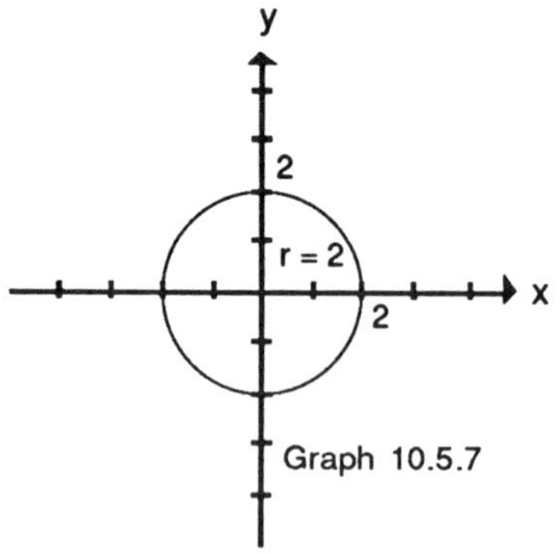

Graph 10.5.7

8.

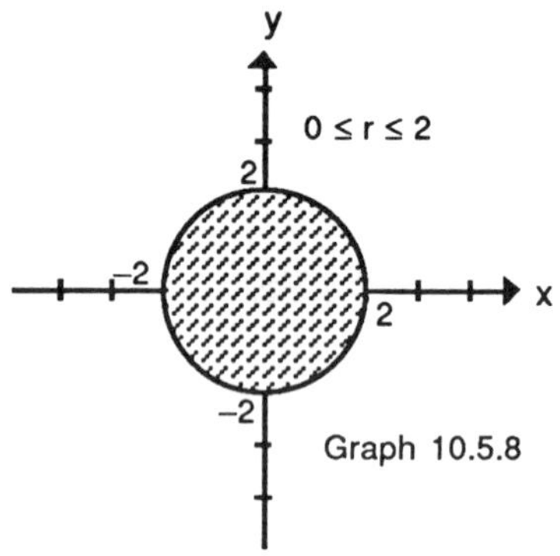

Graph 10.5.8

9.

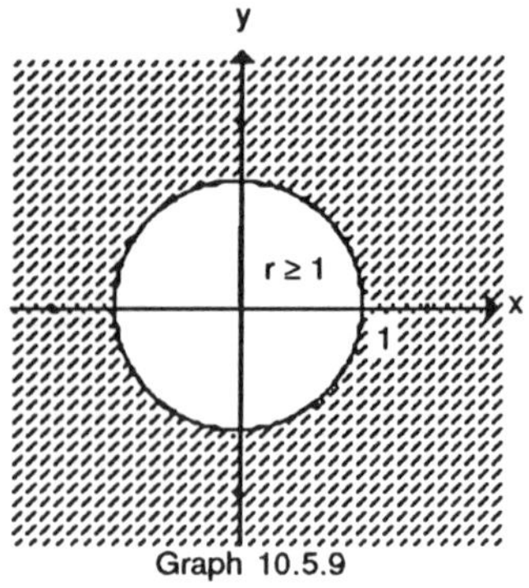

Graph 10.5.9

10.

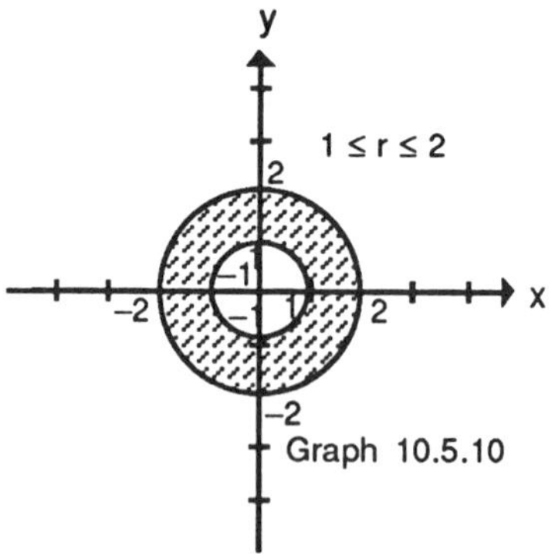

Graph 10.5.10

11.

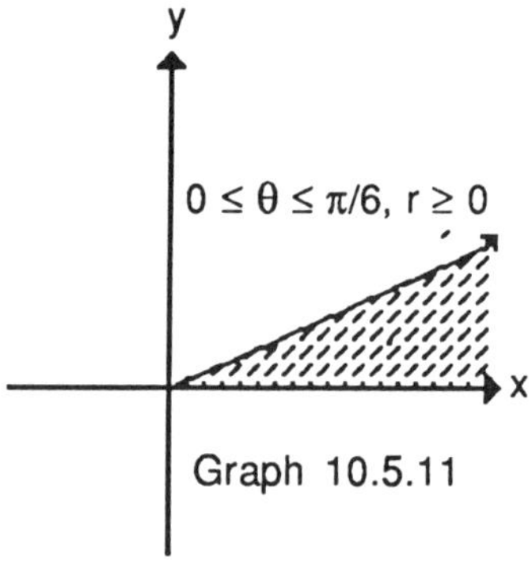

Graph 10.5.11

12.

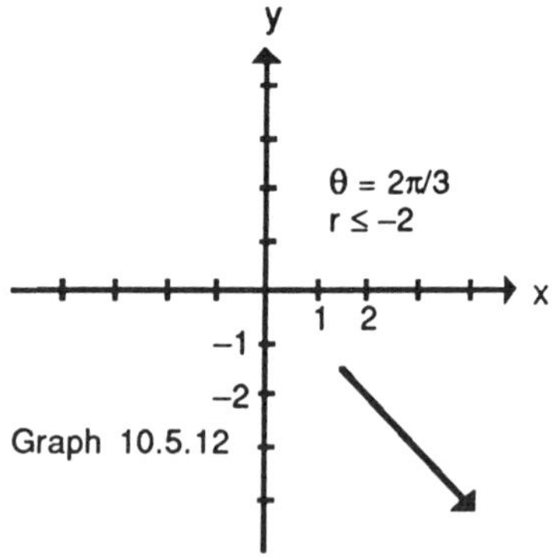

Graph 10.5.12

13.

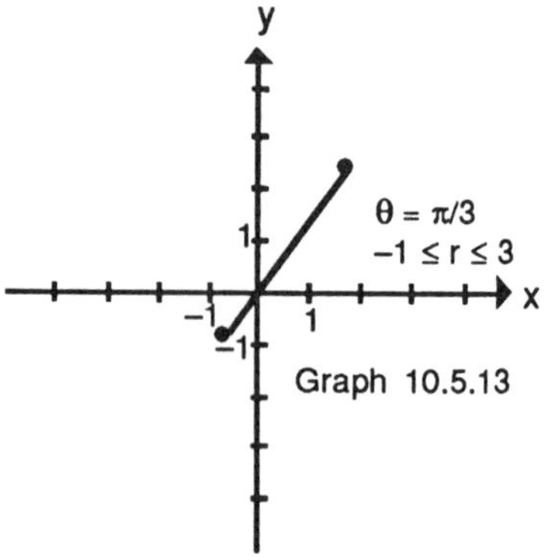

Graph 10.5.13

14.

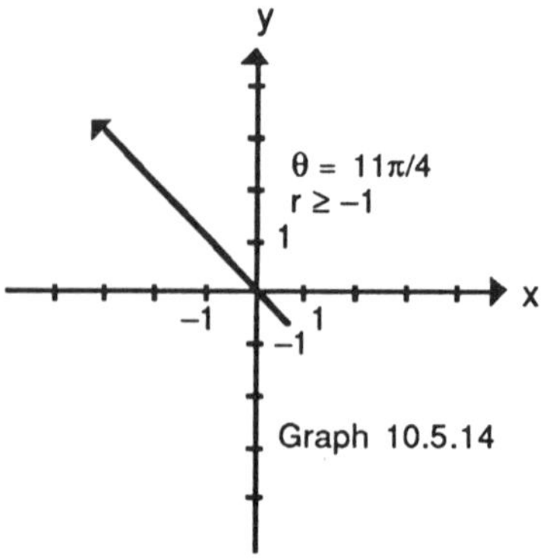

Graph 10.5.14

15.

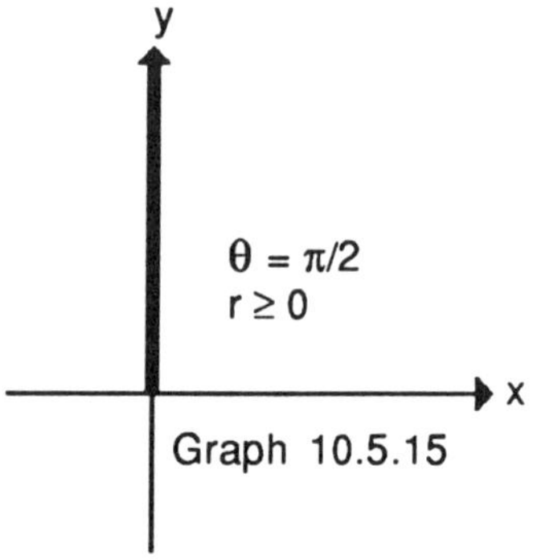

Graph 10.5.15

16.

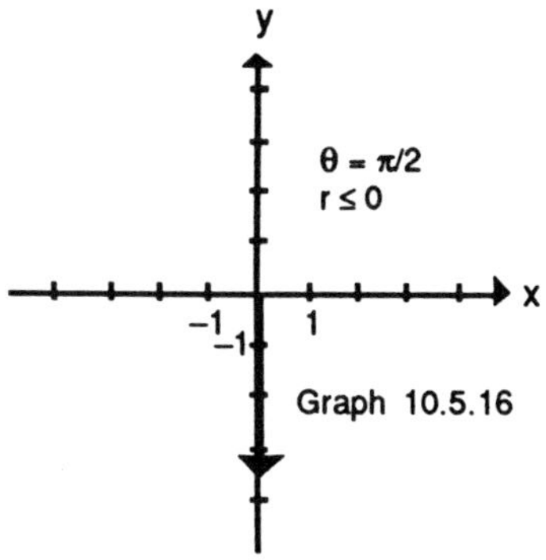

Graph 10.5.16

17.

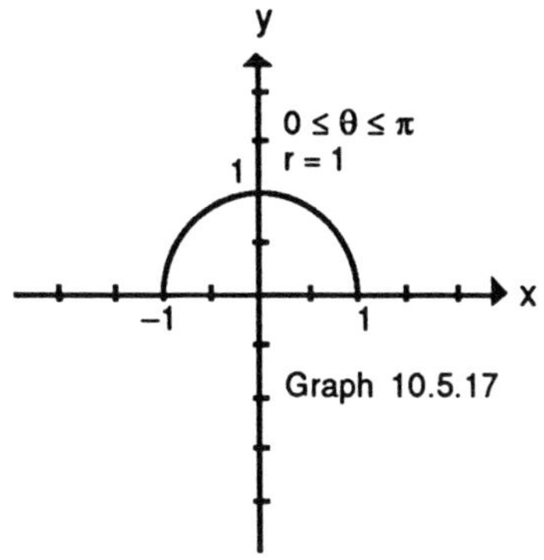

Graph 10.5.17

18.

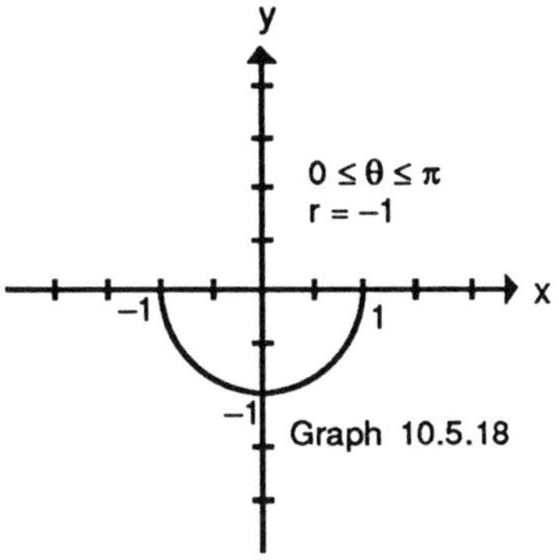

Graph 10.5.18

19.

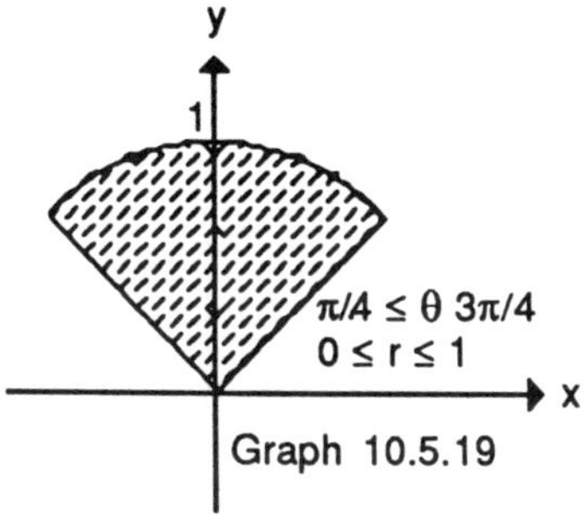

Graph 10.5.19

20.

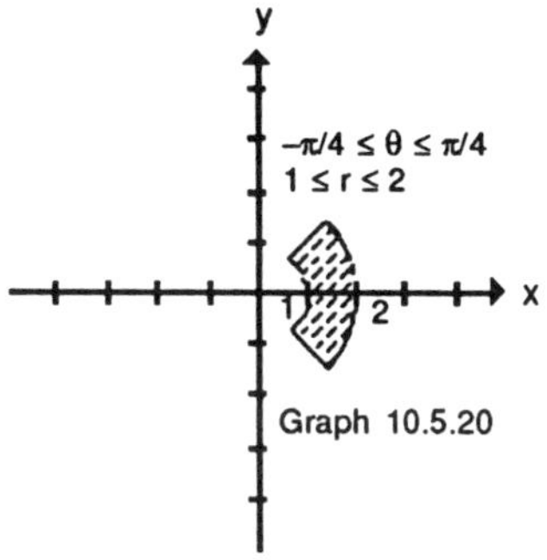

Graph 10.5.20

21.

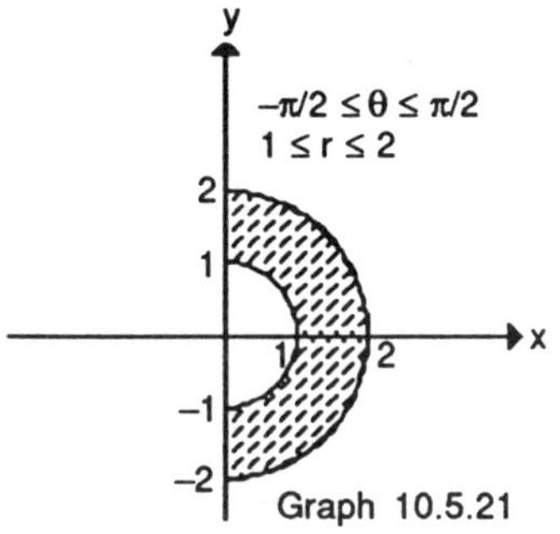

Graph 10.5.21

22.

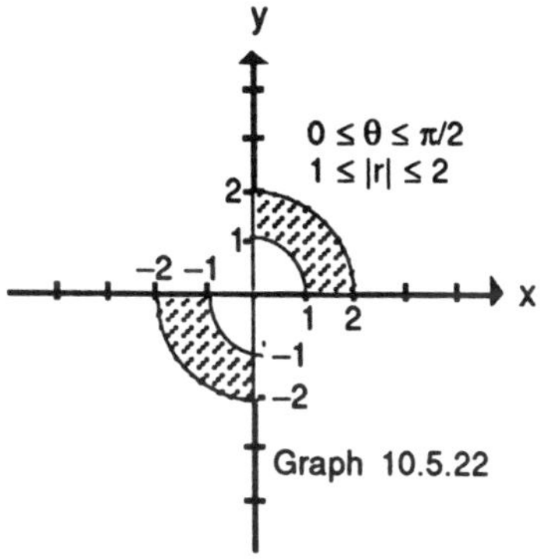

Graph 10.5.22

23. $r \cos \theta = 2 \Rightarrow x = 2$, vertical line through (2,0).

24. $r \sin \theta = -1 \Rightarrow y = -1$, horizontal line through (0,–1).

25. $r \sin \theta = 4 \Rightarrow y = 4$, horizontal line through (0,4).

26. $r \cos \theta = 0 \Rightarrow x = 0$, the y–axis.

27. $r \sin \theta = 0 \Rightarrow y = 0$, the x–axis.

28. $r \cos \theta = -3 \Rightarrow x = -3$, vertical line through (–3,0).

29. $r \cos \theta + r \sin \theta = 1 \Rightarrow x + y = 1$, line, $m = -1$, $b = 1$

30. $r \sin \theta = r \cos \theta \Rightarrow y = x$, line, $m = 1$, $b = 0$

31. $r^2 = 1 \Rightarrow x^2 + y^2 = 1$, circle, $C = (0,0)$, $r = 1$

32. $r^2 = 4r \cos \theta \Rightarrow x^2 + y^2 = 4y$, circle,
$C = (0,2)$, $r = 2$

33. $r = \dfrac{5}{\sin \theta - 2\cos \theta} \Rightarrow r \sin \theta - 2r \cos \theta = 5 \Rightarrow$
$y - 2x = 5$, line, $m = 2$, $b = 5$

34. $r = 4\tan \theta \sec \theta \Rightarrow r = 4\left(\dfrac{\sin \theta}{\cos^2 \theta}\right) \Rightarrow$
$r \cos^2 \theta = 4 \sin \theta \Rightarrow r^2 \cos^2 \theta = 4r \sin \theta$, if $r \neq 0$
$\Rightarrow x^2 = 4y$, parabola, vertex = (0,0)

35. $x = 7 \Rightarrow r \cos \theta = 7$

36. $y = 1 \Rightarrow r \sin \theta = 1$

37. $x = y \Rightarrow r \cos \theta = r \sin \theta \Rightarrow \theta = \frac{\pi}{4}$

38. $x - y = 3 \Rightarrow r\cos\theta - r\sin\theta = 3$

39. $x^2 + y^2 = 4 \Rightarrow r^2 = 4$ or $r = 2$ or $r = -2$

40. $x^2 - y^2 = 1 \Rightarrow r^2\cos^2\theta - r^2\sin^2\theta = 1$
or $r^2\left(\cos^2\theta - \sin^2\theta\right) = 1$
or $r^2\cos 2\theta = 1$

41. $\frac{x^2}{9} + \frac{y^2}{4} = 1 \Rightarrow 4x^2 + 9y^2 = 36 \Rightarrow$
$4r^2\cos^2\theta + 9r^2\sin^2\theta = 36$

42. $xy = 2 \Rightarrow \left(r\cos\theta\right)\left(r\sin\theta\right) = 2 \Rightarrow$
$r^2\cos\theta\sin\theta = 2$ or $2r^2\cos\theta\sin\theta = 4$
or $r^2\sin 2\theta = 4$

43. $y^2 = 4x \Rightarrow r^2\sin^2\theta = 4r\cos\theta$

44. $x^2 - y^2 = 25\sqrt{x^2 + y^2} \Rightarrow r^2\cos^2\theta - r^2\sin^2\theta = 25r \Rightarrow r^2\cos 2\theta = 25r$

SECTION 10.6 GRAPHING IN POLAR COORDINATES

1.

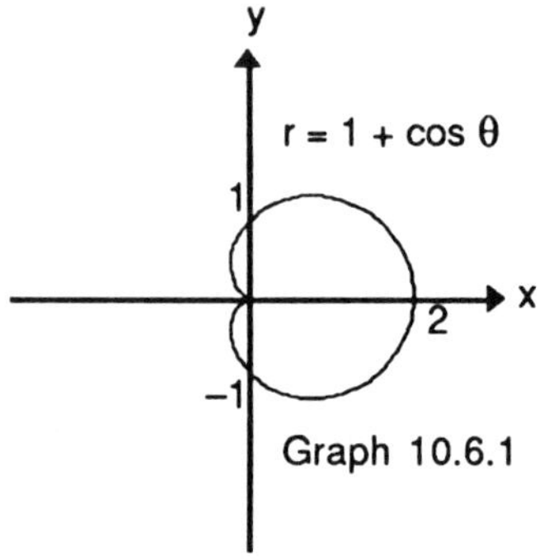

Graph 10.6.1

2.

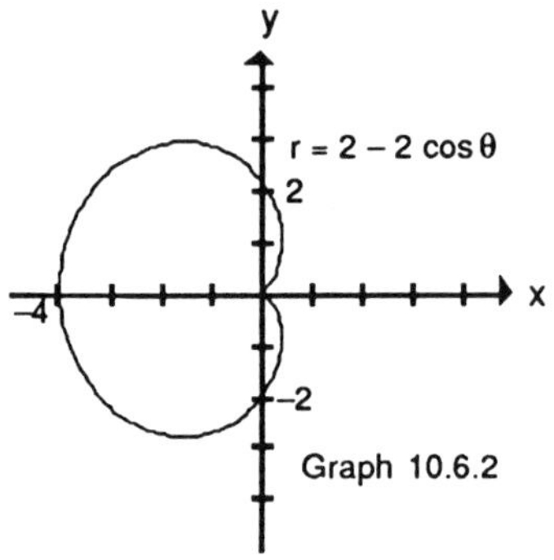

Graph 10.6.2

3.

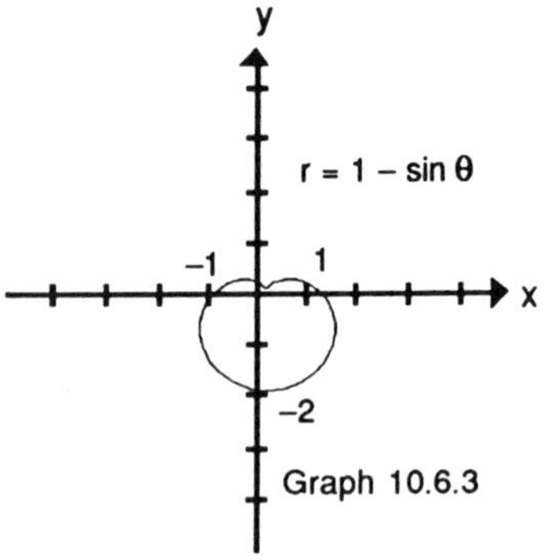

Graph 10.6.3

4.

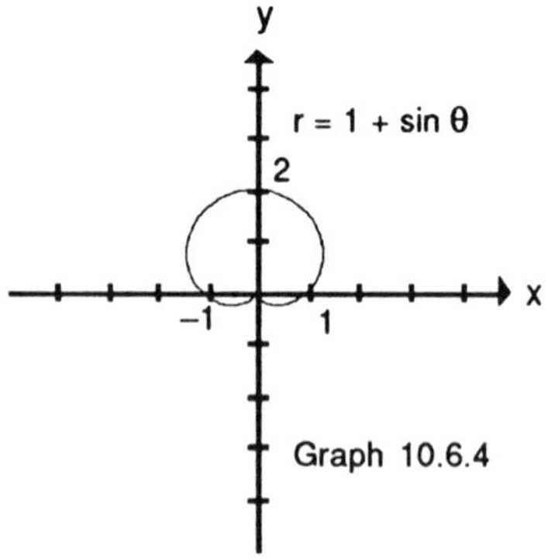

Graph 10.6.4

5.

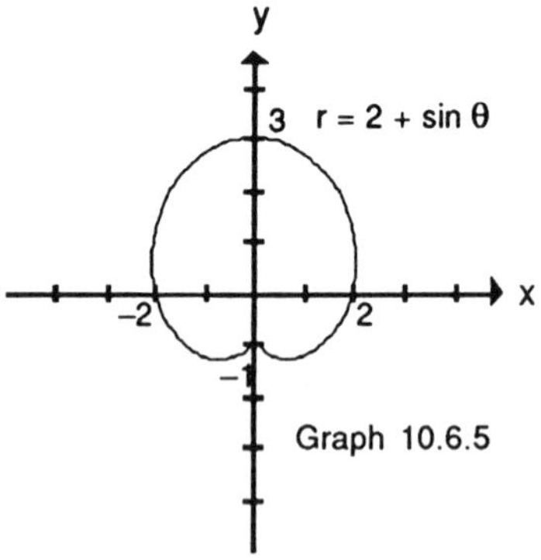

Graph 10.6.5

6.

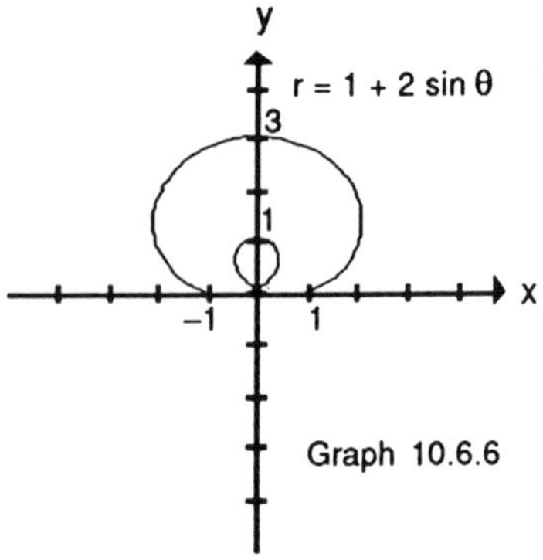

Graph 10.6.6

7.

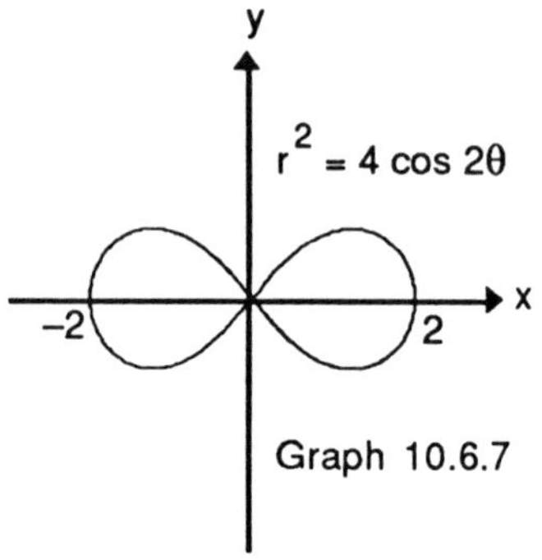

Graph 10.6.7

8.

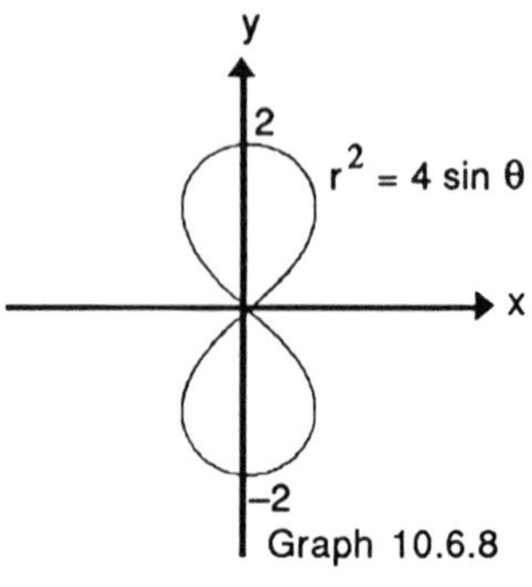

Graph 10.6.8

9.

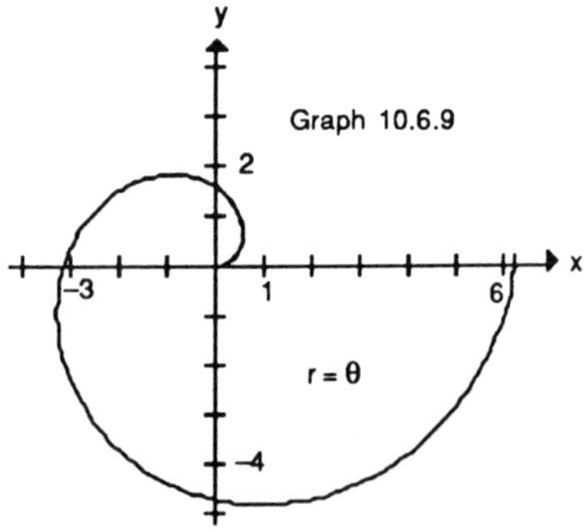

Graph 10.6.9

10.

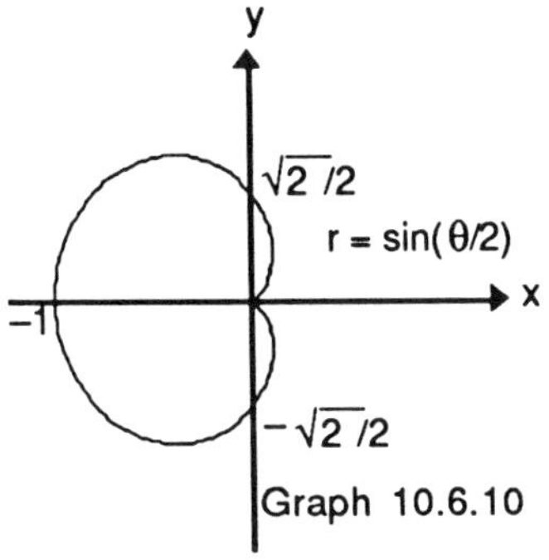

Graph 10.6.10

11. $\theta = \frac{\pi}{2} \Rightarrow r = -1 \Rightarrow \left(-1,\frac{\pi}{2}\right)$, $\theta = -\frac{\pi}{2} \Rightarrow r = -1 \Rightarrow \left(-1,-\frac{\pi}{2}\right)$

$r' = \frac{dr}{d\theta} = -\sin\theta$. Slope $= \frac{r'\sin\theta + r\cos\theta}{r'\cos\theta - r\sin\theta} = \frac{-\sin^2\theta + r\cos\theta}{-\sin\theta\cos\theta - r\sin\theta} \Rightarrow$

Slope at $\left(-1,\frac{\pi}{2}\right) = \frac{-\sin^2\left(\frac{\pi}{2}\right) + (-1)\cos\frac{\pi}{2}}{-\sin\frac{\pi}{2}\cos\frac{\pi}{2} - (-1)\sin\frac{\pi}{2}} = -1$. Slope at $\left(-1,-\frac{\pi}{2}\right) =$

$$\frac{-\sin^2\left(-\frac{\pi}{2}\right) + (-1)\cos\left(-\frac{\pi}{2}\right)}{-\sin\left(-\frac{\pi}{2}\right)\cos\left(-\frac{\pi}{2}\right) - (-1)\sin\left(-\frac{\pi}{2}\right)} = 1$$

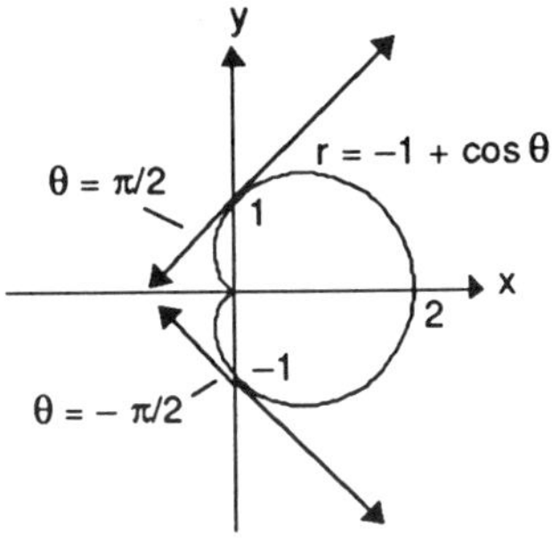

Graph 10.6.11

12. $\theta = 0 \Rightarrow r = -1 \Rightarrow (-1,0)$. $\theta = \frac{\pi}{2} \Rightarrow r = 0 \Rightarrow \left(0,\frac{\pi}{2}\right)$.

$\theta = \pi \Rightarrow r = -1 \Rightarrow (-1,\pi)$. $r' = \frac{dr}{d\theta} = \cos\theta$.

Slope $= \frac{r'\sin\theta + r\cos\theta}{r'\cos\theta - r\sin\theta} = \frac{\cos\theta\sin\theta + r\cos\theta}{\cos\theta\cos\theta - r\sin\theta} =$

$\frac{\cos\theta\sin\theta + r\cos\theta}{\cos^2\theta - r\sin\theta} \Rightarrow$

Slope at $(-1,0) = \frac{\cos 0\sin 0 + (-1)\cos 0}{\cos^2 0 - (-1)\sin 0} = -1$

Slope at $\left(0,\frac{\pi}{2}\right) = \tan\frac{\pi}{2}$ is undefined. Slope at $(-1,\pi) = \frac{\cos\pi\sin\pi + (-1)\cos\pi}{\cos^2\pi - (-1)\sin\pi} = 1$

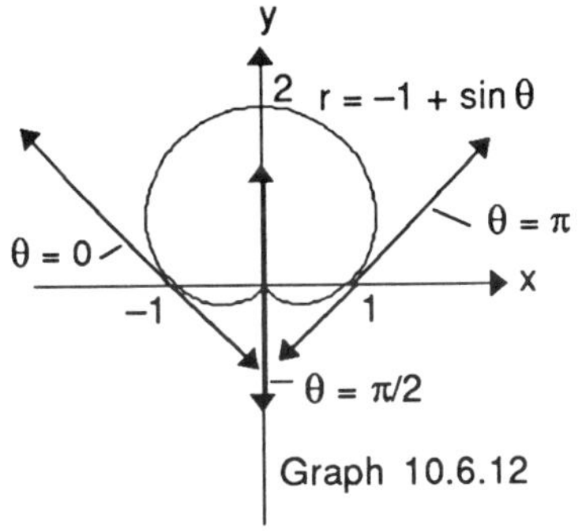

Graph 10.6.12

13. $\theta = \frac{\pi}{4} \Rightarrow r = 1 \Rightarrow \left(1,\frac{\pi}{4}\right)$. $\theta = -\frac{\pi}{4} \Rightarrow r = -1 \Rightarrow \left(-1,-\frac{\pi}{4}\right)$

$\theta = \frac{3\pi}{4} \Rightarrow r = -1 \Rightarrow \left(-1,\frac{3\pi}{4}\right)$. $\theta = -\frac{3\pi}{4} \Rightarrow r = 1 \Rightarrow \left(1,-\frac{3\pi}{4}\right)$

$r = 0 \Rightarrow \theta = 0, \frac{\pi}{2}, \pi, \frac{3\pi}{2}$. $r' = \frac{dr}{d\theta} = 2\cos 2\theta \Rightarrow$

Slope $= \frac{r'\sin\theta + r\cos\theta}{r'\cos\theta - r\sin\theta} = \frac{2\cos 2\theta\sin\theta + r\cos\theta}{2\cos 2\theta\cos\theta - r\sin\theta} \Rightarrow$

Slope at $\left(1,\frac{\pi}{4}\right) = \frac{2\cos\left(\frac{\pi}{2}\right)\sin\frac{\pi}{4} + (1)\cos\frac{\pi}{4}}{2\cos\left(\frac{\pi}{2}\right)\cos\frac{\pi}{4} - (1)\sin\frac{\pi}{4}} = -1$

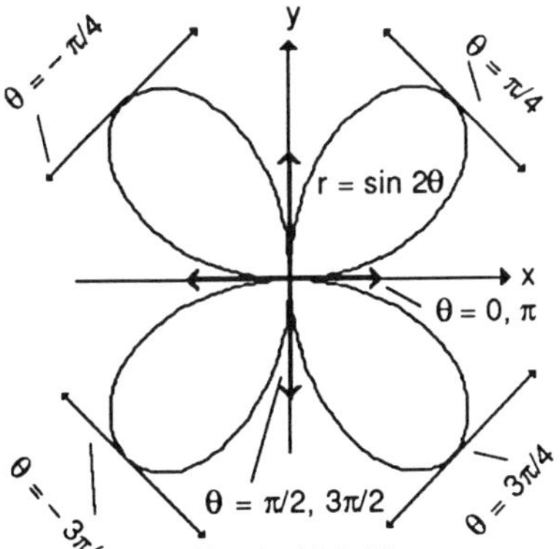

Graph 10.6.13

Slope at $\left(-1,-\frac{\pi}{4}\right) = \dfrac{2\cos\left(-\frac{\pi}{2}\right)\sin\left(-\frac{\pi}{4}\right) + (-1)\cos\left(-\frac{\pi}{4}\right)}{2\cos\left(-\frac{\pi}{2}\right)\cos\left(-\frac{\pi}{4}\right) - (-1)\sin\left(-\frac{\pi}{4}\right)} = 1$ Slope at $(0,0)$, $(0,\pi) = \tan 0 = 0$

Slope at $\left(-1, \frac{3\pi}{4}\right) = \dfrac{2\cos\left(\frac{3\pi}{2}\right)\sin\left(\frac{3\pi}{4}\right) + (-1)\cos\left(\frac{3\pi}{4}\right)}{2\cos\left(\frac{3\pi}{2}\right)\cos\left(\frac{3\pi}{4}\right) - (-1)\sin\left(\frac{3\pi}{4}\right)} = 1$ Slope at $\left(0,\frac{3\pi}{2}\right) = \tan\frac{3\pi}{2}$ is undefined

Slope at $\left(1,-\frac{3\pi}{4}\right) = \dfrac{2\cos\left(-\frac{3\pi}{2}\right)\sin\left(-\frac{3\pi}{4}\right) + (1)\cos\left(-\frac{3\pi}{4}\right)}{2\cos\left(-\frac{3\pi}{2}\right)\cos\left(-\frac{3\pi}{4}\right) - (1)\sin\left(-\frac{3\pi}{4}\right)} = -1$ Slope at $\left(0,\frac{\pi}{2}\right) = \tan\frac{\pi}{2}$ is undefined

14. $\theta = 0 \Rightarrow r = 1 \Rightarrow (1,0)$. $\theta = \frac{\pi}{2} \Rightarrow r = -1 \Rightarrow \left(-1,\frac{\pi}{2}\right)$

$\theta = -\frac{\pi}{2} \Rightarrow r = -1 \Rightarrow \left(-1,-\frac{\pi}{2}\right)$. $\theta = \pi \Rightarrow r = 1 \Rightarrow (1,\pi)$

$r = 0 \Rightarrow \cos 2\theta = 0 \Rightarrow \theta = \frac{\pi}{4}, \frac{3\pi}{4}, \frac{5\pi}{4}, \frac{7\pi}{4}$

$r' = \frac{dr}{d\theta} = -2\sin 2\theta \Rightarrow \text{Slope} = \dfrac{r'\sin\theta + r\cos\theta}{r'\cos\theta - r\sin\theta} =$

$\dfrac{-2\sin 2\theta\sin\theta + r\cos\theta}{-2\sin 2\theta\cos\theta - r\sin\theta} \Rightarrow$

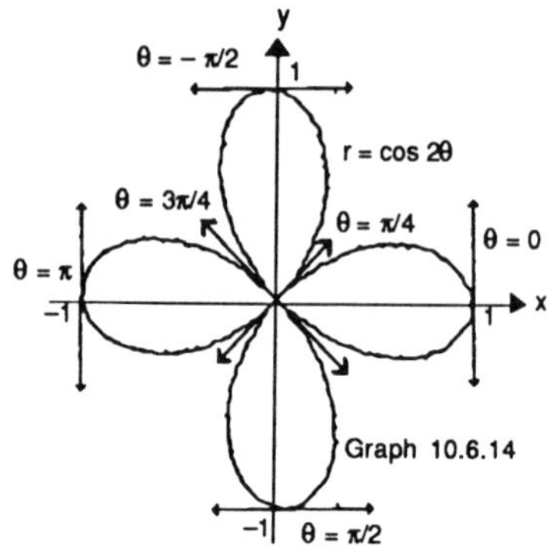

Graph 10.6.14

Slope at $(1,0) = \dfrac{-2\sin 2(0)\sin 0 + (1)\cos 0}{-2\sin 2(0)\cos 0 - (1)\sin 0}$ is undefined.

Slope at $\left(-1,\frac{\pi}{2}\right) = \dfrac{-2\sin 2\left(\frac{\pi}{2}\right)\sin\frac{\pi}{2} + (-1)\cos\frac{\pi}{2}}{-2\sin 2\left(\frac{\pi}{2}\right)\cos\frac{\pi}{2} - (-1)\sin\frac{\pi}{2}} = 0$ Slope at $\left(0,\frac{\pi}{4}\right) = \tan\frac{\pi}{4} = 1$

Slope at $\left(-1,-\frac{\pi}{2}\right) = \dfrac{-2\sin 2\left(-\frac{\pi}{2}\right)\sin\left(-\frac{\pi}{2}\right) + (-1)\cos\left(-\frac{\pi}{2}\right)}{-2\sin 2\left(-\frac{\pi}{2}\right)\cos\left(-\frac{\pi}{2}\right) - (-1)\sin\left(-\frac{\pi}{2}\right)} = 0$ Slope at $\left(0,\frac{3\pi}{4}\right) = \tan\frac{3\pi}{4} = -1$

Slope at $(1,\pi) = \dfrac{-2\sin 2\pi\sin\pi + (1)\cos\pi}{-2\sin 2\pi\cos\pi - (1)\sin\pi}$ is undefined. Slope at $\left(0,\frac{5\pi}{4}\right) = \tan\frac{5\pi}{4} = 1$

Slope at $\left(0,\frac{7\pi}{4}\right) = \tan\frac{7\pi}{4} = -1$

15.

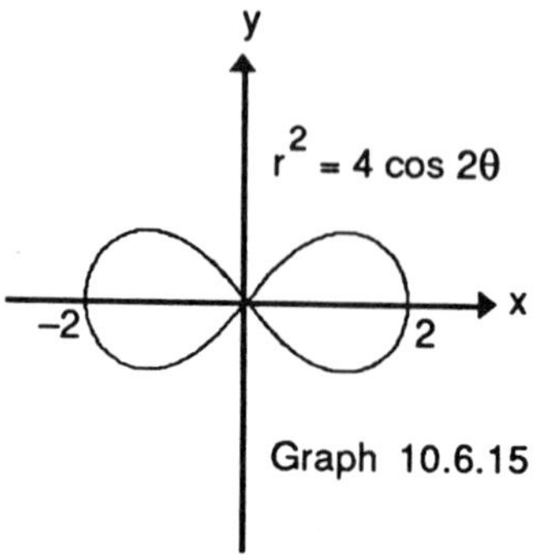

Graph 10.6.15

16.

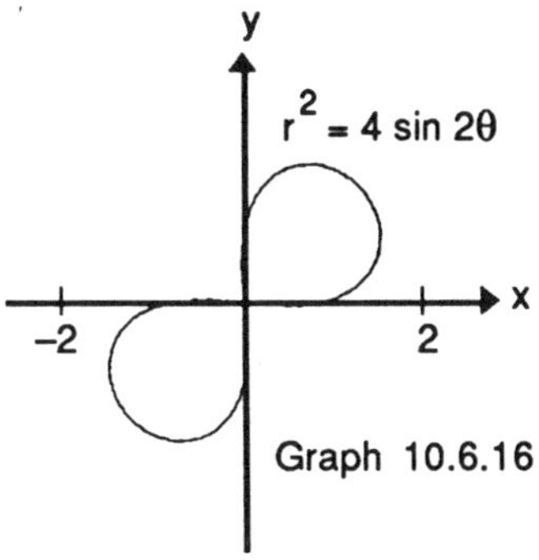

Graph 10.6.16

17. a)

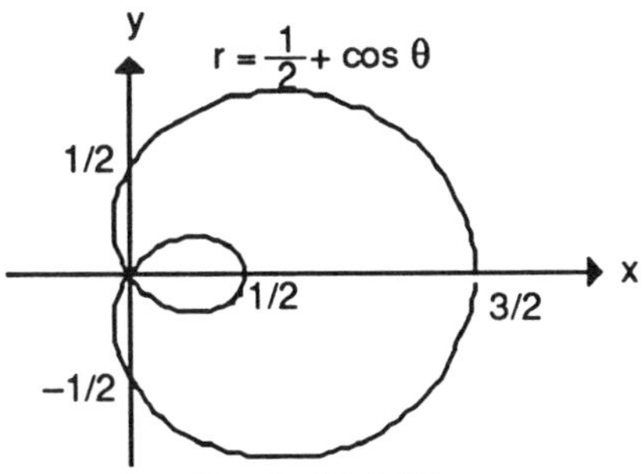

Graph 10.6.17a

17. b)

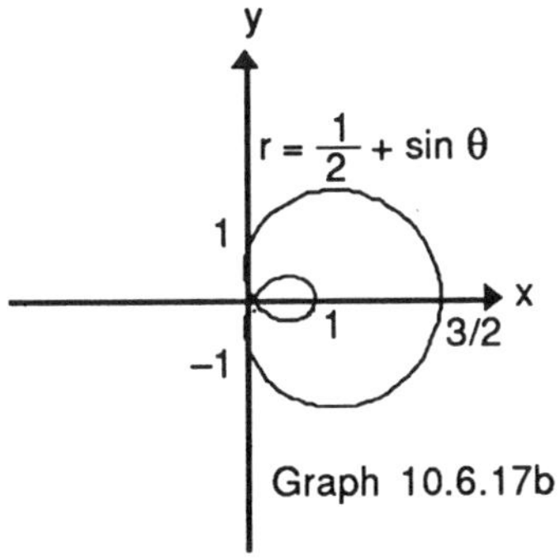

Graph 10.6.17b

18. a)

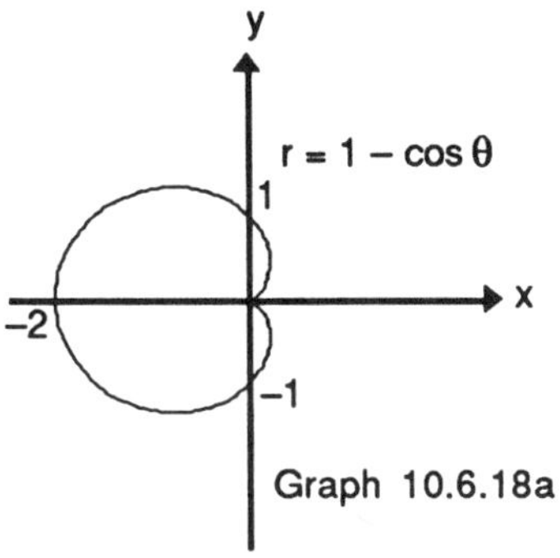

Graph 10.6.18a

18. b)

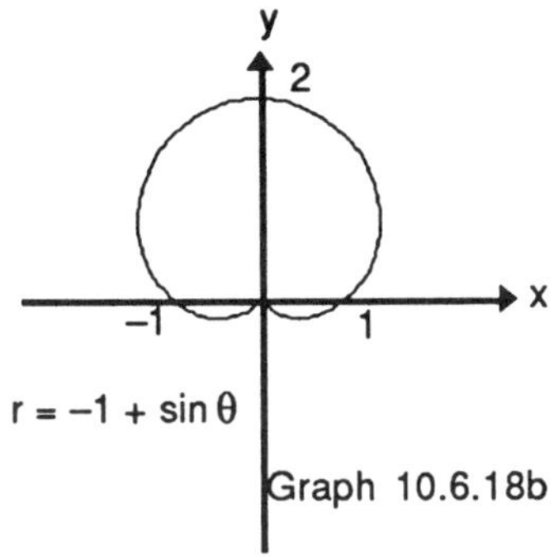

Graph 10.6.18b

19. a)

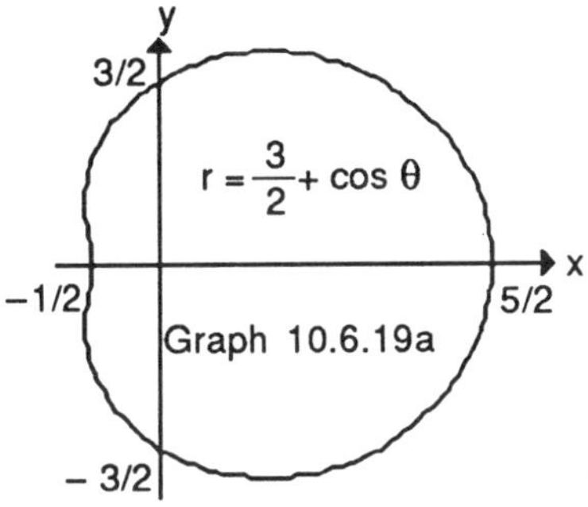

Graph 10.6.19a

19. b)

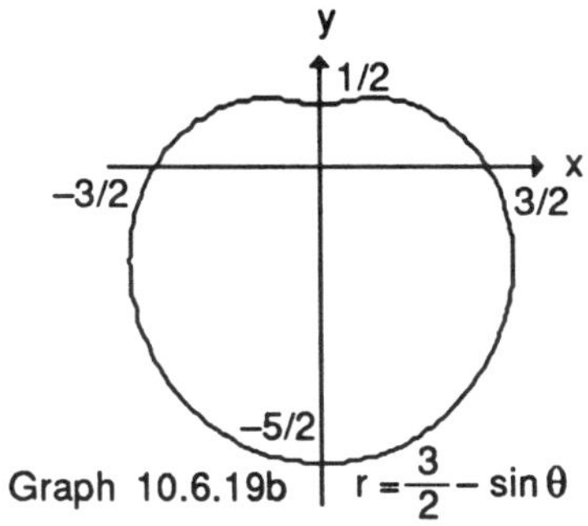

Graph 10.6.19b

20. a)

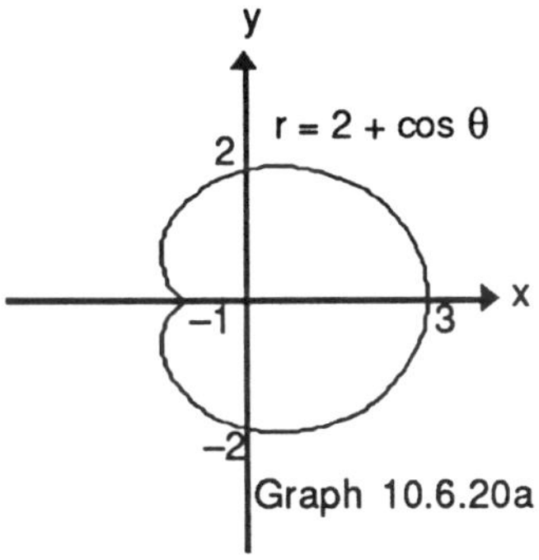

Graph 10.6.20a

20. b)

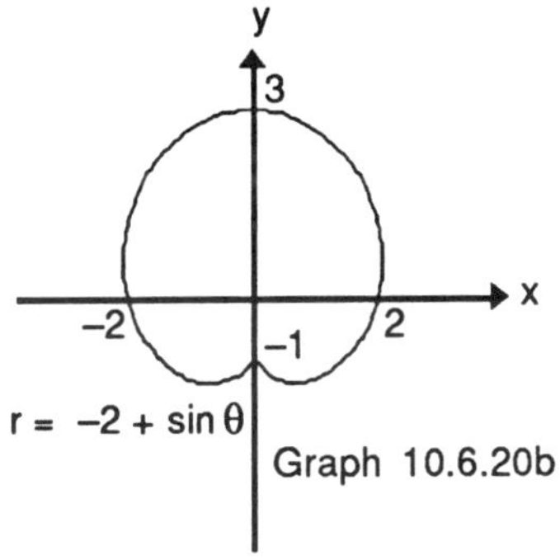

Graph 10.6.20b

21.

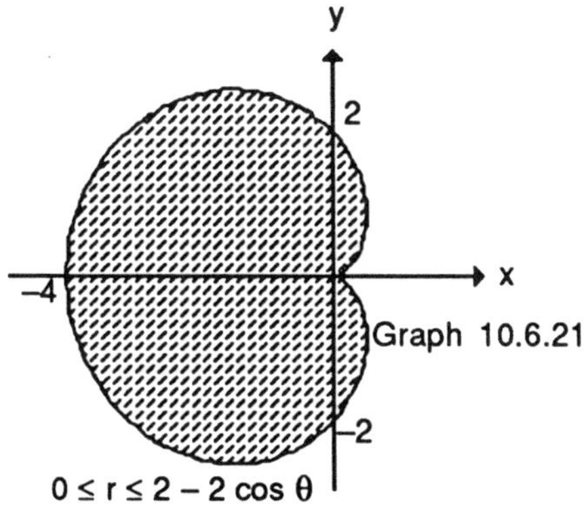

Graph 10.6.21

22.

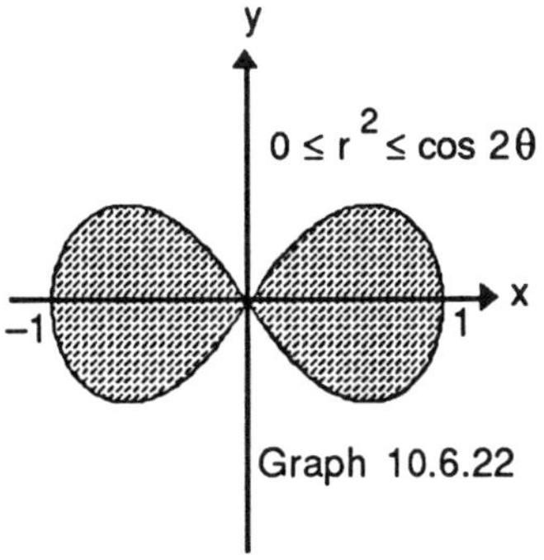

Graph 10.6.22

23. $\left(2,\frac{3\pi}{4}\right)$ is the same point as $\left(-2,-\frac{\pi}{4}\right)$. $r = 2\sin 2\left(-\frac{\pi}{4}\right) = 2\sin\left(-\frac{\pi}{2}\right) = -2 \Rightarrow \left(-2,-\frac{\pi}{4}\right)$ is on the graph $\Rightarrow \left(2,\frac{3\pi}{4}\right)$ is on the graph.

24. $\left(\frac{1}{2},\frac{3\pi}{2}\right)$ is the same point as $\left(-\frac{1}{2},\frac{\pi}{2}\right)$. $r = -\sin\left(\frac{\pi/2}{3}\right) = -\sin\frac{\pi}{6} = -\frac{1}{2} \Rightarrow \left(-\frac{1}{2},\frac{\pi}{2}\right)$ is on the graph $\Rightarrow$ $\left(\frac{1}{2},\frac{3\pi}{2}\right)$ is on the graph.

25. $1 + \cos\theta = 1 - \cos\theta \Rightarrow \cos\theta = 0 \Rightarrow$

$\theta = \frac{\pi}{2}, \frac{3\pi}{2} \Rightarrow r = 1$. Points of intersection are $\left(1,\frac{\pi}{2}\right)$ and $\left(1,\frac{3\pi}{2}\right)$. The point of intersection, (0,0), is found by graphing.

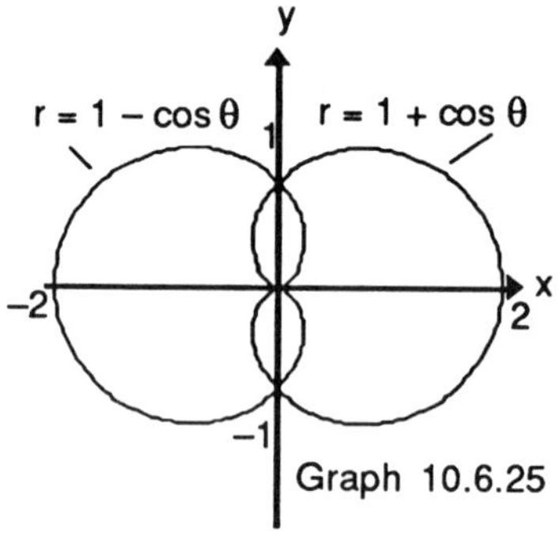

Graph 10.6.25

26. $1 + \sin\theta = 1 - \sin\theta \Rightarrow \sin\theta = 0 \Rightarrow \theta = 0, \pi \Rightarrow$

$r = 1$. Points of intersection are $(1,0)$ and $(1,\pi)$. The point of intersection, (0,0), is found by graphing.

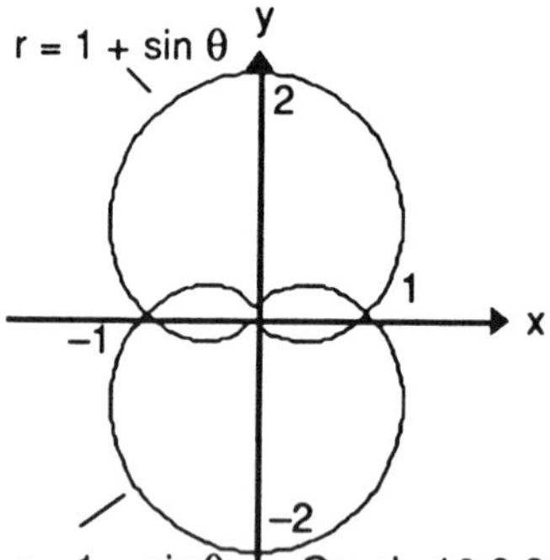

Graph 10.6.26

27. $(1 - \sin\theta)^2 = 4\sin\theta \Rightarrow 1 - 2\sin\theta + \sin^2\theta = 4\sin\theta \Rightarrow$

$1 - 6\sin\theta + \sin^2\theta = 0 \Rightarrow \sin\theta = \frac{6 \pm \sqrt{32}}{2} = 3 \pm 2\sqrt{2}$.

$\sin\theta$ cannot be $3 + 2\sqrt{2}$. $\therefore\ \theta = \sin^{-1}(3 - 2\sqrt{2}) \Rightarrow$

$r^2 = 4\sin\theta = 4(3 - 2\sqrt{2}) \Rightarrow r = \pm 2\sqrt{3 - 2\sqrt{2}} = \pm 2\sqrt{1 - 2\sqrt{2} + 2}$

$\Rightarrow r = \pm 2\left|1 - \sqrt{2}\right| = \pm 2\left(\sqrt{2} - 1\right)$. Points of intersection are

$\left(\pm 2\left(\sqrt{2} - 1\right), \sin^{-1}\left(3 - 2\sqrt{2}\right)\right)$. Points of intersection (0,0) and

$\left(2,\frac{3\pi}{2}\right)$ found by graphing.

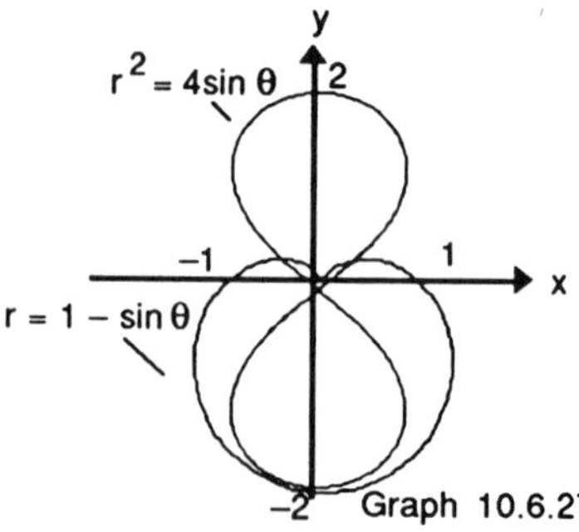

Graph 10.6.27

28. $\sqrt{2}\sin\theta = \sqrt{2}\cos\theta \Rightarrow \sin\theta = \cos\theta \Rightarrow$

$\theta = \frac{\pi}{4}, \frac{5\pi}{4}$. $\theta = \frac{\pi}{4} \Rightarrow r^2 = \sqrt{2}\sin\frac{\pi}{4} = 1 \Rightarrow r = \pm 1.$

$\theta = \frac{5\pi}{4} \Rightarrow r^2 = \sqrt{2}\sin\frac{5\pi}{4} = -1 \Rightarrow$ No solution.

Points of intersection are $\left(\pm 1, \frac{\pi}{4}\right)$. Points of intersection $\left(\pm 1, \frac{3\pi}{4}\right)$ are found by graphing.

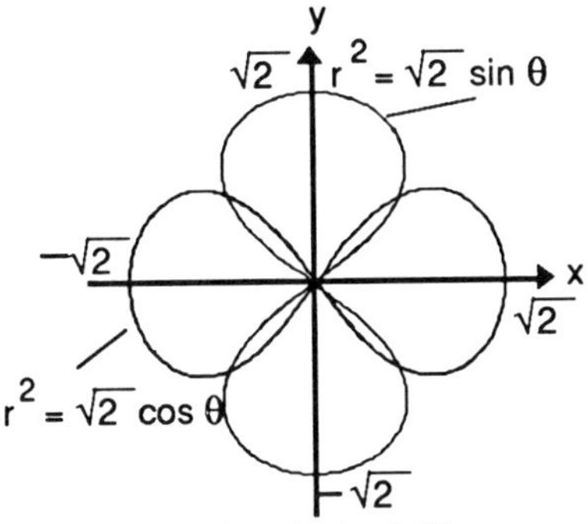

Graph 10.6.28

29. a) $r^2 = -4\cos\theta \Rightarrow \cos\theta = -\frac{r^2}{4}$. $r = 1 - \cos\theta \Rightarrow r = 1 - \left(-\frac{r^2}{4}\right) \Rightarrow 0 = r^2 - 4r + 4 \Rightarrow (r-2)^2 = 0 \Rightarrow$

$r = 2$. $\therefore \cos\theta = -\frac{2^2}{4} = -1 \Rightarrow \theta = \pi$ $\therefore$ $(2,\pi)$ is a point of intersection.

b) $r = 0 \Rightarrow 0^2 = -4\cos\theta \Rightarrow \cos\theta = 0 \Rightarrow \theta = \frac{\pi}{2}, \frac{3\pi}{2} \Rightarrow \left(0, \frac{\pi}{2}\right)$ or $\left(0, \frac{3\pi}{2}\right)$ is on the graph.

$r = 0 \Rightarrow 0 = 1 - \cos\theta \Rightarrow \cos\theta = 1 \Rightarrow \theta = 0 \Rightarrow (0,0)$ is on the graph. Since $(0,0) = \left(0, \frac{\pi}{2}\right)$, the graphs intersect at the origin.

30. a) Let $r = f(\theta)$ be symmetric to the x–axis and the y–axis. Then (r,θ) on the graph $\Rightarrow (r,-\theta)$ is on the graph because of symmetry to the x–axis. Then $\left(-r,-(-\theta)\right) = (-r,\theta)$ is on the graph because of symmetry to the y–axis. $\therefore$ $r = f(\theta)$ is symmetric to the origin.

b) Let $r = f(\theta)$ be symmetric to the x–axis and the origin. Then (r,θ) on the graph $\Rightarrow (r,-\theta)$ is on the graph because of symmetry to the x–axis. Then $(-r,-\theta)$ is on the graph because of symmetry to the origin. $\therefore$ $r = f(\theta)$ is symmetric to the y–axis.

c) Let $r = f(\theta)$ be symmetric to the y–axis and the origin. Then (r,θ) on the graph $\Rightarrow (-r,-\theta)$ is on the graph because of symmetry to the y–axis. Then $(-(-r),-\theta) = (r,-\theta)$ is on the graph because of symmetry to the origin. $\therefore$ $r = f(\theta)$ is symmetric to the x–axis.

31. $r^2 = \sin 2\theta$ and $r^2 = \cos 2\theta$ are generated completely for $0 \le \theta \le \frac{\pi}{2}$. Then $\sin 2\theta = \cos 2\theta$ yields $2\theta = \frac{\pi}{4}$ as the only solution on that interval $\Rightarrow$

$\theta = \frac{\pi}{8} \Rightarrow r^2 = \sin 2\left(\frac{\pi}{8}\right) = \frac{1}{\sqrt{2}} \Rightarrow r = \pm\frac{1}{\sqrt[4]{2}}$.

$\therefore$ Points of intersection are $\left(\pm\frac{1}{\sqrt[4]{2}}, \frac{\pi}{8}\right)$. The point of intersection (0,0) is found by graphing.

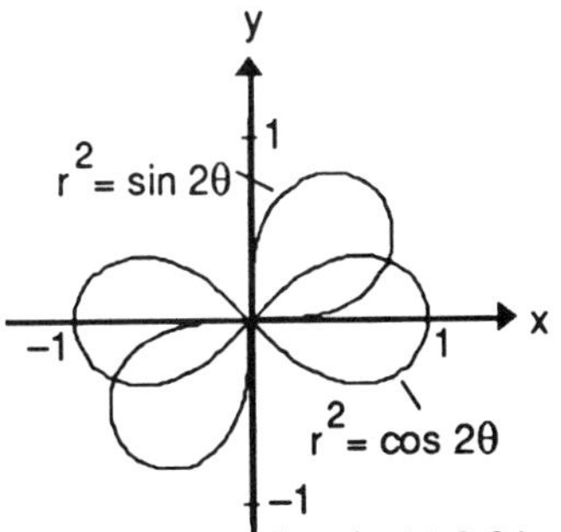

Graph 10.6.31

32. $1 - \sin\frac{\theta}{2} = 1 + \cos\frac{\theta}{2} \Rightarrow -\sin\frac{\theta}{2} = \cos\frac{\theta}{2}$

$\Rightarrow \frac{\theta}{2} = \frac{3\pi}{4} \Rightarrow \theta = \frac{3\pi}{2}$. $\theta = \frac{3\pi}{2} \Rightarrow r = 1 + \cos\frac{3\pi}{4} \Rightarrow$

$r = 1 - \frac{\sqrt{2}}{2}$. Point of intersection is $\left(1 - \frac{\sqrt{2}}{2}, \frac{3\pi}{2}\right)$.

The point of intersection (0,0) is found by graphing.

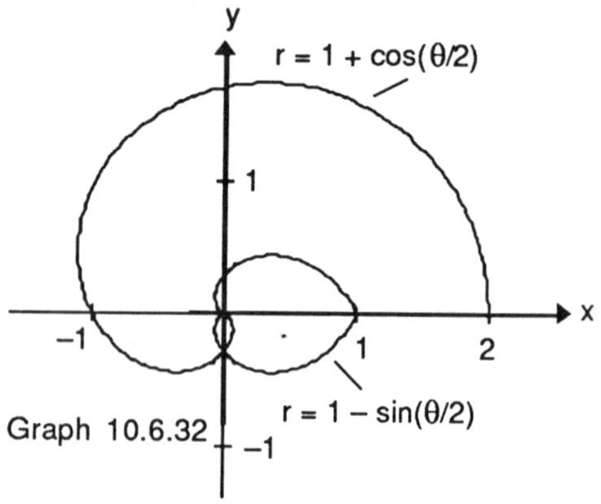

Graph 10.6.32

33. $1 = 2\sin 2\theta \Rightarrow \sin 2\theta = \frac{1}{2} \Rightarrow 2\theta = \frac{\pi}{6}, \frac{5\pi}{6}, \frac{13\pi}{6}, \frac{17\pi}{6}$

$\Rightarrow \theta = \frac{\pi}{12}, \frac{5\pi}{12}, \frac{13\pi}{12}, \frac{17\pi}{12}$. Points of intersection are

$\left(1, \frac{\pi}{12}\right), \left(1, \frac{5\pi}{12}\right), \left(1, \frac{13\pi}{12}\right)$, and $\left(1, \frac{17\pi}{12}\right)$.

Points of intersection $\left(1, \frac{7\pi}{12}\right), \left(1, \frac{11\pi}{12}\right), \left(1, \frac{19\pi}{12}\right)$, and

$\left(1, \frac{23\pi}{12}\right)$ found by graphing and symmetry.

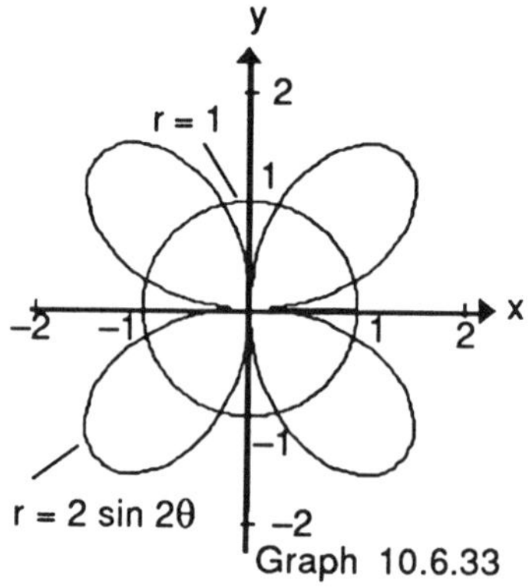

Graph 10.6.33

34. $r^2 = 2\sin 2\theta$ is completely generated on $0 \le \theta \le \frac{\pi}{2}$.

$\therefore$ $1 = 2\sin 2\theta \Rightarrow \sin 2\theta = \frac{1}{2} \Rightarrow 2\theta = \frac{\pi}{6}, \frac{5\pi}{6} \Rightarrow$

$\theta = \frac{\pi}{12}, \frac{5\pi}{12} \Rightarrow r^2 = 2\sin 2\left(\frac{\pi}{12}\right) = 1 \Rightarrow r = \pm 1$ and

$r^2 = 2\sin 2\left(\frac{5\pi}{12}\right) = 1 \Rightarrow r = \pm 1$. The points of

intersection are $\left(\pm 1, \frac{\pi}{12}\right)$ and $\left(\pm 1, \frac{5\pi}{12}\right)$.

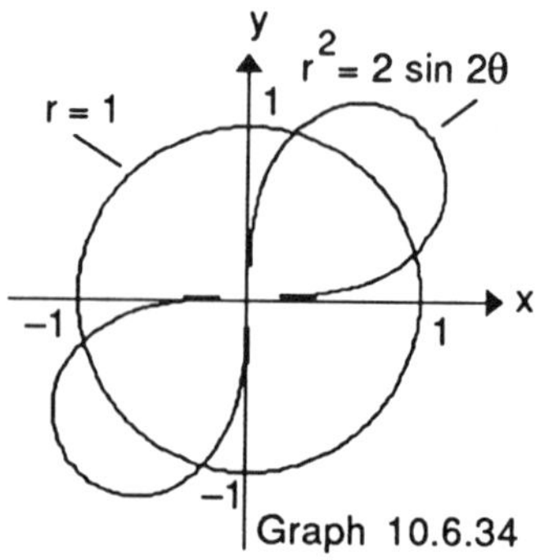

Graph 10.6.34

35.

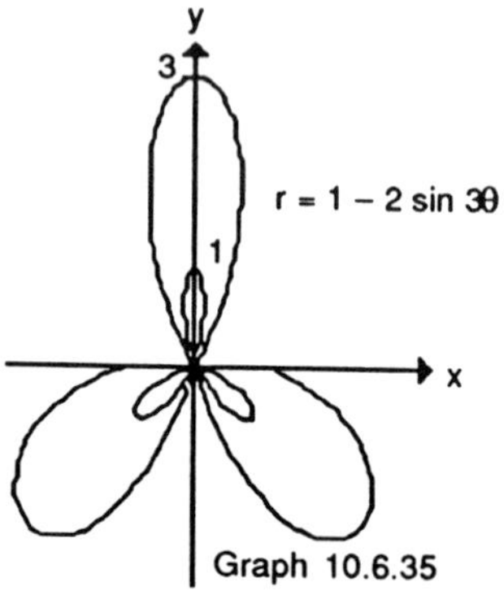

Graph 10.6.35

36.

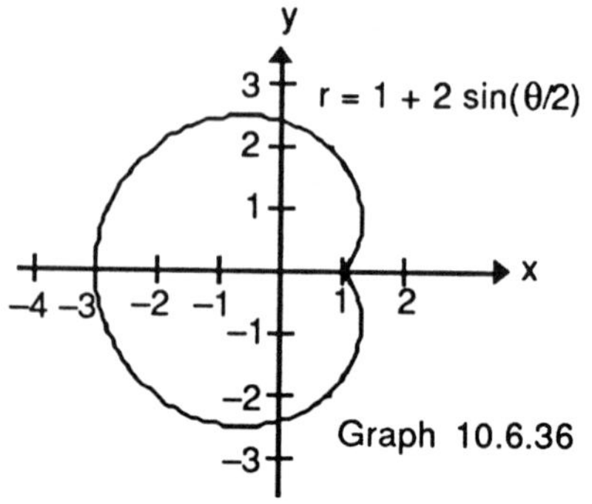

Graph 10.6.36

37. a)

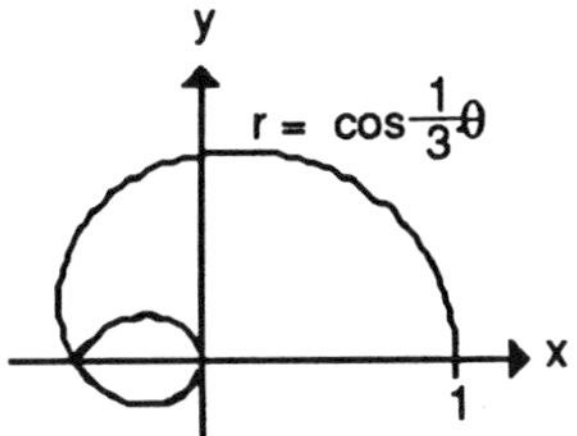

Graph 10.6.37a

37. b)

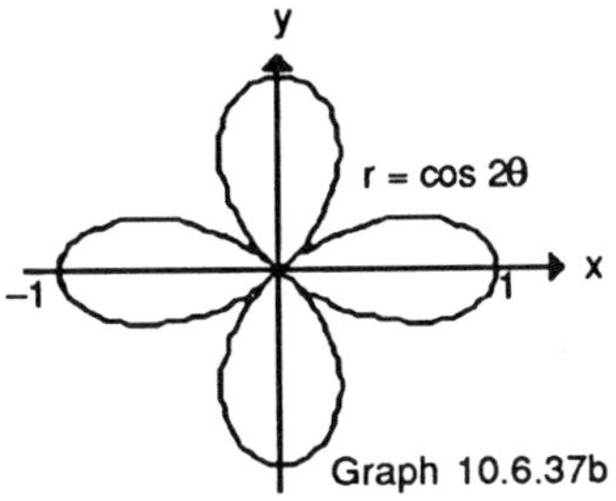

Graph 10.6.37b

37. c)

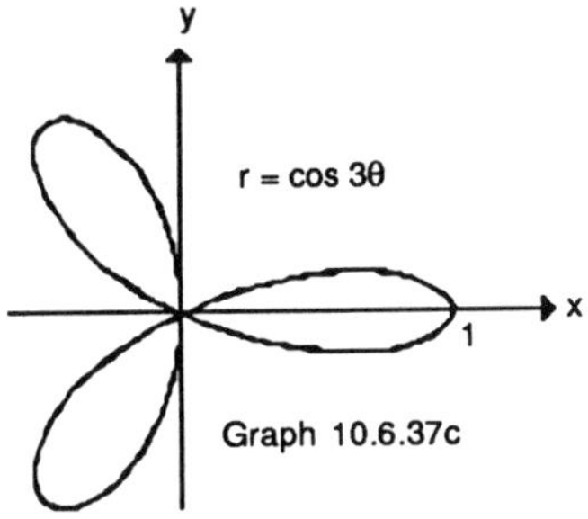

Graph 10.6.37c

37. d)

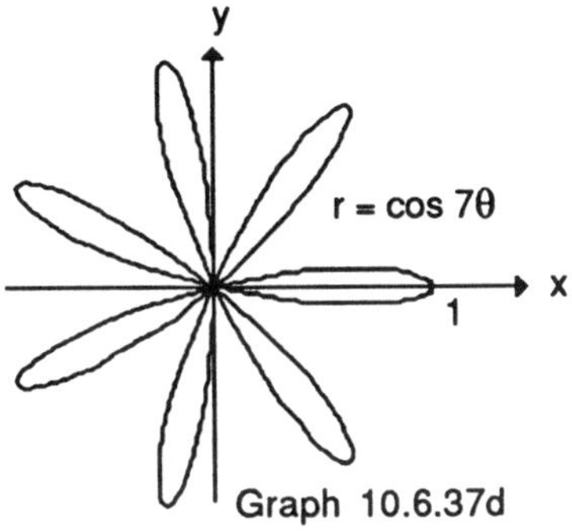

Graph 10.6.37d

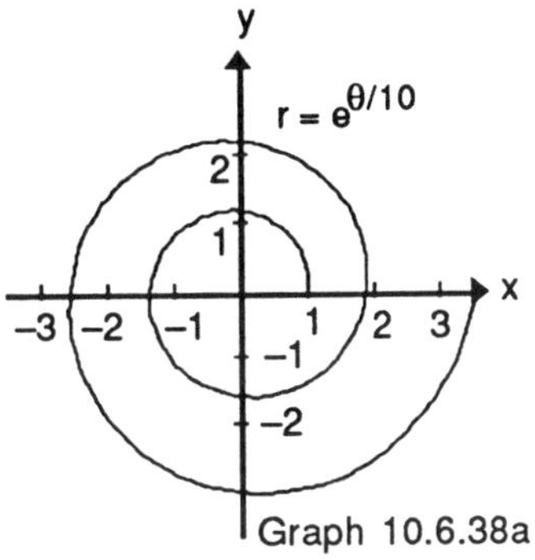

Graph 10.6.38a

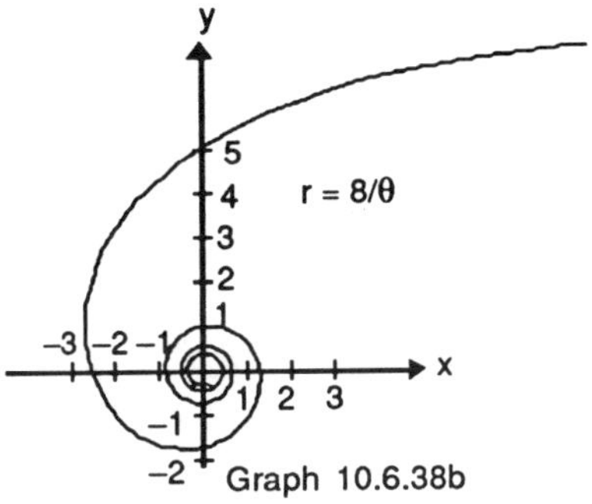

Graph 10.6.38b

38. c)

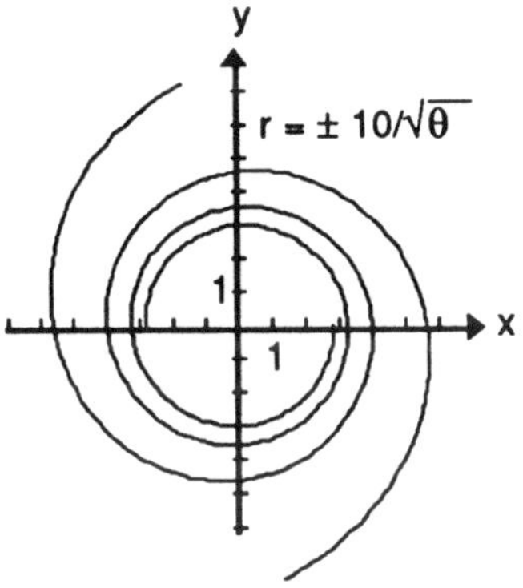

Graph 10.6.38c

SECTION 10.7 POLAR EQUATIONS OF CONIC SECTIONS

1. $r\cos\left(\theta - \frac{\pi}{6}\right) = 5 \Rightarrow r\left(\cos\theta\cos\frac{\pi}{6} + \sin\theta\sin\frac{\pi}{6}\right) = 5 \Rightarrow \frac{\sqrt{3}}{2}r\cos\theta + \frac{1}{2}r\sin\theta = 5 \Rightarrow$

   $\frac{\sqrt{3}}{2}x + \frac{1}{2}y = 5 \Rightarrow \sqrt{3}\,x + y = 10.$

2. $r\cos\left(\theta - \frac{3\pi}{4}\right) = 2 \Rightarrow r\left(\cos\theta\cos\frac{3\pi}{4} + \sin\theta\sin\frac{3\pi}{4}\right) = 2 \Rightarrow -\frac{\sqrt{2}}{2}r\cos\theta + \frac{\sqrt{2}}{2}r\sin\theta = 2 \Rightarrow$

   $-\frac{\sqrt{2}}{2}x + \frac{\sqrt{2}}{2}y = 2 \Rightarrow -\sqrt{2}\,x + \sqrt{2}\,y = 4.$

3. $r\cos\left(\theta - \frac{4\pi}{3}\right) = 3 \Rightarrow r\left(\cos\theta\cos\frac{4\pi}{3} + \sin\theta\sin\frac{4\pi}{3}\right) = 3 \Rightarrow -\frac{1}{2}r\cos\theta - \frac{\sqrt{3}}{2}r\sin\theta = 3 \Rightarrow$

   $-\frac{1}{2}x - \frac{\sqrt{3}}{2}y = 3 \Rightarrow x + \sqrt{3}\,y = -6$

4. $r\cos\left(\theta - \left(-\frac{\pi}{4}\right)\right) = 4 \Rightarrow r\cos\left(\theta + \frac{\pi}{4}\right) = 4 \Rightarrow r\left(\cos\theta\cos\frac{\pi}{4} - \sin\theta\sin\frac{\pi}{4}\right) = 4 \Rightarrow$
$\frac{\sqrt{2}}{2} r\cos\theta - \frac{\sqrt{2}}{2} r\sin\theta = 4 \Rightarrow \frac{\sqrt{2}}{2}x - \frac{\sqrt{2}}{2}y = 4 \Rightarrow \sqrt{2}\,x - \sqrt{2}\,y = 8.$

5. $r\cos\left(\theta - \frac{\pi}{4}\right) = \sqrt{2} \Rightarrow r\left(\cos\theta\cos\frac{\pi}{4} + \sin\theta\sin\frac{\pi}{4}\right) =$
$\sqrt{2} \Rightarrow \frac{1}{\sqrt{2}} r\cos\theta + \frac{1}{\sqrt{2}} r\sin\theta = \sqrt{2} \Rightarrow$
$\frac{1}{\sqrt{2}}x + \frac{1}{\sqrt{2}}y = \sqrt{2} \Rightarrow x + y = 2.$

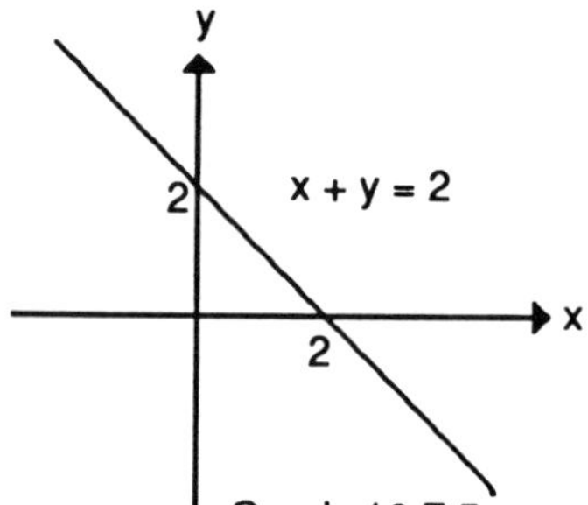

Graph 10.7.5

6. $r\cos\left(\theta - \frac{2\pi}{3}\right) = 3 \Rightarrow r\left(\cos\theta\cos\frac{2\pi}{3} + \sin\theta\sin\frac{2\pi}{3}\right) = 3 \Rightarrow$
$-\frac{1}{2} r\cos\theta + \frac{\sqrt{3}}{2} r\sin\theta = 3 \Rightarrow -\frac{1}{2}x + \frac{\sqrt{3}}{2}y = 3 \Rightarrow$
$x - \sqrt{3}\,y = -6.$

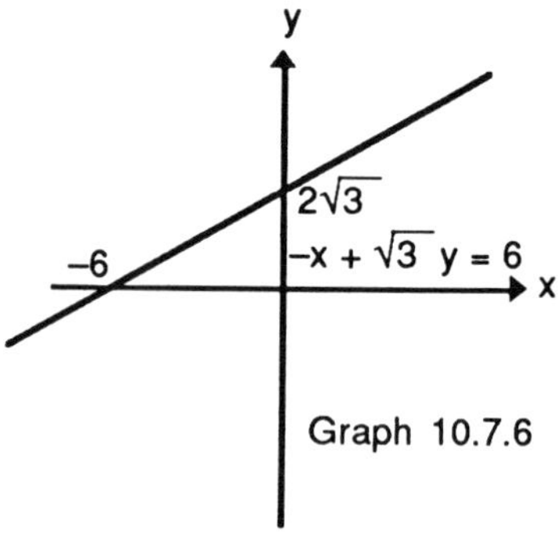

Graph 10.7.6

7. $r\cos\left(\theta - \frac{3\pi}{2}\right) = 1 \Rightarrow r\left(\cos\theta\cos\frac{3\pi}{2} + \sin\theta\sin\frac{3\pi}{2}\right) = 1 \Rightarrow$
$-r\sin\theta = 1 \Rightarrow y = -1.$

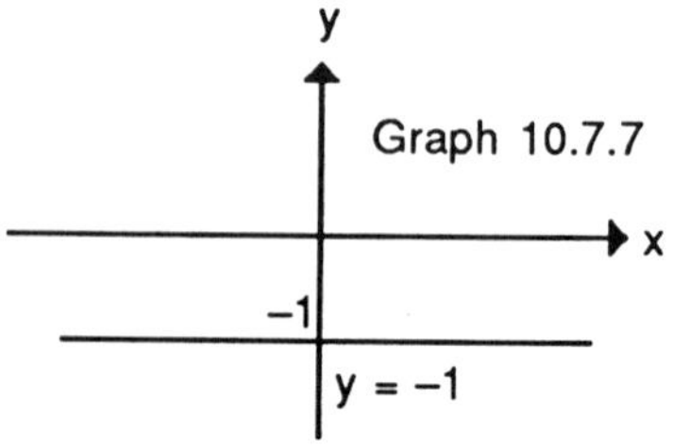

Graph 10.7.7

8. $r\cos\left(\theta + \frac{\pi}{3}\right) = 2 \Rightarrow r\left(\cos\theta\cos\frac{\pi}{3} - \sin\theta\sin\frac{\pi}{3}\right) = 2 \Rightarrow$
$\frac{1}{2} r\cos\theta - \frac{\sqrt{3}}{2} r\sin\theta = 2 \Rightarrow \frac{1}{2}x - \frac{\sqrt{3}}{2}y = 2 \Rightarrow$
$x - \sqrt{3}\,y = 4$

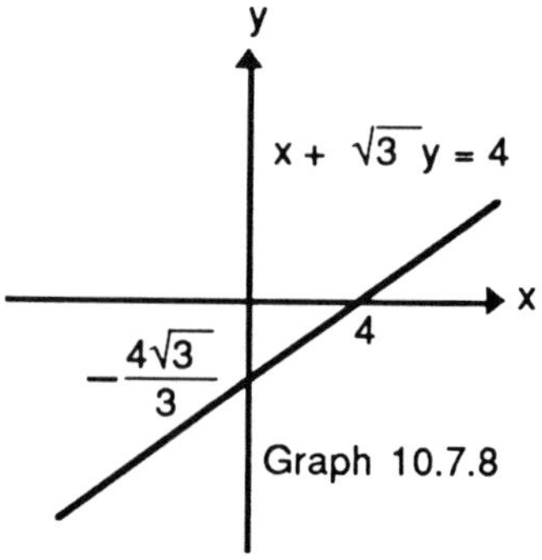

Graph 10.7.8

9. $r = 2(4)\cos\theta = 8\cos\theta$

10. $r = -2(1)\sin\theta = -2\sin\theta$

11. $r = 2\sqrt{2}\,\sin\theta$

12. $r = -2\left(\frac{1}{2}\right)\cos\theta = -\cos\theta$

13.

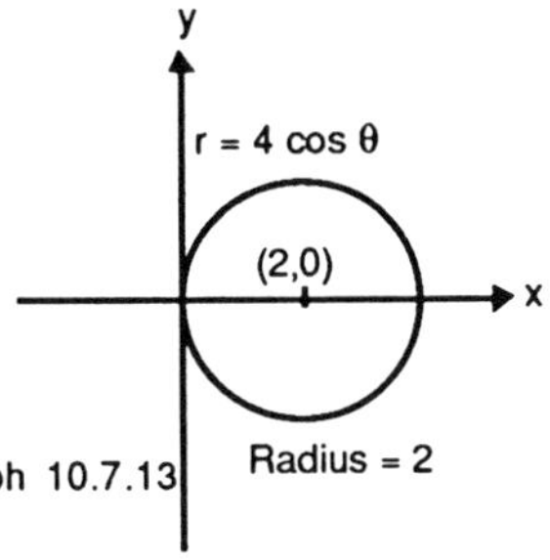

Graph 10.7.13

14.

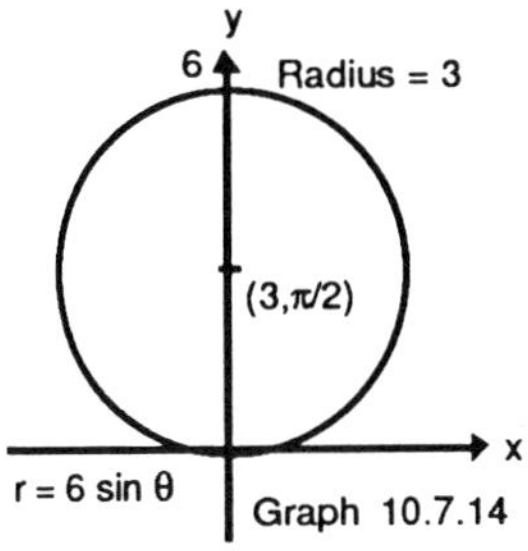

Graph 10.7.14

15.

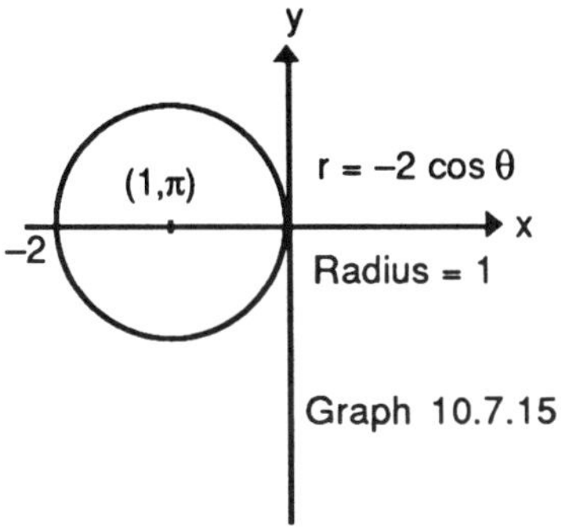

Graph 10.7.15

16.

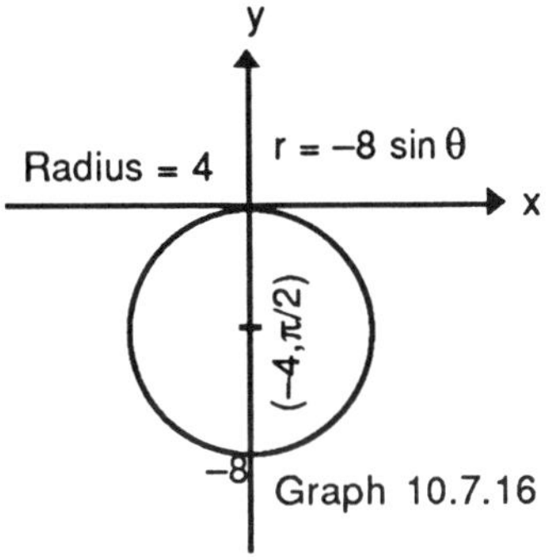

Graph 10.7.16

17. $e = 1,\ x = 2 \Rightarrow k = 2 \Rightarrow$

$$r = \frac{2(1)}{1 + (1)\cos\theta} = \frac{2}{1 + \cos\theta}$$

18. $e = 1,\ y = 2 \Rightarrow k = 2 \Rightarrow$

$$r = \frac{2(1)}{1 + (1)\sin\theta} = \frac{2}{1 + \sin\theta}$$

19. $e = 2,\ x = 4 \Rightarrow k = 4 \Rightarrow r = \dfrac{4(2)}{1 + (2)\cos\theta} = \dfrac{8}{1 + 2\cos\theta}$

20. $e = 5,\ y = -6 \Rightarrow k = 6 \Rightarrow r = \dfrac{6(5)}{1 - (5)\sin\theta} = \dfrac{30}{1 - 5\sin\theta}$

21. $e = \frac{1}{2},\ x = 1 \Rightarrow k = 1 \Rightarrow r = \dfrac{\frac{1}{2}(1)}{1 + (1)\cos\theta} = \dfrac{1}{2 + \cos\theta}$

22. $e = \frac{1}{4},\ x = -2, \Rightarrow k = 2 \Rightarrow r = \dfrac{\frac{1}{4}(2)}{1 - \frac{1}{4}\cos\theta} = \dfrac{2}{4 - \cos\theta}$

23. $e = \frac{1}{5},\ y = -10 \Rightarrow k = 10 \Rightarrow r = \dfrac{\frac{1}{5}(10)}{1 - \frac{1}{5}\sin\theta} = \dfrac{2}{1 - \frac{1}{5}\sin\theta} = \dfrac{10}{5 - \sin\theta}$

24. $e = \frac{1}{3},\ y = 6 \Rightarrow k = 6 \Rightarrow r = \dfrac{\frac{1}{3}(6)}{1 + \frac{1}{3}\sin\theta} = \dfrac{6}{3 + \sin\theta}$

25. $r = \dfrac{1}{1 + \cos\theta} \Rightarrow e = 1,\ k = 1 \Rightarrow x = 1$

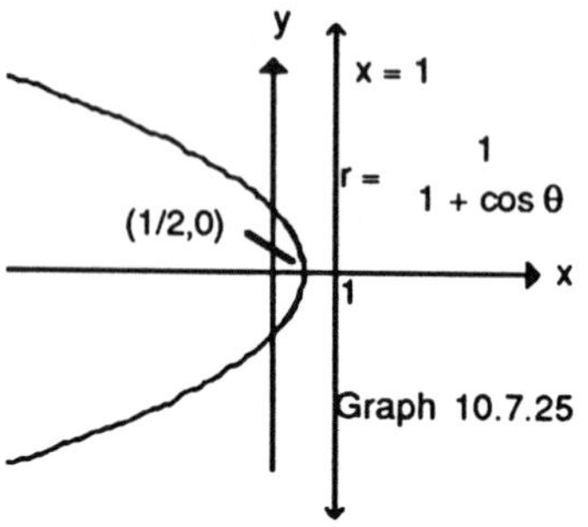

26. $r = \dfrac{6}{2 + \cos\theta} = \dfrac{3}{1 + \frac{1}{2}\cos\theta} \Rightarrow e = \dfrac{1}{2},\ k = 6 \Rightarrow$

$x = 6.$ $\quad a(1 - e^2) = ke \Rightarrow a\left(1 - \left(\dfrac{1}{2}\right)^2\right) = 3 \Rightarrow$

$\dfrac{3}{4}a = 3 \Rightarrow a = 4 \Rightarrow ea = 2.$

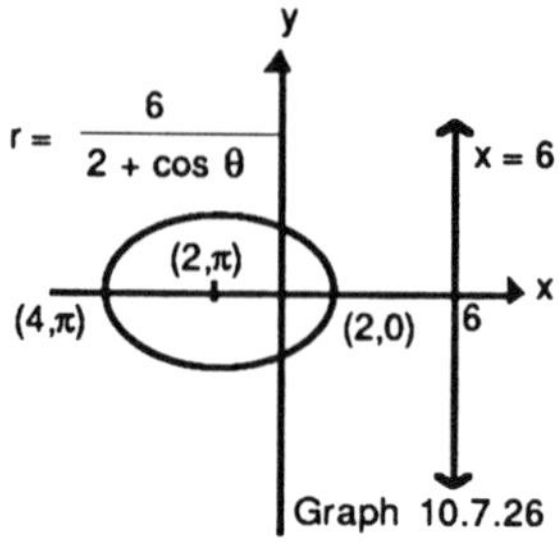

27. $r = \dfrac{25}{10 - 5\cos\theta} \Rightarrow r = \dfrac{\frac{25}{10}}{1 - \frac{5}{10}\cos\theta} = \dfrac{\frac{5}{2}}{1 - \frac{1}{2}\cos\theta} \Rightarrow$

$e = \dfrac{1}{2},\ k = 5.\ \ a(1 - e^2) = ke \Rightarrow a\left(1 - \left(\dfrac{1}{2}\right)^2\right) = \dfrac{5}{2} \Rightarrow$

$\dfrac{3}{4}a = \dfrac{5}{2} \Rightarrow a = \dfrac{10}{3}.\ \ a - ae = \dfrac{10}{3} - \left(\dfrac{10}{3}\right)\dfrac{1}{2} = \dfrac{5}{3}$

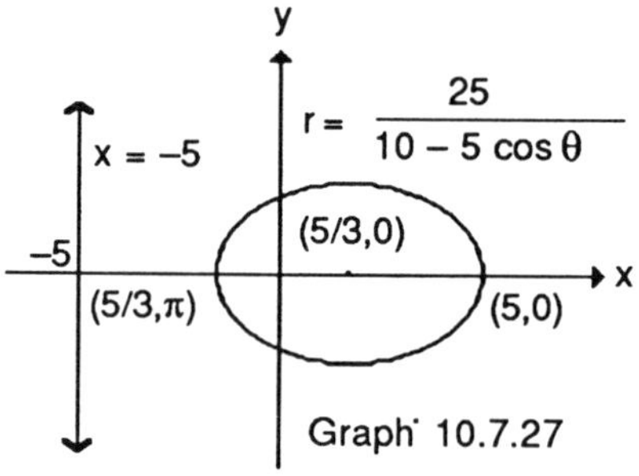

28. $r = \dfrac{4}{2 - 2\cos\theta} \Rightarrow r = \dfrac{2}{1 - \cos\theta} \Rightarrow e = 1,\ k = 2$

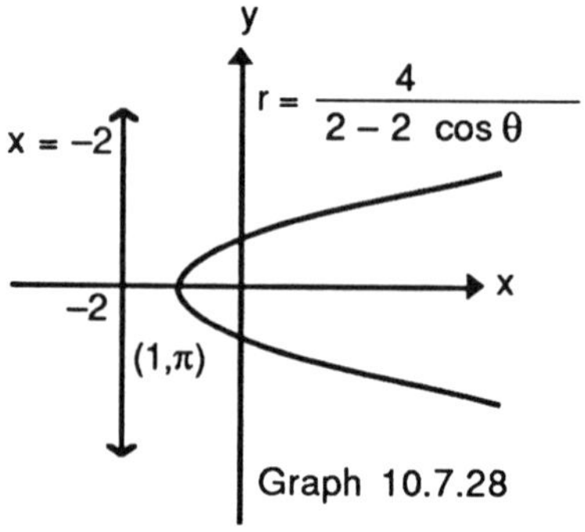

29. $r = \dfrac{400}{16 + 8\sin\theta} \Rightarrow r = \dfrac{\frac{400}{16}}{1 + \frac{8}{16}\sin\theta} \Rightarrow r = \dfrac{25}{1 + \frac{1}{2}\sin\theta} \Rightarrow$

$e = \dfrac{1}{2},\ k = 50.\ \ a(1 - e^2) = ke \Rightarrow a\left(1 - \left(\dfrac{1}{2}\right)^2\right) = 25 \Rightarrow$

$\dfrac{3}{4}a = 25 \Rightarrow\ a = \dfrac{100}{3}.\ \ a - ae = \dfrac{100}{3} - \dfrac{100}{3}\left(\dfrac{1}{2}\right) = \dfrac{50}{3}.$

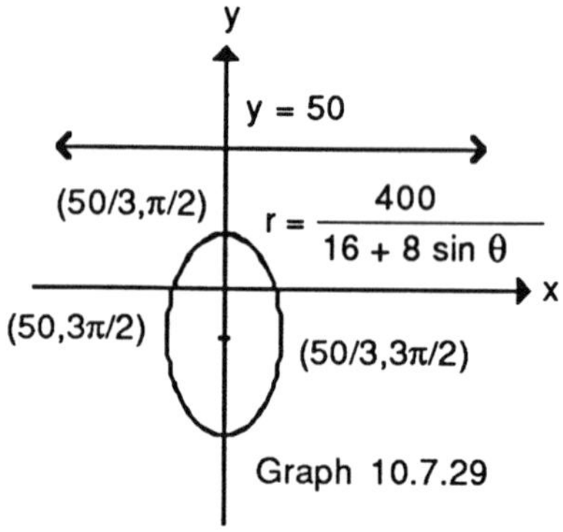

30. $r = \dfrac{12}{3 + 3\sin\theta} \Rightarrow r = \dfrac{4}{1 + \sin\theta} \Rightarrow e = 1, k = 4$

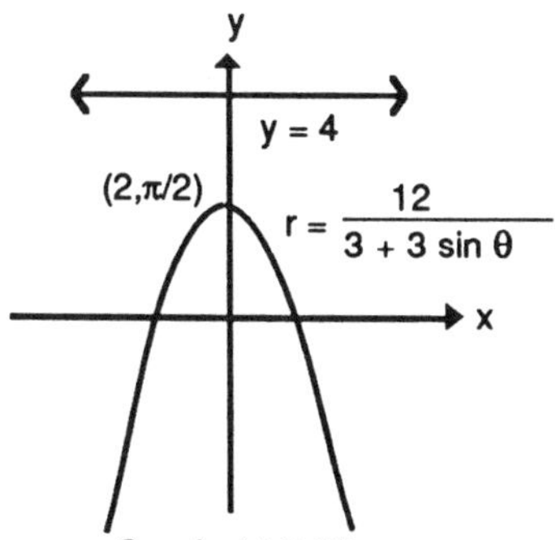

Graph 10.7.30

31. $r = \dfrac{8}{2 - 2\sin\theta} \Rightarrow r = \dfrac{4}{1 - \sin\theta} \Rightarrow e = 1, k = 4$

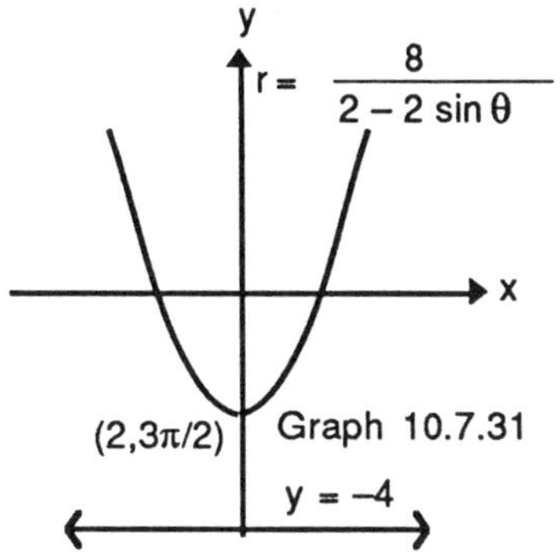

Graph 10.7.31

32. $r = \dfrac{4}{2 - \sin\theta} \Rightarrow r = \dfrac{2}{1 - \frac{1}{2}\sin\theta} \Rightarrow e = \dfrac{1}{2}, k = 4$

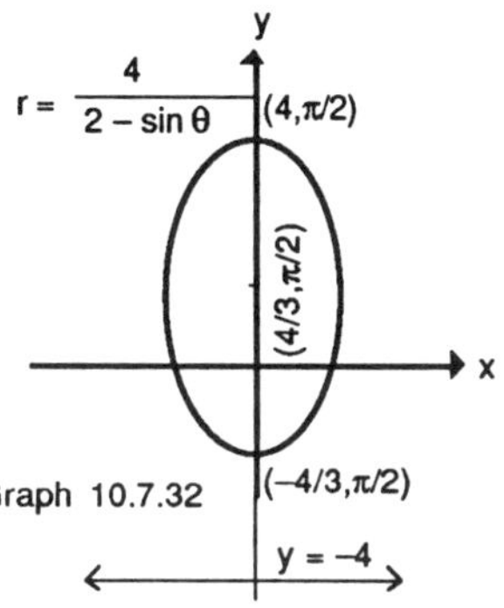

Graph 10.7.32

33.

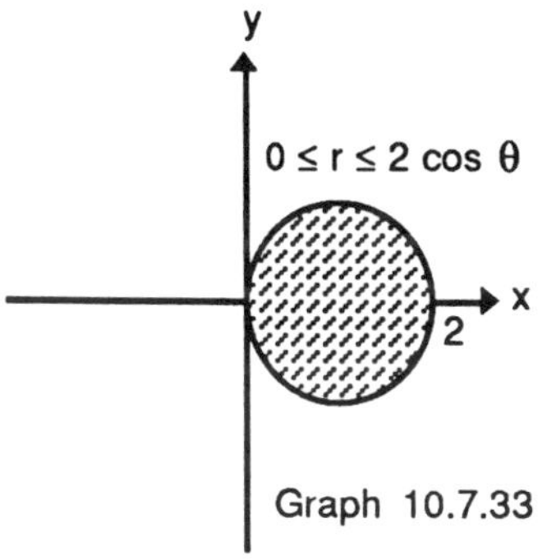

Graph 10.7.33

34.

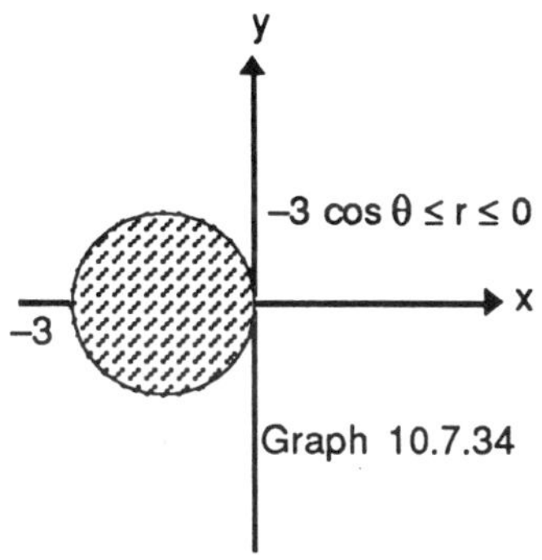

Graph 10.7.34

35. a) Perihelion = $a - ae = a(1 - e)$

Aphelion = $ea + a = a(1 + e)$

b)

Planet	Perihelion	Aphelion
Mercury	0.3075 AU	0.4667 AU
Venus	0.7184 AU	0.7282 AU
Earth	0.9833 AU	1.0167 AU
Mars	1.3817 AU	1.6663 AU
Jupiter	4.9512 AU	5.4548 AU
Saturn	9.0210 AU	10.0570 AU
Uranus	18.2977 AU	20.0623 AU
Neptune	29.8135 AU	30.3065 AU
Pluto	29.6549 AU	49.2251 AU

36. Mercury: $r = \dfrac{0.3871(1 - 0.2056^2)}{1 + 0.2056\cos\theta} = \dfrac{-.3707}{1 + 0.2056\cos\theta}$

Venus: $r = \dfrac{0.7233(1 - 0.0068^2)}{1 + 0.0068\cos\theta} = \dfrac{0.7233}{1 + 0.0068\cos\theta}$

Earth: $r = \dfrac{1(1 - 0.0167^2)}{1 + 0.0167\cos\theta} = \dfrac{0.9997}{1 + 0.0617\cos\theta}$

Mars: $r = \dfrac{1.524(1 - 0.0934^2)}{1 + 0.0934\cos\theta} = \dfrac{1.511}{1 + 0.0934\cos\theta}$

Jupiter: $r = \dfrac{5.203(1 - 0.0484^2)}{1 + 0.0484\cos\theta} = \dfrac{5.191}{1 + 0.0484\cos\theta}$

Saturn: $r = \dfrac{9.539(1 - 0.0543^2)}{1 + 0.0543\cos\theta} = \dfrac{9.511}{1 + 0.0543\cos\theta}$

Uranus: $r = \dfrac{19.18(1 - 0.0460^2)}{1 + 0.0460\cos\theta} = \dfrac{19.14}{1 + 0.0460\cos\theta}$

Neptune: $r = \dfrac{30.06(1 - 0.0082^2)}{1 + 0.0082\cos\theta} = \dfrac{30.06}{1 + 0.0082\cos\theta}$

37. a) $r = 2\sin\theta \Rightarrow a = 1 \Rightarrow$ Center = (0,1)

$\Rightarrow x^2 + (y - 1)^2 = 1.$

$r = \csc\theta \Rightarrow r = \dfrac{1}{\sin\theta} \Rightarrow r\sin\theta = 1$

$\Rightarrow y = 1$

b)

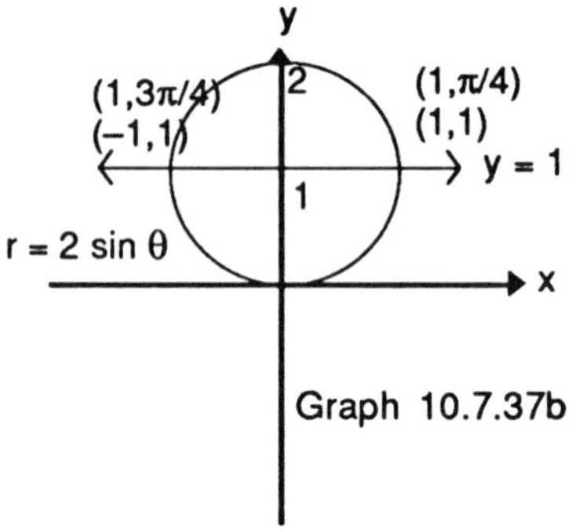

Graph 10.7.37b

38. $r = 2\cos\theta \Rightarrow a = 1 \Rightarrow$ Center = (1,0) $\Rightarrow (x - 1)^2 + y^2 = 1.$ $r = \sec\theta \Rightarrow r = \dfrac{1}{\cos\theta} \Rightarrow r\cos\theta = 1 \Rightarrow x = 1$

39. $r\cos\theta = 4 \Rightarrow x = 4 \Rightarrow k = 4.$ Parabola $\Rightarrow e = 1.$ $\therefore\ r = \dfrac{4}{1 + \cos\theta}$

40. $r\cos\left(\theta - \frac{\pi}{2}\right) = 2 \Rightarrow r\left(\cos\theta\cos\frac{\pi}{2} + \sin\theta\sin\frac{\pi}{2}\right) = 2 \Rightarrow r\sin\theta = 2 \Rightarrow y = 2 \Rightarrow k = 2.$

Parabola $\Rightarrow e = 1.$ $\therefore\ r = \dfrac{2}{1 + \sin\theta}$

41.

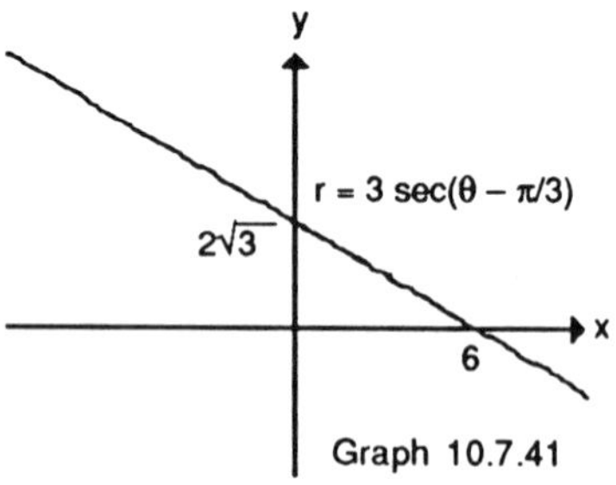

Graph 10.7.41

42.

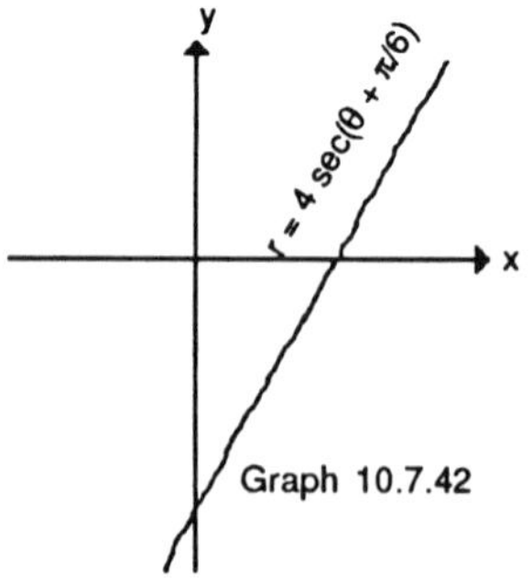

Graph 10.7.42

43.

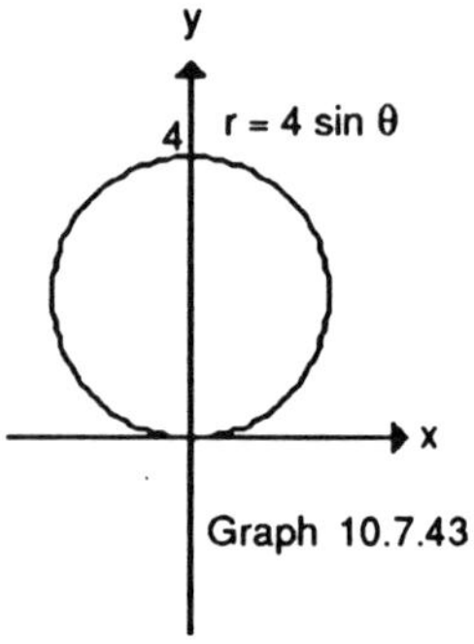

Graph 10.7.43

44.

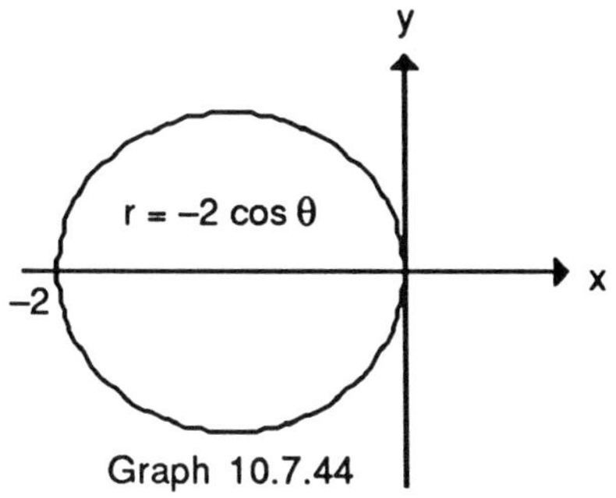

Graph 10.7.44

45.

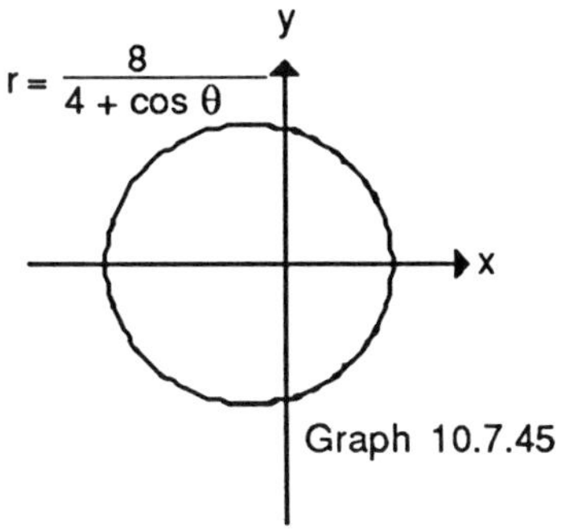

Graph 10.7.45

46.

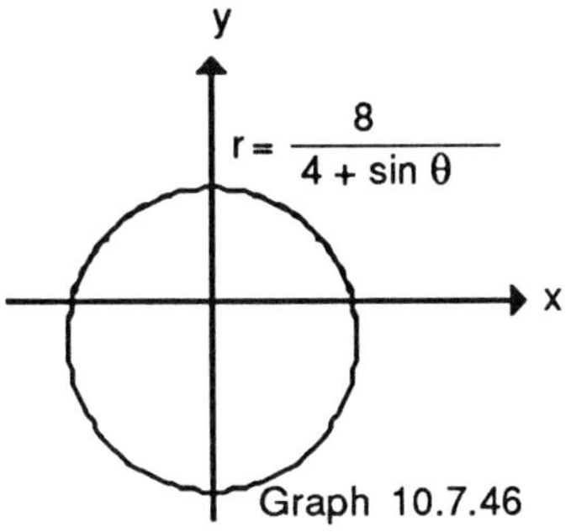

Graph 10.7.46

47.

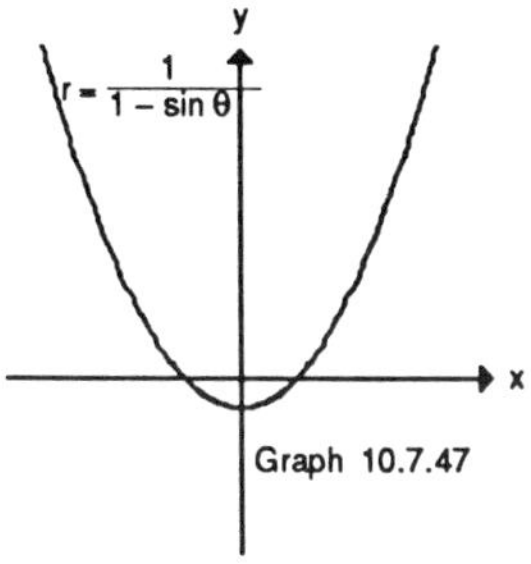

Graph 10.7.47

48.

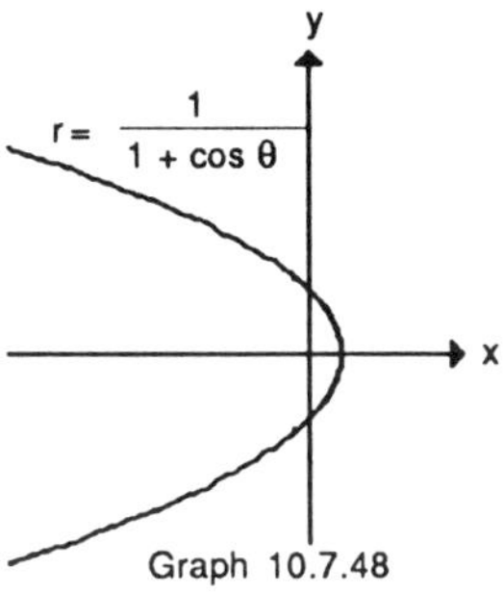

Graph 10.7.48

49.

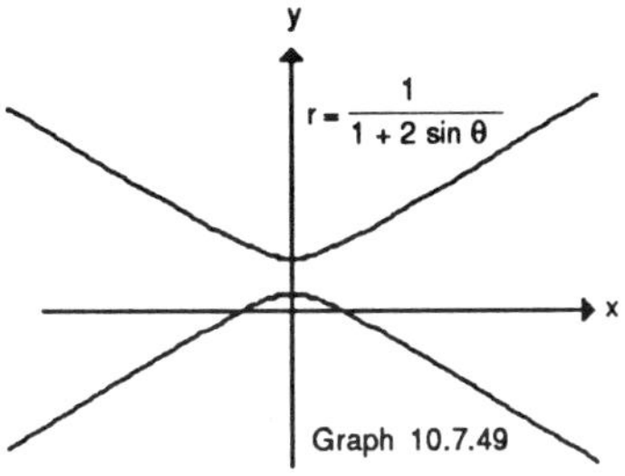

Graph 10.7.49

50.

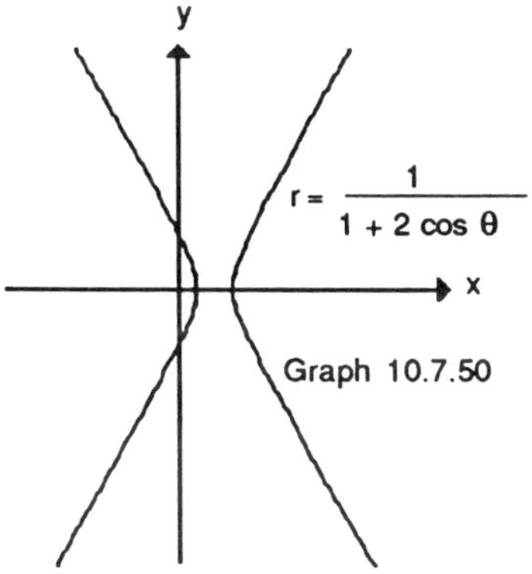

Graph 10.7.50

SECTION 10.8 INTEGRATION IN POLAR COORDINATES

1. $$A = \int_0^{\pi/4} \frac{1}{2} r^2\, d\theta = \int_0^{\pi/4} \frac{1}{2}\cos^2\theta\, d\theta = \int_0^{\pi/4} \frac{1}{2}\left(\frac{1 + \cos 2\theta}{2}\right) d\theta = \frac{1}{4}\int_0^{\pi/4} (1 + \cos 2\theta)\, d\theta$$
$$= \frac{1}{4}\left[\theta + \frac{\sin 2\theta}{2}\right]_0^{\pi/4} = \frac{\pi}{16} + \frac{1}{8}$$

2. $$A = \int_0^{\ln 25} \frac{1}{2} r^2\, d\theta = \int_0^{\ln 25} \frac{1}{2}\left(e^{\theta}\right)^2 d\theta = \int_0^{\ln 25} e^{2\theta}\, d\theta = \left[\frac{1}{4} e^{2\theta}\, d\theta\right]_0^{\ln 25} = 156$$

3. $A = \int_0^{2\pi} \frac{1}{2}(4 + 2\cos\theta)^2 d\theta = \int_0^{2\pi} \frac{1}{2}(16 + 16\cos\theta + 4\cos^2\theta)\, d\theta$

$= \int_0^{2\pi} \left(8 + 8\cos\theta + 2\left(\frac{1 + \cos 2\theta}{2}\right)\right) d\theta = \int_0^{2\pi} (9 + 8\cos\theta + \cos 2\theta)\, d\theta =$

$\left[9\theta + 8\sin\theta + \frac{1}{2}\sin 2\theta\right]_0^{2\pi} = 18\pi$

4. $A = \int_0^{2\pi} \frac{1}{2}(a(1 + \cos\theta))^2 d\theta = \int_0^{2\pi} \frac{1}{2}a^2(1 + 2\cos\theta + \cos^2\theta)\, d\theta =$

$\frac{1}{2}a^2 \int_0^{2\pi} \left(1 + 2\cos\theta + \frac{1 + \cos 2\theta}{2}\right) d\theta = \frac{1}{2}a^2 \int_0^{2\pi} \left(\frac{3}{2} + 2\cos\theta + \frac{1}{2}\cos 2\theta\right) d\theta =$

$\frac{1}{2}a^2\left[\frac{3}{2}\theta + 2\sin\theta + \frac{1}{4}\sin 2\theta\right]_0^{2\pi} = \frac{3}{2}\pi a^2$

5. $A = 2\int_0^{\pi/4} \frac{1}{2}\cos^2 2\theta\, d\theta = \int_0^{\pi/4} \frac{1 + \cos 4\theta}{2} d\theta = \frac{1}{2}\left[\theta + \frac{\sin 4\theta}{4}\right]_0^{\pi/4} = \frac{\pi}{8}$

6. $A = \int_0^{\pi} \frac{1}{2}(2a\sin\theta)^2 d\theta = \int_0^{\pi} \frac{1}{2}(4a^2\sin^2\theta)\, d\theta = \int_0^{\pi} 2a^2\left(\frac{1 - \cos 2\theta}{2}\right) d\theta$

$= \int_0^{\pi} a^2(1 - \cos 2\theta)\, d\theta = \left[a^2(\theta - \frac{1}{2}\sin 2\theta)\right]_0^{\pi} = a^2\pi$

7. $A = 2\int_0^{`\pi/4} \frac{1}{2}\left(2a^2\cos 2\theta\right) d\theta = 2a^2 \int_0^{\pi/4} \cos 2\theta\, d\theta = 2a^2\left[\frac{\sin 2\theta}{2}\right]_0^{\pi/4} = a^2$

8. $A = \int_0^{\pi/2} \frac{1}{2}\left(4\sin 2\theta\right) d\theta = \int_0^{\pi/2} 2\sin 2\theta\, d\theta = \left[-\cos 2\theta\right]_0^{\pi/2} = 2$

9. $A = 6(2)\int_0^{\pi/6} \frac{1}{2}(2\sin 3\theta)\, d\theta = 12\int_0^{\pi/6} \sin 3\theta\, d\theta = 12\left[-\frac{\cos 3\theta}{3}\right]_0^{\pi/6} = 4$

10. $r = 2\cos\theta,\ r = 2\sin\theta \Rightarrow 2\cos\theta = 2\sin\theta \Rightarrow \cos\theta = \sin\theta \Rightarrow \theta = \frac{\pi}{4}$

$$A = 2\int_0^{\pi/4} \frac{1}{2}\left(2\sin\theta\right)^2 d\theta = \int_0^{\pi/4} 4\sin^2\theta\, d\theta = \int_0^{\pi/4} 4\left(\frac{1-\cos 2\theta}{2}\right)d\theta = \int_0^{\pi/4} (2 - 2\cos 2\theta)\, d\theta =$$

$$\Big[2\theta - \sin 2\theta\Big]_0^{\pi/4} = \frac{\pi}{2} - 1$$

11. $r = 1,\ r = 2\sin\theta \Rightarrow\ 2\sin\theta = 1 \Rightarrow\ \sin\theta = \frac{1}{2} \Rightarrow \theta = \frac{\pi}{6}, \frac{5\pi}{6}$

$$A = \pi(1)^2 - \int_{\pi/6}^{5\pi/6} \frac{1}{2}\left((2\sin\theta)^2 - 1^2\right)d\theta = \pi - = \int_{\pi/6}^{5\pi/6}\left(2\sin^2\theta - \frac{1}{2}\right)d\theta = \int_{\pi/6}^{5\pi/6}\left(2(1-\cos^2\theta) - \frac{1}{2}\right)d\theta$$

$$= \pi - \int_{\pi/6}^{5\pi/6}\left(\frac{3}{2} - 2\cos^2\theta\right)d\theta = \pi - \int_{\pi/6}^{5\pi/6}\left(\frac{3}{2} - 2\left(\frac{1+\cos 2\theta}{2}\right)\right)d\theta = \pi - \int_{\pi/6}^{5\pi/6}\left(\frac{1}{2} - \cos 2\theta\right)d\theta =$$

$$\pi - \left[\frac{1}{2}\theta - \frac{\sin 2\theta}{2}\right]_{\pi/6}^{5\pi/6} = \frac{4\pi - 3\sqrt{3}}{6}$$

12. $r = 2,\ r = 2(1-\cos\theta) \Rightarrow 2 = 2(1-\cos\theta) \Rightarrow \cos\theta = 1 \Rightarrow \theta = 0$. Sketch a graph to see the region.

$$A = 2\int_0^{\pi/2} \frac{1}{2}\left(2(1-\cos\theta)\right)^2 d\theta + \frac{1}{2}\text{ of the area of the circle } = \int_0^{\pi/2} 4(1 - 2\cos\theta + \cos^2\theta)d\theta + \frac{1}{2}\pi(2)^2$$

$$= \int_0^{\pi/2}\left(4 - 8\cos\theta + \left(\frac{1+\cos 2\theta}{2}\right)\right)d\theta + 2\pi = \int_0^{\pi/2} (6 - 8\cos\theta + 2\cos 2\theta)d\theta + 2\pi =$$

$$\Big[6\theta - 8\sin\theta + \sin 2\theta\Big]_0^{\pi/2} + 2\pi = 5\pi - 8$$

13. $r = 2(1-\cos\theta),\ r = 2(1+\cos\theta) \Rightarrow 1 - \cos\theta = 1 + \cos\theta \Rightarrow \cos\theta = 0 \Rightarrow \theta = \frac{\pi}{2}, \frac{3\pi}{2}$. The graph gives the point of intersection (0,0).

$$A = 2\int_0^{\pi/2} \frac{1}{2}\left(2(1-\cos\theta)\right)^2 d\theta + 2\int_{\pi/2}^{\pi} \frac{1}{2}\left(2(1+\cos\theta)\right)^2 d\theta =$$

$$\int_0^{\pi/2} 4(1 - 2\cos\theta + \cos^2\theta)d\theta + \int_{\pi/2}^{\pi} 4(1 + 2\cos\theta + \cos^2\theta)d\theta =$$

$$\int_0^{\pi/2}\left(4 - 8\cos\theta\ 4\left(\frac{1+\cos 2\theta}{2}\right)\right)d\theta + \int_{\pi/2}^{\pi}\left(4 + 8\cos\theta + 4\left(\frac{1+\cos 2\theta}{2}\right)\right)d\theta =$$

$$\int_0^{\pi/2} (6 - 8\cos\theta + 2\cos 2\theta)d\theta + \int_{\pi/2}^{\pi} (6 + 8\cos\theta + 2\cos 2\theta)d\theta =$$

$$\Big[6\theta - 8\sin\theta + \sin 2\theta\Big]_0^{\pi/2} + \Big[6\theta + 8\sin\theta + \sin 2\theta\Big]_{\pi/2}^{\pi} = 6\pi - 16$$

14. $r = \sqrt{3}$, $r^2 = 6\cos 2\theta \Rightarrow 3 = 6\cos 2\theta \Rightarrow \cos 2\theta = \frac{1}{2} \Rightarrow \theta = \frac{\pi}{6}$ in the 1st quadrant. Use symmetry to find the area.

$$A = 4\int_0^{\pi/6}\left(\frac{1}{2}(6\cos 2\theta) - \frac{1}{2}\left(\sqrt{3}\right)^2\right)d\theta = 2\int_0^{\pi/6}(6\cos 2\theta - 3)d\theta = 2\Big[3\sin 2\theta - 3\theta\Big]_0^{\pi/6} = 3\sqrt{3} - \pi$$

15. $r = 3a\cos\theta$, $r = a(1 + \cos\theta) \Rightarrow 3a\cos\theta = a(1 + \cos\theta) \Rightarrow 3\cos\theta = 1 + \cos\theta \Rightarrow \cos\theta = \frac{1}{2} \Rightarrow$

$\theta = \frac{\pi}{3}$ or $-\frac{\pi}{3}$. $\therefore$ $A = 2\int_0^{\pi/3}\frac{1}{2}\left((3a\cos\theta)^2 - (a(1+\cos\theta))^2\right)d\theta =$

$$\int_0^{\pi/3}\left(9a^2\cos^2\theta - a^2 - 2a^2\cos\theta - a^2\cos^2\theta\right)d\theta = \int_0^{\pi/3}\left(8a^2\cos^2\theta - 2a^2\cos\theta - a^2\right)d\theta =$$

$$\int_0^{\pi/3}\left(4a^2(1 + \cos 2\theta) - 2a^2\cos\theta - a^2\right)d\theta = \int_0^{\pi/3}\left(3a^2 + 4a^2\cos 2\theta - 2a^2\cos\theta\right)d\theta =$$

$$\Big[3a^2\theta + 2a^2\sin 2\theta - 2a^2\sin\theta\Big]_0^{\pi/3} = \pi a^2$$

16. $r = 1$, $r = -2\cos\theta \Rightarrow 1 = -2\cos\theta \Rightarrow \cos\theta = -\frac{1}{2} \Rightarrow \theta = \frac{2\pi}{3}$ in quadrant II.

$$A = 2\int_{2\pi/3}^{\pi}\frac{1}{2}\left((-2\cos\theta)^2 - 1^2\right)d\theta = \int_{2\pi/3}^{\pi}\left(4\cos^2\theta - 1\right)d\theta = \int_{2\pi/3}^{\pi}\left(2(1 + \cos 2\theta) - 1\right)d\theta =$$

$$\int_{2\pi/3}^{\pi}(1 + 2\cos 2\theta)d\theta = \Big[\theta + \sin 2\theta\Big]_{2\pi/3}^{\pi} = \frac{\pi}{3} + \frac{\sqrt{3}}{2}$$

17. a) $A = 2\int_0^{2\pi/3}\frac{1}{2}(2\cos\theta + 1)^2 d\theta = \int_0^{2\pi/3}(4\cos^2\theta + 4\cos\theta + 1)d\theta =$

$$\int_0^{2\pi/3}\left(2(1 + \cos 2\theta) + 4\cos\theta + 1\right)d\theta = \int_0^{2\pi/3}(3 + 2\cos 2\theta + 4\cos\theta)d\theta =$$

$$\Big[3\theta + 2\sin 2\theta + 4\sin\theta\Big]_0^{2\pi/3} = 2\pi + \frac{3\sqrt{3}}{2}$$

b) $A = \left(2\pi + \frac{3\sqrt{3}}{2}\right) - \left(\pi - \frac{3\sqrt{3}}{2}\right) = \pi + 3\sqrt{3}$ (From 17 a) above and Example 2, page 702, in the text.)

18. $r = 6$, $r\sin\theta = 3 \Rightarrow 6\sin\theta = 3 \Rightarrow \sin\theta = \frac{1}{2} \Rightarrow \theta = \frac{\pi}{4}, \frac{3\pi}{4}$

$$A = \int_{\pi/4}^{3\pi/4}\frac{1}{2}\left(6^2 - \left(\frac{3}{\sin\theta}\right)^2\right)d\theta = \int_{\pi/4}^{3\pi/4}(18 - 9\csc^2\theta)d\theta = \Big[18\theta + 9\cot\theta\Big]_{\pi/4}^{3\pi/4} = 9\pi - 18$$

19. $r = \theta^2,\ 0 \le \theta \le \sqrt{5} \Rightarrow \dfrac{dr}{d\theta} = 2\theta$. $\therefore$ Length $= \displaystyle\int_0^{\sqrt{5}} \sqrt{(\theta)^2 + (2\theta)^2}\, d\theta = \int_0^{\sqrt{5}} \sqrt{\theta^4 + 4\theta^2}\, d\theta =$

$$\int_0^{\sqrt{5}} |\theta|\sqrt{\theta^2 + 4}\, d\theta = \int_0^{\sqrt{5}} \theta\sqrt{\theta^2 + 4}\, d\theta = \int_4^9 \frac{1}{2}\sqrt{u}\, du = \frac{1}{2}\left[\frac{2}{3}u^{3/2}\right]_4^9 = \frac{19}{3}$$

Let $u = \theta^2 + 4 \Rightarrow \frac{1}{2}du = \theta d\theta$; $\theta = 0 \Rightarrow u = 4$, $\theta = \sqrt{5} \Rightarrow u = 9$

20. $r = \dfrac{e^\theta}{\sqrt{2}},\ 0 \le \theta \le \pi \Rightarrow \dfrac{dr}{d\theta} = \dfrac{e^\theta}{\sqrt{2}}$. $\therefore$ Length $\displaystyle\int_0^{\pi} \sqrt{\left(\frac{e^\theta}{\sqrt{2}}\right)^2 + \left(\frac{e^\theta}{\sqrt{2}}\right)^2}\, d\theta = \int_0^{\pi} \sqrt{2\left(\frac{e^{2\theta}}{2}\right)}\, d\theta =$

$$\int_0^{\pi} e^\theta\, d\theta = \left[e^\theta\right]_0^{\pi} = e^\pi - 1$$

21. $r = \sec\theta,\ 0 \le \theta \le \frac{\pi}{4} \Rightarrow \dfrac{dr}{d\theta} = \sec\theta\tan\theta$. $\therefore$ Length $= \displaystyle\int_0^{\pi/4} \sqrt{\sec^2\theta + (\sec\theta\tan\theta)^2}\, d\theta =$

$$\int_0^{\pi/4} \sqrt{\sec^2\theta(1 + \tan^2\theta)}\, d\theta = \int_0^{\pi/4} |\sec\theta|\sqrt{1 + \tan^2\theta}\, d\theta = \int_0^{\pi/4} |\sec\theta|\sqrt{\sec^2\theta}\, d\theta =$$

$$\int_0^{\pi/4} \sec^2\theta\, d\theta = \left[\tan\theta\right]_0^{\pi/4} = 1$$

22. $r = \csc\theta,\ \frac{\pi}{6} \le \theta \le \frac{\pi}{2} \Rightarrow \dfrac{dr}{d\theta} = -\csc\theta\cot\theta$. $\therefore$ Length $= \displaystyle\int_{\pi/6}^{\pi/2} \sqrt{\csc^2\theta + (-\csc\theta\cot\theta)^2}\, d\theta =$

$$\int_{\pi/6}^{\pi/2} \sqrt{\csc^2\theta + \csc^2\theta\cot^2\theta}\, d\theta = \int_{\pi/6}^{\pi/2} |\csc\theta|\sqrt{1 + \cot^2\theta}\, d\theta = \int_{\pi/6}^{\pi/2} |\csc\theta|\sqrt{\csc^2\theta}\, d\theta =$$

$$\int_{\pi/6}^{\pi/2} \csc^2\theta\, d\theta = \left[-\cot^2\theta\right]_{\pi/6}^{\pi/2} = \sqrt{3}$$

23. $r = 1 + \cos\theta \Rightarrow \frac{dr}{d\theta} = -\sin\theta$. $\therefore$ Length $= \int_0^{2\pi} \sqrt{(1+\cos\theta)^2 + (-\sin\theta)^2}\, d\theta =$

$$2\int_0^{\pi} \sqrt{1 + 2\cos\theta + \cos^2\theta + \sin^2\theta}\, d\theta = 2\int_0^{\pi} \sqrt{2 + 2\cos\theta}\, d\theta = 2\int_0^{\pi} \sqrt{\frac{4(1+\cos\theta)}{2}}\, d\theta =$$

$$4\int_0^{\pi} \sqrt{\frac{1+\cos\theta}{2}}\, d\theta = 4\int_0^{\pi} \cos\frac{1}{2}\theta\, d\theta = 4\left[2\sin\frac{1}{2}\theta\right]_0^{\pi} = 8$$

24. a) $r = a \Rightarrow \frac{dr}{d\theta} = 0$. $\therefore$ Length $= \int_0^{2\pi} \sqrt{a^2 + 0^2}\, d\theta = \int_0^{2\pi} |a|\, d\theta = \Big[|a|\theta\Big]_0^{2\pi} = 2\pi|a| = 2\pi a$ if $a \geq 0$

b) $r = a\cos\theta \Rightarrow \frac{dr}{d\theta} = -a\sin\theta$. $\therefore$ Length $= \int_0^{2\pi} \sqrt{(a\cos\theta)^2 + (-a\sin\theta)^2}\, d\theta =$

$$\int_0^{2\pi} \sqrt{a^2(\cos^2\theta + \sin^2\theta)}\, d\theta = \int_0^{2\pi} |a|\, d\theta = \Big[|a|\theta\Big]_0^{2\pi} = 2\pi|a| = 2Pa \text{ if } a \geq 0$$

c) $r = a\sin\theta \Rightarrow \frac{dr}{d\theta} = a\cos\theta$. $\therefore$ Length $= \int_0^{2\pi} \sqrt{(a\cos\theta)^2 + (a\sin\theta)^2}\, d\theta =$

$$\int_0^{2\pi} \sqrt{a^2(\cos^2\theta + \sin^2\theta)}\, d\theta = \int_0^{2\pi} |a|\, d\theta = \Big[|a|\theta\Big]_0^{2\pi} = 2\pi|a| = 2\pi a \text{ if } a \geq 0$$

25. $r = \sqrt{\cos 2\theta}$, $0 \leq \theta \leq \frac{\pi}{4} \Rightarrow \frac{dr}{d\theta} = \frac{1}{2}(\cos 2\theta)^{-1/2}(-\sin 2\theta)(2) = \frac{-\sin 2\theta}{\sqrt{\cos 2\theta}}$

$\therefore$ Surface Area $= \int_0^{\pi/4} 2\pi r\cos\theta \sqrt{\left(\sqrt{\cos 2\theta}\right)^2 + \left(\frac{-\sin 2\theta}{\sqrt{\cos 2\theta}}\right)^2}\, d\theta =$

$$\int_0^{\pi/4} 2\pi\sqrt{\cos 2\theta}\cos\theta \sqrt{\cos 2\theta + \frac{\sin^2 2\theta}{\cos 2\theta}}\, d\theta = \int_0^{\pi/4} 2\pi\sqrt{\cos 2\theta}\cos\theta\sqrt{\frac{1}{\cos 2\theta}}\, d\theta =$$

$$\int_0^{\pi/4} 2\pi\cos\theta\, d\theta = \Big[2\pi\sin\theta\Big]_0^{\pi/4} = \pi\sqrt{2}$$

26. $r = \sqrt{2}\, e^{\theta/2},\ 0 \le \theta \le \frac{\pi}{2} \Rightarrow \frac{dr}{d\theta} = \sqrt{2}\left(\frac{1}{2}\right)e^{\theta/2} = \frac{\sqrt{2}}{2}\, e^{\theta/2}.$

$$\text{Surface Area} = \int_0^{\pi/2} 2\pi\sqrt{2}\, e^{\theta/2}\sin\theta\sqrt{\left(\sqrt{2}e^{\theta/2}\right)^2 + \left(\frac{\sqrt{2}}{2}\, e^{\theta/2}\right)^2}\, d\theta =$$

$$\int_0^{\pi/2} 2\pi\sqrt{2}e^{\theta/2}\sin\theta\sqrt{2e^{\theta} + \frac{1}{2}e^{\theta}}\, d\theta = \int_0^{\pi/2} 2\pi\sqrt{2}e^{\theta/2}\sin\theta\sqrt{\frac{5}{2}e^{\theta}}\, d\theta =$$

$$\int_0^{\pi/2} 2\pi\sqrt{2}e^{\theta/2}\sin\theta\frac{\sqrt{5}}{\sqrt{2}}\, e^{\theta/2}\, d\theta = \int_0^{\pi/2} 2\pi\sqrt{5}\, e^{\theta}\sin\theta\, d\theta = \text{(by parts)}$$

$$2\pi\sqrt{5}\left[\frac{e^{\theta}}{2}(\sin\theta - \cos\theta)\right]_0^{\pi/2} = \pi\sqrt{5}\left(e^{\pi/2} + 1\right)$$

27. $r^2 = \cos 2\theta \Rightarrow r = \pm\sqrt{\cos 2\theta}$. Use $r = \sqrt{\cos 2\theta}$ on $\left[0, \frac{\pi}{4}\right]$. Then $\frac{dr}{d\theta} = \frac{1}{2}(\cos 2\theta)^{-1/2}(-\sin 2\theta)(2) =$

$$\frac{-\sin 2\theta}{\sqrt{\cos 2\theta}}\ \therefore\ \text{Surface Area} = \int_0^{\pi/4} 2\pi\sqrt{\cos 2\theta}\,\sin\theta\sqrt{\left(\sqrt{\cos 2\theta}\right)^2 + \left(\frac{-\sin 2\theta}{\sqrt{\cos 2\theta}}\right)^2}\, d\theta =$$

$$\int_0^{\pi/4} 2\pi\sqrt{\cos 2\theta}\,\sin\theta\sqrt{\cos 2\theta + \frac{\sin^2 2\theta}{\cos 2\theta}}\, d\theta = \int_0^{\pi/4} 2\pi\sqrt{\cos 2\theta}\,\sin\theta\sqrt{\frac{1}{\cos 2\theta}}\, d\theta =$$

$$\int_0^{\pi/4} 2\pi\sin\theta\, d\theta = \left[-2\pi\cos\theta\right]_0^{\pi/4} = \pi\left(2 - \sqrt{2}\right)$$

28. $r = 2a\cos\theta \Rightarrow \frac{dr}{d\theta} = -2a\sin\theta.$

$$\therefore\ \text{Surface Area} = \int_0^{\pi} 2\pi(2a\cos\theta)\cos\theta\sqrt{(2a\cos\theta)^2 + (-2a\sin\theta)^2}\, d\theta =$$

$$\int_0^{\pi} 2\pi(2a\cos^2\theta)\sqrt{4a^2(\cos^2\theta + \sin^2\theta)}\, d\theta = \int_0^{\pi} 4a\pi\cos^2\theta(2|a|)d\theta = \int_0^{\pi} 8a^2\pi\cos^2\theta\, d\theta =$$

$$\int_0^{\pi} 8a^2\pi\left(\frac{1 + \cos 2\theta}{2}\right)d\theta = \int_0^{\pi} 4a^2\pi(1 + \cos 2\theta)d\theta = 4a^2\pi\left[\theta + \frac{1}{2}\sin 2\theta\right]_0^{\pi} = 4a^2\pi^2$$

29. $$\bar{x} = \frac{\frac{2}{3}\int_0^{2\pi} r^3\cos\theta\, d\theta}{\int_0^{2\pi} r^2\, d\theta} = \frac{\frac{2}{3}\int_0^{2\pi} (a(1+\cos\theta))^3\cos\theta\, d\theta}{\int_0^{2\pi} (a(1+\cos\theta))^2\, d\theta} =$$

$$\frac{\frac{2}{3}a^3\int_0^{2\pi} (1 + 3\cos\theta + 3\cos^2\theta + \cos^3\theta)\cos\theta\, d\theta}{a^2\int_0^{2\pi} (1 + 2\cos\theta + \cos^2\theta)d\theta} =$$

$$\frac{\frac{2}{3}a\int_0^{2\pi}\left(\cos\theta + 3\left(\frac{1+\cos 2\theta}{2}\right) + 3(1-\sin^2\theta)\cos\theta + \left(\frac{1+\cos 2\theta}{2}\right)^2\right)d\theta}{\int_0^{2\pi}\left(1 + 2\cos\theta + \left(\frac{1+\cos 2\theta}{2}\right)\right)d\theta} =$$

(After much work using the identity $\cos A = \frac{1+\cos 2A}{2}$)

$$\frac{a\int_0^{2\pi}\left(\frac{15}{12} + \frac{8}{3}\cos\theta + \frac{4}{3}\cos 2\theta - 2\cos\theta\sin^2\theta + \frac{1}{12}\cos 4\theta\right)d\theta}{\int_0^{2\pi}\left(\frac{3}{2} + 2\cos\theta + \frac{1}{2}\cos 2\theta\right)d\theta} =$$

$$\frac{a\left[\frac{15}{12}\theta + \frac{8}{3}\sin\theta + \frac{2}{3}\sin 2\theta - \frac{2}{3}\sin^3\theta + \frac{1}{48}\sin 4\theta\right]_0^{2\pi}}{\left[\frac{3}{2}\theta + 2\sin\theta + \frac{1}{4}\sin 2\theta\right]_0^{2\pi}} = \frac{a\left(\frac{15}{6}\pi\right)}{3\pi} = \frac{5}{6}a$$

$$\bar{y} = \frac{\frac{2}{3}\int_0^{2\pi} r^3\sin\theta\, d\theta}{\int_0^{2\pi} r^2\, d\theta} = \frac{\frac{2}{3}\int_0^{2\pi} (a(1+\cos\theta))^3\sin\theta\, d\theta}{3\pi} = \frac{\frac{2}{3}\int_{2a}^{2a} -\frac{1}{a}u^3 du}{3\pi} = \frac{0}{3\pi} = 0$$ Centroid is $\left(\frac{5}{6}a, 0\right)$

Let $u = a(1+\cos\theta) \Rightarrow -\frac{1}{a}du = \sin\theta\, d\theta$; $\theta = 0 \Rightarrow u = 2a$,n $\theta = 2\pi \Rightarrow u = 2a$

30. $\int_0^{\pi} r^2\, d\theta = \int_0^{\pi} a^2\, d\theta = \left[a^2\theta\right]_0^{\pi} = a^2\pi.$

$$\bar{x} = \frac{\frac{2}{3}\int_0^{\pi} r^3\cos\theta\, d\theta}{\int_0^{\pi} r^2\, d\theta} = \frac{\frac{2}{3}\int_0^{\pi} a^3\cos\theta\, d\theta}{a^2\pi} = \frac{\frac{2}{3}a^3\left[-\sin\theta\right]_0^{\pi}}{a^2\pi} = \frac{0}{a^2\pi} = 0$$

$$\bar{y} = \frac{\frac{2}{3}\int_0^{\pi} r^3\sin\theta\, d\theta}{\int_0^{\pi} r^2\, d\theta} = \frac{\frac{2}{3}\int_0^{\pi} a^3\sin\theta\, d\theta}{a^2\pi} = \frac{\frac{2}{3}a^3\left[-\cos\theta\right]_0^{\pi}}{a^2\pi} =$$

$\frac{\frac{4}{3}a^3}{a^2\pi} = \frac{4a}{3\pi}$. Centroid is $\left(0, \frac{4a}{3\pi}\right)$

PRACTICE EXERCISES

1. $x = \frac{y^2}{8} \Rightarrow 4p = 8 \Rightarrow p = 2$

   ∴ Focus: (2,0); Directrix: $x = -2$

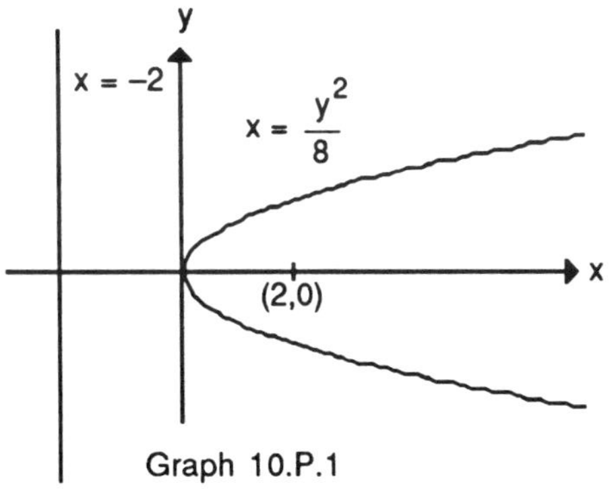

Graph 10.P.1

2. $y = -\frac{x^2}{4} \Rightarrow 4p = 4 \Rightarrow p = 1$

   ∴ Focus: (0,−1); Directrix: $y = 1$

Graph 10.P.2

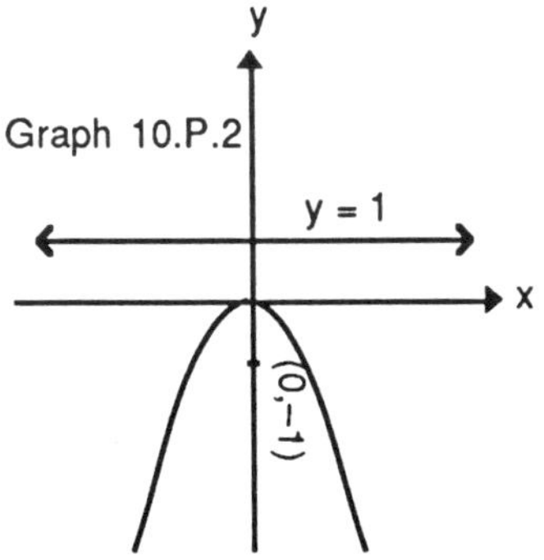

3. $16x^2 + 7y^2 = 112 \Rightarrow \frac{x^2}{7} + \frac{y^2}{16} = 1 \Rightarrow$

$c^2 = 16 - 7 = 9 \Rightarrow c = 3.$ $e = \frac{c}{a} = \frac{3}{4} \Rightarrow$

$y = \frac{a}{e} = \frac{4}{\frac{3}{4}} = \frac{16}{3}$ and $y = -\frac{16}{3}$

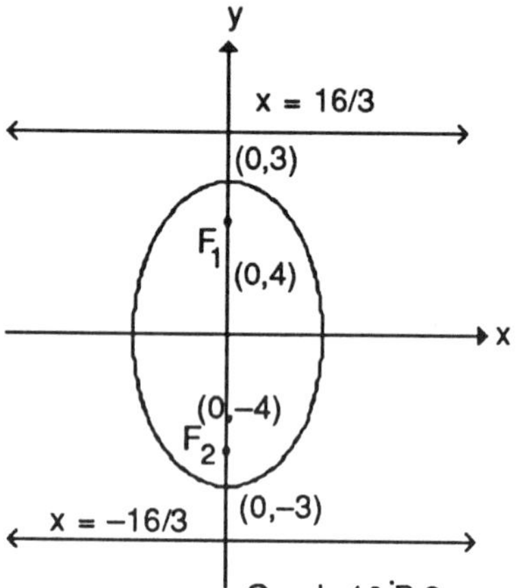

Graph 10.P.3

4. $x^2 + 2y^2 = 4 \Rightarrow \frac{x^2}{4} + \frac{y^2}{2} = 1 \Rightarrow c^2 = 4 - 2 = 2 \Rightarrow$

$c = \sqrt{2}$. $e = \frac{c}{a} = \frac{\sqrt{2}}{2} \Rightarrow x = \frac{2}{\frac{\sqrt{2}}{2}} = 2\sqrt{2}$ and

$x = -2\sqrt{2}$

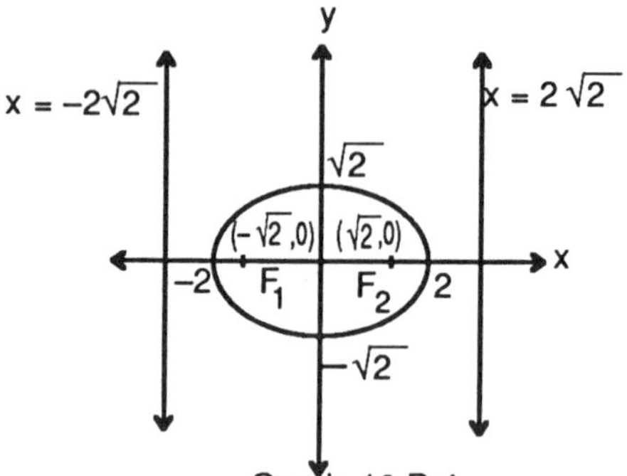

Graph 10.P.4

5. $3x^2 - y^2 = 3 \Rightarrow x^2 - \frac{y^2}{3} = 1 \Rightarrow$

$c^2 = 1 + 3 = 4 \Rightarrow c = 2.$ $e = \frac{c}{a} = \frac{2}{1} = 2 \Rightarrow$

$x = \frac{a}{e} = \frac{1}{2}$ and $x = -\frac{1}{2}$

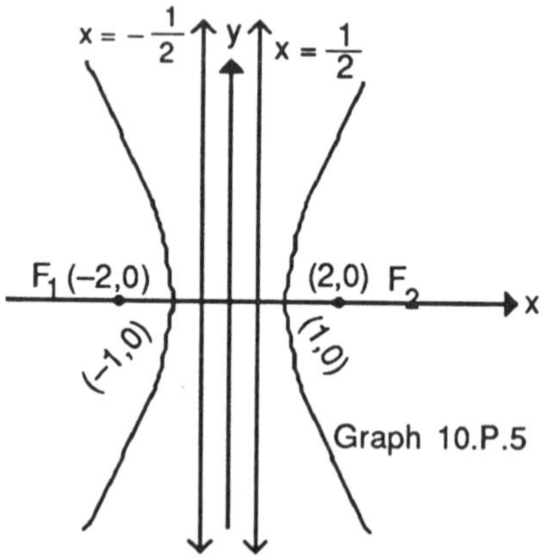

Graph 10.P.5

6. $2y^2 - 8x^2 = 16 \Rightarrow \frac{y^2}{8} - \frac{x^2}{2} = 1 \Rightarrow c^2 = 8 + 2 = 10$

$\Rightarrow c = \sqrt{10}.$ $e = \frac{c}{a} = \frac{\sqrt{10}}{\sqrt{8}} = \frac{\sqrt{5}}{2} \Rightarrow$

$y = \frac{a}{e} = \frac{\sqrt{8}}{\frac{\sqrt{5}}{2}} = \frac{4\sqrt{2}}{\sqrt{5}} = \frac{4\sqrt{10}}{5}$ and $y = -\frac{4\sqrt{10}}{5}$

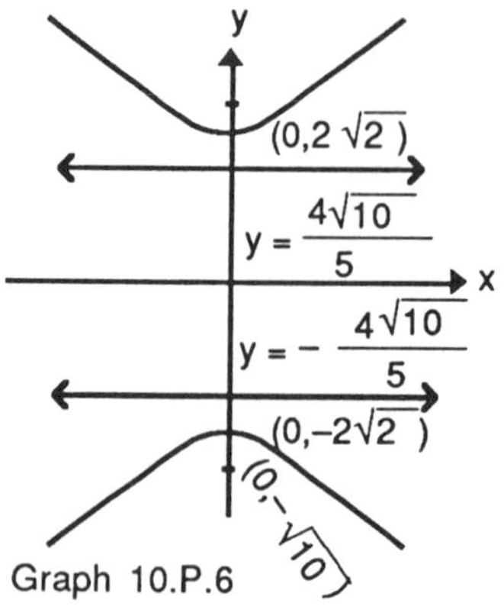

Graph 10.P.6

7. $B^2 - 4AC = 1 - 4(1)(1) = -3 < 0 \Rightarrow$ Ellipse

8. $B^2 - 4AC = 3^2 - 4(1)(2) = 1 > 0 \Rightarrow$ Hyperbola

9. $B^2 - 4AC = 4^2 - 4(1)(4) = 0 \Rightarrow$ Parabola

10. $B^2 - 4AC = 2^2 - 4(1)(-2) = 12 > 0 \Rightarrow$ Hyperbola

11. $B^2 - 4AC = 1^2 - 4(2)(2) = -15 < 0 \Rightarrow$ Ellipse. $\cot 2\alpha = \frac{A-C}{B} = 0 \Rightarrow 2\alpha = \frac{\pi}{2} \Rightarrow \alpha = \frac{\pi}{4}$

$x = \frac{\sqrt{2}}{2}x' - \frac{\sqrt{2}}{2}y',\ y = \frac{\sqrt{2}}{2}x' + \frac{\sqrt{2}}{2}y' \Rightarrow 2\left(\frac{\sqrt{2}}{2}x' - \frac{\sqrt{2}}{2}y'\right)^2 + \left(\frac{\sqrt{2}}{2}x' - \frac{\sqrt{2}}{2}y'\right)\left(\frac{\sqrt{2}}{2}x' + \frac{\sqrt{2}}{2}y'\right) +$

$2\left(\frac{\sqrt{2}}{2}x' + \frac{\sqrt{2}}{2}y'\right)^2 - 15 = 0 \Rightarrow 5x'^2 + 3y'^2 - 30 = 0.$

12. $B^2 - 4AC = \left(2\sqrt{3}\right)^2 - 4(1)(-1) = 16 \Rightarrow$ Hyperbola. $\cot 2\alpha = \frac{A-C}{B} = \frac{1}{\sqrt{3}} \Rightarrow 2\alpha = \frac{\pi}{3} \Rightarrow \alpha = \frac{\pi}{6}$

$x = \frac{\sqrt{3}}{2}x' - \frac{1}{2}y',\ y = \frac{1}{2}x' + \frac{\sqrt{3}}{2}y' \Rightarrow \left(\frac{\sqrt{3}}{2}x' - \frac{1}{2}y'\right)^2 + 2\sqrt{3}\left(\frac{\sqrt{3}}{2}x' - \frac{1}{2}y'\right)\left(\frac{1}{2}x' + \frac{\sqrt{3}}{2}y'\right) - \left(\frac{1}{2}x' + \frac{\sqrt{3}}{2}y'\right)^2 = 4$

$\Rightarrow 2x'^2 - 2y'^2 = 4 \Rightarrow x'^2 - y'^2 = 2$

13. $xy = 2 \Rightarrow \alpha = \frac{\pi}{4}$ and $x'^2 - y'^2 = 4$ (See Exercise 17, Section 10.2). $a = 1,\ b = 1 \Rightarrow c = \sqrt{2} \Rightarrow e = \frac{c}{a} = \sqrt{2}$

14. $e = 2$, vertices $(2,0)$, $(-2,0) \Rightarrow a = 2$. $e = \frac{c}{a} \Rightarrow 2 = \frac{c}{2} \Rightarrow c = 4$. $c^2 = a^2 + b^2 \Rightarrow 4^2 = 2^2 + b^2 \Rightarrow$

$b^2 = 12 \Rightarrow b = \sqrt{12} = 2\sqrt{3}$. $\therefore \frac{x^2}{4} - \frac{y^2}{12} = 1$ or $3x^2 - y^2 = 12$.

15. a) Around the x–axis: $9x^2 + 4y^2 = 36 \Rightarrow y^2 = 9 - \frac{9}{4}x^2 \Rightarrow y = \pm\sqrt{9 - \frac{9}{4}x^2}$, use the positive root.

$$V = 2\int_0^2 \pi\left(\sqrt{9 - \frac{9}{4}x^2}\right)^2 dx = 2\int_0^2 \pi\left(9 - \frac{9}{4}x^2\right) dx = 2\pi\left[9x - \frac{3}{4}x^3\right]_0^2 = 24\pi$$

b) Around the y–axis: $9x^2 + 4y^2 = 36 \Rightarrow x^2 = 4 - \frac{4}{9}y^2 \Rightarrow x = \pm\sqrt{4 - \frac{4}{9}y^2}$, use the positive root.

$$V = 2\int_0^3 \pi\left(\sqrt{4 - \frac{4}{9}y^2}\right)^2 dy = 2\int_0^3 \pi\left(4 - \frac{4}{9}y^2\right) dy = 2\pi\left[4y - \frac{4}{27}y^3\right]_0^3 = 16\pi$$

16. $9x^2 - 4y^2 = 36,\ x = 4 \Rightarrow y^2 = \frac{9x^2 - 36}{4} \Rightarrow y = \frac{3}{2}\sqrt{x^2 - 4}$

$$V = \int_2^4 \pi\left(\frac{3}{2}\sqrt{x^2 - 4}\right)^2 dx = \frac{3\pi}{2}\int_2^4 \left(x^2 - 4\right) dx = \frac{3\pi}{2}\left[\frac{x^3}{3} - 4x\right]_2^4 = 16\pi$$

17. $x = 2t,\ y = t^2 \Rightarrow d = \sqrt{(2t - 0)^2 + (t^2 - 3)^2} = \sqrt{t^4 - 2t^2 + 9}$. Minimize $d^2 = r = t^4 - 2t^2 + 9$ to minimize $d(t)$. $\frac{dr}{dt} = 4t^3 - 4t$. $\frac{dr}{dt} = 0 \Rightarrow 4t^3 - 4t = 0 \Rightarrow 4t(t^2 - 1) = 0 \Rightarrow t = 0$ or $t = \pm 1$. $\frac{d^2r}{dt^2} = 12t^2 - 4 \Rightarrow$ $\frac{d^2r}{dt^2}(0) = -4 \Rightarrow$ Relative Max. $\frac{d^2r}{dt^2}(1) = 8 \Rightarrow$ Relative Min. $r(1) = 8$. $\frac{d^2r}{dt^2}(-1) = 8 \Rightarrow$ Relative Min. $r(-1) = 8$. $\therefore$ $t = \pm 1$ both give points on the parabola that are closest to $(0,3)$. $t = 1 \Rightarrow x = 2,\ y = 1 \Rightarrow$ $(2,1)$ is one of the points. $t = -1 \Rightarrow x = -2,\ y = 1 \Rightarrow (-2,1)$ is the other point.

18. $A = 2\displaystyle\int_0^{b/2}\left(h - \frac{4h}{b^2}x^2\right)dx = 2\left[hx - \frac{4h}{b^2}\left(\frac{x^3}{3}\right)\right]_0^{b/2} = 2\left[h\left(\frac{b}{2}\right) - \frac{4h}{b^2}\left(\frac{\left(\frac{b}{2}\right)^3}{3}\right) - 0\right] = \frac{2}{3}\,bh$

19. $x = \frac{t}{2}$, $y = t + 1 \Rightarrow 2x = t \Rightarrow y = 2x + 1$

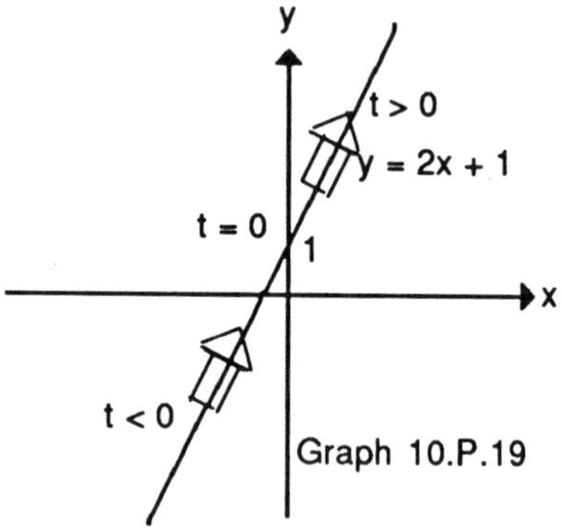

Graph 10.P.19

20. $x = \sqrt{t}$, $y = 1 - \sqrt{t} \Rightarrow y = 1 - x$

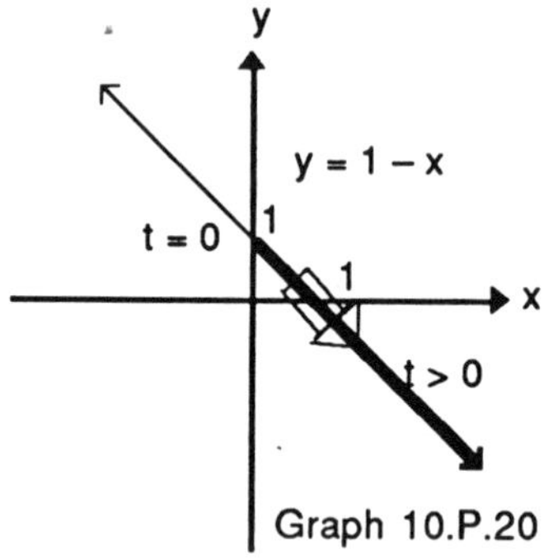

Graph 10.P.20

21. $x = \frac{1}{2}\tan t$, $y = \frac{1}{2}\sec t \Rightarrow x^2 = \frac{1}{4}\tan^2 t$,
$y = \frac{1}{4}\sec^2 t \Rightarrow 4x^2 = \tan^2 t$, $4y^2 = \sec^2 t \Rightarrow$
$4x^2 + 1 = 4y^2 \Rightarrow 1 = 4y^2 - 4x^2$

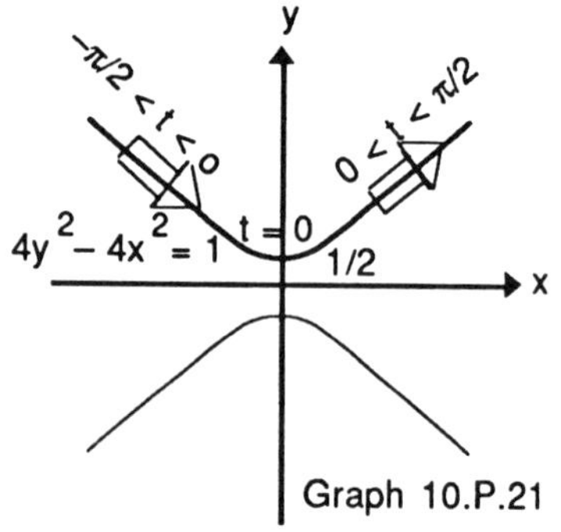

Graph 10.P.21

22. $x = -2\cos t$, $y = 2\sin t \Rightarrow x^2 = 4\cos^2 t$, $y^2 =$
$4\sin^2 t \Rightarrow x^2 + y^2 = 1$

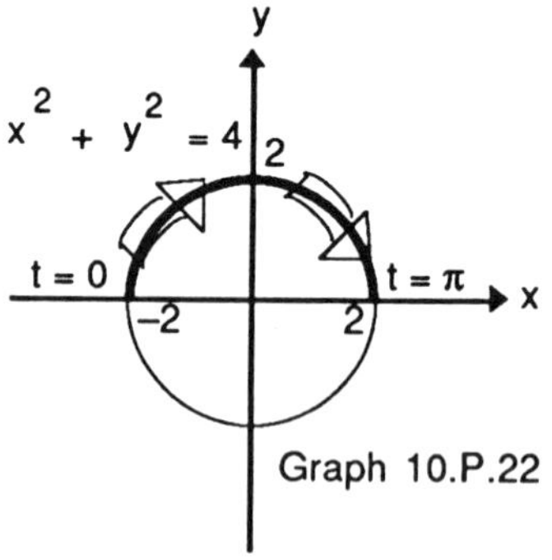

Graph 10.P.22

23. $x = -\cos t$, $y = \cos^2 t \Rightarrow y = (-x)^2 = x^2$

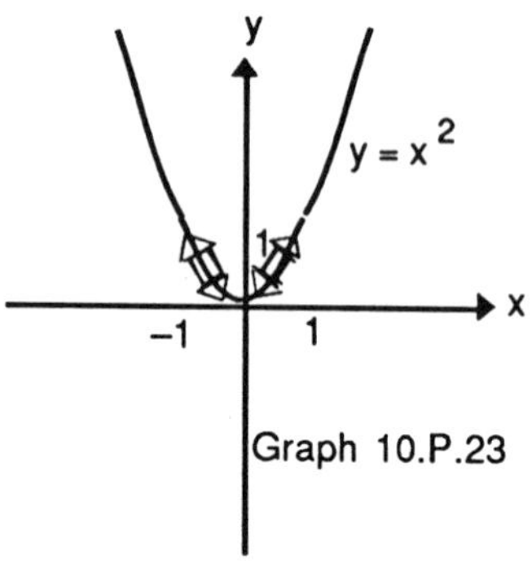

Graph 10.P.23

24. $x = 4\cos t$, $y = 9\sin t \Rightarrow x^2 = 16\cos^2 t$,
$y^2 = 81\sin^2 t \Rightarrow \frac{x^2}{16} + \frac{y^2}{81} = 1$

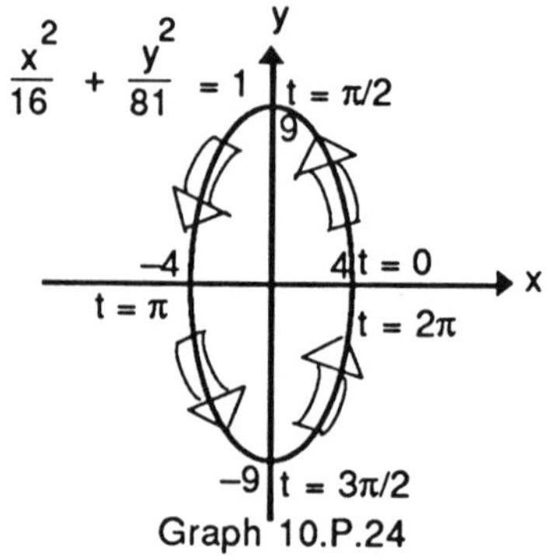

Graph 10.P.24

25. $16x^2 + 9y^2 = 144 \Rightarrow \frac{x^2}{9} + \frac{y^2}{16} = 1 \Rightarrow a = 3,\ b = 4 \Rightarrow x = 3\cos t,\ y = 4\sin t,\ 0 \le t \le 2\pi$

26. $x^2 + y^2 = 4 \Rightarrow\ x = -2\cos t,\ y = 2\sin t,\ 0 \le t \le 6\pi$

27. $x = \frac{1}{2}\tan t,\ y = \frac{1}{2}\sec t \Rightarrow \frac{dy}{dx} = \frac{dy/dt}{dx/dt} = \frac{\frac{1}{2}\sec t\tan t}{\frac{1}{2}\sec^2 t} = \frac{\tan t}{\sec t} = \sin t \Rightarrow \frac{dy}{dx}\left(\frac{\pi}{3}\right) = \sin\frac{\pi}{3} = \frac{\sqrt{3}}{2}.\quad t = \frac{\pi}{3} \Rightarrow$

$x = \frac{1}{2}\tan\frac{\pi}{3} = \frac{\sqrt{3}}{2}$ and $y = \frac{1}{2}\sec t = 1 \Rightarrow y = \frac{\sqrt{3}}{2}x + \frac{1}{4}.\ \frac{d^2y}{dx^2} = \frac{\cos t}{\frac{1}{2}\sec^2 t} = 2\cos^3 t \Rightarrow \frac{d^2y}{dx^2}\left(\frac{\pi}{3}\right)\ 2\cos^3\frac{\pi}{3}$

$= \frac{1}{4}$

28. $x = 1 + \frac{1}{t^2},\ y = 1 - \frac{3}{t} \Rightarrow \frac{dy}{dx} = \frac{dy/dt}{dx/dt} = \frac{3/t^2}{-2/t^3} = -\frac{3}{2}t \Rightarrow \frac{dy}{dx}(2) = -\frac{3}{2}(2) = -3.\ t = 2 \Rightarrow x = 1 + \frac{1}{2^2} = \frac{5}{4}$ and

$y = 1 - \frac{3}{2} = -\frac{1}{2} \Rightarrow y = -3x + \frac{13}{4}.\ \frac{d^2y}{dx^2} = \frac{-3/2}{-2/t^3} \Rightarrow \frac{d^2y}{dx^2}(2) = \frac{3/2}{2/2^3} = 6$

29. $x = e^{2t} - \frac{t}{8},\ y = e^t,\ 0 \le t \le \ln 2 \Rightarrow \frac{dx}{dt} = 2e^{2t} - \frac{1}{8},\ \frac{dy}{dt} = e^t \Rightarrow \text{Length} = \int_0^{\ln 2} \sqrt{\left(2e^{2t} - \frac{1}{8}\right)^2 + \left(e^t\right)^2}\,dt =$

$\int_0^{\ln 2} \sqrt{4e^{4t} + \frac{1}{2}e^{2t} + \frac{1}{16}}\,dt = \int_0^{\ln 2} \sqrt{\left(2e^{2t} + \frac{1}{8}\right)^2}\,dt = \int_0^{\ln 2} \left(2e^{2t} + \frac{1}{8}\right)dt =$

$\left[e^{2t} + \frac{t}{8}\right]_0^{\ln 2} = 3 + \frac{\ln 2}{8}$

30. $x = t^2,\ y = \frac{t^3}{3} - t,\ -\sqrt{3} \le t \le \sqrt{3} \Rightarrow \frac{dx}{dt} = 2t,\ \frac{dy}{dt} = t^2 - 1 \Rightarrow$

$\text{Length} = \int_{-\sqrt{3}}^{\sqrt{3}} \sqrt{(2t)^2 + (t^2 - 1)^2}\,dt = \int_{-\sqrt{3}}^{\sqrt{3}} \sqrt{t^4 + 2t^2 + 1}\,dt = \int_{-\sqrt{3}}^{\sqrt{3}} \sqrt{(t^2 + 1)^2}\,dt =$

$\int_{-\sqrt{3}}^{\sqrt{3}} (t^2 + 1)dt = \left[\frac{t^3}{3} + t\right]_{-\sqrt{3}}^{\sqrt{3}} = 4\sqrt{3}$

31. $x = \frac{t^2}{2},\ y = 2t,\ 0 \le t \le \sqrt{5} \Rightarrow \frac{dx}{dt} = t^2,\ \frac{dy}{dt} = 2 \Rightarrow \text{Area} = \int_0^{\sqrt{5}} 2\pi(2t)\sqrt{t^2 + 4}\,dt = \int_4^9 2\pi\,u^{1/2}\,du =$

$2\pi\left[\frac{2}{3}u^{3/2}\right]_4^9 = \frac{76\pi}{3}$ Let $u = t^2 + 4 \Rightarrow du = 2t\,dt.\ x = 0 \Rightarrow u = 4,\ x = \sqrt{5} \Rightarrow u = 9$

32. $x = t^2 + \frac{1}{2t}$, $y = 4\sqrt{t}$, $\frac{1}{\sqrt{2}} \le t \le 1 \Rightarrow \frac{dx}{dt} = 2t - \frac{1}{2t^2}$, $\frac{dy}{dt} = \frac{2}{\sqrt{t}} \Rightarrow$

$$\text{Area} = \int_{1/\sqrt{2}}^{1} 2\pi\left(t^2 + \frac{1}{2t}\right)\sqrt{\left(2t - \frac{1}{2t^2}\right)^2 + \left(\frac{2}{\sqrt{t}}\right)^2}\, dt = \int_{1/\sqrt{2}}^{1} 2\pi\left(t^2 + \frac{1}{2t}\right)\sqrt{\left(2t + \frac{1}{2t^2}\right)^2}\, dt =$$

$$\int_{1/\sqrt{2}}^{1} 2\pi\left(t^2 + \frac{1}{2t}\right)\left(2t + \frac{1}{2t^2}\right)dt = \int_{1/\sqrt{2}}^{1} 2\pi\left(2t^3 + \frac{3}{2} + \frac{1}{4}t^{-3}\right) dt = 2\pi\left[\frac{1}{2}t^4 + \frac{3}{2}t - \frac{1}{8}t^{-2}\right]_{1/\sqrt{2}}^{1} = 2\pi\left(2 - \frac{3\sqrt{2}}{4}\right.$$

33. d 34. e 35. l 36. f 37. k 38. h 39. i 40. j

41. $r\cos\left(\theta - \frac{\pi}{3}\right) = 2\sqrt{3} \Rightarrow r\left(\cos\theta\cos\frac{\pi}{3} + \sin\theta\sin\frac{\pi}{3}\right) =$
$2\sqrt{3} \Rightarrow \frac{1}{2}r\cos\theta + \frac{\sqrt{3}}{2}r\sin\theta = 2\sqrt{3} \Rightarrow$
$r\cos\theta + \sqrt{3}\,r\sin\theta = 4\sqrt{3} \Rightarrow x + \sqrt{3}\,y = 4\sqrt{3}$

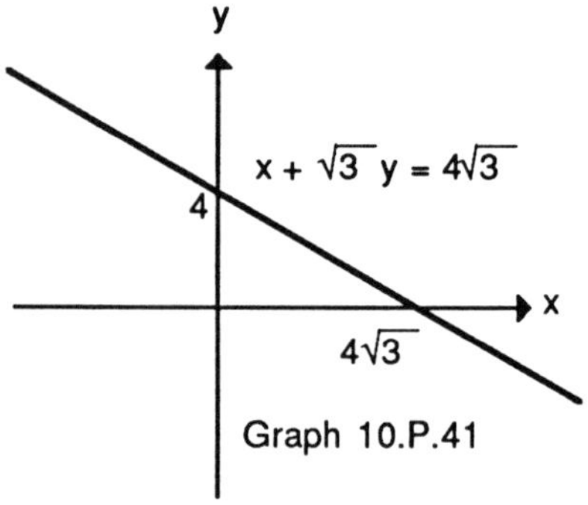

Graph 10.P.41

42. $r\cos\left(\theta - \frac{3\pi}{4}\right) = \frac{\sqrt{2}}{2} \Rightarrow r\left(\cos\theta\cos\frac{3\pi}{4} + \sin\theta\sin\frac{3\pi}{4}\right) = \frac{\sqrt{2}}{2} \Rightarrow$
$-\frac{\sqrt{2}}{2}r\cos\theta + \frac{\sqrt{2}}{2}r\sin\theta = \frac{\sqrt{2}}{2} \Rightarrow -x + y = 1$

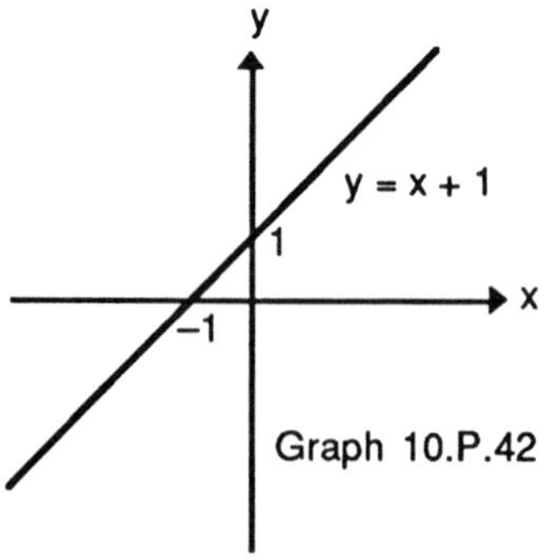

Graph 10.P.42

43. C: (0,1), r = 1

44. C: (–2,0), r = 2

45.

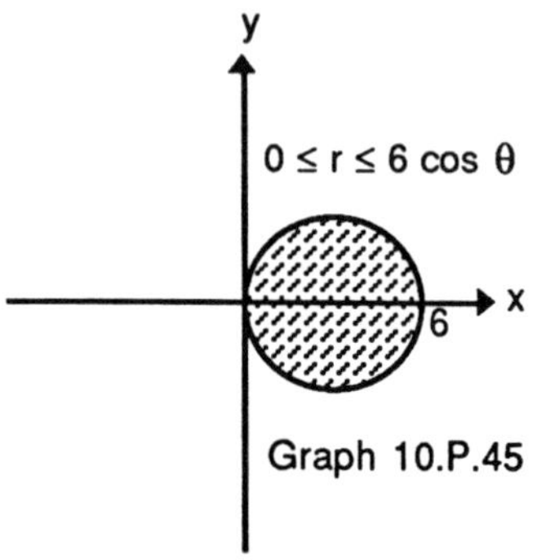

Graph 10.P.45

46.

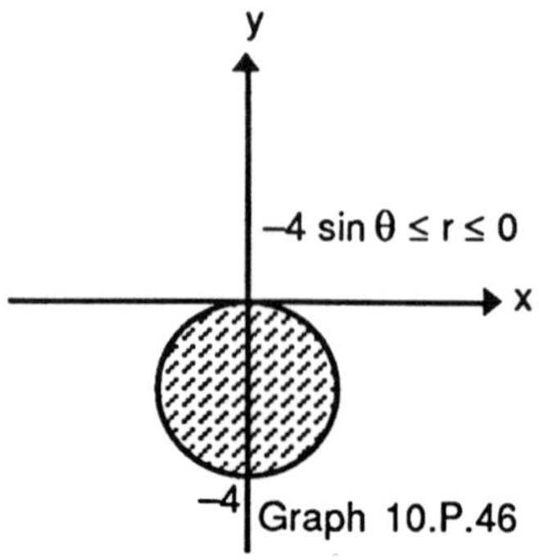

Graph 10.P.46

47. $r = \sin\theta$, $r = 1 + \sin\theta \Rightarrow \sin\theta = 1 + \sin\theta \Rightarrow \varnothing$. There are no points of intersection found by solving the system. The point of intersection (0,0) is found by graphing.

48. $r = \cos\theta$, $r = 1 - \cos\theta \Rightarrow \cos\theta = 1 - \cos\theta \Rightarrow \cos\theta = \frac{1}{2} \Rightarrow \theta = \frac{\pi}{3}, -\frac{\pi}{3}$. $\theta = \frac{\pi}{3} \Rightarrow r = \frac{1}{2}$, $\theta = -\frac{\pi}{3} \Rightarrow r = \frac{1}{2}$. The points of intersection $\left(\frac{1}{2}, \frac{\pi}{3}\right)$ and $\left(\frac{1}{2}, -\frac{\pi}{3}\right)$ are found by solving the system. The point of intersection (0,0) is found by graphing.

49. $r = 1 + \sin\theta$ and $r = -1 + \sin\theta$ intersect at all points of $r = 1 + \sin\theta$. This can be seen by graphing them.

50. $r = 1 + \cos\theta$, $r = -1 - \cos\theta \Rightarrow 1 + \cos\theta = -1 - \cos\theta \Rightarrow \cos\theta = -1 \Rightarrow \theta = \pi$. $\theta = \pi \Rightarrow r = 1$. The point of intersection $(1,\pi)$ is found by solving the system. The points of intersection $\left(1, \frac{\pi}{2}\right)$ and $\left(1, \frac{3\pi}{2}\right)$ are found by graphing.

51. $r = \dfrac{2}{1 + \cos\theta} \Rightarrow e = 1 \Rightarrow$ Parabola

Vertex = (1,0)

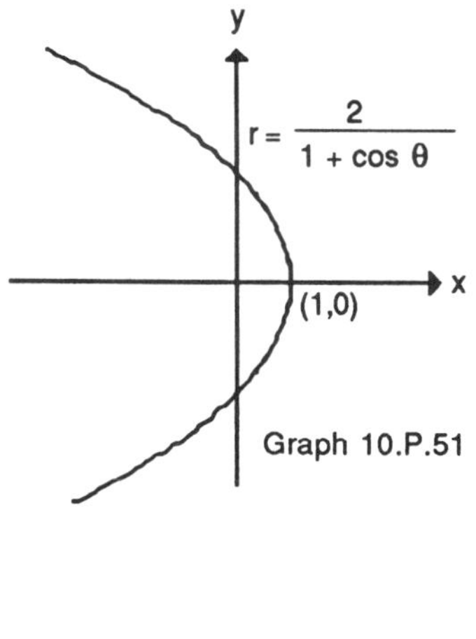

Graph 10.P.51

52. $r = \dfrac{8}{2 + \cos\theta} \Rightarrow r = \dfrac{4}{1 + \frac{1}{2}\cos\theta} \Rightarrow e = \frac{1}{2} \Rightarrow$ Ellipse. $ke = 4 \Rightarrow \frac{1}{2}k = 4 \Rightarrow k = 8$. $k = \frac{a}{e} - ea$

$8 = \frac{a}{1/2} - \frac{1}{2}a \Rightarrow a = \frac{16}{3} \Rightarrow ea = \frac{1}{2}\left(\frac{16}{3}\right) = \frac{8}{3}$

$\therefore$ Center is $\left(\frac{8}{3}, \pi\right)$; Vertices are $(8,\pi)$ and $\left(\frac{8}{3}, 0\right)$

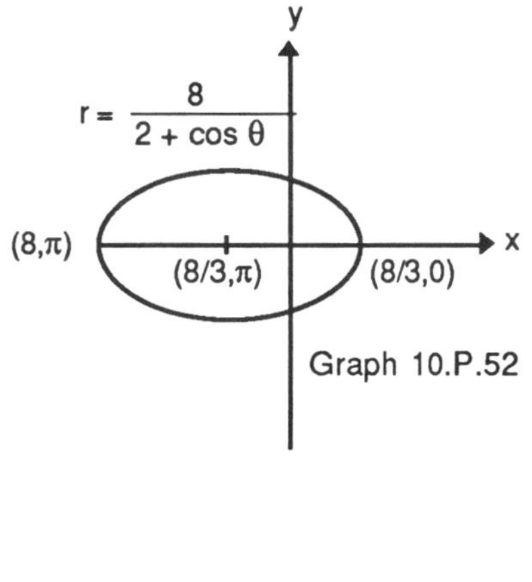

Graph 10.P.52

53. $r = \dfrac{6}{1 - 2\cos\theta} \Rightarrow e = 2 \Rightarrow$ Hyperbola

$ke = 6 \Rightarrow 2k = 6 \Rightarrow k = 3 \Rightarrow$ Vertices are

$(2,\pi)$ and $(6,\pi)$.

54. $r = \dfrac{12}{3 + \sin\theta} \Rightarrow r = \dfrac{4}{1 + \frac{1}{3}\sin\theta} \Rightarrow e = \dfrac{1}{3}.\ ke = 4$

$\Rightarrow \dfrac{1}{3}k = 4 \Rightarrow k = 12.\ a(1 - e^2) = 4 \Rightarrow$

$a\left(1 - \left(\dfrac{1}{3}\right)^2\right) = 4 \Rightarrow a = \dfrac{9}{2}.\ ea = \dfrac{1}{3}\left(\dfrac{9}{2}\right) = \dfrac{3}{2}.$

$\therefore$ Center is $\left(\dfrac{3}{2}, \dfrac{3\pi}{2}\right)$; Vertices are $\left(3, \dfrac{\pi}{2}\right)$ and $\left(6, \dfrac{3\pi}{2}\right)$

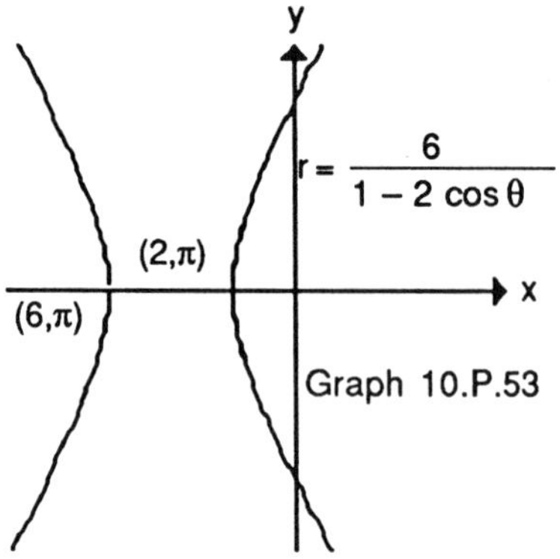

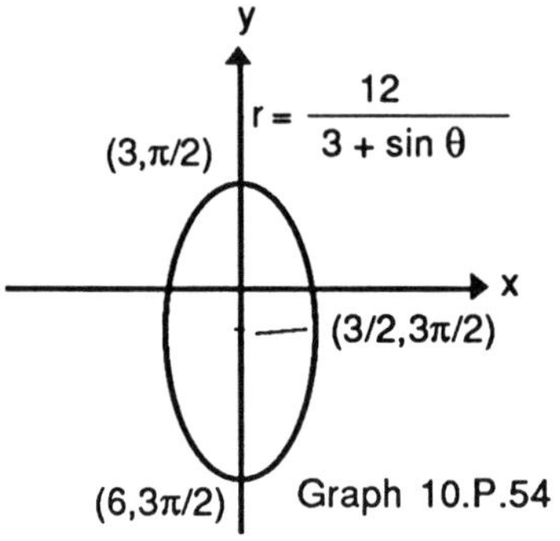

55. $e = 2$, $r\cos\theta = 2 \Rightarrow x = 2$ is directrix $\Rightarrow k = 2$ The conic is a hyperbola. $r = \dfrac{ke}{1 + e\cos\theta} \Rightarrow$

$r = \dfrac{2(2)}{1 + 2\cos\theta} \Rightarrow r = \dfrac{4}{1 + 2\cos\theta}$

56. $e = 1$, $r\cos\theta = -4 \Rightarrow x = -4$ is directrix $\Rightarrow k = 4$. The conic is a parabola.. $r = \dfrac{ke}{1 - e\cos\theta} \Rightarrow$

$r = \dfrac{4(1)}{1 - \cos\theta} \Rightarrow r = \dfrac{4}{1 - \cos\theta}$

57. $e = \dfrac{1}{2}$, $r\sin\theta = 2$ Y $y = 2$ is directrix $\Rightarrow k = 2$. The conic is an ellipse. $r = \dfrac{ke}{1 + e\sin\theta} \Rightarrow$

$r = \dfrac{2\left(\frac{1}{2}\right)}{1 + \frac{1}{2}\sin\theta} \Rightarrow r = \dfrac{2}{2 + \sin\theta}$

58. $e = \dfrac{1}{3}$, $r\sin\theta = -6 \Rightarrow y = -6$ is directrix $\Rightarrow k = 6$. The conic is an ellipse. $r = \dfrac{ke}{1 - e\sin\theta} \Rightarrow$

$r = \dfrac{6\left(\frac{1}{3}\right)}{1 - \frac{1}{3}\sin\theta} \Rightarrow r = \dfrac{6}{3 - \sin\theta}$

59. $r^2 = \cos 2\theta \Rightarrow r = 0$ when $\cos 2\theta = 0 \Rightarrow 2\theta = \dfrac{\pi}{2}, \dfrac{3\pi}{2} \Rightarrow \theta = \dfrac{\pi}{4}, \dfrac{3\pi}{4}$. $\theta_1 = \dfrac{\pi}{4} \Rightarrow m_1 = \tan\dfrac{\pi}{4} = 1 \Rightarrow$

$y = x$ is one tangent line. $\theta_2 = \dfrac{3\pi}{4} \Rightarrow m_2 = \tan\dfrac{3\pi}{4} = -1 \Rightarrow y = -x$ is other tangent line.

60. $r = 2\cos\theta + 1 \Rightarrow r = 0$ when $2\cos\theta + 1 = 0 \Rightarrow \cos\theta = -\dfrac{1}{2} \Rightarrow \theta = \dfrac{2\pi}{3}, \dfrac{4\pi}{3}$. $\theta_1 = \dfrac{2\pi}{3} \Rightarrow m_1 = \tan\dfrac{2\pi}{3}$

$= -\sqrt{3} \Rightarrow y = -\sqrt{3}\,x$ is one tangent line. $\theta_2 = \dfrac{4\pi}{3} \Rightarrow m_2 = \tan\dfrac{4\pi}{3} = \sqrt{3} \Rightarrow y = \sqrt{3}\,x$ is other tangent line.

61. Tips of the petals are at $\theta = \frac{\pi}{4}, \frac{3\pi}{4}, \frac{5\pi}{4}, \frac{7\pi}{4}$, r = 1 at those values of π. Then for $\theta = \frac{\pi}{4}$, the line is $r\cos\left(\theta - \frac{\pi}{4}\right) = 1$; for $\theta = \frac{3\pi}{4}$, $r\cos\left(\theta - \frac{3\pi}{4}\right) = 1$; for $\theta = \frac{5\pi}{4}$, $r\cos\left(\theta - \frac{5\pi}{4}\right) = 1$; and for $\theta = \frac{7\pi}{4}$, $r\cos\left(\theta - \frac{7\pi}{4}\right) = 1$.

62. $r = 1 + \sin\theta$ crosses the x–axis at $(1,0)$, $(1,\pi)$, and $\left(0,\frac{3\pi}{2}\right)$. At $\left(0,\frac{3\pi}{2}\right)$ the tangent line is $r\cos\theta = 0$.

$$\frac{dr}{d\theta} = \cos\theta \Rightarrow \text{Slope} = \frac{r'\sin\theta + r\cos\theta}{r'\cos\theta - r\sin\theta} = \frac{\cos\theta(\sin\theta) + (1+\sin\theta)\cos\theta}{\cos\theta\cos\theta - (1+\sin\theta)\sin\theta} = \frac{2\cos\theta\sin\theta + \cos\theta}{\cos^2\theta - \sin^2\theta - \sin\theta}.$$

$$\text{Slope at } (1,0) = \frac{2\cos 0\sin 0 + \cos 0}{\cos^2 0 - \sin^2 0 - \sin 0} = 1 \Rightarrow \text{tangent line is } y = x - 1 \Rightarrow r\sin\theta = r\cos\theta - 1.$$

$$\text{Slope at } (1,\pi) = \frac{2\cos\pi\sin\pi + \cos\pi}{\cos^2\pi - \sin^2\pi - \sin\pi} = -1 \Rightarrow \text{tangent line is } y = -x - 1 \Rightarrow r\sin\theta = -r\cos\theta - 1.$$

63. $$A = 2\int_0^\pi \frac{1}{2}r^2\,d\theta = \int_0^\pi (2-\cos\theta)^2\,d\theta = \int_0^\pi \left(4 - 2\cos\theta + \cos^2\theta\right)d\theta =$$

$$\int_0^\pi \left(4 - 2\cos\theta + \frac{1+\cos 2\theta}{2}\right)d\theta = \int_0^\pi \left(\frac{9}{2} - 2\cos\theta + \frac{\cos 2\theta}{2}\right)d\theta = \left[\frac{9}{2}\theta - 2\sin\theta + \frac{\sin 2\theta}{4}\right]_0^\pi = \frac{9}{2}\pi$$

64. $$A = \int_0^{\pi/3} \frac{1}{2}\left(\sin^2 3\theta\right)d\theta = \int_0^{\pi/3} \frac{1}{2}\left(\frac{1-\cos 6\theta}{2}\right)d\theta = \frac{1}{4}\left[\theta - \frac{1}{6}\sin 6\theta\right]_0^{\pi/3} = \frac{\pi}{12}$$

65. $r = 1 + \cos 2\theta$, $r = 1 \Rightarrow 1 = 1 + \cos 2\theta \Rightarrow 0 = \cos 2\theta \Rightarrow 2\theta = \frac{\pi}{2} \Rightarrow \theta = \frac{\pi}{4}$.

$$\therefore\ A = 4\int_0^{\pi/4} \frac{1}{2}\left((1+\cos 2\theta)^2 - 1^2\right)d\theta = 2\int_0^{\pi/4} (1 + 2\cos 2\theta + \cos^2 2\theta - 1)\,d\theta =$$

$$2\int_0^{\pi/4} \left(2\cos 2\theta + \frac{1}{2} + \frac{\cos 4\theta}{2}\right)d\theta = 2\left[\sin 2\theta + \frac{1}{2}\theta + \frac{\sin 4\theta}{8}\right]_0^{\pi/4} = 2 + \frac{\pi}{4}$$

66. $$A = 2\int_{-\pi/2}^{\pi/2} \frac{1}{2}\left((2(1+\sin\theta))^2 - (2\sin\theta)^2\right)d\theta = \int_{-\pi/2}^{\pi/2} \left(4(1 + 2\sin\theta + \sin^2\theta) - 4\sin^2\theta\right)d\theta =$$

$$\int_{-\pi/2}^{\pi/2} (4 + 8\sin\theta)\,d\theta = \left[4\theta - 8\cos\theta\right]_{-\pi/2}^{\pi/2} = 4\pi$$

67. $r = \sqrt{\cos 2\theta} \Rightarrow \frac{dr}{d\theta} = \frac{-\sin 2\theta}{\sqrt{\cos 2\theta}}$

$$\text{Surface Area} = \int_0^{\pi/4} 2\pi r \sin\theta \sqrt{r^2 + \left(\frac{dr}{d\theta}\right)^2}\, d\theta =$$

$$\int_0^{\pi/4} 2\pi\sqrt{\cos 2\theta}\, \sin\theta \sqrt{\left(\sqrt{\cos 2\theta}\right)^2 + \left(\frac{-\sin 2\theta}{\sqrt{\cos 2\theta}}\right)^2}\, d\theta =$$

$$\int_0^{\pi/4} 2\pi\sqrt{\cos 2\theta}\, \sin\theta \sqrt{\cos 2\theta + \frac{\sin^2 2\theta}{\cos 2\theta}}\, d\theta = \int_0^{\pi/4} 2\pi\sqrt{\cos 2\theta}\, \sin\theta \sqrt{\frac{1}{\cos 2\theta}}\, d\theta =$$

$$\int_0^{\pi/4} 2\pi \sin\theta\, d\theta = \Big[2\pi(-\cos\theta)\Big]_0^{\pi/4} = 2\pi\left(1 - \frac{\sqrt{2}}{2}\right)$$

68. $r^2 = \sin 2\theta \Rightarrow 2r\frac{dr}{d\theta} = 2\cos 2\theta \Rightarrow r\frac{dr}{d\theta} = \cos 2\theta$

$$\text{Surface Area} = \int_0^{\pi/2} 2\pi\, r \cos\theta \sqrt{r^2 + \left(\frac{dr}{d\theta}\right)^2}\, d\theta = \int_0^{\pi/2} 2\pi \cos\theta \sqrt{r^4 + \left(r\frac{dr}{d\theta}\right)^2}\, d\theta =$$

$$\int_0^{\pi/2} 2\pi \cos\theta \sqrt{(\sin 2\theta)^2 + (\cos 2\theta)^2}\, d\theta = \int_0^{\pi/2} 2\pi \cos\theta\, d\theta = \Big[2\pi \sin\theta\Big]_0^{\pi/2} = 2\pi$$

69. $r = -1 + \cos\theta \Rightarrow \frac{dr}{d\theta} = -\sin\theta$

$$\text{Length} = \int_0^{2\pi} \sqrt{(-1 + \cos\theta) + \left(-\sin\theta\right)^2}\, d\theta = \int_0^{2\pi} \sqrt{2 - 2\cos\theta}\, d\theta =$$

$$\int_0^{2\pi} \sqrt{\frac{4\left(1 - \cos\theta\right)}{2}}\, d\theta = \int_0^{2\pi} 2\sin\frac{1}{2}\theta\, d\theta = \left[-4\cos\frac{1}{2}\theta\right]_0^{2\pi} = 8$$

70. $r = \sin\theta \Rightarrow \frac{dr}{d\theta} = \cos\theta$

$$\text{Length} = \int_0^{\pi} \sqrt{(\sin\theta)^2 + (\cos\theta)^2}\, d\theta = \int_0^{\pi} d\theta = \Big[\theta\Big]_0^{\pi} = \pi$$

71. $r = \cos^3\left(\frac{\theta}{3}\right) \Rightarrow \frac{dr}{d\theta} = -\cos^2\left(\frac{\theta}{3}\right)\sin\left(\frac{\theta}{3}\right)$

$$\text{Length} = \int_0^{\pi/4} \sqrt{\left(\cos^2\left(\frac{\theta}{3}\right)\right)^2 + \left(\cos^2\left(\frac{\theta}{3}\right)\sin\left(\frac{\theta}{3}\right)\right)^2}\, d\theta = \int_0^{\pi/4} \sqrt{\cos^6\left(\frac{\theta}{3}\right) + \cos^4\left(\frac{\theta}{3}\right)\sin^2\left(\frac{\theta}{3}\right)}\, d\theta$$

$$= \int_0^{\pi/4} \sqrt{\cos^4\left(\frac{\theta}{3}\right)}\, d\theta = \int_0^{\pi/4} \cos^2\left(\frac{\theta}{3}\right) d\theta = \int_0^{\pi/4} \frac{1 + \cos\left(\frac{2\theta}{3}\right)}{2}\, d\theta = \left[\frac{\theta + \frac{3}{2}\sin\left(\frac{2\theta}{3}\right)}{2}\right]_0^{\pi/4} = \frac{\pi + 3}{8}$$

72. $r = \sqrt{1 + \sin 2\theta} \Rightarrow \frac{dr}{d\theta} = \cos 2\theta(1 + \sin 2\theta)^{-1/2}$

$$\text{Length} = \int_0^{\pi/2} \sqrt{\left(\sqrt{1 + \sin 2\theta}\right)^2 + \left((1 + \sin 2\theta)^{-1/2}(\cos 2\theta)\right)^2}\, d\theta =$$

$$\int_0^{\pi/2} \sqrt{1 + \sin 2\theta + \frac{\cos^2 2\theta}{1 + \sin 2\theta}}\, d\theta = \int_0^{\pi/2} \sqrt{1 + \sin 2\theta + 1 - \sin 2\theta}\, d\theta =$$

$$\int_0^{\pi/4} \sqrt{2}\, d\theta = \left[\sqrt{2}\,\theta\right]_0^{\pi/4} = \frac{\pi\sqrt{2}}{2}$$

73. a) $r_{av} = \frac{1}{2\pi - 0} \int_0^{2\pi} a(1 - \cos\theta)\, d\theta = \frac{1}{2\pi}\left[a(\theta - \sin\theta)\right]_0^{2\pi} = a$

b) $r_{av} = \frac{1}{2\pi - 0} \int_0^{2\pi} a\, d\theta = \frac{1}{2\pi}\left[a\theta\right]_0^{2\pi} = a$

c) $r_{av} = \frac{1}{\frac{\pi}{2} - \left(-\frac{\pi}{2}\right)} \int_{-\pi/2}^{\pi/2} a\cos\theta\, d\theta = \frac{1}{\pi}\left[a\sin\theta\right]_{-\pi/2}^{\pi/2} = \frac{2a}{\pi}$

74. Let (r_1, θ_1) be on the graph where $r_1 = a\,\theta_1$. Let (r_2, θ_2) be on the graph where $r_2 = a\,\theta_2$ and $\theta_2 = \theta_1 + 2\pi$. Then r_1 and r_2 lie on the same ray on consecutive turns of the spiral and the distance between the two points is $r_2 - r_1 = a\,\theta_2 - a\,\theta_1 = a(\theta_2 - \theta_1) = a(2\pi)$, a constant.

75. $r_{max} + r_{min} = 2a$. $r_{max} - r_{min} = a + ea - (a - ea) = 2ea$. $\therefore \dfrac{r_{max} + r_{min}}{r_{max} - r_{min}} = \dfrac{2ea}{2a} = e$, the eccentricity.

76. a) The length of the major axis is 300 miles + 8000 miles + 1000 miles = 2a $\Rightarrow$ a = 4650 miles. If the center of the earth is one focus and the distance from the center of the earth to the satellite's low point is 4300 miles (half the diameter plus the distance above the North Pole), then the distance from the center of the ellipse to the focus (center of the earth) is 4650 miles – 4300 miles = 350 miles = c. $\therefore e = \dfrac{c}{a} = \dfrac{350 \text{ miles}}{4650 \text{ miles}} = \dfrac{7}{93}$.

b) $r = \dfrac{a(1 - e^2)}{1 + e\cos\theta} \Rightarrow r = \dfrac{4650\left(1 - \left(\frac{7}{93}\right)^2\right)}{1 + \frac{7}{93}\cos\theta} = \dfrac{430\,000}{93 + 7\cos\theta}$

77. Length of $r = f(\theta) = \displaystyle\int_\alpha^\beta \sqrt{\left(f(\theta)\right)^2 + \left(f'(\theta)\right)^2}\, d\theta$.

Length of $r = 2f(\theta) = \displaystyle\int_\alpha^\beta \sqrt{\left(2f(\theta)\right)^2 + \left(2f'(\theta)\right)^2}\, d\theta = \int_\alpha^\beta 2\sqrt{\left(f(\theta)\right)^2 + \left(f'(\theta)\right)^2}\, d\theta =$

2(Length of $r = f(\theta)$)

78. Surface Area of the solid formed by revolving $r = 2f(\theta)$ about the x–axis =

$$\int_\alpha^\beta 2\pi r \sin\theta \sqrt{r^2 + \left(\frac{dr}{d\theta}\right)^2}\, d\theta = \int_\alpha^\beta 2\pi(2f(\theta))\sin\theta \sqrt{\left(2f(\theta)\right)^2 + \left(2f'(\theta)\right)^2}\, d\theta =$$

$$\int_\alpha^\beta 2(2\pi f(\theta)\sin\theta)\sqrt{4\left(\left(f(\theta)\right)^2 + \left(f'(\theta)\right)^2\right)}\, d\theta = 4\int_\alpha^\beta 2\pi \sin\theta \sqrt{\left(f(\theta)\right)^2 + \left(f'(\theta)\right)^2}\, d\theta =$$

4 times the area of the solid generated by revolving $r = f(\theta)$ about the x–axis.

CHAPTER 11

VECTORS AND ANALYTIC GEOMETRY IN SPACE

11.1 VECTORS IN THE PLANE

1.

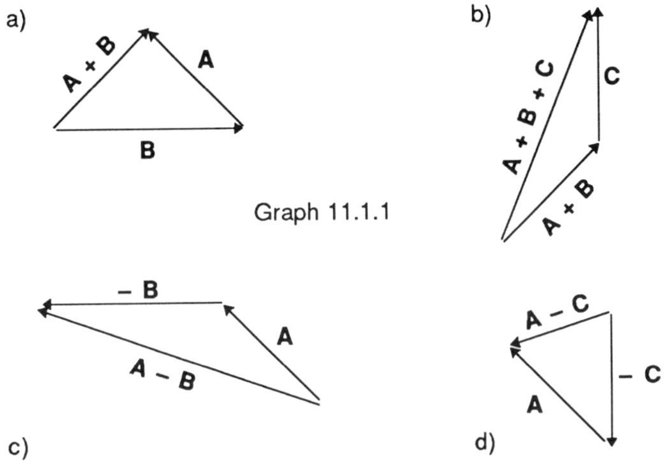

Graph 11.1.1

2.

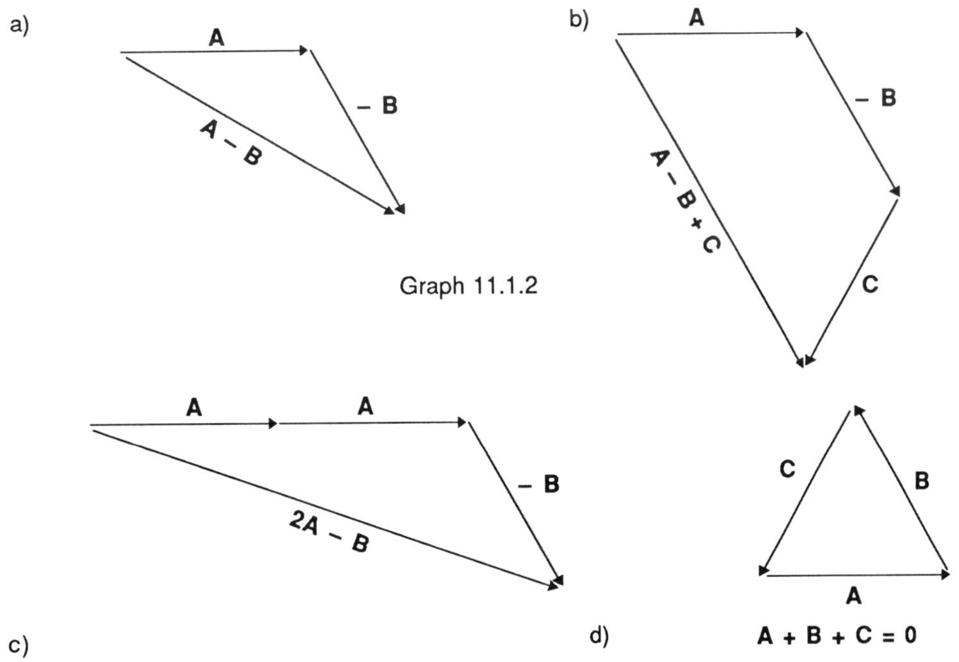

Graph 11.1.2

3. $\overrightarrow{P_1P_2} = \mathbf{i} - 4\mathbf{j}$

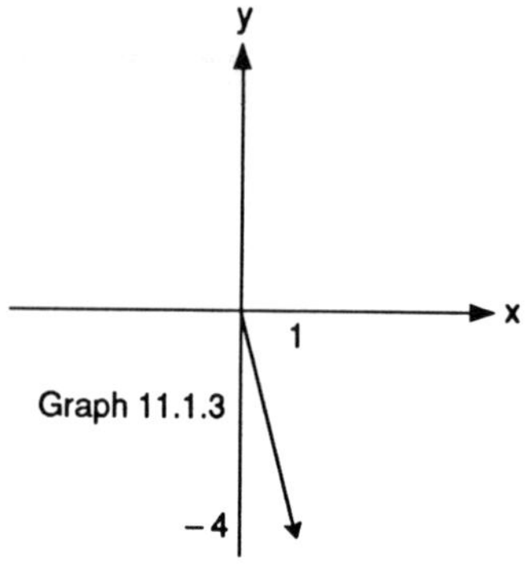

Graph 11.1.3

4. $\overrightarrow{OP_3} = -\mathbf{i} + \mathbf{j}$

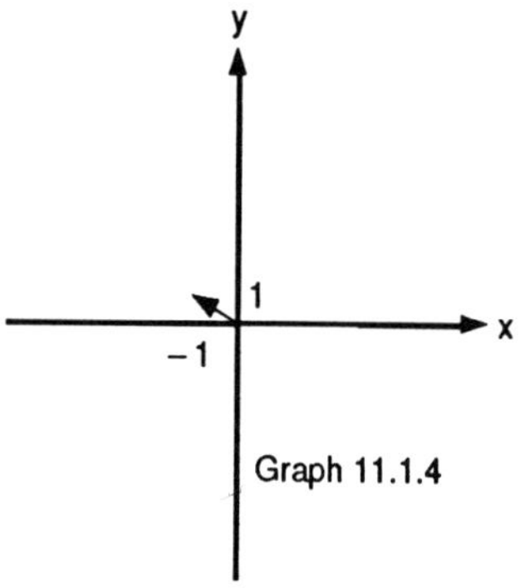

Graph 11.1.4

5. $\overrightarrow{AO} = -2\mathbf{i} - 3\mathbf{j}$

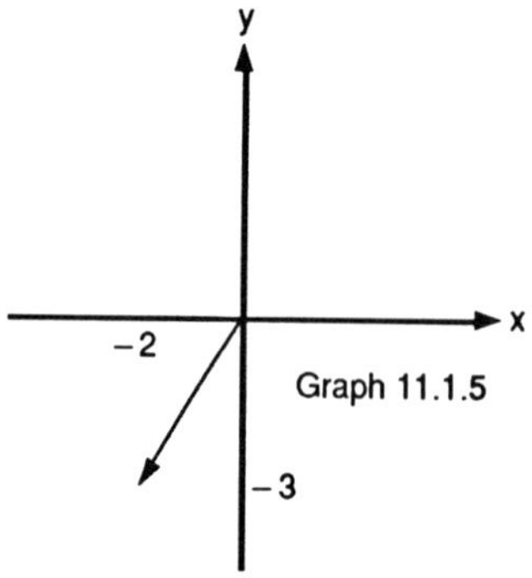

Graph 11.1.5

6. $\overrightarrow{AB} + \overrightarrow{CD} = \vec{0}$

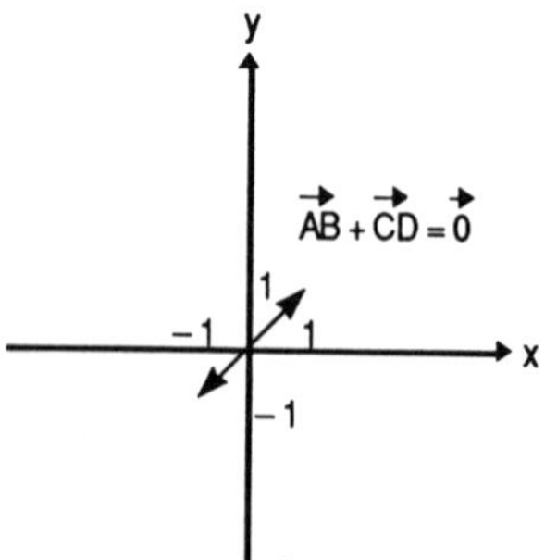

Graph 11.1.6

7. $\mathbf{U} = \frac{\sqrt{3}}{2}\mathbf{i} + \frac{1}{2}\mathbf{j}$, when $\theta = \frac{\pi}{6}$

$\mathbf{U} = -\frac{1}{2}\mathbf{i} + \frac{\sqrt{3}}{2}\mathbf{j}$, when $\theta = \frac{2\pi}{3}$

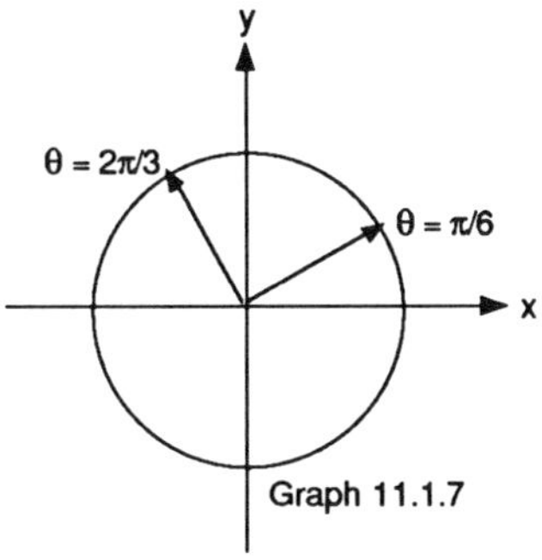

Graph 11.1.7

8. $\mathbf{U} = \frac{1}{\sqrt{2}}\mathbf{i} - \frac{1}{\sqrt{2}}\mathbf{j}$, when $\theta = -\frac{\pi}{4}$

$\mathbf{U} = -\frac{1}{\sqrt{2}}\mathbf{i} - \frac{1}{\sqrt{2}}\mathbf{i}$, when $\theta = -\frac{3\pi}{4}$

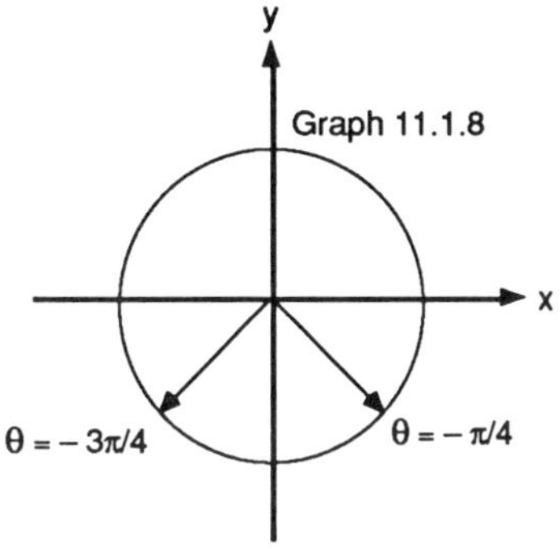

Graph 11.1.8

9. $\mathbf{U} = \frac{\sqrt{3}}{2}\mathbf{i} - \frac{1}{2}\mathbf{j}$

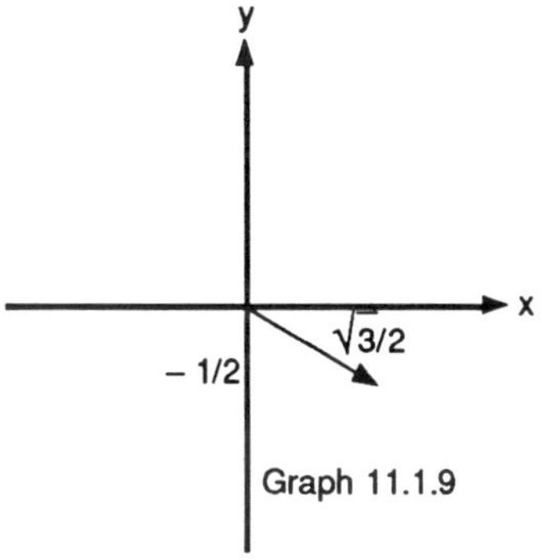

Graph 11.1.9

10. $\mathbf{U} = -\frac{1}{\sqrt{2}}\mathbf{i} + \frac{1}{\sqrt{2}}\mathbf{j}$

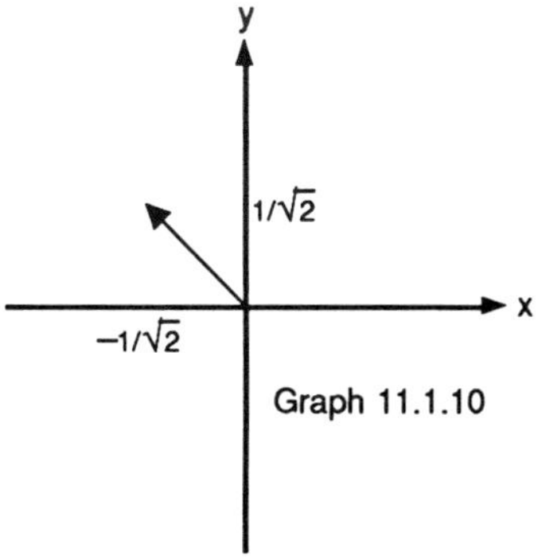

Graph 11.1.10

11. $\mathbf{U} = \frac{1}{\sqrt{17}}\mathbf{i} + \frac{4}{\sqrt{17}}\mathbf{j}, -\mathbf{U} = -\frac{1}{\sqrt{17}}\mathbf{i} - \frac{4}{\sqrt{17}}\mathbf{j},$
$\mathbf{V} = \frac{4}{\sqrt{17}}\mathbf{i} - \frac{1}{\sqrt{17}}\mathbf{j}, -\mathbf{V} = -\frac{4}{\sqrt{17}}\mathbf{i} + \frac{1}{\sqrt{17}}\mathbf{j}$

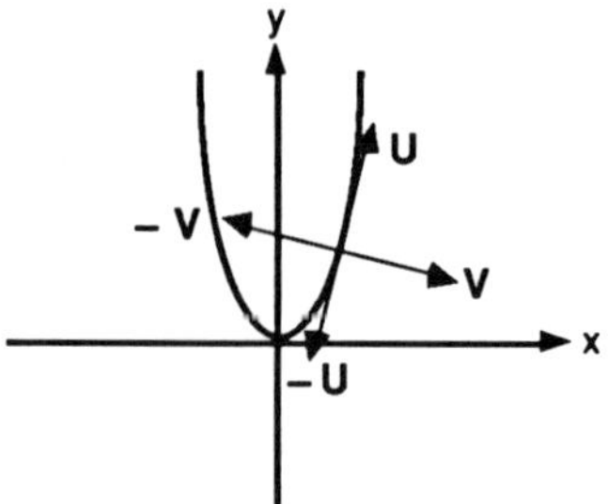

Graph 11.1.11

12. $\mathbf{U} = \frac{1}{\sqrt{2}}(\mathbf{i} - \mathbf{j}), -\mathbf{U} = \frac{1}{\sqrt{2}}(-\mathbf{i} + \mathbf{j})$
$\mathbf{V} = \frac{1}{\sqrt{2}}(\mathbf{i} + \mathbf{i}), -\mathbf{V} = \frac{1}{\sqrt{2}}(-\mathbf{i} - \mathbf{j})$

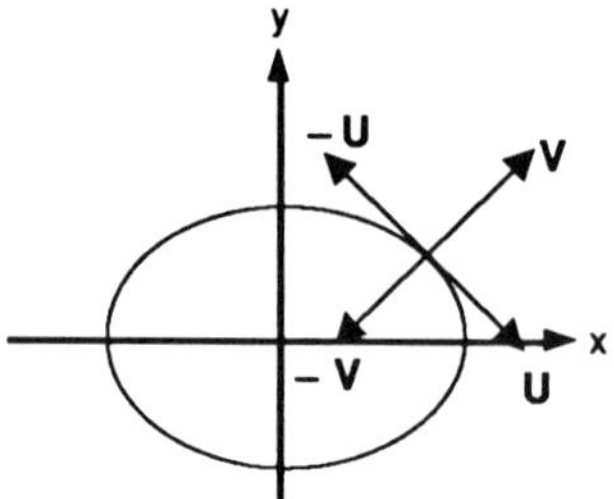

Graph 11.12

13. $|\mathbf{i} + \mathbf{j}| = \sqrt{1^2 + 1^2} = \sqrt{2}, \mathbf{i} + \mathbf{j} = \sqrt{2}\left[\frac{1}{\sqrt{2}}\mathbf{i} + \frac{1}{\sqrt{2}}\mathbf{j}\right]$

14. $|2\mathbf{i} - 3\mathbf{j}| = \sqrt{2^2 + 3^2} = \sqrt{13}, 2\mathbf{i} - 3\mathbf{j} = \sqrt{13}\left[\frac{2}{\sqrt{13}}\mathbf{i} - \frac{3}{\sqrt{13}}\mathbf{j}\right]$

15. $\left|\sqrt{3}\,\mathbf{i} + \mathbf{j}\right| = \sqrt{\left(\sqrt{3}\right)^2 + 1^2} = 2, 2\left[\frac{\sqrt{3}}{2}\mathbf{i} + \frac{1}{2}\mathbf{j}\right]$

16. $|-2\mathbf{i} + 3\mathbf{j}| = \sqrt{13}, \sqrt{13}\left[-\frac{2}{\sqrt{13}}\mathbf{i} + \frac{3}{\sqrt{13}}\mathbf{j}\right]$

17. $|5\mathbf{i} + 12\mathbf{j}| = 13, 13\left[\frac{5}{13}\mathbf{i} + \frac{12}{13}\mathbf{j}\right]$

18. $|-5\mathbf{i} - 12\mathbf{j}| = 13, 13\left[-\frac{5}{13}\mathbf{i} - \frac{12}{13}\mathbf{j}\right]$

19. $|\mathbf{A}| = |3\mathbf{i} + 6\mathbf{j}| = \sqrt{3^2 + 6^2} = 3\sqrt{5} \Rightarrow \mathbf{A} = 3\sqrt{5}\left[\frac{1}{\sqrt{5}}\mathbf{i} + \frac{2}{\sqrt{5}}\mathbf{j}\right];$
$|\mathbf{B}| = |-\mathbf{i} - 2\mathbf{j}| = \sqrt{5} \Rightarrow \mathbf{B} = \sqrt{5}\left[-\frac{1}{\sqrt{5}}\mathbf{i} - \frac{2}{\sqrt{5}}\mathbf{j}\right]$

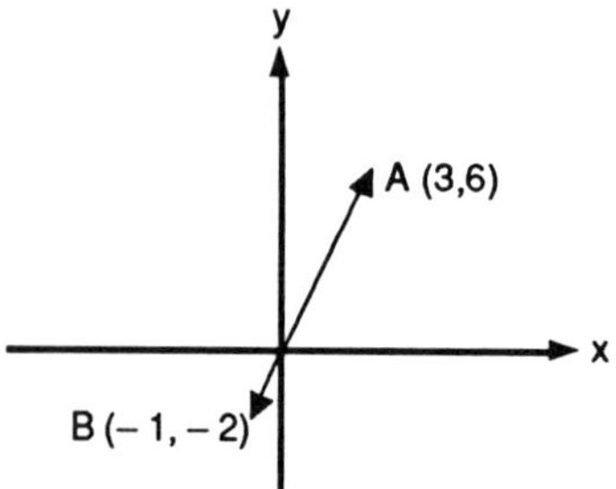

Graph 11.1.19

20. $\mathbf{A} = 3\mathbf{i} + 6\mathbf{j} = 3\sqrt{5}\left[\frac{1}{\sqrt{5}}\mathbf{i} + \frac{2}{\sqrt{5}}\mathbf{j}\right]$ and $\mathbf{B} = \frac{1}{2}\mathbf{i} + \mathbf{j} = \frac{\sqrt{5}}{2}\left[\frac{1}{\sqrt{5}}\mathbf{i} + \frac{2}{\sqrt{5}}\mathbf{j}\right] \Rightarrow \frac{1}{6}\mathbf{A} = \mathbf{B} \Rightarrow$ **A** and **B** have the same direction

11.2 CARTESIAN (RECTANGULAR) COORDINATES AND VECTORS IN SPACE

1. a line throught the point (2,3,0) parallel to the z–axis
2. a line through the point (– 1,0,0) parallel to the y–axis
3. the x–axis
4. a line through the point (1,0,0) parallel to the z–axis
5. the circle, $x^2 + y^2 = 4$ in the xy–plane
6. the circle, $x^2 + y^2 = 4$ in the $z = -2$ plane
7. the circle, $x^2 + z^2 = 4$ in the xz–plane
8. the circle, $y^2 + z^2 = 1$ in the yz–plane
9. the circle, $y^2 + z^2 = 1$ in the yz–plane
10. the circle, $x^2 + z^2 = 9$ in the $y = -4$ plane
11. the circle, $x^2 + y^2 = 16$ in the xy–plane
12. the circle, $x^2 + z^2 = 3$ in the xz–plane
13. a) the first quadrant of the xy–plane b) the fourth quadrant of the xy–plane
14. a) a slab 1 unit thick containing the yz–plane
    b) a square column one unit wide containing the z–plane
    c) a unit cube having one vertex at the origin
15. a) a solid sphere of radius 1 centered at the origin
    b) all points which are greater than 1 unit from the origin
16. a) the circumference and interior of the circle $x^2 + y^2 = 1$ in the xy–plane
    b) the circumference and interior of the circle $x^2 + y^2 = 1$ in the $z = 3$ plane
    c) a solid column having a radius of 1 centered about the z–axis
17. a) the upper hemisphere of radius 1 centered at the origin
    b) the solid upper hemisphere of raduis 1 centered at rhe origin
18. a) the line $y = x$ in the xy–plane b) the plane $y = x$
19. a) $x = 3$ b) $y = -1$ c) $z = -2$
20. a) $x = 3$ b) $y = -1$ c) $z = 2$
21. a) $z = 1$ b) $x = 3$ c) $y = -1$
22. a) $x^2 + y^2 = 4, z = 0$ b) $y^2 + z^2 = 4, x = 0$ c) $x^2 + z^2 = 4, y = 0$
23. a) $x^2 + (y-2)^2 = 4$ b) $(y-2)^2 + z^2 = 4$ c) $x^2 + z^2 = 4$
24. a) $(x+3)^2 + (y-4)^2 = 1, z = 1$ b) $(y-4)^2 + (z-1)^2 = 1, x = -3$
    c) $(x+3)^2 + (z-1)^2 = 1, y = 4$
25. a) $y = 3, z = -1$ b) $x = 1, z = -1$ c) $x = 1, y = 3$

26. $\sqrt{x^2+y^2+z^2} = \sqrt{x^2+(y-2)^2+z^2} \Rightarrow x^2+y^2+z^2 = x^2+(y-2)^2+z^2 \Rightarrow y^2 = y^2-4y+4 \Rightarrow y = 1$

27. $x^2+y^2+z^2 = 25,\ z = 3$

28. solve $x^2+y^2+(z-1)^2 = 4$ and $x^2+y^2+(z+1)^2 = 4 \Rightarrow x^2+y^2+(z-1)^2 = x^2+y^2+(z+1)^2 \Rightarrow$
$z = 0,\ x^2+y^2 = 3$

29. $0 \le z \le 1$

30. $0 \le x \le 2,\ 0 \le y \le 2,\ 0 \le z \le 2$

31. $z \le 0$

32. $z = \sqrt{1-x^2-y^2}$

33. a) $(x-1)^2+(y-1)^2+(z-1)^2 < 1$ b) $(x-1)^2+(y-1)^2+(z-1)^2 > 1$

34. $1 < x^2+y^2+z^2 \le 4$

35. length $= |2\mathbf{i}+\mathbf{j}-2\mathbf{k}| = \sqrt{2^2+1^2+(-2)^2} = 3,\ 2\mathbf{i}+\mathbf{j}-2\mathbf{k} = 3\left[\frac{2}{3}\mathbf{i}+\frac{1}{3}\mathbf{j}-\frac{2}{3}\mathbf{k}\right] \Rightarrow$
the direction is $\frac{2}{3}\mathbf{i}+\frac{1}{3}\mathbf{j}-\frac{2}{3}\mathbf{k}$

36. length $= |3\mathbf{i}-6\mathbf{j}+2\mathbf{k}| = \sqrt{3^2+(-6)^2+2^2} = 7,\ 3\mathbf{i}-6\mathbf{j}+2\mathbf{k} = 7\left[\frac{3}{7}\mathbf{i}-\frac{6}{7}\mathbf{j}+\frac{2}{7}\mathbf{k}\right] \Rightarrow$
the direction is $\frac{3}{7}\mathbf{i}-\frac{6}{7}\mathbf{j}+\frac{2}{7}\mathbf{k}$

37. length $= |\mathbf{i}+4\mathbf{j}-8\mathbf{k}| = \sqrt{1+16+64} = 9,\ \mathbf{i}+4\mathbf{j}-8\mathbf{k} = 9\left[\frac{1}{9}\mathbf{i}+\frac{4}{9}\mathbf{j}-\frac{8}{9}\mathbf{k}\right] \Rightarrow$
the direction is $\frac{1}{9}\mathbf{i}+\frac{4}{9}\mathbf{j}-\frac{8}{9}\mathbf{k}$

38. length $= |9\mathbf{i}-2\mathbf{j}+6\mathbf{k}| = \sqrt{81+4+36} = 11,\ 9\mathbf{i}-2\mathbf{j}+6\mathbf{k} = 11\left[\frac{9}{11}\mathbf{i}-\frac{2}{11}\mathbf{j}+\frac{6}{11}\mathbf{k}\right] \Rightarrow$
the direction is $\frac{9}{11}\mathbf{i}-\frac{2}{11}\mathbf{j}+\frac{6}{11}\mathbf{k}$

39. length $= |5\mathbf{k}| = \sqrt{25} = 5,\ 5\mathbf{k} = 5[\mathbf{k}] \Rightarrow$ the direction is $\mathbf{k}$

40. length $= |6\mathbf{i}| = \sqrt{36} = 6,\ 6\mathbf{i} = 6[\mathbf{i}] \Rightarrow$ the direction is $\mathbf{i}$

41. length $= |-4\mathbf{j}| = \sqrt{(-4)^2} = \sqrt{16} = 4,\ 4[-\mathbf{j}] \Rightarrow$ the direction is $-\mathbf{j}$

42. length $= \left|\frac{3}{5}\mathbf{i}+\frac{4}{5}\mathbf{k}\right| = \sqrt{\frac{9}{25}+\frac{16}{26}} = 1,\ \frac{3}{4}\mathbf{i}+\frac{4}{5}\mathbf{k} = 1\left[\frac{3}{4}\mathbf{i}+\frac{4}{5}\mathbf{k}\right] \Rightarrow$ the direction is $\frac{3}{4}\mathbf{i}+\frac{4}{5}\mathbf{k}$

43. length $= \left|-\frac{1}{3}\mathbf{j}+\frac{1}{4}\mathbf{k}\right| = \sqrt{\left(-\frac{1}{3}\right)^2+\left(\frac{1}{4}\right)} = \frac{5}{12},\ -\frac{1}{3}\mathbf{j}+\frac{1}{4}\mathbf{k} = \frac{5}{12}\left[-\frac{4}{5}\mathbf{j}+\frac{3}{5}\mathbf{k}\right] \Rightarrow$
the direction is $-\frac{4}{5}\mathbf{j}+\frac{3}{5}\mathbf{k}$

44. length $= \left|\frac{1}{\sqrt{2}}\mathbf{i}-\frac{1}{\sqrt{2}}\mathbf{k}\right| = \sqrt{\frac{1}{2}+\frac{1}{2}} = 1,\ \frac{1}{\sqrt{2}}\mathbf{i}-\frac{1}{\sqrt{2}}\mathbf{k} = 1\left[\frac{1}{\sqrt{2}}\mathbf{i}-\frac{1}{\sqrt{2}}\mathbf{k}\right]$

45. length $= \left|\frac{1}{\sqrt{6}}\mathbf{i}-\frac{1}{\sqrt{6}}\mathbf{j}-\frac{1}{\sqrt{6}}\mathbf{k}\right| = \sqrt{3\left(\frac{1}{\sqrt{6}}\right)^2} = \sqrt{\frac{1}{2}},\ \frac{1}{\sqrt{6}}\mathbf{i}-\frac{1}{\sqrt{6}}\mathbf{j}-\frac{1}{\sqrt{6}}\mathbf{k} =$
$\sqrt{\frac{1}{2}}\left[\frac{1}{\sqrt{3}}\mathbf{i}-\frac{1}{\sqrt{3}}\mathbf{j}-\frac{1}{\sqrt{3}}\mathbf{k}\right] \Rightarrow$ the direction is $\frac{1}{\sqrt{3}}\mathbf{i}-\frac{1}{\sqrt{3}}\mathbf{j}-\frac{1}{\sqrt{3}}\mathbf{k}$

46. the length $= \left[\frac{1}{\sqrt{3}}\mathbf{i}+\frac{1}{\sqrt{3}}\mathbf{j}+\frac{1}{\sqrt{3}}\mathbf{k}\right] = \sqrt{3\left(\frac{1}{\sqrt{3}}\right)^2} = 1,\ \frac{1}{\sqrt{3}}\mathbf{i}+\frac{1}{\sqrt{3}}\mathbf{j}+\frac{1}{\sqrt{3}}\mathbf{k} = 1\left[\frac{1}{\sqrt{3}}\mathbf{i}+\frac{1}{\sqrt{3}}\mathbf{j}+\frac{1}{\sqrt{3}}\mathbf{k}\right] \Rightarrow$
the direction is $\frac{1}{\sqrt{3}}\mathbf{i}+\frac{1}{\sqrt{3}}\mathbf{j}+\frac{1}{\sqrt{3}}\mathbf{k}$

47. the length = $\left|\overrightarrow{P_1P_2}\right| = |2\mathbf{i} + 2\mathbf{j} - \mathbf{k}| = \sqrt{2^2 + 2^2 + (-1)^2} = 3$, $2\mathbf{i} + 2\mathbf{j} - \mathbf{k} = 3\left[\frac{2}{3}\mathbf{i} + \frac{2}{3}\mathbf{j} - \frac{1}{3}\mathbf{k}\right] \Rightarrow$ the direction is $\frac{2}{3}\mathbf{i} + \frac{2}{3}\mathbf{j} - \frac{1}{3}\mathbf{k}$

48. the length = $\left|\overrightarrow{P_1P_2}\right| = |3\mathbf{i} + 4\mathbf{j} - 5\mathbf{k}| = \sqrt{9 + 16 + 25} = 5\sqrt{2}$, $3\mathbf{i} + 4\mathbf{j} - 5\mathbf{k} = 5\sqrt{2}\left[\frac{3}{5\sqrt{2}}\mathbf{i} + \frac{4}{5\sqrt{2}}\mathbf{j} - \frac{1}{\sqrt{2}}\mathbf{k}\right] \Rightarrow$ the direction is $\frac{3}{5\sqrt{2}}\mathbf{i} + \frac{4}{5\sqrt{2}}\mathbf{j} - \frac{1}{\sqrt{2}}\mathbf{k}$

49. the length = $\left|\overrightarrow{P_1P_2}\right| = |(3\mathbf{i} - 6\mathbf{j} + 2\mathbf{k}) = \sqrt{9 + 36 + 4} = 7$, $3\mathbf{i} - 6\mathbf{j} + 2\mathbf{k} = 7\left[\frac{3}{7}\mathbf{i} - \frac{6}{7}\mathbf{j} + \frac{2}{7}\mathbf{k}\right] \Rightarrow$ the direction is $\frac{3}{7}\mathbf{i} - \frac{6}{7}\mathbf{j} + \frac{2}{7}\mathbf{k}$

50. the length = $\left|\overrightarrow{P_1P_2}\right| = |-\mathbf{i} - \mathbf{j} - \mathbf{k}| = \sqrt{3}$, $-\mathbf{i} - \mathbf{j} - \mathbf{k} = \sqrt{3}\left[-\frac{1}{\sqrt{3}}\mathbf{i} - \frac{1}{\sqrt{3}}\mathbf{j} - \frac{1}{\sqrt{3}}\mathbf{k}\right] \Rightarrow$ the direction is $-\frac{1}{\sqrt{3}}\mathbf{i} - \frac{1}{\sqrt{3}}\mathbf{j} - \frac{1}{\sqrt{3}}\mathbf{k}$

51. the length = $\left|\overrightarrow{P_1P_2}\right| = |2\mathbf{i} - 2\mathbf{j} - 2\mathbf{k}| = \sqrt{3 \cdot 2^2} = 2\sqrt{3}$, $2\mathbf{i} - 2\mathbf{j} - 2\mathbf{k} = 2\sqrt{3}\left[\frac{1}{\sqrt{3}}\mathbf{i} - \frac{1}{\sqrt{3}}\mathbf{j} - \frac{1}{\sqrt{3}}\mathbf{k}\right] \Rightarrow$ the direction is $\frac{1}{\sqrt{3}}\mathbf{i} - \frac{1}{\sqrt{3}}\mathbf{j} - \frac{1}{\sqrt{3}}\mathbf{k}$

52. the length = $\left|\overrightarrow{P_1P_2}\right| = |-5\mathbf{i} - 3\mathbf{j} + 2\mathbf{k}| = \sqrt{38}$, $-5\mathbf{i} - 3\mathbf{j} + 2\mathbf{k} = \sqrt{38}\left[-\frac{5}{\sqrt{38}}\mathbf{i} - \frac{3}{\sqrt{38}}\mathbf{j} + \frac{2}{\sqrt{38}}\mathbf{k}\right] \Rightarrow$ the direction is $-\frac{5}{\sqrt{38}}\mathbf{i} - \frac{3}{\sqrt{38}}\mathbf{j} + \frac{2}{\sqrt{38}}\mathbf{k}$

53. a) $2\mathbf{i}$ b) $4\mathbf{j}$ c) $\sqrt{3}\mathbf{k}$
d) $\frac{3}{10}\mathbf{j} + \frac{2}{5}\mathbf{k}$ e) $6\mathbf{i} + 2\mathbf{j} + 3\mathbf{k}$ f) $au_1\mathbf{i} + au_2\mathbf{j} + au_3\mathbf{k}$

54. $\frac{\mathbf{A}}{|\mathbf{A}|} = \frac{1}{13}\mathbf{A} = \frac{1}{13}[12\mathbf{i} - 5\mathbf{k}] = \frac{12}{13}\mathbf{i} - \frac{5}{13}\mathbf{j}$, where $|\mathbf{A}| = \sqrt{12^2 + 5^2} = \sqrt{169} = 13$

55. $|\mathbf{A}| = |\mathbf{i} + \mathbf{j} + \mathbf{k}| = \sqrt{3}$, $\mathbf{A} = \sqrt{3}\left[\frac{1}{\sqrt{3}}\mathbf{i} + \frac{1}{\sqrt{3}}\mathbf{j} + \frac{1}{\sqrt{3}}\mathbf{k}\right] \Rightarrow$ the desired vector is $5\left[\frac{1}{\sqrt{3}}\mathbf{i} + \frac{1}{\sqrt{3}}\mathbf{j} + \frac{1}{\sqrt{3}}\mathbf{k}\right]$

56. $|\mathbf{A}| = |\mathbf{i} - \mathbf{j}| = \sqrt{2}$, $\mathbf{A} = \sqrt{2}\left[\frac{1}{\sqrt{2}}\mathbf{i} - \frac{1}{\sqrt{2}}\mathbf{j}\right] \Rightarrow$ the desired vector is $\frac{1}{2}\left[\frac{1}{\sqrt{2}}\mathbf{i} - \frac{1}{\sqrt{2}}\mathbf{j}\right]$

57. a) center $(-2,0,2)$, radius $2\sqrt{2}$ b) center $(-1/2,-1/2,-1/2)$, $\frac{\sqrt{21}}{2}$
c) center $(\sqrt{2},\sqrt{2},-\sqrt{2})$, radius $\sqrt{2}$ c) center $(0,-1/3,1/3)$, radius $\frac{\sqrt{29}}{3}$

58. a) $(x-1)^2 + (y-2)^2 + (z-3)^2 = 14$ b) $x^2 + (y+1)^2 + (z-5)^2 = 4$
c) $(x+2)^2 + y^2 + z^2 = 3$ d) $x^2 + (y+7)^2 + z^2 = 49$

59. a) $\sqrt{y^2 + z^2}$ b) $\sqrt{x^2 + z^2}$ c) $\sqrt{x^2 + y^2}$

60. a) z b) x c) y

11.3 DOT PRODUCTS

	$\mathbf{A}\cdot\mathbf{B}$	$\lvert\mathbf{A}\rvert$	$\lvert\mathbf{B}\rvert$	$\cos\theta$	$\lvert\mathbf{B}\rvert\cos\theta$	$\text{Proj}_{\mathbf{A}}\,\mathbf{B}$
1.	10	$\sqrt{13}$	$\sqrt{26}$	$\frac{10}{13\sqrt{2}}$	$\frac{10}{\sqrt{13}}$	$\frac{10}{13}[3\mathbf{i}+2\mathbf{j}]$
2.	0	1	$\sqrt{34}$	0	0	$\mathbf{0}$
3.	4	$\sqrt{14}$	2	$\frac{2}{\sqrt{14}}$	$\frac{4}{\sqrt{14}}$	$\frac{2}{7}[3\mathbf{i}-2\mathbf{j}-\mathbf{k}]$
4.	0	$\sqrt{53}$	1	0	0	$\mathbf{0}$
5.	2	$\sqrt{34}$	$\sqrt{3}$	$\frac{2}{\sqrt{3}\sqrt{34}}$	$\frac{2}{\sqrt{34}}$	$\frac{1}{17}[5\mathbf{j}-3\mathbf{k}]$
6.	0	1	$\sqrt{\frac{3}{2}}$	0	0	$\mathbf{0}$
7.	$\sqrt{3}-\sqrt{2}$	$\sqrt{2}$	3	$\frac{\sqrt{3}-\sqrt{2}}{\sqrt{2}}$	$\frac{\sqrt{3}-\sqrt{2}}{\sqrt{2}}$	$\frac{\sqrt{3}-\sqrt{2}}{2}[-\mathbf{i}+\mathbf{j}]$
8.	2	$\sqrt{2}$	$\sqrt{3}$	$\sqrt{\frac{2}{3}}$	$\sqrt{2}$	$\mathbf{i}+\mathbf{k}$
9.	-25	5	5	-1	-5	$-2\mathbf{i}+4\mathbf{j}-\sqrt{5}\mathbf{k}$
10.	$-10+\sqrt{17}$	$\sqrt{26}$	11	$\frac{\sqrt{17}-10}{11\sqrt{26}}$	$\frac{\sqrt{17}-10}{\sqrt{26}}$	$\frac{\sqrt{17}-10}{26}[-5\mathbf{i}+\mathbf{j}]$
11.	25	15	5	$\frac{1}{3}$	$\frac{5}{3}$	$\frac{1}{9}[10\mathbf{i}+11\mathbf{j}-2\mathbf{k}]$
12.	13	15	3	$\frac{13}{45}$	$\frac{13}{15}$	$\frac{13}{225}[2\mathbf{i}+10\mathbf{j}-11\mathbf{k}]$

13. $\mathbf{B}=\left(\frac{\mathbf{A}\cdot\mathbf{B}}{\mathbf{A}\cdot\mathbf{A}}\mathbf{A}\right)+\left(\mathbf{B}-\frac{\mathbf{A}\cdot\mathbf{B}}{\mathbf{A}\cdot\mathbf{A}}\mathbf{A}\right)=\frac{3}{2}[\mathbf{i}+\mathbf{j}]+\left[(3\mathbf{j}+4\mathbf{k})-\frac{3}{2}(\mathbf{i}+\mathbf{j})\right]=\left[\frac{3}{2}\mathbf{i}+\frac{3}{2}\mathbf{j}\right]+\left[-\frac{3}{2}\mathbf{i}+\frac{3}{2}\mathbf{j}+4\mathbf{k}\right]$, where $\mathbf{A}\cdot\mathbf{B}=3$ and $\mathbf{A}\cdot\mathbf{A}=2$

14. $\mathbf{B}=\left(\frac{\mathbf{A}\cdot\mathbf{B}}{\mathbf{A}\cdot\mathbf{A}}\mathbf{A}\right)+\left(\mathbf{B}-\frac{\mathbf{A}\cdot\mathbf{B}}{\mathbf{A}\cdot\mathbf{A}}\mathbf{A}\right)=\frac{1}{2}\mathbf{A}+\mathbf{B}-\frac{1}{2}\mathbf{A}=\frac{1}{2}[\mathbf{i}+\mathbf{j}]+\left[(\mathbf{j}+\mathbf{k})-\frac{1}{2}(\mathbf{i}+\mathbf{j})\right]=$ $\left[\frac{1}{2}\mathbf{i}+\frac{1}{2}\mathbf{j}\right]+\left[-\frac{1}{2}\mathbf{i}+\frac{1}{2}\mathbf{j}+\mathbf{k}\right]$, where $\mathbf{A}\cdot\mathbf{B}=1$ and $\mathbf{A}\cdot\mathbf{A}=2$

15. $\mathbf{B}=\left(\frac{\mathbf{A}\cdot\mathbf{B}}{\mathbf{A}\cdot\mathbf{A}}\mathbf{A}\right)+\left(\mathbf{B}-\frac{\mathbf{A}\cdot\mathbf{B}}{\mathbf{A}\cdot\mathbf{A}}\mathbf{A}\right)=\frac{14}{3}[\mathbf{i}+2\mathbf{j}-\mathbf{k}]+\left[(8\mathbf{i}+4\mathbf{j}-12\mathbf{k})-\left(\frac{14}{3}\mathbf{i}+\frac{28}{3}\mathbf{j}-\frac{14}{3}\mathbf{k}\right)\right]=$ $\left[\frac{14}{3}\mathbf{i}+\frac{28}{3}\mathbf{j}-\frac{14}{3}\mathbf{k}\right]+\left[\frac{10}{3}\mathbf{i}-\frac{16}{3}\mathbf{j}-\frac{22}{3}\mathbf{k}\right]$, where $\mathbf{A}\cdot\mathbf{B}=28$ and $\mathbf{A}\cdot\mathbf{A}=6$

16. $\mathbf{B}=\left(\frac{\mathbf{A}\cdot\mathbf{B}}{\mathbf{A}\cdot\mathbf{A}}\mathbf{A}\right)+\left(\mathbf{B}-\frac{\mathbf{A}\cdot\mathbf{B}}{\mathbf{A}\cdot\mathbf{A}}\mathbf{A}\right)=\frac{1}{1}[\mathbf{A}]+\left[(\mathbf{i}+\mathbf{j}+\mathbf{k})-\frac{1}{1}\mathbf{A}\right]=[\mathbf{i}]+[\mathbf{j}+\mathbf{k}]$, where $\mathbf{A}\cdot\mathbf{B}=1$ and $\mathbf{A}\cdot\mathbf{A}=1$; yes

17. $(\mathbf{i} + 2\mathbf{j}) \cdot \big((x-2)\mathbf{i} + (y-1)\mathbf{j}\big) = 0 \Rightarrow x + 2y = 4$

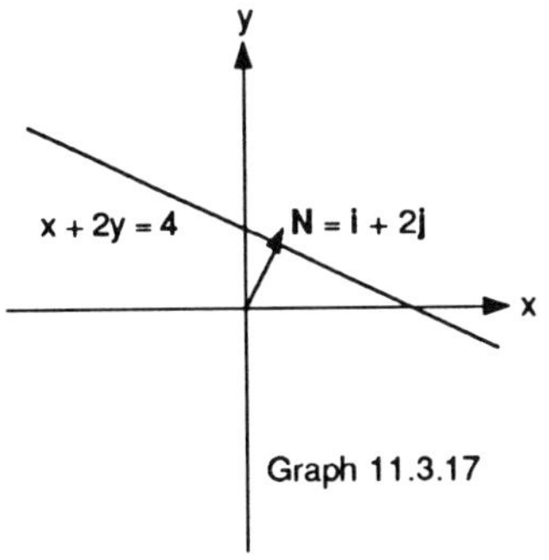

Graph 11.3.17

18. $(\mathbf{i} - \mathbf{j}) \cdot \big((x+2)\mathbf{i} + (y-1)\mathbf{j}\big) = 0 \Rightarrow x - y = -3$

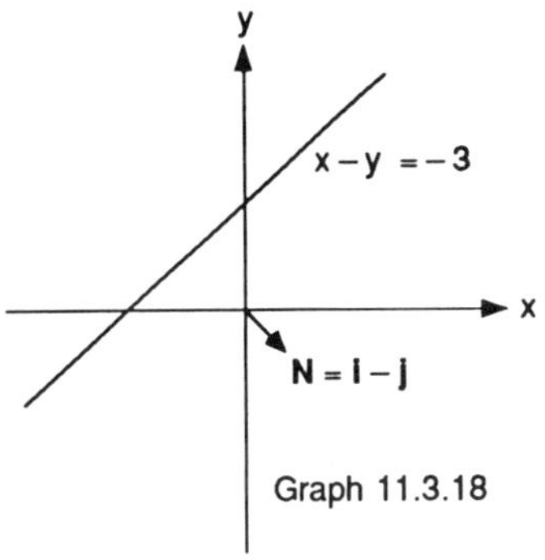

Graph 11.3.18

19. $(-2\mathbf{i} - \mathbf{j}) \cdot \big((x+1)\mathbf{i} + (y-2)\mathbf{j}\big) = 0 \Rightarrow -2x - y = 0$

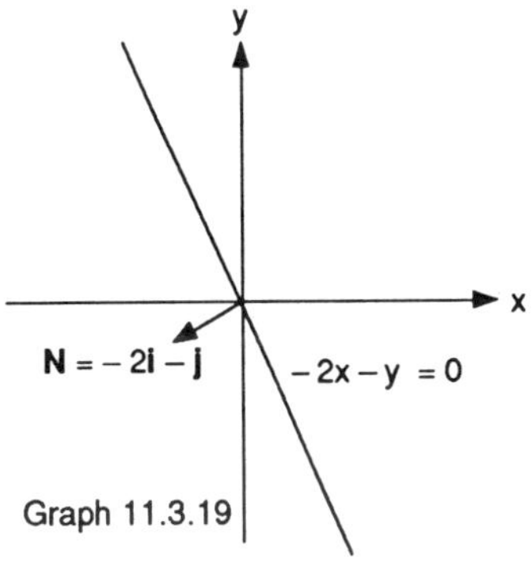

Graph 11.3.19

20. $(2\mathbf{j} - 3\mathbf{j}) \cdot \big((x+1)\mathbf{i} + (y-2)\mathbf{j}\big) = 0 \Rightarrow 2x - 3y = -8$

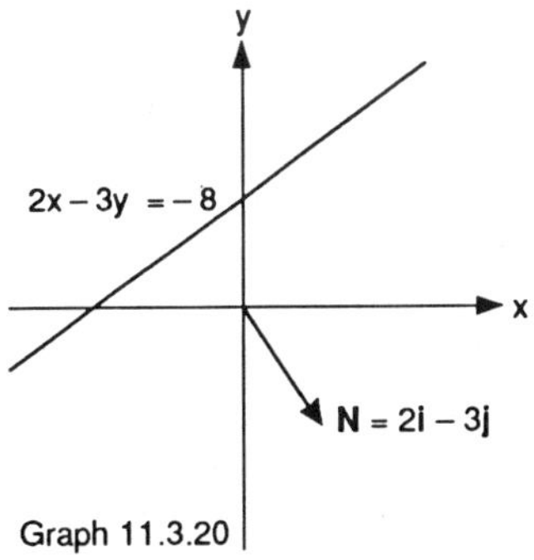

Graph 11.3.20

21. distance $= \left|\text{proj}_{\mathbf{N}}\overrightarrow{PS}\right| = \left|\dfrac{\mathbf{N}\cdot\overrightarrow{PS}}{|\mathbf{N}|}\right| = \left|\dfrac{(\mathbf{i}+3\mathbf{j})\cdot(2\mathbf{i}+6\mathbf{j})}{\sqrt{1^2+3^2}}\right| = \left|\dfrac{20}{\sqrt{10}}\right| = 2\sqrt{10}$, where S(2,8), P(0,2)

and $\mathbf{N} + \mathbf{i} + 3\mathbf{j}$

22. distance $= \left|\text{proj}_{\mathbf{N}}\overrightarrow{PS}\right| = \left|\dfrac{\mathbf{N}\cdot\overrightarrow{PS}}{|\mathbf{N}|}\right| = \left|\dfrac{(\mathbf{i}+3\mathbf{j})\cdot(-2\mathbf{j})}{\sqrt{1+9}}\right| = \left|\dfrac{-6}{\sqrt{10}}\right| = \dfrac{6}{\sqrt{10}}$, where $S(0,0)$, $P(0,2)$ and $\mathbf{N} = \mathbf{i}+3\mathbf{j}$

23. distance $= \left|\text{proj}_{\mathbf{N}}\overrightarrow{PS}\right| = \left|\dfrac{\mathbf{N}\cdot\overrightarrow{PS}}{|\mathbf{N}|}\right| = \left|\dfrac{(\mathbf{i}+\mathbf{j})\cdot(\mathbf{i}+\mathbf{j})}{\sqrt{1+1}}\right| = \sqrt{2}$, where $S(2,1)$, $P(1,0)$ and $\mathbf{N} = \mathbf{i}+\mathbf{j}$

24. distance $= \left|\text{proj}_{\mathbf{N}}\overrightarrow{PS}\right| = \left|\dfrac{\mathbf{N}\cdot\overrightarrow{PS}}{|\mathbf{N}|}\right| = \left|\dfrac{(2\mathbf{i}+\mathbf{j})\cdot(\mathbf{i}+3\mathbf{j})}{\sqrt{4+1}}\right| = \dfrac{5}{\sqrt{5}} = \sqrt{5}$, where $S(1,3)$, $P(0,0)$ and $\mathbf{N} = 2\mathbf{i}+\mathbf{j}$

25. $\mathbf{A}\cdot\mathbf{B} = \left(\frac{1}{\sqrt{3}}\right)(0) + \left(-\frac{1}{\sqrt{3}}\right)\left(\frac{1}{\sqrt{2}}\right) + \left(\frac{1}{\sqrt{3}}\right)\left(\frac{1}{\sqrt{2}}\right) = 0$, $\mathbf{A}\cdot\mathbf{C} = \left(\frac{1}{\sqrt{3}}\right)\left(-\frac{2}{\sqrt{6}}\right) + \left(-\frac{1}{\sqrt{3}}\right)\left(-\frac{1}{\sqrt{6}}\right) + \left(\frac{1}{\sqrt{3}}\right)\left(\frac{1}{\sqrt{6}}\right) = 0$ and $\mathbf{B}\cdot\mathbf{C} = (0)\left(-\frac{2}{\sqrt{6}}\right) + \left(\frac{1}{\sqrt{2}}\right)\left(-\frac{1}{\sqrt{6}}\right) + \left(\frac{1}{\sqrt{2}}\right)\left(\frac{1}{\sqrt{6}}\right) = 0 \Rightarrow$ that $\mathbf{A}$, $\mathbf{B}$ and $\mathbf{C}$ are orthogonal

26. $\text{proj}_{\mathbf{A}}\mathbf{D} = \dfrac{\mathbf{A}\cdot\mathbf{D}}{\mathbf{A}\cdot\mathbf{A}}\mathbf{A} = \dfrac{\frac{1}{\sqrt{3}} - \frac{1}{\sqrt{3}} + \frac{1}{\sqrt{3}}}{1}\mathbf{A} = \dfrac{1}{\sqrt{3}}\mathbf{A} = \dfrac{1}{3}(\mathbf{i}-\mathbf{j}+\mathbf{k})$, $\text{proj}_{\mathbf{B}}\mathbf{D} = \dfrac{\mathbf{B}\cdot\mathbf{D}}{\mathbf{B}\cdot\mathbf{B}}\mathbf{B} = \dfrac{\sqrt{2}}{1}\mathbf{B} = (\mathbf{j}+\mathbf{k})$

$\text{proj}_{\mathbf{C}}\mathbf{D} = \dfrac{\mathbf{C}\cdot\mathbf{D}}{\mathbf{C}\cdot\mathbf{C}}\mathbf{C} = \dfrac{-2/\sqrt{6}}{1}\mathbf{C} = \dfrac{2}{3}\mathbf{i} + \dfrac{1}{3}\mathbf{j} - \dfrac{1}{3}\mathbf{k} \Rightarrow \mathbf{D} = \text{proj}_{\mathbf{A}}\mathbf{D} + \text{proj}_{\mathbf{B}}\mathbf{D} + \text{proj}_{\mathbf{C}}\mathbf{D}$

27. If $\mathbf{A} = (\mathbf{i}+\mathbf{j}+\mathbf{k})$, $\mathbf{B}_1 = \mathbf{i}$, and $\mathbf{B}_2 = \mathbf{j}$, then $\mathbf{A}\cdot\mathbf{B}_1 = \mathbf{A}\cdot\mathbf{B}_2$, but $\mathbf{B}_1 \neq \mathbf{B}_2$

28. $(\mathbf{V}_1+\mathbf{V}_2)\cdot(\mathbf{V}_1-\mathbf{V}_2) = 0 \Rightarrow \mathbf{V}_1\cdot\mathbf{V}_1 + \mathbf{V}_1\cdot\mathbf{V}_2 - \mathbf{V}_1\cdot\mathbf{V}_2 - \mathbf{V}_2\cdot\mathbf{V}_2 = 0 \Rightarrow |\mathbf{V}_1|^2 - |\mathbf{V}_2|^2 = 0 \Rightarrow |\mathbf{V}_1| = |\mathbf{V}_2|$ or $|\mathbf{V}_1| = -|\mathbf{V}_2| \Rightarrow \mathbf{V}_1$ and $\mathbf{V}_2$ must have the same length or both must be zero vectors

29. $\overrightarrow{AB} = 3\mathbf{i}+\mathbf{j}-3\mathbf{k}$, $\overrightarrow{AC} = 2\mathbf{i}-2\mathbf{j}$, $\overrightarrow{BA} = -3\mathbf{i}-\mathbf{j}+3\mathbf{k}$, $\overrightarrow{CA} = -2\mathbf{i}+2\mathbf{j}$, $\overrightarrow{CB} = \mathbf{i}+3\mathbf{j}-3\mathbf{k}$, $\overrightarrow{BC} = -\mathbf{i}-3\mathbf{j}+3\mathbf{k} \Rightarrow \angle A = \cos^{-1}\left(\dfrac{\overrightarrow{AB}\cdot\overrightarrow{AC}}{\left|\overrightarrow{AB}\right|\left|\overrightarrow{AC}\right|}\right) = \cos^{-1}\left(\dfrac{4}{\sqrt{152}}\right) \approx 71.1°$,

$\angle B = \cos^{-1}\left(\dfrac{\overrightarrow{BA}\cdot\overrightarrow{BC}}{\left|\overrightarrow{BA}\right|\left|\overrightarrow{BC}\right|}\right) = \cos^{-1}\left(\dfrac{15}{19}\right) \approx 37.9°$, $\angle C = \cos^{-1}\left(\dfrac{\overrightarrow{CA}\cdot\overrightarrow{CB}}{\left|\overrightarrow{CA}\right|\left|\overrightarrow{CB}\right|}\right) = \cos^{-1}\left(\dfrac{4}{\sqrt{152}}\right) \approx 71.1°$

30. $\theta = \cos^{-1}\left(\dfrac{\mathbf{A}\cdot\mathbf{B}}{|\mathbf{A}|\,|\mathbf{B}|}\right) = \cos^{-1}\left(\dfrac{13}{45}\right) \approx 73.2°$

31. $\theta = \cos^{-1}\left(\dfrac{\mathbf{A}\cdot\mathbf{B}}{|\mathbf{A}|\,|\mathbf{B}|}\right) = \cos^{-1}\left(\dfrac{2}{\sqrt{2}\sqrt{3}}\right) \approx 35.3°$, where $\mathbf{A} = \mathbf{i}+\mathbf{k}$ and $\mathbf{B} = \mathbf{i}+\mathbf{j}+\mathbf{k}$

32. $\theta = \cos^{-1}\left(\dfrac{\mathbf{A}\cdot\mathbf{B}}{|\mathbf{A}|\,|\mathbf{B}|}\right) = \cos^{-1}\left(\dfrac{1}{1\cdot\sqrt{3}}\right) \approx 54.7°$, where $\mathbf{A} = \mathbf{i}$ and $\mathbf{B} = \mathbf{i}+\mathbf{j}+\mathbf{k}$

33. $P(0,0,0)$, $Q(1,1,1)$ and $\mathbf{F} = -5\mathbf{k} \Rightarrow \overrightarrow{PQ} = \mathbf{i}+\mathbf{j}+\mathbf{k}$, $W = \mathbf{F}\cdot\overrightarrow{PQ} = (-5\mathbf{k})\cdot(\mathbf{i}+\mathbf{j}+\mathbf{k}) = -5\ \text{N}\cdot\text{m}$

34. $W = |\mathbf{F}|\left|\overrightarrow{PQ}\right| \cos\theta = (60000)(1000)(\cos 0) = 6 \times 10^7 \text{ N}\cdot\text{m}$

35. $W = |\mathbf{F}|\left|\overrightarrow{PQ}\right| \cos\theta = (200)(20)\left(\cos\frac{\pi}{6}\right) = 20000\sqrt{3} \approx 3464.10 \text{ N}\cdot\text{m}$

36. $W = |\mathbf{F}|\left|\overrightarrow{PQ}\right| \cos\theta = (1000)(5280)(\cos 60°) = 2640000 \text{ ft}\cdot\text{lb}$

37. The angle between the corresponding normals is equal to the angle between the corresponding tangents. $\theta = \cos^{-1}\left(\frac{\mathbf{N}_1\cdot\mathbf{N}_2}{|\mathbf{N}_1|\,|\mathbf{N}_2|}\right) = \cos^{-1}\left(\frac{1}{\sqrt{2}}\right) = 45°$, where $\mathbf{N}_1 = 3\mathbf{i} + \mathbf{j}$, $\mathbf{N}_2 = 2\mathbf{i} - \mathbf{j}$; the angle is either 45° or 135°

38. $\theta = \cos^{-1}\left(\frac{\mathbf{N}_1\cdot\mathbf{N}_2}{|\mathbf{N}_1|\,|\mathbf{N}_2|}\right) = \cos^{-1}\left(-\frac{1}{2}\right) = \frac{2\pi}{3}$; the angle is either $\frac{\pi}{3}$ or $\frac{2\pi}{3}$. See exercise 37 for additional details.

39. The curve $y = \sqrt{\frac{3}{4} + x}$ has $y = \frac{1}{\sqrt{3}}x + \frac{\sqrt{3}}{2}$ as its tangent line at $(0,\sqrt{3}/2)$. The normal vector for $y = \sqrt{\frac{3}{4} + x}$ at $(0,\sqrt{3}/2)$ is $\mathbf{N}_1 = \frac{1}{\sqrt{3}}\mathbf{i} - \mathbf{j}$. The curve $y = \sqrt{\frac{3}{4} - x}$ has $y = -\frac{1}{\sqrt{3}}x + \frac{\sqrt{3}}{2}$ as its tangent line at $(0,\sqrt{3}/2)$. The normal vector for $y = \sqrt{\frac{3}{4} + x}$ at $(0,\sqrt{3}/2)$ is $\mathbf{N}_2 = -\frac{1}{\sqrt{3}}\mathbf{i} - \mathbf{j}$.

$\theta = \cos^{-1}\left(\frac{\mathbf{N}_1\cdot\mathbf{N}_2}{|\mathbf{N}_1|\,|\mathbf{N}_2|}\right) = \cos^{-1}\left(\frac{1}{2}\right) = \frac{\pi}{3}$; the angle is either $\frac{\pi}{3}$ or $\frac{2\pi}{3}$. See exercise 37 for additional details.

40. The points of intersection for the curves $y = x^3$ and $y = \sqrt{x}$ are (0,0) and (1,1). At (0,0) the tangent line for $y = x^3$ is $y = 0$ and the tangent line for $y = \sqrt{x}$ is $x = 0$. Therefore, the angle of intersection at (0,0) is $\frac{\pi}{2}$. At (1,1) the tangent line for $y = x^3$ is $y = 3x - 2$ and the tangent line for $y = \sqrt{x}$ is $y = \frac{1}{2}x + \frac{1}{2}$. The corresponding normal vectors are: $\mathbf{N}_1 = -3\mathbf{i} + \mathbf{j}$, $\mathbf{N}_2 = -\frac{1}{2}\mathbf{i} + \mathbf{j}$.

$\theta = \cos^{-1}\left(\frac{\mathbf{N}_1\cdot\mathbf{N}_2}{|\mathbf{N}_1|\,|\mathbf{N}_2|}\right) = \cos^{-1}\frac{1}{\sqrt{2}} = \frac{\pi}{4}$; the angle is either 45° or 135°

41. The points of intersection for the curves $y = x^2$ and $y = \sqrt[3]{x}$ are (0,0) and (1,1). At (0,0) the tangent line for $y = x^2$ is $y = 0$ and the tangent line for $y = \sqrt{x}$ is $x = 0$. Therefore, the angle of intersection at (0,0) is $\frac{\pi}{2}$. At (1,1) the tangent line for $y = x^2$ is $y = 2x - 1$ and the tangent line for $y = \sqrt[3]{x}$ is $y = \frac{1}{3}x + \frac{2}{3}$. The corresponding normal vectors are: $\mathbf{N}_1 = 2\mathbf{i} - \mathbf{j}$, $\mathbf{N}_2 = \frac{1}{3}\mathbf{i} - \mathbf{j}$.

$\theta = \cos^{-1}\left(\frac{\mathbf{N}_1\cdot\mathbf{N}_2}{|\mathbf{N}_1|\,|\mathbf{N}_2|}\right) = \cos^{-1}\frac{1}{\sqrt{2}} = \frac{\pi}{4}$; the angle is either 45° or 135°

11.4 CROSS PRODUCTS

1. $\mathbf{A}\times\mathbf{B} = \begin{vmatrix} \mathbf{i} & \mathbf{j} & \mathbf{k} \\ 2 & -2 & -1 \\ 1 & 0 & -1 \end{vmatrix} = 3\left[\frac{2}{3}\mathbf{i}+\frac{1}{3}\mathbf{j}+\frac{2}{3}\mathbf{k}\right] \Rightarrow$ length = 3 and the direction is $\frac{2}{3}\mathbf{i}+\frac{1}{3}\mathbf{j}+\frac{2}{3}\mathbf{k}$

   $\mathbf{B}\times\mathbf{A} = \begin{vmatrix} \mathbf{i} & \mathbf{j} & \mathbf{k} \\ 1 & 0 & -1 \\ 2 & -2 & -1 \end{vmatrix} = -3\left[\frac{2}{3}\mathbf{i}+\frac{1}{3}\mathbf{j}+\frac{2}{3}\mathbf{k}\right] \Rightarrow$ length = 3 and the direction is $-\frac{2}{3}\mathbf{i}-\frac{1}{3}\mathbf{j}-\frac{2}{3}\mathbf{k}$

2. $\mathbf{A}\times\mathbf{B} = \begin{vmatrix} \mathbf{i} & \mathbf{j} & \mathbf{k} \\ 2 & 3 & 0 \\ -1 & 1 & 0 \end{vmatrix} = 5\,[\mathbf{k}] \Rightarrow$ length = 5 and the direction is $\mathbf{k}$

   $\mathbf{B}\times\mathbf{A} = \begin{vmatrix} \mathbf{i} & \mathbf{j} & \mathbf{k} \\ -1 & 1 & 0 \\ 2 & 3 & 0 \end{vmatrix} = -5\,[\mathbf{k}] \Rightarrow$ length = 5 and the direction is $-\mathbf{k}$

3. $\mathbf{A}\times\mathbf{B} = \begin{vmatrix} \mathbf{i} & \mathbf{j} & \mathbf{k} \\ 2 & -2 & 4 \\ -1 & 1 & -2 \end{vmatrix} = \mathbf{0} \Rightarrow$ length = 0 and has no direction

   $\mathbf{B}\times\mathbf{A} = \begin{vmatrix} \mathbf{i} & \mathbf{j} & \mathbf{k} \\ -1 & 1 & -2 \\ 2 & -2 & 4 \end{vmatrix} = \mathbf{0} \Rightarrow$ length = 0 and has no direction

4. $\mathbf{A}\times\mathbf{B} = \begin{vmatrix} \mathbf{i} & \mathbf{j} & \mathbf{k} \\ 1 & 1 & -1 \\ 0 & 0 & 0 \end{vmatrix} = \mathbf{0} \Rightarrow$ length = 0 and has no direction

   $\mathbf{B}\times\mathbf{A} = \begin{vmatrix} \mathbf{i} & \mathbf{j} & \mathbf{k} \\ 0 & 0 & 0 \\ 1 & 1 & -1 \end{vmatrix} = \mathbf{0} \Rightarrow$ length = 0 and has no direction

5. $\mathbf{A}\times\mathbf{B} = \begin{vmatrix} \mathbf{i} & \mathbf{j} & \mathbf{k} \\ 2 & 0 & 0 \\ 0 & -3 & 0 \end{vmatrix} = -6\,[\mathbf{k}] \Rightarrow$ length = 6 and the direction is $-\mathbf{k}$

   $\mathbf{B}\times\mathbf{A} = \begin{vmatrix} \mathbf{i} & \mathbf{j} & \mathbf{k} \\ 0 & -3 & 0 \\ 2 & 0 & 0 \end{vmatrix} = 6\,[\mathbf{k}] \Rightarrow$ length = 6 and the direction is $\mathbf{k}$

6. $\mathbf{A}\times\mathbf{B} = \begin{vmatrix} \mathbf{i} & \mathbf{j} & \mathbf{k} \\ 0 & 0 & 1 \\ 1 & 0 & 0 \end{vmatrix} = [\mathbf{j}] \Rightarrow$ length = 1 and the direction is $\mathbf{j}$

   $\mathbf{B}\times\mathbf{A} = \begin{vmatrix} \mathbf{i} & \mathbf{j} & \mathbf{k} \\ 1 & 0 & 0 \\ 0 & 0 & 1 \end{vmatrix} = -[\mathbf{j}] \Rightarrow$ length = 1 and the direction is $-\mathbf{j}$

7. $\mathbf{A}\times\mathbf{B} = \begin{vmatrix} \mathbf{i} & \mathbf{j} & \mathbf{k} \\ -8 & -2 & -4 \\ 2 & 2 & 1 \end{vmatrix} = [6\mathbf{i}-12\mathbf{k}] \Rightarrow$ length = $6\sqrt{5}$ and the direction is $\frac{1}{\sqrt{5}}\mathbf{i}-\frac{2}{\sqrt{5}}\mathbf{k}$

   $\mathbf{B}\times\mathbf{A} = \begin{vmatrix} \mathbf{i} & \mathbf{j} & \mathbf{k} \\ 2 & 2 & 1 \\ -8 & -2 & -4 \end{vmatrix} = -[6\mathbf{i}-12\mathbf{k}] \Rightarrow$ length = $6\sqrt{5}$ and the direction is $-\frac{1}{\sqrt{5}}\mathbf{i}+\frac{2}{\sqrt{5}}\mathbf{k}$

8. $\mathbf{A}\times\mathbf{B} = \begin{vmatrix} \mathbf{i} & \mathbf{j} & \mathbf{k} \\ 3/2 & -1/2 & -1 \\ 1 & 1 & 2 \end{vmatrix} = [-2\mathbf{i}-2\mathbf{j}+2\mathbf{k}] \Rightarrow$ length = $2\sqrt{3}$ and the direction is $-\frac{1}{\sqrt{3}}\mathbf{i}-\frac{1}{\sqrt{3}}\mathbf{j}+\frac{1}{\sqrt{3}}\mathbf{k}$

   $\mathbf{B}\times\mathbf{A} = \begin{vmatrix} \mathbf{i} & \mathbf{j} & \mathbf{k} \\ 1 & 1 & 2 \\ 3/2 & -1/2 & -1 \end{vmatrix} = -[-2\mathbf{i}-2\mathbf{j}+2\mathbf{k}] \Rightarrow$ length = $2\sqrt{3}$ and the direction is $\frac{1}{\sqrt{3}}\mathbf{i}+\frac{1}{\sqrt{3}}\mathbf{j}-\frac{1}{\sqrt{3}}\mathbf{k}$

9. $\mathbf{A} \times \mathbf{B} = \begin{vmatrix} \mathbf{i} & \mathbf{j} & \mathbf{k} \\ 1 & 0 & 0 \\ 0 & 1 & 0 \end{vmatrix} = \mathbf{k}$

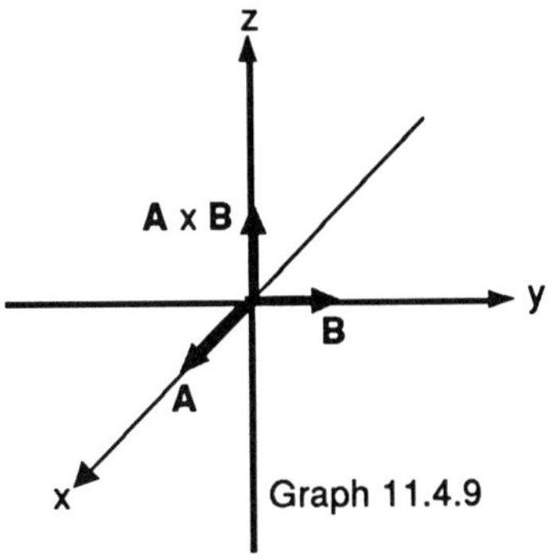

Graph 11.4.9

10. $\mathbf{A} \times \mathbf{B} = \begin{vmatrix} \mathbf{i} & \mathbf{j} & \mathbf{k} \\ 1 & 0 & 1 \\ 0 & 1 & 0 \end{vmatrix} = -\mathbf{i} + \mathbf{k}$

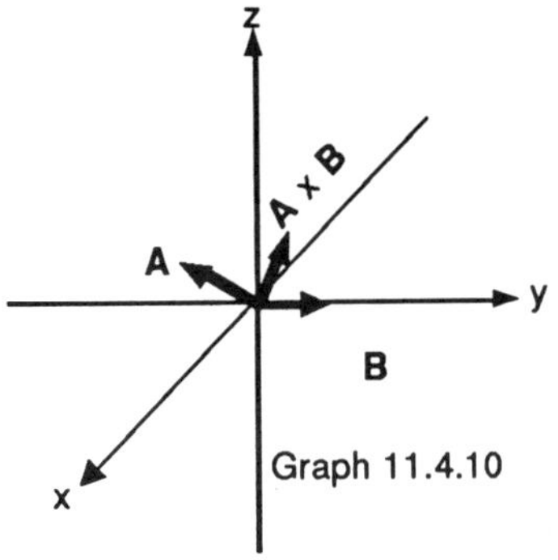

Graph 11.4.10

11. $\mathbf{A} \times \mathbf{B} = \begin{vmatrix} \mathbf{i} & \mathbf{j} & \mathbf{k} \\ 1 & 0 & -1 \\ 0 & 1 & 1 \end{vmatrix} = \mathbf{i} - \mathbf{j} + \mathbf{k}$

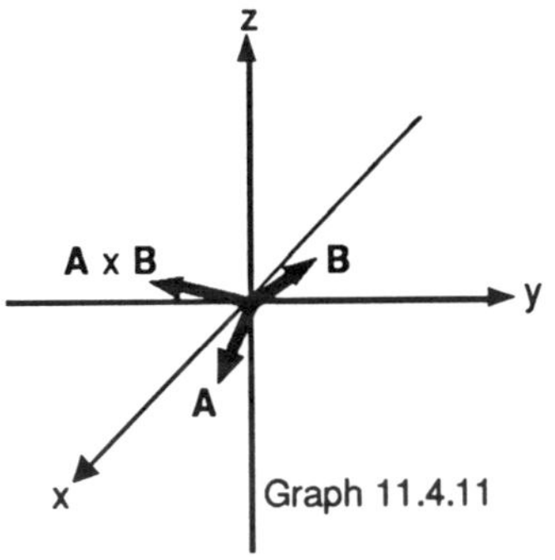

Graph 11.4.11

12. $\mathbf{A} \times \mathbf{B} = \begin{vmatrix} \mathbf{i} & \mathbf{j} & \mathbf{k} \\ 2 & -1 & 0 \\ 1 & 2 & 0 \end{vmatrix} = 5\mathbf{k}$

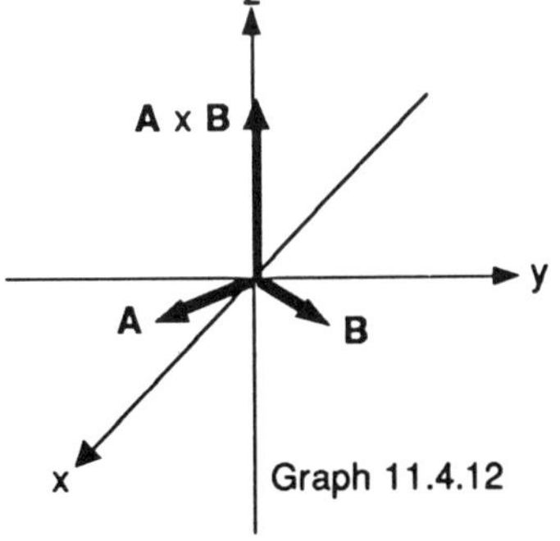

Graph 11.4.12

13. $\mathbf{A} \times \mathbf{B} = \begin{vmatrix} \mathbf{i} & \mathbf{j} & \mathbf{k} \\ 1 & 3 & 2 \\ 0 & 0 & 1 \end{vmatrix} = 3\mathbf{i} - \mathbf{j}$

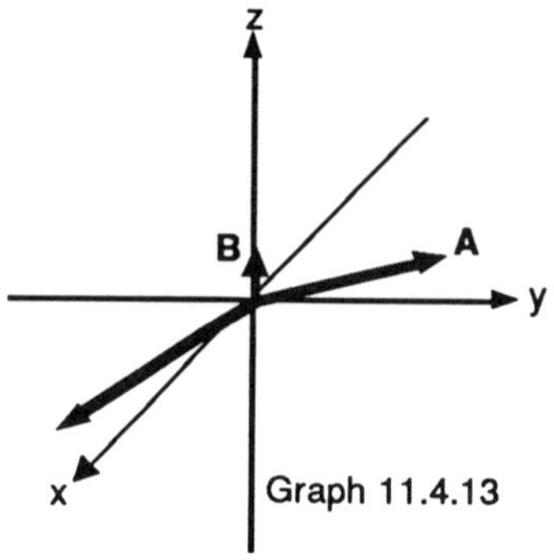

Graph 11.4.13

14. $\mathbf{A} \times \mathbf{B} = \begin{vmatrix} \mathbf{i} & \mathbf{j} & \mathbf{k} \\ 1 & 2 & 0 \\ 0 & 2 & 1 \end{vmatrix} = 2\mathbf{i} - \mathbf{j} + 2\mathbf{k}$

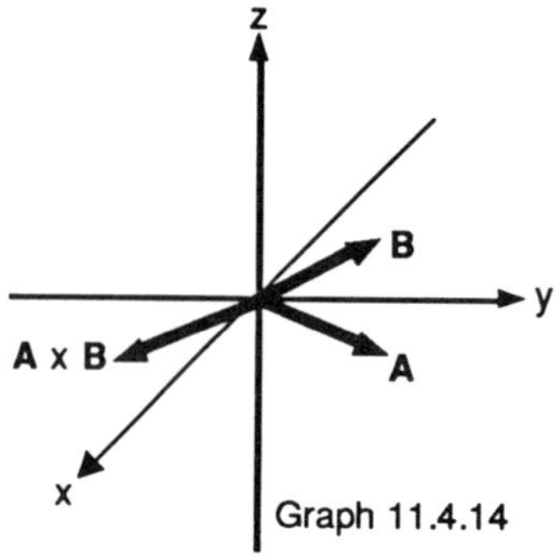

Graph 11.4.14

15. a) $\overrightarrow{PQ} = \mathbf{i} + \mathbf{j} - 3\mathbf{k}, \overrightarrow{PR} = -\mathbf{i} + 3\mathbf{j} - \mathbf{k} \Rightarrow \pm\left(\overrightarrow{PQ} \times \overrightarrow{PR}\right) = \pm\begin{vmatrix} \mathbf{i} & \mathbf{j} & \mathbf{k} \\ 1 & 1 & -3 \\ -1 & 3 & -1 \end{vmatrix} = \pm(8\mathbf{i} + 4\mathbf{j} + 4\mathbf{k})$

b) $\frac{\left|\overrightarrow{PQ} \times \overrightarrow{PR}\right|}{2} = \frac{\sqrt{64 + 16 + 16}}{2} = 2\sqrt{6}$

c) $\pm\left[\frac{2}{\sqrt{6}}\mathbf{i} + \frac{1}{\sqrt{6}}\mathbf{j} + \frac{1}{\sqrt{6}}\mathbf{k}\right]$

16. a) $\overrightarrow{PQ} = \mathbf{i} + 2\mathbf{k}, \overrightarrow{PR} = 2\mathbf{i} - 2\mathbf{j} \Rightarrow \pm\left(\overrightarrow{PQ} \times \overrightarrow{PR}\right) = \pm\begin{vmatrix} \mathbf{i} & \mathbf{j} & \mathbf{k} \\ 1 & 0 & 2 \\ 2 & -2 & 0 \end{vmatrix} = \pm(4\mathbf{i} + 4\mathbf{j} - 2\mathbf{k})$

b) $\frac{\left|\overrightarrow{PQ} \times \overrightarrow{PR}\right|}{2} = \frac{6}{2} = 3$

c) $\pm\left[\frac{2}{3}\mathbf{i} + \frac{2}{3}\mathbf{j} - \frac{1}{3}\mathbf{k}\right]$

17. a) $\overrightarrow{PQ} = \mathbf{i} + \mathbf{j} + \mathbf{k}, \overrightarrow{PR} = \mathbf{i} + \mathbf{j} \Rightarrow \pm\left(\overrightarrow{PQ} \times \overrightarrow{PR}\right) = \pm\begin{vmatrix} \mathbf{i} & \mathbf{j} & \mathbf{k} \\ 1 & 1 & 1 \\ 1 & 1 & 0 \end{vmatrix} = \pm(-\mathbf{i} + \mathbf{j})$

b) $\frac{\left|\overrightarrow{PQ} \times \overrightarrow{PR}\right|}{2} = \frac{\sqrt{2}}{2}$

c) $\pm\left[\frac{-1}{\sqrt{2}}\mathbf{i} + \frac{1}{\sqrt{2}}\mathbf{j}\right]$

18. a) $\overrightarrow{PQ} = 2\mathbf{i} - \mathbf{j} - \mathbf{k},\ \overrightarrow{PR} = \mathbf{i} - 2\mathbf{k} \Rightarrow \pm\left(\overrightarrow{PQ} \times \overrightarrow{PR}\right) = \pm\begin{vmatrix} \mathbf{i} & \mathbf{j} & \mathbf{k} \\ 2 & -1 & -1 \\ 1 & 0 & -2 \end{vmatrix} = \pm(2\mathbf{i} + 3\mathbf{j} + \mathbf{k})$

b) $\dfrac{\left|\overrightarrow{PQ} \times \overrightarrow{PR}\right|}{2} = \dfrac{\sqrt{14}}{2}$

c) $\pm\left[\dfrac{2}{\sqrt{14}}\mathbf{i} + \dfrac{3}{\sqrt{14}}\mathbf{j} + \dfrac{1}{\sqrt{14}}\mathbf{k}\right]$

19. a) $\mathbf{A} \cdot \mathbf{B} = -6,\ \mathbf{A} \cdot \mathbf{C} = -81,\ \mathbf{B} \cdot \mathbf{C} = 18 \Rightarrow$ none

b) $\mathbf{A} \times \mathbf{B} = \begin{vmatrix} \mathbf{i} & \mathbf{j} & \mathbf{k} \\ 5 & -1 & 1 \\ 0 & 1 & -5 \end{vmatrix} = 4\mathbf{i} \neq \mathbf{0},\ \mathbf{A} \times \mathbf{C} = \begin{vmatrix} \mathbf{i} & \mathbf{j} & \mathbf{k} \\ 5 & -1 & 1 \\ -15 & 3 & -3 \end{vmatrix} = \mathbf{0},$

$\mathbf{B} \times \mathbf{C} = \begin{vmatrix} \mathbf{i} & \mathbf{j} & \mathbf{k} \\ 0 & 1 & -5 \\ -15 & 3 & -3 \end{vmatrix} = 11\mathbf{i} \neq \mathbf{0} \Rightarrow$ **A** and **C** are parallel

20. a) $\mathbf{A} \cdot \mathbf{B} = -1,\ \mathbf{A} \cdot \mathbf{C} = -1,\ \mathbf{A} \cdot \mathbf{D} = 7,\ \mathbf{B} \cdot \mathbf{C} = 0,\ \mathbf{B} \cdot \mathbf{D} = 0,\ \mathbf{C} \cdot \mathbf{D} = 0 \Rightarrow \mathbf{B} \perp \mathbf{C},\ \mathbf{B} \perp \mathbf{D}$ and $\mathbf{C} \perp \mathbf{D}$

b) $\mathbf{A} \times \mathbf{B} = \begin{vmatrix} \mathbf{i} & \mathbf{j} & \mathbf{k} \\ 1 & 2 & -2 \\ -1 & 1 & 1 \end{vmatrix} \neq \mathbf{0},\ \mathbf{A} \times \mathbf{C} = \begin{vmatrix} \mathbf{i} & \mathbf{j} & \mathbf{k} \\ 1 & 2 & -2 \\ 1 & 0 & 1 \end{vmatrix} \neq \mathbf{0},\ \mathbf{A} \times \mathbf{D} = \begin{vmatrix} \mathbf{i} & \mathbf{j} & \mathbf{k} \\ 1 & 2 & -2 \\ 1 & 2 & -1 \end{vmatrix} \neq \mathbf{0},$

$\mathbf{B} \times \mathbf{C} = \begin{vmatrix} \mathbf{i} & \mathbf{j} & \mathbf{k} \\ -1 & 1 & 1 \\ 1 & 0 & 1 \end{vmatrix} \neq \mathbf{0},\ \mathbf{B} \times \mathbf{D} = \begin{vmatrix} \mathbf{i} & \mathbf{j} & \mathbf{k} \\ -1 & 1 & 1 \\ 1 & 2 & -1 \end{vmatrix} \neq \mathbf{0},\ \mathbf{C} \times \mathbf{D} = \begin{vmatrix} \mathbf{i} & \mathbf{j} & \mathbf{k} \\ 1 & 0 & 1 \\ 1 & 2 & -1 \end{vmatrix} \neq \mathbf{0} \Rightarrow$ none

21. $\mathbf{A} \times \mathbf{B} = \begin{vmatrix} \mathbf{i} & \mathbf{j} & \mathbf{k} \\ 2 & -1 & 0 \\ 1 & 3 & -2 \end{vmatrix} = 2\mathbf{i} + 4\mathbf{j} + 7\mathbf{k},\ (\mathbf{A} \times \mathbf{B}) \cdot \mathbf{A} = (2\mathbf{i} + 4\mathbf{j} + 7\mathbf{k}) \cdot (2\mathbf{i} - \mathbf{j}) = 0,$

$(\mathbf{A} \times \mathbf{B}) \cdot \mathbf{B} = (2\mathbf{i} + 4\mathbf{j} + 7\mathbf{k}) \cdot (\mathbf{i} + 3\mathbf{j} - 2\mathbf{k}) = 0$

22. **A** x **B** is perpendicular to the plane containing A and B. Consequently, both $(\mathbf{A} \times \mathbf{B}) \cdot \mathbf{A}$ and $(\mathbf{A} \times \mathbf{B}) \cdot \mathbf{B}$ are always zero.

23. a) $\text{proj}_{\mathbf{B}}\, \mathbf{A} = \dfrac{\mathbf{A} \cdot \mathbf{B}}{\mathbf{B} \cdot \mathbf{B}}\mathbf{B}$ b) $(\pm)(\mathbf{A} \times \mathbf{B})$ c) $\sqrt{\mathbf{A} \cdot \mathbf{A}}\dfrac{\mathbf{B}}{\sqrt{\mathbf{B} \cdot \mathbf{B}}}$

d) $(\pm)(\mathbf{A} \times \mathbf{B}) \times \mathbf{C}$ e) $(\pm)(\mathbf{B} \times \mathbf{C}) \times \mathbf{A}$

24. $\mathbf{i} \times \mathbf{j} = \mathbf{k}$ and $\mathbf{i} \times (\mathbf{i} + \mathbf{j}) = \mathbf{k}$, but $\mathbf{j} \neq \mathbf{i} + \mathbf{j}$

25. $\left|\overrightarrow{PQ} \times \mathbf{F}\right| = \left|\overrightarrow{PQ}\right| |\mathbf{F}| \sin(60°) = 10\sqrt{3}$ ft · lb

26. $\left|\overrightarrow{PQ} \times \mathbf{F}\right| = \left|\overrightarrow{PQ}\right| |\mathbf{F}| \sin(135°) = 120\sqrt{2}$ ft · lb

27. If $\mathbf{A} = a_1\mathbf{i} + a_2\mathbf{j} + a_3\mathbf{k}$, $\mathbf{B} = b_1\mathbf{i} + b_2\mathbf{j} + b_3\mathbf{k}$, and $\mathbf{C} = c_1\mathbf{i} + c_2\mathbf{j} + c_3\mathbf{k}$, then $\mathbf{A} \cdot (\mathbf{B} \times \mathbf{C}) = \begin{vmatrix} a_1 & a_2 & a_3 \\ b_1 & b_2 & b_3 \\ c_1 & c_2 & c_3 \end{vmatrix}$,

$\mathbf{B} \cdot (\mathbf{C} \times \mathbf{A}) = \begin{vmatrix} b_1 & b_2 & b_3 \\ c_1 & c_2 & c_3 \\ a_1 & a_2 & a_3 \end{vmatrix}$ and $\mathbf{C} \cdot (\mathbf{A} \times \mathbf{B}) = \begin{vmatrix} c_1 & c_2 & c_3 \\ a_1 & a_2 & a_3 \\ b_1 & b_2 & b_3 \end{vmatrix}$ which all have the same value, since the interchanging of two rows, in a determinant, does not change its value. The volume is $\left|(\mathbf{A} \times \mathbf{B}) \cdot \mathbf{C}\right| = \begin{vmatrix} 2 & 0 & 0 \\ 0 & 2 & 0 \\ 0 & 0 & 2 \end{vmatrix} = 8.$

28. $\left|(\mathbf{A} \times \mathbf{B}) \cdot \mathbf{C}\right| = \begin{vmatrix} 1 & -1 & 1 \\ 2 & 1 & -2 \\ -1 & 2 & -1 \end{vmatrix} = 5$. For details about verification, see exercise 27.

29. $\left|(\mathbf{A} \times \mathbf{B}) \cdot \mathbf{C}\right| = \begin{vmatrix} 2 & 1 & 0 \\ 2 & -1 & 1 \\ 1 & 0 & 2 \end{vmatrix} = 7$. For details about verification, see exercise 27.

30. $\left|(\mathbf{A} \times \mathbf{B}) \cdot \mathbf{C}\right| = \begin{vmatrix} 1 & 1 & -2 \\ -1 & 0 & -1 \\ 2 & 4 & -2 \end{vmatrix} = 6$. For details about verification, see exercise 27.

11.5 LINES AND PLANES IN SPACE

1. the direction $\mathbf{i} + \mathbf{j} + \mathbf{k}$ and $P(3,-4,-1) \Rightarrow x = 3 + t,\ y = -4 + t,\ z = -1 + t$
2. the direction $\overrightarrow{PQ} = -2\mathbf{i} - 2\mathbf{j} + 2\mathbf{k}$ and $P(1,2,-1) \Rightarrow x = 1 - 2t,\ y = 2 - 2t,\ z = -1 + 2t$
3. the direction $\overrightarrow{PQ} = 5\mathbf{i} + 5\mathbf{j} - 5\mathbf{k}$ and$P(-2,0,3) \Rightarrow x = -2 + 5t,\ y = 5t,\ z = 3 - 5t$
4. the direction $\overrightarrow{PQ} = -\mathbf{j} - \mathbf{k}$ and $P(1,2,0) \Rightarrow x = 1,\ y = 2 - t,\ z = -t$
5. the direction $2\mathbf{j} + \mathbf{k}$ and $P(0,0,0) \Rightarrow x = 0,\ y = 2t,\ z = t$
6. the direction is $2\mathbf{i} - \mathbf{j} + 3\mathbf{k}$ and $P(3,-2,1) \Rightarrow x = 3 + 2t,\ y = -2 - t,\ z = 1 + 3t$
7. the direction $\mathbf{k}$ and $P(1,1,1) \Rightarrow x = 1,\ y = 1,\ z = 1 + t$
8. the direction $3\mathbf{i} + 7\mathbf{j} - 5\mathbf{k}$ and $(2,4,5) \Rightarrow x = 2 + 3t,\ y = 4 + 7t,\ z = 5 - 5t$
9. the direction $\mathbf{i} + 2\mathbf{j} + 2\mathbf{k}$ and $(0,-7,0) \Rightarrow x = t,\ y = -7 + 2t,\ z = 2t$
10. the direction is $\mathbf{A} \times \mathbf{B} = \begin{vmatrix} \mathbf{i} & \mathbf{j} & \mathbf{k} \\ 1 & 2 & 3 \\ 3 & 4 & 5 \end{vmatrix} = -2\mathbf{i} + 4\mathbf{j} - 2\mathbf{k}$ and $P(2,3,0) \Rightarrow x = 2 - 2t,\ y = 3 + 4t,\ z = -2t$
11. the direction $\mathbf{i}$ and $P(0,0,0) \Rightarrow x = t,\ y = 0,\ z = 0$
12. the direction $\mathbf{k}$ and $P(0,0,0) \Rightarrow x = 0,\ y = 0,\ z = t$
13. the direction $\overrightarrow{PQ} = \mathbf{i} + \mathbf{j} + \mathbf{k}$ and $P(0,0,0) \Rightarrow$
    $x = t,\ y = t,\ z = t$, where $0 \le t \le 1$

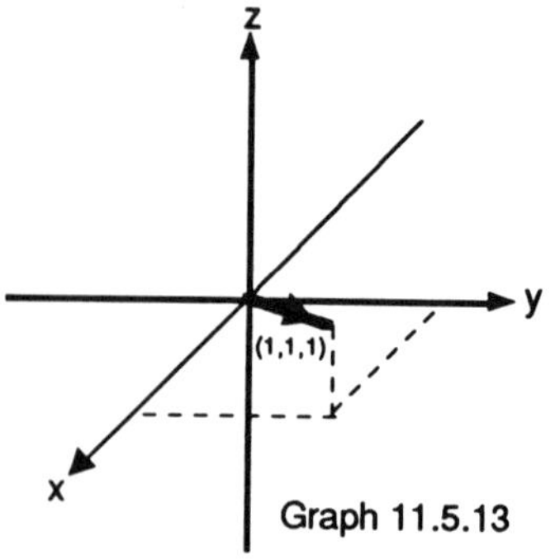

Graph 11.5.13

14. the direction $\overrightarrow{PQ} = \mathbf{i}$ and P(0,0,0) ⇒
$x = t$, $y = 0$, $z = 0$, where $0 \le t \le 1$

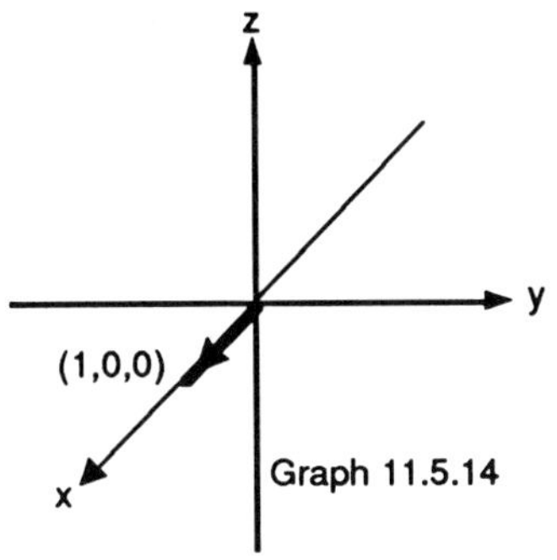

Graph 11.5.14

15. the direction $\overrightarrow{PQ} = \mathbf{j}$ and P(1,0,0) ⇒
$x = 1$, $y = 1 + t$, $z = 0$, where $-1 \le t \le 0$

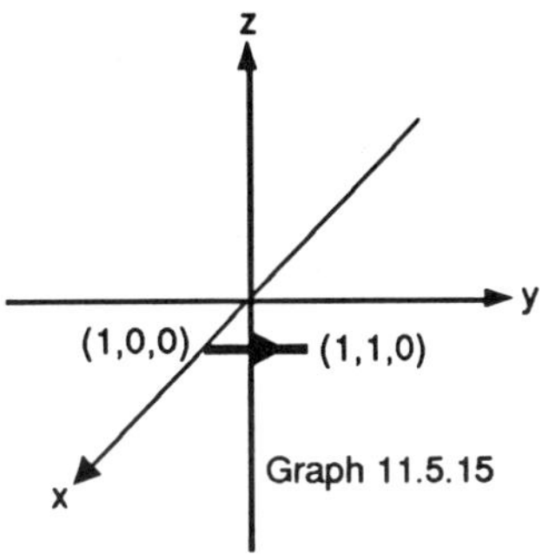

Graph 11.5.15

16. the direction $\overrightarrow{PQ} = \mathbf{k}$ and P(1,1,1) ⇒
$x = 1$, $y = 1$, $z = t$, where $0 \le t \le 1$

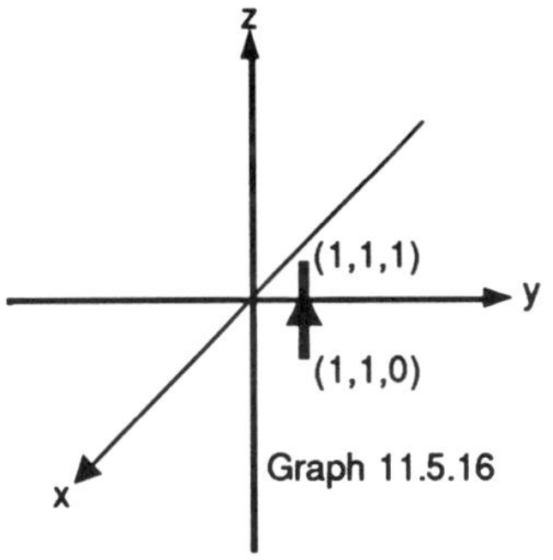

Graph 11.5.16

17. the direction $\overrightarrow{PQ} = 2\mathbf{j}$ and P(0,– 1,1) ⇒
$x = 0$, $y = -1 + 2t$, $z = 1$, where $0 \le t \le 1$

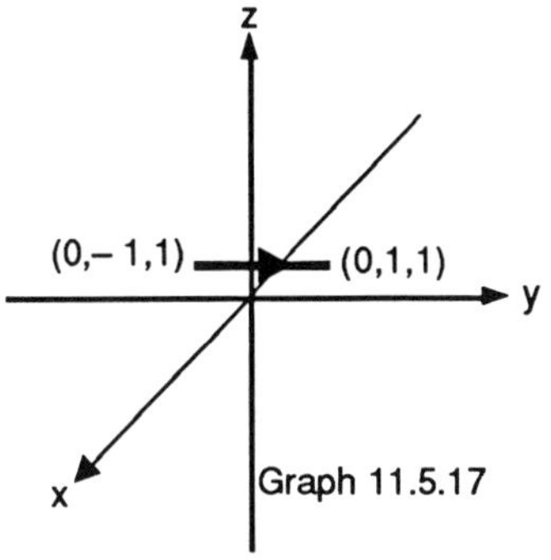

Graph 11.5.17

18. the direction $\overrightarrow{PQ} = -3\mathbf{i} + 2\mathbf{j}$ and $P(3,0,0) \Rightarrow$
$x = 3 - 3t,\ y = 2t,\ z = 0,\ 0 \le t \le 1$

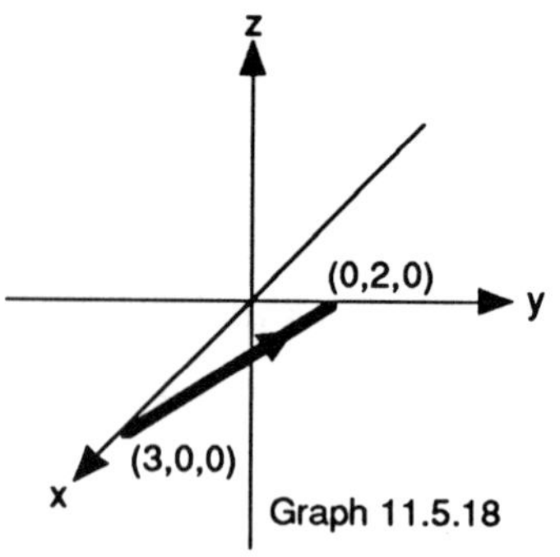

Graph 11.5.18

19. the direction $\overrightarrow{PQ} = -\mathbf{i} - 2\mathbf{k}$ and $P(2,2,0) \Rightarrow$
$x = 2 - t,\ y = 2,\ z = -2t$, where $0 \le t \le 1$

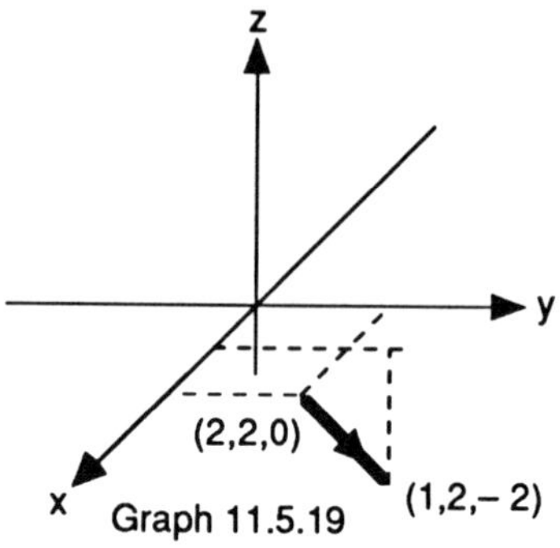

Graph 11.5.19

20. the direction $\overrightarrow{PQ} = -\mathbf{i} + 3\mathbf{j} + 3\mathbf{k}$ and $P(1, -1, -2) \Rightarrow$
$x = 1 - t,\ y = -1 + 3t,\ z = -2 + 3t$, where $0 \le t \le 1$

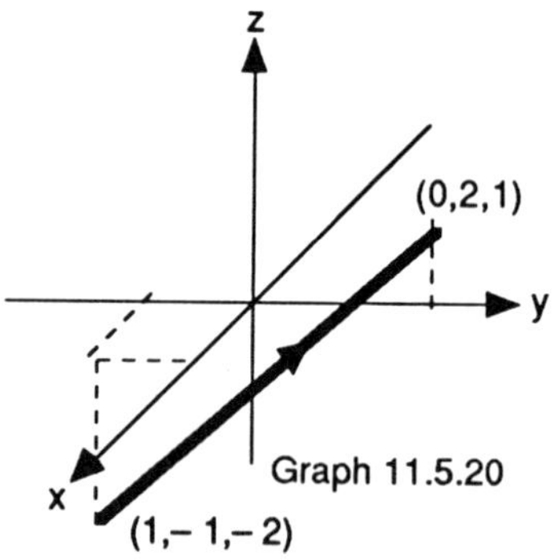

Graph 11.5.20

21. $3(x) + (-2)(y-2) + (-1)(z+1) = 0 \Rightarrow 3x - 2y - z = -3$

22. $(x-1)(3) + (y+1)(1) + (z-3)(1) = 0 \Rightarrow 3x + y + z = 5$

23. $\overrightarrow{PQ} = \mathbf{i} - \mathbf{j} + 3\mathbf{k},\ \overrightarrow{PS} = -\mathbf{i} - 3\mathbf{j} + 2\mathbf{k} \Rightarrow \overrightarrow{PQ} \times \overrightarrow{PS} = \begin{vmatrix} \mathbf{i} & \mathbf{j} & \mathbf{k} \\ 1 & -1 & 3 \\ -1 & -3 & 2 \end{vmatrix} = 7\mathbf{i} - 5\mathbf{j} - 4\mathbf{k}$, the normal;

$(x-2)(7) + (y-0)(-5) + (z-2)(-4) = 0 \Rightarrow 7x - 5y - 4z + 6$

24. $\overrightarrow{PQ} = -\mathbf{i} + \mathbf{j} + 2\mathbf{k},\ \overrightarrow{PS} = -3\mathbf{i} + 2\mathbf{j} + 3\mathbf{k} \Rightarrow \overrightarrow{PQ} \times \overrightarrow{PS} = \begin{vmatrix} \mathbf{i} & \mathbf{j} & \mathbf{k} \\ -1 & 1 & 2 \\ -3 & 2 & 3 \end{vmatrix} = -\mathbf{i} - 3\mathbf{j} + \mathbf{k}$, the normal;

$(x-1)(-1) + (y-5)(-3) + (z-7)(1) = 0 \Rightarrow x + 3y - z = 9$

25. $\mathbf{N} = \mathbf{i} + 3\mathbf{j} + 4\mathbf{k}$, $P(2,4,5) \Rightarrow (x-2)(1) + (y-4)(3) + (z-5)(4) = 0 \Rightarrow x + 3y + 4z = 34$

26. $\mathbf{N} = \mathbf{i} - 2\mathbf{j} + \mathbf{k}$, $P(1,-2,1) \Rightarrow (x-1)(1) + (y+2)(-2) + (z-1)(1) = 0 \Rightarrow x - 2y + z = 6$

27. The distance between $(4t,-2t,2t)$ and $(0,0,12)$ is $d = \sqrt{(4t)^2 + (-2t)^2 + (2t-12)^2}$. If $f(t) = (4t)^2 + (-2t)^2 + (2t-12)^2$ is minimized, then d is minimized. $f'(t) = 0 \Rightarrow t = 1 \Rightarrow$ $d = \sqrt{16 + 4 + 100} = 2\sqrt{30}$

28. The distance between $(5+3t, 5+4t, -3-5t)$ and $(0,0,0)$ is $d = \sqrt{(5+3t)^2 + (5+4t)^2 + (-3-5t)^2}$. If $f(t) = (5+3t)^2 + (5+4t)^2 + (-3-5t)^2$ is minimized, then d is minimized. $f'(t) = 0 \Rightarrow t = -1 \Rightarrow$ $d = \sqrt{2^2 + 1^2 + 2^2} = 3$

29. The distance between $(2+2t, 1+6t, 3)$ and $(2,1,3)$ is $d = \sqrt{(2t)^2 + (6t)^2}$. If $f(t) = (2t)^2 + (6t)^2$ is minimized, then d is minimized. $f'(t) = 0 \Rightarrow t = 0 \Rightarrow d = 0$

30. The distance between $(2t, 1+2t, 2t)$ and $(2,1,-1)$ is $d = \sqrt{(2t-2)^2 + (2t)^2 + (2t+1)^2}$. If $f(t) = (2t-2)^2 + (2t)^2 + (2t+1)^2$ is minimized, then d is minimized. $f'(t) = 0 \Rightarrow t = \frac{1}{6} \Rightarrow d = \sqrt{\frac{14}{3}}$

31. $S(2,-3,4)$, $x + 2y + 2z = 13$ and $P(13,0,0)$ is on the plane $\Rightarrow \overrightarrow{PS} = -11\mathbf{i} - 3\mathbf{j} + 4\mathbf{k}$, $\mathbf{N} = \mathbf{i} + 2\mathbf{j} + 2\mathbf{k}$;
$$d = \left|\overrightarrow{PS} \cdot \frac{\mathbf{N}}{|\mathbf{N}|}\right| = \left|\frac{-11-6+8}{\sqrt{1+4+4}}\right| = 3$$

32. $S(0,0,0)$, $3x + 2y + 6z = 6$ and $P(2,0,0)$ is on the plane $\Rightarrow \overrightarrow{PS} = 2\mathbf{i}$, $\mathbf{N} = 3\mathbf{i} + 2\mathbf{j} + 6\mathbf{k}$;
$$d = \left|\overrightarrow{PS} \cdot \frac{\mathbf{N}}{|\mathbf{N}|}\right| = \left|\frac{6}{\sqrt{9+4+36}}\right| = \frac{6}{7}$$

33. $S(0,1,1)$, $4y + 3z = -12$ and $P(0,-3,0)$ is on the plane $\Rightarrow \overrightarrow{PS} = 4\mathbf{j} + \mathbf{k}$, $\mathbf{N} = 4\mathbf{j} + 3\mathbf{k}$;
$$d = \left|\overrightarrow{PS} \cdot \frac{\mathbf{N}}{|\mathbf{N}|}\right| = \left|\frac{16+3}{\sqrt{16+9}}\right| = \frac{19}{5}$$

34. $S(2,2,3)$, $2x + y + 2z = 4$ and $P(2,0,0)$ is on the plane $\Rightarrow \overrightarrow{PS} = 2\mathbf{j} + 3\mathbf{k}$, $\mathbf{N} = 2\mathbf{i} + \mathbf{j} + 2\mathbf{k}$;
$$d = \left|\overrightarrow{PS} \cdot \frac{\mathbf{N}}{|\mathbf{N}|}\right| = \frac{2+6}{\sqrt{4+1+4}} = \frac{8}{3}$$

35. $S(0,-1,0)$, $2x + y + 2z = 4$ and $P(2,0,0)$ is on the plane $\Rightarrow \overrightarrow{PS} = -2\mathbf{i} - \mathbf{j}$, $\mathbf{N} = 2\mathbf{i} + \mathbf{j} + 2\mathbf{k}$;
$$d = \left|\overrightarrow{PS} \cdot \frac{\mathbf{N}}{|\mathbf{N}|}\right| = \left|\frac{-4-1+0}{\sqrt{4+1+4}}\right| = \frac{5}{3}$$

36. $S(1,0,-1)$, $-4 + y + z = 4$ and $P(-1,0,0)$ is on the plane $\Rightarrow \overrightarrow{PS} = 2\mathbf{i} - \mathbf{k}$, $\mathbf{N} = -4\mathbf{i} + \mathbf{j} + \mathbf{k}$;
$$d = \left|\overrightarrow{PS} \cdot \frac{\mathbf{N}}{|\mathbf{N}|}\right| = \left|\frac{-8-1}{\sqrt{16+1+1}}\right| = \frac{9}{\sqrt{18}} = \frac{3\sqrt{2}}{2}$$

37. $2(1-t) - (3t) + 3(1+t) = 6 \Rightarrow t = -\ 1/2 \Rightarrow (3/2, -3/2, 1/2)$

38. $6(2) + 3(3+2t) - 4(-2-2t) = -12 \Rightarrow t = -41/14 \Rightarrow (2, -20/7, 27/7)$

39. $1(1+2t)+1(1+5t)+1(3t)=2 \Rightarrow t=0 \Rightarrow (1,1,0)$

40. $2(-1+3t)+0(-2)-3(5t)=7 \Rightarrow t=-1 \Rightarrow (-4,-2,-5)$

41. $\mathbf{N}_1=\mathbf{i}+\mathbf{j},\ \mathbf{N}_2=2\mathbf{i}+\mathbf{j}-2\mathbf{k} \Rightarrow \theta=\cos^{-1}\left(\frac{\mathbf{N}_1\cdot\mathbf{N}_2}{|\mathbf{N}_1|\,|\mathbf{N}_2|}\right)=\cos^{-1}\left(\frac{2+1}{\sqrt{2}\sqrt{9}}\right)=\cos^{-1}\frac{1}{\sqrt{2}}=\frac{\pi}{4}$

42. $\mathbf{N}_1=5\mathbf{i}+\mathbf{j}-\mathbf{k},\ \mathbf{N}_2=\mathbf{i}-2\mathbf{j}+3\mathbf{k} \Rightarrow \theta=\cos^{-1}\left(\frac{\mathbf{N}_1\cdot\mathbf{N}_2}{|\mathbf{N}_1|\,|\mathbf{N}_2|}\right)=\cos^{-1}\left(\frac{5-2-3}{\sqrt{26}\sqrt{14}}\right)=\cos^{-1}0=\frac{\pi}{2}$

43. $\mathbf{N}_1=2\mathbf{i}+2\mathbf{j}+2\mathbf{k},\ \mathbf{N}_2=2\mathbf{i}-2\mathbf{j}-\mathbf{k} \Rightarrow \theta=\cos^{-1}\left(\frac{\mathbf{N}_1\cdot\mathbf{N}_2}{|\mathbf{N}_1|\,|\mathbf{N}_2|}\right)=\cos^{-1}\left(\frac{4-4-2}{\sqrt{12}\sqrt{9}}\right)=$
$\cos^{-1}\left(\frac{-1}{3\sqrt{3}}\right)\approx 101.1°$

44. $\mathbf{N}_1=\mathbf{i}+\mathbf{j}+\mathbf{k},\ \mathbf{N}_2=\mathbf{k} \Rightarrow \theta=\cos^{-1}\left(\frac{\mathbf{N}_1\cdot\mathbf{N}_2}{|\mathbf{N}_1|\,|\mathbf{N}_2|}\right)=\cos^{-1}\left(\frac{1}{\sqrt{3}\sqrt{1}}\right)\approx 54.7°$

45. $\mathbf{N}_1=2\mathbf{i}+2\mathbf{j}-\mathbf{k},\ \mathbf{N}_2=\mathbf{i}+2\mathbf{j}+\mathbf{k} \Rightarrow \theta=\cos^{-1}\left(\frac{\mathbf{N}_1\cdot\mathbf{N}_2}{|\mathbf{N}_1|\,|\mathbf{N}_2|}\right)=\cos^{-1}\left(\frac{2+4-1}{\sqrt{9}\sqrt{6}}\right)=$
$\cos^{-1}\left(\frac{5}{3\sqrt{6}}\right)\approx 47.1°$

46. $\mathbf{N}_1=4\mathbf{j}+3\mathbf{k},\ \mathbf{N}_2=3\mathbf{i}+2\mathbf{j}+6\mathbf{k} \Rightarrow \theta=\cos^{-1}\left(\frac{\mathbf{N}_1\cdot\mathbf{N}_2}{|\mathbf{N}_1|\,|\mathbf{N}_2|}\right)=\cos^{-1}\left(\frac{12+18}{\sqrt{25}\sqrt{49}}\right)=$
$\cos^{-1}\left(\frac{6}{7}\right)\approx 31°$

47. $\mathbf{N}_1=\mathbf{i}+\mathbf{j}+\mathbf{k},\ \mathbf{N}_2=\mathbf{i}+\mathbf{j} \Rightarrow \mathbf{N}_1\times\mathbf{N}_2=\begin{vmatrix}\mathbf{i}&\mathbf{j}&\mathbf{k}\\1&1&1\\1&1&0\end{vmatrix}=-\mathbf{i}+\mathbf{j}$, the direction of the desired line; $(1,1,-1)$ is on both planes; the desired line is $x=1-t,\ y=1+t,\ z=-1$

48. $\mathbf{N}_1=3\mathbf{i}-6\mathbf{j}-2\mathbf{k},\ \mathbf{N}_2=2\mathbf{i}+\mathbf{j}-2\mathbf{k} \Rightarrow \mathbf{N}_1\times\mathbf{N}_2=\begin{vmatrix}\mathbf{i}&\mathbf{j}&\mathbf{k}\\3&-6&-2\\2&1&-2\end{vmatrix}=14\mathbf{i}+2\mathbf{j}+15\mathbf{k}$, the direction of the desired line; $(1,0,0)$ is on both planes; the desired line is $x=1+14t,\ y=2t,\ z=15t$

49. $\mathbf{N}_1=\mathbf{i}-2\mathbf{j}+4\mathbf{k},\ \mathbf{N}_2=\mathbf{i}+\mathbf{j}-2\mathbf{k} \Rightarrow \mathbf{N}_1\times\mathbf{N}_2=\begin{vmatrix}\mathbf{i}&\mathbf{j}&\mathbf{k}\\1&-2&4\\1&1&-2\end{vmatrix}=6\mathbf{j}+3\mathbf{k}$, the direction of the desired line; $(4,3,1)$ is on both planes; the desired line is $x=4,\ y=3+6t,\ z=1+3t$

50. $\mathbf{N}_1=5\mathbf{i}-2\mathbf{j},\ \mathbf{N}_2=4\mathbf{j}-5\mathbf{k} \Rightarrow \mathbf{N}_1\times\mathbf{N}_2=\begin{vmatrix}\mathbf{i}&\mathbf{j}&\mathbf{k}\\5&-2&0\\0&4&-5\end{vmatrix}=10\mathbf{i}+25\mathbf{j}+20\mathbf{k}$, the direction of the desired line; $(1,-3,1)$ is on both planes; the desired line is $x=1+10t,\ y=-3+25t,\ z=1+20t$

11.6 SURFACES IN SPACE

1. $x^2 + y^2 = 4$

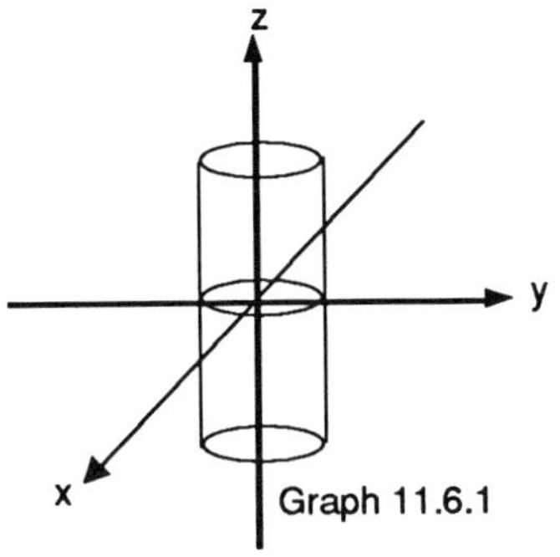

Graph 11.6.1

2. $x^2 + z^2 = 4$

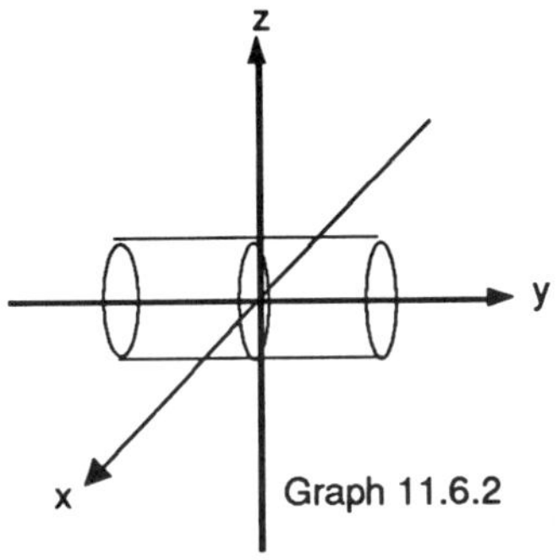

Graph 11.6.2

3. $y^2 + z^2 = 1$

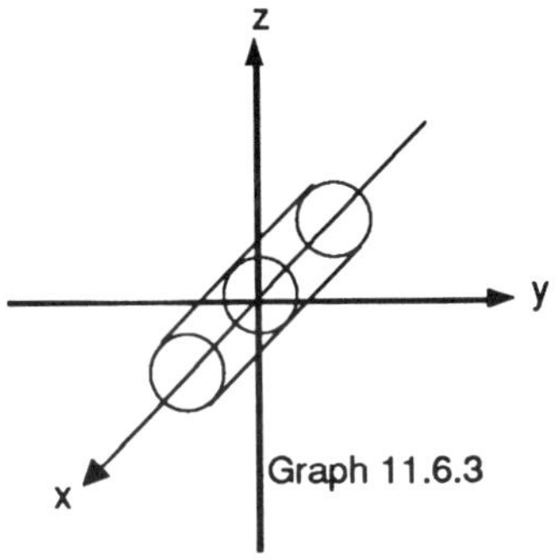

Graph 11.6.3

4. $z = y^2/4$

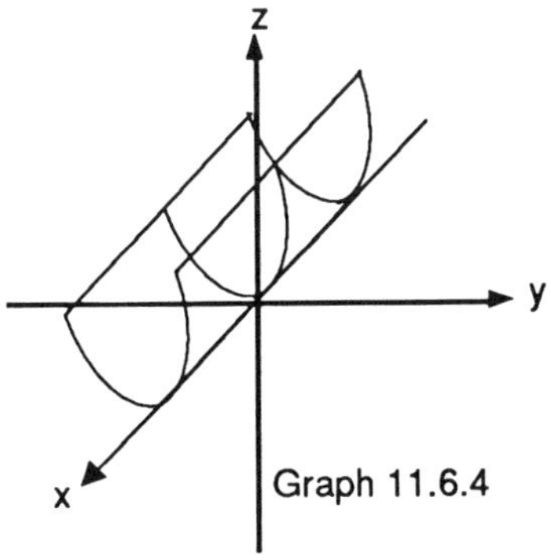

Graph 11.6.4

5. $z = y^2 - 1$

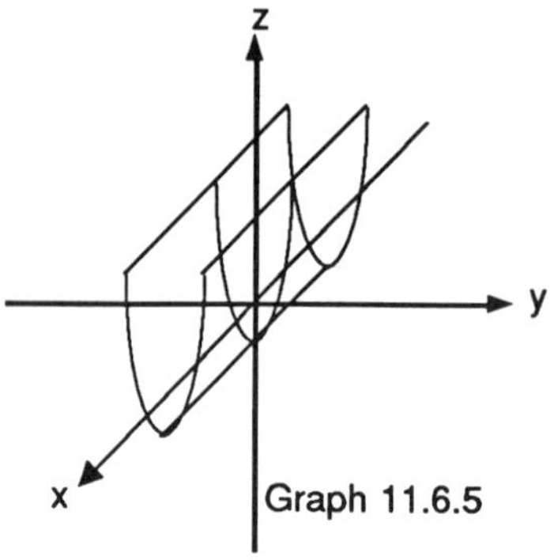

6. $x = y^2$

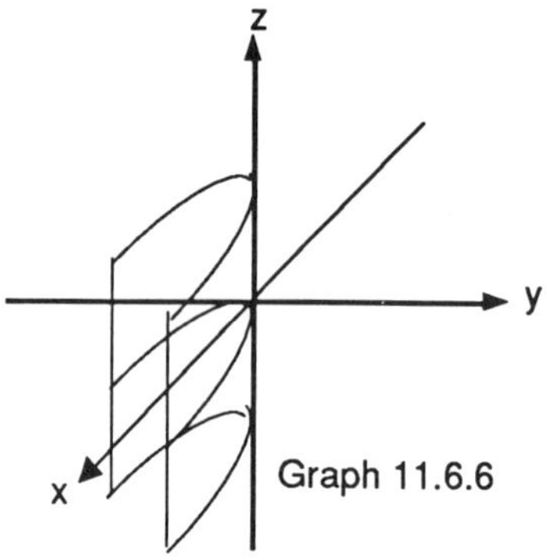

7. $z = 4 - x^2$

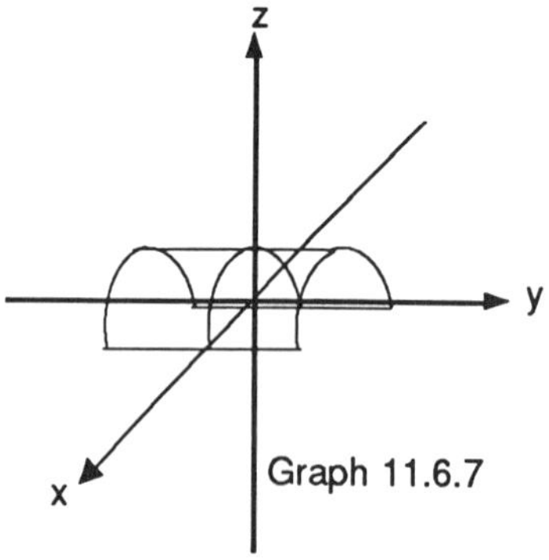

8. $x = 4 - y^2$

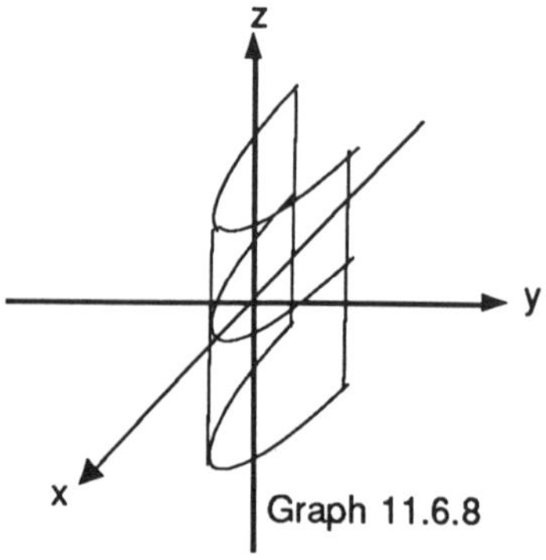

9. $y = x^2$

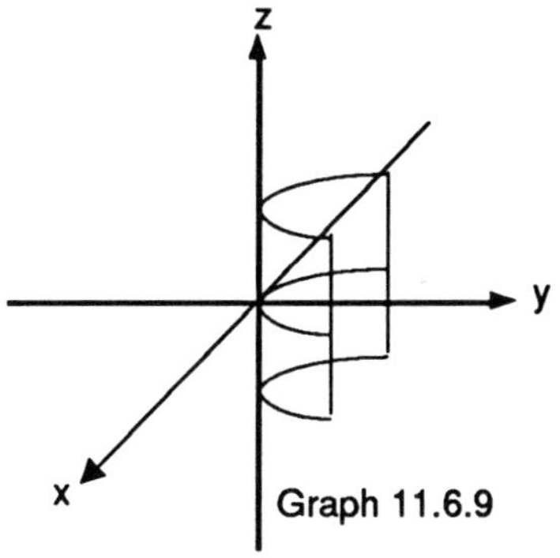

Graph 11.6.9

10. $y = x^2 - 2$

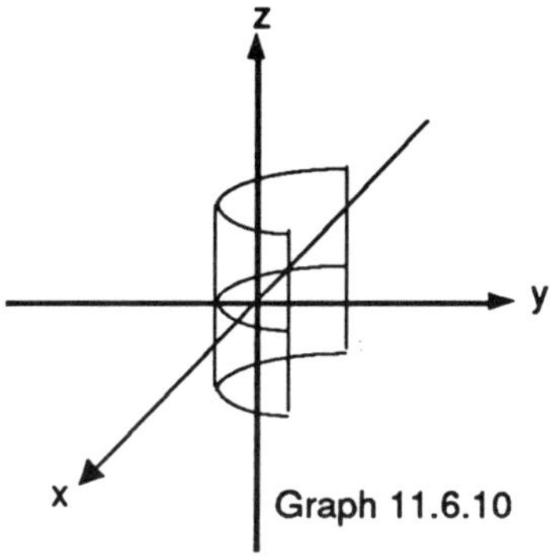

Graph 11.6.10

11. $y^2 + 4z^2 = 16$

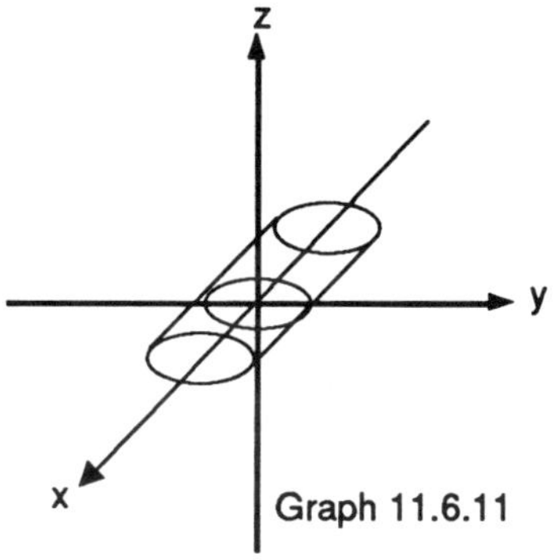

Graph 11.6.11

12. $4x^2 + y^2 = 36$

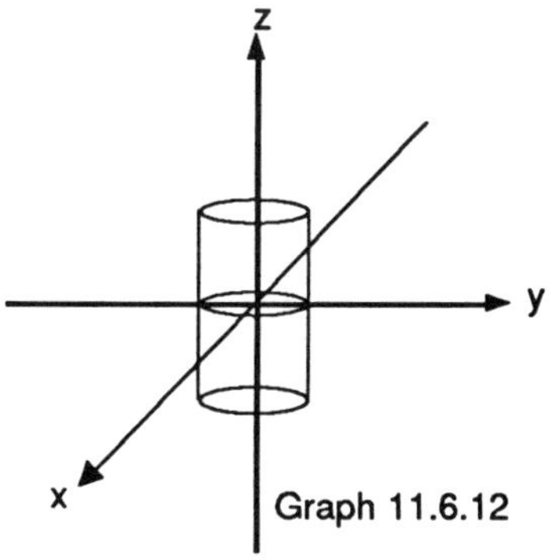

Graph 11.6.12

13. $z^2 + 4y^2 = 9$

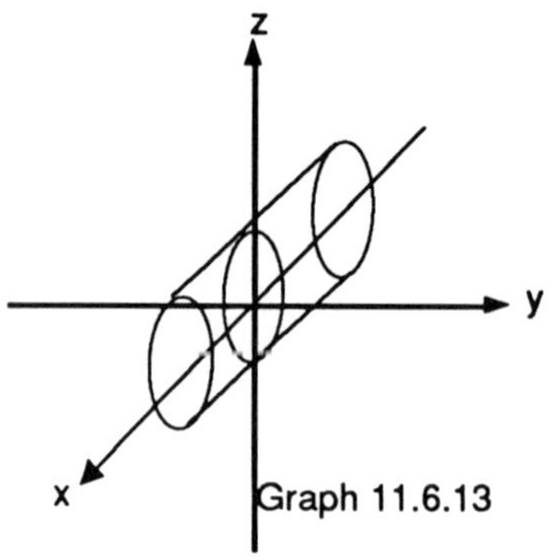

Graph 11.6.13

14. $y^2 - z^2 = 4$

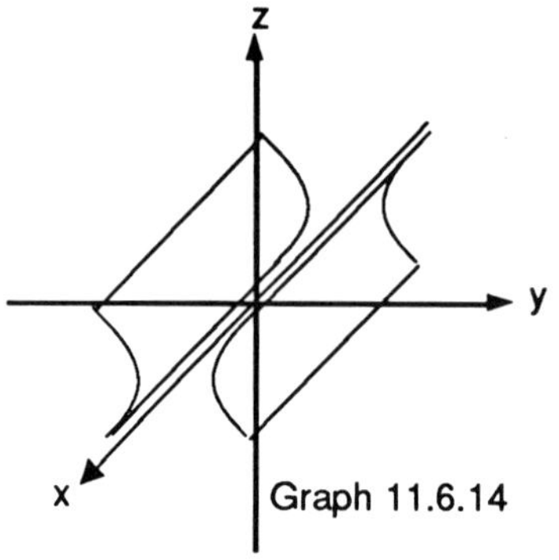

Graph 11.6.14

15. $z^2 - y^2 = 1$

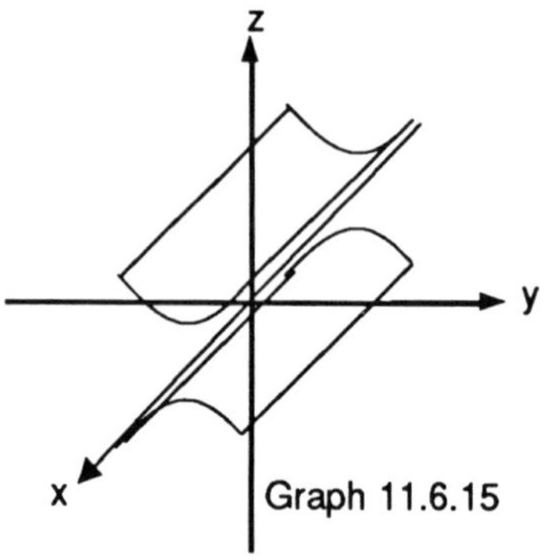

Graph 11.6.15

16. $yz = 1$

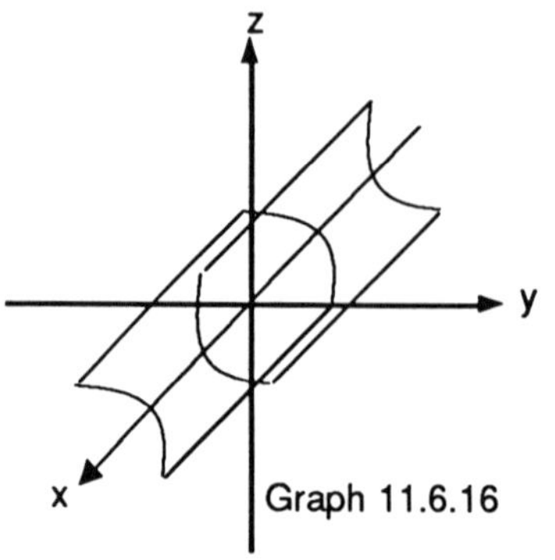

Graph 11.6.16

17. $9x^2 + y^2 + z^2 = 9$

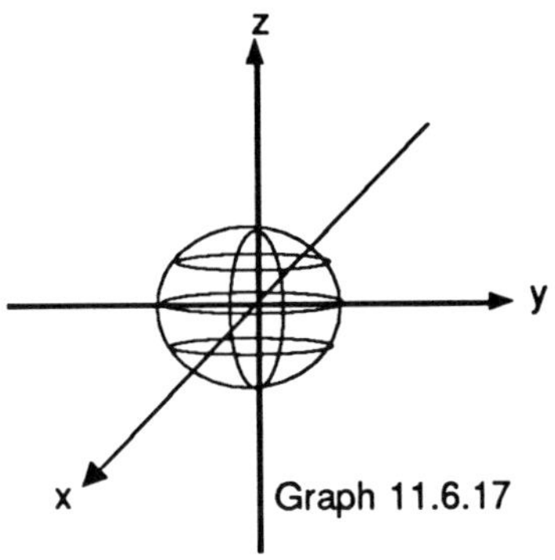

Graph 11.6.17

18. $4x^2 + 4y^2 + z^2 = 16$

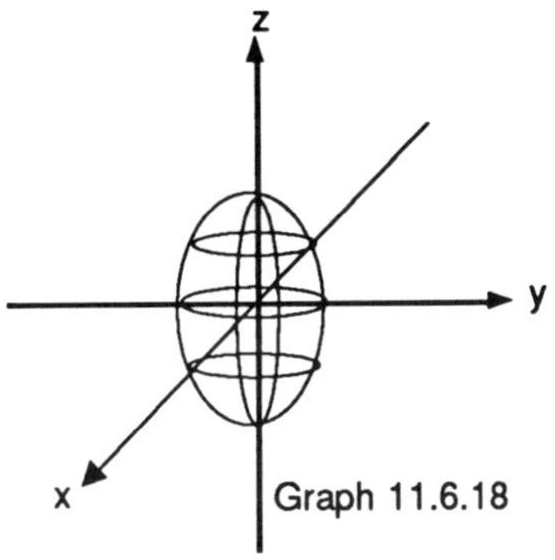

Graph 11.6.18

19. $x^2 + y^2 + z^2 = 4$

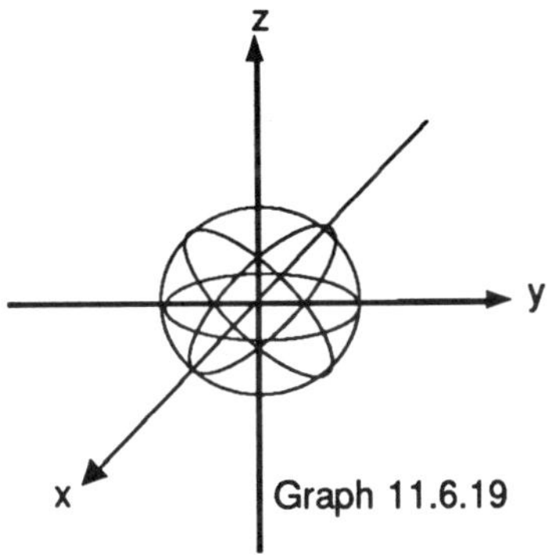

Graph 11.6.19

20. $9x^2 + 4y^2 + z^2 = 36$

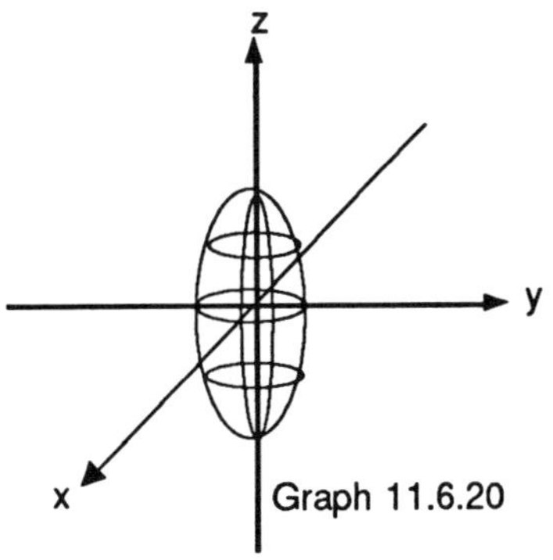

Graph 11.6.20

21. $4x^2 + 9y^2 + 4z^2 = 36$

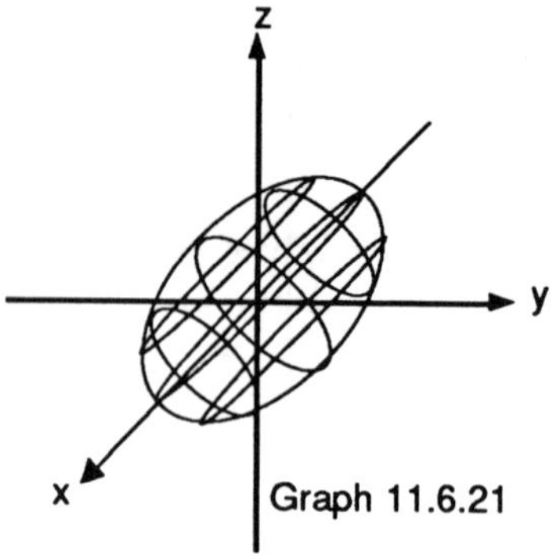

Graph 11.6.21

22. $9x^2 + 4y^2 + 36z^2 = 36$

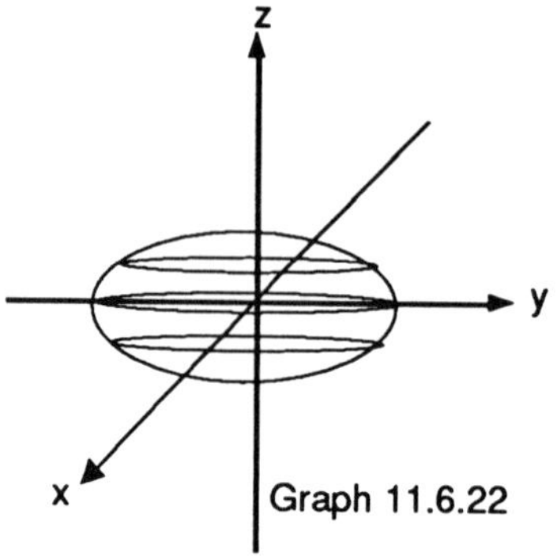

Graph 11.6.22

23. $x^2 + y^2 = z$

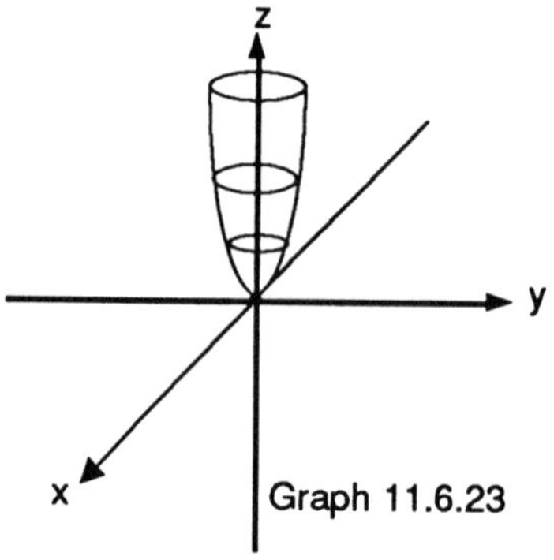

Graph 11.6.23

24. $x^2 + z^2 = y$

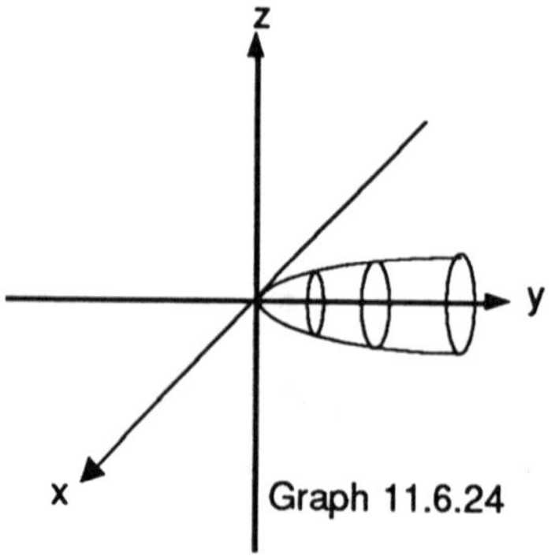

Graph 11.6.24

25. $x^2 + 4y^2 = z$

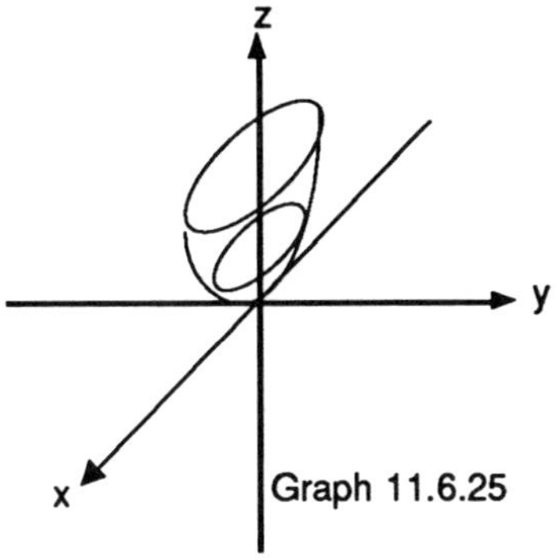

Graph 11.6.25

26. $z = x^2 + 9y^2$

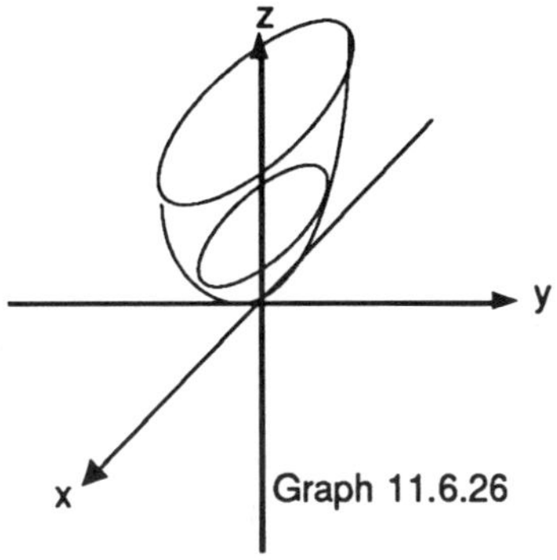

Graph 11.6.26

27. $z = 8 - x^2 - y^2$

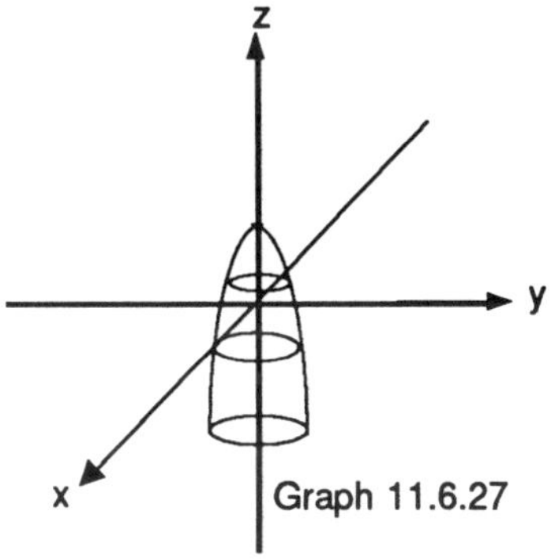

Graph 11.6.27

28. $z = 18 - x^2 - 9y^2$

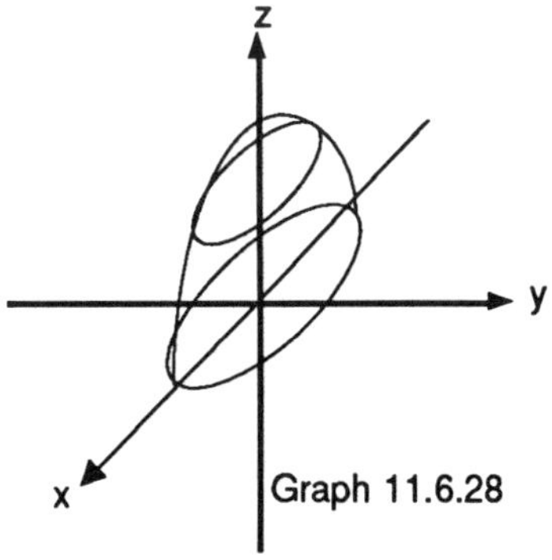

Graph 11.6.28

29. $x = 4 - 4y^2 - z^2$

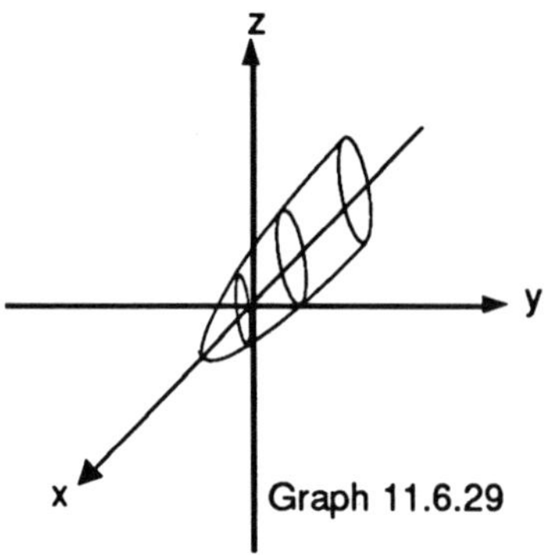

Graph 11.6.29

30. $y = 1 - x^2 - z^2$

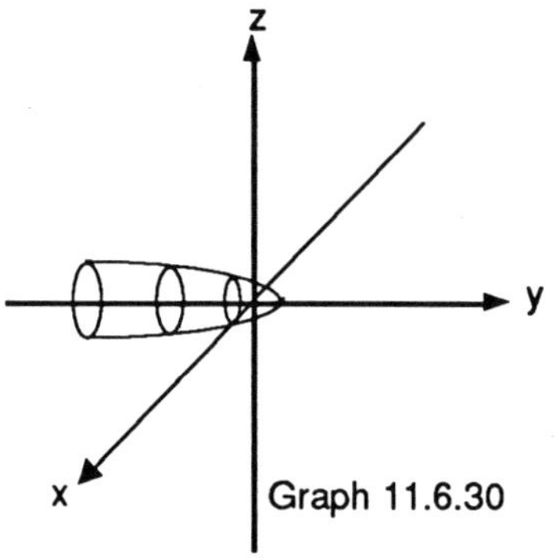

Graph 11.6.30

31. $z = x^2 + y^2 + 1$

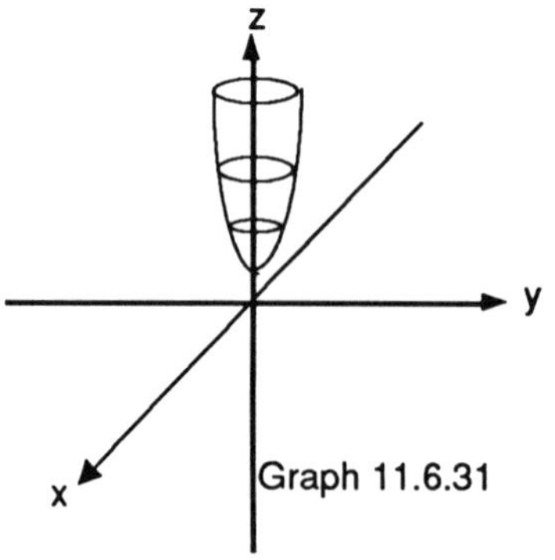

Graph 11.6.31

32. $z = 4x^2 + y^2 - 4$

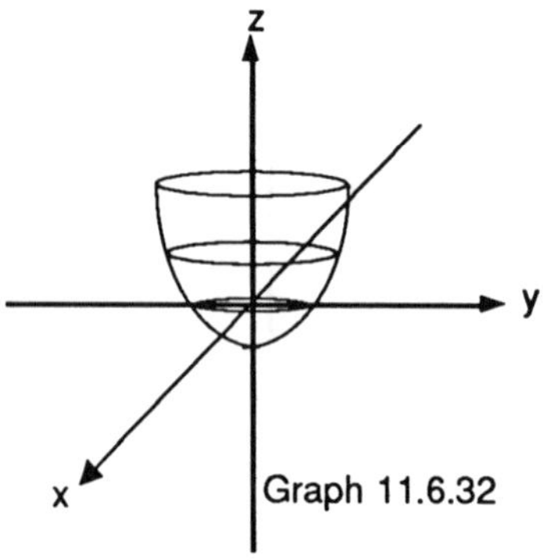

Graph 11.6.32

33. $x^2 + y^2 = z^2$

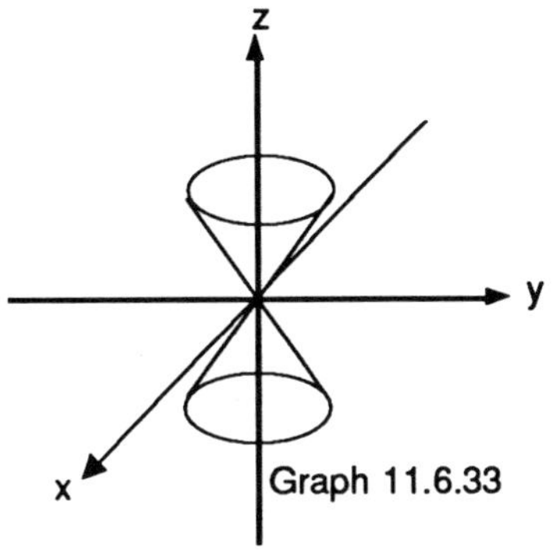

Graph 11.6.33

34. $y^2 + z^2 = x^2$

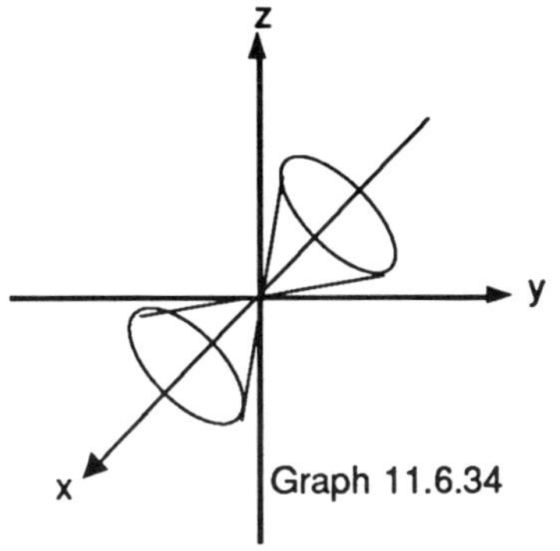

Graph 11.6.34

35. $x^2 + z^2 = y^2$

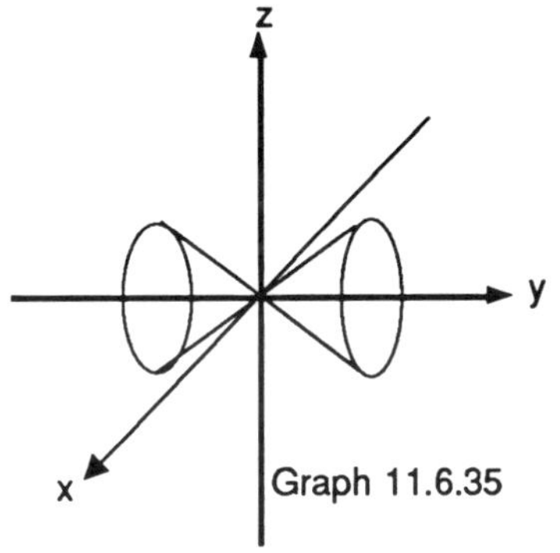

Graph 11.6.35

36. $4x^2 + 9y^2 = z^2$

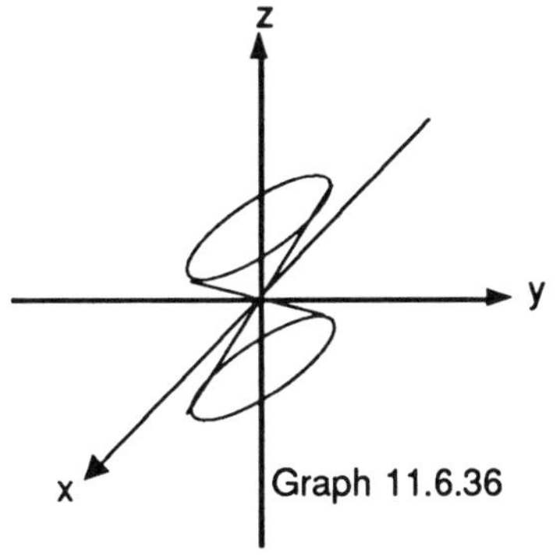

Graph 11.6.36

37. $9x^2 + 4y^2 = 36z^2$

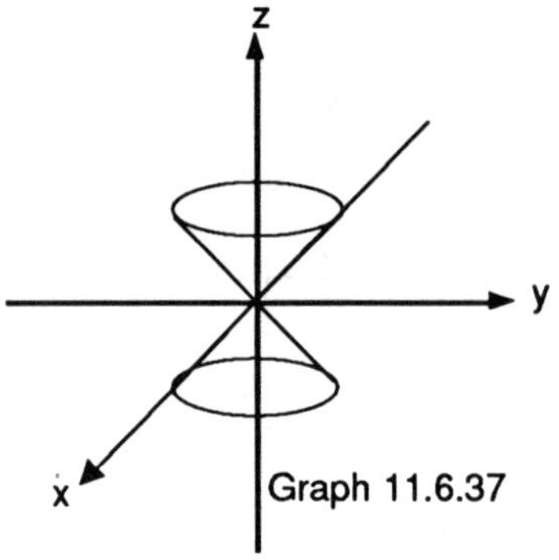

Graph 11.6.37

38. $4x^2 + 9z^2 = 9y^2$

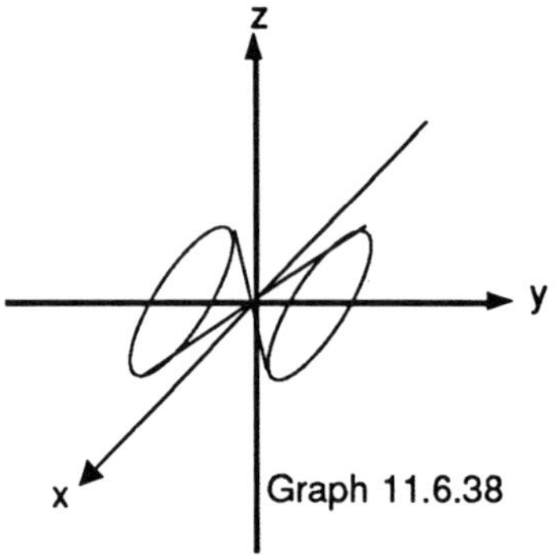

Graph 11.6.38

39. $x^2 + y^2 - z^2 = 1$

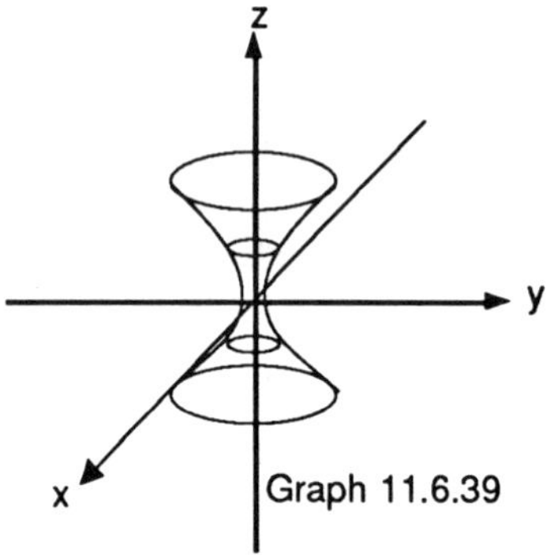

Graph 11.6.39

40. $y^2 + z^2 - x^2 = 1$

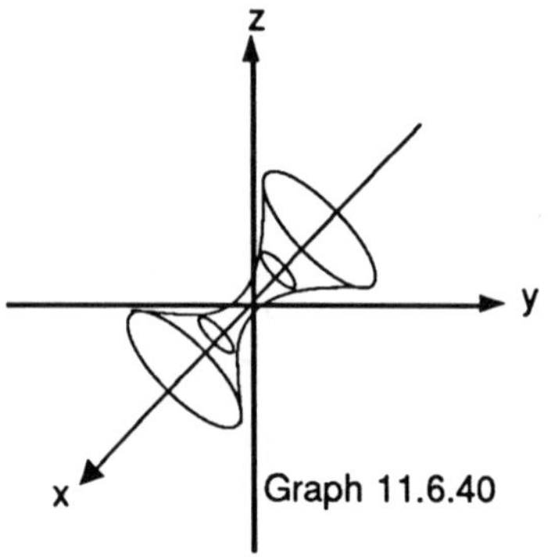

Graph 11.6.40

41. $\left(y^2/4\right) + \left(z^2/9\right) - \left(x^2/4\right) = 1$

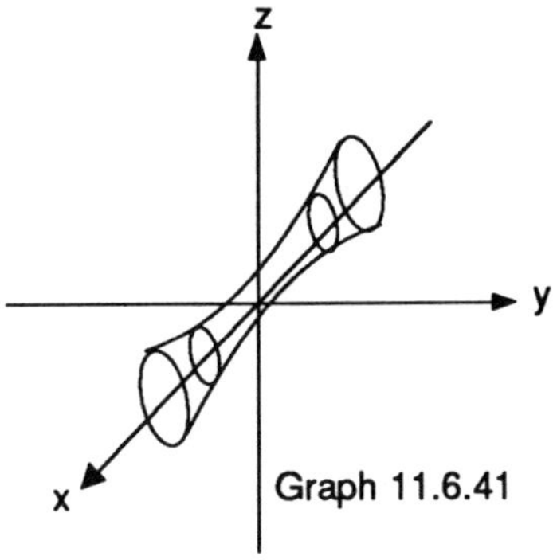

Graph 11.6.41

42. $\left(x^2/4\right) + \left(y^2/4\right) - \left(z^2/9\right) = 1$

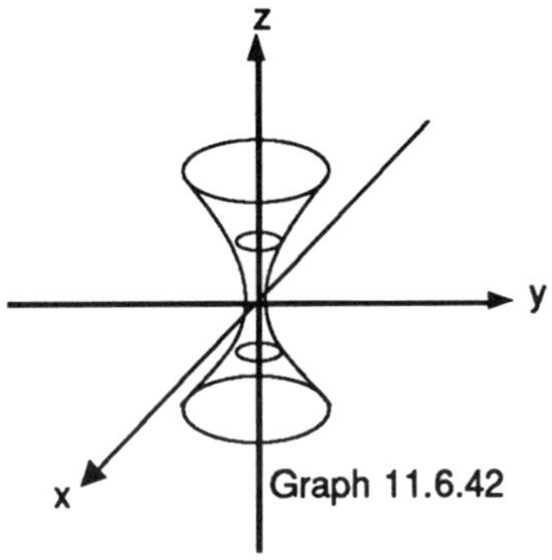

Graph 11.6.42

43. $\left(x^2/4\right) + y^2 - z^2 = 1$

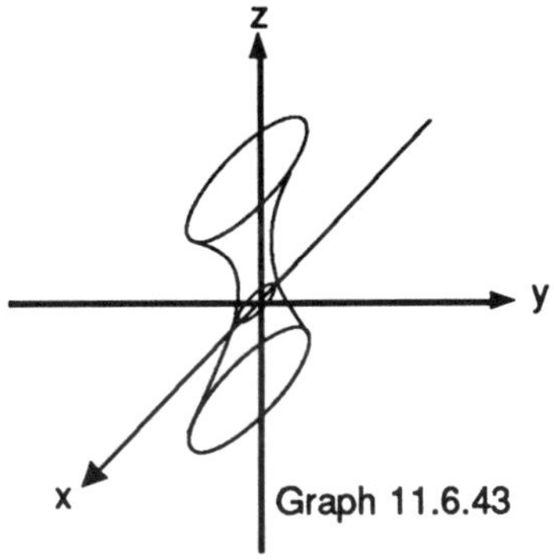

Graph 11.6.43

44. $z^2 - x^2 - y^2 = 1$

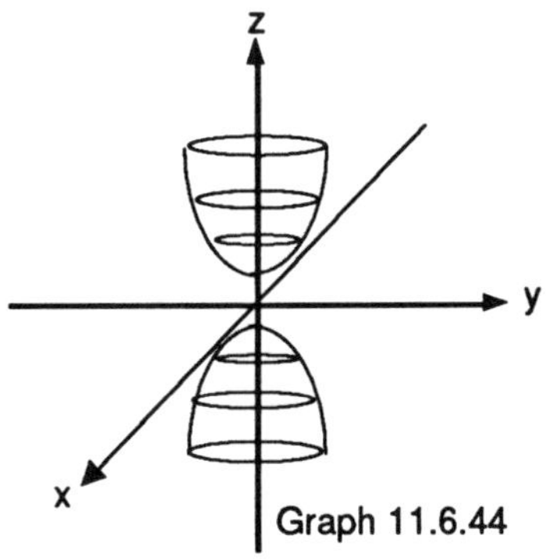

Graph 11.6.44

45. $z^2 - (x^2/4) - y^2 = 1$

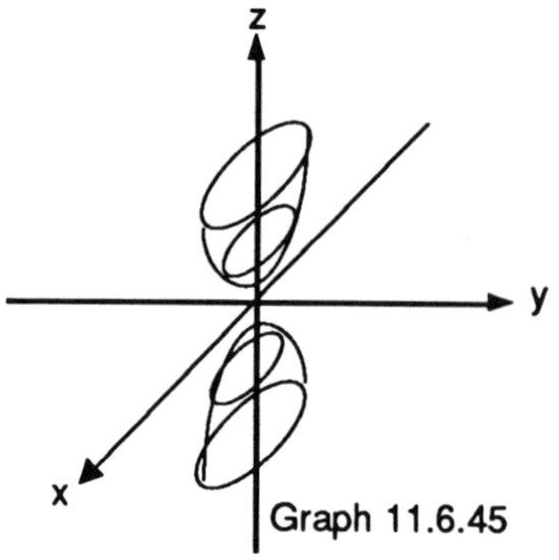

Graph 11.6.45

46. $(y^2/4) - (x^2/4) - z^2 = 1$

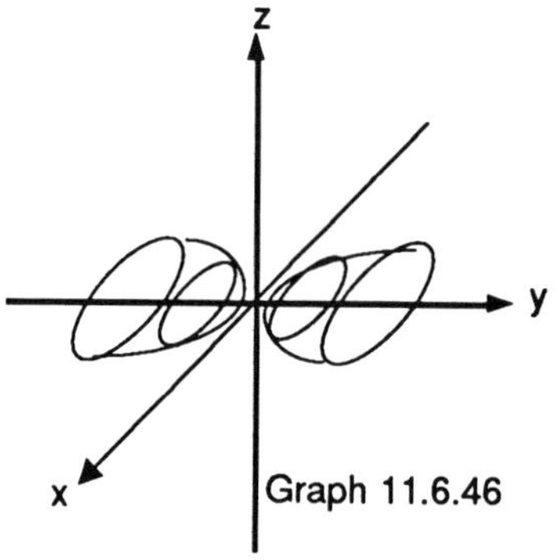

Graph 11.6.46

47. $x^2 - y^2 - (z^2/4) = 1$

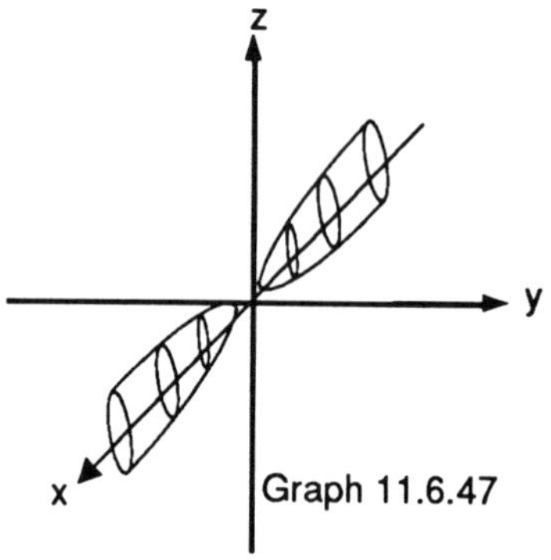

Graph 11.6.47

48. $(x^2/4) - (z^2/4) - y^2 = 1$

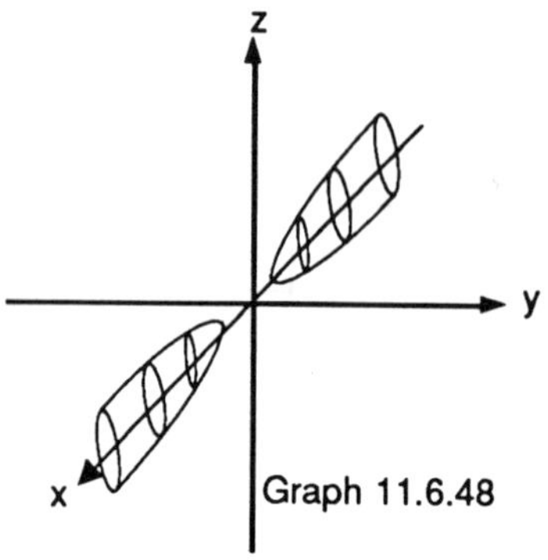

Graph 11.6.48

49. $y^2 - x^2 = z$

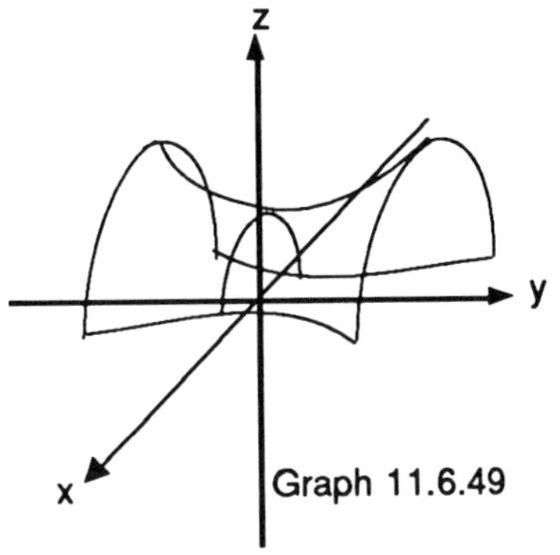

Graph 11.6.49

50. $x^2 - y^2 = z$

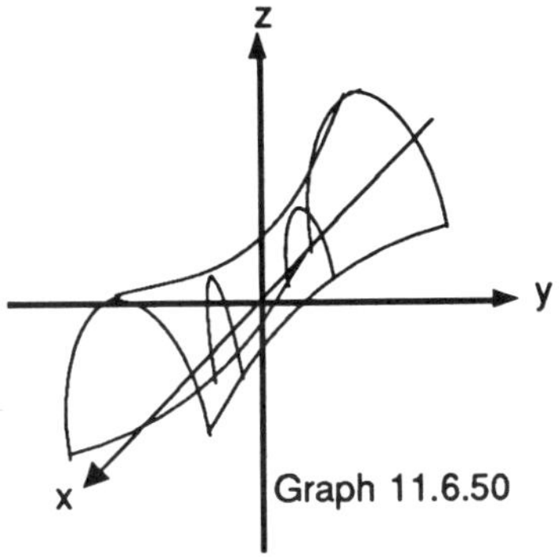

Graph 11.6.50

51. $z = y^2$

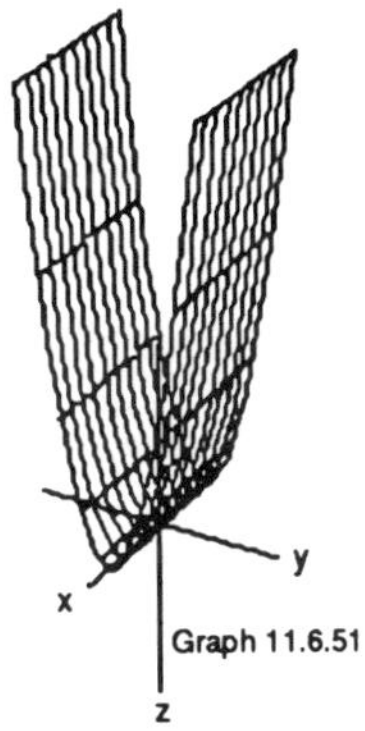

Graph 11.6.51

52. $z = 1 - y^2$

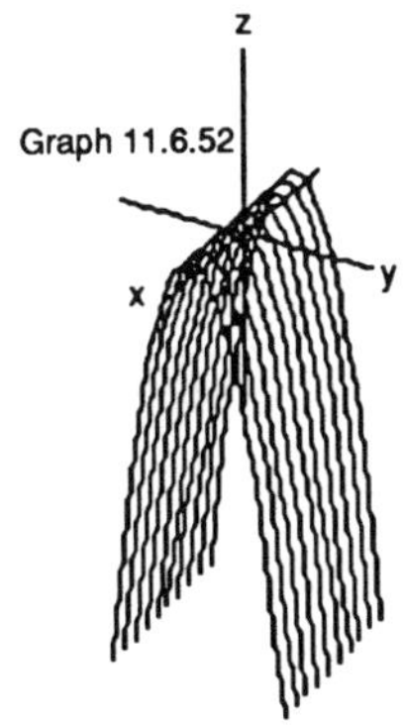

Graph 11.6.52

53. $z = x^2 + y^2$

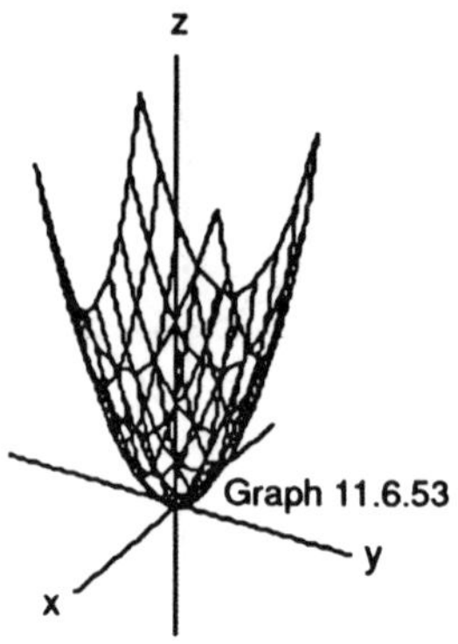

Graph 11.6.53

54. $z = x^2 + 2y^2$

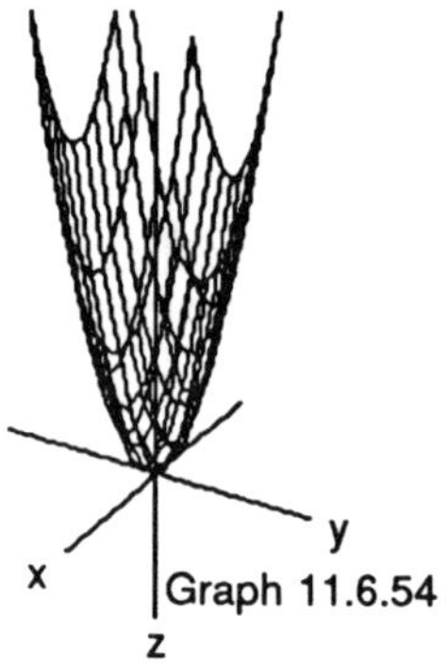

Graph 11.6.54

55. $z = \sqrt{1 - x^2}$

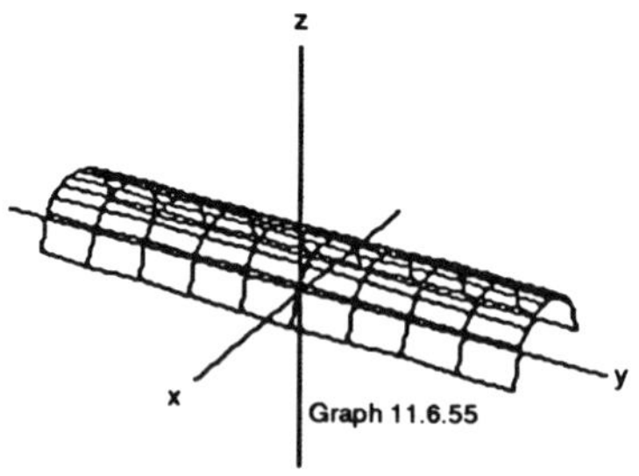

Graph 11.6.55

56. $z = \sqrt{1 - (y^2/4)}$

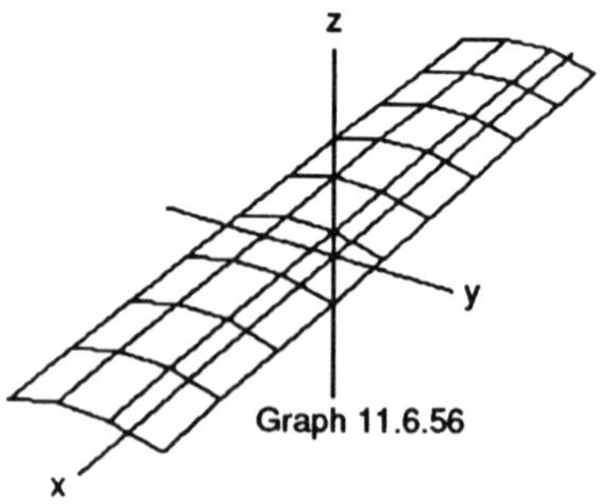

Graph 11.6.56

57. $z = \sqrt{x^2 + 2y^2 + 4}$

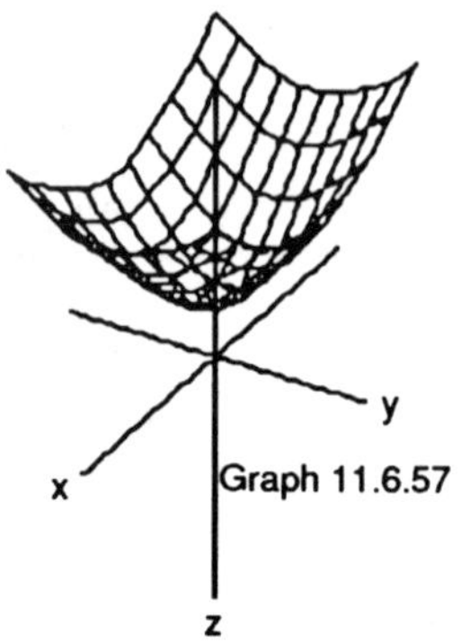

Graph 11.6.57

11.7 CYLINDRICAL AND SPHERICAL COORDINATES

	Rectangular	Cylindrical	Spherical
1.	$(0,0,0)$	$(0,0,0)$	$(0,0,0)$
2.	$(1,0,0)$	$(1,0,0)$	$(1,\pi/2,0)$
3.	$(0,1,0)$	$(1,\pi/2,0)$	$(1,\pi/2,\pi/2)$
4.	$(0,0,1)$	$(0,*,1)$	$(1,0,*)$
5.	$(1,0,0)$	$(1,0,0)$	$(1,\pi/2,0)$
6.	$(\sqrt{2},0,1)$	$(\sqrt{2},0,1)$	$(\sqrt{3},\cos^{-1}\left(1/\sqrt{3}\right),0)$
7.	$(0,1,1)$	$(1,\pi/2,1)$	$(\sqrt{2},\pi/4,\pi/2)$
8.	$(0,-3/2,\sqrt{3}/2)$	$(3/2,3\pi/2,\sqrt{3}/2)$	$(\sqrt{3},\pi/3,-\pi/2)$
9.	$(0,-2\sqrt{2},0)$	$(2\sqrt{2},3\pi/2,0)$	$(2\sqrt{2},\pi/2,3\pi/2)$
10.	$(0,0,-\sqrt{2})$	$(0,*,-\sqrt{2})$	$(\sqrt{2},\pi,3\pi/2)$

* any angle may be used

11. $r = 0 \Rightarrow$ rectangular, $x^2 + y^2 = 0$; spherical, $\phi = 0$ and $\phi = \pi$; the z–axis
12. $x^2 + y^2 = 5 \Rightarrow$ cylindrical, $r = \sqrt{5}$; spherical, $\rho \sin \phi = \sqrt{5}$; a cylinder
13. $z = 0 \Rightarrow$ cylindrical, $z = 0$; spherical, $\phi = \frac{\pi}{2}$; the xy–plane
14. $z = -2 \Rightarrow$ cylindrical, $z = -2$; spherical, $\rho \cos \phi = -2$; the plane $z = -2$
15. $\rho \cos \phi = 3 \Rightarrow$ rectangular, $z = 3$; cylindrical, $z = 3$; the plane $z = 3$
16. $\sqrt{x^2 + y^2} = z \Rightarrow$ cylindrical,$z = r$; spherical, $\phi = \frac{\pi}{4}$; a cone
17. $\rho \sin \phi \cos \phi = 0 \Rightarrow$ rectangular, $x = 0$; cylindrical $\theta = \frac{\pi}{2}$; the yz–plane
18. $\tan^2 \phi = 1 \Rightarrow$ rectangular $x^2 + y^2 = z^2$; cylindrical, $r^2 = z^2$; a circular cone symmetric to the z–axis
19. $x^2 + y^2 + z^2 = 4 \Rightarrow$ cylindrical, $r^2 + z^2 = 4$; spherical, $\rho = 2$; a sphere centered at the origin with a radius of 2
20. $\rho = 6 \cos \phi \Rightarrow$ rectangular, $x^2 + y^2 + (z-3)^2 = 9$; cylindrical, $r^2 + z^2 = 6z$; a sphere
21. a right circular cylinder whose generating curve is a circle of radius 2 in the $r\theta$–plane

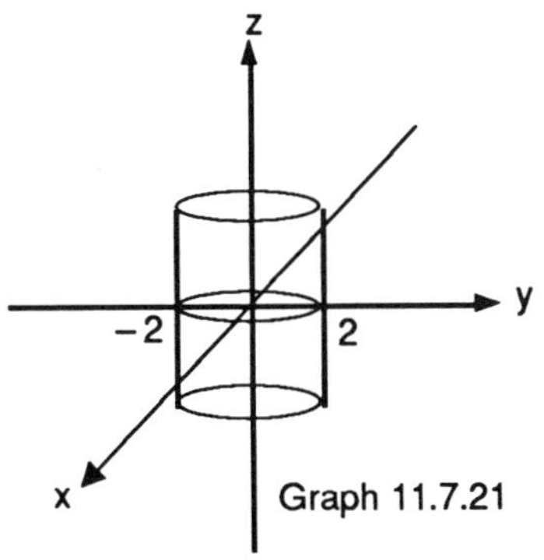

Graph 11.7.21

22. a sphere with center (0,0,0) and radius 1

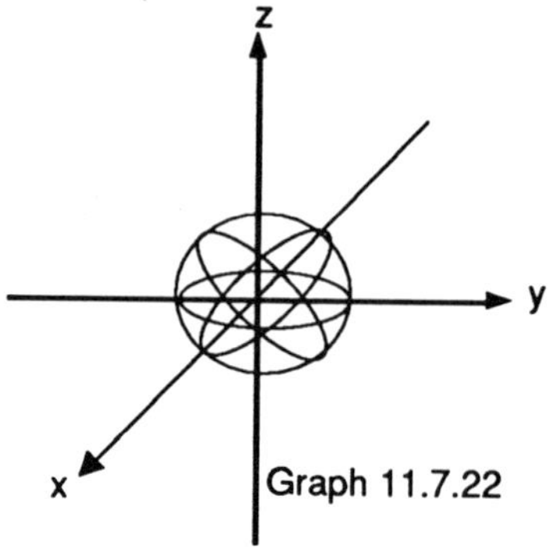

Graph 11.7.22

23. a cylinder whose generation curve is a cardioid in the $r\theta$–plane

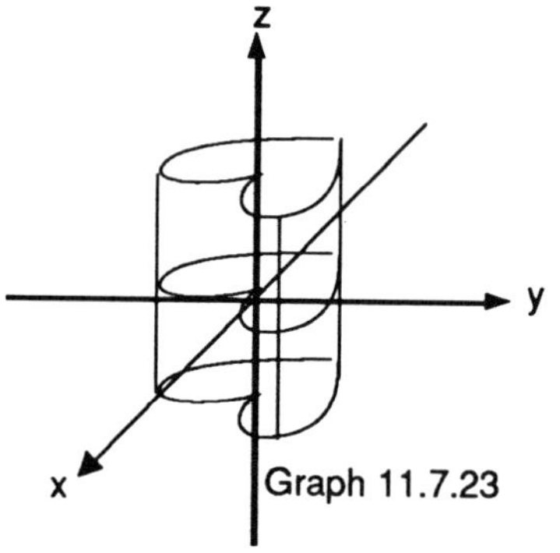

Graph 11.7.23

24. a cylinder whose generating curve is the circle $(x-1)^2+y^2=1$ in the xy–plane

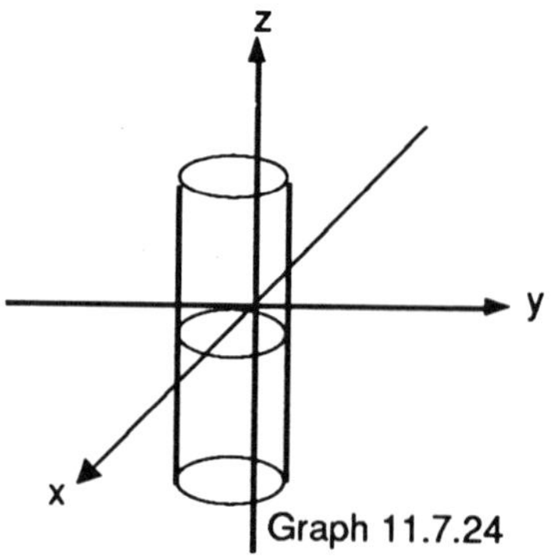

Graph 11.7.24

25. a circle contained in the plane $z = 3$ having a radius of $\sqrt{2}$ and center at (0,0,3)

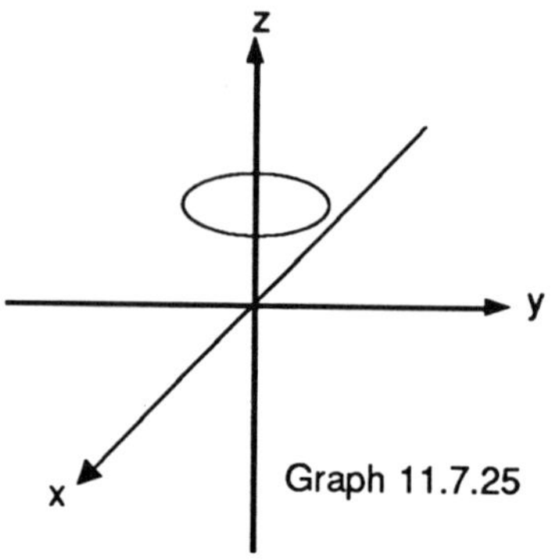

Graph 11.7.25

26. the intersection of the plane $\theta = \frac{\pi}{6}$ and the upper nappe of the cone $z = \sqrt{x^2 + y^2}$

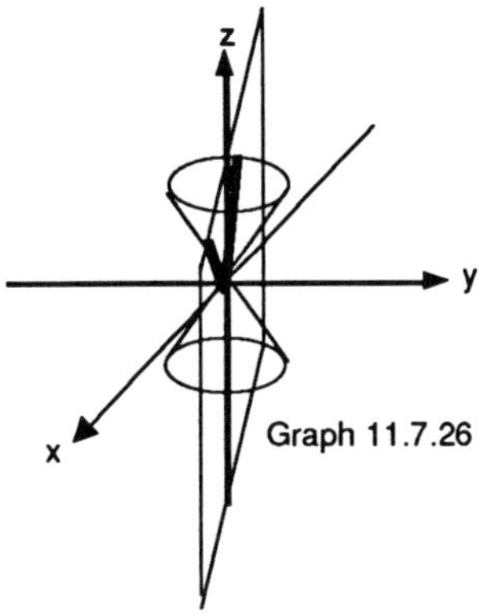

Graph 11.7.26

27. a space curve called a helix

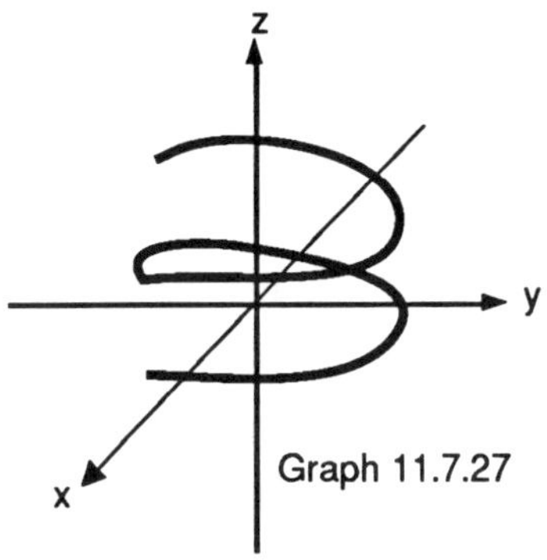

Graph 11.7.27

28. a cylinder whose generating curve is a lemniscate in the rθ–plane; the generating curve a lemniscate similar to exercise 15 in section 10.6.

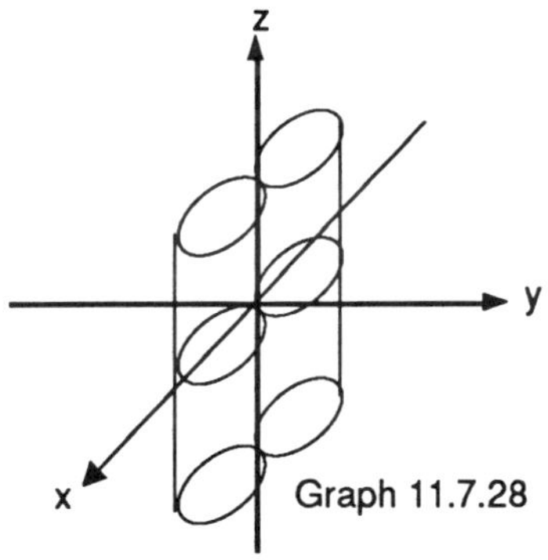

Graph 11.7.28

29. the upper nappe of a cone

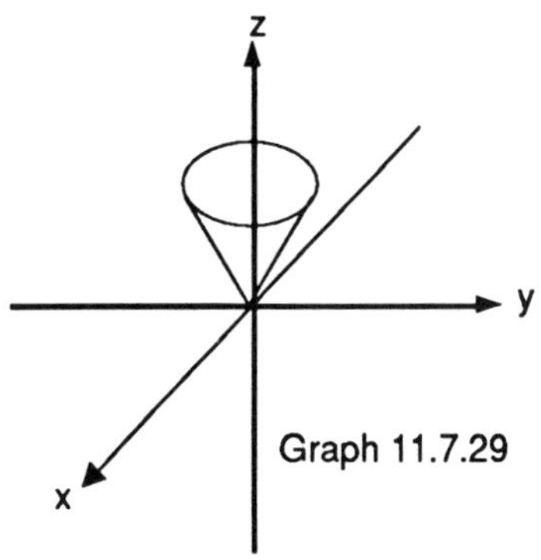

Graph 11.7.29

30. the intersection of the sphere $\rho = 6$ and cone $\phi = \frac{\pi}{6}$

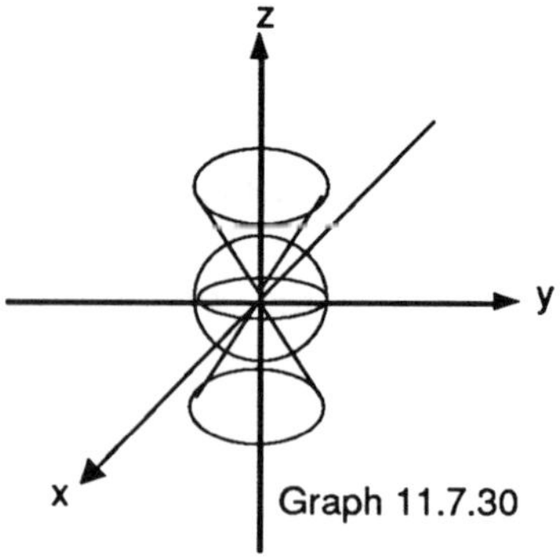

Graph 11.7.30

31. a " vertical semicircle " in the y = x plane

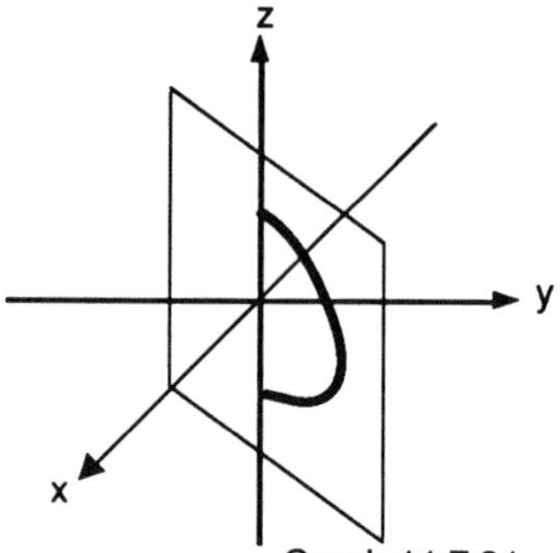

Graph 11.7.31

32. the intersection of the cone $\phi = \frac{\pi}{4}$ and the plane $\theta = \frac{\pi}{4}$

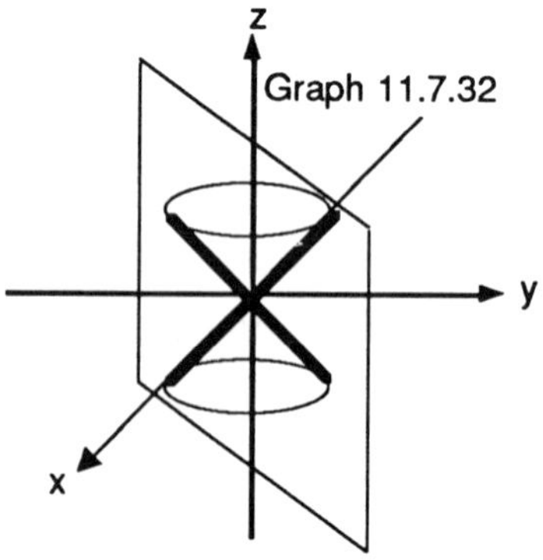

Graph 11.7.32

33. the intersection of the sphere, $x^2 + y^2 + (z-2)^2 = 4$, and the yz–plane when $y \geq 0$

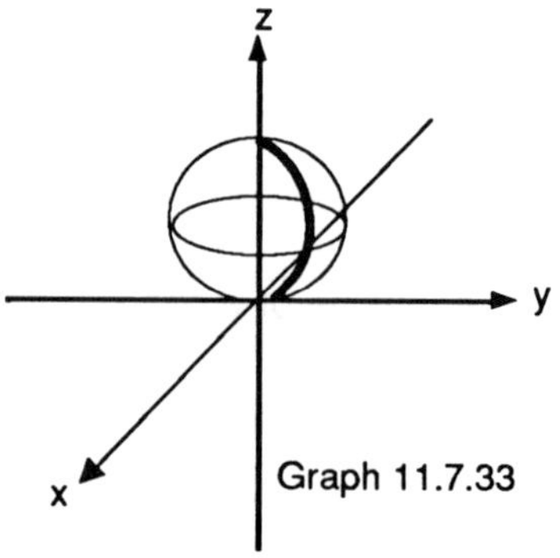

Graph 11.7.33

34. two concentric spheres centered at (0,0,– 1/2) having radii of $\frac{1}{4}$ and $\frac{9}{4}$; all points contained on and between these spheres

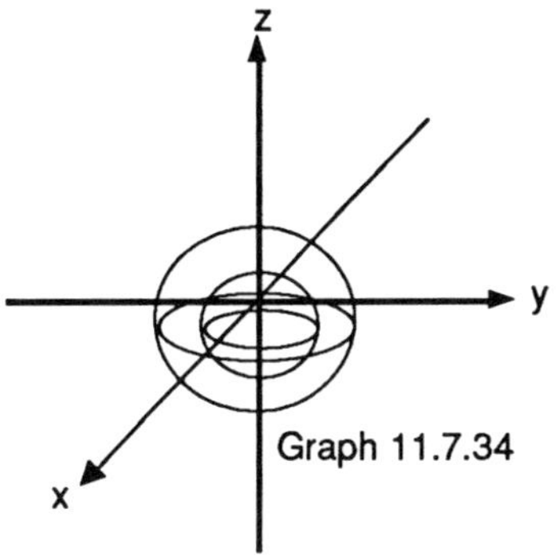

Graph 11.7.34

PRACTICE EXERCISES

1. $\theta = 0 \Rightarrow \mathbf{u} = \mathbf{i}$; $\theta = \frac{\pi}{2} \Rightarrow \mathbf{u} = \mathbf{j}$; $\theta = \frac{2\pi}{3} \Rightarrow \mathbf{u} = -\frac{1}{2}\mathbf{i} + \frac{\sqrt{3}}{2}\mathbf{j}$;
$\theta = \frac{5\pi}{4} \Rightarrow \mathbf{u} = -\frac{1}{\sqrt{2}}\mathbf{i} - \frac{1}{\sqrt{2}}\mathbf{j}$; $\theta = \frac{5\pi}{3} \Rightarrow \mathbf{u} = \frac{1}{2}\mathbf{i} - \frac{\sqrt{3}}{2}\mathbf{j}$

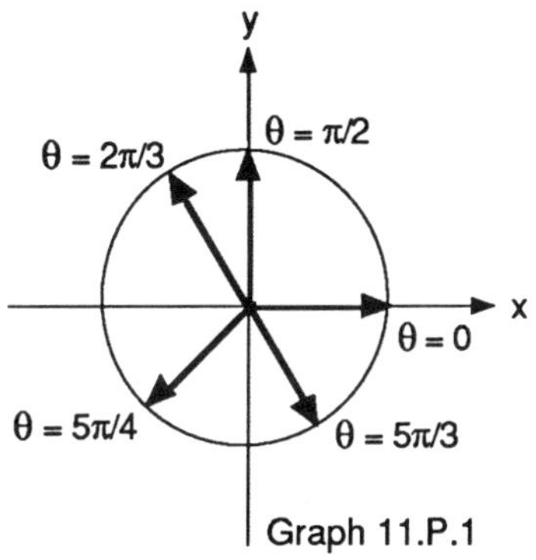

Graph 11.P.1

2. a) Rotating $\mathbf{i}$ clockwise 45° yields $\frac{1}{\sqrt{2}}\mathbf{i} + \frac{1}{\sqrt{2}}\mathbf{j}$.

   b) Rotating $\mathbf{j}$ counterclockwise 120° yields $-\frac{\sqrt{3}}{2}\mathbf{i} - \frac{1}{2}\mathbf{j}$.

3. $y = \tan x \Rightarrow [y']_{\pi/4} = \left[\sec^2 x\right]_{\pi/4} = 2 = \frac{2}{1} \Rightarrow \mathbf{T} = \mathbf{i} + 2\mathbf{j} \Rightarrow$ the unit tangents are $\pm\left(\frac{1}{\sqrt{5}}\mathbf{i} + \frac{2}{\sqrt{5}}\mathbf{j}\right)$ and the unit normals are $\pm\left(-\frac{2}{\sqrt{5}}\mathbf{i} + \frac{1}{\sqrt{5}}\mathbf{j}\right)$

4. $x^2 + y^2 = 5 \Rightarrow [y']_{(3,4)} = \left[-\frac{x}{y}\right]_{(3,4)} = \frac{-3}{4} \Rightarrow \mathbf{T} = 4\mathbf{i} - 3\mathbf{j} \Rightarrow$ the unit tangents are $\pm\frac{1}{5}(4\mathbf{i} - 3\mathbf{j})$ and the unit normnal are $\pm\frac{1}{5}(3\mathbf{i} + 4\mathbf{j})$

5. length $= \left|\sqrt{2}\mathbf{i} + \sqrt{2}\mathbf{j}\right| = \sqrt{2+2} = 2$, $\sqrt{2}\mathbf{i} + \sqrt{2}\mathbf{j} = 2\left[\frac{1}{\sqrt{2}}\mathbf{i} + \frac{1}{\sqrt{2}}\mathbf{j}\right] \Rightarrow$
the direction is $= \frac{1}{\sqrt{2}}\mathbf{i} + \frac{1}{\sqrt{2}}\mathbf{j}$

6. length $= |-\mathbf{i} - \mathbf{j}| = \sqrt{1+1} = \sqrt{2}$, $-\mathbf{i} - \mathbf{j} = \sqrt{2}\left[-\frac{1}{\sqrt{2}}\mathbf{i} - \frac{1}{\sqrt{2}}\mathbf{j}\right] \Rightarrow$
the direction is $= -\frac{1}{\sqrt{2}}\mathbf{i} - \frac{1}{\sqrt{2}}\mathbf{j}$

7. length = $|2\mathbf{i} - 3\mathbf{j} + 6\mathbf{k}| = \sqrt{4 + 9 + 36} = 7$, $2\mathbf{i} - 3\mathbf{j} + 6\mathbf{k} = 7\left[\frac{2}{7}\mathbf{i} - \frac{3}{7}\mathbf{j} + \frac{6}{7}\mathbf{k}\right] \Rightarrow$

the direction is = $\frac{2}{7}\mathbf{i} - \frac{3}{7}\mathbf{j} + \frac{6}{7}\mathbf{k}$

8. length = $|\mathbf{i} + 2\mathbf{j} - \mathbf{k}| = \sqrt{1 + 4 + 1} = \sqrt{6}$, $\mathbf{i} + 2\mathbf{j} - \mathbf{k} = \sqrt{6}\left[\frac{1}{\sqrt{6}}\mathbf{i} + \frac{2}{\sqrt{6}}\mathbf{j} - \frac{1}{\sqrt{6}}\mathbf{k}\right] \Rightarrow$

the direction is = $\frac{1}{\sqrt{6}}\mathbf{i} + \frac{2}{\sqrt{6}}\mathbf{j} - \frac{1}{\sqrt{6}}\mathbf{k}$

9. $\frac{2\,\mathbf{A}}{|\mathbf{A}|} = \frac{2}{\sqrt{33}}[4\mathbf{i} - \mathbf{j} + 4\mathbf{k}]$

10. $-\frac{5\,\mathbf{A}}{|\mathbf{A}|} = -5\left[\frac{3}{5}\mathbf{i} + \frac{4}{5}\mathbf{j}\right] = -3\mathbf{i} - 4\mathbf{j}$

11. $|\mathbf{A}| = \sqrt{2}$, $|\mathbf{B}| = 3$, $\mathbf{A}\cdot\mathbf{B} = 3$, $\mathbf{B}\cdot\mathbf{A} = 3$, $\mathbf{A}\times\mathbf{B} = \begin{vmatrix} \mathbf{i} & \mathbf{j} & \mathbf{k} \\ 1 & 1 & 0 \\ 2 & 1 & -2 \end{vmatrix} = -2\mathbf{i} + 2\mathbf{j} - \mathbf{k}$,

$\mathbf{B}\times\mathbf{A} = \begin{vmatrix} \mathbf{i} & \mathbf{j} & \mathbf{k} \\ 2 & 1 & -2 \\ 1 & 1 & 0 \end{vmatrix} = 2\mathbf{i} - 2\mathbf{j} + \mathbf{k}$, $|\mathbf{A}\times\mathbf{B}| = \sqrt{4 + 4 + 1} = 3$, $\theta = \cos^{-1}\left(\frac{\mathbf{A}\cdot\mathbf{B}}{|\mathbf{A}|\,|\mathbf{B}|}\right) = \cos^{-1}\left(\frac{1}{\sqrt{2}}\right) = \frac{\pi}{4}$,

$\text{comp}_{\mathbf{A}}\,\mathbf{B} = \frac{\mathbf{A}\cdot\mathbf{B}}{|\mathbf{A}|} = \frac{3}{\sqrt{2}}$, $\text{proj}_{\mathbf{A}}\,\mathbf{B} = \frac{\mathbf{A}\cdot\mathbf{B}}{\mathbf{A}\cdot\mathbf{A}}\,\mathbf{A} = \frac{3}{2}[\mathbf{i} + \mathbf{j}]$

12. $|\mathbf{A}| = \sqrt{27}$, $|\mathbf{B}| = \sqrt{14}$, $\mathbf{A}\cdot\mathbf{B} = 6$, $\mathbf{B}\cdot\mathbf{A} = 6$, $\mathbf{A}\times\mathbf{B} = \begin{vmatrix} \mathbf{i} & \mathbf{j} & \mathbf{k} \\ 5 & 1 & 1 \\ 1 & -2 & 3 \end{vmatrix} = 5\mathbf{i} - 14\mathbf{j} - 11\mathbf{k}$,

$\mathbf{B}\times\mathbf{A} = \begin{vmatrix} \mathbf{i} & \mathbf{j} & \mathbf{k} \\ 1 & -2 & 3 \\ 5 & 1 & 1 \end{vmatrix} = -5\mathbf{i} + 14\mathbf{j} + 11\mathbf{k}$, $|\mathbf{A}\times\mathbf{B}| = \sqrt{25 + 196 + 121} = \sqrt{342}$, $\theta = \cos^{-1}\left(\frac{\mathbf{A}\cdot\mathbf{B}}{|\mathbf{A}|\,|\mathbf{B}|}\right) =$

$\cos^{-1}\left(\frac{6}{\sqrt{27}\sqrt{14}}\right) \approx 72.02°$,

$\text{comp}_{\mathbf{A}}\,\mathbf{B} = \frac{\mathbf{A}\cdot\mathbf{B}}{|\mathbf{A}|} = \frac{6}{\sqrt{27}} = \frac{2}{\sqrt{3}}$, $\text{proj}_{\mathbf{A}}\,\mathbf{B} = \frac{\mathbf{A}\cdot\mathbf{B}}{\mathbf{A}\cdot\mathbf{A}}\,\mathbf{A} = \frac{2}{\sqrt{3}}[5\mathbf{i} + \mathbf{j} + \mathbf{k}]$

13. $\mathbf{B} = \left(\frac{\mathbf{A}\cdot\mathbf{B}}{\mathbf{A}\cdot\mathbf{A}}\,\mathbf{A}\right) + \left(\mathbf{B} - \frac{\mathbf{A}\cdot\mathbf{B}}{\mathbf{A}\cdot\mathbf{A}}\,\mathbf{A}\right) = \frac{4}{3}[2\mathbf{i} + \mathbf{j} - \mathbf{k}] + \left[(\mathbf{i} + \mathbf{j} - 5\mathbf{k}) - \frac{4}{3}(2\mathbf{i} + \mathbf{j} - \mathbf{k})\right] =$

$\frac{4}{3}[2\mathbf{i} + \mathbf{j} - \mathbf{k}] - \frac{1}{3}[5\mathbf{i} + \mathbf{j} + 11\mathbf{k}]$, where $\mathbf{A}\cdot\mathbf{B} = 8$ and $\mathbf{A}\cdot\mathbf{A} = 6$

14. $\mathbf{B} = \left(\frac{\mathbf{A}\cdot\mathbf{B}}{\mathbf{A}\cdot\mathbf{A}}\,\mathbf{A}\right) + \left(\mathbf{B} - \frac{\mathbf{A}\cdot\mathbf{B}}{\mathbf{A}\cdot\mathbf{A}}\,\mathbf{A}\right) = -\frac{1}{5}[\mathbf{i} - 2\mathbf{j}] + \left[(\mathbf{i} + \mathbf{j} + \mathbf{k}) - \frac{-1}{5}(\mathbf{i} - 2\mathbf{j})\right] =$

$-\frac{1}{5}[\mathbf{i} - 2\mathbf{j}] + \left[\frac{6}{5}\mathbf{i} + \frac{3}{5}\mathbf{j} + \mathbf{k}\right]$, where $\mathbf{A}\cdot\mathbf{B} = -1$ and $\mathbf{A}\cdot\mathbf{A} = 5$

15. $\mathbf{A}\times\mathbf{B} = \begin{vmatrix} \mathbf{i} & \mathbf{j} & \mathbf{k} \\ 1 & 0 & 0 \\ 1 & 1 & 0 \end{vmatrix} = \mathbf{k}$

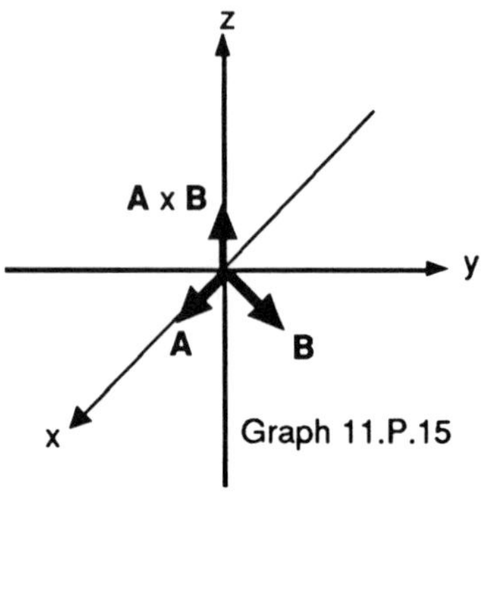

Graph 11.P.15

16. $\mathbf{A} \times \mathbf{B} = \begin{vmatrix} \mathbf{i} & \mathbf{j} & \mathbf{k} \\ 1 & -1 & 0 \\ 1 & 1 & 0 \end{vmatrix} = 2\mathbf{k}$

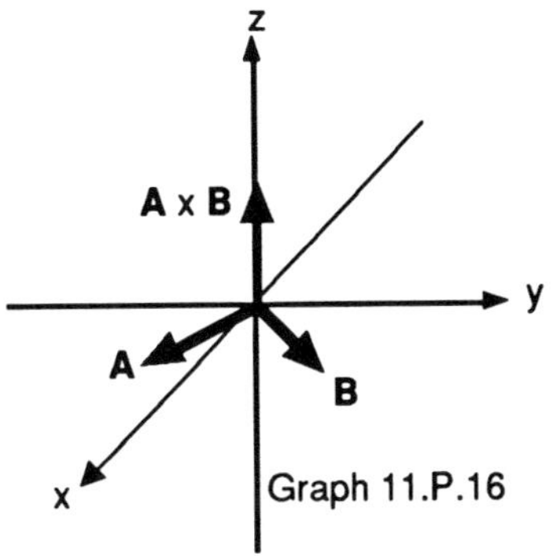

Graph 11.P.16

17. S(3,2), P(2,– 1) is on 3x + 4y = 2 $\Rightarrow \overrightarrow{PS} = \mathbf{j} + 3\mathbf{k}$ and $\mathbf{N} = 3\mathbf{i} + 4\mathbf{j} \Rightarrow d = \left|\text{proj}_{\mathbf{N}}\overrightarrow{PS}\right| =$

$\left|\dfrac{\mathbf{N}\cdot\overrightarrow{PS}}{|\mathbf{N}|}\right| = \left|\dfrac{3+12}{\sqrt{9+16}}\right| = 3$

18. S(– 1,1), P(0,–3/4) is on 5x – 12y = 9 $\Rightarrow \overrightarrow{PS} = \mathbf{i} - \frac{7}{4}\mathbf{j}$ and $\mathbf{N} = 5\mathbf{i} - 12\mathbf{j} \Rightarrow d = \left|\text{proj}_{\mathbf{N}}\overrightarrow{PS}\right| =$

$\left|\dfrac{\mathbf{N}\cdot\overrightarrow{PS}}{|\mathbf{N}|}\right| = \left|\dfrac{5+21}{\sqrt{25+144}}\right| = 2$

19. S(6,0,– 6), P(4,0,0) is on x – y = 4 $\Rightarrow \overrightarrow{PS} = 2\mathbf{i} - 6\mathbf{k}$ and $\mathbf{N} = \mathbf{i} - \mathbf{j} \Rightarrow d = \left|\dfrac{\mathbf{N}\cdot\overrightarrow{PS}}{|\mathbf{N}|}\right| = \left|\dfrac{2}{\sqrt{2}}\right| = \sqrt{2}$

20. S(3,0,10), P(0,0,2) is on 2x + 3y + z = 2 $\Rightarrow \overrightarrow{PS} = 3\mathbf{i} + 8\mathbf{k}$ and $\mathbf{N} = 2\mathbf{i} + 3\mathbf{j} + \mathbf{k} \Rightarrow d = \left|\dfrac{\mathbf{N}\cdot\overrightarrow{PS}}{|\mathbf{N}|}\right| =$

$\dfrac{6+8}{\sqrt{4+9+1}} = \sqrt{14}$

21. (3,– 2,1) and $\mathbf{N} = 2\mathbf{i} + \mathbf{j} - \mathbf{k} \Rightarrow 2(x-3) + 1(y+2) + (-1)(z-1) = 0 \Rightarrow 2x + y - z = 3$

22. (– 1,6,0), x = – 1 + t, y = 6 – 2t and z = 3t $\Rightarrow 1(x+1) + (-2)(y-6) + 3(z) = 0 \Rightarrow x - 2y + 3z = -13$

23. P(1,– 1,2), Q(2,1,3) and R(– 1,2,– 1) $\Rightarrow \overrightarrow{PQ} = \mathbf{i} + 2\mathbf{j} + \mathbf{k}$, $\overrightarrow{PR} = -2\mathbf{i} + 3\mathbf{j} - 3\mathbf{k}$ and $\overrightarrow{PQ} \times \overrightarrow{PR} =$

$\begin{vmatrix} \mathbf{i} & \mathbf{j} & \mathbf{k} \\ 1 & 2 & 1 \\ -2 & 3 & -3 \end{vmatrix} = -9\mathbf{i} + \mathbf{j} + 7\mathbf{k}$, the normal of the plane; $-9(x-1) + 1(y+1) + 7(z-2) = 0 \Rightarrow$

$-9x + y + 7z = 4$

24. P(1,0,0), Q(0,1,0) and R(0,0,1) $\Rightarrow \overrightarrow{PQ} = \mathbf{i} - \mathbf{j}$, $\overrightarrow{PR} = \mathbf{i} - \mathbf{k}$ and $\overrightarrow{PQ} \times \overrightarrow{PR} =$

$\begin{vmatrix} \mathbf{i} & \mathbf{j} & \mathbf{k} \\ 1 & -1 & 0 \\ 1 & 0 & -1 \end{vmatrix} = \mathbf{i} + \mathbf{j} + \mathbf{k}$, the normal of the plane; $1(x-1) + 1(y) + 1(z) = 0 \Rightarrow x + y + z = 1$

25. (1,2,3) and $\mathbf{v} = -3\mathbf{i} + 7\mathbf{k} \Rightarrow x = 1 - 3t,\ y = 2,\ z = 3 + 7t$

26. $(0,-\frac{1}{2},-\frac{3}{2})$, since $t=-\frac{1}{2}$, $y=-\frac{1}{2}$ and $z=-\frac{3}{2}$ when $x=0$; $(-1,0,-3)$, since $t=-1$, $x=-1$ and $z=-3$ when $y=0$; $(1,-1,0)$, since $t=0$, $x=1$ and $y=-1$ when $z=0$

27. $x=2t$, $y=-t$, $z=-t$ a line containing the origin and perpendicular to the plane $2x-y-z=4$; this line intersects the plane $3x-5y+2z=6$ when t is the solution of $3(2t)-5(-t)+2(-t)=6 \Rightarrow t=\frac{2}{3} \Rightarrow (4/3,-2/3,-2/3)$ is the point of intersection

28. $P(1,2,0)$ and $Q(1,3,-1) \Rightarrow \overrightarrow{PQ}=-\mathbf{j}+\mathbf{k} \Rightarrow x=-1$, $y=2-t$, $z=t$ where $-1\le t\le 0$

29. the desired line contains $P(-5,3,0)$ and $Q(10,0,-9)$; $\overrightarrow{PQ}=15\mathbf{i}-3\mathbf{j}-9\mathbf{k}$, the direction of this line $\therefore x=-5+15t$, $y=3-3t$, $z=-9t$

30. $\theta=\cos^{-1}\left(\frac{\mathbf{N}_1\cdot\mathbf{N}_2}{|\mathbf{N}_1|\,|\mathbf{N}_2|}\right)=\cos^{-1}\left(\frac{1}{\sqrt{2}\sqrt{2}}\right)=60°$, where $\mathbf{N}_1=\mathbf{i}+\mathbf{j}$ and $\mathbf{N}_2=\mathbf{j}+\mathbf{k}$

31. The distance $d=\sqrt{2^2+(5+2t)^2+(4t+10)^2}$ is minimized when $f(t)=2^2+(5+2t)^2+(4t+10)^2$ is minimized. $f'(t)=0 \Rightarrow t=-\frac{5}{2} \Rightarrow d=\sqrt{4+0+0}=2$

32. The distance $d=\sqrt{(11+4t)^2+49+(4t-3)^2}$ is minimized when $f(t)=(11+4t)^2+49+(4t-3)^2$ is minimized. $f'(t)=0 \Rightarrow t=-1 \Rightarrow d=\sqrt{49+49+49}=7\sqrt{3}$

33. $W=\mathbf{F}\cdot\overrightarrow{PQ}=|\mathbf{F}|\left|\overrightarrow{PQ}\right|\cos\theta=[40\text{ lb}](800\text{ ft})\cos 30°=16000\sqrt{3}\text{ ft}\cdot\text{lb}$

34. $\left|\overrightarrow{PQ}\times\mathbf{F}\right|=\left|\overrightarrow{PQ}\right||\mathbf{F}|\sin 90° \Rightarrow 15\text{ ft}\cdot\text{lb}=\frac{3}{4}\text{ ft}\,|\mathbf{F}|\,1 \Rightarrow |\mathbf{F}|=20\text{ lb}$; for additional details, see example 1 on page 740 in the text

35. a) area $=|\mathbf{A}\times\mathbf{B}|=\left|\begin{vmatrix}\mathbf{i}&\mathbf{j}&\mathbf{k}\\1&1&-1\\2&1&1\end{vmatrix}\right|=|2\mathbf{i}-3\mathbf{j}-\mathbf{k}|=\sqrt{4+9+1}=\sqrt{14}$

b) volume $=\mathbf{A}\cdot(\mathbf{B}\times\mathbf{C})=\begin{vmatrix}1&1&-1\\2&1&1\\-1&-2&3\end{vmatrix}=1$

36. a) area $=|\mathbf{A}\times\mathbf{B}|=\left|\begin{vmatrix}\mathbf{i}&\mathbf{j}&\mathbf{k}\\1&1&0\\0&1&0\end{vmatrix}\right|=|\mathbf{k}|=1$

b) volume $=\mathbf{A}\cdot(\mathbf{B}\times\mathbf{C})=\begin{vmatrix}1&1&0\\0&1&0\\1&1&1\end{vmatrix}=\begin{vmatrix}1&0&0\\0&1&0\\0&0&1\end{vmatrix}=1$

37. a) true b) false, $(2\mathbf{i})\cdot(2\mathbf{i})=4$ while $|2\mathbf{i}|=\sqrt{2^2}=2$ c) true

d) true e) false, they have opposite directions f) true

g) true h) true

38. All parts are true.

39. the y–axis in the xy–plane and the yz–plane in three dimensional space

40. a line in the xy–plane and a plane in three dimensional space

41. a circle centered at (0,0) with a radius of 2 in the xy–plane; a cylinder parallel with the z–axis in three dimensional space with the circle as its generating curve

42. an ellipse in the xy–plane; a cylinder in three dimensional space with the ellipse as the generating curve

43. a horizontal parabola opening to the right with its vertex at (0,0) in the xy–plane; a cylinder parallel with the z–axis in three dimensional space with the parabola as the generating curve

44. a vertical hyperbola centered at (0,0) in the xy–plane; a cylinder parallel with the z–axis in three dimensional space with the hyperbola as the generating curve

45. a horizontal cardioid in the $r\theta$–plane; a cylinder parallel with the z–axis in three dimensional space with the cardioid as the generating curve

46. a circle centered at (0,1/2) having a radius of 1/2 in the xy–plane; a cylinder parallel with the z–axis in three dimensional space with the circle as the generating curve

47. a horizontal lemniscate of length $2\sqrt{2}$ in the $r\theta$–plane; a cylinder parallel with the z–axis in three dimensional space with the lemniscate as the generating curve

48. a rose in the $r\theta$–plane; a cylinder in three dimensional space with the rose as the generating curve

49. a sphere with a radius of 2 centered at the origin

50. a plane which intersects the xy–plane at a right angle on the line y = x

51. the upper nappe of a cone whose surface makes a $\frac{\pi}{6}$ angle with the z–axis

52. the circle $x^2 + y^2 = 1$ in the xy–plane

53. the upper hemisphere of a sphere with a radius 1 centered at the origin

54. two concentric spheres centered at the origin having radii of 1 and 2; all points contained on and between the spheres

	Rectangular	Cylindrical	Spherical
55.	(1,0,0)	(1,0,0)	$(1,\pi/2,0)$
56.	(0,1,0)	$(1,\pi/2,0)$	$(1,\pi/2,\pi/2)$
57.	(0,1,1)	$(1,\pi/2,1)$	$(\sqrt{2},\pi/4,\pi/2)$
58.	$(1,0,-\sqrt{3})$	$(1,0,-\sqrt{3})$	$(2,5\pi/6,0)$
59.	(1,1,1)	$(\sqrt{2},\pi/4,1)$	$(\sqrt{3},\tan^{-1}\sqrt{2},\pi/4)$
60.	(0,– 1,1)	$(1,3\pi/2,1)$	$(\sqrt{2},\pi/4,3\pi/2)$

61. $z = 2 \Rightarrow$ cylindrical, $z = 2$; spherical, $\rho \cos \phi = 2$; a plane parallel with the xy–plane

62. $z = \sqrt{x^2 + y^2} \Rightarrow$ cylindrical, $z = r$; spherical, $\phi = \frac{\pi}{4}$; a cone

63. $z = r^2 \Rightarrow$ rectangular, $z = x^2 + y^2$; spherical, $\rho = 0$ or $\rho = \frac{\cos \phi}{\sin^2 \phi}$ when $0 < \phi < \frac{\pi}{2}$; a paraboloid symmetric to the z–axis opening upward

64. $r = \cos \theta \Rightarrow$ rectangular, $\left(x - \frac{1}{2}\right)^2 + y^2 = \frac{1}{4}$; spherical, $\rho \sin \phi = \cos \theta$; a cylinder whose generating curve is the circle

65. $\rho = 4 \Rightarrow$ rectangular, $x^2 + y^2 + z^2 = 16$; cylindrical, $r^2 + z^2 = 16$; a sphere with a radius of 4 centered at the origin

66. $\rho \cos \phi = 1 \Rightarrow$ rectangular $z = 1$; cylindrical, $z = 1$; the plane $z = 1$

67. $x^2 + y^2 + z^2 = 4$

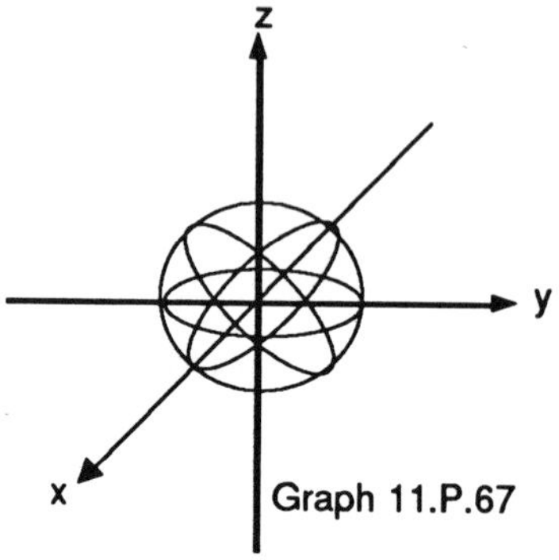

Graph 11.P.67

68. $x^2 + y^2 - 1 + z^2 = 1$

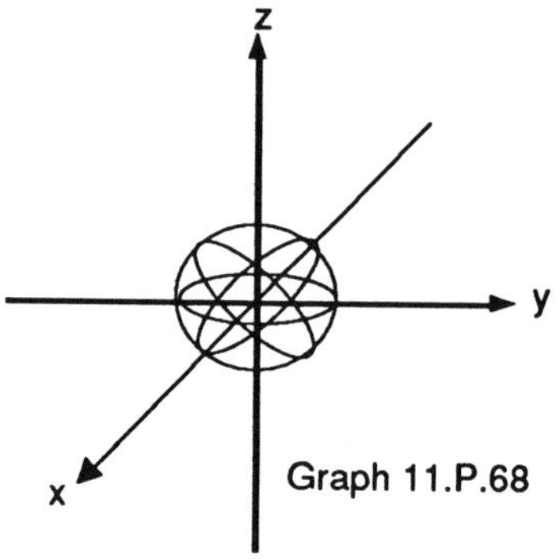

Graph 11.P.68

69. $4x^2 + 4y^2 + z^2 = 4$

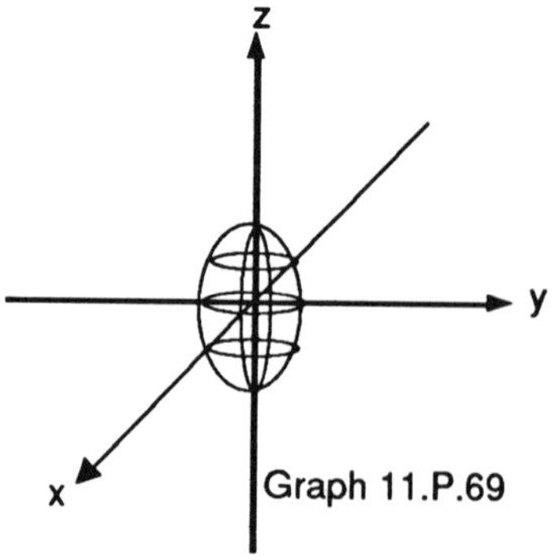

Graph 11.P.69

70. $36x^2 + 9y^2 + 4z^2 = 36$

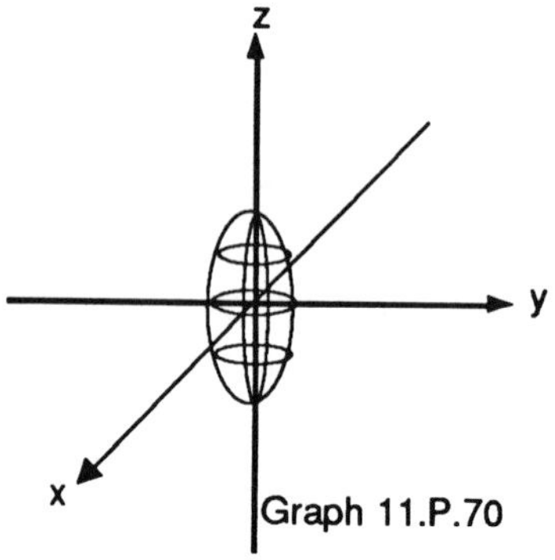

Graph 11.P.70

71. $z = -\left(x^2 + y^2\right)$

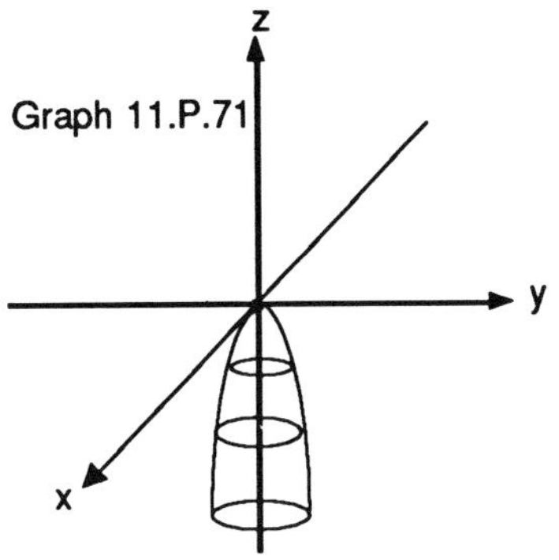

72. $y = -\left(x^2 + z^2\right)$

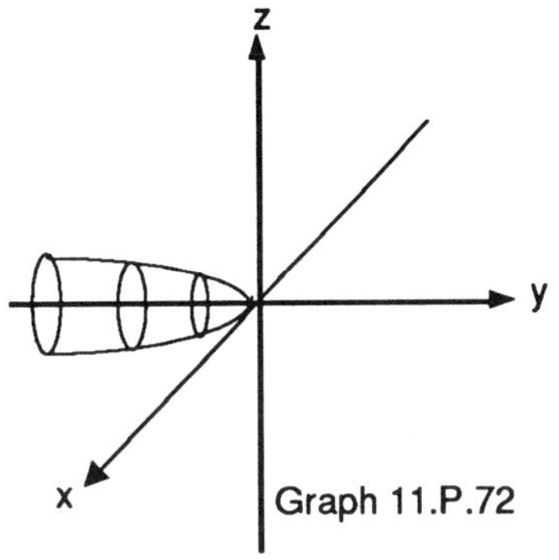

73. $x^2 + y^2 = z^2$

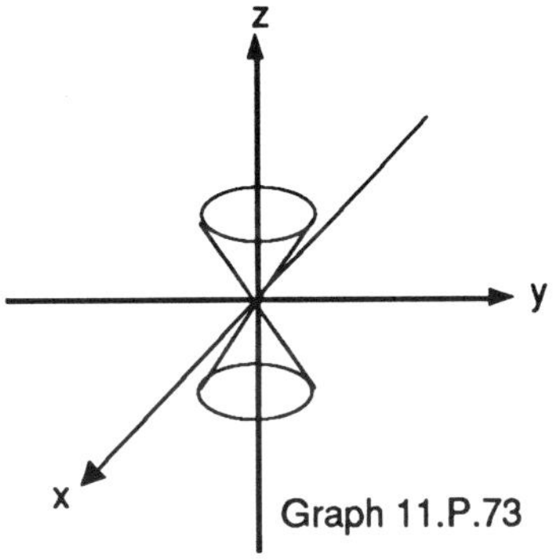

74. $x^2 + z^2 = y^2$

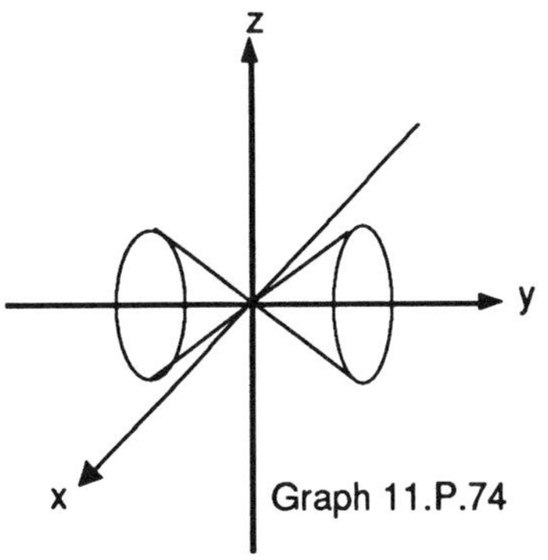

75. $x^2 + y^2 - z^2 = 4$

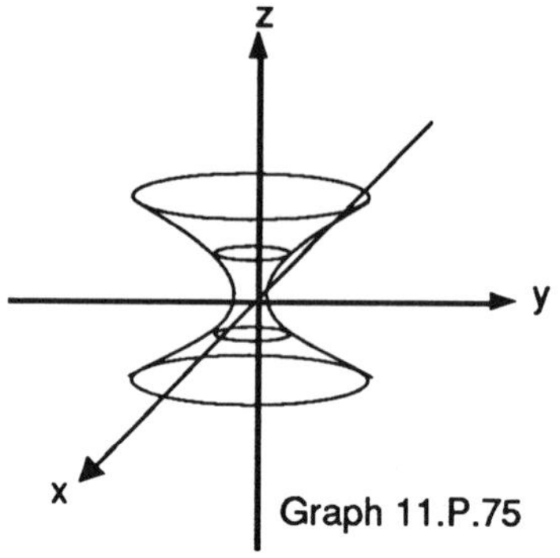

Graph 11.P.75

76. $4y^2 + z^2 - 4x^2 = 4$

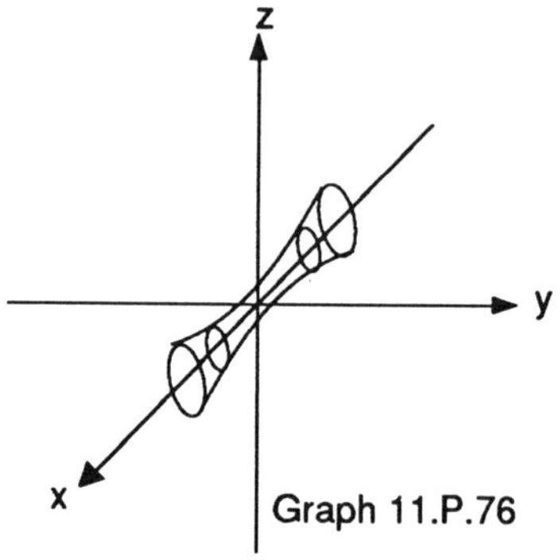

Graph 11.P.76

77. $y^2 - x^2 - z^2 = 1$

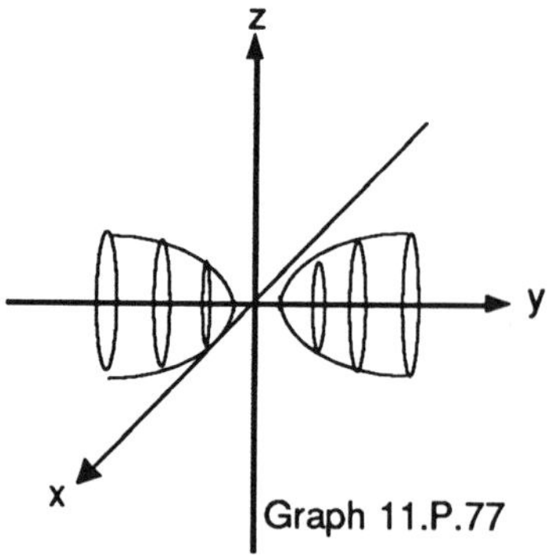

Graph 11.P.77

78. $z^2 - x^2 - y^2 = 1$

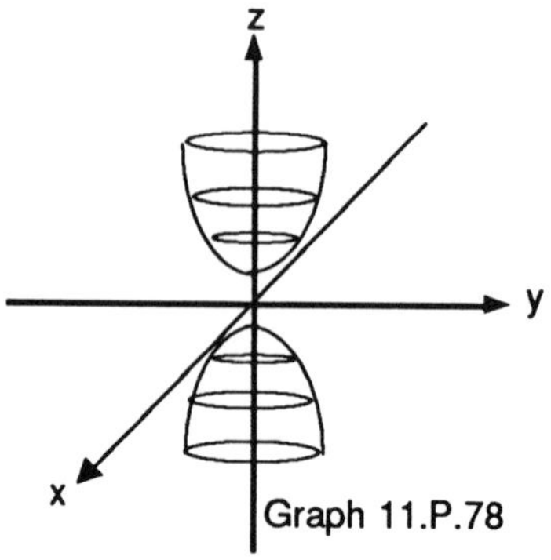

Graph 11.P.78

CHAPTER 12

VECTOR-VALUED FUNCTIONS AND MOTION IN SPACE

12.1 VECTOR-VALUED FUNCTIONS AND CURVES IN SPACE. DERIVATIVES AND INTEGRALS

1. $\mathbf{r} = (2\cos t)\,\mathbf{i} + (3\sin t)\,\mathbf{j} + 4t\,\mathbf{k} \Rightarrow \mathbf{v} = \frac{d\mathbf{r}}{dt} = (-2\sin t)\,\mathbf{i} + (3\cos t)\,\mathbf{j} + 4\,\mathbf{k}$

   $\mathbf{a} = \frac{d^2\mathbf{r}}{dt^2} = (-2\cos t)\,\mathbf{i} - (3\sin t)\,\mathbf{j}$. Speed: $\left|\mathbf{v}\left(\frac{\pi}{2}\right)\right| = \sqrt{\left(-2\sin\frac{\pi}{2}\right)^2 + \left(3\cos\frac{\pi}{2}\right)^2 + 4^2} = 2\sqrt{5}$

   Direction: $\frac{\mathbf{v}\left(\frac{\pi}{2}\right)}{\left|\mathbf{v}\left(\frac{\pi}{2}\right)\right|} = \left(-\frac{2}{2\sqrt{5}}\sin\frac{\pi}{2}\right)\mathbf{i} + \left(\frac{3}{2\sqrt{5}}\cos\frac{\pi}{2}\right)\mathbf{j} + \frac{4}{2\sqrt{5}}\mathbf{k} = -\frac{1}{\sqrt{5}}\mathbf{i} + \frac{2}{\sqrt{5}}\mathbf{k}$

   $\mathbf{v}\left(\frac{\pi}{2}\right) = 2\sqrt{5}\left[-\frac{1}{\sqrt{5}}\mathbf{i} + \frac{2}{\sqrt{5}}\mathbf{k}\right]$

2. $\mathbf{r} = (t+1)\,\mathbf{i} + (t^2-1)\,\mathbf{j} + 2t\,\mathbf{k} \Rightarrow \mathbf{v} = \frac{d\mathbf{r}}{dt} = \mathbf{i} + 2t\,\mathbf{j} + 2\,\mathbf{k}$, $\mathbf{a} = \frac{d^2\mathbf{r}}{dt^2} = 2\,\mathbf{j}$. Speed: $|\mathbf{v}(1)| = \sqrt{1^2 + (2(1))^2 + 2^2}$ $= 3$. Direction: $\frac{\mathbf{v}(1)}{|\mathbf{v}(1)|} = \frac{\mathbf{i} + 2(1)\,\mathbf{j} + 2\,\mathbf{k}}{3} = \frac{1}{3}\mathbf{i} + \frac{2}{3}\mathbf{j} + \frac{2}{3}\mathbf{k}$. $\mathbf{v}(1) = 3\left[\frac{1}{3}\mathbf{i} + \frac{2}{3}\mathbf{j} + \frac{2}{3}\mathbf{k}\right]$

3. $\mathbf{r} = (\cos 2t)\,\mathbf{j} + (2\sin t)\,\mathbf{k} \Rightarrow \mathbf{v} = (-2\sin 2t)\,\mathbf{j} + (2\cos t)\,\mathbf{k}$, $\mathbf{a} = (-4\cos 2t)\,\mathbf{j} - (2\sin t)\,\mathbf{k}$. Speed: $|\mathbf{v}(0)| = \sqrt{(-2\sin 2(0))^2 + (2\cos 0)^2} = 2$ Direction: $\frac{\mathbf{v}(0)}{|\mathbf{v}(0)|} = \frac{-2\sin 2(0)\,\mathbf{j} + 2\cos 0\,\mathbf{k}}{2} = \mathbf{k}$. $\mathbf{v}(0) = 2\,\mathbf{k}$

4. $\mathbf{r} = e^t\,\mathbf{i} + \frac{2}{9}e^{2t}\,\mathbf{j} \Rightarrow \mathbf{v} = \frac{d\mathbf{r}}{dt} = e^t\,\mathbf{i} + \frac{4}{9}e^{2t}\,\mathbf{j}$, $\mathbf{a} = \frac{d^2\mathbf{r}}{dt^2} = e^t\,\mathbf{i} + \frac{8}{9}e^{2t}\,\mathbf{j}$. Speed: $|\mathbf{v}(\ln 3)| =$

   $\sqrt{\left(e^{\ln 3}\right)^2 + \left(\frac{4}{9}e^{2\ln 3}\right)^2} = 5$ Direction: $\frac{\mathbf{v}(\ln 3)}{|\mathbf{v}(\ln 3)|} = \frac{e^{\ln 3}\,\mathbf{i} + \frac{4}{9}n\,e^{2\ln 3}\,\mathbf{j}}{5} = \frac{3}{5}\mathbf{i} + \frac{4}{5}\mathbf{j}$.

   $\mathbf{v}(\ln 3) = 5\left[\frac{3}{5}\mathbf{i} + \frac{4}{5}\mathbf{j}\right]$

5. $\mathbf{r} = (\sec t)\,\mathbf{i} + (\tan t)\,\mathbf{j} + \frac{4}{3}t\,\mathbf{k} \Rightarrow \mathbf{v} = \frac{d\mathbf{r}}{dt} = (\sec t\tan t)\,\mathbf{i} + (\sec^2 t)\,\mathbf{j} + \frac{4}{3}\mathbf{k}$, $\mathbf{a} = \frac{d^2\mathbf{r}}{dt^2} = (\sec t\tan^2 t + \sec^3 t)\,\mathbf{i} +$

   $(2\sec^2 t\tan t)\,\mathbf{j}$. Speed: $\left|\mathbf{v}\left(\frac{\pi}{6}\right)\right| = \sqrt{\left(\sec\frac{\pi}{6}\tan\frac{\pi}{6}\right)^2 + \left(\sec^2\frac{\pi}{6}\right)^2 + \left(\frac{4}{3}\right)^2} = 2$. Direction: $\frac{\mathbf{v}\left(\frac{\pi}{6}\right)}{\left|\mathbf{v}\left(\frac{\pi}{6}\right)\right|}$

   $= \frac{\sec\frac{\pi}{6}\tan\frac{\pi}{6}\,\mathbf{i} + \sec^2\frac{\pi}{6}\,\mathbf{j} + \frac{4}{3}\mathbf{k}}{2} = \frac{1}{3}\mathbf{i} + \frac{2}{3}\mathbf{j} + \frac{2}{3}\mathbf{k}$. $\mathbf{v}\left(\frac{\pi}{6}\right) = 2\left(\frac{1}{3}\mathbf{i} + \frac{2}{3}\mathbf{j} + \frac{2}{3}\mathbf{k}\right)$

6. $\mathbf{r} = (2\ln(t+1))\,\mathbf{i} + t^2\,\mathbf{j} + \frac{t^2}{2}\mathbf{k} \Rightarrow \mathbf{v} = \frac{d\mathbf{r}}{dt} = \left(\frac{2}{t+1}\right)\mathbf{i} + 2t\,\mathbf{j} + t\,\mathbf{k},\ \mathbf{a} = \frac{d^2\mathbf{r}}{dt^2} = \left(\frac{-2}{(t+1)^2}\right)\mathbf{i} + 2\,\mathbf{j} + \mathbf{k}$

Speed: $|\mathbf{v}(1)| = \sqrt{\left(\frac{2}{1+1}\right)^2 + (2(1))^2 + 1^2} = \sqrt{6}$. Direction: $\frac{\mathbf{v}(1)}{|\mathbf{v}(1)|} = \frac{\left(\frac{2}{1+1}\right)\mathbf{i} + 2(1)\,\mathbf{j} + (1)\,\mathbf{k}}{\sqrt{6}} =$
$\frac{1}{\sqrt{6}}\mathbf{i} + \frac{2}{\sqrt{6}}\mathbf{j} + \frac{1}{\sqrt{6}}\mathbf{k}$. $\mathbf{v}(1) = \sqrt{6}\left[\frac{1}{\sqrt{6}}\mathbf{i} + \frac{2}{\sqrt{6}}\mathbf{j} + \frac{1}{\sqrt{6}}\mathbf{k}.\right]$

7. $\mathbf{r} = \left(e^{-t}\right)\mathbf{i} + (2\cos 3t)\,\mathbf{j} + (2\sin 3t)\,\mathbf{k} \Rightarrow \mathbf{v} = \frac{d\mathbf{r}}{dt} = \left(-e^{-t}\right)\mathbf{i} - (6\sin 3t)\,\mathbf{j} + (6\cos 3t)\,\mathbf{k},\ \mathbf{a} = \frac{d^2\mathbf{r}}{dt^2} =$
$\left(e^{-t}\right)\mathbf{i} - (18\cos 3t)\,\mathbf{j} - (18\sin 3t)\,\mathbf{k}$. Speed: $|\mathbf{v}(0)| = \sqrt{\left(-e^0\right)^2 + (-6\sin 3(0))^2 + (6\cos 3(0))^2} = \sqrt{37}$
Direction: $\frac{\mathbf{v}(0)}{|\mathbf{v}(0)|} = \frac{\left(-e^0\right)\mathbf{i} - 6\sin 3(0)\,\mathbf{j} + 6\cos 3(0)\,\mathbf{k}}{\sqrt{37}} = -\frac{1}{\sqrt{37}}\mathbf{i} + \frac{6}{\sqrt{37}}\mathbf{k}$.
$\mathbf{v}(0) = \sqrt{37}\left[-\frac{1}{\sqrt{37}}\mathbf{i} + \frac{6}{\sqrt{37}}\mathbf{k}.\right]$

8. $\mathbf{r} = (1+t)\,\mathbf{i} + \frac{t^2}{\sqrt{2}}\mathbf{j} + \frac{t^3}{3}\mathbf{k} \Rightarrow \mathbf{v} = \frac{d\mathbf{r}}{dt} = \mathbf{i} + \frac{2t}{\sqrt{2}}\mathbf{j} + t^2\,\mathbf{k},\ \mathbf{a} = \frac{d^2\mathbf{r}}{dt^2} = \frac{2}{\sqrt{2}}\mathbf{j} + 2t\,\mathbf{k}$. Speed: $|\mathbf{v}(1)| =$
$\sqrt{1^2 + \left(\frac{2(1)}{\sqrt{2}}\right)^2 + \left(1^2\right)^2} = 2$. Direction: $\frac{\mathbf{v}(1)}{|\mathbf{v}(1)|} = \frac{\mathbf{i} + \frac{2(1)}{\sqrt{2}}\mathbf{j} + \left(1^2\right)\mathbf{k}}{2} = \frac{1}{2}\mathbf{i} + \frac{1}{\sqrt{2}}\mathbf{j} + \frac{1}{2}\mathbf{k}$
$\mathbf{v}(1) = 2\left[\frac{1}{2}\mathbf{i} + \frac{1}{\sqrt{2}}\mathbf{j} + \frac{1}{2}\mathbf{k}\right]$

9. $\mathbf{v} = 3\,\mathbf{i} + \sqrt{3}\,\mathbf{j} + 2t\,\mathbf{k},\ \mathbf{a} = 2\,\mathbf{k}$. $\mathbf{v}(0) = 3\,\mathbf{i} + \sqrt{3}\,\mathbf{j},\ \mathbf{a}(0) = 2\,\mathbf{k} \Rightarrow |\mathbf{v}(0)| = \sqrt{3^2 + \left(\sqrt{3}\right)^2 + 0^2} = \sqrt{12}$,
$|\mathbf{a}(0)| = \sqrt{2^2} = 2$. $\mathbf{v}(0)\cdot\mathbf{a}(0) = 0 \Rightarrow \cos\theta = \frac{0}{2\sqrt{12}} = 0 \Rightarrow \theta = \frac{\pi}{2}$

10. $\mathbf{v} = \frac{\sqrt{2}}{2}\mathbf{i} + \left(\frac{\sqrt{2}}{2} - 32t\right)\mathbf{j},\ \mathbf{a} = -32\,\mathbf{j}$. $\mathbf{v}(0) = \frac{\sqrt{2}}{2}\mathbf{i} + \frac{\sqrt{2}}{2}\mathbf{j},\ \mathbf{a}(0) = -32\,\mathbf{j} \Rightarrow |\mathbf{v}(0)| = \sqrt{\left(\frac{\sqrt{2}}{2}\right)^2 + \left(\frac{\sqrt{2}}{2}\right)^2} = 1$
$|\mathbf{a}(0)| = \sqrt{(-32)^2} = 32$. $\mathbf{v}(0)\cdot\mathbf{a}(0) = \left(\frac{\sqrt{2}}{2}\right)(-32) = -16\sqrt{2} \Rightarrow \cos\theta = \frac{-16\sqrt{2}}{1(32)} = -\frac{\sqrt{2}}{2} \Rightarrow \theta = \frac{3\pi}{4}$

11. $\mathbf{v} = \frac{2t}{t^2+1}\mathbf{i} + \frac{1}{t^2+1}\mathbf{j} + t\left(t^2+1\right)^{-1/2}\mathbf{k},\ \mathbf{a} = \frac{-2t^2+2}{\left(t^2+1\right)^2}\mathbf{i} - \frac{2t}{\left(t^2+1\right)^2}\mathbf{j} + \frac{1}{\left(t^2+1\right)^{3/2}}\mathbf{k}$. $\mathbf{v}(0) = \mathbf{j},\ \mathbf{a}(0) =$
$2\,\mathbf{i} + \mathbf{k} \Rightarrow |\mathbf{v}(0)| = 1,\ |\mathbf{a}(0)| = \sqrt{2^2 + 1^2} = \sqrt{5}$. $\mathbf{v}(0)\cdot\mathbf{a}(0) = 0 \Rightarrow \cos\theta = \frac{0}{1\sqrt{5}} = 0 \Rightarrow \theta = \frac{\pi}{2}$

12. $\mathbf{v} = \frac{2}{3}(1+t)^{1/2}\,\mathbf{i} - \frac{2}{3}(1-t)^{1/2}\,\mathbf{j} + \frac{1}{3}\mathbf{k},\ \mathbf{a} = \frac{1}{3}(1+t)^{-1/2}\,\mathbf{i} + \frac{1}{3}(1-t)^{-1/2}\,\mathbf{j}$. $\mathbf{v}(0) = \frac{2}{3}\mathbf{i} - \frac{2}{3}\mathbf{j} + \frac{1}{3}\mathbf{k},\ \mathbf{a}(0) =$
$\frac{1}{3}\mathbf{i} + \frac{1}{3}\mathbf{j} \Rightarrow |\mathbf{v}(0)| = \sqrt{\left(\frac{2}{3}\right)^2 + \left(-\frac{2}{3}\right)^2 + \left(\frac{1}{3}\right)^2} = 1,\ |\mathbf{a}(0)| = \sqrt{\left(\frac{1}{3}\right)^2 + \left(\frac{1}{3}\right)^2} = \frac{\sqrt{2}}{3}$.
$\mathbf{v}(0)\cdot\mathbf{a}(0) = \frac{2}{9} - \frac{2}{9} = 0 \Rightarrow \cos\theta = \frac{0}{1\left(\frac{\sqrt{2}}{3}\right)} = 0 \Rightarrow \theta = \frac{\pi}{2}$

13. $\mathbf{v} = (1 - \cos t)\,\mathbf{i} + (\sin t)\,\mathbf{j}$, $\mathbf{a} = (\sin t)\,\mathbf{i} + (\cos t)\,\mathbf{j} \Rightarrow \mathbf{v} \cdot \mathbf{a} = \sin t(1 - \cos t) + \sin t(\cos t) = \sin t$.
$\mathbf{v} \cdot \mathbf{a} = 0 \Rightarrow \sin t = 0 \Rightarrow t = 0, \pi, 2\pi$

14. $\mathbf{v} = (\cos t)\,\mathbf{i} + \mathbf{j} - (\sin t)\,\mathbf{k}$, $\mathbf{a} = (-\sin t)\,\mathbf{i} - (\cos t)\,\mathbf{k} \Rightarrow \mathbf{v} \cdot \mathbf{a} = -\sin t \cos t + \sin t \cos t = 0$ for all $t \geq 0$

15. $$\int_0^1 \left(t^3\,\mathbf{i} + 7\,\mathbf{j} + (t+1)\,\mathbf{k}\right) dt = \left[\frac{t^4}{4}\right]_0^1 \mathbf{i} + [7t]_0^1\,\mathbf{j} + \left[\frac{t^2}{2} + t\right]_0^1 \mathbf{k} = \frac{1}{4}\,\mathbf{i} + 7\,\mathbf{j} + \frac{3}{2}\,\mathbf{k}$$

16. $$\int_1^2 \left((6 - 6t)\,\mathbf{i} + 3\sqrt{t}\,\mathbf{j} + \left(\frac{4}{t^2}\right)\mathbf{k}\right) dt = \left[6t - 3t^2\right]_1^2 \mathbf{i} + \left[2t^{3/2}\right]_1^2 \mathbf{j} + \left[-4t^{-1}\right]_1^2 \mathbf{k} = -3\,\mathbf{i} + \left(4\sqrt{2} - 2\right)\mathbf{j} + 2\,\mathbf{k}$$

17. $$\int_{-\pi/4}^{\pi/4} \left((\sin t)\,\mathbf{i} + (1 + \cos t)\,\mathbf{j} + (\sec^2 t)\,\mathbf{k}\right) dt = [-\cos t]_{-\pi/4}^{\pi/4}\,\mathbf{i} + [t + \sin t]_{-\pi/4}^{\pi/4}\,\mathbf{j} + [\tan t]_{-\pi/4}^{\pi/4}\,\mathbf{k} =$$
$$\left(\frac{\pi + 2\sqrt{2}}{2}\right)\mathbf{j} + 2\,\mathbf{k}$$

18. $$\int_0^{\pi/3} \left((\sec t \tan t)\,\mathbf{i} + (\tan t)\,\mathbf{j} + (2 \sin t \cos t)\,\mathbf{k}\right) dt = \int_0^{\pi/3} \left((\sec t \tan t)\,\mathbf{i} + \frac{\sin t}{\cos t}\,\mathbf{j} + (\sin 2t)\,\mathbf{k}\right) dt =$$
$$[\sec t]_0^{\pi/3}\,\mathbf{i} + [-\ln(\cos t)]_0^{\pi/3}\,\mathbf{j} + \left[-\frac{1}{2}\cos 2t\right]_0^{\pi/3} \mathbf{k} = \mathbf{i} + (\ln 2)\,\mathbf{j} + \frac{3}{4}\,\mathbf{k}$$

19. $$\int_1^4 \left(\frac{1}{t}\,\mathbf{i} + \frac{1}{5 - t}\,\mathbf{j} + \frac{1}{2t}\,\mathbf{k}\right) dt = [\ln t]_1^4\,\mathbf{i} + [-\ln(5 - t)]_1^4\,\mathbf{j} + \left[\frac{1}{2}\ln 2t\right]_1^4 \mathbf{k} = (\ln 4)\,\mathbf{i} + (\ln 4)\,\mathbf{j} + (\ln 2)\,\mathbf{k}$$

20. $$\int_0^1 \left(\frac{2}{\sqrt{1 - t^2}}\,\mathbf{i} + \frac{\sqrt{3}}{1 + t^2}\,\mathbf{k}\right) dt = \left[2 \sin^{-1} t\right]_0^1 \mathbf{i} + \left[\sqrt{3} \tan^{-1} t\right]_0^1 \mathbf{k} = \pi\,\mathbf{i} + \frac{\pi\sqrt{3}}{4}\,\mathbf{k}$$

21. $\mathbf{v} = (\cos t)\,\mathbf{i} - (\sin t)\,\mathbf{j}$, $\mathbf{a} = -(\sin t)\,\mathbf{i} - (\cos t)\,\mathbf{j} \Rightarrow$ For $t = \frac{\pi}{4}$, $\mathbf{v}\left(\frac{\pi}{4}\right) = \frac{\sqrt{2}}{2}\,\mathbf{i} - \frac{\sqrt{2}}{2}\,\mathbf{j}$, $\mathbf{a}\left(\frac{\pi}{4}\right) = -\frac{\sqrt{2}}{2}\,\mathbf{i} - \frac{\sqrt{2}}{2}\,\mathbf{j}$;

For $t = \frac{\pi}{2}$, $\mathbf{v}\left(\frac{\pi}{2}\right) = -\mathbf{j}$, $\mathbf{a}\left(\frac{\pi}{2}\right) = -\mathbf{i}$

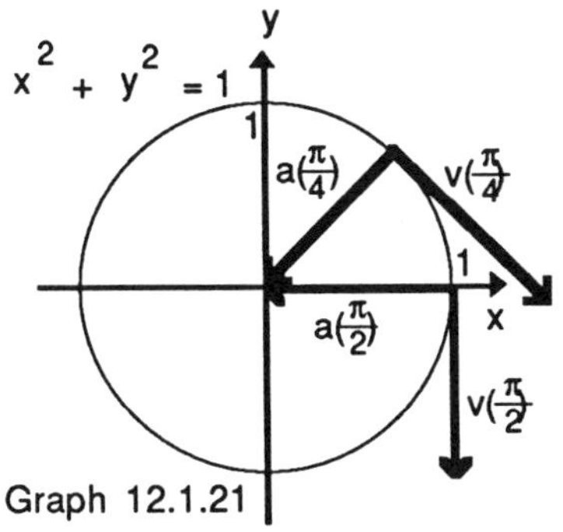

Graph 12.1.21

22. $\mathbf{v} = \left(4\cos\frac{t}{2}\right)\mathbf{i} + \left(4\sin\frac{t}{2}\right)\mathbf{j}$, $\mathbf{a} = \left(-\cos\frac{t}{2}\right)\mathbf{i} + \left(-\sin\frac{t}{2}\right)\mathbf{j} \Rightarrow$ For $t = \pi$, $\mathbf{v}(\pi) = -2\,\mathbf{i}$, $\mathbf{a}(\pi) = -\mathbf{j}$;
For $t = \frac{3\pi}{2}$, $\mathbf{v}\left(\frac{3\pi}{2}\right) = -\sqrt{2}\,\mathbf{i} - \sqrt{2}\,\mathbf{j}$, $\mathbf{a}\left(\frac{3\pi}{2}\right) = \frac{\sqrt{2}}{2}\mathbf{i} - \frac{\sqrt{2}}{2}\mathbf{j}$.

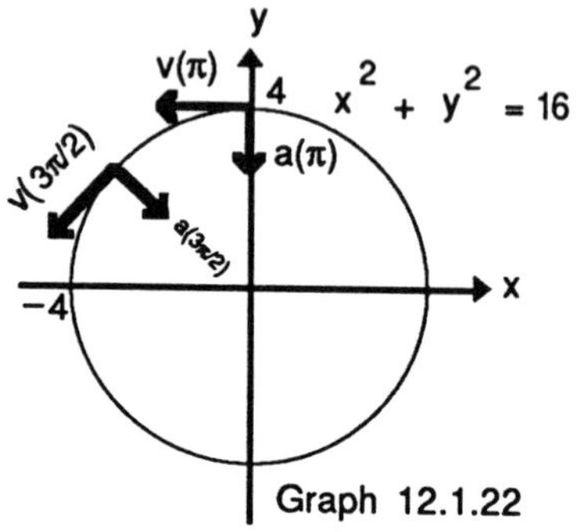

Graph 12.1.22

23. $\mathbf{v} = (1 - \cos t)\,\mathbf{i} + (\sin t)\,\mathbf{j}$, $\mathbf{a} = (\sin t)\,\mathbf{i} + (\cos t)\,\mathbf{j} \Rightarrow$ For $t = \pi$, $\mathbf{v}(\pi) = 2\,\mathbf{i}$, $\mathbf{a}(\pi) = -\mathbf{j}$; For $t = \frac{3\pi}{2}$,
$\mathbf{v}\left(\frac{3\pi}{2}\right) = \mathbf{i} - \mathbf{j}$, $\mathbf{a}\left(\frac{3\pi}{2}\right) = -\mathbf{i}$

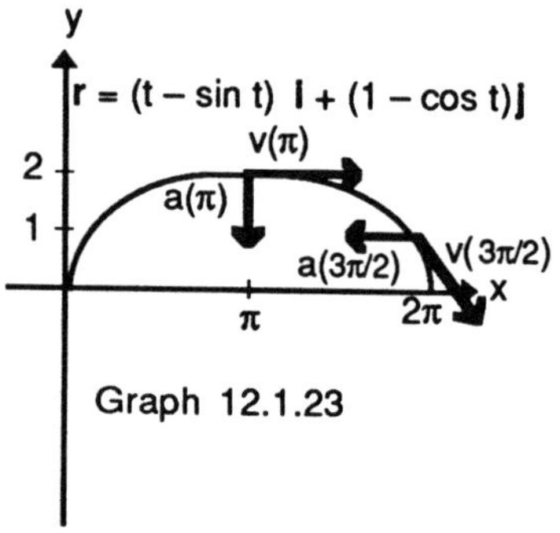

Graph 12.1.23

24. $\mathbf{v} = \mathbf{i} + 2t\,\mathbf{j}$, $\mathbf{a} = 2\,\mathbf{j} \Rightarrow$ For $t = -1$, $\mathbf{v}(-1) = \mathbf{i} - 2\,\mathbf{j}$, $\mathbf{a}(-1) = 2\,\mathbf{j}$; For $t = 0$, $\mathbf{v}(0) = \mathbf{i}$, $\mathbf{a}(0) = 2\,\mathbf{j}$; For $t = 1$,
$\mathbf{v}(1) = \mathbf{i} + 2\,\mathbf{j}$, $\mathbf{a}(1) = 2\,\mathbf{j}$

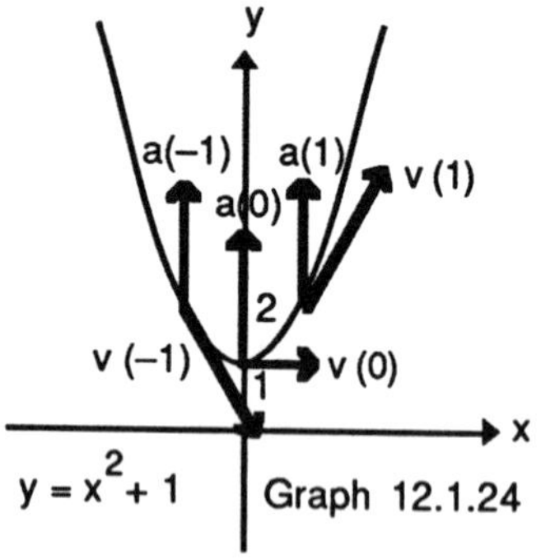

Graph 12.1.24

25. $\mathbf{r} = \int (-t\,\mathbf{i} - t\,\mathbf{j} - t\,\mathbf{k})\,dt = -\frac{t^2}{2}\mathbf{i} - \frac{t^2}{2}\mathbf{j} - \frac{t^2}{2}\mathbf{k} + \mathbf{C}$. $\mathbf{r}(0) = 0\,\mathbf{i} - 0\,\mathbf{j} - 0\,\mathbf{k} + \mathbf{C} = \mathbf{i} + 2\,\mathbf{j} + 3\,\mathbf{k} \Rightarrow$
$\mathbf{C} = \mathbf{i} + 2\,\mathbf{j} + 3\,\mathbf{k}$. $\therefore$ $\mathbf{r} = \left(-\frac{t^2}{2} + 1\right)\mathbf{i} + \left(-\frac{t^2}{2} + 2\right)\mathbf{j} + \left(-\frac{t^2}{2} + 3\right)\mathbf{k}$

26. $\mathbf{r} = \int \left((180\,t)\,\mathbf{i} + (180t - 16t^2)\mathbf{j}\right) dt = 90t^2\,\mathbf{i} + \left(90t^2 - \frac{16}{3}t^3\right)\mathbf{j} + \mathbf{C}$. $\mathbf{r}(0) = 100\,\mathbf{j} \Rightarrow$
$90(0)^2\,\mathbf{i} + \left(90(0)^2 - \frac{16}{3}(0)^3\right)\mathbf{j} + \mathbf{C} = 100\,\mathbf{j} \Rightarrow \mathbf{C} = 100\,\mathbf{j}$. $\therefore$ $\mathbf{r} = 90t^2\,\mathbf{i} + \left(90t^2 - \frac{16}{3}t^3 + 100\right)\mathbf{j}$

27. $\mathbf{r} = \int \left(\frac{3}{2}(t+1)^{1/2}\,\mathbf{i} + e^{-t}\,\mathbf{j} + \frac{1}{t+1}\mathbf{k}\right) dt = (t+1)^{3/2}\,\mathbf{i} - e^{-t}\,\mathbf{j} + \ln(t+1)\,\mathbf{k} + \mathbf{C}$. $\mathbf{r}(0) = \mathbf{k} \Rightarrow$
$(0+1)^{3/2}\,\mathbf{i} - e^{-0}\,\mathbf{j} + \ln(0+1)\,\mathbf{k} + \mathbf{C} = \mathbf{k} \Rightarrow \mathbf{C} = -\mathbf{i} + \mathbf{j} + \mathbf{k}$.
$\therefore$ $\mathbf{r} = \left((t+1)^{3/2} - 1\right)\mathbf{i} + \left(1 - e^{-t}\right)\mathbf{j} + (1 + \ln(t+1))\,\mathbf{k}$

28. $\mathbf{r} = \int \left((t^3 + 4t)\,\mathbf{i} + t\,\mathbf{j} + 2t^2\,\mathbf{k}\right) dt = \left(\frac{t^4}{4} + 2t^2\right)\mathbf{i} + \frac{t^2}{2}\mathbf{j} + \frac{2t^3}{3}\mathbf{k} + \mathbf{C}$. $\mathbf{r}(0) = \mathbf{i} + \mathbf{j} \Rightarrow$
$\left(\frac{0^4}{4} + 2(0)^2\right)\mathbf{i} + \frac{0^2}{2}\mathbf{j} + \frac{2(0)^3}{3}\mathbf{k} + \mathbf{C} = \mathbf{i} + \mathbf{j} \Rightarrow \mathbf{C} = \mathbf{i} + \mathbf{j}$. $\therefore$ $\mathbf{r} = \left(\frac{t^4}{4} + 2t^2 + 1\right)\mathbf{i} + \left(\frac{t^2}{2} + 1\right)\mathbf{j} + \frac{2t^3}{3}\mathbf{k}$

29. $\frac{d\mathbf{r}}{dt} = \int (-32\,\mathbf{k})\,dt = -32t\,\mathbf{k} + \mathbf{C}_1$. $\frac{d\mathbf{r}}{dt}(0) = 8\,\mathbf{i} + 8\,\mathbf{j} \Rightarrow -32(0)\,\mathbf{k} + \mathbf{C}_1 = 8\,\mathbf{i} + 8\,\mathbf{j} \Rightarrow \mathbf{C}_1 = 8\,\mathbf{i} + 8\,\mathbf{j}$
$\therefore$ $\frac{d\mathbf{r}}{dt} = 8\,\mathbf{i} + 8\,\mathbf{j} - 32t\,\mathbf{k}$. $\mathbf{r} = \int (8\,\mathbf{i} + 8\,\mathbf{j} - 32t\,\mathbf{k})\,dt = 8t\,\mathbf{i} + 8t\,\mathbf{j} - 16t^2\,\mathbf{k} + \mathbf{C}_2$. $\mathbf{r}(0) = 100\,\mathbf{k} \Rightarrow$
$8(0)\,\mathbf{i} + 8(0)\,\mathbf{j} - 16(0)^2\,\mathbf{k} + \mathbf{C}_2 = 100\,\mathbf{k} \Rightarrow \mathbf{C}_2 = 100\,\mathbf{k}$. $\therefore$ $\mathbf{r} = 8t\,\mathbf{i} + 8t\,\mathbf{j} + (100 - 16t^2)\,\mathbf{k}$

30. $\frac{d\mathbf{r}}{dt} = \int -(\mathbf{i} + \mathbf{j} + \mathbf{k})\,dt = -(t\,\mathbf{i} + t\,\mathbf{j} + t\,\mathbf{k}) + \mathbf{C}_1$. $\frac{d\mathbf{r}}{dt}(0) = \mathbf{0} \Rightarrow -(0\,\mathbf{i} + 0\,\mathbf{j} + 0\,\mathbf{k}) + \mathbf{C}_1 = \mathbf{0} \Rightarrow \mathbf{C}_1 = \mathbf{0}$
$\therefore$ $\frac{d\mathbf{r}}{dt} = -(t\,\mathbf{i} + t\,\mathbf{j} + t\,\mathbf{k})$. $\mathbf{r} = \int -(t\,\mathbf{i} + t\,\mathbf{j} + t\,\mathbf{k})\,dt = -\left(\frac{t^2}{2}\mathbf{i} + \frac{t^2}{2}\mathbf{j} + \frac{t^2}{2}\mathbf{k}\right) + \mathbf{C}_2$ $\mathbf{r}(0) = 10\,\mathbf{i} + 10\,\mathbf{j} + 10\,\mathbf{k}$
$\Rightarrow -\left(\frac{0^2}{2}\mathbf{i} + \frac{0^2}{2}\mathbf{j} + \frac{0^2}{2}\mathbf{k}\right) + \mathbf{C}_2 = 10\,\mathbf{i} + 10\,\mathbf{j} + 10\,\mathbf{k} \Rightarrow \mathbf{C}_2 = 10\,\mathbf{i} + 10\,\mathbf{j} + 10\,\mathbf{k}$
$\therefore$ $\mathbf{r} = \left(-\frac{t^2}{2} + 10\right)\mathbf{i} + \left(-\frac{t^2}{2} + 10\right)\mathbf{j} + \left(-\frac{t^2}{2} + 10\right)\mathbf{k}$

31. $\mathbf{v} = (1 - \cos t)\,\mathbf{i} + (\sin t)\,\mathbf{j}$, $\mathbf{a} = (\sin t)\,\mathbf{i} + (\cos t)\,\mathbf{j}$. $|\mathbf{v}|^2 = (1 - \cos t)^2 + \sin^2 t = 2 - 2\cos t$. $|\mathbf{v}|^2$ is at a max when $\cos t = -1 \Rightarrow t = \pi, 3\pi, 5\pi$, etc. At these values of t, $|\mathbf{v}|^2 = 4 \Rightarrow \max|\mathbf{v}| = \sqrt{4} = 2$. $|\mathbf{v}|^2$ is at a min when $\cos t = 1 \Rightarrow t = 0, 2\pi, 4\pi$, etc. At these values of t, $|\mathbf{v}|^2 = 0 \Rightarrow \min|\mathbf{v}| = 0$.
$|\mathbf{a}|^2 = \sin^2 t + \cos^2 t = 1$ for every $t \Rightarrow \max|\mathbf{a}| = \min|\mathbf{a}| = \sqrt{1} = 1$.

32. $\mathbf{v} = (-3\sin t)\,\mathbf{j} + (2\cos t)\,\mathbf{k}$, $\mathbf{a} = (-3\cos t)\,\mathbf{j} - (2\sin t)\,\mathbf{k}$. $|\mathbf{v}|^2 = 9\sin^2 t + 4\cos^2 t \Rightarrow \frac{d}{dt}\left(|\mathbf{v}|^2\right) =$ $18\sin t\cos t - 8\cos t\sin t = 10\sin t\cos t$. Let $\frac{d}{dt}\left(|\mathbf{v}|^2\right) = 0 \Rightarrow 10\sin t\cos t = 0 \Rightarrow \sin t = 0$ or $\cos t = 0 \Rightarrow t = 0, \pi$ or $t = \frac{\pi}{2}, \frac{3\pi}{2}$. When $t = 0, \pi$, $|\mathbf{v}|^2 = 4 \Rightarrow |\mathbf{v}| = \sqrt{4} = 2$; when $t = \frac{\pi}{2}, \frac{3\pi}{2}$, $|\mathbf{v}| = \sqrt{9} = 3$. $\therefore$ $\max|\mathbf{v}|$ is 3 when $t = \frac{\pi}{2}, \frac{3\pi}{2}$; $\min|\mathbf{v}| = 2$ when $t = 0, \pi$. $|\mathbf{a}|^2 = 9\cos^2 t + 4\sin^2 t \Rightarrow \frac{d}{dt}\left(|\mathbf{a}|^2\right) =$ $-18\cos t\sin t + 8\sin t\cos t = -10\sin t\cos t$. Let $\frac{d}{dt}\left(|\mathbf{a}|^2\right) = 0 \Rightarrow -10\sin t\cos t = 0 \Rightarrow \sin t = 0$ or $\cos t = 0 \Rightarrow t = 0, \pi$ or $t = \frac{\pi}{2}, \frac{3\pi}{2}$. When $t = 0, \pi$, $|\mathbf{a}|^2 = 9 \Rightarrow |\mathbf{a}| = 3$; when $t = \frac{\pi}{2}, \frac{3\pi}{2}$, $|\mathbf{a}|^2 = 4 \Rightarrow$ $|\mathbf{a}| = 2$. $\therefore$ $\max|\mathbf{a}| = 3$ when $t = 0, \pi$; $\min|\mathbf{a}| = 2$ when $t = \frac{\pi}{2}, \frac{3\pi}{2}$.

33. Let **f** = **C**, a constant vector. Then **f** = c_1 **i** + c_2 **j** + c_3 **k** where c_1, c_2, c_3 are Real Numbers.
$\frac{d\mathbf{f}}{dt} = 0\,\mathbf{i} + 0\,\mathbf{j} + 0\,\mathbf{k} = \mathbf{0}$

34. Let **f** be a differentiable function of t $\Rightarrow$ **f**(t) = f(t) **i** + g(t) **j** + h(t) **k** where f, g, h are differentiable functions of t. Then $\frac{d}{dt}(c\mathbf{f}(t)) = \frac{d}{dt}\left(c(f(t)\,\mathbf{i} + g(t)\,\mathbf{j} + h(t)\,\mathbf{k})\right) = \frac{d}{dt}\left((cf(t))\,\mathbf{i} + (cg(t))\,\mathbf{j} + (ch(t))\,\mathbf{k}\right) =$
$\left(\frac{d}{dt}(cf(t))\right)\mathbf{i} + \left(\frac{d}{dt}(cg(t))\right)\mathbf{j} + \left(\frac{d}{dt}(ch(t))\right)\mathbf{k} = \left(c\frac{d}{dt}(f(t))\right)\mathbf{i} + \left(c\frac{d}{dt}(g(t))\right)\mathbf{j} + \left(c\frac{d}{dt}(h(t))\right)\mathbf{k} =$
$c\left(\frac{d}{dt}(f(t))\,\mathbf{i} + \frac{d}{dt}(g(t))\,\mathbf{j} + \frac{d}{dt}(h(t))\,\mathbf{k}\right) = c\left(\frac{d}{dt}(f(t)\,\mathbf{i} + g(t)\,\mathbf{j} + h(t)\,\mathbf{k})\right) = c\frac{d\mathbf{f}}{dt}$

35. Let **u** = $f_1(t)\,\mathbf{i} + f_2(t)\,\mathbf{j} + f_3(t)\,\mathbf{k}$, **v** = $g_1(t)\,\mathbf{i} + g_2(t)\,\mathbf{j} + g_3(t)\,\mathbf{k}$. Then $\mathbf{u} + \mathbf{v} = (f_1(t) + g_1(t))\,\mathbf{i} + (f_2(t) + g_2(t))\,\mathbf{j} +$
$(f_3(t) + g_3(t))\,\mathbf{k} \Rightarrow \frac{d}{dt}(\mathbf{u} + \mathbf{v}) = (f'_1(t) + g'_1(t))\,\mathbf{i} + (f'_2(t) + g'_2(t))\,\mathbf{j} + (f'_3(t) + g'_3(t))\,\mathbf{k} =$
$(f'_1(t)\,\mathbf{i} + f'_2(t)\,\mathbf{j} + f'_3(t)\,\mathbf{k}) + (g'_1(t)\,\mathbf{i} + g'_2(t)\,\mathbf{j} + g'_3(t)\,\mathbf{k} = \frac{d\mathbf{u}}{dt} + \frac{d\mathbf{v}}{dt}$.
$\mathbf{u} - \mathbf{v} = (f_1(t) - g_1(t))\,\mathbf{i} + (f_2(t) - g_2(t))\,\mathbf{j} + (f_3(t) - g_3(t))\,\mathbf{k} \Rightarrow \frac{d}{dt}(\mathbf{u} - \mathbf{v}) = (f'_1(t) - g'_1(t))\,\mathbf{i} + (f'_2(t) - g'_2(t))\,\mathbf{j} +$
$(f'_3(t) - g'_3(t))\,\mathbf{k} = (f'_1(t)\,\mathbf{i} + f'_2(t)\,\mathbf{j} + f'_3(t)\,\mathbf{k}) - (g'_1(t)\,\mathbf{i} + g'_2(t)\,\mathbf{j} + g'_3(t)\,\mathbf{k}) = \frac{d\mathbf{u}}{dt} - \frac{d\mathbf{v}}{dt}$.

SECTION 12.2 MODELING PROJECTILE MOTION

1. $x = (v_0 \cos\alpha)t \Rightarrow (21\text{ km})\left(\frac{1000\text{ m}}{1\text{ km}}\right) = 840\text{ m/s}(\cos 60°)t \Rightarrow t = \frac{21\,000\text{ m}}{(840\text{ km/s})(\cos 60°)} = 50$ seconds

2. Maximum R = 24.5 km $\Rightarrow \alpha = 45°$. $\therefore$ $24.5\text{ km} = \frac{{v_0}^2}{9.8\text{ m/s}^2} \Rightarrow {v_0}^2 = (9.8)(24\,500)\text{ m}^2/\text{s}^2 \Rightarrow v_0 =$
$\sqrt{(9.8)(24\,500)\text{ m}^2/\text{s}^2} = 490$ m/s

3. a) $t = \frac{2v_0 \sin\alpha}{g} = \frac{2(500\text{ m/s})\sin 45°}{9.8\text{ m/s}^2} = 72.2$ seconds. $R = \frac{{v_0}^2}{g}\sin 2\alpha = \frac{(500\text{ m/s})^2}{9.8\text{ m/s}^2}(\sin 2(45°)) =$
25 510.2 m
b) $x = (v_0 \cos\alpha)t \Rightarrow 5000\text{ m} = (500\text{ m/s})(\cos 45°)t \Rightarrow t = \frac{5000\text{ m}}{(500\text{ m/s})\cos 45°} = 14.14$ s
$y = (v_0 \sin\alpha)t - \frac{1}{2}gt^2 \Rightarrow y = (500\text{ m/s})(\sin 45°)(14.14\text{ s}) - \frac{1}{2}(9.8\text{ m/s}^2)(14.14\text{ s})^2 = 4020.3$ m
c) $y_{max} = \frac{(v_0 \sin\alpha)^2}{2g} = \frac{((500\text{ m/s})\sin 45°)^2}{2(9.8\text{ m/s}^2)} = 6377.6$ m

4. $x = (v_0 \cos\alpha)t \Rightarrow x = (32\text{ ft/sec})(\cos 30°)t = 32\left(\frac{\sqrt{3}}{2}\right)t\text{ ft} = 16\sqrt{3}\,t$ ft. $y = y_0 + (v_0 \sin\alpha)t - \frac{1}{2}gt^2 \Rightarrow$
$y = 32\text{ ft} + (32\text{ ft/sec})(\sin 30°)t - \frac{1}{2}(32\text{ ft/sec}^2)t^2 \Rightarrow y = 32 + 16t - 16t^2$. Let $y = 0 \Rightarrow$
$0 = 32 + 16t - 16t^2 \Rightarrow t = -1$ or $t = 2$. Since $t \geq 0$, $t = 2$ is when the ball hits the ground $\Rightarrow$
$x = 16\sqrt{3}(2)\text{ ft} = 55.4$ ft

5. $R = \frac{v_0^2}{g}\sin 2\alpha = \frac{v_0^2}{g}\left(2\sin\alpha\cos\alpha\right) = \frac{v_0^2}{g}\left(2\cos(90° - \alpha)\sin(90° - \alpha)\right) = \frac{v_0^2}{g}\left(\sin 2(90° - \alpha)\right)$

6. $R = \frac{v_0^2}{g}\sin 2\alpha \Rightarrow 16\,000\text{ m} = \frac{(400\text{ m/s})^2}{9.8\text{ m/s}^2}\sin 2\alpha \Rightarrow \sin 2\alpha = 0.98 \Rightarrow 2\alpha = 78.5°$ or $2\alpha = 101.5° \Rightarrow$ $\alpha = 39.25°$ or $50.75°$

7. $R = \frac{v_0^2}{g}\sin 2\alpha \Rightarrow 10\text{ m} = \frac{v_0^2}{9.8\text{ m/s}^2}\sin 2(45°) \Rightarrow v_0^2 = 98\text{ m}^2/\text{s}^2 \Rightarrow v_0 = 9.9\text{ m/s}.$

$6\text{ m} = \frac{(9.9\text{ m/s})^2}{9.8\text{ m/s}^2}\sin 2\alpha \Rightarrow \sin 2\alpha = 0.59999 \Rightarrow 2\alpha = 36.87°$ or $143.12° \Rightarrow \alpha = 18.44°$ or $71.56°$

8. $v_0 = 5 \times 10^6$ m/s and $x = 40$ cm $= 0.4$ m. Then $x = \left(v_0\cos\alpha\right)t \Rightarrow 0.4\text{ m} = \left(5 \times 10^6\text{ m/s}\right)(\cos 0°)t \Rightarrow$ $t = 0.08 \times 10^{-6}\text{ s} = 8 \times 10^{-8}$ s. $y = y_0 + \left(v_0\sin\alpha\right)t - \frac{1}{2}gt^2 \Rightarrow y = \left(5 \times 10^6\text{ m/s}\right)(\sin 0°)(8 \times 10^{-8}\text{ s}) -$ $\frac{1}{2}\left(9.8\text{ m/s}^2\right)(8 \times 10^{-8}\text{ s})^2 = -3.92 \times 10^{-7}$ m or -3.92×10^{-5} cm. It drops -3.92×10^{-5} cm.

9. $R = \frac{v_0^2}{g}\sin 2\alpha \Rightarrow (746.4\text{ ft}) = \frac{v_0^2}{32\text{ ft/sec}^2}\sin 2(9°) \Rightarrow v_0^2 = 77\,292.84\text{ ft}^2/\text{sec}^2 \Rightarrow v_0 = 278.01$ ft/sec = 189.6 mph

10. If the initial speed is $2v_0$, where v_0 is the original speed, then $y_{max} = \frac{\left(2v_0\sin\alpha\right)^2}{2g} = \frac{4\left(v_0\sin\alpha\right)^2}{2g} =$ $4\left(\frac{\left(v_0\sin\alpha\right)^2}{2g}\right)$ or 4 times the original maximum height.

$R = \frac{\left(2v_0\right)^2}{g}\sin 2\alpha = \frac{4v_0^2}{g}\sin 2\alpha = 4\left(\frac{v_0^2}{g}\sin 2\alpha\right)$ or 4 times the original range.

Let $R = \frac{v_0^2}{g}\sin 2\alpha$. Then we want the factor, p, so that pv_0 will double the range $\Rightarrow \frac{\left(pv_0\right)^2}{g}\sin 2\alpha = 2R$ $= 2\left(\frac{v_0^2}{g}\sin 2\alpha\right) \Rightarrow p^2v_0^2 = 2v_0^2 \Rightarrow p^2 = 2 \Rightarrow p = \sqrt{2}$ or 1.41 or 141%. The same percent will approximately double the height.

11. $y_{max} = \frac{\left(v_0\sin\alpha\right)^2}{2g} \Rightarrow \frac{3}{4}y_{max} = \frac{3\left(v_0\sin\alpha\right)^2}{8g}$. $y = \left(v_0\sin\alpha\right)t - \frac{1}{2}gt^2 \Rightarrow \frac{3\left(v_0\sin\alpha\right)^2}{8g} =$ $\left(v_0\sin\alpha\right)t - \frac{1}{2}gt^2 \Rightarrow 3\left(v_0\sin\alpha\right)^2 = \left(8gv_0\sin\alpha\right)t - 4g^2t^2 \Rightarrow$

$4g^2t^2 - \left(8gv_0\sin\alpha\right)t + 3\left(v_0\sin\alpha\right)^2 = 0 \Rightarrow 2gt - 3v_0\sin\alpha = 0$ or $2gt - v_0\sin\alpha = 0 \Rightarrow t = \frac{3v_0\sin\alpha}{2g}$ or $t = \frac{v_0\sin\alpha}{2g}$. Since the time it takes to reach y_{max} is $t_{max} = \frac{v_0\sin\alpha}{g}$, then the time it takes the projectile to reach $\frac{3}{4}$ of y_{max} is $t = \frac{v_0\sin\alpha}{2g}$ or $\frac{1}{2}t_{max}$.

12. $v_0 = \frac{80\sqrt{10}}{3}$ ft/sec. We want R = 200 ft $\Rightarrow 200 = \frac{(80\sqrt{10}/3)^2}{32}\sin 2\alpha \Rightarrow \sin 2\alpha = .9 \Rightarrow 2\alpha = 64.2° \Rightarrow$ $\alpha = 32.1°$. $y_{max} = \frac{((80\sqrt{10}/3)\sin 32.1°)^2}{2(32)} = 31.4$ ft. In order to reach the cushion, the angle of elevation will need to be 32.1°. At this angle, Hugo will go 31.4 ft into the air at maximum height. He will make it with room to spare.

13. $x = (v_0 \cos\alpha)t \Rightarrow 135$ ft = (90 ft/sec)(cos 30°)t $\Rightarrow$ t = 1.732 sec. $y = (v_0 \sin\alpha)t - \frac{1}{2}gt^2 \Rightarrow$ $y = (90 \text{ ft/sec})(\sin 30°)(1.732 \text{ sec}) - \frac{1}{2}(32 \text{ ft/sec}^2)(1.732 \text{ sec})^2 \Rightarrow y = 29.94$ ft. The golf ball will clip the leaves at the top.

14. $v_0 = 116$ ft/sec, $\alpha = 45° \Rightarrow R = \frac{116^2}{32}\sin 2(45°) = 420.5$ ft. Since the tee is 369 ft from the pin, the ball will have enough range. $x = (v_0\cos\alpha)t \Rightarrow 369 = (116\cos 45°)t \Rightarrow t = 4.50$ sec. $y = (v_0\sin\alpha)t - \frac{1}{2}gt^2 \Rightarrow y = (116\sin 45°)(4.50) - \frac{1}{2}(32)(4.50)^2 = 45.11$ ft. It will take the ball 4.50 sec to travel 369 ft. At that time the ball will be 45.11 ft in the air. It should reach the green.

15. $x = (v_0\cos\alpha)\,t \Rightarrow 315$ ft = $(v_0 \cos 20°)t \Rightarrow v_0 = \frac{315}{t\cos 20°}$. $y = (v_0\sin\alpha)t - \frac{1}{2}gt^2 \Rightarrow$ $34 \text{ ft} = \frac{315}{t\cos 20°}(t\sin 20°) - \frac{1}{2}(32)t^2 \Rightarrow 34 = 315\tan 20° - 16t^2 \Rightarrow t^2 = 5.04 \text{ sec}^2 \Rightarrow t = 2.25$ sec $t = 2.25 \text{ sec} \Rightarrow v_0 = \frac{315}{(2.25)\cos 20°} = 148.98$ ft/sec

16. Flight time = 1 sec. The measure of the angle of elevation is 64° (using a protractor). $\therefore\ t = \frac{2v_0\sin\alpha}{g}$ $\Rightarrow 1 = \frac{2v_0\sin 64°}{32} \Rightarrow v_0 = 17.80$ ft/sec. Then $y_{max} = \frac{(17.80\sin 64°)^2}{2(32)} = 4.00$ ft. $R = \frac{v_0^2}{g}\sin 2\alpha \Rightarrow$ $R = \frac{(17.80)^2}{32}\sin 2(64°) = 7.80$ ft $\Rightarrow$ Engine traveled 7.80 ft in 1 sec $\Rightarrow$ Engine velocity was 7.80 ft/sec.

17. Height of the Marble A, R units downrange: $x = (v_0\cos\alpha)\,t$ and $x = R \Rightarrow R = (v_0\cos\alpha)\,t \Rightarrow$ $t = \frac{R}{v_0\cos\alpha}$. $y = (v_0\sin\alpha)t - \frac{1}{2}gt^2 \Rightarrow y = (v_0\sin\alpha)\left(\frac{R}{v_0\cos\alpha}\right) - \frac{1}{2}g\left(\frac{R}{v_0\cos\alpha}\right)^2 \Rightarrow$

$y = R\tan\alpha - \frac{1}{2}g\left(\frac{R^2}{v_0^2\cos^2\alpha}\right)$ is the height of Marble A after $t = \frac{R}{v_0\cos\alpha}$ seconds.

Height of the Marble B, at $t = \frac{R}{v_0\cos\alpha}$ seconds: $y = R\tan\alpha - \frac{1}{2}gt^2 = R\tan\alpha - \frac{1}{2}g\left(\frac{R}{v_0\cos\alpha}\right)^2 =$ $R\tan\alpha - \frac{1}{2}g\left(\frac{R^2}{v_0^2\cos^2\alpha}\right)$ which is the height of Marble A. $\therefore$ They collide regardless of the initial velocity.

18. a)

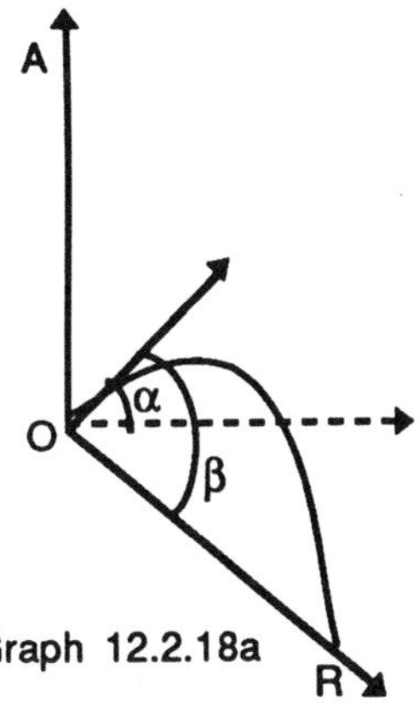

Graph 12.2.18a

$\tan\beta = \frac{y}{x}$. $x = (v_0 \cos\alpha)\,t$ and $y = (v_0 \sin\alpha)t - \frac{1}{2}gt^2 \Rightarrow \tan\beta =$
$\frac{(v_0 \sin\alpha)t - \frac{1}{2}gt^2}{(v_0\cos\alpha)\,t} = \frac{(v_0 \sin\alpha) - \frac{1}{2}gt}{(v_0\cos\alpha)} \Rightarrow v_0 \cos\alpha \tan\beta = v_0 \sin\alpha - \frac{1}{2}gt \Rightarrow$
$t = \frac{2v_0 \sin\alpha - 2v_0 \cos\alpha \tan\beta}{g}$. This is the time when the projectile will hit the downhill slope. $\therefore x = v_0 \cos\alpha\left(\frac{2v_0 \sin\alpha - 2v_0\cos\alpha\tan\beta}{g}\right) =$
$\frac{2v_0^2}{g}\left(\sin\alpha\cos\alpha - \cos^2\alpha\tan\beta\right)$. If x is maximized, then OR is maximized.

$\frac{dx}{d\alpha} = \frac{2v_0^2}{g}(\cos 2\alpha + \sin 2\alpha \tan\beta)$. $\frac{dx}{d\alpha} = 0 \Rightarrow \cos 2\alpha + \sin 2\alpha\tan\beta = 0 \Rightarrow \cot 2\alpha + \tan\beta = 0 \Rightarrow$
$\cot 2\alpha = -\tan\beta = \tan(-\beta) \Rightarrow 2\alpha = 90° - (-\beta) = 90° + \beta \Rightarrow \alpha = \frac{1}{2}(90° + \beta) = \frac{1}{2}$ of $\angle$ AOR.

b)

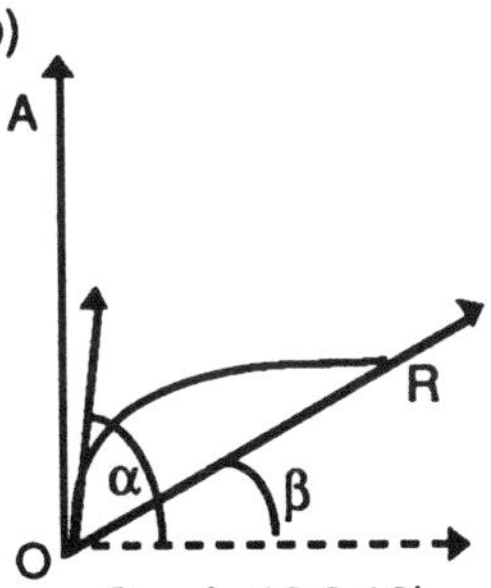

Graph 12.2.18b

By the same reasoning in 18 a), $\tan\beta = \frac{y}{x}$. $x = (v_0\cos\alpha)\,t$ and
$y = (v_0\sin\alpha)t - \frac{1}{2}gt^2 \Rightarrow \tan\beta = \frac{(v_0 \sin\alpha)t - \frac{1}{2}gt^2}{(v_0\cos\alpha)\,t} = \frac{(v_0 \sin\alpha) - \frac{1}{2}gt}{(v_0\cos\alpha)} \Rightarrow$
$v_0\cos\alpha\tan\beta = v_0\sin\alpha - \frac{1}{2}gt \Rightarrow t = \frac{2v_0 \sin\alpha - 2v_0 \cos\alpha \tan\beta}{g}$.
This is the time when the projectile will hit the uphill slope.
$\therefore x = v_0 \cos\alpha\left(\frac{2v_0 \sin\alpha - 2v_0\cos\alpha\tan\beta}{g}\right) =$
$\frac{2v_0^2}{g}\left(\sin\alpha\cos\alpha - \cos^2\alpha\tan\beta\right)$. If x is maximized, then OR is maximized.
$\frac{dx}{d\alpha} = \frac{2v_0^2}{g}(\cos 2\alpha + \sin 2\alpha \tan\beta)$. $\frac{dx}{d\alpha} = 0 \Rightarrow \cos 2\alpha + \sin 2\alpha\tan\beta = 0 \Rightarrow \cot 2\alpha + \tan\beta = 0 \Rightarrow$
$\cot 2\alpha = -\tan\beta = \tan(-\beta) \Rightarrow 2\alpha = 90° - (-\beta) = 90° + \beta \Rightarrow \alpha = \frac{1}{2}(90° + \beta) = \frac{1}{2}(90°) + \frac{1}{2}\beta =$
$\frac{1}{2}(90°) + \beta - \frac{1}{2}\beta = \beta + \frac{1}{2}(90° - \beta)$. Now $\frac{1}{2}(90° - \beta) =$
$\frac{1}{2}$ of $\angle$ AOR. $\therefore v_0$ would bisect $\angle$ AOR for maximum range uphill.

19. $\frac{d\mathbf{r}}{dt} = \int (-g\,\mathbf{j})\,dt = -gt\,\mathbf{j} + \mathbf{C}_1$. $\frac{d\mathbf{r}}{dt}(0) = (v_0\cos\alpha)\,\mathbf{i} + (v_0 \sin\alpha)\,\mathbf{j} \Rightarrow -g(0)\,\mathbf{j} + \mathbf{C}_1 = (v_0\cos\alpha)\,\mathbf{i} +$
$(v_0\sin\alpha)\,\mathbf{j} \Rightarrow \mathbf{C}_1 = (v_0\cos\alpha)\,\mathbf{i} + (v_0\sin\alpha)\,\mathbf{j} \Rightarrow \frac{d\mathbf{r}}{dt} = (v_0\cos\alpha)\,\mathbf{i} + (v_0\sin\alpha - gt)\,\mathbf{j}$.
$\mathbf{r} = \int \left((v_0\cos\alpha)\,\mathbf{i} + (v_0\sin\alpha - gt)\,\mathbf{j}.\right)dt = (v_0t\cos\alpha)\,\mathbf{i} + (v_0t\sin\alpha - \frac{1}{2}gt^2)\,\mathbf{j} + \mathbf{C}_2$.
$\mathbf{r}(0) = x_0\,\mathbf{i} + y_0\,\mathbf{j} \Rightarrow (v_0(0)\cos\alpha)\,\mathbf{i} + (v_0(0)\sin\alpha - \frac{1}{2}g(0)^2)\,\mathbf{j} + \mathbf{C}_2 = x_0\,\mathbf{i} + y_0\,\mathbf{j} \Rightarrow$
$\mathbf{C}_2 = x_0\,\mathbf{i} + y_0\,\mathbf{j}$. $\therefore$ $\mathbf{r} = (x_0 + v_0t\cos\alpha)\,\mathbf{i} + (y_0 + v_0t\sin\alpha - \frac{1}{2}gt^2)\,\mathbf{j} \Rightarrow x = x_0 + v_0t\cos\alpha$,
$y = y_0 + v_0t\sin\alpha - \frac{1}{2}gt^2$

20. See Exercise 39, Section 10.3

SECTION 12.3 DIRECTED DISTANCE AND THE UNIT TANGENT VECTOR T

1. $\mathbf{r} = (2\cos t)\,\mathbf{i} + (2\sin t)\,\mathbf{j} + \sqrt{5}\,t\,\mathbf{k} \Rightarrow \mathbf{v} = (-2\sin t)\,\mathbf{i} + (2\cos t)\,\mathbf{j} + \sqrt{5}\,\mathbf{k} \Rightarrow$
$|\mathbf{v}| = \sqrt{(-2\sin t)^2 + (2\cos t)^2 + \left(\sqrt{5}\right)^2} = \sqrt{4\sin^2 t + 4\cos^2 t + 5} = 3$. $\mathbf{T} = \frac{\mathbf{v}}{|\mathbf{v}|} = \left(-\frac{2}{3}\sin t\right)\mathbf{i} + \left(\frac{2}{3}\cos t\right)\mathbf{j} + \frac{\sqrt{5}}{3}\mathbf{k}$. Length $= \int_0^{\pi} |\mathbf{v}|\,dt = \int_0^{\pi} 3\,dt = \left[3t\right]_0^{\pi} = 3\pi$

2. $\mathbf{r} = (6\sin 2t)\,\mathbf{i} + (6\cos 2t)\,\mathbf{j} + 5t\,\mathbf{k} \Rightarrow \mathbf{v} = (12\cos 2t)\,\mathbf{i} + (-12\sin 2t)\,\mathbf{j} + 5\,\mathbf{k} \Rightarrow$
$|\mathbf{v}| = \sqrt{(12\cos 2t)^2 + (-12\sin 2t)^2 + 5^2} = \sqrt{144\cos^2 2t + 144\sin^2 2t + 25} = 13$. $\mathbf{T} = \frac{\mathbf{v}}{|\mathbf{v}|} = \left(\frac{12}{13}\cos 2t\right)\mathbf{i} - \left(\frac{12}{13}\sin 2t\right)\mathbf{j} + \frac{5}{13}\mathbf{k}$. Length $= \int_0^{\pi} |\mathbf{v}|\,dt = \int_0^{\pi} 13\,dt = \left[13t\right]_0^{\pi} = 13\pi$

3. $\mathbf{r} = t\,\mathbf{i} + \frac{2}{3}t^{3/2}\,\mathbf{k} \Rightarrow \mathbf{v} = \mathbf{i} + t^{1/2}\,\mathbf{k} \Rightarrow |\mathbf{v}| = \sqrt{1^2 + \left(t^{1/2}\right)^2} = \sqrt{1+t}$. $\mathbf{T} = \frac{\mathbf{v}}{|\mathbf{v}|} = \frac{1}{\sqrt{1+t}}\mathbf{i} + \frac{\sqrt{t}}{\sqrt{1+t}}\mathbf{k}$
Length $= \int_0^{8} \sqrt{1+t}\,dt = \left[\frac{2}{3}(1+t)^{3/2}\right]_0^{8} = \frac{52}{3}$

4. $\mathbf{r} = (\cos^3 t)\,\mathbf{j} + (\sin^3 t)\,\mathbf{k} \Rightarrow \mathbf{v} = (-3\cos^2 t\sin t)\,\mathbf{j} + (3\sin^2 t\cos t)\,\mathbf{k} \Rightarrow$
$|\mathbf{v}| = \sqrt{(-3\cos^2 t\sin t)^2 + (3\sin^2 t\cos t)^2} = \sqrt{9\cos^2 t\sin^2 t(\cos^2 t + \sin^2 t)} = 3|\cos t\sin t|$
$\mathbf{T} = \frac{\mathbf{v}}{|\mathbf{v}|} = \frac{-3\cos^2 t\sin t}{3|\cos t\sin t|}\mathbf{j} + \frac{3\sin^2 t\cos t}{3|\cos t\sin t|}\mathbf{k}$. $\mathbf{T} = -\cos t\,\mathbf{j} + \sin t\,\mathbf{k}$ if $0 \le t \le \frac{\pi}{2}$.
Length $= \int_0^{\pi/2} 3|\cos t\sin t|\,dt = \int_0^{\pi/2} 3\cos t\sin t\,dt = \int_0^{\pi/2} \frac{3}{2}\sin 2t\,dt = \left[-\frac{3}{4}\cos 2t\right]_0^{\pi/2} = \frac{3}{4}$

5. $\mathbf{r} = (2+t)\,\mathbf{i} - (t+1)\,\mathbf{j} + t\,\mathbf{k} \Rightarrow \mathbf{v} = \mathbf{i} - \mathbf{j} + \mathbf{k} \Rightarrow |\mathbf{v}| = \sqrt{1^2 + (-1)^2 + 1^2} = \sqrt{3}$. $\mathbf{T} = \frac{\mathbf{v}}{|\mathbf{v}|} = \frac{1}{\sqrt{3}}\mathbf{i} - \frac{1}{\sqrt{3}}\mathbf{j} + \frac{1}{\sqrt{3}}\mathbf{k}$
Length $= \int_0^{3} \sqrt{3}\,dt = \left[\sqrt{3}\,t\right]_0^{3} = 3\sqrt{3}$

6. $\mathbf{r} = 6t^3\,\mathbf{i} - 2t^3\,\mathbf{j} - 3t^3\,\mathbf{k} \Rightarrow \mathbf{v} = 18t^2\,\mathbf{i} - 6t^2\,\mathbf{j} - 9t^2\,\mathbf{k} \Rightarrow |\mathbf{v}| = \sqrt{(18t^2)^2 + (-6t^2)^2 + (-9t^2)^2} = \sqrt{441t^4} = 21t^2$ $\mathbf{T} = \frac{\mathbf{v}}{|\mathbf{v}|} = \frac{18t^2}{21t^2}\mathbf{i} - \frac{6t^2}{21t^2}\mathbf{j} - \frac{9t^2}{21t^2}\mathbf{k} = \frac{6}{7}\mathbf{i} - \frac{2}{7}\mathbf{j} - \frac{3}{7}\mathbf{k}$. Length $= \int_{-1}^{1} 21t^2\,dt = \left[7t^3\right]_{-1}^{1} = 14$

7. $\mathbf{r} = (t\cos t)\,\mathbf{i} + (t\sin t)\,\mathbf{j} + \frac{2\sqrt{2}}{3}t^{3/2}\,\mathbf{k} \Rightarrow \mathbf{v} = (\cos t - t\sin t)\,\mathbf{i} + (\sin t + t\cos t)\,\mathbf{j} + (\sqrt{2}\ t^{1/2})\,\mathbf{k} \Rightarrow$ $|\mathbf{v}| = \sqrt{(\cos t - t\sin t)^2 + (\sin t + t\cos t)^2 + (\sqrt{2}\ t^{1/2})^2} = \sqrt{1 + t^2 + 2t} = \sqrt{(t+1)^2} = |t+1| = t + 1$ since $t \ge 0$. $\mathbf{T} = \frac{\mathbf{v}}{|\mathbf{v}|} = \left(\frac{\cos t - t\sin t}{t+1}\right)\mathbf{i} + \left(\frac{\sin t + t\cos t}{t+1}\right)\mathbf{j} + \left(\frac{\sqrt{2}\ t^{1/2}}{t+1}\right)\mathbf{k}$. Length $= \int_0^\pi (t+1)\,dt =$ $\left[\frac{t^2}{2} + t\right]_0^\pi = \frac{\pi^2}{2} + \pi$

8. $\mathbf{r} = (t\sin t + \cos t)\,\mathbf{i} + (t\cos t - \sin t)\,\mathbf{j} \Rightarrow \mathbf{v} = (\sin t + t\cos t - \sin t)\,\mathbf{i} + (\cos t - t\sin t - \cos t)\,\mathbf{j} =$ $(t\cos t)\,\mathbf{i} - (t\sin t)\,\mathbf{j} \Rightarrow |\mathbf{v}| = \sqrt{(t\cos t)^2 + (-t\sin t)^2} = \sqrt{t^2} = |t| = t$ since $0 \le t \le \sqrt{2}$.
$\mathbf{T} = \frac{\mathbf{v}}{|\mathbf{v}|} = \left(\frac{t\cos t}{t}\right)\mathbf{i} - \left(\frac{t\sin t}{t}\right)\mathbf{j} = \cos t\,\mathbf{i} - \sin t\,\mathbf{j}$. Length $= \int_0^{\sqrt{2}} t\,dt = \left[\frac{t^2}{2}\right]_0^{\sqrt{2}} = 1$

9. $\mathbf{r} = (4\cos t)\,\mathbf{i} + (4\sin t)\,\mathbf{j} + 3t\,\mathbf{k} \Rightarrow \mathbf{v} = (-4\sin t)\,\mathbf{i} + (4\cos t)\,\mathbf{j} + 3\,\mathbf{k} \Rightarrow |\mathbf{v}| = \sqrt{(-4\sin t)^2 + (4\cos t)^2 + 3^2}$ $= \sqrt{25} = 5$. $s(t) = \int_0^t 5\,d\tau = 5t$ Length $= s\left(\frac{\pi}{2}\right) = \frac{5\pi}{2}$

10. $\mathbf{r} = (\cos t + t\sin t)\,\mathbf{i} + (\sin t - t\cos t)\,\mathbf{j} \Rightarrow \mathbf{v} = (-\sin t + \sin t + t\cos t)\,\mathbf{i} + (\cos t - \cos t + t\sin t)\,\mathbf{j} =$ $(t\cos t)\,\mathbf{i} + (t\sin t)\,\mathbf{j} \Rightarrow |\mathbf{v}| = \sqrt{(t\cos t)^2 + (t\sin t)^2} = \sqrt{t^2} = |t| = t$ since $\frac{\pi}{2} \le t \le \pi$.
$s(t) = \int_0^t \tau\,d\tau = \frac{t^2}{2}$. Length $= s(\pi) - s\left(\frac{\pi}{2}\right) = \frac{\pi^2}{2} - \frac{(\pi/2)^2}{2} = \frac{3\pi^2}{8}$

11. $\mathbf{r} = (e^t\cos t)\,\mathbf{i} + (e^t\sin t)\,\mathbf{j} + e^t\,\mathbf{k} \Rightarrow \mathbf{v} = (e^t\cos t - e^t\sin t)\,\mathbf{i} + (e^t\sin t + e^t\cos t)\,\mathbf{j} + e^t\,\mathbf{k} \Rightarrow$ $|\mathbf{v}| = \sqrt{(e^t\cos t - e^t\sin t)^2 + (e^t\sin t + e^t\cos t)^2 + (e^t)^2} = \sqrt{3e^{2t}} = \sqrt{3}\,e^t$
$s(t) = \int_0^t \sqrt{3}\,e^\tau\,d\tau = \sqrt{3}\,e^t - \sqrt{3}$. Length $= s(\ln 4) = \sqrt{3}\,e^{\ln 4} - \sqrt{3} = 3\sqrt{3}$

12. $\mathbf{r} = (1 + 2t)\,\mathbf{i} + (1 + 3t)\,\mathbf{j} + (6 - 6t)\,\mathbf{k} \Rightarrow \mathbf{v} = 2\,\mathbf{i} + 3\,\mathbf{j} - 6\,\mathbf{k} \Rightarrow |\mathbf{v}| = \sqrt{2^2 + 3^2 + (-6)^2} = 7$
$s(t) = \int_0^t 7\,d\tau = 7t$. Length $= s(1) = 7$

13. $\mathbf{r} = (\sqrt{2}\,t)\,\mathbf{i} + (\sqrt{2}\,t)\,\mathbf{j} + (1 - t^2)\,\mathbf{k} \Rightarrow \mathbf{v} = \sqrt{2}\,\mathbf{i} + \sqrt{2}\,\mathbf{j} - 2t\,\mathbf{k} \Rightarrow |\mathbf{v}| = \sqrt{(\sqrt{2})^2 + (\sqrt{2})^2 + (-2t)^2} = \sqrt{4 + 4t^2}$ $= 2\sqrt{1 + t^2}$ Length $= \int_0^1 2\sqrt{1 + t^2}\,dt = \left[2\left(\frac{t}{2}\sqrt{1 + t^2} + \frac{1}{2}\ln\left(t + \sqrt{1 + t^2}\right)\right)\right]_0^1 = \sqrt{2} + \ln\left(1 + \sqrt{2}\right)$

14.

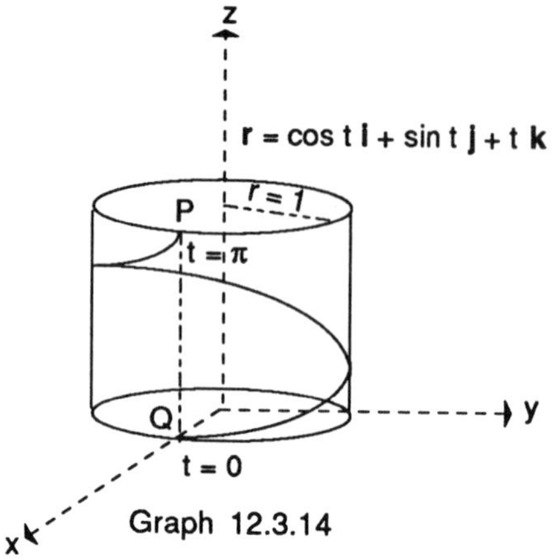

Graph 12.3.14

Let the helix make one complete turn from t = 0 to t = 2π. Note that the radius of the cylinder is 1 ⇒ the circumference of the base is 2π. When t = 2π, the point P is (cos 2π,sin 2π,2π) = (1,0,2π) ⇒ the cylinder is 2π units high. Cut the cylinder along PQ and flatten. The resulting rectangle has a width equalt to the circumference of the cylinder = 2π and a height equal to 2π, the height of the cylinder. ∴ the rectangle is a square and the portion of the helix from t = 0 to t = 2π is the diagonal.

15. a) $\mathbf{r} = (\cos 4t)\,\mathbf{i} + (\sin 4t)\,\mathbf{j} + 4t\,\mathbf{k} \Rightarrow \mathbf{v} = (-4\sin 4t)\,\mathbf{i} + (4\cos 4t)\,\mathbf{j} + 4\,\mathbf{k} \Rightarrow$

$|\mathbf{v}| = \sqrt{(-4\sin 4t)^2 + (4\cos 4t)^2 + 4^2} = \sqrt{32} = 4\sqrt{2}$. Length $= \int_0^{\pi/2} 4\sqrt{2}\,dt = \left[4\sqrt{2}\,t\right]_0^{\pi/2} = 2\pi\sqrt{2}$

b) $\mathbf{r} = (\cos \frac{t}{2})\,\mathbf{i} + (\sin \frac{t}{2})\,\mathbf{j} + \frac{t}{2}\mathbf{k} \Rightarrow \mathbf{v} = \left(-\frac{1}{2}\sin\frac{t}{2}\right)\mathbf{i} + \left(\frac{1}{2}\cos\frac{t}{2}\right)\mathbf{j} + \frac{1}{2}\mathbf{k} \Rightarrow$

$|\mathbf{v}| = \sqrt{\left(-\frac{1}{2}\sin\frac{t}{2}\right)^2 + \left(\frac{1}{2}\cos\frac{t}{2}\right)^2 + \left(\frac{1}{2}\right)^2} = \sqrt{\frac{1}{4} + \frac{1}{4}} = \frac{\sqrt{2}}{2}$. Length $= \int_0^{4\pi} \frac{\sqrt{2}}{2}\,dt = \left[\frac{\sqrt{2}}{2}\,t\right]_0^{4\pi} =$ $2\pi\sqrt{2}$

c) $\mathbf{r} = (\cos t)\,\mathbf{i} - (\sin t)\,\mathbf{j} - t\,\mathbf{k} \Rightarrow \mathbf{v} = (-\sin t)\,\mathbf{i} - (\cos t)\,\mathbf{j} - \mathbf{k} \Rightarrow |\mathbf{v}| = \sqrt{(-\sin t)^2 + (-\cos t)^2 + (-1)^2} =$ $\sqrt{1+1} = \sqrt{2}$. Length $= \int_{-2\pi}^{0} \sqrt{2}\,dt = \left[\sqrt{2}\,t\right]_{-2\pi}^{0} = 2\pi\sqrt{2}$

SECTION 12.4 CURVATURE, TORSION AND THE TNB FRAME

1. $\mathbf{r} = t\,\mathbf{i} + \ln(\cos t)\,\mathbf{j} \Rightarrow \mathbf{v} = \mathbf{i} + \frac{-\sin t}{\cos t}\mathbf{j} = \mathbf{i} - \tan t\,\mathbf{j} \Rightarrow |\mathbf{v}| = \sqrt{1^2 + (-\tan t)^2} = \sqrt{\sec^2 t} = |\sec t| = \sec t$ since $-\frac{\pi}{2} < t < \frac{\pi}{2}$. $\mathbf{T} = \frac{\mathbf{v}}{|\mathbf{v}|} = \frac{1}{\sec t}\mathbf{i} - \frac{\tan t}{\sec t}\mathbf{j} = \cos t\,\mathbf{i} - \sin t\,\mathbf{j}$. $\frac{d\mathbf{T}}{dt} = -\sin t\,\mathbf{i} - \cos t\,\mathbf{j} \Rightarrow$ $\left|\frac{d\mathbf{T}}{dt}\right| = \sqrt{(-\sin t)^2 + (-\cos t)^2} = 1$. $\mathbf{N} = \frac{d\mathbf{T}/dt}{|d\mathbf{T}/dt|} = (-\sin t)\,\mathbf{i} - (\cos t)\,\mathbf{j}$. $\mathbf{a} = (-\sec^2 t)\,\mathbf{j} \Rightarrow$

$$\mathbf{v} \times \mathbf{a} = \begin{vmatrix} \mathbf{i} & \mathbf{j} & \mathbf{k} \\ 1 & -\tan t & 0 \\ 0 & -\sec^2 t & 0 \end{vmatrix} = (-\sec^2 t)\,\mathbf{k}.\ |\mathbf{v} \times \mathbf{a}| = \sqrt{(-\sec^2 t)^2} = \sec^2 t \Rightarrow \kappa = \frac{|\mathbf{v} \times \mathbf{a}|}{|\mathbf{v}|^3} = \frac{\sec^2 t}{\sec^3 t} = \cos t$$

2. $\mathbf{r} = (\ln(\sec t))\,\mathbf{i} + t\,\mathbf{j} \Rightarrow \mathbf{v} = \frac{1}{\sec t}(\sec t \tan t)\,\mathbf{i} + \mathbf{j} = (\tan t)\,\mathbf{i} + \mathbf{j} \Rightarrow |\mathbf{v}| = \sqrt{(\tan t)^2 + 1^2} = \sqrt{\sec^2 t} =$ $|\sec t| = \sec t$ since $-\frac{\pi}{2} < t < \frac{\pi}{2}$. $\mathbf{T} = \frac{\mathbf{v}}{|\mathbf{v}|} = \frac{\tan t}{\sec t}\mathbf{i} + \frac{1}{\sec t}\mathbf{j} = (\sin t)\,\mathbf{i} + (\cos t)\,\mathbf{j} \Rightarrow \frac{d\mathbf{T}}{dt} = (\cos t)\,\mathbf{i} - (\sin t)\,\mathbf{j}$ $\Rightarrow \left|\frac{d\mathbf{T}}{dt}\right| = \sqrt{(\cos t)^2 + (-\sin t)^2} = 1$. $\mathbf{N} = \frac{d\mathbf{T}/dt}{|d\mathbf{T}/dt|} = (\cos t)\,\mathbf{i} - (\sin t)\,\mathbf{j}$. $\mathbf{a} = (\sec^2 t)\,\mathbf{i} \Rightarrow$

2. (Continued)

$$\mathbf{v} \times \mathbf{a} = \begin{vmatrix} \mathbf{i} & \mathbf{j} & \mathbf{k} \\ \tan t & 1 & 0 \\ \sec^2 t & 0 & 0 \end{vmatrix} = (-\sec^2 t)\,\mathbf{k} \Rightarrow |\mathbf{v} \times \mathbf{a}| = \sqrt{(-\sec^2 t)^2} = \sec^2 t \Rightarrow \kappa = \frac{|\mathbf{v} \times \mathbf{a}|}{|\mathbf{v}|^3} = \frac{\sec^2 t}{\sec^3 t} = \cos t$$

3. $\mathbf{r} = (2t + 3)\,\mathbf{i} + \left(5 - t^2\right)\mathbf{j} \Rightarrow \mathbf{v} = 2\,\mathbf{i} - 2t\,\mathbf{j} \Rightarrow |\mathbf{v}| = \sqrt{2^2 + (-2t)^2} = 2\sqrt{1 + t^2}$. $\mathbf{T} = \frac{\mathbf{v}}{|\mathbf{v}|} = \frac{2}{2\sqrt{1 + t^2}}\mathbf{i} + \frac{-2t}{2\sqrt{1 + t^2}}\mathbf{j} = \frac{1}{\sqrt{1 + t^2}}\mathbf{i} - \frac{t}{\sqrt{1 + t^2}}\mathbf{j}$. $\frac{d\mathbf{T}}{dt} = = \frac{-t}{\left(\sqrt{1 + t^2}\right)^3}\mathbf{i} - \frac{1}{\left(\sqrt{1 + t^2}\right)^3}\mathbf{j} \Rightarrow$

$\left|\frac{d\mathbf{T}}{dt}\right| = \sqrt{\left(\frac{-t}{\left(\sqrt{1 + t^2}\right)^3}\right)^2 + \left(-\frac{1}{\left(\sqrt{1 + t^2}\right)^3}\right)^2} = \sqrt{\frac{1}{(1 + t^2)^2}} = \frac{1}{1 + t^2}$. $\mathbf{N} = \frac{d\mathbf{T}/dt}{|d\mathbf{T}/dt|} = \frac{-t}{\sqrt{1 + t^2}}\mathbf{i} - \frac{1}{\sqrt{1 + t^2}}\mathbf{j}$. $\mathbf{a} = -2\,\mathbf{j} \Rightarrow \mathbf{v} \times \mathbf{a} = \begin{vmatrix} \mathbf{i} & \mathbf{j} & \mathbf{k} \\ 2 & -2t & 0 \\ 0 & -2 & 0 \end{vmatrix} = -4\,\mathbf{k} \Rightarrow |\mathbf{v} \times \mathbf{a}| = \sqrt{(-4)^2} = 4$. $\kappa = \frac{|\mathbf{v} \times \mathbf{a}|}{|\mathbf{v}|^3} = \frac{4}{\left(2\sqrt{1 + t^2}\right)^3} = \frac{1}{2\left(\sqrt{1 + t^2}\right)^3}$

4. $\mathbf{r} = (\cos t + t \sin t)\,\mathbf{i} + (\sin t - t \cos t)\,\mathbf{j} \Rightarrow \mathbf{v} = (t \cos t)\,\mathbf{i} + (t \sin t)\,\mathbf{j} \Rightarrow |\mathbf{v}| = \sqrt{(t \cos t)^2 + (t \sin t)^2} = \sqrt{t^2} = |t| = t$ since $t > 0$. $\mathbf{T} = \frac{\mathbf{v}}{|\mathbf{v}|} = \frac{(t \cos t)\,\mathbf{i} + (t \sin t)\,\mathbf{j}}{t} = (\cos t)\,\mathbf{i} + (\sin t)\,\mathbf{j}$. $\frac{d\mathbf{T}}{dt} = (-\sin t)\,\mathbf{i} + (\cos t)\,\mathbf{j} \Rightarrow$ $\left|\frac{d\mathbf{T}}{dt}\right| = \sqrt{(-\sin t)^2 + (\cos t)^2} = 1$. $\mathbf{N} = \frac{d\mathbf{T}/dt}{|d\mathbf{T}/dt|} = (-\sin t)\,\mathbf{i} + (\cos t)\,\mathbf{j}$. $\mathbf{a} = (\cos t - t \sin t)\,\mathbf{i} + (\sin t + t \cos t)\,\mathbf{j}$

$$\Rightarrow \mathbf{v} \times \mathbf{a} = \begin{vmatrix} \mathbf{i} & \mathbf{j} & \mathbf{k} \\ t\cos t & t\sin t & 0 \\ \cos t - t\sin t & \sin t + t\cos t & 0 \end{vmatrix} = \left[t \cos t(\sin t + t \cos t) - t \sin t(\cos t - t \sin t)\right]\mathbf{k} =$$

$t^2\,\mathbf{k} \Rightarrow |\mathbf{v} \times \mathbf{a}| = \sqrt{\left(t^2\right)^2} = t^2$. $\kappa = \frac{|\mathbf{v} \times \mathbf{a}|}{|\mathbf{v}|^3} = \frac{t^2}{t^3} = \frac{1}{t}$

5. $\mathbf{r} = (3 \sin t)\,\mathbf{i} + (3 \cos t)\,\mathbf{j} + 4t\,\mathbf{k} \Rightarrow \mathbf{v} = (3 \cos t)\,\mathbf{i} + (-3 \sin t)\,\mathbf{j} + 4\,\mathbf{k} \Rightarrow |\mathbf{v}| = \sqrt{(3 \cos t)^2 + (-3 \sin t)^2 + 4^2} = \sqrt{25} = 5$. $\mathbf{T} = \frac{\mathbf{v}}{|\mathbf{v}|} = \frac{3 \cos t}{5}\mathbf{i} - \frac{3 \sin t}{5}\mathbf{j} + \frac{4}{5}\mathbf{k} \Rightarrow \frac{d\mathbf{T}}{dt} = \left(-\frac{3}{5}\sin t\right)\mathbf{i} - \left(\frac{3}{5}\cos t\right)\mathbf{j}$

$\left|\frac{d\mathbf{T}}{dt}\right| = \sqrt{\left(-\frac{3}{5}\sin t\right)^2 + \left(-\frac{3}{5}\cos t\right)^2} = \frac{3}{5}$. $\mathbf{N} = \frac{d\mathbf{T}/dt}{|d\mathbf{T}/dt|} = (-\sin t)\,\mathbf{i} - (\cos t)\,\mathbf{j}$

$$\mathbf{a} = (-3 \sin t)\,\mathbf{i} + (-3 \cos t)\,\mathbf{j} \Rightarrow \mathbf{v} \times \mathbf{a} = \begin{vmatrix} \mathbf{i} & \mathbf{j} & \mathbf{k} \\ 3\cos t & -3\sin t & 4 \\ -3\sin t & -3\cos t & 0 \end{vmatrix} = (12 \cos t)\,\mathbf{i} - (12 \sin t)\,\mathbf{j} - 9\,\mathbf{k} \Rightarrow$$

$|\mathbf{v} \times \mathbf{a}| = \sqrt{(12 \cos t)^2 + (-12 \sin t)^2 + (-9)^2} = \sqrt{225} = 15$. $\kappa = \frac{|\mathbf{v} \times \mathbf{a}|}{|\mathbf{v}|^3} = \frac{15}{5^3} = \frac{3}{25}$

$$\mathbf{B} = \mathbf{T} \times \mathbf{N} = \begin{vmatrix} \mathbf{i} & \mathbf{j} & \mathbf{k} \\ \frac{3}{5}\cos t & -\frac{3}{5}\sin t & \frac{4}{5} \\ -\sin t & -\cos t & 0 \end{vmatrix} = \left(\frac{4}{5}\cos t\right)\mathbf{i} - \left(\frac{4}{5}\sin t\right)\mathbf{j} + \left(-\frac{3}{5}\cos^2 t - \frac{3}{5}\sin^2 t\right)\mathbf{k} =$$

5. (Continued)

$\left(\frac{4}{5}\cos t\right)\mathbf{i} - \left(\frac{4}{5}\sin t\right)\mathbf{j} - \frac{3}{5}\mathbf{k}$. $\dot{\mathbf{a}} = (-3\cos t)\,\mathbf{i} + (3\sin t)\,\mathbf{j} \Rightarrow \tau = \dfrac{\begin{vmatrix} 3\cos t & -3\sin t & 4 \\ -3\sin t & -3\cos t & 0 \\ -3\cos t & 3\sin t & 0 \end{vmatrix}}{|\mathbf{v} \times \mathbf{a}|^2} =$

$\dfrac{|-36\sin^2 t - 36\cos^2 t|}{15^2} = \dfrac{4}{25}$

6. $\mathbf{r} = (\cos t + t\sin t)\,\mathbf{i} + (\sin t - t\cos t)\,\mathbf{j} + 3\,\mathbf{k} \Rightarrow \mathbf{v} = (t\cos t)\,\mathbf{i} + (t\sin t)\,\mathbf{j} \Rightarrow |\mathbf{v}| = \sqrt{(t\cos t)^2 + (t\sin t)^2}$
$= \sqrt{t^2} = |t| = t$ if $t > 0$. $\mathbf{T} = \dfrac{\mathbf{v}}{|\mathbf{v}|} = \dfrac{(t\cos t)\,\mathbf{i} + (t\sin t)\,\mathbf{j}}{t} = (\cos t)\,\mathbf{i} + (\sin t)\,\mathbf{j},\ t > 0, \Rightarrow$
$\dfrac{d\mathbf{T}}{dt} = (-\sin t)\,\mathbf{i} + (\cos t)\,\mathbf{j} \Rightarrow \left|\dfrac{d\mathbf{T}}{dt}\right| = \sqrt{(-\sin t)^2 + (\cos t)^2} = 1$. $\mathbf{N} = \dfrac{d\mathbf{T}/dt}{|d\mathbf{T}/dt|} = (-\sin t)\,\mathbf{i} + (\cos t)\,\mathbf{j}$.

$\mathbf{a} = (\cos t - t\sin t)\,\mathbf{i} + (\sin t + t\cos t)\,\mathbf{j} \Rightarrow \mathbf{v} \times \mathbf{a} = \begin{vmatrix} \mathbf{i} & \mathbf{j} & \mathbf{k} \\ t\cos t & t\sin t & 0 \\ \cos t - t\sin t & \sin t + t\cos t & 0 \end{vmatrix} =$

$[t\cos t(\sin t + t\cos t) - t\sin t(\cos t - t\sin t)]\,\mathbf{k} = t^2\,\mathbf{k} \Rightarrow |\mathbf{v} \times \mathbf{a}| = \sqrt{\left(t^2\right)^2} = t^2 \Rightarrow$

$\kappa = \dfrac{|\mathbf{v} \times \mathbf{a}|}{|\mathbf{v}|^3} = \dfrac{t^2}{t^3} = \dfrac{1}{t}$. $\mathbf{B} = \mathbf{T} \times \mathbf{N} = \begin{vmatrix} \mathbf{i} & \mathbf{j} & \mathbf{k} \\ \cos t & \sin t & 0 \\ -\sin t & \cos t & 0 \end{vmatrix} = (\cos^2 t + \sin^2 t)\,\mathbf{k} = \mathbf{k}$.

$\dot{\mathbf{a}} = (-2\sin t - t\cos t)\,\mathbf{i} + (2\cos t - t\sin t)\,\mathbf{j} \Rightarrow \tau = \dfrac{\begin{vmatrix} t\cos t & t\sin t & 0 \\ \cos t - t\sin t & \sin t + t\cos t & 0 \\ -2\sin t - t\cos t & 2\cos t - t\sin t & 0 \end{vmatrix}}{|\mathbf{v} \times \mathbf{a}|^2} = \dfrac{0}{|\mathbf{v} \times \mathbf{a}|^2} = 0$

7. $\mathbf{r} = (e^t\cos t)\,\mathbf{i} + (e^t\sin t)\,\mathbf{j} + 2\,\mathbf{k} \Rightarrow \mathbf{v} = (e^t\cos t - e^t\sin t)\,\mathbf{i} + (e^t\sin t + e^t\cos t)\,\mathbf{j} \Rightarrow$
$|\mathbf{v}| = \sqrt{(e^t\cos t - e^t\sin t)^2 + (e^t\sin t + e^t\cos t)^2} = \sqrt{2e^{2t}} = e^t\sqrt{2}$.
$\mathbf{a} = \left(e^t(\cos t - \sin t) + e^t(-\sin t - \cos t)\right)\mathbf{i} + \left(e^t(\sin t + \cos t) + e^t(\cos t - \sin t)\right)\mathbf{j} =$

$(-2e^t\sin t)\,\mathbf{i} + (2e^t\cos t)\,\mathbf{j} \Rightarrow \mathbf{v} \times \mathbf{a} = \begin{vmatrix} \mathbf{i} & \mathbf{j} & \mathbf{k} \\ e^t\cos t - e^t\sin t & e^t\sin t + e^t\cos t & 0 \\ -2e^t\sin t & 2e^t\cos t & 0 \end{vmatrix} = (2e^{2t})\,\mathbf{k} \Rightarrow$

$|\mathbf{v} \times \mathbf{a}| = \sqrt{(2e^{2t})^2} = 2e^{2t}$. $\kappa = \dfrac{|\mathbf{v} \times \mathbf{a}|}{|\mathbf{v}|^3} = \dfrac{2e^{2t}}{\left(e^t\sqrt{2}\right)^3} = \dfrac{1}{e^t\sqrt{2}}$.

$\mathbf{T} = \dfrac{\mathbf{v}}{|\mathbf{v}|} = \left(\dfrac{e^t\cos t - e^t\sin t}{e^t\sqrt{2}}\right)\mathbf{i} + \left(\dfrac{e^t\cos t + e^t\sin t}{e^t\sqrt{2}}\right)\mathbf{j} \Rightarrow \dfrac{d\mathbf{T}}{dt} = \left(\dfrac{-\sin t - \cos t}{\sqrt{2}}\right)\mathbf{i} + \left(\dfrac{\cos t - \sin t}{\sqrt{2}}\right)\mathbf{j} \Rightarrow$

$\left|\dfrac{d\mathbf{T}}{dt}\right| = \sqrt{\left(\dfrac{-\sin t - \cos t}{\sqrt{2}}\right)^2 + \left(\dfrac{\cos t - \sin t}{\sqrt{2}}\right)^2} = 1$. $\mathbf{N} = \dfrac{d\mathbf{T}/dt}{|d\mathbf{T}/dt|} = \left(\dfrac{-\cos t - \sin t}{\sqrt{2}}\right)\mathbf{i} + \left(\dfrac{-\sin t + \cos t}{\sqrt{2}}\right)\mathbf{j}$

$\dot{\mathbf{a}} = (-2e^t\sin t - 2e^t\cos t)\,\mathbf{i} + (2e^t\cos t - 2e^t\sin t)\,\mathbf{j} \Rightarrow$

7. (Continued)

$$\tau = \frac{\begin{vmatrix} e^t\cos t - e^t\sin t & e^t\sin t + e^t\cos t & 0 \\ -2e^t\sin t & 2e^t\cos t & 0 \\ -2e^t\sin t - 2e^t\cos t & 2e^t\cos t - 2e^t\sin t & 0 \end{vmatrix}}{|\mathbf{v} \times \mathbf{a}|^2} = 0.$$

$$\mathbf{B} = \mathbf{T} \times \mathbf{N} = \begin{vmatrix} \mathbf{i} & \mathbf{j} & \mathbf{k} \\ \frac{\cos t - \sin t}{\sqrt{2}} & \frac{\sin t + \cos t}{\sqrt{2}} & 0 \\ \frac{-\cos t - \sin t}{\sqrt{2}} & \frac{-\sin t + \cos t}{\sqrt{2}} & 0 \end{vmatrix} = \mathbf{k}$$

8. $\mathbf{r} = (6 \sin 2t)\,\mathbf{i} + (6 \cos 2t)\,\mathbf{j} + 5t\,\mathbf{k} \Rightarrow \mathbf{v} = (12 \cos 2t)\,\mathbf{i} - (12 \sin 2t)\,\mathbf{j} + 5\,\mathbf{k} \Rightarrow$ $|\mathbf{v}| = \sqrt{(12 \cos 2t)^2 + (-12 \sin 2t)^2 + 5^2} = \sqrt{169} = 13.$ $\mathbf{T} = \frac{\mathbf{v}}{|\mathbf{v}|} = \frac{(12 \cos 2t)\,\mathbf{i} - (12 \sin 2t)\,\mathbf{j} + 5\,\mathbf{k}}{13} =$ $\left(\frac{12}{13}\cos 2t\right)\mathbf{i} - \left(\frac{12}{13}\sin 2t\right)\mathbf{j} + \frac{5}{13}\mathbf{k}.$ $\frac{d\mathbf{T}}{dt} = \left(-\frac{24}{13}\sin 2t\right)\mathbf{i} - \left(\frac{24}{13}\sin 2t\right)\mathbf{j} \Rightarrow$ $\left|\frac{d\mathbf{T}}{dt}\right| = \sqrt{\left(-\frac{24}{13}\sin 2t\right)^2 + \left(-\frac{24}{13}\sin 2t\right)^2} = \frac{24}{13}.$ $\mathbf{N} = \frac{d\mathbf{T}/dt}{|d\mathbf{T}/dt|} = \frac{\left(-\frac{24}{13}\sin 2t\right)\mathbf{i} - \left(\frac{24}{13}\cos 2t\right)}{\frac{24}{13}} =$ $(-\sin 2t)\,\mathbf{i} - (\cos 2t)\,\mathbf{j}.$ $\mathbf{B} = \mathbf{T} \times \mathbf{N} = \begin{vmatrix} \mathbf{i} & \mathbf{j} & \mathbf{k} \\ \frac{12}{13}\cos 2t & -\frac{12}{13}\sin 2t & \frac{5}{13} \\ -\sin 2t & -\cos 2t & 0 \end{vmatrix} =$ $\left(\frac{5}{13}\cos 2t\right)\mathbf{i} - \left(\frac{5}{13}\sin 2t\right)\mathbf{j} - \frac{12}{13}\mathbf{k}.$ $\mathbf{a} = (-24 \sin 2t)\,\mathbf{i} - (24 \cos 2t)\,\mathbf{j} \Rightarrow$

$$\mathbf{v} \times \mathbf{a} = \begin{vmatrix} \mathbf{i} & \mathbf{j} & \mathbf{k} \\ 12\cos 2t & -12\sin 2t & 5 \\ -24\sin 2t & -24\cos 2t & 0 \end{vmatrix} = (120 \cos 2t)\,\mathbf{i} - (120 \sin 2t)\,\mathbf{j} - 1440\,\mathbf{k}$$

$|\mathbf{v} \times \mathbf{a}| = \sqrt{(120 \cos 2t)^2 + (-120 \sin 2t)^2 + (-1440)^2} = 120\sqrt{145} \Rightarrow \kappa = \frac{|\mathbf{v} \times \mathbf{a}|}{|\mathbf{v}|^3} = \frac{120\sqrt{145}}{13^3} = \frac{120\sqrt{145}}{2197}$

$$\dot{\mathbf{a}} = (-48 \cos 2t)\,\mathbf{i} + (48 \sin 2t)\,\mathbf{j} \Rightarrow \tau = \frac{\begin{vmatrix} 12\cos 2t & -12\sin 2t & 5 \\ -24\sin 2t & -24\cos 2t & 0 \\ -48\cos 2t & 48\sin 2t & 0 \end{vmatrix}}{|\mathbf{v} \times \mathbf{a}|^2} = \frac{5760}{(120\sqrt{145})^2} = 0.00276$$

9. $\mathbf{r} = (2t + 3)\,\mathbf{i} + (t^2 - 1)\,\mathbf{j} \Rightarrow \mathbf{v} = 2\,\mathbf{i} + 2t\,\mathbf{j} \Rightarrow |\mathbf{v}| = \sqrt{2^2 + (2t)^2} = 2\sqrt{1 + t^2}$ $a_T = 2\left(\frac{1}{2}\right)(1 + t^2)^{-1/2}(2t) = \frac{2t}{\sqrt{1 + t^2}}$; $\mathbf{a} = 2\,\mathbf{j} \Rightarrow |\mathbf{a}| = 2 \Rightarrow a_N = \sqrt{|\mathbf{a}|^2 - a_T^2} = \sqrt{2^2 - \left(\frac{2t}{\sqrt{1 + t^2}}\right)^2}$ $= \frac{2}{\sqrt{1 + t^2}}.$ $\therefore \mathbf{a} = \frac{2t}{\sqrt{1 + t^2}}\mathbf{T} + \frac{2}{\sqrt{1 + t^2}}\mathbf{N}.$

10. $\mathbf{r} = (\ln(t^2+1))\,\mathbf{i} + (t - 2\tan^{-1}t)\,\mathbf{j} \Rightarrow \mathbf{v} = \frac{2t}{t^2+1}\mathbf{i} + \left(1 - \frac{2}{t^2+1}\right)\mathbf{j} \Rightarrow$

$|\mathbf{v}| = \sqrt{\left(\frac{2t}{t^2+1}\right)^2 + \left(1 - \frac{2}{t^2+1}\right)^2} = 1.$ $a_T = 0.$ $\mathbf{a} = \frac{2-2t^2}{(t^2+1)^2}\mathbf{i} + \frac{4t}{(t^2+1)^2}\mathbf{j} \Rightarrow$

$|\mathbf{a}| = \sqrt{\left(\frac{2-2t^2}{(t^2+1)^2}\right)^2 + \left(\frac{4t}{(t^2+1)^2}\right)^2} = \frac{2}{(t^2+1)^2}.$ $a_N = \sqrt{|\mathbf{a}|^2 - a_T^2} = \sqrt{|\mathbf{a}|^2 - 0^2} = |\mathbf{a}| = \frac{2}{(t^2+1)^2}$

$\therefore\ \mathbf{a} = (0)\,\mathbf{T} + \frac{2}{(t^2+1)^2}\mathbf{N} = \frac{2}{(t^2+1)^2}\mathbf{N}$

11. $\mathbf{r} = (a\cos t)\,\mathbf{i} + (a\sin t)\,\mathbf{j} + bt\,\mathbf{k} \Rightarrow \mathbf{v} = (-a\sin t)\,\mathbf{i} + (a\cos t)\,\mathbf{j} + b\,\mathbf{k} \Rightarrow |\mathbf{v}| = \sqrt{(-a\sin t)^2 + (a\cos t)^2 + b^2}$ $= \sqrt{a^2+b^2}.$ $a_T = 0.$ $\mathbf{a} = (-a\cos t)\,\mathbf{i} + (-a\sin t)\,\mathbf{j} \Rightarrow |\mathbf{a}| = \sqrt{(-a\cos t)^2 + (a\sin t)^2} = \sqrt{a^2} = |a|.$

$a_N = \sqrt{|\mathbf{a}|^2 - a_T^2} = \sqrt{|\mathbf{a}|^2 - 0^2} = |\mathbf{a}| = |a|.$ $\therefore\ \mathbf{a} = (0)\,\mathbf{T} + |a|\,\mathbf{N} = |a|\,\mathbf{N}$

12. $\mathbf{r} = (1+3t)\,\mathbf{i} + (t-2)\,\mathbf{j} - 3t\,\mathbf{k} \Rightarrow \mathbf{v} = 3\,\mathbf{i} + \mathbf{j} - 3\,\mathbf{k} \Rightarrow |\mathbf{v}| = \sqrt{3^2 + 1^2 + (-3)^2} = \sqrt{19}.$

$a_T = 0.$ $\mathbf{a} = \mathbf{0}.$ $a_N = \sqrt{|\mathbf{a}|^2 - a_T^2} = 0.$ $\therefore\ \mathbf{a} = (0)\,\mathbf{T} + (0)\,\mathbf{N} = \mathbf{0}$

13. $\mathbf{r} = (t+1)\,\mathbf{i} + 2t\,\mathbf{j} + t^2\,\mathbf{k} \Rightarrow \mathbf{v} = \mathbf{i} + 2\,\mathbf{j} + 2t\,\mathbf{k} \Rightarrow \mathbf{v}(1) = \mathbf{i} + 2\,\mathbf{j} + 2\,\mathbf{k} \Rightarrow |\mathbf{v}| = \sqrt{1^2 + 2^2 + (2t)^2} =$ $\sqrt{5+4t^2}.$ $a_T = \frac{1}{2}\left(5+4t^2\right)^{-1/2}(8t) = 4t\left(5+4t^2\right)^{-1/2} \Rightarrow a_T(1) = \frac{4}{\sqrt{9}} = \frac{4}{3}.$ $\mathbf{a} = 2\,\mathbf{k} \Rightarrow \mathbf{a}(1) = 2\,\mathbf{k} \Rightarrow$

$|\mathbf{a}(1)| = 2.$ $a_N = \sqrt{|\mathbf{a}|^2 - a_T^2} = \sqrt{2^2 - \left(\frac{4}{3}\right)^2} = \frac{2\sqrt{5}}{3}.$ $\therefore\ \mathbf{a}(1) = \frac{4}{3}\mathbf{T} + \frac{2\sqrt{5}}{3}\mathbf{N}$

14. $\mathbf{r} = (t\cos t)\,\mathbf{i} + (t\sin t)\,\mathbf{j} + t^2\,\mathbf{k} \Rightarrow \mathbf{v} = (\cos t - t\sin t)\,\mathbf{i} + (\sin t + t\cos t)\,\mathbf{j} + 2t\,\mathbf{k} \Rightarrow$

$|\mathbf{v}| = \sqrt{(\cos t - t\sin t)^2 + (\sin t + t\cos t)^2 + (2t)^2} = \sqrt{5t^2+1}.$ $a_T = \frac{1}{2}(5t^2+1)^{-1/2}(10t) = \frac{5t}{\sqrt{5t^2+1}}.$

$a_T(0) = 0.$ $\mathbf{a} = (-\sin t - \sin t - t\cos t)\,\mathbf{i} + (\cos t + \cos t - t\sin t)\,\mathbf{j} + 2\,\mathbf{k} =$

$(-2\sin t - t\cos t)\,\mathbf{i} + (2\cos t - t\sin t)\,\mathbf{j} + 2\,\mathbf{k}.$ $\mathbf{a}(0) = 2\,\mathbf{j} + 2\,\mathbf{k} \Rightarrow |\mathbf{a}(0)| = \sqrt{2^2 + 2^2} = 2\sqrt{2}.$

$a_N = \sqrt{|\mathbf{a}|^2 - a_T^2} = \sqrt{(2\sqrt{2})^2 - 0^2} = 2\sqrt{2}.$ $\therefore\ \mathbf{a}(0) = (0)\,\mathbf{T} + 2\sqrt{2}\,\mathbf{N} = 2\sqrt{2}\,\mathbf{N}$

15. $\mathbf{r} = t^2\,\mathbf{i} + (t + \frac{1}{3}t^3)\,\mathbf{j} + (t - \frac{1}{3}t^3)\,\mathbf{k} \Rightarrow \mathbf{v} = 2t\,\mathbf{i} + (1+t^2)\,\mathbf{j} + (1-t^2)\,\mathbf{k} \Rightarrow |\mathbf{v}| = \sqrt{(2t)^2 + (1+t^2)^2 + (1-t^2)^2}$ $= \sqrt{t^4 + 4t^2 + 2}.$ $a_T = \frac{1}{2}(t^4 + 4t^2 + 2)^{-1/2}(4t^3 + 8t) = \frac{2t^3 + 4t}{\sqrt{t^4 + 4t^2 + 2}} \Rightarrow a_T(0) = 0.$ $\mathbf{a} = 2\,\mathbf{i} + 2t\,\mathbf{j} - 2t\,\mathbf{k} \Rightarrow$

$\mathbf{a}(0) = 2\,\mathbf{i} \Rightarrow |\mathbf{a}(0)| = 2.$ $a_N = \sqrt{|\mathbf{a}|^2 - a_T^2} = \sqrt{2^2 - 0^2} = 2.$ $\therefore\ \mathbf{a}(0) = (0)\,\mathbf{T} + 2\,\mathbf{N} = 2\,\mathbf{N}$

16. $\mathbf{r} = (e^t\cos t)\,\mathbf{i} + (e^t\sin t)\,\mathbf{j} + \sqrt{2}\,e^t\,\mathbf{k} \Rightarrow \mathbf{v} = (e^t\cos t - e^t\sin t)\,\mathbf{i} + (e^t\sin t + e^t\cos t)\,\mathbf{j} + \sqrt{2}\,e^t\,\mathbf{k} \Rightarrow$

$|\mathbf{v}| = \sqrt{(e^t\cos t - e^t\sin t)^2 + (e^t\sin t + e^t\cos t)^2 + (\sqrt{2}\,e^t)^2} = \sqrt{4e^{2t}} = 2e^t.$ $a_T = 2e^t \Rightarrow a_T(0) = 2.$

$\mathbf{a} = (e^t\cos t - e^t\sin t - e^t\sin t - e^t\cos t)\,\mathbf{i} + (e^t\sin t + e^t\cos t + e^t\cos t - e^t\sin t)\,\mathbf{j} + \sqrt{2}\,e^t\,\mathbf{k} =$

$(-2e^t\sin t)\,\mathbf{i} + (2e^t\cos t)\,\mathbf{j} + \sqrt{2}\,e^t\,\mathbf{k} \Rightarrow \mathbf{a}(0) = 2\,\mathbf{j} + \sqrt{2}\,\mathbf{k} \Rightarrow |\mathbf{a}(0)| = \sqrt{2^2 + (\sqrt{2})^2} = \sqrt{6}.$

$a_N = \sqrt{|\mathbf{a}|^2 - a_T^2} = \sqrt{(\sqrt{6})^2 - 2^2} = \sqrt{2}.$ $\therefore \mathbf{a}(0) = 2\,\mathbf{T} + \sqrt{2}\,\mathbf{N}$

17. $\mathbf{r} = (\cos t)\,\mathbf{i} + (\sin t)\,\mathbf{j} - \mathbf{k} \Rightarrow \mathbf{v} = (-\sin t)\,\mathbf{i} + (\cos t)\,\mathbf{j} \Rightarrow |\mathbf{v}| = \sqrt{(-\sin t)^2 + (\cos t)^2} = 1.$ $\mathbf{T} = \frac{\mathbf{v}}{|\mathbf{v}|} =$ $(-\sin t)\,\mathbf{i} + (\cos t)\,\mathbf{j} \Rightarrow \mathbf{T}\left(\frac{\pi}{4}\right) = -\frac{\sqrt{2}}{2}\mathbf{i} + \frac{\sqrt{2}}{2}\mathbf{j}.$ $\frac{d\mathbf{T}}{dt} = (-\cos t)\,\mathbf{i} - (\sin t)\,\mathbf{j} \Rightarrow \left|\frac{d\mathbf{T}}{dt}\right| = \sqrt{(-\cos t)^2 + (-\sin t)^2}$ $= 1.$ $\mathbf{N} = \frac{d\mathbf{T}/dt}{|d\mathbf{T}/dt|} = (-\cos t)\,\mathbf{i} - (\sin t)\,\mathbf{j} \Rightarrow \mathbf{N}\left(\frac{\pi}{4}\right) = -\frac{\sqrt{2}}{2}\mathbf{i} - \frac{\sqrt{2}}{2}\mathbf{j}.$ $\mathbf{r}\left(\frac{\pi}{4}\right) = \frac{\sqrt{2}}{2}\mathbf{i} + \frac{\sqrt{2}}{2}\mathbf{j} - \mathbf{k}$

$$\mathbf{B} = \mathbf{T} \times \mathbf{N} = \begin{vmatrix} \mathbf{i} & \mathbf{j} & \mathbf{k} \\ -\sin t & \cos t & 0 \\ -\cos t & -\sin t & 0 \end{vmatrix} = \mathbf{k} \Rightarrow \mathbf{B}\left(\frac{\pi}{4}\right) = \mathbf{k}.$$

$P = \left(\frac{\sqrt{2}}{2}, \frac{\sqrt{2}}{2}, -1\right)$ (see $\mathbf{r}\left(\frac{\pi}{4}\right)$), the osculating plane is $z = -1$ since $\mathbf{B}$ is the normal vector and $(0)x + (0)y + (1)z = (0)\left(\frac{\sqrt{2}}{2}\right) + (0)\left(\frac{\sqrt{2}}{2}\right) + (1)(-1).$

The normal plane is $-x + y = 0$ since $\mathbf{T}$ is the normal vector and $-\frac{\sqrt{2}}{2}x + \frac{\sqrt{2}}{2}y + (0)z = \left(-\frac{\sqrt{2}}{2}\right)\left(\frac{\sqrt{2}}{2}\right) + \left(\frac{\sqrt{2}}{2}\right)\left(\frac{\sqrt{2}}{2}\right) + (-1)(0) \Rightarrow -\frac{\sqrt{2}}{2}x + \frac{\sqrt{2}}{2}y = 0.$ The rectifying plane is $x + y = \sqrt{2}$ since $\mathbf{N}$ is the normal vector and $-\frac{\sqrt{2}}{2}x - \frac{\sqrt{2}}{2}y + (0)z = \left(-\frac{\sqrt{2}}{2}\right)\left(\frac{\sqrt{2}}{2}\right) - \left(\frac{\sqrt{2}}{2}\right)\left(\frac{\sqrt{2}}{2}\right) + (-1)(0) \Rightarrow -\frac{\sqrt{2}}{2}x - \frac{\sqrt{2}}{2}y = -1.$

18. $\mathbf{r} = (\cos t)\,\mathbf{i} + (\sin t)\,\mathbf{j} + t\,\mathbf{k} \Rightarrow \mathbf{r}(0) = \mathbf{i}$ and $\mathbf{v} = (-\sin t)\,\mathbf{i} + (\cos t)\,\mathbf{j} + \mathbf{k} \Rightarrow |\mathbf{v}| = \sqrt{(-\sin t)^2 + (\cos t)^2 + 1^2}$ $= \sqrt{2}.$ Then $\mathbf{v}(0) = \mathbf{j} + \mathbf{k}$ and $|\mathbf{v}(0)| = \sqrt{2}.$ $\therefore$ $\mathbf{T}(0) = \frac{\mathbf{j}+\mathbf{k}}{\sqrt{2}} = \frac{1}{\sqrt{2}}\mathbf{j} + \frac{1}{\sqrt{2}}\mathbf{k}.$

$\mathbf{T} = \frac{\mathbf{v}}{|\mathbf{v}|} = \frac{(-\sin t)\,\mathbf{i} + (\cos t)\,\mathbf{j} + \mathbf{k}}{\sqrt{2}} \Rightarrow \frac{d\mathbf{T}}{dt} = \frac{-\cos t}{\sqrt{2}}\mathbf{i} - \frac{\sin t}{\sqrt{2}}\mathbf{j} \Rightarrow \left|\frac{d\mathbf{T}}{dt}\right| = \sqrt{\left(\frac{-\cos t}{\sqrt{2}}\right)^2 + \left(\frac{-\sin t}{\sqrt{2}}\right)^2} = \sqrt{\frac{1}{2}}$ $= \frac{1}{\sqrt{2}}.$ $\mathbf{N} = \frac{d\mathbf{T}/dt}{|d\mathbf{T}/dt|} = \frac{(-\cos t/\sqrt{2})\,\mathbf{i} - (\sin t/\sqrt{2})\,\mathbf{j}}{1/\sqrt{2}} = (-\cos t)\,\mathbf{i} - (\sin t)\,\mathbf{j} \Rightarrow \mathbf{N}(0) = -\mathbf{i}$

$$\mathbf{B}(0) = \mathbf{T}(0) \times \mathbf{N}(0) = \begin{vmatrix} \mathbf{i} & \mathbf{j} & \mathbf{k} \\ 0 & \frac{1}{\sqrt{2}} & \frac{1}{\sqrt{2}} \\ -1 & 0 & 0 \end{vmatrix} = -\frac{1}{\sqrt{2}}\mathbf{j} + \frac{1}{\sqrt{2}}\mathbf{k}.$$

P is the point $(1,0,0)$ (see $\mathbf{r}(0)$).

The osculating plane is $y - z = 0$ since $\mathbf{B}$ is the normal vector and $(0)x - \frac{1}{\sqrt{2}}y + \frac{1}{\sqrt{2}}z = (0)(1) - \frac{1}{\sqrt{2}}(0) + \frac{1}{\sqrt{2}}(0).$

The normal plane is $y + z = 0$ since $\mathbf{T}$ is the normal vector and $(0)x + \frac{1}{\sqrt{2}}y + \frac{1}{\sqrt{2}}z = (0)(1) + \frac{1}{\sqrt{2}}(0) + \frac{1}{\sqrt{2}}(0).$ The rectifying plane is $x = 1$ since $\mathbf{N}$ is the normal vector and $(-1)x + (0)y + (0)z = (-1)(1).$

19. If $|\mathbf{v}|$ is constant, then $a_T = 0 \Rightarrow \mathbf{a} = a_N\,\mathbf{N} \Rightarrow \mathbf{a}$ is normal to $\mathbf{T} \Rightarrow$ acceleration is normal to the path.

20. a) $\mathbf{r} = x\,\mathbf{i} + f(x)\,\mathbf{j} \Rightarrow \mathbf{v} = \mathbf{i} + f'(x)\,\mathbf{j} \Rightarrow \mathbf{a} = f''(x)\,\mathbf{j}.$ $\mathbf{v} \times \mathbf{a} = \begin{vmatrix} \mathbf{i} & \mathbf{j} & \mathbf{k} \\ 1 & f'(x) & 0 \\ 0 & f''(x) & 0 \end{vmatrix} = f''(x)\,\mathbf{k} \Rightarrow$

$|\mathbf{v} \times \mathbf{a}| = \sqrt{(f''(x))^2} = |f''(x)|.$ $|\mathbf{v}| = \sqrt{1^2 + (f'(x))^2} = \sqrt{1 + (f'(x))^2}.$ Then $\kappa = \frac{|\mathbf{v} \times \mathbf{a}|}{|\mathbf{v}|^3} =$

20. (Continued)

$$\frac{|f''(x)|}{\left(\sqrt{1+(f'(x))^2}\right)^3} = \frac{|f''|}{\left(1+(f')^2\right)^3}$$

b) $y = \ln(\cos x) \Rightarrow \frac{dy}{dx} = \frac{1}{\cos x}(-\sin x) = -\tan x \Rightarrow \frac{d^2y}{dx^2} = -\sec^2 x$. Then $\kappa = \frac{|-\sec^2 x|}{\left(\sqrt{1+(-\tan x)^2}\right)^3} =$

$\frac{\sec^2 x}{|\sec^3 x|} = \frac{\sec^2 x}{\text{sex}^3 x} = \frac{1}{\sec x} = \cos x$ since $-\frac{\pi}{2} < x < \frac{\pi}{2}$.

21. $\mathbf{r} = t\,\mathbf{i} + (\sin t)\,\mathbf{j} \Rightarrow \mathbf{v} = \mathbf{i} + (\cos t)\,\mathbf{j} \Rightarrow |\mathbf{v}| = \sqrt{1^2 + (\cos t)^2} = \sqrt{1+\cos^2 t} \Rightarrow \left|\mathbf{v}\left(\frac{\pi}{2}\right)\right| = \sqrt{1+\cos^2\left(\frac{\pi}{2}\right)}$

$= 1$. $\mathbf{a} = (-\sin t)\,\mathbf{j} \Rightarrow \mathbf{v} \times \mathbf{a} = \begin{vmatrix} \mathbf{i} & \mathbf{j} & \mathbf{k} \\ 1 & \cos t & 0 \\ 0 & -\sin t & 0 \end{vmatrix} = (-\sin t)\,\mathbf{k} \Rightarrow |\mathbf{v} \times \mathbf{a}| = \sqrt{(-\sin t)^2} = |\sin t| \Rightarrow$

$|\mathbf{v} \times \mathbf{a}|\left(\frac{\pi}{2}\right) = \left|\sin\left(\frac{\pi}{2}\right)\right| = 1 \Rightarrow \kappa = \frac{|\mathbf{v} \times \mathbf{a}|}{|\mathbf{v}|^3} = \frac{1}{1^3} = 1$. $\therefore\ \rho = \frac{1}{1} = 1 \Rightarrow$ center is $\left(\frac{\pi}{2}, 0\right)$, $r = 1 \Rightarrow$

$\left(x - \frac{\pi}{2}\right)^2 + y^2 = 1$

22. $\mathbf{r} = (2\ln t)\,\mathbf{i} - \left(t + \frac{1}{t}\right)\mathbf{j} \Rightarrow \mathbf{v} = \left(\frac{2}{t}\right)\mathbf{i} - \left(1 - \frac{1}{t^2}\right)\mathbf{j} \Rightarrow \mathbf{v}(1) = 2\,\mathbf{i}$. $\mathbf{a} = \left(-\frac{2}{t^2}\right)\mathbf{i} - \left(\frac{2}{t^3}\right)\mathbf{j} \Rightarrow \mathbf{a}(1) = -2\,\mathbf{i} - 2\,\mathbf{j}$

$\mathbf{v} \times \mathbf{a} = \begin{vmatrix} \mathbf{i} & \mathbf{j} & \mathbf{k} \\ 2 & 0 & 0 \\ -2 & -2 & 0 \end{vmatrix} = -4\,\mathbf{k}$. $|\mathbf{v}(1)| = \sqrt{2^2} = 2$. $|\mathbf{v} \times \mathbf{a}| = \sqrt{(-4)^2} = 4$. $\kappa = \frac{|\mathbf{v} \times \mathbf{a}|}{|\mathbf{v}|^3} = \frac{4}{2^3} = \frac{1}{2} \Rightarrow$

$\rho = \frac{1}{\frac{1}{2}} = 2$. The circle of curvature is tangent to the curve at P(0,–2) $\Rightarrow$ Circle has same tangent as the curve $\Rightarrow$ $\mathbf{v}(1)$ is tangent to the curve $\Rightarrow \mathbf{v}(1) = 2\,\mathbf{i}$ is tangent to the circle $\Rightarrow$ the center lies on the y–axis. If $t \neq 1$ $(t > 0)$, then $(t-1)^2 > 0 \Rightarrow t^2 - 2t + 1 > 0 \Rightarrow t^2 + 1 > 2t \Rightarrow \frac{t^2+1}{t} > 2$ since $t > 0 \Rightarrow$ $t + \frac{1}{t} > 2 \Rightarrow -\left(t + \frac{1}{t}\right) < -2 \Rightarrow$ $y < -2$ on both sides of (0,–2) $\Rightarrow$ Curve is concave down $\Rightarrow$ center of circle of curvature is (0,–4) $\Rightarrow$ $x^2 + (y+4)^2 = 4$ is the equation of the circle of curvature.

23. $\mathbf{v} = \mathbf{T}(ds/dt)$ and $\mathbf{a}$ $(d^2s/dt^2)\mathbf{T} + \kappa(ds/dt)^2\mathbf{N} \Rightarrow |\mathbf{v} \times \mathbf{a}| = \left|(ds/dt)\mathbf{T} \times \left((d^2s/dt^2)\mathbf{T} + \kappa(ds/dt)^2\mathbf{N}\right)\right| =$

$\left|\kappa(ds/dt)^3(\mathbf{T} \times \mathbf{N})\right| = \kappa\left|(ds/dt)^3\mathbf{B}\right| = \kappa\left||\mathbf{v}|^3\mathbf{B}\right| = \kappa\,|\mathbf{v}|^3|\mathbf{B}| = \kappa\,|\mathbf{v}|^3$. $\therefore\ |\mathbf{v} \times \mathbf{a}| = \kappa\,|\mathbf{v}|^3 \Rightarrow \kappa = \frac{|\mathbf{v} \times \mathbf{a}|}{|\mathbf{v}|^3}$

24. $\mathbf{v} = (-\sin t)\,\mathbf{i} + (\cos t)\,\mathbf{j} + \mathbf{k} \Rightarrow |\mathbf{v}| = \sqrt{(-\sin t)^2 + (\cos t)^2 + 1^2} = \sqrt{2}$. $\mathbf{T} = \frac{\mathbf{v}}{|\mathbf{v}|} = \frac{-\sin t}{\sqrt{2}}\mathbf{i} + \frac{\cos t}{\sqrt{2}}\mathbf{j} + \frac{1}{\sqrt{2}}\mathbf{k}$

$\Rightarrow \frac{d\mathbf{T}}{dt} = \frac{-\cos t}{\sqrt{2}}\mathbf{i} - \frac{\sin t}{\sqrt{2}}\mathbf{j} \Rightarrow \left|\frac{d\mathbf{T}}{dt}\right| = \sqrt{\left(\frac{-\cos t}{\sqrt{2}}\right)^2 + \left(\frac{-\sin t}{\sqrt{2}}\right)^2} = \sqrt{\frac{1}{2}} = \frac{1}{\sqrt{2}}$. $\mathbf{N} = \frac{d\mathbf{T}/dt}{|d\mathbf{T}/dt|} =$

$\frac{\frac{-\cos t}{\sqrt{2}}\mathbf{i} - \frac{\sin t}{\sqrt{2}}\mathbf{j}}{\frac{1}{\sqrt{2}}} = (-\cos t)\,\mathbf{i} - (\sin t)\,\mathbf{j}$. $\mathbf{B} = \mathbf{T} \times \mathbf{N} = \begin{vmatrix} \mathbf{i} & \mathbf{j} & \mathbf{k} \\ \frac{-\sin t}{\sqrt{2}} & \frac{\cos t}{\sqrt{2}} & \frac{1}{\sqrt{2}} \\ -\cos t & -\sin t & 0 \end{vmatrix} = \frac{\sin t}{\sqrt{2}}\mathbf{i} - \frac{\cos t}{\sqrt{2}}\mathbf{j} + \frac{1}{\sqrt{2}}\mathbf{k}$

24. (Continued)

$$\frac{d\mathbf{B}}{dt} = \frac{\cos t}{\sqrt{2}}\mathbf{i} + \frac{\sin t}{\sqrt{2}}\mathbf{j} \Rightarrow \left|\frac{d\mathbf{B}}{dt}\right| = \sqrt{\left(\frac{\cos t}{\sqrt{2}}\right)^2 + \left(\frac{\sin t}{\sqrt{2}}\right)^2} = \frac{1}{\sqrt{2}} \Rightarrow \tau = \frac{|d\mathbf{B}/dt|}{|\mathbf{v}|} = \frac{1/\sqrt{2}}{\sqrt{2}} = \frac{1}{2}$$

25. $\kappa = \frac{a}{a^2 + b^2} \Rightarrow \frac{d\kappa}{da} = \frac{-a^2 + b^2}{(a^2 + b^2)^2}$. If $\frac{d\kappa}{da} = 0$, then $-a^2 + b^2 = 0 \Rightarrow a = \pm b$. When $a = b$ $(b > 0)$, $\frac{d\kappa}{da} > 0$ if $a < b$ and $\frac{d\kappa}{da} < 0$ if $a > b$. $\therefore$ κ is at a maximum when $a = b \Rightarrow \kappa(b) = \frac{b}{b^2 + b^2} = \frac{1}{2b}$, the maximum value of κ.

SECTION 12.5 PLANETARY MOTION AND SATELLITES

1. $\frac{T^2}{a^3} = \frac{4\pi^2}{GM} \Rightarrow T^2 = \frac{4\pi^2}{GM}a^3 \Rightarrow T^2 = \frac{4\pi^2}{(6.6720 \text{ X } 10^{-11}\text{Nm}^2\text{kg}^{-2})(5.975 \text{ X } 10^{24} \text{ kg})}(6\ 808\ 000 \text{ m})^3 =$ $3.125 \text{ X } 10^7 \text{ sec}^2 \Rightarrow T = \sqrt{3125 \text{ X } 10^4 \text{ sec}^2} = 55.90 \text{ X } 10^2 \text{ sec} = 93.17$ minutes.

2. $e = 0.0167$ and perihelion distance $= 149\ 577\ 000$ km and $e = \frac{r_0 v_0^2}{GM} - 1 \Rightarrow$

$$0.0167 = \frac{(149\ 577\ 000\ 000 \text{ m})v_0^2}{(6.6720 \text{ X } 10^{-11}\text{Nm}^2\text{kg}^{-2})(1.99 \text{ X } 10^{30}\text{kg})} - 1$$

$\Rightarrow v_0^2 = 9.02 \text{ X } 10^8 \text{ m}^2/\text{s}^2 \Rightarrow v_0 = \sqrt{9.02 \text{ X } 10^8 \text{ m}^2/\text{s}^2} = 3.00 \text{ X } 10^4$ m/s

3. 92.25 minutes = 5535 seconds. $\frac{T^2}{a^3} = \frac{4\pi^2}{GM} \Rightarrow a^3 = \frac{GM}{4\pi^2}T^2 \Rightarrow a^3 =$ $\frac{(6.6720 \text{ X } 10^{-11}\text{Nm}^2\text{kg}^{-2})(5.975 \text{ X } 10^{24}\text{kg})}{4\pi^2}(5535 \text{ s})^2 = 3.094 \text{ X } 10^{20} \text{ m}^3 \Rightarrow a = \sqrt[3]{3.094 \text{ X } 10^{20} \text{ m}^3}$ $= 6.764 \text{ X } 10^6$ m = 6764 km. Mean distance from center of the Earth $= \frac{12758 \text{ km} + 183 \text{ km} + 589 \text{ km}}{2}$ = 6765 km

4. a) p = 1639 minutes = 98340 seconds and mass of Mars $= 6.418 \text{ X } 10^{23}$ kg $\Rightarrow a^3 = \frac{GM}{4\pi^2}T^2 =$

$$\frac{(6.6720 \text{ X } 10^{-11} \text{ Nm}^2\text{kg}^{-2})(6.418 \text{ X } 10^{23} \text{ kg})(98340 \text{ s})^2}{4\pi^2} = 1.049 \text{ X } 10^{22} \text{ m}^3 \Rightarrow$$

$a = \sqrt[3]{1.049 \text{ X } 10^{22} \text{ m}^3} = 2.19 \text{ X } 10^7$ m = 21900 km

b) 2a = Diameter of Mars + Perigee Height + Apogee Height = D + 1499 km + 35800 km $\Rightarrow$ 2(21900)km = D + 37800 km $\Rightarrow$ D = 6501 km.

5. a = 22030 km $= 2.203 \text{ X } 10^7$ m. $T^2 = \frac{4\pi^2}{GM}a^3 \Rightarrow T^2 =$ $\frac{4\pi^2}{(6.670 \text{ X } 10^{-11}\text{Nm}^2\text{kg}^{-2})(6.418 \text{ X } 10^{23} \text{ kg})}(2.203 \text{ X } 10^7\text{s})^3 = 9.857\text{X } 10^8 \text{ sec}^2 \Rightarrow T =$ $\sqrt{9.857 \text{ X } 10^8 \text{ sec}^2} = 9.928 \text{ X } 10^4$ sec = 1655 minutes.

6. Period of the satellite = Rotational period of the Earth $\Rightarrow$ Period of the satellite = 1436.1 minutes = 86166 seconds. $a^3 = \frac{GMT^2}{4\pi^2} \Rightarrow a^3 = \frac{(6.6720 \text{ X } 10^{-11} \text{ Nm}^2\text{kg}^{-2})(5.975 \text{ X } 10^{24} \text{ kg})(86166 \text{ s})^2}{4\pi^2} =$ $74.973 \text{ X } 10^{21} \text{ m}^3 \Rightarrow a = \sqrt[3]{74.973 \text{ X } 10^{21} \text{ m}^3} = 4.2166 \text{ X } 10^7$ m = 42166 km

7. T = 1477.4 minutes = 88644 seconds. $a^3 = \frac{GMT^2}{4\pi^2} =$

$$\frac{(6.6720 \times 10^{-11} Nm^2kg^{-2})(6.418 \times 10^{23} kg)(88644\ s)^2}{4\pi^2} = 8.523 \times 10^{21}\ m^3 \Rightarrow$$

$$a = \sqrt[3]{8.523 \times 10^{21}\ m^3} = 2.043 \times 10^7\ m = 20430\ km$$

8. Period of the Moon = 2.36055 X 10^6 seconds. $a^3 = \frac{GMT^2}{4\pi^2} =$

$$\frac{(6.6720 \times 10^{-11} Nm^2kg^{-2})(5.975 \times 10^{24} kg)(2.36055 \times 10^6\ s)^2}{4\pi^2} = 5.627 \times 10^{25}\ m^3 \Rightarrow$$

$a = \sqrt[3]{5.627 \times 10^{25}\ m^3} = 3.832 \times 10^8\ m$ = 383200 km from the center of the Earth or about 376821 km from the surface.

9. $r = \frac{GM}{v^2} \Rightarrow v^2 = \frac{GM}{r} \Rightarrow |v| = \sqrt{\frac{GM}{r}} = \sqrt{\frac{(6.6720 \times 10^{-11}\ Nm^2kg^{-2})(5.975 \times 10^{24}\ kg)}{r}} =$ $1.9966 \times 10^7\ r^{-1/2}$ m/s

10. Solar System: $\frac{T^3}{a^3} = \frac{4\pi^2}{(6.6720 \times 10^{-11} Nm^2kg^{-2})(1.99 \times 10^{30}\ kg)} = 2.97 \times 10^{-19}\ s^2/m^3$

Earth: $\frac{T^3}{a^3} = \frac{4\pi^2}{(6.6720 \times 10^{-11} Nm^2kg^{-2})(5.975 \times 10^{24}\ kg)} = 9.832 \times 10^{-14} s^2/m^3$

Moon: $\frac{T^3}{a^3} = \frac{4\pi^2}{(6.6720 \times 10^{-11}\ Nm^2kg^{-2})(7.354 \times 10^{22}\ kg)} = 8.046 \times 10^{-12}\ s^2/m^3$

11. $e = \frac{r_0 v_0^2}{GM} - 1 \Rightarrow v_0^2 = \frac{GM(e+1)}{r_0} \Rightarrow v_0 = \sqrt{\frac{GM(e+1)}{r_0}}$

Circle: $e = 0 \Rightarrow v_0 = \sqrt{\frac{GM}{r_0}}$

Ellipse: $0 < e < 1 \Rightarrow \sqrt{\frac{GM}{r_0}} < v_0 < \sqrt{\frac{2GM}{r_0}}$

Parabola: $e = 1 \Rightarrow v_0 = \sqrt{\frac{2GM}{r_0}}$

Hyperbola: $e > 1 \Rightarrow v_0 > \sqrt{\frac{2GM}{r_0}}$

12. $r = \frac{GM}{v^2} \Rightarrow v^2 = \frac{GM}{r} \Rightarrow v = \sqrt{\frac{GM}{r}}$ which is constant since G, M, and r (the radius of orbit) are constant.

13. $\frac{d\theta}{dt} = 2$, $\theta = 2t$, $r = \cosh\theta = \cosh(2t)$. $\mathbf{a} = \left(\frac{d^2r}{dt^2} - r\left(\frac{d\theta}{dt}\right)^2\right)\mathbf{u}_r + \left(r\left(\frac{d^2\theta}{dt^2}\right) + 2\left(\frac{dr}{dt}\right)\left(\frac{d\theta}{dt}\right)\right)\mathbf{u}_\theta \Rightarrow$ the $\mathbf{u}_r$ component of acceleration $= \left(\frac{d^2r}{dt^2} - r\left(\frac{d\theta}{dt}\right)^2\right)$ $\frac{dr}{dt} = 2\sinh(2t) \Rightarrow \frac{d^2r}{dt^2} = 4\cosh(2t)$.

$\therefore \left(\frac{d^2r}{dt^2} - r\left(\frac{d\theta}{dt}\right)^2\right) = 4\cosh(2t) - \cosh(2t)(2^2) = 0$

PRACTICE EXERCISES

1. $\mathbf{r} = (4\cos t)\,\mathbf{i} + (\sqrt{2}\sin t)\,\mathbf{j} \Rightarrow x = 4\cos t \Rightarrow x^2 = 16\cos^2 t$
$y = \sqrt{2}\sin t \Rightarrow y^2 = 2\sin^2 t \Rightarrow 8y^2 = 16\sin^2 t \Rightarrow$
$x^2 + 8y^2 = 16 \Rightarrow \frac{x^2}{16} + \frac{y^2}{2} = 1$. $t = 0 \Rightarrow x = 4,\ y = 0$;
$t = \frac{\pi}{4} \Rightarrow x = 2\sqrt{2}$, $y = 1$. $\mathbf{v} = (-4\sin t)\,\mathbf{i} + (\sqrt{2}\cos t)\,\mathbf{j}$
$\Rightarrow \mathbf{v}(0) = \sqrt{2}\,\mathbf{j},\ \mathbf{v}\left(\frac{\pi}{4}\right) = -2\sqrt{2}\,\mathbf{i} + \mathbf{j}$.
$\mathbf{a} = (-4\cos t)\,\mathbf{i} + (-\sqrt{2}\sin t)\,\mathbf{j} \Rightarrow \mathbf{a}(0) = -4\,\mathbf{i},\ \mathbf{a}\left(\frac{\pi}{4}\right) =$
$-2\sqrt{2}\,\mathbf{i} - \mathbf{j}$.

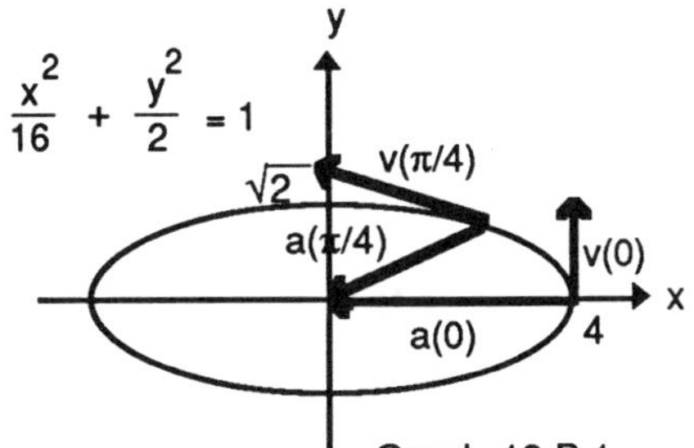

Graph 12.P.1

2. $\mathbf{r} = (\sqrt{3}\sec t)\,\mathbf{i} + (\sqrt{3}\tan t)\,\mathbf{j} \Rightarrow x = \sqrt{3}\sec t,\ y = \sqrt{3}\tan t$
$1 + \tan^2 t = \sec^2 t \Rightarrow 3 + 3\tan^2 t = 3\sec^2 t \Rightarrow$
$3 + y^2 = x^2 \Rightarrow x^2 - y^2 = 3$. $\mathbf{v} = (\sqrt{3}\sec t\tan t)\,\mathbf{i} +$
$\left(\sqrt{3}\sec^2 t\right)\mathbf{j} \Rightarrow \mathbf{v}(0) = \sqrt{3}\,\mathbf{j},\ \mathbf{v}\left(\frac{\pi}{6}\right) = \frac{2}{3}\sqrt{3}\,\mathbf{i} + \frac{4}{3}\sqrt{3}\,\mathbf{j}$.
$\mathbf{a} = (\sqrt{3}\sec t\tan^2 t + \sqrt{3}\sec^3 t)\,\mathbf{i} + (2\sqrt{3}\sec^2 t\tan t)\,\mathbf{j} \Rightarrow$
$\mathbf{a}(0) = \sqrt{3}\,\mathbf{i},\ \mathbf{a}\left(\frac{\pi}{6}\right) = \frac{10}{3}\,\mathbf{i} + \frac{8}{3}\,\mathbf{j}$

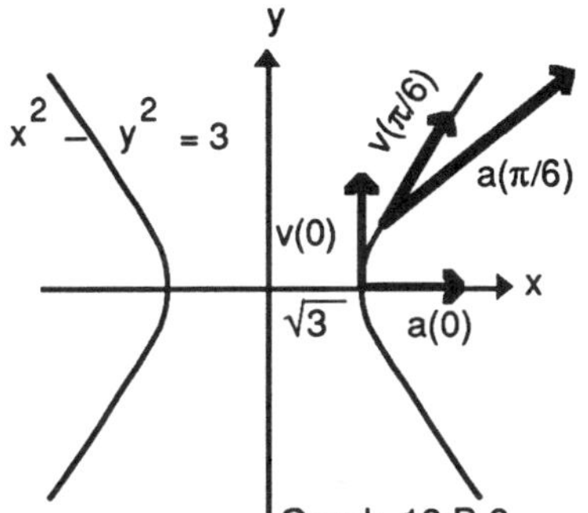

Graph 12.P.2

3. $\int_0^1 [(3 + 6t)\,\mathbf{i} + (4 + 8t)\,\mathbf{j} + (6\pi\cos \pi t)\,\mathbf{k}]\,dt = \left[3t + 3t^2\right]_0^1 \mathbf{i} + \left[4t + 4t^2\right]_0^1 \mathbf{j} + \left[-6\sin \pi t\right]_0^1 \mathbf{j} = 6\,\mathbf{i} + 8\,\mathbf{j}$

4. $\int_e^{e^2} \left[\frac{2\ln t}{t}\,\mathbf{i} + \frac{1}{t\ln t}\,\mathbf{j} + \frac{1}{t}\,\mathbf{k}\right] dt = \left[(\ln t)^2\right]_e^{e^2} \mathbf{i} + \left[\ln(\ln t)\right]_e^{e^2} \mathbf{j} + \left[\ln t\right]_e^{e^2} \mathbf{k} = 3\,\mathbf{i} + (\ln 2)\,\mathbf{j} + \mathbf{k}$

5. $\mathbf{r} = \int ((-\sin t)\,\mathbf{i} + (\cos t)\,\mathbf{j} + \mathbf{k})\,dt = (\cos t)\,\mathbf{i} + (\sin t)\,\mathbf{j} + t\,\mathbf{k} + \mathbf{C}$. $\mathbf{r}(0) = \mathbf{j} \Rightarrow (\cos 0)\,\mathbf{i} + (\sin 0)\,\mathbf{j} + (0)\,\mathbf{k}$
$+ \mathbf{C} = \mathbf{j} \Rightarrow \mathbf{C} = \mathbf{j} - \mathbf{i} \Rightarrow \mathbf{r} = ((\cos t) - 1)\,\mathbf{i} + ((\sin t) + 1)\,\mathbf{j} + t\,\mathbf{k}$

6. $\mathbf{r} = \int \left[\frac{1}{t^2 + 1}\,\mathbf{i} - \frac{1}{\sqrt{1 - t^2}}\,\mathbf{j} + \frac{t}{\sqrt{t^2 + 1}}\,\mathbf{k}\right] dt = (\tan^{-1} t)\,\mathbf{i} - (\sin^{-1} t)\,\mathbf{j} + (t^2 + 1)^{1/2}\,\mathbf{k} + \mathbf{C}$. $\mathbf{r}(0) = \mathbf{j} + \mathbf{k} \Rightarrow$
$(\tan^{-1} 0)\,\mathbf{i} - (\sin^{-1} 0)\,\mathbf{j} + (0^2 + 1)^{1/2}\,\mathbf{k} + \mathbf{C} = \mathbf{j} + \mathbf{k} \Rightarrow \mathbf{C} = \mathbf{j} \Rightarrow \mathbf{r} = (\tan^{-1} t)\,\mathbf{i} - ((\sin^{-1} t) - 1)\,\mathbf{j} +$
$(t^2 + 1)^{1/2}\,\mathbf{k}$

7. $\frac{d\mathbf{r}}{dt} = \int 2\,\mathbf{j}\,dt = 2t\,\mathbf{j} + \mathbf{C}_1$. $\frac{d\mathbf{r}}{dt}(0) = \mathbf{k} \Rightarrow 2(0)\,\mathbf{j} + \mathbf{C}_1 = \mathbf{k} \Rightarrow \mathbf{C}_1 = \mathbf{k}$. $\therefore \frac{d\mathbf{r}}{dt} = 2t\,\mathbf{j} + \mathbf{k}$.
$\mathbf{r} = \int (2t\,\mathbf{j} + \mathbf{k})\,dt = t^2\,\mathbf{j} + t\,\mathbf{k} + \mathbf{C}_2$. $\mathbf{r}(0) = \mathbf{i} \Rightarrow (0^2)\,\mathbf{j} + (0)\,\mathbf{k} + \mathbf{C}_2 = \mathbf{i} \Rightarrow \mathbf{C}_2 = \mathbf{i}$
$\therefore \mathbf{r} = \mathbf{i} + t^2\,\mathbf{j} + t\,\mathbf{k}$

8. $\frac{d\mathbf{r}}{dt} = \int (-2\,\mathbf{i} - 4\,\mathbf{j})\,dt = -2t\,\mathbf{i} - 4t\,\mathbf{j} + \mathbf{C_1}$. $\frac{d\mathbf{r}}{dt}(1) = 4\,\mathbf{i} \Rightarrow -2(1)\,\mathbf{i} - 4(1)\,\mathbf{j} + \mathbf{C_1} = 4\,\mathbf{i} \Rightarrow \mathbf{C_1} = 6\,\mathbf{i} + 4\,\mathbf{j}$

$\therefore \frac{d\mathbf{r}}{dt} = (6 - 2t)\,\mathbf{i} + (4 - 4t)\,\mathbf{j}$. $\mathbf{r} = \int ((6 - 2t)\,\mathbf{i} + (4 - 4t)\,\mathbf{j})\,dt = (6t - t^2)\,\mathbf{i} + (4t - 2t^2)\,\mathbf{j} + \mathbf{C_2}$. $\mathbf{r}(1) =$ $3\,\mathbf{i} + 3\,\mathbf{j} \Rightarrow (6(1) - 1^2)\,\mathbf{i} + \left(4(1) - 2(1)^2\right)\mathbf{j} + \mathbf{C_2} = 3\,\mathbf{i} + 3\,\mathbf{j} \Rightarrow \mathbf{C_2} = -2\,\mathbf{i} + \mathbf{j}$.

$\therefore\ \mathbf{r} = (6t - t^2 - 2)\,\mathbf{i} + (4t - 2t^2 + 1)\,\mathbf{j}$

9. $\mathbf{r} = (2\cos t)\,\mathbf{i} + (2\sin t)\,\mathbf{j} + t^2\,\mathbf{k} \Rightarrow \mathbf{v} = (-2\sin t)\,\mathbf{i} + (2\cos t)\,\mathbf{j} + 2t\,\mathbf{k} \Rightarrow$

$|\mathbf{v}| = \sqrt{(-2\sin t)^2 + (2\cos t)^2 + (2t)^2} = 2\sqrt{1 + t^2}$. Length $= \int_0^{\pi/4} 2\sqrt{1 + t^2}\,dt =$

$\left[t\sqrt{1 + t^2} + \ln\left|t + \sqrt{1 + t^2}\right|\right]_0^{\pi/4} = \frac{\pi}{4}\sqrt{1 + \frac{\pi^2}{16}} + \ln\left(\frac{\pi}{4} + \sqrt{1 + \frac{\pi^2}{16}}\right)$

10. $\mathbf{r} = (3\cos t)\,\mathbf{i} + (3\sin t)\,\mathbf{j} + 2t^{3/2}\,\mathbf{k} \Rightarrow \mathbf{v} = (-3\sin t)\,\mathbf{i} + (3\cos t)\,\mathbf{j} + 3t^{1/2}\,\mathbf{k} \Rightarrow$

$|\mathbf{v}| = \sqrt{(-3\sin t)^2 + (3\cos t)^2 + (3t^{1/2})^2} = \sqrt{9 + 9t} = 3\sqrt{1 + t}$.

Length $= \int_0^3 \sqrt{1 + t}\,dt = \left[\frac{2}{3}(1 + t)^{3/2}\right]_0^3 = \frac{16}{3}$

11. $\mathbf{r} = \frac{4}{9}(1 + t)^{3/2}\,\mathbf{i} + \frac{4}{9}(1 - t)^{3/2}\,\mathbf{j} + \frac{1}{3}t\,\mathbf{k} \Rightarrow \mathbf{v} = \frac{2}{3}(1 + t)^{1/2}\,\mathbf{i} - \frac{2}{3}(1 - t)^{1/2}\,\mathbf{j} + \frac{1}{3}\mathbf{k} \Rightarrow$

$|\mathbf{v}| = \sqrt{\left(\frac{2}{3}(1 + t)^{1/2}\right)^2 + \left(-\frac{2}{3}(1 - t)^{1/2}\right)^2 + \left(\frac{1}{3}\right)^2} = 1$. $\mathbf{T} = \frac{2}{3}(1 + t)^{1/2}\,\mathbf{i} - \frac{2}{3}(1 - t)^{1/2} + \frac{1}{3}\mathbf{k} \Rightarrow$

$\mathbf{T}(0) = \frac{2}{3}\mathbf{i} - \frac{2}{3}\mathbf{j} + \frac{1}{3}\mathbf{k}$. $\frac{d\mathbf{T}}{dt} = \frac{1}{3}(1 + t)^{-1/2}\,\mathbf{i} + \frac{1}{3}(1 - t)^{-1/2}\,\mathbf{j} \Rightarrow \left|\frac{d\mathbf{T}}{dt}\right| =$

$\sqrt{\left(\frac{1}{3}(1 + t)^{-1/2}\right)^2 + \left(\frac{1}{3}(1 - t)^{-1/2}\right)^2} = \frac{1}{3}\sqrt{\frac{2}{1 - t^2}}$. $\frac{d\mathbf{T}}{dt}(0) = \frac{1}{3}\mathbf{i} + \frac{1}{3}\mathbf{j} \Rightarrow \left|\frac{d\mathbf{T}}{dt}(0)\right| = \frac{\sqrt{2}}{3}$.

$\therefore\ \mathbf{N}(0) = \frac{1}{\sqrt{2}}\mathbf{i} + \frac{1}{\sqrt{2}}\mathbf{j}$. $\mathbf{B}(0) = \mathbf{T}(0) \times \mathbf{N}(0) = \begin{vmatrix} \mathbf{i} & \mathbf{j} & \mathbf{k} \\ \frac{2}{3} & -\frac{2}{3} & \frac{1}{3} \\ \frac{1}{\sqrt{2}} & \frac{1}{\sqrt{2}} & 0 \end{vmatrix} = -\frac{1}{3\sqrt{2}}\mathbf{i} + \frac{1}{3\sqrt{2}}\mathbf{j} + \frac{4}{3\sqrt{2}}\mathbf{k}$

$\mathbf{a} = \frac{1}{3}(1 + t)^{-1/2}\,\mathbf{i} + \frac{1}{3}(1 - t)^{-1/2}\,\mathbf{j} \Rightarrow \mathbf{a}(0) = \frac{1}{3}\mathbf{i} + \frac{1}{3}\mathbf{j}$. $\mathbf{v}(0) = \frac{2}{3}\mathbf{i} - \frac{2}{3}\mathbf{j} + \frac{1}{3}\mathbf{k}$.

$\therefore\ \mathbf{v}(0) \times \mathbf{a}(0) = \begin{vmatrix} \mathbf{i} & \mathbf{j} & \mathbf{k} \\ \frac{2}{3} & -\frac{2}{3} & \frac{1}{3} \\ \frac{1}{3} & \frac{1}{3} & 0 \end{vmatrix} = -\frac{1}{9}\mathbf{i} + \frac{1}{9}\mathbf{j} + \frac{4}{9}\mathbf{k} \Rightarrow |\mathbf{v} \times \mathbf{a}| = \frac{\sqrt{2}}{3} \Rightarrow \kappa = \frac{|\mathbf{v} \times \mathbf{a}|}{|\mathbf{v}|^3} = \frac{\sqrt{2}/3}{1^3} = \frac{\sqrt{2}}{3}$.

11. (Continued)

$$\dot{\mathbf{a}} = -\frac{1}{6}(1+t)^{-3/2}\,\mathbf{i} + \frac{1}{6}(1-t)^{-3/2}\,\mathbf{j} \Rightarrow \dot{\mathbf{a}}(0) = -\frac{1}{6}\mathbf{i} + \frac{1}{6}\mathbf{j} \Rightarrow \tau = \frac{\begin{vmatrix} \frac{2}{3} & -\frac{2}{3} & \frac{1}{3} \\ \frac{1}{3} & \frac{1}{3} & 0 \\ -\frac{1}{6} & \frac{1}{6} & 0 \end{vmatrix}}{|\mathbf{v} \times \mathbf{a}|^2} = \frac{1/27}{(\sqrt{2}/3)^2} = \frac{1}{6}$$

12. $\mathbf{r} = (\sin t)\,\mathbf{i} + (\sqrt{2}\cos t)\,\mathbf{j} + (\sin t)\,\mathbf{k} \Rightarrow \mathbf{v} = (\cos t)\,\mathbf{i} - (\sqrt{2}\sin t)\,\mathbf{j} + (\cos t)\,\mathbf{k} \Rightarrow$

$|\mathbf{v}| = \sqrt{(\cos t)^2 + (-\sqrt{2}\sin t)^2 + (\cos t)^2} = \sqrt{2} \Rightarrow \mathbf{T} = \frac{\mathbf{v}}{|\mathbf{v}|} = \frac{(\cos t)\,\mathbf{i} - (\sqrt{2}\sin t)\,\mathbf{j} + (\cos t)\,\mathbf{k}}{\sqrt{2}} =$

$\left(\frac{1}{\sqrt{2}}\cos t\right)\mathbf{i} - (\sin t)\,\mathbf{j} + \left(\frac{1}{\sqrt{2}}\cos t\right)\mathbf{k}$. $\frac{d\mathbf{T}}{dt} = \left(-\frac{1}{\sqrt{2}}\sin t\right)\mathbf{i} - (\cos t)\,\mathbf{j} - \left(\frac{1}{\sqrt{2}}\sin t\right)\mathbf{k} \Rightarrow$

$\left|\frac{d\mathbf{T}}{dt}\right| = \sqrt{\left(-\frac{1}{\sqrt{2}}\sin t\right)^2 + (-\cos t)^2 + \left(-\frac{1}{\sqrt{2}}\sin t\right)^2} = 1$. $\mathbf{N} = \frac{d\mathbf{T}/dt}{|d\mathbf{T}/dt|} = \left(-\frac{1}{\sqrt{2}}\sin t\right)\mathbf{i} - (\cos t)\,\mathbf{j} -$

$\left(\frac{1}{\sqrt{2}}\sin t\right)\mathbf{k}$. $\mathbf{B} = \mathbf{T} \times \mathbf{N} = \begin{vmatrix} \mathbf{i} & \mathbf{j} & \mathbf{k} \\ \frac{1}{\sqrt{2}}\cos t & -\sin t & \frac{1}{\sqrt{2}}\cos t \\ -\frac{1}{\sqrt{2}}\sin t & -\cos t & -\frac{1}{\sqrt{2}}\sin t \end{vmatrix} = \frac{1}{\sqrt{2}}\mathbf{i} - \frac{1}{\sqrt{2}}\mathbf{k}$

$\mathbf{a} = (-\sin t)\,\mathbf{i} - (\sqrt{2}\cos t)\,\mathbf{j} - (\sin t)\,\mathbf{k} \Rightarrow \mathbf{v} \times \mathbf{a} = \begin{vmatrix} \mathbf{i} & \mathbf{j} & \mathbf{k} \\ \cos t & -\sqrt{2}\sin t & \cos t \\ -\sin t & -\sqrt{2}\cos t & -\sin t \end{vmatrix} = \sqrt{2}\,\mathbf{i} - \sqrt{2}\,\mathbf{k}$

$|\mathbf{v} \times \mathbf{a}| = \sqrt{(\sqrt{2})^2 + (-\sqrt{2})^2} = \sqrt{4} = 2 \Rightarrow \kappa = \frac{|\mathbf{v} \times \mathbf{a}|}{|\mathbf{v}|^3} = \frac{2}{(\sqrt{2})^3} = \frac{1}{\sqrt{2}}$.

$$\dot{\mathbf{a}} = (-\cos t)\,\mathbf{i} + (\sqrt{2}\sin t)\,\mathbf{j} - (\cos t)\,\mathbf{k} \Rightarrow \tau = \frac{\begin{vmatrix} \cos t & -\sqrt{2}\sin t & \cos t \\ -\sin t & -\sqrt{2}\cos t & -\sin t \\ -\cos t & \sqrt{2}\sin t & -\cos t \end{vmatrix}}{|\mathbf{v} \times \mathbf{a}|^2} = = \frac{0}{|\mathbf{v} \times \mathbf{a}|^2} = 0$$

13. $\mathbf{r} = (2 + 3t + 3t^2)\,\mathbf{i} + (4t + 4t^2)\,\mathbf{j} - (6\cos t)\,\mathbf{k} \Rightarrow \mathbf{v} = (3 + 6t)\,\mathbf{i} + (4 + 8t)\,\mathbf{j} + (6\sin t)\,\mathbf{k} \Rightarrow$

$|\mathbf{v}| = \sqrt{(3 + 6t)^2 + (4 + 8t)^2 + (6\sin t)^2} = \sqrt{25 + 100t + 100t^2 + 36\sin^2 t}$

$\frac{d|\mathbf{v}|}{dt} = \frac{1}{2}(25 + 100t + 100t^2 + 36\sin^2 t)^{-1/2}(100 + 200t + 72\sin t\cos t) \Rightarrow a_T(0) = \frac{d|\mathbf{v}|}{dt}(0) = 10.$

$\mathbf{a} = 6\,\mathbf{i} + 8\,\mathbf{j} + (6\cos t)\,\mathbf{k} \Rightarrow |\mathbf{a}| = \sqrt{6^2 + 8^2 + (6\cos t)^2} = \sqrt{100 + 36\cos^2 t} \Rightarrow |\mathbf{a}(0)| = \sqrt{136} = 2\sqrt{34}$

$a_{\mathbf{N}} = \sqrt{|\mathbf{a}|^2 - a_{\mathbf{T}}^2} = \sqrt{(2\sqrt{34})^2 - 10^2} = \sqrt{36} = 6.$ $\therefore$ $\mathbf{a}(0) = 10\,\mathbf{T} + 6\,\mathbf{N}$

14. $\mathbf{r} = (2 + t)\,\mathbf{i} + (t + 2t^2)\,\mathbf{j} + (1 + t^2)\,\mathbf{k} \Rightarrow \mathbf{v} = \mathbf{i} + (1 + 4t)\,\mathbf{j} + 2t\,\mathbf{k} \Rightarrow |\mathbf{v}| = \sqrt{1^2 + (1 + 4t)^2 + (2t)^2} =$
$\sqrt{2(1 + 4t + 10t^2)}$ $\frac{d|\mathbf{v}|}{dt} = \frac{1}{2}\left(2(1 + 4t + 10t^2)\right)^{-1/2}(2(4 + 20t)) \Rightarrow a_T = \frac{d|\mathbf{v}|}{dt}(0) = 2\sqrt{2}$
$\mathbf{a} = 4\,\mathbf{j} + 2\,\mathbf{k} \Rightarrow |\mathbf{a}| = \sqrt{4^2 + 2^2} = \sqrt{20} = 2\sqrt{5}$. $a_N = \sqrt{|\mathbf{a}|^2 - a_T{}^2} = \sqrt{(2\sqrt{5})^2 - (2\sqrt{2})^2} = \sqrt{12} = 2\sqrt{3}$
$\therefore\ \mathbf{a}(0) = 2\sqrt{2}\,\mathbf{T} + 2\sqrt{3}\,\mathbf{N}$

15. $\mathbf{r} = \frac{1}{\sqrt{1 + t^2}}\mathbf{i} + \frac{t}{\sqrt{1 + t^2}}\mathbf{j} \Rightarrow \mathbf{v} = -t\left(1 + t^2\right)^{-3/2}\,\mathbf{i} + (1 + t^2)^{-3/2}\,\mathbf{j} \Rightarrow$
$|\mathbf{v}| = \sqrt{\left(-t\left(1 + t^2\right)^{-3/2}\right)^2 + \left((1 + t^2)^{-3/2}\right)^2} = \frac{1}{1 + t^2}$. Want to maximize $|\mathbf{v}|$: $\frac{d|\mathbf{v}|}{dt} = \frac{-2t}{(1 + t^2)^2}$
When $\frac{d|\mathbf{v}|}{dt} = 0$, $\frac{-2t}{(1 + t^2)^2} = 0 \Rightarrow -2t = 0 \Rightarrow t = 0$. For $t < 0$, $\frac{-2t}{(1 + t^2)^2} > 0$; for $t > 0$, $\frac{-2t}{(1 + t^2)^2} < 0 \Rightarrow$
$|\mathbf{v}|_{max}$ occurs when $t = 0 \Rightarrow |\mathbf{v}|_{max} = 1$

16. $\mathbf{r} = \left(e^t \cos t\right)\mathbf{i} + \left(e^t \sin t\right)\mathbf{j} \Rightarrow \mathbf{v} = \left(e^t \cos t - e^t \sin t\right)\mathbf{i} + \left(e^t \sin t + e^t \cos t\right)\mathbf{j} \Rightarrow$
$\mathbf{a} = \left(e^t \cos t - e^t \sin t - e^t \sin t - e^t \cos t\right)\mathbf{i} + \left(e^t \sin t + e^t \cos t + e^t \cos t - e^t \sin t\right)\mathbf{j}$
$= \left(-2e^t \sin t\right)\mathbf{i} + \left(2e^t \cos t\right)\mathbf{j}$. Let θ be the angle between $\mathbf{r}$ and $\mathbf{a}$.
Then $\theta = \cos^{-1}\left(\frac{\mathbf{r}\cdot\mathbf{a}}{|\mathbf{r}||\mathbf{a}|}\right) = \cos^{-1}\left(\frac{-2e^{2t}\sin t \cos t + 2e^{2t} \sin t \cos t}{\sqrt{\left(e^t \cos t\right)^2 + \left(e^t \sin t\right)^2}\sqrt{\left(-2e^t \sin t\right)^2 + \left(2e^t \cos t\right)^2}}\right) =$
$\cos^{-1}\left(\frac{0}{e^{2t}}\right) = \cos^{-1} 0 = \frac{\pi}{2}$ for all t.

17. $\mathbf{r} = \mathbf{i} + (5 \cos t)\,\mathbf{j} + (3 \sin t)\,\mathbf{k} \Rightarrow \mathbf{v} = (-5 \sin t)\,\mathbf{j} + (3 \cos t)\,\mathbf{k} \Rightarrow \mathbf{a} = (-5 \cos t)\,\mathbf{j} - (3 \sin t)\,\mathbf{k} \Rightarrow$
$\mathbf{v}\cdot\mathbf{a} = 25 \sin t \cos t - 9 \sin t \cos t = 16 \sin t \cos t$. $\mathbf{v}\cdot\mathbf{a} = 0 \Rightarrow 16 \sin t \cos t = 0 \Rightarrow \sin t = 0$ or $\cos t = 0$
$\Rightarrow t = 0, \pi$ or $t = \frac{\pi}{2}$

18. $\mathbf{r} = 2\,\mathbf{i} + \left(4 \sin\frac{t}{2}\right)\mathbf{j} + \left(3 - \frac{t}{\pi}\right)\mathbf{k} \Rightarrow \mathbf{r}\cdot(\mathbf{i} - \mathbf{j}) = 2(1) + \left(4 \sin\frac{t}{2}\right)(-1)$. $\mathbf{r}\cdot(\mathbf{i} - \mathbf{j}) = 0 \Rightarrow 2 - 4 \sin\frac{t}{2} = 0 \Rightarrow$
$\sin\frac{t}{2} = \frac{1}{2} \Rightarrow \frac{t}{2} = \frac{\pi}{6} \Rightarrow t = \frac{\pi}{3}$ (for the first time).

19. $y = \left(v_0 \sin\alpha\right)t - \frac{1}{2}gt^2 \Rightarrow y = (44\text{ ft/sec})(\sin 45°) - \frac{1}{2}(32\text{ ft/sec}^2)(3\text{ sec})^2 = 66\sqrt{2} - 144\text{ ft} = -50.66\text{ ft}$
$y + 8\text{ ft} = -42.66\text{ ft}$. The shot put is on the ground.

20. $y_{max} = y_0 + \frac{(v_0 \sin\alpha)^2}{2g} = 7\text{ ft} + \frac{((80\text{ ft/sec})(\sin 45°))^2}{2(32\text{ ft/sec}^2)} = 57\text{ ft}$

21. $R = \frac{v_0{}^2}{g}\sin 2\alpha \Rightarrow v_0 = \sqrt{\frac{Rg}{\sin 2\alpha}}$

For 4325 yards: 4325 yards = 12975 ft $\Rightarrow v_0 = \sqrt{\frac{(12975\text{ ft})(32\text{ ft/sec}^2)}{\sin 90°}} = 644.36\text{ ft/sec}$

For 4752 yards: 4752 yards = 14256 ft $\Rightarrow v_0 = \sqrt{\frac{(14256\text{ ft})(32\text{ ft/sec}^2)}{\sin 90°}} = 675.42\text{ ft/sec}$

22. $R = \frac{v_0^2}{g} \sin 2\alpha \Rightarrow 109.5 \text{ ft} = \frac{v_0^2}{32 \text{ ft/sec}^2} = \sin 2(45°) \Rightarrow v_0^2 = 3504 \text{ ft}^2/\text{sec}^2 \Rightarrow v_0 = \sqrt{3504 \text{ ft}^2/\text{sec}^2} =$ 59.19 ft/sec

23. $\mathbf{r} = e^t\,\mathbf{i} + (\sin t)\,\mathbf{j} + (\ln(1-t))\,\mathbf{k} \Rightarrow \mathbf{v} = e^t\,\mathbf{i} + (\cos t)\,\mathbf{j} - \frac{1}{1-t}\mathbf{k} \Rightarrow \mathbf{v}(0) = \mathbf{i} + \mathbf{j} - \mathbf{k}$. $\mathbf{r}(0) = \mathbf{i} \Rightarrow (1,0,0)$ is on the line. $\therefore$ $x = 1 + t,\ y = t,\ z = -t$ are the parametric equations of the line.

24. $\mathbf{r} = (\sqrt{2}\cos t)\,\mathbf{i} + (\sqrt{2}\sin t)\,\mathbf{j} + t\,\mathbf{k} \Rightarrow \mathbf{v} = (-\sqrt{2}\sin t)\,\mathbf{i} + (\sqrt{2}\cos t)\,\mathbf{j} + \mathbf{k}$. $\mathbf{v}\left(\frac{\pi}{4}\right) = \left(-\sqrt{2}\sin\frac{\pi}{4}\right)\mathbf{i} + \left(\sqrt{2}\cos\frac{\pi}{4}\right)\mathbf{j} + \mathbf{k} = -\mathbf{i} + \mathbf{j} + \mathbf{k}$ is a vector tangent to the helix when $t = \frac{\pi}{4} \Rightarrow$ the tangent line is parallel to $\mathbf{v}\left(\frac{\pi}{4}\right)$. $\mathbf{r}\left(\frac{\pi}{4}\right) = (\sqrt{2}\cos\frac{\pi}{4})\,\mathbf{i} + \left(\sqrt{2}\sin\frac{\pi}{4}\right)\mathbf{j} + \frac{\pi}{4}\mathbf{k} \Rightarrow$ the point $\left(1, 1, \frac{\pi}{4}\right)$ is on the line.

$\therefore$ $x = 1 - t,\ y = 1 + t,\ z = \frac{\pi}{4} + t$ are the parametric equations of the line.

25. $\mathbf{v} = 3\,\mathbf{i} + 4\,\mathbf{j},\ \mathbf{a} = 5\,\mathbf{i} + 15\,\mathbf{j} \Rightarrow \mathbf{v} \times \mathbf{a} = \begin{vmatrix} \mathbf{i} & \mathbf{j} & \mathbf{k} \\ 3 & 4 & 0 \\ 5 & 15 & 0 \end{vmatrix} = 25\,\mathbf{k} \Rightarrow |\mathbf{v} \times \mathbf{a}| = \sqrt{25^2} = 25$. $|\mathbf{v}| = \sqrt{3^2 + 4^2}$ $= 5$. $\therefore$ $\kappa = \frac{|\mathbf{v} \times \mathbf{a}|}{|\mathbf{v}|^3} = \frac{25}{5} = 5$.

26. $\mathbf{r} = t\,\mathbf{i} + t\,\mathbf{j} + \frac{2}{3}t^{3/2}\,\mathbf{k} \Rightarrow \mathbf{v} = \mathbf{i} + \mathbf{j} + t^{1/2}\,\mathbf{k} \Rightarrow |\mathbf{v}| = \sqrt{1^2 + 1^2 + \left(t^{1/2}\right)^2} = \sqrt{2 + t} \Rightarrow |\mathbf{v}(1)| = \sqrt{3}$.
$\mathbf{T} = \frac{\mathbf{v}}{|\mathbf{v}|} = \frac{\mathbf{i} + \mathbf{j} + t^{1/2}\,\mathbf{k}}{\sqrt{2+t}} \Rightarrow \mathbf{T}(1) = \frac{\mathbf{i} + \mathbf{j} + \mathbf{k}}{\sqrt{3}} = \frac{1}{\sqrt{3}}\mathbf{i} + \frac{1}{\sqrt{3}}\mathbf{j} + \frac{1}{\sqrt{3}}\mathbf{k}$. $\mathbf{T}(1)$ is orthogonal to the normal plane $\Rightarrow \frac{1}{\sqrt{3}}x + \frac{1}{\sqrt{3}}y + \frac{1}{\sqrt{3}}z = \frac{1}{\sqrt{3}}(1) + \frac{1}{\sqrt{3}}(1) + \frac{1}{\sqrt{3}}\left(\frac{2}{3}\right) \Rightarrow x + y + z = \frac{8}{3}$ is the equation of the normal plane.

27. $\mathbf{r} = (a\cos t)\,\mathbf{i} + (a\sin t)\,\mathbf{j} + bt\,\mathbf{k} \Rightarrow \mathbf{v} = (-a\sin t)\,\mathbf{i} + (a\cos t)\,\mathbf{j} + b\,\mathbf{k} \Rightarrow \mathbf{a} = (-a\cos t)\,\mathbf{i} - (a\sin t)\,\mathbf{j}$

$|\mathbf{v}| = \sqrt{(-a\sin t)^2 + (a\cos t)^2 + b^2} = \sqrt{a^2 + b^2}$. $\mathbf{v} \times \mathbf{a} = \begin{vmatrix} \mathbf{i} & \mathbf{j} & \mathbf{k} \\ -a\sin t & a\cos t & b \\ -a\cos t & -a\sin t & 0 \end{vmatrix} =$

$(ab\sin t)\,\mathbf{i} + (ab\cos t)\,\mathbf{j} + a^2\,\mathbf{k} \Rightarrow |\mathbf{v} \times \mathbf{a}| = \sqrt{(ab\sin t)^2 + (ab\cos t)^2 + (a^2)^2} = \sqrt{a^2b^2 + a^4}$

$\dot{\mathbf{a}} = (a\sin t)\,\mathbf{i} - (a\cos t)\,\mathbf{j} \Rightarrow \tau = \frac{\begin{vmatrix} -a\sin t & a\cos t & b \\ -a\cos t & -a\sin t & 0 \\ a\sin t & -a\cos t & 0 \end{vmatrix}}{|\mathbf{v} \times \mathbf{a}|^2} = \frac{b(a^2\cos^2 t + a^2\sin^2 t)}{a^2b^2 + a^4} = \frac{b}{a^2 + b^2}$

$\tau'(b) = \frac{a^2 - b^2}{(a^2 + b^2)^2}$. $\tau'(b) = 0 \Rightarrow \frac{a^2 - b^2}{(a^2 + b^2)^2} = 0 \Rightarrow a^2 - b^2 = 0 \Rightarrow b = \pm a$. Use $b = a$. Then $b < a \Rightarrow$ $\tau' > 0$; $b > a \Rightarrow \tau' < 0$. $\therefore$ τ_{max} occurs when $b = a$. $\tau_{max} = \frac{a}{a^2 + a^2} = \frac{1}{2a}$

28. $\mathbf{r} = (\ln(\sec t))\,\mathbf{i} + t\,\mathbf{j} \Rightarrow x = \ln(\sec t) \Rightarrow \frac{dx}{dt} = \tan t \Rightarrow \frac{d^2x}{dt^2} = \sec^2 t$ and $y = t \Rightarrow \frac{dy}{dt} = 1 \Rightarrow \frac{d^2y}{dt^2} = 0.$

$|\dot{x}\ddot{y} - \dot{y}\ddot{x}| = |(\tan t)(0) - (1)(\sec^2 t)| = \sec^2 t.$ $\left(\dot{x}^2 + \dot{y}^2\right)^{3/2} = \left((\tan t)^2 + 1^2\right)^{3/2} = \sec^3 t$ since $-\frac{\pi}{2} < t < \frac{\pi}{2}$. $\therefore\ \kappa = \frac{\sec^2 t}{\sec^3 t} = \frac{1}{\sec t} = \cos t$

29. $\Delta\, ACB \approx \Delta\, BCD \Rightarrow \frac{DC}{BC} = \frac{BC}{AC} \Rightarrow$

$\frac{y_0}{6380 \text{ miles}} = \frac{6380 \text{ miles}}{(6380 + 437) \text{ miles}} \Rightarrow$

$y_0 = \frac{6380^2 \text{ miles}^2}{6817 \text{ miles}} = 5971 \text{ miles}$

$VA = \int_{5971}^{6817} 2\pi x\, ds =$

$2\pi \int_{5971}^{6817} \sqrt{6380^2 - y^2}\, \frac{6380}{\sqrt{6380^2 - y^2}}\, dy =$

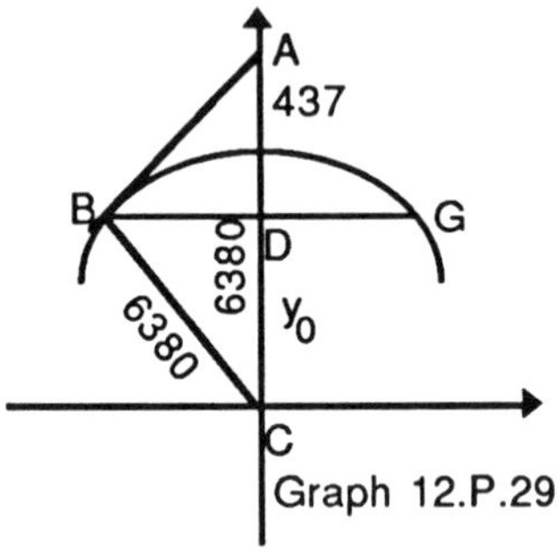

Graph 12.P.29

$2\pi \int_{5971}^{6817} 6380\, dy = 2\pi\Big[6380y\Big]_{5971}^{6817} = 16\,395\,469 \text{ km}^2$. Percentage visible $= \frac{16\,395\,469 \text{ km}^2}{4\pi(6380 \text{ km})^2} = 3.21\%$

30. $r = \frac{(1 + e)r_0}{1 + e\cos\theta} \Rightarrow \frac{dr}{d\theta} = \frac{(1 + e)r_0(e\sin\theta)}{(1 + e\cos\theta)^2}$. $\frac{dr}{d\theta} = 0 \Rightarrow \frac{(1 + e)r_0(e\sin\theta)}{(1 + e\cos\theta)^2} = 0 \Rightarrow (1 + e)r_0(e\sin\theta) = 0 \Rightarrow$

$\sin\theta = 0 \Rightarrow \theta = 0$ or π. $\frac{d^2r}{d\theta^2} = \frac{er_0(1 + e)(1 + e\cos\theta)^2(\cos\theta) - 2(1 + e\cos\theta)(-e\sin\theta)(\sin\theta)}{\left((1 + e\cos\theta)^2\right)^2} =$

$\frac{er_0(1 + e)(\cos\theta + e\cos^2\theta + 2e\sin^2\theta)}{(1 + e\cos\theta)^3}$. $\frac{d^2r}{d\theta^2}(0) = \frac{er_0}{1 + e} > 0$ since $e > 0$, $r_0 > 0 \Rightarrow$ minimum.

$\frac{d^2r}{d\theta^2}(\pi) = \frac{er_0(1 + e)(-1 + e)}{(1 - e)^3} < 0$ since $0 < e < 1$ (for an ellipse) and $r_0 > 0 \Rightarrow$ maximum.

$\therefore$ r is a minimum when $\theta = 0$. Then $r(0) = \frac{(1 + e)r_0}{1 + e\cos 0} = r_0$.

31. a) Given $f(x) = x - 1 - \frac{1}{2}\sin x = 0$, $f(0) = -1$ and $f(2) = 2 - 1 - \frac{1}{2}\sin 2 = 0.54535$ Since f is continuous on [0,2], the intermediate value theorem implies there is a root between 0 and 2.

b) Root ≈ 1.49870113

CHAPTER 13

FUNCTIONS OF TWO OR MORE VARIABLES AND THEIR DERIVATIVES

13.1 FUNCTIONS OF TWO OR MORE INDEPENDENT VARIABLES

1. Domain: All points in the xy–plane; Range: All Real Numbers
   Level curves are straight lines parallel to the line $y = x$.

2. Domain: Set of all (x,y) so that $y - x \geq 0 \Rightarrow y \geq x$; Range: $z \geq 0$
   Level curves are straight lines of the form $y - x = c$ where $c \geq 0$.

3. Domain: Set of all (x,y) so that $(x,y) \neq (0,0)$; Range: All Real Numbers
   Level curves are circles with center $(0,0)$ and radii ≥ 0.

4. Domain: All points in the xy–plane; Range: $-1 \leq z \leq 1$
   Level curves are circles with center $(0,0)$ and radii ≥ 0.

5. Domain: All point in the xy–plane; Range: All Real Numbers
   Level curves are hyperbolas with the x and y axes as asymptotes when $f(x,y) \neq 0$, and the x and y axes when $f(x,y) = 0$.

6. Domain: All $(x,y) \neq (0,y)$; Range: All Real Numbers
   Level curves are parabolas with vertex $(0,0)$ and the y–axis as axis of the parabola.

7. Domain: All points in the xy–plane; Range: $z \geq 0$
   Level curves are ellipses with center $(0,0)$ and major and minor axes along the x and y axes.

8. Domain: All points on the xy–plane; Range: All Real Numbers
   Level curves are hyperbolas centered at $(0,0)$ and transverse and conjugate axes along the x and y axes.

9. a) b)

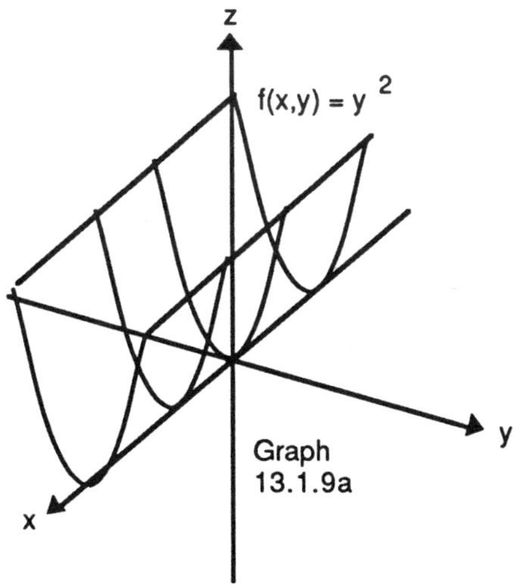

Graph 13.1.9a

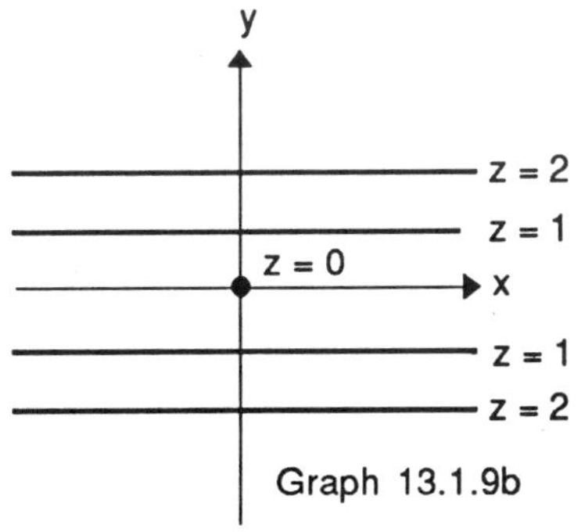

Graph 13.1.9b

10. a)

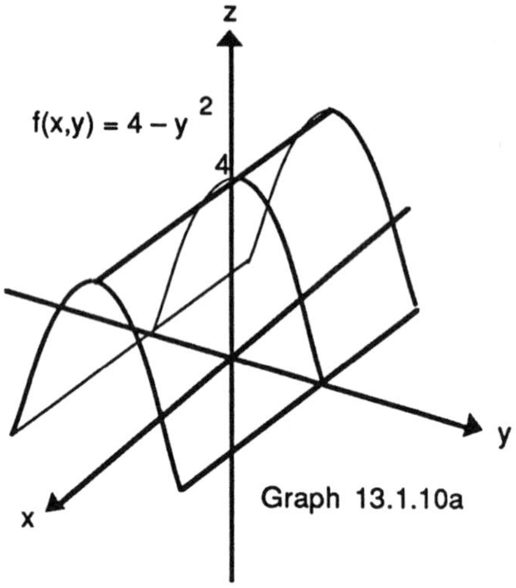

Graph 13.1.10a

b)

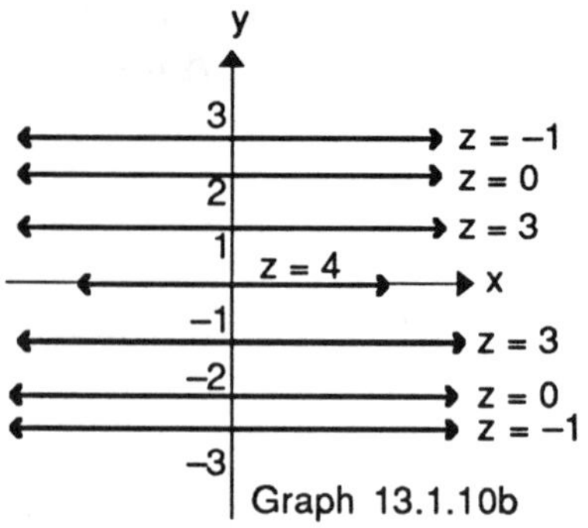

Graph 13.1.10b

11. a)

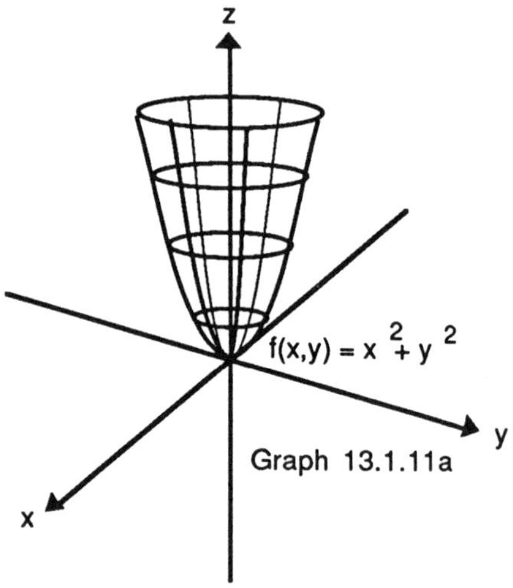

Graph 13.1.11a

b)

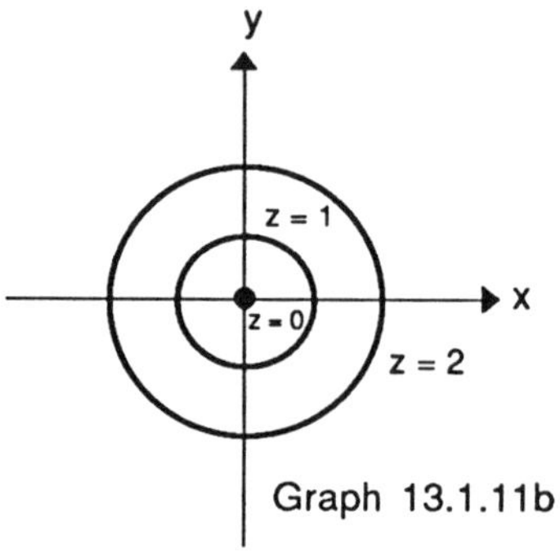

Graph 13.1.11b

12. a)

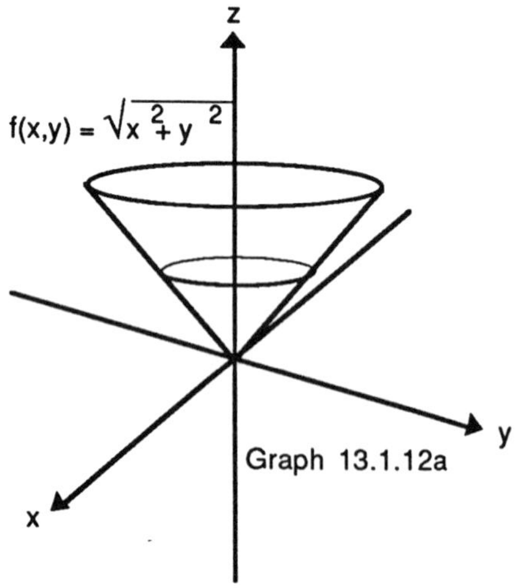

Graph 13.1.12a

b)

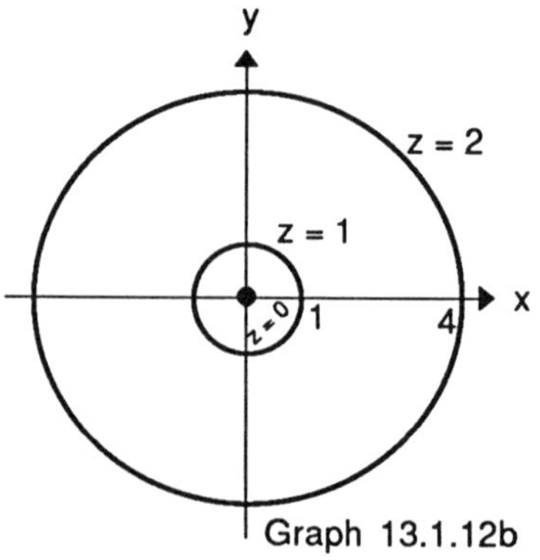

Graph 13.1.12b

13. a)

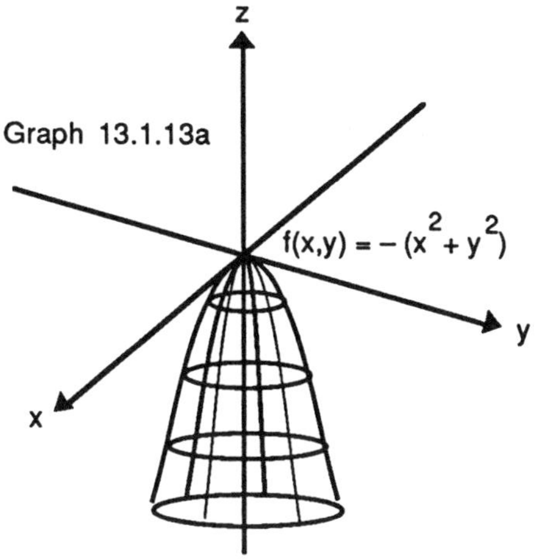

Graph 13.1.13a

b)

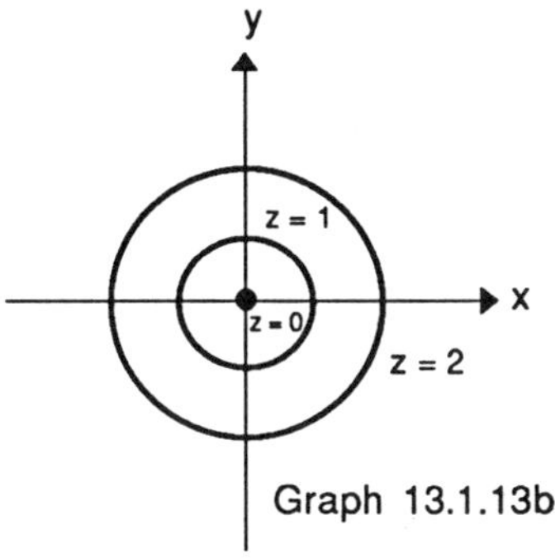

Graph 13.1.13b

14. a)

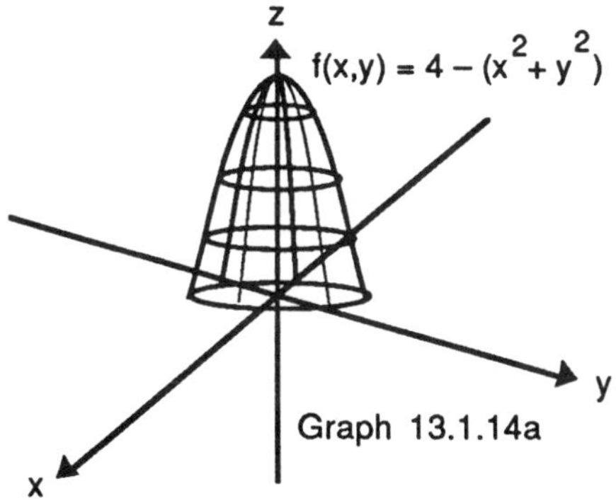

Graph 13.1.14a

b)

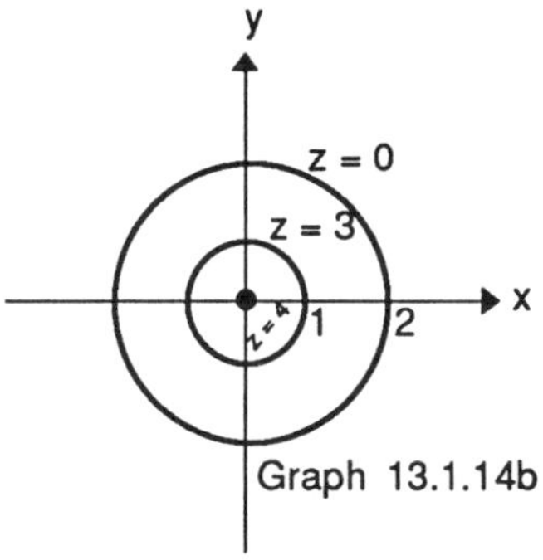

Graph 13.1.14b

15. a)

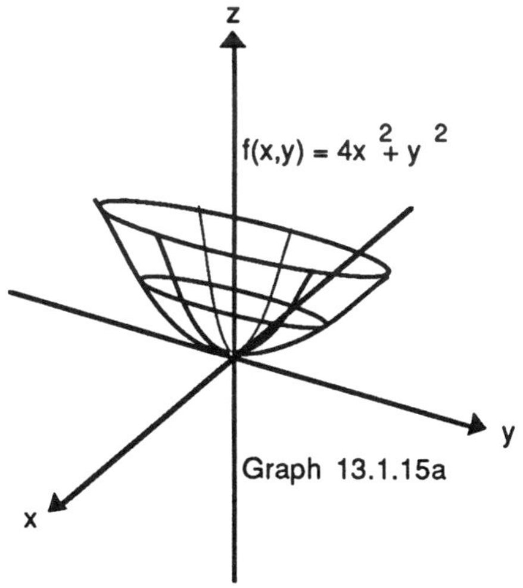

Graph 13.1.15a

b)

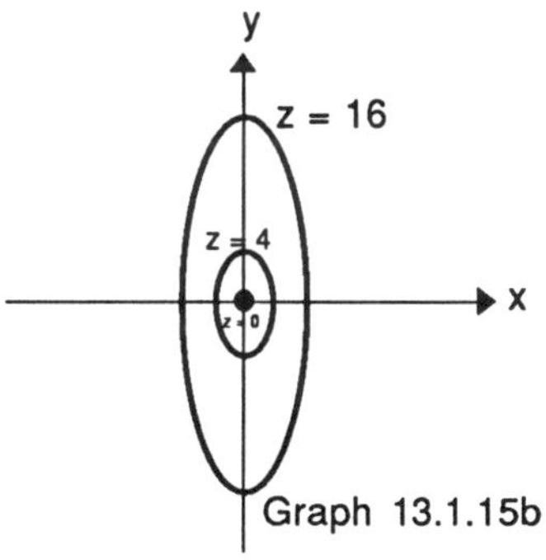

Graph 13.1.15b

16. a)

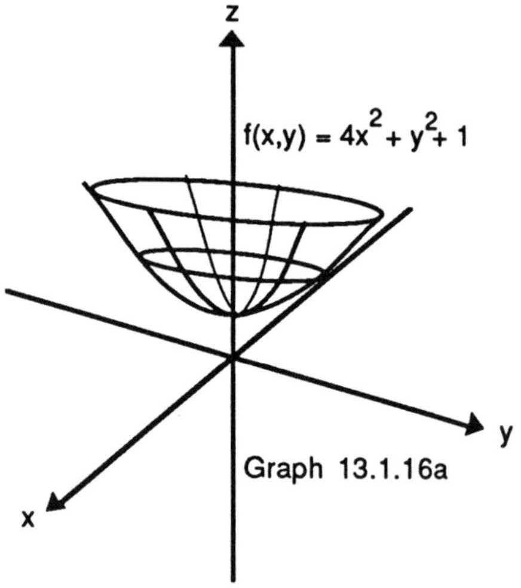

Graph 13.1.16a

b)

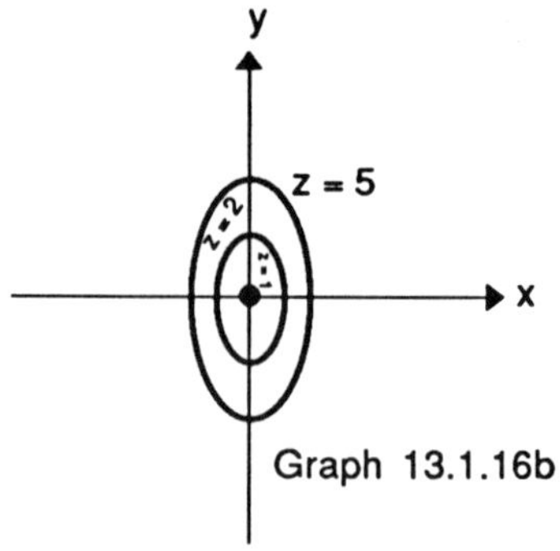

Graph 13.1.16b

17. f 18. e 19. a 20. c 21. d 22. b

23.

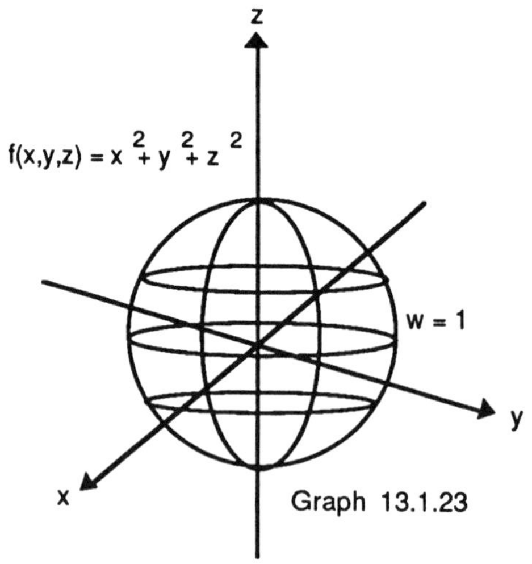

Graph 13.1.23

24.

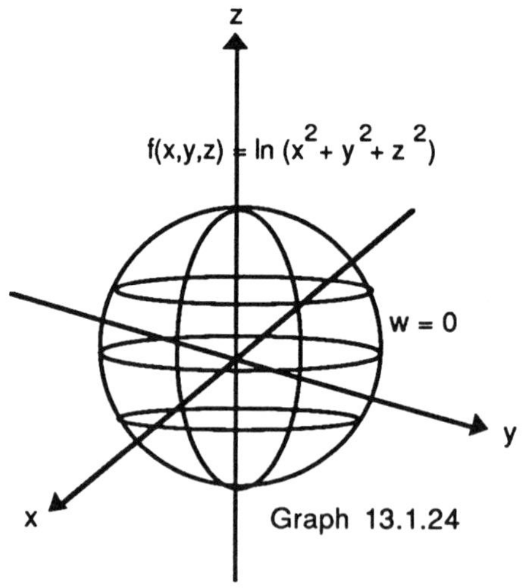

Graph 13.1.24

25.

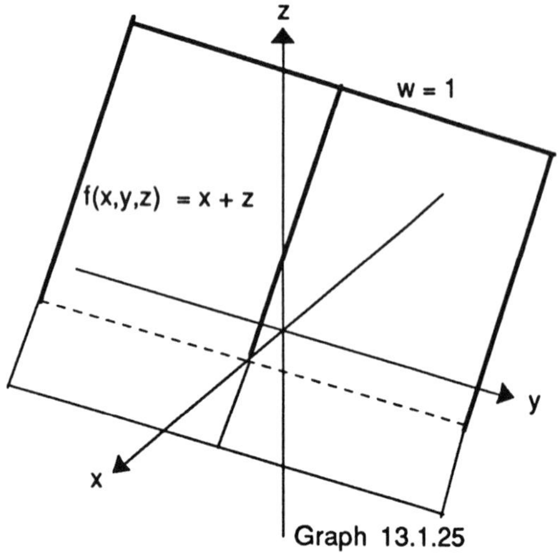

Graph 13.1.25

26.

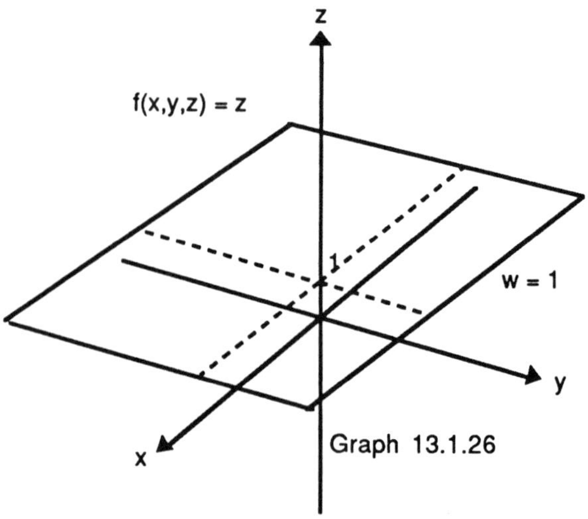

Graph 13.1.26

27.

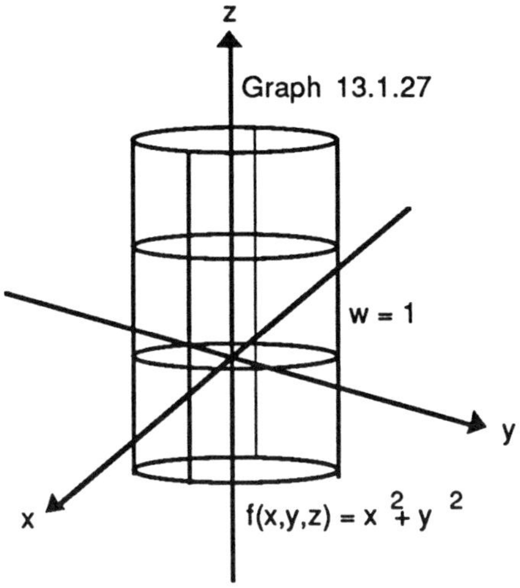

28.

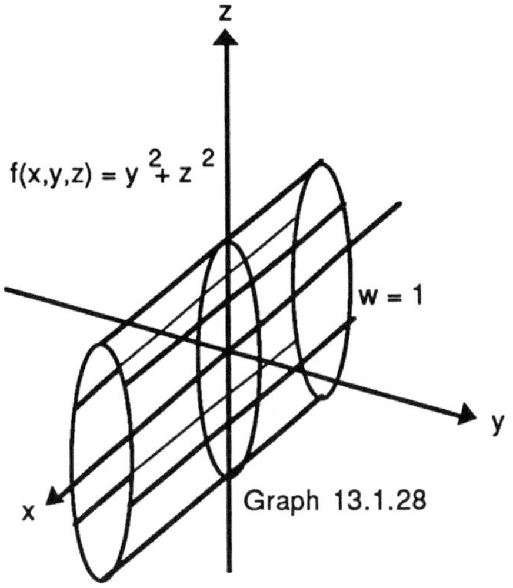

29.

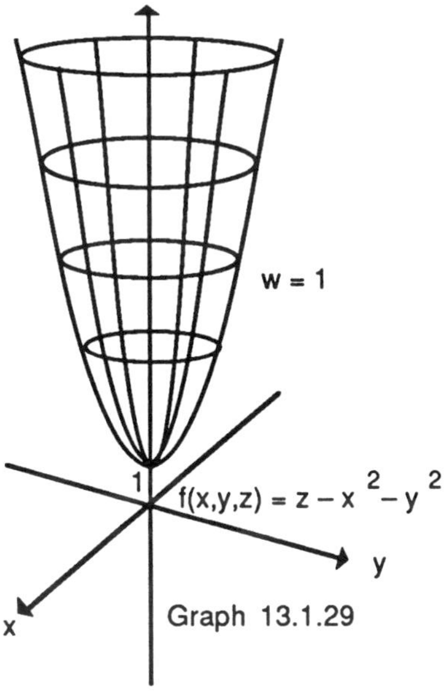

30.

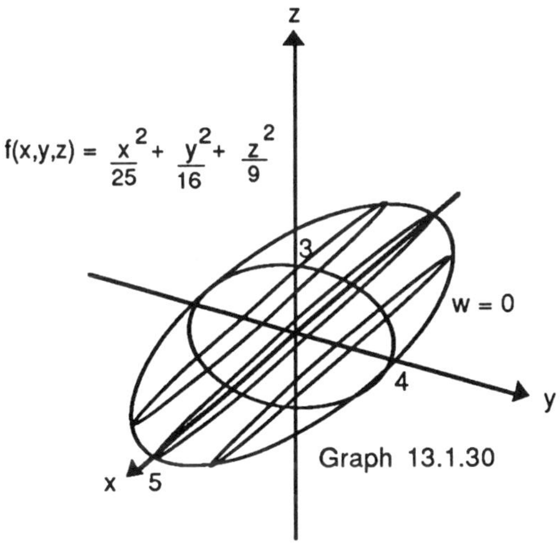

31. $f(x,y) = 16 - x^2 - y^2$ and $(2\sqrt{2}, \sqrt{2}) \Rightarrow z = 16 - \left(2\sqrt{2}\right)^2 - \left(\sqrt{2}\right)^2 = 6 \Rightarrow 6 = 16 - x^2 - y^2 \Rightarrow x^2 + y^2 = 10$

32. $f(x,y) = \sqrt{x^2 - 1}$ and $(1,0) \Rightarrow z = \sqrt{1^2 - 1} = 0 \Rightarrow x^2 - 1 = 0 \Rightarrow x = 1$ or $x = -1$

33. $w = 4\left(\frac{Th}{d}\right)^{1/2} = 4\left(\frac{(290\text{ k})(16.8\text{ km})}{5\text{ k/km}}\right)^{1/2} = 124.86$ km. $\therefore$ must be 62.43 km south of Nantucket.

34. $f(x,y,z) = xyz$ and $x = 20 - t$, $y = t$, $z = 20 \Rightarrow w = (20 - t)(t)(20)$ along the line $\Rightarrow w = 400t - 20t^2 \Rightarrow$ $\frac{dw}{dt} = 400 - 40t$. Let $\frac{dw}{dt} = 0 \Rightarrow 400 - 40t = 0 \Rightarrow t = 10$. $\frac{d^2w}{dt^2} = -40$ for all $t \Rightarrow$ maximum at $t = 10 \Rightarrow$ $x = 20 - 10 = 10$, $y = 10$, $z = 20 \Rightarrow$ maximum of f along the line is $f(10,10,20) = (10)(10)(20) = 2000$.

SECTION 13.2 LIMITS AND CONTINUITY

1. $\lim\limits_{(x,y)\to(0,0)} \dfrac{3x^2 - y^2 + 5}{x^2 + y^2 + 2} = \dfrac{5}{2}$

2. $\lim\limits_{(x,y)\to(1,1)} \ln|1 + x^2y^2| = \ln 2$

3. $\lim\limits_{(x,y)\to(0,\ln 2)} e^{x-y} = \dfrac{1}{2}$

4. $\lim\limits_{(x,y)\to(0,4)} \dfrac{x}{\sqrt{y}} = 0$

5. $\lim\limits_{P\to(1,3,4)} \sqrt{x^2 + y^2 + z^2 - 1} = 5$

6. $\lim\limits_{P\to(1,2,6)} \left(\dfrac{1}{x} + \dfrac{1}{y} + \dfrac{1}{z}\right) = \dfrac{5}{3}$

7. $\lim\limits_{(x,y)\to(0,\pi/4)} \sec x \tan y = 1$

8. $\lim\limits_{(x,y)\to(0,0)} \cos\left(\dfrac{x^2 + y^2}{x + y + 1}\right) = 1$

9. $\lim\limits_{(x,y)\to(1,1)} \cos\left(\sqrt[3]{|xy| - 1}\right) = 1$

10. $\lim\limits_{(x,y)\to(1,0)} \dfrac{x \sin y}{x^2 + 1} = 0$

11. $\lim\limits_{(x,y)\to(0,0)} \dfrac{e^y \sin x}{x} = 1$

12. $\lim\limits_{(x,y)\to(0,0)} \tan^{-1}\left(\dfrac{1}{\sqrt{x^2 + y^2}}\right) = \dfrac{\pi}{2}$

13. $\lim\limits_{\substack{(x,y)\to(1,1)\\ x \neq y}} \dfrac{x^2 - 2xy + y^2}{x - y} = \lim\limits_{(x,y)\to(1,1)} \dfrac{(x - y)^2}{x - y} = \lim\limits_{(x,y)\to(1,1)} (x - y) = 0$

14. $\lim\limits_{\substack{(x,y)\to(1,1)\\ x \neq y}} \dfrac{x^2 - y^2}{x - y} = \lim\limits_{(x,y)\to(1,1)} (x + y) = 2$

15. $\lim\limits_{\substack{(x,y)\to(1,1)\\ x \neq 1}} \dfrac{xy - y - 2x + 2}{x - 1} = \lim\limits_{(x,y)\to(1,1)} \dfrac{(x - 1)(y - 2)}{x - 1} = \lim\limits_{(x,y)\to(1,1)} (y - 2) = -1$

16. $\lim\limits_{\substack{(x,y)\to(2,-4)\\ y \neq 4,\ x \neq x^2}} \dfrac{y + 4}{x^2y - xy + 4x^2 - 4x} = \lim\limits_{(x,y)\to(2,-4)} \dfrac{y + 4}{x(x - 1)(y + 4)} = \lim\limits_{(x,y)\to(2,-4)} \dfrac{1}{x(x - 1)} = \dfrac{1}{2}$

17. $\lim\limits_{P\to(2,3,-6)} \sqrt{x^2 + y^2 + z^2} = 7$

18. $\lim\limits_{P\to(0,-2,0)} \ln\sqrt{x^2 + y^2 + z^2} = \ln 2$

19. $\lim\limits_{P\to(3,3,0)} \left(\sin^2 x + \cos^2 y + \sec^2 z\right) = \sin^2 3 + \cos^2 3 + \sec^2 0 = 2$

20. $\lim\limits_{P\to(\pi,0,3)} ze^{-2y} \cos 2x = (3)e^{-2(0)} \cos 2\pi = 3$

21. $\lim\limits_{P\to(-1/4,\pi/2,2)} \tan^{-1}(xyz) = \tan^{-1}\left(-\dfrac{\pi}{4}\right)$

22. $\lim\limits_{P\to(1,-1,-1)} \dfrac{2xy + yz}{x^2 + z^2} = -\dfrac{1}{2}$

23. a) Continuous at all (x,y)
    b) Continuous at all (x,y) except (0,0)

24. a) Continuous at all (x,y) so that $x \neq y$
    b) Continuous at all (x,y)

25. a) Continuous at all (x,y) except where $x = 0$ or $y = 0$
    b) Continuous at all (x,y)

26. a) Continuous at all (x,y) so that $x^2 - 3x + 2 \neq 0 \Rightarrow (x-2)(x-1) \neq 0 \Rightarrow x \neq 2$ and $x \neq 1$

b) Continuous at all (x,y) so that $y \neq x^2$

27. a) Continuous at all (x,y,z)

b) Continuous at all (x,y,z) except the interior of the cylinder $x^2 + y^2 = 1$

28. a) Continuous at all (x,y,z) so that $xyz > 0$

b) Continuous at all (x,y,z)

29. a) Continuous at all (x,y,z) so that $(x,y,z) \neq (x,y,0)$

b) Continuous at all (x,y,z) except those on the sphere $x^2 + y^2 + z^2 = 1$

30. a) Continuous at all (x,y,z) except (0,0,0)

b) Continuous at all (x,y,z) except those of the form (0,y,0) or (x,0,0), that is all (x,y,z) except those on the x and y axes

31. $$\lim_{\substack{(x,y)\to(0,0)\\ \text{along } y=x,\\ x>0}} \frac{x}{\sqrt{x^2+y^2}} = \lim_{(x,y)\to(0,0)} \frac{x}{\sqrt{x^2+x^2}} = \lim_{(x,y)\to(0,0)} \frac{x}{\sqrt{2}\,|x|} = \lim_{(x,y)\to(0,0)} \frac{x}{\sqrt{2}\,x} = \lim_{(x,y)\to(0,0)} \frac{1}{\sqrt{2}} = \frac{1}{\sqrt{2}}$$

$$\lim_{\substack{(x,y)\to(0,0)\\ \text{along } y=x,\\ x<0}} \frac{x}{\sqrt{x^2+y^2}} = \lim_{(x,y)\to(0,0)} \frac{x}{\sqrt{2}\,|x|} = \lim_{(x,y)\to(0,0)} \frac{x}{\sqrt{2}(-x)} = \lim_{(x,y)\to(0,0)} -\frac{1}{\sqrt{2}} = -\frac{1}{\sqrt{2}}$$

$\therefore$ consider paths along $y = x$ where $x > 0$ or $x < 0$.

32. $$\lim_{\substack{(x,y)\to(0,0)\\ \text{along } y=0}} \frac{x^4}{x^4+y^2} = \lim_{(x,y)\to(0,0)} \frac{x^4}{x^4+0^2} = 1. \quad \lim_{\substack{(x,y)\to(0,0)\\ \text{along } y=x^2}} \frac{x^4}{x^4+y^2} = \lim_{(x,y)\to(0,0)} \frac{x^4}{x^4+\left(x^2\right)^2} = \frac{1}{2}$$

$\therefore$ consider paths along $y = 0$ and $y = x^2$

33. $$\lim_{\substack{(x,y)\to(0,0)\\ \text{along } y=kx^2}} \frac{x^4-y^2}{x^4+y^2} = \lim_{(x,y)\to(0,0)} \frac{x^4-\left(kx^2\right)^2}{x^4+\left(kx^2\right)^2} = \lim_{(x,y)\to(0,0)} \frac{x^4-k^2x^4}{x^4+k^2x^4} = \frac{1-k^2}{1+k^2} \Rightarrow$$ different limits for different values of k. $\therefore$ consider paths along $y = kx^2$, k a constant.

34. $$\lim_{\substack{(x,y)\to(0,0)\\ \text{along } y=kx,\ k\neq 0}} \frac{xy}{|xy|} = \lim_{(x,y)\to(0,0)} \frac{x(kx)}{|x(kx)|} = \lim_{(x,y)\to(0,0)} \frac{kx^2}{|kx^2|} = \lim_{(x,y)\to(0,0)} \frac{k}{|k|}.$$ If $k > 0$, the limit is 1; if $k < 0$, the limit is -1. $\therefore$ consider paths along $y = kx$, $k \neq 0$, k a constant.

35. $$\lim_{\substack{(x,y)\to(0,0)\\ \text{along } y=kx,\ k\neq -1}} \frac{x-y}{x+y} = \lim_{(x,y)\to(0,0)} \frac{x-kx}{x+kx} = \frac{1-k}{1+k} \Rightarrow$$ different limits for different values of k. $\therefore$ consider paths along $y = kx$, k a constant, $k \neq -1$.

36. $$\lim_{\substack{(x,y)\to(0,0)\\ \text{along } y=kx,\ k\neq 1}} \frac{x+y}{x-y} = \lim_{(x,y)\to(0,0)} \frac{x+kx}{x-kx} = \frac{1+k}{1-k} \Rightarrow$$ different limits for different values of k. $\therefore$ consider paths along $y = kx$, k a constant, $k \neq 1$

37. $\lim_{(x,y)\to(0,0)} \frac{x^2+y}{y} = \lim_{(x,y)\to(0,0)} \frac{x^2+kx^2}{kx^2} = \frac{1+k}{k} \Rightarrow$ different limits for different values of k. ∴ consider along $y = kx^2$, $k \neq 0$

paths along $y = kx^2$, k a constant, $k \neq 0$.

38. $\lim_{(x,y)\to(0,0)} \frac{x^2}{x^2-y} = \lim_{(x,y)\to(0,0)} \frac{x^2}{x^2-kx^2} = \frac{1}{1-k} \Rightarrow$ different limits for different values of k. ∴ consider along $y = kx^2$, $k \neq 1$

paths along $y = kx^2$, k a constant, $k \neq 0$.

39. a) $f(x,y)\big|_{y=mx} = \frac{2m}{1+m^2} = \frac{2\tan\theta}{1+\tan^2\theta} = \sin 2\theta$. The value of f(x,y) is $\sin 2\theta$ where $\tan\theta = m$ along $y = mx$.

b) Since $f(x,y)\big|_{y=mx} = \sin 2\theta$ and since $-1 \leq \sin 2\theta \leq 1$ for every θ, $\lim_{(x,y)\to(0,0)} f(x,y)$ varies from -1 to 1 along $y = mx$.

SECTION 13.3 PARTIAL DERIVATIVES

1. $\frac{\partial f}{\partial x} = 2, \frac{\partial f}{\partial y} = 0$

2. $\frac{\partial f}{\partial x} = 0, \frac{\partial f}{\partial y} = -3$

3. $\frac{\partial f}{\partial x} = 0, \frac{\partial f}{\partial y} = 0$

4. $\frac{\partial f}{\partial x} = 2, \frac{\partial f}{\partial y} = -3$

5. $\frac{\partial f}{\partial x} = y - 1, \frac{\partial f}{\partial y} = x$

6. $\frac{\partial f}{\partial x} = 2x, \frac{\partial f}{\partial y} = 2y$

7. $\frac{\partial f}{\partial x} = 2x - y, \frac{\partial f}{\partial y} = -x + 2y$

8. $\frac{\partial f}{\partial x} = y + 3, \frac{\partial f}{\partial y} = x + 2$

9. $\frac{\partial f}{\partial x} = 5y - 14x + 3,$ $\frac{\partial f}{\partial y} = 5x - 2y - 6$

10. $\frac{\partial f}{\partial x} = \frac{-1}{(x+y)^2}, \frac{\partial f}{\partial y} = \frac{-1}{(x+y)^2}$

11. $\frac{\partial f}{\partial x} = x(x^2+y^2)^{-1/2}, \frac{\partial f}{\partial y} = y(x^2+y^2)^{-1/2}$

12. $\frac{\partial f}{\partial x} = \frac{y^2-x^2}{(x^2+y^2)^2}, \frac{\partial f}{\partial y} = \frac{-2xy}{(x^2+y^2)^2}$

13. $\frac{\partial f}{\partial x} = \frac{-y^2-1}{(xy-1)^2}, \frac{\partial f}{\partial y} = \frac{-x^2-1}{(xy-1)^2}$

14. $\frac{\partial f}{\partial x} = -e^x + e^x(y-x), \frac{\partial f}{\partial y} = e^x$

15. $\frac{\partial f}{\partial x} = e^x \ln y, \frac{\partial f}{\partial y} = \frac{e^x}{y}$

16. $\frac{\partial f}{\partial x} = \cos(x+y), \frac{\partial f}{\partial y} = \cos(x+y)$

17. $\frac{\partial f}{\partial x} = e^x \sin(y+1), \frac{\partial f}{\partial y} = e^x \cos(y+1)$

18. $\frac{\partial f}{\partial x} = \frac{-y}{y^2+x^2}, \frac{\partial f}{\partial y} = \frac{x}{y^2+x^2}$

19. $f_x(x,y,z) = y + z, f_y(x,y,z) = x + z, f_z(x,y,z) = y + x$

20. $f_x(x,y,z) = -x(x^2+y^2+z^2)^{-3/2}, f_y(x,y,z) = -y(x^2+y^2+z^2)^{-3/2}, f_z(x,y,z) = -z(x^2+y^2+z^2)^{-3/2}$

21. $f_x(x,y,z) = 0, f_y(x,y,z) = 2y, f_z(x,y,z) = 4z$

22. $f_x(x,y,z) = 1, f_y(x,y,z) = -y(y^2+z^2)^{-1/2}, f_z(x,y,z) = -z(y^2+z^2)^{-1/2}$

23. $f_x(x,y,z) = \cos(x + yz)$, $f_y(x,y,z) = z\cos(x + yz)$, $f_z(x,y,z) = y\cos(x + yz)$

24. $f_x(x,y,z) = \sec^2(x + y + z)$, $f_y(x,y,z) = \sec^2(x + y + z)$, $f_z(x,y,z) = \sec^2(x + y + z)$

25. $\frac{\partial f}{\partial t} = -2\pi\sin(2\pi t - \alpha)$, $\frac{\partial f}{\partial \alpha} = \sin(2\pi t - \alpha)$

26. $\frac{\partial g}{\partial u} = 2v^2e^{2u/v}$, $\frac{\partial g}{\partial v} = 2ve^{2u/v} - 2ue^{2u/v}$

27. $\frac{\partial h}{\partial \rho} = \sin\phi\cos\theta$, $\frac{\partial h}{\partial \phi} = \rho\cos\phi\cos\theta$, $\frac{\partial h}{\partial \theta} = -\rho\sin\phi\sin\theta$

28. $\frac{\partial g}{\partial r} = 1 - \cos\theta$, $\frac{\partial g}{\partial \theta} = r\sin\theta$, $\frac{\partial g}{\partial z} = -1$

29. $\frac{\partial W}{\partial P} = v$, $\frac{\partial W}{\partial V} = P + \frac{\rho v^2}{2g}$, $\frac{\partial W}{\partial \rho} = \frac{Vv^2}{2g}$, $\frac{\partial W}{\partial v} = \frac{V\rho v}{g}$, $\frac{\partial W}{\partial g} = -\frac{V\rho v^2}{2g^2}$

30. $\frac{\partial A}{\partial c} = m$, $\frac{\partial A}{\partial h} = \frac{q}{2}$, $\frac{\partial A}{\partial k} = \frac{m}{q}$, $\frac{\partial A}{\partial m} = \frac{k}{q} + c$, $\frac{\partial A}{\partial q} = -\frac{km}{q^2} + \frac{h}{2}$

31. $\frac{\partial f}{\partial x} = y + z$, $\frac{\partial f}{\partial y} = x + z$, $\frac{\partial^2 f}{\partial x^2} = 0$, $\frac{\partial^2 f}{\partial y^2} = 0$, $\frac{\partial^2 f}{\partial y\partial x} = \frac{\partial^2 f}{\partial x\partial y} = 1$

32. $\frac{\partial f}{\partial x} = y\cos xy$, $\frac{\partial f}{\partial y} = x\cos xy$, $\frac{\partial^2 f}{\partial x^2} = -y^2\sin xy$, $\frac{\partial^2 f}{\partial y^2} = -x^2\sin xy$, $\frac{\partial^2 f}{\partial y\partial x} = \frac{\partial^2 f}{\partial x\partial y} = \cos xy - xy\sin xy$

33. $\frac{\partial g}{\partial x} = 2xy + y\cos x$, $\frac{\partial g}{\partial y} = x^2 - \sin y + \sin x$, $\frac{\partial^2 g}{\partial x^2} = 2y - y\sin x$, $\frac{\partial^2 g}{\partial y^2} = -\cos y$, $\frac{\partial^2 g}{\partial y\partial x} = \frac{\partial^2 g}{\partial x\partial y} = 2x + \cos x$

34. $\frac{\partial h}{\partial x} = e^y$, $\frac{\partial h}{\partial y} = xe^y + 1$, $\frac{\partial^2 h}{\partial x^2} = 0$, $\frac{\partial^2 h}{\partial y^2} = xe^y$, $\frac{\partial^2 h}{\partial y\partial x} = \frac{\partial^2 h}{\partial x\partial y} = e^y$

35. $\frac{\partial r}{\partial x} = \frac{1}{x + y}$, $\frac{\partial r}{\partial y} = \frac{1}{x + y}$, $\frac{\partial^2 r}{\partial x^2} = \frac{-1}{(x + y)^2}$, $\frac{\partial^2 r}{\partial y^2} = \frac{-1}{(x + y)^2}$, $\frac{\partial^2 r}{\partial y\partial x} = \frac{\partial^2 r}{\partial x\partial y} = \frac{-1}{(x + y)^2}$

36. $\frac{\partial s}{\partial x} = \frac{-y}{y^2 + x^2}$, $\frac{\partial s}{\partial y} = \frac{x}{y^2 + x^2}$, $\frac{\partial^2 s}{\partial x^2} = \frac{2xy}{(y^2 + x^2)^2}$, $\frac{\partial^2 s}{\partial y^2} = \frac{-2xy}{(y^2 + x^2)^2}$, $\frac{\partial^2 s}{\partial y\partial x} = \frac{\partial^2 s}{\partial x\partial y} = \frac{y^2 - x^2}{(y^2 + x^2)^2}$

37. $\frac{\partial w}{\partial x} = \frac{2}{2x + 3y}$, $\frac{\partial w}{\partial y} = \frac{3}{2x + 3y}$, $\frac{\partial^2 w}{\partial y\partial x} = \frac{-6}{(2x + 3y)^2}$ and $\frac{\partial^2 w}{\partial x\partial y} = \frac{-6}{(2x + 3y)^2}$

38. $\frac{\partial w}{\partial x} = e^x + \ln y + \frac{y}{x}$, $\frac{\partial w}{\partial y} = \frac{x}{y} + \ln x$, $\frac{\partial^2 w}{\partial y\partial x} = \frac{1}{y} + \frac{1}{x}$ and $\frac{\partial^2 w}{\partial x\partial y} = \frac{1}{y} + \frac{1}{x}$

39. $\frac{\partial w}{\partial x} = y^2 + 2xy^3 + 3x^2y^4$, $\frac{\partial w}{\partial y} = 2xy + 3x^2y^2 + 4x^3y^3$, $\frac{\partial^2 w}{\partial y\partial x} = 2y + 6xy^2 + 12x^2y^3$ and
$\frac{\partial^2 w}{\partial x\partial y} = 2y + 6xy^2 + 12x^2y^3$

40. $\frac{\partial w}{\partial x} = \sin y + y\cos x + y$, $\frac{\partial w}{\partial y} = x\cos y + \sin x + x$, $\frac{\partial^2 w}{\partial y\partial x} = \cos y + \cos x + 1$ and $\frac{\partial^2 w}{\partial x\partial y} = \cos y + \cos x + 1$

41. a) x first b) y first c) x first d) x first e) y first f) y first

42. a) y first three times b) y first three times c) y first twice d) x first twice

43. $xy + z^3x - 2yz = 0 \Rightarrow y + 3z^2x\frac{\partial z}{\partial x} - 2y\frac{\partial z}{\partial x} = 0 \Rightarrow (3z^2x - 2y)\frac{\partial z}{\partial x} = -y - z^3 \Rightarrow \frac{\partial z}{\partial x} = \frac{-y - z^3}{3z^2x - 2y} \Rightarrow$
$\frac{\partial z}{\partial x}(1,1,1) = -2$

44. $xz + y\ln x + x^2 + 4 = 0 \Rightarrow z\frac{\partial x}{\partial z} + x + y\left(\frac{1}{x}\right)\frac{\partial x}{\partial z} + 2x\frac{\partial x}{\partial z} = 0 \Rightarrow \left(z + \frac{y}{x} + 2x\right)\frac{\partial x}{\partial z} = -x \Rightarrow$
$\frac{\partial x}{\partial z} = \frac{-x}{z + \frac{y}{x} + 2x} \Rightarrow \frac{\partial x}{\partial z}(1,-1,-3) = \frac{1}{2}$

45. $\frac{\partial f}{\partial x} = 2x, \frac{\partial f}{\partial y} = 2y, \frac{\partial f}{\partial z} = -4z \Rightarrow \frac{\partial^2 f}{\partial x^2} = 2, \frac{\partial^2 f}{\partial y^2} = 2, \frac{\partial^2 f}{\partial z^2} = -4 \Rightarrow \frac{\partial^2 f}{\partial x^2} + \frac{\partial^2 f}{\partial y^2} + \frac{\partial^2 f}{\partial z^2} = 2 + 2 + (-4) = 0$

46. $\frac{\partial f}{\partial x} = -6xz, \frac{\partial f}{\partial y} = -6yz, \frac{\partial f}{\partial z} = 6z^2 - 3(x^2 + y^2), \frac{\partial^2 f}{\partial x^2} = -6z, \frac{\partial^2 f}{\partial y^2} = -6z, \frac{\partial^2 f}{\partial z^2} = 12z$
$\therefore \frac{\partial^2 f}{\partial x^2} + \frac{\partial^2 f}{\partial y^2} + \frac{\partial^2 f}{\partial z^2} = -6z - 6z + 12z = 0$

47. $\frac{\partial f}{\partial x} = -2e^{-2y}\sin 2x, \frac{\partial f}{\partial y} = -2e^{-2y}\cos 2x, \frac{\partial^2 f}{\partial x^2} = -4e^{-2y}\cos 2x, \frac{\partial^2 f}{\partial y^2} = 4e^{-2y}\cos 2x$
$\therefore \frac{\partial^2 f}{\partial x^2} + \frac{\partial^2 f}{\partial y^2} = -4e^{-2y}\cos 2x + 4e^{-2y}\cos 2x = 0$

48. $\frac{\partial f}{\partial x} = \frac{x}{x^2 + y^2}, \frac{\partial f}{\partial y} = \frac{y}{x^2 + y^2}, \frac{\partial^2 f}{\partial x^2} = \frac{y^2 - x^2}{(x^2 + y^2)^2}, \frac{\partial^2 f}{\partial y^2} = \frac{x^2 - y^2}{(x^2 + y^2)^2}$
$\therefore \frac{\partial^2 f}{\partial x^2} + \frac{\partial^2 f}{\partial y^2} = \frac{y^2 - x^2}{(x^2 + y^2)^2} + \frac{x^2 - y^2}{(x^2 + y^2)^2} = 0$

49. $\frac{\partial f}{\partial x} = -\frac{1}{2}(x^2 + y^2 + z^2)^{-3/2}(2x) = -x(x^2 + y^2 + x^2)^{-3/2}, \frac{\partial f}{\partial y} = -\frac{1}{2}(x^2 + y^2 + z^2)^{-3/2}(2y) = -y(x^2 + y^2 + z^2)^{-3/2}$
$\frac{\partial f}{\partial z} = -\frac{1}{2}(x^2 + y^2 + z^2)^{-3/2}(2z) = -z(x^2 + y^2 + z^2)^{-3/2}. \frac{\partial^2 f}{\partial x^2} = -(x^2 + y^2 + z^2)^{-3/2} + 3x^2(x^2 + y^2 + z^2)^{-5/2}$
$\frac{\partial^2 f}{\partial y^2} = -(x^2 + y^2 + z^2)^{-3/2} + 3y^2(x^2 + y^2 + z^2)^{-5/2}, \frac{\partial^2 f}{\partial z^2} = -(x^2 + y^2 + z^2)^{-3/2} + 3z^2(x^2 + y^2 + z^2)^{-5/2}$
$\therefore \frac{\partial^2 f}{\partial x^2} + \frac{\partial^2 f}{\partial y^2} + \frac{\partial^2 f}{\partial z^2} = \left(-(x^2 + y^2 + z^2)^{-3/2} + 3x^2(x^2 + y^2 + z^2)^{-5/2}\right) +$
$\left(-(x^2 + y^2 + z^2)^{-3/2} + 3y^2(x^2 + y^2 + z^2)^{-5/2}\right) + \left(-(x^2 + y^2 + z^2)^{-3/2} + 3z^2(x^2 + y^2 + z^2)^{-5/2}\right) =$
$-3(x^2 + y^2 + z^2)^{-3/2} + (3x^2 + 3y^2 + 3z^2)(x^2 + y^2 + z^2)^{-5/2} = 0$

50. $\frac{\partial f}{\partial x} = 3e^{3x+4y}\cos 5z, \frac{\partial f}{\partial y} = 4e^{3x+4y}\cos 5z, \frac{\partial f}{\partial z} = -5e^{3x+4y}\sin 5z. \frac{\partial^2 f}{\partial x^2} = 9e^{3x+4y}\cos 5z,$
$\frac{\partial^2 f}{\partial y^2} = 16e^{3x+4y}\cos 5z, \frac{\partial^2 f}{\partial z^2} = -25e^{3x+4y}\cos 5z \Rightarrow \frac{\partial^2 f}{\partial x^2} + \frac{\partial^2 f}{\partial y^2} + \frac{\partial^2 f}{\partial z^2} = 9e^{3x+4y}\cos 5z + 16e^{3x+4y}\cos 5z -$
$25e^{3x+4y}\cos 5z = 0$

51. $\frac{\partial w}{\partial x} = \cos(x + ct)$, $\frac{\partial w}{\partial t} = c\cos(x + ct)$. $\frac{\partial^2 w}{\partial x^2} = -\sin(x + ct)$, $\frac{\partial^2 w}{\partial t^2} = -c^2 \sin(x + ct)$

$\therefore \frac{\partial^2 w}{\partial t^2} = c^2(-\sin(x + ct)) = c^2 \frac{\partial^2 w}{\partial x^2}$

52. $\frac{\partial w}{\partial x} = -2\sin(2x + 2ct)$, $\frac{\partial w}{\partial t} = -2c\sin(2x + 2ct)$. $\frac{\partial^2 w}{\partial x^2} = -4\cos(2x + 2ct)$, $\frac{\partial^2 w}{\partial t^2} = -4c^2\cos(2x + 2ct)$

$\therefore \frac{\partial^2 w}{\partial t^2} = c^2(-4\cos(2x + 2ct)) = c^2 \frac{\partial^2 w}{\partial x^2}$

53. $\frac{\partial w}{\partial x} = \cos(x + ct) - 2\sin(2x + 2ct)$, $\frac{\partial w}{\partial t} = c\cos(x + ct) - 2c\sin(2x + 2ct)$. $\frac{\partial^2 w}{\partial x^2} = -\sin(x + ct) -$ $4\cos(2x + 2ct)$, $\frac{\partial^2 w}{\partial t^2} = -c^2\sin(x + ct) - 4c^2\cos(2x + 2ct)$ $\therefore \frac{\partial^2 w}{\partial t^2} = c^2(-\sin(x + ct) - 4\cos(2x + 2ct)) =$ $c^2 \frac{\partial^2 w}{\partial x^2}$

54. $\frac{\partial w}{\partial x} = \frac{1}{x + ct}$, $\frac{\partial w}{\partial t} = \frac{c}{x + ct}$. $\frac{\partial^2 w}{\partial x^2} = \frac{-1}{(x + ct)^2}$, $\frac{\partial^2 w}{\partial t^2} = \frac{-c^2}{(x + ct)^2}$. $\therefore \frac{\partial^2 w}{\partial t^2} = c^2\left(\frac{-1}{(x + ct)^2}\right) = c^2 \frac{\partial^2 w}{\partial x^2}$

55. $\frac{\partial w}{\partial x} = 2\sec^2(2x - 2ct)$, $\frac{\partial w}{\partial t} = -2c\sec^2(2x - 2ct)$. $\frac{\partial^2 w}{\partial x^2} = 8\sec^2(2x - 2ct)\tan(2x - 2ct)$, $\frac{\partial^2 w}{\partial t^2} = 8c^2\sec^2(2x - 2ct)\tan(2x - 2ct)$ $\therefore \frac{\partial^2 w}{\partial t^2} = c^2\left(8\sec^2(2x - 2ct)\tan(2x - 2ct)\right) = c^2 \frac{\partial^2 w}{\partial x^2}$

56. $\frac{\partial w}{\partial x} = -15\sin(3x + 3ct) + e^{x+ct}$, $\frac{\partial w}{\partial t} = -15c\sin(3x + 3ct) + ce^{x+ct}$. $\frac{\partial^2 w}{\partial x^2} = -45\cos(3x + 3ct) + e^{x+ct}$ $\frac{\partial^2 w}{\partial t^2} = -45c^2\cos(3x + 3ct) + c^2e^{x+ct}$ $\therefore \frac{\partial^2 w}{\partial t^2} = c^2\left(-45\cos(3x + 3ct) + e^{x+ct}\right) = c^2 \frac{\partial^2 w}{\partial x^2}$

SECTION 13.4 THE CHAIN RULE

1. $\frac{\partial w}{\partial x} = 2x$, $\frac{\partial w}{\partial y} = 2y$, $\frac{dx}{dt} = -\sin t$, $\frac{dy}{dt} = \cos t \Rightarrow \frac{dw}{dt} = -2x\sin t + 2y\cos t = -2\cos t\sin t + 2\sin t\cos t = 0$ $\Rightarrow \frac{dw}{dt}(0) = 0$

2. $\frac{\partial w}{\partial x} = 2x$, $\frac{\partial w}{\partial y} = 2y$, $\frac{dx}{dt} = -\sin t + \cos t$, $\frac{dy}{dt} = -\sin t - \cos t \Rightarrow \frac{dw}{dt} = 2x(-\sin t + \cos t) +$ $2y(-\sin t - \cos t) = 2(\cos t + \sin t)(\cos t - \sin t) - 2(\cos t - \sin t)(\sin t + \cos t) = 0 \Rightarrow \frac{dw}{dt}(0) = 0.$

3. $\frac{\partial w}{\partial x} = \frac{1}{z}$, $\frac{\partial w}{\partial y} = \frac{1}{z}$, $\frac{\partial w}{\partial z} = \frac{-(x + y)}{z^2}$, $\frac{dx}{dt} = -2\cos t\sin t$, $\frac{dy}{dt} = 2\sin t\cos t$, $\frac{dz}{dt} = -\frac{1}{t^2} \Rightarrow$ $\frac{dw}{dt} = -\frac{2}{z}\cos t\sin t + \frac{2}{z}\sin t\cos t + \frac{x + y}{z^2t^2} = \frac{\cos^2 t + \sin^2 t}{\frac{1}{t^2}(t^2)} = 1 \Rightarrow \frac{dw}{dt}(3) = 1$

4. $\frac{\partial w}{\partial x} = \frac{2x}{x^2 + y^2 + z^2}$, $\frac{\partial w}{\partial y} = \frac{2y}{x^2 + y^2 + z^2}$, $\frac{\partial w}{\partial z} = \frac{2z}{x^2 + y^2 + z^2}$, $\frac{dx}{dt} = -\sin t$, $\frac{dy}{dt} = \cos t$, $\frac{dz}{dt} = 2t^{-1/2} \Rightarrow$
$\frac{dw}{dt} = \frac{-2x \sin t}{x^2 + y^2 + z^2} + \frac{2y \cos t}{x^2 + y^2 + z^2} + \frac{4zt^{-1/2}}{x^2 + y^2 + z^2} = \frac{-2 \sin t \cos t + 2 \sin t \cos t + 4\left(4t^{1/2}\right)t^{-1/2}}{\cos^2 t + \sin^2 t + 16t} = \frac{16}{1 + 16t} \Rightarrow$
$\frac{dw}{dt}(3) = \frac{16}{49}$

5. $\frac{\partial w}{\partial x} = 2ye^x$, $\frac{\partial w}{\partial y} = 2e^x$, $\frac{\partial w}{\partial z} = -\frac{1}{z}$, $\frac{dx}{dt} = \frac{2t}{t^2 + 1}$, $\frac{dy}{dt} = \frac{1}{t^2 + 1}$, $\frac{dz}{dt} = e^t \Rightarrow \frac{dw}{dt} = \frac{4yte^x}{t^2 + 1} + \frac{2e^x}{t^2 + 1} - \frac{e^t}{z} =$
$\frac{4t \tan^{-1} t\, e^{\ln(t^2+1)}}{t^2 + 1} + \frac{2(t^2 + 1)}{t^2 + 1} - \frac{e^t}{e^t} = 4t \tan^{-1} t + 1 \Rightarrow \frac{dw}{dt}(1) = \pi + 1$

6. $\frac{\partial w}{\partial x} = -y \cos xy$, $\frac{\partial w}{\partial y} = -x \cos xy$, $\frac{\partial w}{\partial z} = 1$, $\frac{dx}{dt} = 1$, $\frac{dy}{dt} = \frac{1}{t}$, $\frac{dz}{dt} = e^{t-1} \Rightarrow \frac{dw}{dt} = -y \cos xy - \frac{x \cos xy}{t} + e^{t-1} =$
$-\ln t(\cos(t \ln t)) - \frac{t \cos(t \ln t)}{t} + e^{t-1} = -\ln t(\cos(t \ln t)) - \cos(t \ln t) + e^{t-1} \Rightarrow \frac{dw}{dt}(1) = 0$

7. $\frac{dz}{dt} = \frac{\partial z}{\partial x}\frac{dx}{dt} + \frac{\partial z}{\partial y}\frac{dy}{dt}$

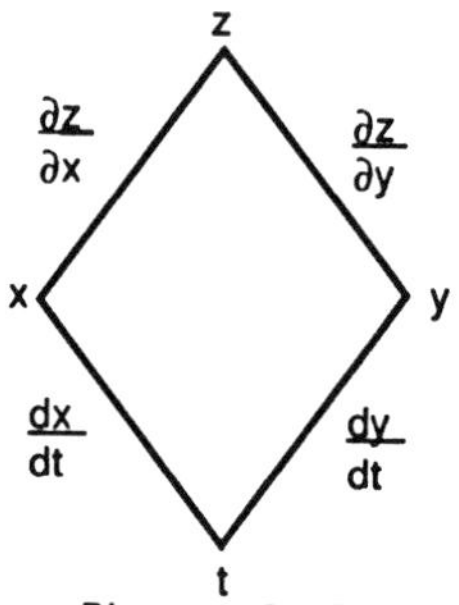

Diagram 13.4.7

8. $\frac{dz}{dt} = \frac{\partial z}{\partial u}\frac{du}{dt} + \frac{\partial z}{\partial v}\frac{dv}{dt} + \frac{\partial z}{\partial w}\frac{dw}{dt}$

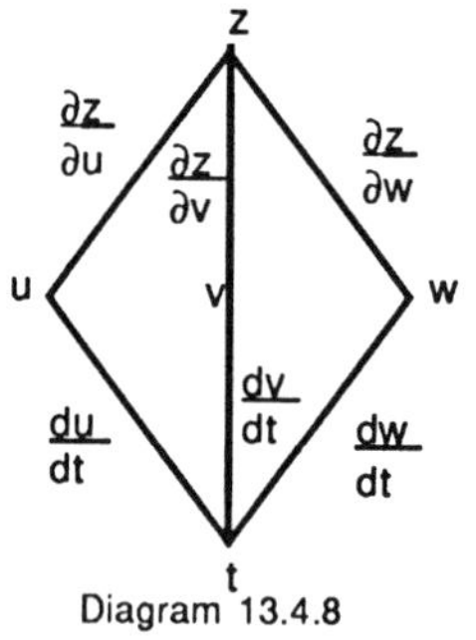

Diagram 13.4.8

9. $\frac{\partial w}{\partial u} = \frac{\partial w}{\partial x}\frac{\partial x}{\partial u} + \frac{\partial w}{\partial y}\frac{\partial y}{\partial u} + \frac{\partial w}{\partial z}\frac{\partial z}{\partial u}$

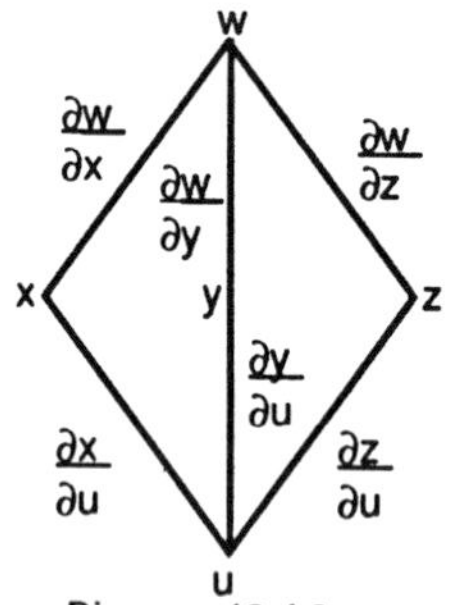

Diagram 13.4.9 a

$\frac{\partial w}{\partial v} = \frac{\partial w}{\partial x}\frac{\partial x}{\partial v} + \frac{\partial w}{\partial y}\frac{\partial y}{\partial v} + \frac{\partial w}{\partial z}\frac{\partial z}{\partial v}$

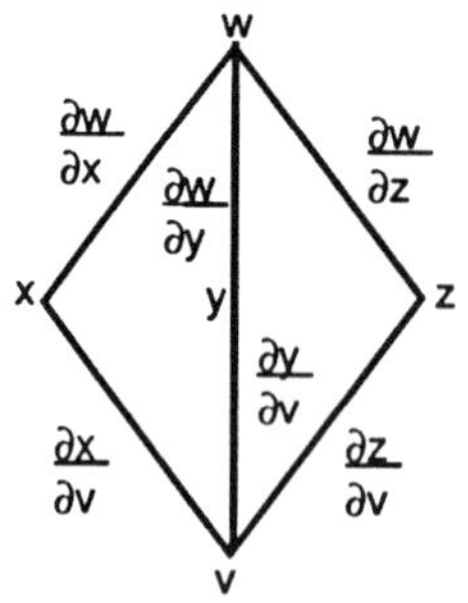

Diagram 13.4.9 b

10. $\frac{\partial w}{\partial x} = \frac{\partial w}{\partial r}\frac{\partial r}{\partial x} + \frac{\partial w}{\partial s}\frac{\partial s}{\partial x} + \frac{\partial w}{\partial t}\frac{\partial t}{\partial x}$

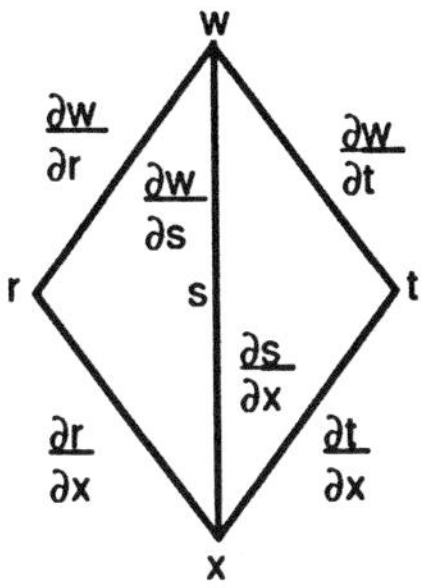

Diagram 13.4.10 a

$\frac{\partial w}{\partial y} = \frac{\partial w}{\partial r}\frac{\partial r}{\partial y} + \frac{\partial w}{\partial s}\frac{\partial s}{\partial y} + \frac{\partial w}{\partial t}\frac{\partial t}{\partial y}$

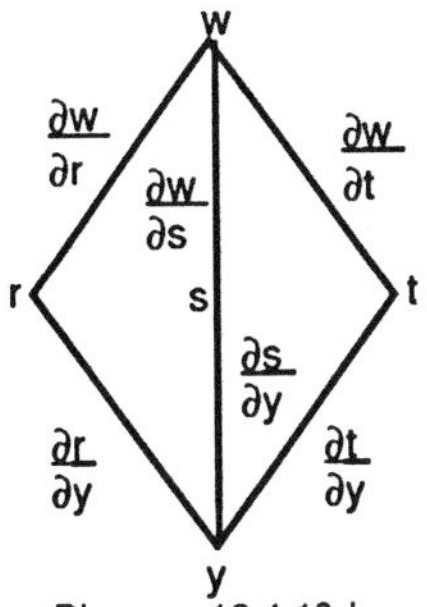

Diagram 13.4.10 b

11. $\frac{\partial w}{\partial u} = \frac{\partial w}{\partial x}\frac{\partial x}{\partial u} + \frac{\partial w}{\partial y}\frac{\partial y}{\partial u}$

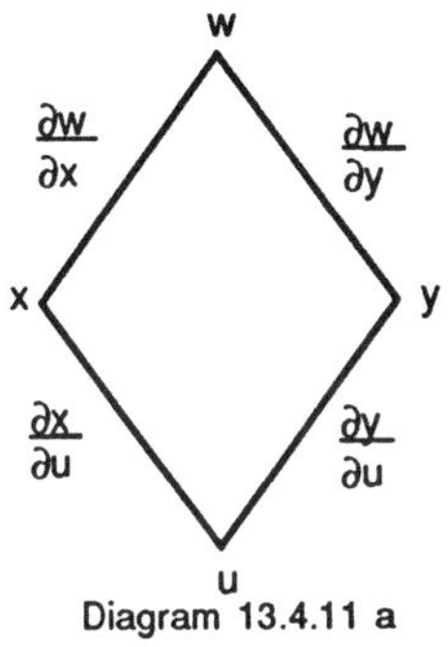

Diagram 13.4.11 a

$\frac{\partial w}{\partial v} = \frac{\partial w}{\partial x}\frac{\partial x}{\partial v} + \frac{\partial w}{\partial y}\frac{\partial y}{\partial v}$

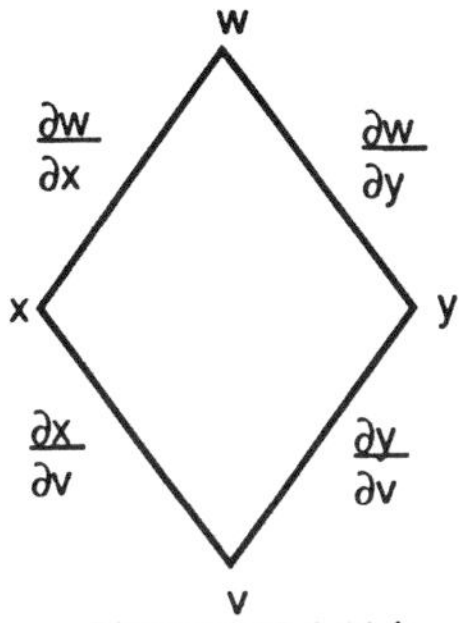

Diagram 13.4.11 b

12. $\frac{\partial w}{\partial x} = \frac{\partial w}{\partial u}\frac{\partial u}{\partial x} + \frac{\partial w}{\partial v}\frac{\partial v}{\partial x}$

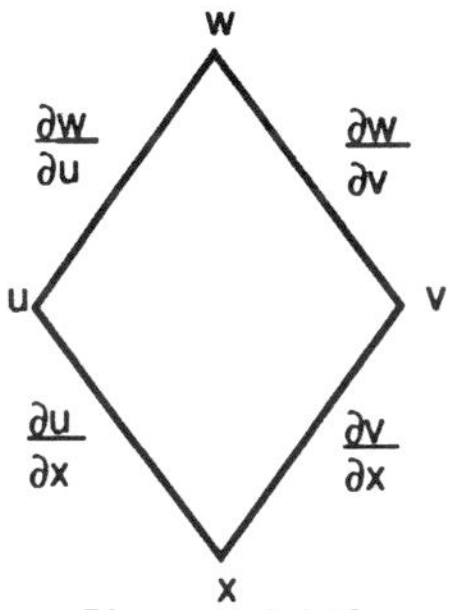

Diagram 13.4.12 a

$\frac{\partial w}{\partial y} = \frac{\partial w}{\partial u}\frac{\partial u}{\partial y} + \frac{\partial w}{\partial v}\frac{\partial v}{\partial y}$

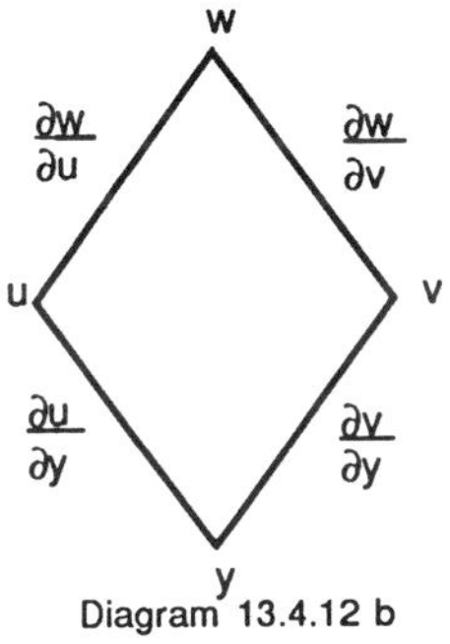

Diagram 13.4.12 b

13. $\frac{\partial z}{\partial t} = \frac{\partial z}{\partial x}\frac{\partial x}{\partial t} + \frac{\partial z}{\partial y}\frac{\partial y}{\partial t}$

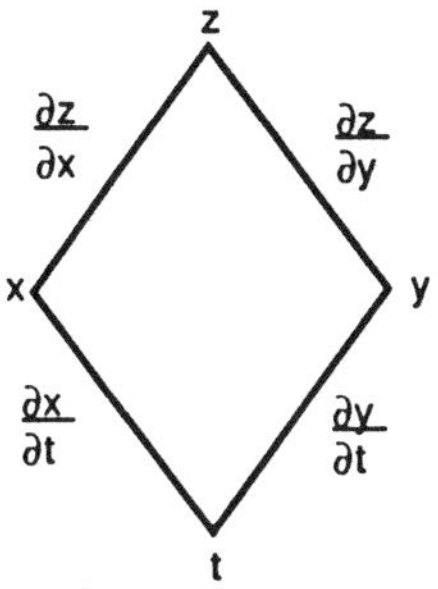

Diagram 13.4.13 a

$\frac{\partial z}{\partial s} = \frac{\partial z}{\partial x}\frac{\partial x}{\partial s} + \frac{\partial z}{\partial y}\frac{\partial y}{\partial s}$

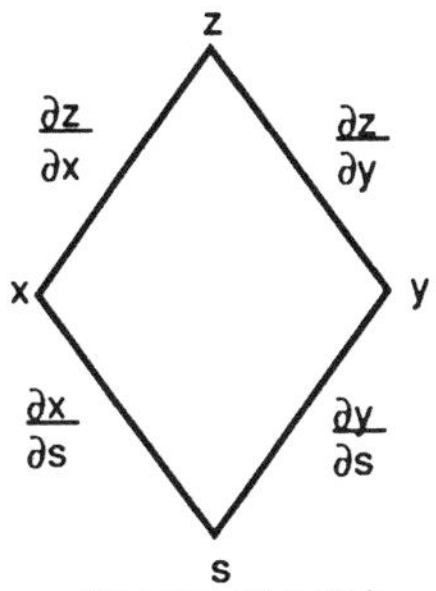

Diagram 13.4.13 b

14. $\frac{\partial y}{\partial r} = \frac{dy}{du}\frac{\partial u}{\partial r}$

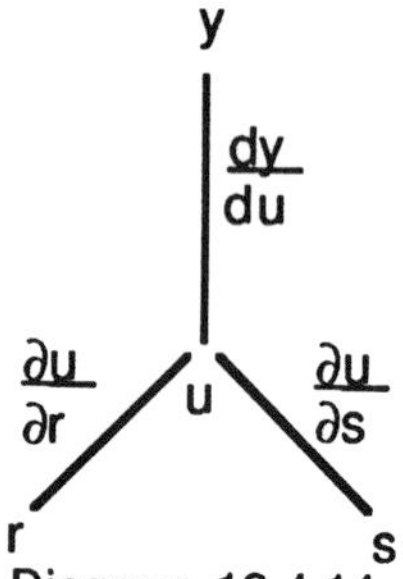

Diagram 13.4.14

15. $\frac{\partial w}{\partial s} = \frac{dw}{du}\frac{\partial u}{\partial s}$

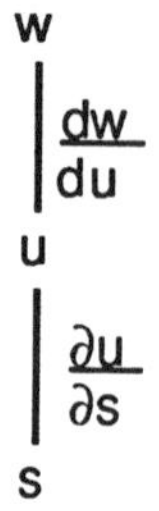

Diagram 13.4.15 a

$\frac{\partial w}{\partial t} = \frac{dw}{du}\frac{\partial u}{\partial t}$

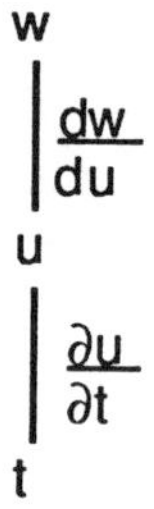

Diagram 13.4.15 b

16. $\frac{\partial w}{\partial p} = \frac{\partial w}{\partial x}\frac{\partial x}{\partial p} + \frac{\partial w}{\partial y}\frac{\partial y}{\partial p} + \frac{\partial w}{\partial z}\frac{\partial z}{\partial p} + \frac{\partial w}{\partial v}\frac{\partial v}{\partial p}$

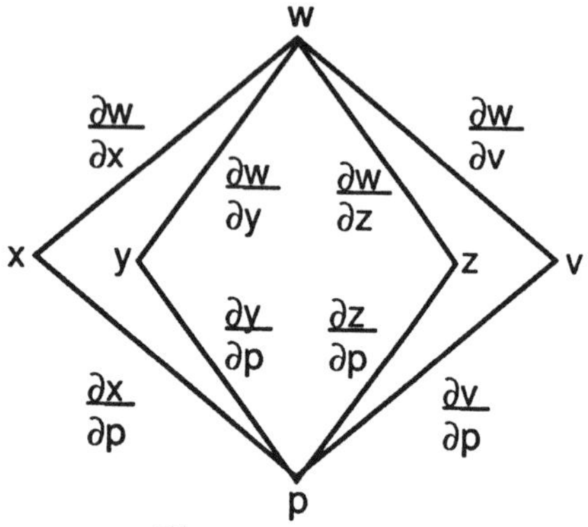

Diagram 13.4.16

17. $\frac{\partial w}{\partial r} = \frac{\partial w}{\partial x}\frac{dx}{dr} + \frac{\partial w}{\partial y}\frac{dy}{dr} = \frac{\partial w}{\partial x}\frac{dx}{dr}$

since $\frac{dy}{dr} = 0$

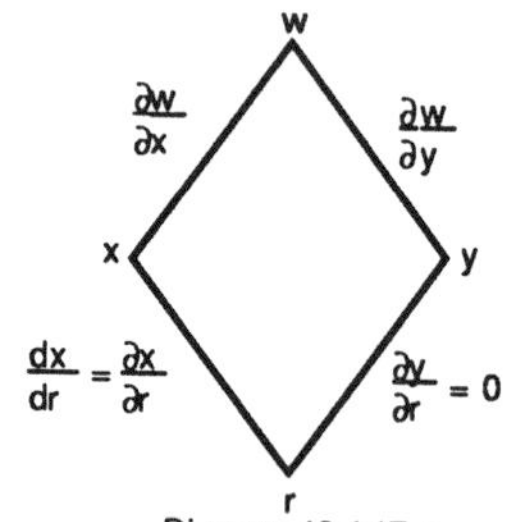

Diagram 13.4.17 a

$\frac{\partial w}{\partial s} = \frac{\partial w}{\partial x}\frac{dx}{ds} + \frac{\partial w}{\partial y}\frac{dy}{ds} = \frac{\partial w}{\partial y}\frac{dy}{ds}$

since $\frac{dx}{ds} = 0$

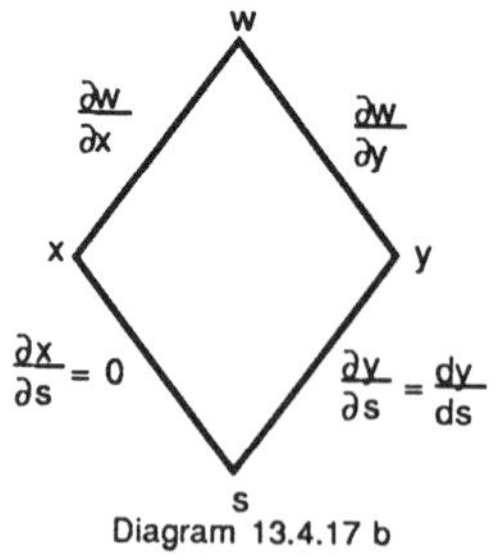

Diagram 13.4.17 b

18. $\frac{\partial w}{\partial s} = \frac{\partial w}{\partial x}\frac{\partial x}{\partial s} + \frac{\partial w}{\partial y}\frac{\partial y}{\partial s}$

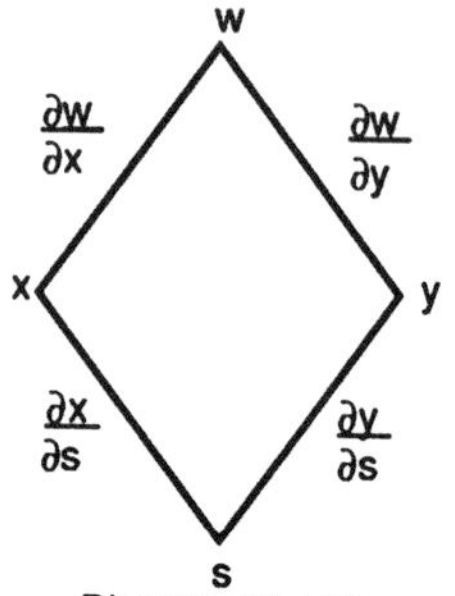

Diagram 13.4.18

19. Let $F(x,y) = x^3 - 2y^2 + xy = 0 \Rightarrow F_x(x,y) = 3x^2 + y$,

$F_y(x,y) = -4y + x \Rightarrow \frac{dy}{dx} = -\frac{F_x}{F_y} = -\frac{3x^2 + y}{-4y + x} \Rightarrow$

$\frac{dy}{dx}(1,1) = \frac{4}{3}$

20. Let $F(x,y) = xy + y^2 - 3x - 3 = 0 \Rightarrow F_x(x,y) = y - 3,\ F_y(x,y) = x + 2y \Rightarrow \frac{dy}{dx} = -\frac{F_x}{F_y} = -\frac{y-3}{x+2y} \Rightarrow$
$\frac{dy}{dx}(-1,1) = 2$

21. Let $F(x,y) = x^2 + xy + y^2 - 7 = 0 \Rightarrow F_x(x,y) = 2x + y,\ F_y(x,y) = x + 2y \Rightarrow \frac{dy}{dx} = -\frac{F_x}{F_y} = -\frac{2x+y}{x+2y} \Rightarrow$
$\frac{dy}{dx}(1,2) = -\frac{4}{5}$

22. Let $F(x,y) = xe^y + \sin xy + y - \ln 2 = 0 \Rightarrow F_x(x,y) = e^y + y\cos xy,\ F_y(x,y) = xe^y + x\sin xy + 1 \Rightarrow$
$\frac{dy}{dx} = -\frac{F_x}{F_y} = -\frac{e^y + y\cos xy}{xe^y + x\sin xy + 1} \Rightarrow \frac{dy}{dx}(0,\ln 2) = -(2 + \ln 2)$

23. Let $F(x,y,z)) = z^3 - xy + yz + y^3 - 2 = 0 \Rightarrow F_x(x,y,z) = -y,\ F_y(x,y,z) = -x + z + 3y^2,\ F_z(x,y,z) = 3z^2 + y \Rightarrow$
$\frac{\partial z}{\partial x} = -\frac{F_x}{F_z} = -\frac{-y}{3z^2+y} = \frac{y}{3z^2+y} \Rightarrow \frac{\partial z}{\partial x}(1,1,1) = \frac{1}{4}.\ \frac{\partial z}{\partial y} = -\frac{F_y}{F_z} = -\frac{-x+z+3y^2}{3z^2+y} = \frac{x-z-3y^2}{3z^2+y} \Rightarrow$
$\frac{\partial z}{\partial y}(1,1,1) = -\frac{3}{4}$

24. Let $F(x,y,z) = \frac{1}{x} + \frac{1}{y} + \frac{1}{z} - 1 = 0 \Rightarrow F_x(x,y,z) = -\frac{1}{x^2},\ F_y(x,y,z) = -\frac{1}{y^2},\ F_z(x,y,z) = -\frac{1}{z^2} \Rightarrow$
$\frac{\partial z}{\partial x} = -\frac{F_x}{F_z} = -\frac{-\frac{1}{x^2}}{-\frac{1}{z^2}} = -\frac{z^2}{x^2} \Rightarrow \frac{\partial z}{\partial x}(2,3,6) = -9.\ \frac{\partial z}{\partial y} = -\frac{F_y}{F_z} = -\frac{-\frac{1}{y^2}}{-\frac{1}{z^2}} = -\frac{z^2}{y^2} \Rightarrow \frac{\partial z}{\partial y}(2,3,6) = -4$

25. Let $F(x,y,z) = \sin(x+y) + \sin(y+z) + \sin(x+z) = 0 \Rightarrow F_x(x,y,z) = \cos(x+y) + \cos(x+z)$,
$F_y(x,y,z) = \cos(x+y) + \cos(y+z),\ F_z(x,y,z) = \cos(y+z) + \cos(x+z) \Rightarrow \frac{\partial z}{\partial x} = -\frac{F_x}{F_z} = -\frac{\cos(x+y)+\cos(x+z)}{\cos(y+z)+\cos(x+z)}$
$\Rightarrow \frac{\partial z}{\partial x}(\pi,\pi,\pi) = -1.\ \frac{\partial z}{\partial y} = -\frac{F_y}{F_z} = -\frac{\cos(x+y)+\cos(y+z)}{\cos(y+z)+\cos(x+z)} \Rightarrow \frac{\partial z}{\partial y}(\pi,\pi,\pi) = -1$

26. Let $F(x,y,z) = xe^y + ye^z + 2\ln x - 2 - 3\ln 2 = 0 \Rightarrow F_x(x,y,z) = e^y + \frac{2}{x},\ F_y(x,y,z) = xe^y + e^z,\ F_z(x,y,z) = y$
$\Rightarrow \frac{\partial z}{\partial x} = -\frac{F_x}{F_z} = -\frac{e^y + \frac{2}{x}}{y} \Rightarrow \frac{\partial z}{\partial x}(1,\ln 2,\ln 3) = -\frac{4}{\ln 2}.\ \frac{\partial z}{\partial y} = -\frac{F_y}{F_z} = -\frac{xe^y + e^z}{y} \Rightarrow \frac{\partial z}{\partial y}(1,\ln 2,\ln 3) = -\frac{5}{\ln 2}$

27. $\frac{\partial w}{\partial r} = \frac{\partial w}{\partial x}\frac{\partial x}{\partial r} + \frac{\partial w}{\partial y}\frac{\partial y}{\partial r} + \frac{\partial w}{\partial z}\frac{\partial z}{\partial r} = 2(x+y+z)(1) + 2(x+y+z)(-\sin(r+s)) + 2(x+y+z)(\cos(r+s)) =$
$2(x+y+z)(1 - \sin(r+s) + \cos(r+s)) = 2(r - s + \cos(r+s) + \sin(r+s))(1 - \sin(r+s) + \cos(r+s)) \Rightarrow$
$\left.\frac{\partial w}{\partial r}\right|_{r=1,s=1} = 12$

28. $\frac{\partial w}{\partial v} = \frac{\partial w}{\partial x}\frac{\partial x}{\partial v} + \frac{\partial w}{\partial y}\frac{\partial y}{\partial v} + \frac{\partial w}{\partial z}\frac{\partial z}{\partial v} = y\left(\frac{2v}{u}\right) + x(1) + \frac{1}{z}(0) = (u+v)\left(\frac{2v}{u}\right) + \frac{v^2}{u} \Rightarrow \left.\frac{\partial w}{\partial v}\right|_{u=-1,v=2} = -5$

29. $\frac{\partial w}{\partial v} = \frac{\partial w}{\partial x}\frac{\partial x}{\partial v} + \frac{\partial w}{\partial y}\frac{\partial y}{\partial v} = \left(2x - \frac{y}{x^2}\right)(-2) + \frac{1}{x}(1) = \left(2(u-2v+1) - \frac{2u+v-2}{(u-2v+1)^2}\right)(-2) + \frac{1}{u-2v+1} \Rightarrow$
$\left.\frac{\partial w}{\partial v}\right|_{u=0,v=0} = -7$

30. $\frac{\partial z}{\partial u} = \frac{\partial z}{\partial x}\frac{\partial x}{\partial u} + \frac{\partial z}{\partial y}\frac{\partial y}{\partial u} = (y\cos xy + \sin y)2u + (x\cos xy + x\cos y)v = \left(uv\cos\left(\left(u^2+v^2\right)uv\right) + \sin uv\right)2u + \left(\left(u^2+v^2\right)\cos\left(u^2+v^2\right)uv + \left(u^2+v^2\right)\cos uv\right)v \Rightarrow \left.\frac{\partial z}{\partial u}\right|_{u=0,v=1} = 2$

31. $\frac{\partial z}{\partial u} = \frac{dz}{dx}\frac{\partial x}{\partial u} = \frac{5}{1+x^2}\left(e^u + \ln v\right) = \frac{5}{1+\left(e^u+\ln v\right)^2}\left(e^u + \ln v\right) \Rightarrow \left.\frac{\partial z}{\partial u}\right|_{u=\ln 2,v=1} = 2$

$\frac{\partial z}{\partial v} = \frac{dz}{dx}\frac{\partial x}{\partial v} = \frac{5}{1+x^2}\left(\frac{1}{v}\right) = \frac{5}{1+\left(e^u+\ln v\right)^2}\left(\frac{1}{v}\right) \Rightarrow \left.\frac{\partial z}{\partial v}\right|_{u=\ln 2,v=1} = 5$

32. $\frac{\partial w}{\partial q} = \frac{\partial w}{\partial x}\frac{\partial x}{\partial q} + \frac{\partial w}{\partial y}\frac{\partial y}{\partial q} + \frac{\partial w}{\partial z}\frac{\partial z}{\partial q} + \frac{\partial w}{\partial u}\frac{\partial u}{\partial q} + \frac{\partial w}{\partial v}\frac{\partial v}{\partial q} = yp + (x+z)\tan^{-1}p + (y+u)\frac{1}{q-3p} + z\left(\frac{1}{2}\right)(q+5p)^{-1/2} + u(1) = qp\tan^{-1}p + (pq + \ln(q-3p))\tan^{-1}p + \frac{q\tan^{-1}p + \sqrt{q+5p}}{q-3p} + \frac{\ln(q-3p)}{2(q+5p)^{1/2}} + \sqrt{q+5p} \Rightarrow$
$\left.\frac{\partial w}{\partial q}\right|_{p=1,q=4} = 3\pi + 6$

33. $\frac{dV}{dt} = \frac{\partial V}{\partial I}\frac{dI}{dt} + \frac{\partial V}{\partial R}\frac{dR}{dt}$. $V = IR \Rightarrow \frac{\partial V}{\partial I} = R, \frac{\partial V}{\partial R} = I \Rightarrow \frac{dV}{dt} = R\frac{dI}{dt} + I\frac{dR}{dt} \Rightarrow -0.01$ volts/sec = (600 ohms)$\frac{dI}{dt}$ + (0.04 amps)(0.5 ohms/sec) $\Rightarrow \frac{dI}{dt} = -0.00005$ amps/sec.

34. a) $V = abc \Rightarrow \frac{dV}{dt} = \frac{\partial V}{\partial a}\frac{da}{dt} + \frac{\partial V}{\partial b}\frac{db}{dt} + \frac{\partial V}{\partial c}\frac{dc}{dt} = bc\left(\frac{da}{dt}\right) + ac\left(\frac{db}{dt}\right) + ab\left(\frac{dc}{dt}\right) \Rightarrow$
$\left.\frac{dV}{dt}\right|_{a=13,b=9,c=5}$ = (9 cm)(5 cm)(2 cm/sec) + (13 cm)(5 cm)(−5 cm/sec) + (13 cm)(9 cm)(2 cm/sec) = -1 cm^3/sec $\Rightarrow$ Volume is decreasing.

$S = 4ac + 2bc \Rightarrow \frac{dS}{dt} = \frac{\partial S}{\partial a}\frac{da}{dt} + \frac{\partial S}{\partial b}\frac{db}{dt} + \frac{\partial S}{\partial c}\frac{dc}{dt} = 4c\left(\frac{da}{dt}\right) + 2c\left(\frac{db}{dt}\right) + (4a+2b)\left(\frac{dc}{dt}\right) \Rightarrow$
$\left.\frac{dS}{dt}\right|_{a=13,b=9,c=5}$ = 4(5 cm)(2 cm/sec) + 2(5 cm)(−5 cm/sec) + (4(13 cm) + 2(9 cm))(2 cm/sec) = 130 cm^2/sec. S is increasing.

b) The diagonal is $D = \sqrt{a^2+b^2+c^2}$. Then $\frac{dD}{dt} = \frac{\partial D}{\partial a}\frac{da}{dt} + \frac{\partial D}{\partial b}\frac{db}{dt} + \frac{\partial D}{\partial c}\frac{dc}{dt} = \frac{1}{2}(a^2+b^2+c^2)^{-1/2}(2a)\frac{da}{dt} + \frac{1}{2}(a^2+b^2+c^2)^{-1/2}(2b)\frac{db}{dt} + \frac{1}{2}(a^2+b^2+c^2)^{-1/2}(2c)\frac{dc}{dt} = (a^2+b^2+c^2)^{-1/2}\left(a\frac{da}{dt} + b\frac{db}{dt} + c\frac{dc}{dt}\right)$
$\Rightarrow \left.\frac{dD}{dt}\right|_{a=13,b=9,c=5} =$
$\left(13^2\text{ cm}^2 + 9^2\text{ cm}^2 + 5^2\text{ cm}^2\right)^{-1/2}$((13 cm)(2 cm/sec) + (9 cm)(−5 cm/sec) + (5 cm)(2 cm/sec))
$= \frac{-9}{\sqrt{275}}$ cm/sec. The diagonal is decreasing.

35. $f_x(x,y,z) = \cos t$, $f_y(x,y,z) = \sin t$, $f_z(x,y,z) = t^2 + t - 2$. $\frac{df}{dt} = \frac{\partial f}{\partial x}\frac{dx}{dt} + \frac{\partial f}{\partial y}\frac{dy}{dt} + \frac{\partial f}{\partial z}\frac{dz}{dt} = (\cos t)(-\sin t) + (\sin t)(\cos t) + (t^2+t-2)(1) = t^2 + t - 2$. $\frac{df}{dt} = 0 \Rightarrow t^2 + t - 2 = 0 \Rightarrow t = -2$ or $t = 1$
$t = -2 \Rightarrow x = \cos(-2), y = \sin(-2), z = -2$; $t = 1 \Rightarrow x = \cos 1, y = \sin 1, z = 1$

36. $\frac{dw}{dt} = \frac{\partial w}{\partial x}\frac{dx}{dt} + \frac{\partial w}{\partial y}\frac{dy}{dt} + \frac{\partial w}{\partial z}\frac{dz}{dt} = \left(2xe^{2y}\cos 3z\right)(-\sin t) + \left(2x^2e^{2y}\cos 3z\right)\left(\frac{1}{t+2}\right) +$
$\left(-3x^2e^{2y}\sin 3z\right)(1) = -2xe^{2y}\cos 3z \sin t + \frac{2x^2e^{2y}\cos 3z}{t+2} - 3x^2e^{2y}\sin 3z$. Now $z = 0$ and $z = t \Rightarrow t = 0$
$\therefore \left.\frac{dw}{dt}\right|_{(1,\ln 2,0);t=0} = 4$

37. a) $\frac{\partial T}{\partial x} = 8x - 4y$, $\frac{\partial T}{\partial y} = 8y - 4x$. $\frac{dT}{dt} = \frac{\partial T}{\partial x}\frac{dx}{dt} + \frac{\partial T}{\partial y}\frac{dy}{dt} = (8x - 4y)(-\sin t) + (8y - 4x)(\cos t) =$
$(8\cos t - 4\sin t)(-\sin t) + (8\sin t - 4\cos t)(\cos t) = 4\sin^2 t - 4\cos^2 t \Rightarrow \frac{d^2T}{dt^2} = 16 \sin t \cos t$
$\frac{dT}{dt} = 0 \Rightarrow 4\sin^2 t - 4\cos^2 t = 0 \Rightarrow \sin^2 t = \cos^2 t \Rightarrow \sin t = \cos t$ or $\sin t = -\cos t \Rightarrow$
$t = \frac{\pi}{4}, \frac{5\pi}{4}$ or $\frac{3\pi}{4}, \frac{7\pi}{4}$ on the interval $0 \le t \le 2\pi$.
$\left.\frac{d^2T}{dt^2}\right|_{t=\pi/4} = 16 \sin\frac{\pi}{4}\cos\frac{\pi}{4} > 0 \Rightarrow$ T has a minimum at $t = \frac{\pi}{4}$
$\left.\frac{d^2T}{dt^2}\right|_{t=3\pi/4} = 16 \sin\frac{3\pi}{4}\cos\frac{3\pi}{4} < 0 \Rightarrow$ T has a maximum at $t = \frac{3\pi}{4}$
$\left.\frac{d^2T}{dt^2}\right|_{t=5\pi/4} = 16 \sin\frac{5\pi}{4}\cos\frac{5\pi}{4} > 0 \Rightarrow$ T has a minimum at $t = \frac{5\pi}{4}$
$\left.\frac{d^2T}{dt^2}\right|_{t=7\pi/4} = 16 \sin\frac{7\pi}{4}\cos\frac{7\pi}{4} < 0 \Rightarrow$ T has a maximum at $t = \frac{7\pi}{4}$

b) $T = 4x^2 - 4xy + 4y^2 \Rightarrow \frac{\partial T}{\partial x} = 8x - 4y$, $\frac{\partial T}{\partial y} = 8y - 4x$ (See part a above.)
$t = \frac{\pi}{4} \Rightarrow x = \cos\frac{\pi}{4} = \frac{\sqrt{2}}{2}$, $y = \sin\frac{\pi}{4} = \frac{\sqrt{2}}{2} \Rightarrow T\left(\frac{\pi}{4}\right) = 2$
$t = \frac{3\pi}{4} \Rightarrow x = \cos\frac{3\pi}{4} = -\frac{\sqrt{2}}{2}$, $y = \sin\frac{3\pi}{4} = \frac{\sqrt{2}}{2} \Rightarrow T\left(\frac{3\pi}{4}\right) = 6$
$t = \frac{5\pi}{4} \Rightarrow x = \cos\frac{5\pi}{4} = -\frac{\sqrt{2}}{2}$, $y = \sin\frac{5\pi}{4} = -\frac{\sqrt{2}}{2} \Rightarrow T\left(\frac{5\pi}{4}\right) = 2$
$t = \frac{7\pi}{4} \Rightarrow x = \cos\frac{7\pi}{4} = \frac{\sqrt{2}}{2}$, $y = \sin\frac{7\pi}{4} = -\frac{\sqrt{2}}{2} \Rightarrow T\left(\frac{7\pi}{4}\right) = 6$
$\therefore T_{max} = 6$ and $T_{min} = 2$.

SECTION 13.5 DIRECTIONAL DERIVATIVES AND GRADIENT VECTORS

1. $\frac{\partial f}{\partial x} = 2x \Rightarrow \frac{\partial f}{\partial x}(1,1,1) = 2$. $\frac{\partial f}{\partial y} = 2y \Rightarrow \frac{\partial f}{\partial y}(1,1,1) = 2$. $\frac{\partial f}{\partial z} = -4z \Rightarrow \frac{\partial f}{\partial z}(1,1,1) = -4 \Rightarrow \nabla f = 2\,\mathbf{i} + 2\,\mathbf{j} - 4\,\mathbf{k}$

2. $\frac{\partial f}{\partial x} = -6xz \Rightarrow \frac{\partial f}{\partial x}(1,1,1) = -6$. $\frac{\partial f}{\partial y} = -6yz \Rightarrow \frac{\partial f}{\partial y}(1,1,1) = -6$. $\frac{\partial f}{\partial z} = 6z^2 - 3(x^2 + y^2) \Rightarrow \frac{\partial f}{\partial z}(1,1,1) = 0 \Rightarrow$
$\nabla f = -6\,\mathbf{i} - 6\,\mathbf{j}$

3. $\frac{\partial f}{\partial x} = -x(x^2 + y^2 + z^2)^{-3/2} \Rightarrow \frac{\partial f}{\partial x}(1,2,-2) = -\frac{1}{27}$. $\frac{\partial f}{\partial y} = -y(x^2 + y^2 + z^2)^{-3/2} \Rightarrow \frac{\partial f}{\partial y}(1,2,-2) = -\frac{2}{27}$.
$\frac{\partial f}{\partial z} = -z(x^2 + y^2 + z^2)^{-3/2} \Rightarrow \frac{\partial f}{\partial z}(1,2,-2) = \frac{2}{27} \Rightarrow \nabla f = -\frac{1}{27}\mathbf{i} - \frac{2}{27}\mathbf{j} + \frac{2}{27}\mathbf{k}$

4. $\frac{\partial f}{\partial x} = e^{x+y}\cos z \Rightarrow \frac{\partial f}{\partial x}\left(0,0,\frac{\pi}{6}\right) = \frac{\sqrt{3}}{2}$. $\frac{\partial f}{\partial y} = e^{x+y}\cos z \Rightarrow \frac{\partial f}{\partial y}\left(0,0,\frac{\pi}{6}\right) = \frac{\sqrt{3}}{2}$. $\frac{\partial f}{\partial z} = -e^{x+y}\sin z \Rightarrow$ $\frac{\partial f}{\partial z}\left(0,0,\frac{\pi}{6}\right) = -\frac{1}{2} \Rightarrow \nabla f = \frac{\sqrt{3}}{2}\mathbf{i} + \frac{\sqrt{3}}{2}\mathbf{j} - \frac{1}{2}\mathbf{k}$

5. $\frac{\partial f}{\partial x} = -1$, $\frac{\partial f}{\partial y} = 1 \Rightarrow \nabla f = -\mathbf{i} + \mathbf{j}$

$-1 = y - x$ is the level curve.

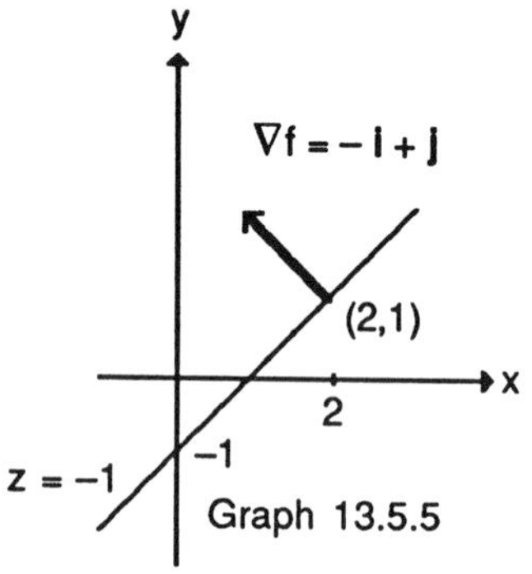

Graph 13.5.5

6. $\frac{\partial f}{\partial x} = \frac{2x}{x^2 + y^2} \Rightarrow \frac{\partial f}{\partial x}(1,1) = 1$. $\frac{\partial f}{\partial y} = \frac{2y}{x^2 + y^2} \Rightarrow$ $\frac{\partial f}{\partial y}(1,1) = 1 \Rightarrow \nabla f = \mathbf{i} + \mathbf{j}$. $f(1,1) = \ln 2 \Rightarrow \ln 2 =$ $\ln(x^2 + y^2) \Rightarrow 2 = x^2 + y^2$ is the level curve.

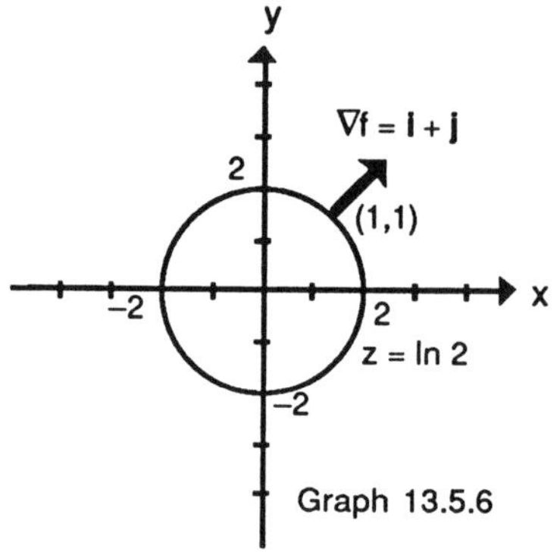

Graph 13.5.6

7. $\frac{\partial f}{\partial x} = -2x \Rightarrow \frac{\partial f}{\partial x}(-1,0) = 2$. $\frac{\partial f}{\partial y} = 1 \Rightarrow$ $\nabla f = 2\mathbf{i} + \mathbf{j}$. $-1 = y - x^2$ is the level curve.

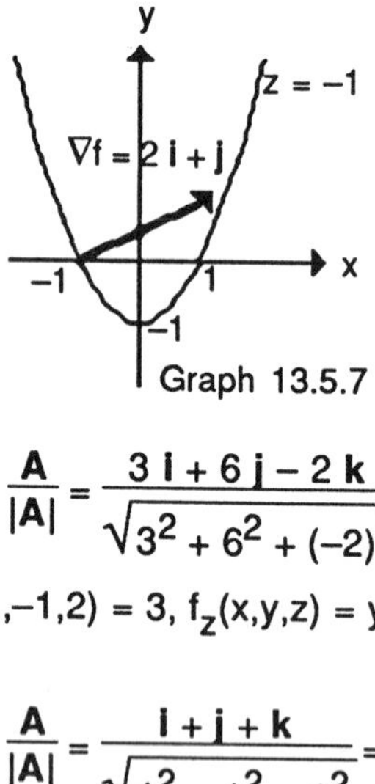

Graph 13.5.7

8. $\frac{\partial f}{\partial x} = 2x \Rightarrow \frac{\partial f}{\partial x}(2,\sqrt{3}) = 4$. $\frac{\partial f}{\partial y} = -2y \Rightarrow \frac{\partial f}{\partial y}(2,\sqrt{3}) =$ $-2\sqrt{3} \Rightarrow \nabla f = 4\mathbf{i} - 2\sqrt{3}\,\mathbf{j}$. $1 = x^2 - y^2$ is the level curve.

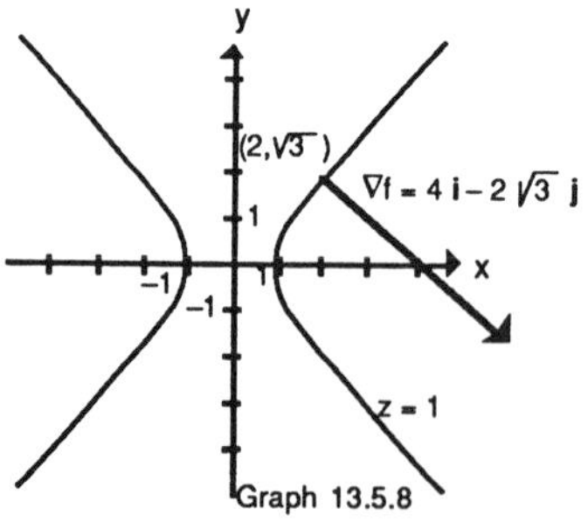

Graph 13.5.8

9. $\mathbf{u} = \frac{\mathbf{A}}{|\mathbf{A}|} = \frac{3\mathbf{i} + 6\mathbf{j} - 2\mathbf{k}}{\sqrt{3^2 + 6^2 + (-2)^2}} = \frac{3}{7}\mathbf{i} + \frac{6}{7}\mathbf{j} - \frac{2}{7}\mathbf{k}$. $f_x(x,y,z) = y + z \Rightarrow f_x(1,-1,2) = 1$, $f_y(x,y,z) = x + z \Rightarrow$ $f_y(1,-1,2) = 3$, $f_z(x,y,z) = y + x \Rightarrow f_z(1,-1,2) = 0 \Rightarrow \nabla f = \mathbf{i} + 3\mathbf{j} \Rightarrow \left(D_{\mathbf{u}}f\right)_{P_0} = \nabla f \cdot \mathbf{u} = \frac{3}{7} + \frac{18}{7} = 3$

10. $\mathbf{u} = \frac{\mathbf{A}}{|\mathbf{A}|} = \frac{\mathbf{i} + \mathbf{j} + \mathbf{k}}{\sqrt{1^2 + 1^2 + 1^2}} = \frac{1}{\sqrt{3}}\mathbf{i} + \frac{1}{\sqrt{3}}\mathbf{j} + \frac{1}{\sqrt{3}}\mathbf{k}$. $f_x(x,y,z) = 2x \Rightarrow f_x(1,1,1) = 2$, $f_y(x,y,z) = 4y \Rightarrow$ $f_y(1,1,1) = 4$, $f_z(x,y,z) = -6z \Rightarrow f_z(1,1,1) = -6 \Rightarrow \nabla f = 2\mathbf{i} + 4\mathbf{j} - 6\mathbf{k} \Rightarrow \left(D_{\mathbf{u}}f\right)_{P_0} = \nabla f \cdot \mathbf{u} = 2\left(\frac{1}{\sqrt{3}}\right) +$ $4\left(\frac{1}{\sqrt{3}}\right) - 6\left(\frac{1}{\sqrt{3}}\right) = 0$

11. $\mathbf{u} = \frac{\mathbf{A}}{|\mathbf{A}|} = \frac{2\mathbf{i} + \mathbf{j} - 2\mathbf{k}}{\sqrt{2^2 + 1^2 + (-2)^2}} = \frac{2}{3}\mathbf{i} + \frac{1}{3}\mathbf{j} - \frac{2}{3}\mathbf{k}$. $f_x(x,y,z) = 3e^x \cos yz \Rightarrow f_x(0,0,0) = 3$, $f_y(x,y,z) =$ $-3ze^x \sin yz \Rightarrow f_y(0,0,0) = 0$, $f_z(x,y,z) = -3ye^x \sin yz \Rightarrow f_z(0,0,0) = 0 \Rightarrow \nabla f = 3\mathbf{i} \Rightarrow \left(D_{\mathbf{u}}f\right)_{P_0} = \nabla f \cdot \mathbf{u} = 2$

12. $\mathbf{u} = \frac{\mathbf{A}}{|\mathbf{A}|} = \frac{\mathbf{i} + 2\mathbf{j} + 2\mathbf{k}}{\sqrt{1^2 + 2^2 + 2^2}} = \frac{1}{3}\mathbf{i} + \frac{2}{3}\mathbf{j} + \frac{2}{3}\mathbf{k}$. $f_x(x,y,z) = y \cos xy + \frac{1}{x} \Rightarrow f_x\left(1,0,\frac{1}{2}\right) = 1$, $f_y(x,y,z) = x \cos xy + ze^{yz} \Rightarrow f_y\left(1,0,\frac{1}{2}\right) = \frac{3}{2}$, $f_z(x,y,z) = y\, e^{yz} + \frac{1}{z} \Rightarrow f_z\left(1,0,\frac{1}{2}\right) = 2 \Rightarrow \nabla f = \mathbf{i} + \frac{3}{2}\mathbf{j} + 2\mathbf{k} \Rightarrow \left(D_{\mathbf{u}}f\right)_{P_0} = \nabla f \cdot \mathbf{u} = \frac{1}{3} + 1 + \frac{4}{3} = \frac{8}{3}$

13. $\mathbf{u} = \frac{\mathbf{A}}{|\mathbf{A}|} = \frac{12\mathbf{i} + 5\mathbf{j}}{\sqrt{12^2 + 5^2}} = \frac{12}{13}\mathbf{i} + \frac{5}{13}\mathbf{j}$. $f_x(x,y,z) = 1 + \frac{y^2}{x^2} \Rightarrow f_x(1,1) = 2$, $f_y(x,y,z) = -\frac{2y}{x} \Rightarrow f_y(1,1) = -2 \Rightarrow \nabla f = 2\mathbf{i} - 2\mathbf{j} \Rightarrow \left(D_{\mathbf{u}}f\right)_{P_0} = \nabla f \cdot \mathbf{u} = \frac{24}{13} - \frac{10}{13} = \frac{14}{13}$

14. $\mathbf{u} = \frac{\mathbf{A}}{|\mathbf{A}|} = \frac{3\mathbf{i} - 4\mathbf{j}}{\sqrt{3^2 + (-4)^2}} = \frac{3}{5}\mathbf{i} - \frac{4}{5}\mathbf{j}$. $f_x(x,y,z) = 4x \Rightarrow f_x(-1,1) = -4$, $f_y(x,y,z) = 2y \Rightarrow f_y(-1,1) = 2 \Rightarrow \nabla f = -4\mathbf{i} + 2\mathbf{j} \Rightarrow \left(D_{\mathbf{u}}f\right)_{P_0} = \nabla f \cdot \mathbf{u} = -\frac{12}{5} - \frac{8}{5} = -4$

15. $\mathbf{u} = \frac{\mathbf{A}}{|\mathbf{A}|} = \frac{3\mathbf{i} + 4\mathbf{j}}{\sqrt{3^2 + 4^2}} = \frac{3}{5}\mathbf{i} + \frac{4}{5}\mathbf{j}$. $f_x(x,y,z) = 2x + 2y \Rightarrow f_x(1,1) = 4$, $f_y(x,y,z) = 2x = 6y \Rightarrow f_y(1,1) = -4 \Rightarrow \nabla f = 4\mathbf{i} - 4\mathbf{j} \Rightarrow \left(D_{\mathbf{u}}f\right)_{P_0} = \nabla f \cdot \mathbf{u} = \frac{12}{5} - \frac{16}{5} = -\frac{4}{5}$

16. $\mathbf{u} = \frac{\mathbf{A}}{|\mathbf{A}|} = \frac{2\mathbf{i} - \mathbf{j}}{\sqrt{2^2 + (-1)^2}} = \frac{2}{\sqrt{5}}\mathbf{i} - \frac{1}{\sqrt{5}}\mathbf{j}$. $f_x(x,y,z) = \tan^{-1}\left(\frac{y}{x}\right) - \frac{xy}{x^2 + y^2} \Rightarrow f_x(1,1) = \frac{\pi}{4} - \frac{1}{2}$, $f_y(x,y,z) = \frac{1}{\left(\frac{y}{x}\right)^2 + 1} \Rightarrow f_y(1,1) = \frac{1}{2} \Rightarrow \nabla f = \left(\frac{\pi}{4} - \frac{1}{2}\right)\mathbf{i} + \frac{1}{2}\mathbf{j} \Rightarrow \left(D_{\mathbf{u}}f\right)_{P_0} = \nabla f \cdot \mathbf{u} = \left(\frac{\pi}{4} - \frac{1}{2}\right)\frac{2}{\sqrt{5}} + \frac{1}{2}\left(-\frac{1}{\sqrt{5}}\right) = \frac{\pi - 3}{2\sqrt{5}}$

17. $\nabla f = 2x\,\mathbf{i} + \cos y\,\mathbf{j} \Rightarrow \nabla f(1,0) = 2\mathbf{i} + \mathbf{j} \Rightarrow \mathbf{u} = \frac{\nabla f}{|\nabla f|} = \frac{2\mathbf{i} + \mathbf{j}}{\sqrt{2^2 + 1^2}} = \frac{2}{\sqrt{5}}\mathbf{i} + \frac{1}{\sqrt{5}}\mathbf{j}$. f increases most rapidly in the direction of $\mathbf{u} = \frac{2}{\sqrt{5}}\mathbf{i} + \frac{1}{\sqrt{5}}\mathbf{j}$; decreases most rapidly in the direction $-\mathbf{u} = -\frac{2}{\sqrt{5}}\mathbf{i} - \frac{1}{\sqrt{5}}\mathbf{j}$. $\left(D_{\mathbf{u}}f\right)_{P_0} = \nabla f \cdot \mathbf{u} = \sqrt{5}$, $\left(D_{-\mathbf{u}}f\right)_{P_0} = -\sqrt{5}$.

18. $\nabla f = (2x + y)\,\mathbf{i} + (x + 2y)\,\mathbf{j} \Rightarrow \nabla f(-1,1) = -\mathbf{i} + \mathbf{j} \Rightarrow \mathbf{u} = \frac{\nabla f}{|\nabla f|} = \frac{-\mathbf{i} + \mathbf{j}}{\sqrt{(-1)^2 + 1^2}} = -\frac{1}{\sqrt{2}}\mathbf{i} + \frac{1}{\sqrt{2}}\mathbf{j}$. f increases most rapidly in the direction $\mathbf{u} = -\frac{1}{\sqrt{2}}\mathbf{i} + \frac{1}{\sqrt{2}}\mathbf{j}$ and decreases most rapidly in the direction $-\mathbf{u} = \frac{1}{\sqrt{2}}\mathbf{i} - \frac{1}{\sqrt{2}}\mathbf{j}$. $\left(D_{\mathbf{u}}f\right)_{P_0} = \nabla f \cdot \mathbf{u} = \sqrt{2}$, $\left(D_{-\mathbf{u}}f\right)_{P_0} = -\sqrt{2}$

19. $\nabla f = e^y\,\mathbf{i} + xe^y\,\mathbf{j} + 2z\,\mathbf{k} \Rightarrow \nabla f\left(1, \ln 2, \frac{1}{2}\right) = 2\mathbf{i} + 2\mathbf{j} + \mathbf{k} \Rightarrow \mathbf{u} = \frac{\nabla f}{|\nabla f|} = \frac{2\mathbf{i} + 2\mathbf{j} + \mathbf{k}}{\sqrt{2^2 + 2^2 + 1^2}} = \frac{2}{3}\mathbf{i} + \frac{2}{3}\mathbf{j} + \frac{1}{3}\mathbf{k}$. f increases most rapidly in the direction $\mathbf{u} = \frac{2}{3}\mathbf{i} + \frac{2}{3}\mathbf{j} + \frac{1}{3}\mathbf{k}$; decreases most rapidly in the direction $-\mathbf{u} = -\frac{2}{3}\mathbf{i} - \frac{2}{3}\mathbf{j} - \frac{1}{3}\mathbf{k}$. $\left(D_{\mathbf{u}}f\right)_{P_0} = \nabla f \cdot \mathbf{u} = 3$; $\left(D_{-\mathbf{u}}f\right)_{P_0} = -3$

20. $\nabla f = \frac{1}{y}\mathbf{i} - \left(\frac{x}{y^2} + z\right)\mathbf{j} - y\,\mathbf{k} \Rightarrow \nabla f(4,1,1) = \mathbf{i} - 5\,\mathbf{j} - \mathbf{k}$. $\mathbf{u} = \frac{\nabla f}{|\nabla f|} = \frac{\mathbf{i} - 5\,\mathbf{j} - \mathbf{k}}{\sqrt{1^2 + (-5)^2 + (-1)^2}} = \frac{1}{3\sqrt{3}}\mathbf{i} - \frac{5}{3\sqrt{3}}\mathbf{j} - \frac{1}{3\sqrt{3}}\mathbf{k}$. f increases most rapidly in the direction of $\mathbf{u} = \frac{1}{3\sqrt{3}}\mathbf{i} - \frac{5}{3\sqrt{3}}\mathbf{j} - \frac{1}{3\sqrt{3}}\mathbf{k}$; decreases most rapidly in the direction $-\mathbf{u} = -\frac{1}{3\sqrt{3}}\mathbf{i} + \frac{5}{3\sqrt{3}}\mathbf{j} + \frac{1}{3\sqrt{3}}\mathbf{k}$. $\left(D_{\mathbf{u}}f\right)_{P_0} = \nabla f \cdot \mathbf{u} = 3\sqrt{3}$; $\left(D_{-\mathbf{u}}f\right)_{P_0} = -3\sqrt{3}$

21. $\nabla f = (\cos x)\,\mathbf{i} + (\cos y)\,\mathbf{j} + (\cos z)\,\mathbf{k} \Rightarrow \nabla f(0,0,0) = \mathbf{i} + \mathbf{j} + \mathbf{k} \Rightarrow \mathbf{u} = \frac{\nabla f}{|\nabla f|} = \frac{\mathbf{i} + \mathbf{j} + \mathbf{k}}{\sqrt{1^2 + 1^2 + 1^2}} = \frac{1}{\sqrt{3}}\mathbf{i} + \frac{1}{\sqrt{3}}\mathbf{j} + \frac{1}{\sqrt{3}}\mathbf{k}$. f increases most rapidly in the direction $\mathbf{u} = \frac{1}{\sqrt{3}}\mathbf{i} + \frac{1}{\sqrt{3}}\mathbf{j} + \frac{1}{\sqrt{3}}\mathbf{k}$; decreases most rapidly in the direction $-\mathbf{u} = -\frac{1}{\sqrt{3}}\mathbf{i} - \frac{1}{\sqrt{3}}\mathbf{j} - \frac{1}{\sqrt{3}}\mathbf{k}$. $\left(D_{\mathbf{u}}f\right)_{P_0} = \nabla f \cdot \mathbf{u} = \sqrt{3}$, $\left(D_{-\mathbf{u}}f\right)_{P_0} = -\sqrt{3}$

22. $\nabla f = \frac{2x}{x^2 + y^2 - 1}\mathbf{i} + \left(\frac{2y}{x^2 + y^2 - 1} + 1\right)\mathbf{j} + 6\,\mathbf{k} \Rightarrow \nabla f(1,1,0) = 2\,\mathbf{i} + 3\,\mathbf{j} + 6\,\mathbf{k}. \Rightarrow \mathbf{u} = \frac{\nabla f}{|\nabla f|} = \frac{2\,\mathbf{i} + 3\,\mathbf{j} + 6\,\mathbf{k}}{\sqrt{2^2 + 3^2 + 6^2}} = \frac{2}{7}\mathbf{i} + \frac{3}{7}\mathbf{j} + \frac{6}{7}\mathbf{k}$. f increases most rapidly in the direction $\mathbf{u} = \frac{2}{7}\mathbf{i} + \frac{3}{7}\mathbf{j} + \frac{6}{7}\mathbf{k}$; decreases most rapidly in the direction $-\mathbf{u} = -\frac{2}{7}\mathbf{i} - \frac{3}{7}\mathbf{j} - \frac{6}{7}\mathbf{k}$. $\left(D_{\mathbf{u}}f\right)_{P_0} = \nabla f \cdot \mathbf{u} = 7$; $\left(D_{-\mathbf{u}}f\right)_{P_0} = -7$

23. $\nabla f = (-\pi y \sin(\pi xy) + y^2)\,\mathbf{i} + (-\pi x \sin(\pi xy) + 2xy)\,\mathbf{j} \Rightarrow \nabla f(-1,-1) = \mathbf{i} + 2\,\mathbf{j}$. $\mathbf{u} = \frac{\mathbf{A}}{|\mathbf{A}|} = \frac{\mathbf{i} + \mathbf{j}}{\sqrt{1^2 + 1^2}} = \frac{1}{\sqrt{2}}\mathbf{i} + \frac{1}{\sqrt{2}}\mathbf{j}$

$\nabla f \cdot \mathbf{u} = \frac{1}{\sqrt{2}} + \frac{2}{\sqrt{2}} = \frac{3}{\sqrt{2}}$. $\therefore$ $df = (\nabla f \cdot \mathbf{u})ds = \frac{3}{\sqrt{2}}(0.1) = 0{,}15\sqrt{2} \approx 0.212132$

24. $\nabla f = \frac{x}{x^2 + y^2 + z^2}\mathbf{i} + \frac{y}{x^2 + y^2 + z^2}\mathbf{j} + \frac{z}{x^2 + y^2 + z^2}\mathbf{k} \Rightarrow \nabla f(3,4,12) = \frac{3}{169}\mathbf{i} + \frac{4}{169}\mathbf{j} + \frac{12}{169}\mathbf{k}$. $\mathbf{u} = \frac{\mathbf{A}}{|\mathbf{A}|} = \frac{3\,\mathbf{i} + 6\,\mathbf{j} - 2\,\mathbf{k}}{\sqrt{3^2 + 6^2 + (-2)^2}} = \frac{3}{7}\mathbf{i} + \frac{6}{7}\mathbf{j} - \frac{2}{7}\mathbf{k} \Rightarrow \nabla f \cdot \mathbf{u} = \frac{9}{1183}$. $\therefore$ $df = (\nabla f \cdot \mathbf{u})ds = \frac{9}{1183}(0.1) \approx 0.000760$

25. $\nabla f = \left(e^x \cos yz\right)\mathbf{i} - \left(ze^x \sin yz\right)\mathbf{j} - \left(ye^x \sin yz\right)\mathbf{k} \Rightarrow \nabla f(0,0,0) = \mathbf{i}$. $\mathbf{u} = \frac{\mathbf{A}}{|\mathbf{A}|} = \frac{2\,\mathbf{i} + \mathbf{j} - 2\,\mathbf{k}}{\sqrt{2^2 + 1^2 + (-2)^2}} = \frac{2}{3}\mathbf{i} + \frac{1}{3}\mathbf{j} - \frac{2}{3}\mathbf{k} \Rightarrow \nabla f \cdot \mathbf{u} = \frac{2}{3}$. $\therefore$ $df = (\nabla f \cdot \mathbf{u})ds = \frac{2}{3}(0.1) = \frac{1}{15}$ or 0.067

26. $\mathbf{A} = \overrightarrow{P_0P_1} = -2\,\mathbf{i} + 2\,\mathbf{j} + 2\,\mathbf{k}$. $\nabla f = (1 + \cos z)\,\mathbf{i} + (1 - \sin z)\,\mathbf{j} + (-x \sin z - y \cos z)\,\mathbf{k} \Rightarrow \nabla f(2,-1,0) = 2\,\mathbf{i} + \mathbf{j} + \mathbf{k}$. $\mathbf{u} = \frac{\mathbf{A}}{|\mathbf{A}|} = \frac{-2\,\mathbf{i} + 2\,\mathbf{j} + 2\,\mathbf{k}}{\sqrt{(-2)^2 + 2^2 + 2^2}} = -\frac{1}{\sqrt{3}}\mathbf{i} + \frac{1}{\sqrt{3}}\mathbf{j} + \frac{1}{\sqrt{3}}\mathbf{k} \Rightarrow \nabla f \cdot \mathbf{u} = 0$. $\therefore$ $df = (\nabla f \cdot \mathbf{u})ds = 0(0.2) = 0$

27. $\nabla f = y\,\mathbf{i} + (x + 2y)\,\mathbf{j} \Rightarrow \nabla f(3,2) = 2\,\mathbf{i} + 7\,\mathbf{j}$. **A**, orthogonal to ∇f, is $\mathbf{A} = 7\,\mathbf{i} - 2\,\mathbf{j} \Rightarrow \mathbf{u} = \frac{\mathbf{A}}{|\mathbf{A}|} = \frac{7\,\mathbf{i} - 2\,\mathbf{j}}{\sqrt{7^2 + (-2)^2}} = \frac{7}{\sqrt{53}}\mathbf{i} - \frac{2}{\sqrt{53}}\mathbf{j} \Rightarrow -\mathbf{u} = -\frac{7}{\sqrt{53}}\mathbf{i} + \frac{2}{\sqrt{53}}\mathbf{j}$

28. $\nabla f = \frac{4xy^2}{(x^2 + y^2)^2}\mathbf{i} - \frac{4x^2y}{(x^2 + y^2)^2}\mathbf{j} \Rightarrow \nabla f(1,1) = \mathbf{i} - \mathbf{j}$. **A**, orthogonal to ∇f, is $\mathbf{A} = \mathbf{i} + \mathbf{j} \Rightarrow$
$\mathbf{u} = \frac{\mathbf{A}}{|\mathbf{A}|} = \frac{\mathbf{i} + \mathbf{j}}{\sqrt{1^2 + 1^2}} = \frac{1}{\sqrt{2}}\mathbf{i} + \frac{1}{\sqrt{2}}\mathbf{j} \Rightarrow -\mathbf{u} = -\frac{1}{\sqrt{2}}\mathbf{i} - \frac{1}{\sqrt{2}}\mathbf{j}$

29. $\nabla f = 2x\,\mathbf{i} + 2y\,\mathbf{j} + 2z\,\mathbf{k} = (2\cos t)\,\mathbf{i} + (2\sin t)\,\mathbf{j} + 2t\,\mathbf{k}$. $\mathbf{v} = (-\sin t)\,\mathbf{i} + (\cos t)\,\mathbf{j} + \mathbf{k} \Rightarrow \mathbf{T} = \frac{\mathbf{v}}{|\mathbf{v}|} =$

$$\frac{(-\sin t)\,\mathbf{i} + (\cos t)\,\mathbf{j} + \mathbf{k}}{\sqrt{(\sin t)^2 + (\cos t)^2 + 1^2}} = \left(\frac{-\sin t}{\sqrt{2}}\right)\mathbf{i} + \left(\frac{\cos t}{\sqrt{2}}\right)\mathbf{j} + \frac{1}{\sqrt{2}}\mathbf{k}.$$

$$\left(D_{\mathbf{T}}f\right)_{P_0} = \nabla f \cdot \mathbf{T} = (2\cos t)\left(\frac{-\sin t}{\sqrt{2}}\right) + (2\sin t)\left(\frac{\cos t}{\sqrt{2}}\right) + 2t\left(\frac{1}{\sqrt{2}}\right) = \frac{2t}{\sqrt{2}} \Rightarrow \left(D_{\mathbf{T}}f\right)\left(\frac{-\pi}{4}\right) = \frac{-\pi}{2\sqrt{2}},$$

$$\left(D_{\mathbf{T}}f\right)(0) = 0,\ \left(D_{\mathbf{T}}f\right)\left(\frac{\pi}{4}\right) = \frac{\pi}{2\sqrt{2}}$$

30. $\nabla f = 2x\,\mathbf{i} + 2y\,\mathbf{j} = 2(\cos t + t\sin t)\,\mathbf{i} + 2(\sin t - t\cos t)\,\mathbf{j}$. $\mathbf{v} = (t\cos t)\,\mathbf{i} + (t\sin t)\,\mathbf{j} \Rightarrow$

$$\mathbf{T} = \frac{\mathbf{v}}{|\mathbf{v}|} = \frac{(t\cos t)\,\mathbf{i} + (t\sin t)\,\mathbf{j}}{\sqrt{(t\cos t)^2 + (t\sin t)^2}} = (\cos t)\,\mathbf{i} + (\sin t)\,\mathbf{j} \text{ since } t > 0.\ \left(D_{\mathbf{T}}f\right)_{P_0} = \nabla f \cdot \mathbf{T} = 2(\cos t + t\sin t)\cos t +$$
$$2(\sin t - t\cos t)\sin t = 2$$

31. $\nabla f = f_x(1,2)\,\mathbf{i} + f_y(1,2)\,\mathbf{j}$. $\mathbf{u_1} = \frac{\mathbf{i}+\mathbf{j}}{\sqrt{1^2+1^2}} = \frac{1}{\sqrt{2}}\mathbf{i} + \frac{1}{\sqrt{2}}\mathbf{j}$. $\left(D_{\mathbf{u_1}}f\right)(1,2) =$

$f_x(1,2)\left(\frac{1}{\sqrt{2}}\right) + f_y(1,2)\left(\frac{1}{\sqrt{2}}\right) = 2\sqrt{2} \Rightarrow f_x(1,2) + f_y(1,2) = 4$. $\mathbf{u_2} = -\mathbf{j}$. $\left(D_{\mathbf{u_2}}f\right)(1,2) = f_x(1,2)(0) + f_y(1,2)(-1)$
$= -3 \Rightarrow -f_y(1,2) = -3 \Rightarrow f_y(1,2) = 3$. $\therefore$ $f_x(1,2) + 3 = 4 \Rightarrow f_x(1,2) = 1$. Then $\nabla f(1,2) = \mathbf{i} + 3\,\mathbf{j}$.

$$\mathbf{u} = \frac{\mathbf{A}}{|\mathbf{A}|} = \frac{-\mathbf{i} - 2\,\mathbf{j}}{\sqrt{(-1)^2 + (-2)^2}} = -\frac{1}{\sqrt{5}}\mathbf{i} - \frac{2}{\sqrt{5}}\mathbf{j} \Rightarrow \left(D_{\mathbf{u}}f\right)_{P_0} = \nabla f \cdot \mathbf{u} = -\frac{1}{\sqrt{5}} - \frac{6}{\sqrt{5}} = -\frac{7}{\sqrt{5}}$$

32. a) $\left(D_{\mathbf{u}}f\right)_P = 2\sqrt{3} \Rightarrow |\nabla f| = 2\sqrt{3}$. $\mathbf{u} = \frac{\mathbf{A}}{|\mathbf{A}|} = \frac{\mathbf{i}+\mathbf{j}-\mathbf{k}}{\sqrt{1^2+1^2+(-1)^2}} = \frac{1}{\sqrt{3}}\mathbf{i} + \frac{1}{\sqrt{3}}\mathbf{j} - \frac{1}{\sqrt{3}}\mathbf{k}$. $\mathbf{u} = \frac{\nabla f}{|\nabla f|} \Rightarrow$

$$\nabla f = |\nabla f|\mathbf{u} \Rightarrow \nabla f = 2\sqrt{3}\left(\frac{1}{\sqrt{3}}\mathbf{i} + \frac{1}{\sqrt{3}}\mathbf{j} - \frac{1}{\sqrt{3}}\mathbf{k}\right) = 2\,\mathbf{i} + 2\,\mathbf{j} - 2\,\mathbf{k}$$

b) $\mathbf{A} = \mathbf{i} + \mathbf{j} \Rightarrow \mathbf{u} = \frac{\mathbf{A}}{|\mathbf{A}|} = \frac{\mathbf{i}+\mathbf{j}}{\sqrt{1^2+1^2}} = \frac{1}{\sqrt{2}}\,\mathbf{i} + \frac{1}{\sqrt{2}}\mathbf{j}$. $\left(D_{\mathbf{u}}f\right)_P = \nabla f \cdot \mathbf{u} = 2\left(\frac{1}{\sqrt{2}}\right)\mathbf{i} + 2\left(\frac{1}{\sqrt{2}}\right)\mathbf{j} = \sqrt{2}\,\mathbf{i} + \sqrt{2}\,\mathbf{j}$

33. $x = g(t),\ y = h(t) \Rightarrow \mathbf{r} = g(t)\,\mathbf{i} + h(t)\,\mathbf{j} \Rightarrow \mathbf{v} = g'(t)\,\mathbf{i} + h'(t)\,\mathbf{j} \Rightarrow \mathbf{T} = \frac{\mathbf{v}}{|\mathbf{v}|} = \frac{g'(t)\,\mathbf{i} + h'(t)\,\mathbf{j}}{\sqrt{(g'(t))^2 + (h'(t))^2}}$

$z = f(x,y) \Rightarrow \frac{df}{dt} = \frac{\partial f}{\partial x}\frac{dx}{dt} + \frac{\partial f}{\partial y}\frac{dy}{dt} = \frac{\partial f}{\partial x}g'(t) + \frac{\partial f}{\partial y}h'(t)$. If $f(g(t),h(t)) = c$, then $\frac{df}{dt} = 0 \Rightarrow \frac{\partial f}{\partial x}g'(t) + \frac{\partial f}{\partial y}h'(t) = 0$.

Now $\nabla f = \frac{\partial f}{\partial x}\mathbf{i} + \frac{\partial f}{\partial y}\mathbf{j}$. $\therefore$ $\left(D_{\mathbf{T}}f\right) = \nabla f \cdot \mathbf{T} = \dfrac{\frac{\partial f}{\partial x}g'(t) + \frac{\partial f}{\partial y}h'(t)}{\sqrt{(g'(t))^2 + (h'(t))^2}} = 0 \Rightarrow \nabla f$ is normal to $\mathbf{T}$

34. a) $$\nabla(kf) = \frac{\partial(kf)}{\partial x}\mathbf{i} + \frac{\partial(kf)}{\partial y}\mathbf{j} + \frac{\partial(kf)}{\partial z}\mathbf{k} = k\left(\frac{\partial f}{\partial x}\right)\mathbf{i} + k\left(\frac{\partial f}{\partial y}\right)\mathbf{j} + k\left(\frac{\partial f}{\partial z}\right)\mathbf{k} = k\left(\frac{\partial f}{\partial x}\mathbf{i} + \frac{\partial f}{\partial y}\mathbf{j} + \frac{\partial f}{\partial z}\mathbf{k}\right) = k\nabla f$$

b) $$\nabla(f+g) = \frac{\partial(f+g)}{\partial x}\mathbf{i} + \frac{\partial(f+g)}{\partial y}\mathbf{j} + \frac{\partial(f+g)}{\partial z}\mathbf{k} = \left(\frac{\partial f}{\partial x} + \frac{\partial g}{\partial x}\right)\mathbf{i} + \left(\frac{\partial f}{\partial y} + \frac{\partial g}{\partial y}\right)\mathbf{j} + \left(\frac{\partial f}{\partial z} + \frac{\partial g}{\partial z}\right)\mathbf{k} =$$
$$\frac{\partial f}{\partial x}\mathbf{i} + \frac{\partial g}{\partial x}\mathbf{i} + \frac{\partial f}{\partial y}\mathbf{j} + \frac{\partial g}{\partial y}\mathbf{j} + \frac{\partial f}{\partial z}\mathbf{k} + \frac{\partial g}{\partial z}\mathbf{k} = \left(\frac{\partial f}{\partial x}\mathbf{i} + \frac{\partial f}{\partial y}\mathbf{j} + \frac{\partial f}{\partial z}\mathbf{k}\right) + \left(\frac{\partial g}{\partial x}\mathbf{i} + \frac{\partial g}{\partial y}\mathbf{j} + \frac{\partial g}{\partial z}\mathbf{k}\right) = \nabla f + \nabla g$$

c) $\nabla(f - g) = \nabla f - \nabla g$ (Substitute $-g$ in for g in part b above)

d) $$\nabla(fg) = \frac{\partial(fg)}{\partial x}\mathbf{i} + \frac{\partial(fg)}{\partial y}\mathbf{j} + \frac{\partial(fg)}{\partial z}\mathbf{k} = \left(\frac{\partial f}{\partial x}g + \frac{\partial g}{\partial x}f\right)\mathbf{i} + \left(\frac{\partial f}{\partial y}g + \frac{\partial g}{\partial y}f\right)\mathbf{j} + \left(\frac{\partial f}{\partial z}g + \frac{\partial g}{\partial z}f\right)\mathbf{k} =$$
$$\frac{\partial f}{\partial x}g\,\mathbf{i} + \frac{\partial g}{\partial x}f\,\mathbf{i} + \frac{\partial f}{\partial y}g\,\mathbf{j} + \frac{\partial g}{\partial y}f\,\mathbf{j} + \frac{\partial f}{\partial z}g\,\mathbf{k} + \frac{\partial g}{\partial z}f\,\mathbf{k} = f\left(\frac{\partial g}{\partial x}\mathbf{i} + \frac{\partial g}{\partial y}\mathbf{j} + \frac{\partial g}{\partial z}\mathbf{k}\right) + g\left(\frac{\partial f}{\partial x}\mathbf{i} + \frac{\partial f}{\partial y}\mathbf{j} + \frac{\partial f}{\partial z}\mathbf{k}\right) =$$
$$f\,\nabla g + g\,\nabla f$$

SECTION 13.6 TANGENT PLANES AND NORMAL LINES

1. $\nabla f = 2x\,\mathbf{i} + 2y\,\mathbf{j} + 2z\,\mathbf{k} \Rightarrow \nabla f(1,1,1) = 2\,\mathbf{i} + 2\,\mathbf{j} + 2\,\mathbf{k} \Rightarrow$ Tangent plane: $2(x-1) + 2(y-1) + 2(z-1) = 0$ $\Rightarrow x + y + z = 3$; Normal line: $x = 1 + 2t,\ y = 1 + 2t,\ z = 1 + 2t$

2. $\nabla f = 2x\,\mathbf{i} + 2y\,\mathbf{j} - 2z\,\mathbf{k} \Rightarrow \nabla f(3,4,-5) = 6\,\mathbf{i} + 8\,\mathbf{j} + 10\,\mathbf{k} \Rightarrow$ Tangent plane: $6(x-3) + 8(y-4) + 10(z+5) = 0 \Rightarrow 6x + 8y + 10z = 0$; Normal line: $x = 3 + 6t,\ y = 4 + 8t,\ z = -5 + 10t$

3. $\nabla f = -2x\,\mathbf{i} - 2y\,\mathbf{j} + 2z\,\mathbf{k} \Rightarrow \nabla f(3,4,-5) = -6\,\mathbf{i} - 8\,\mathbf{j} - 10\,\mathbf{k} \Rightarrow$ Tangent plane: $-6(x-3) - 8(y-4) - 10(z+5) = 0 \Rightarrow 3x + 4y + 5z = 0$; Normal line: $x = 3 - 6t,\ y = 4 - 8t,\ z = -5 - 10t$

4. $\nabla f = -2x\,\mathbf{i} + 2\,\mathbf{k} \Rightarrow \nabla f(2,0,2) = -4\,\mathbf{i} + 2\,\mathbf{k} \Rightarrow$ Tangent plane: $-4(x-2) + 2(z-2) = 0 \Rightarrow -4x + 2z + 4 = 0$; Normal line: $x = 2 - 4t,\ y = 0,\ z = 2 + 2t$

5. $\nabla f = \frac{-2x}{x^2+y^2}\,\mathbf{i} - \frac{2y}{x^2+y^2}\,\mathbf{j} \Rightarrow \nabla f(1,0,0) = -2\,\mathbf{i} + \mathbf{k} \Rightarrow$ Tangent plane: $-2(x-1) + z = 0 \Rightarrow -2x + z = -2$; Normal line: $x = 1 - 2t,\ y = 0,\ z = t$

6. $\nabla f = 2(x+y)\,\mathbf{i} + 2(x+y)\,\mathbf{j} + 2z\,\mathbf{k} \Rightarrow \nabla f(1,2,4) = 6\,\mathbf{i} + 6\,\mathbf{j} + 8\,\mathbf{k} \Rightarrow$ Tangent plane: $6(x-1) + 6(y-2) + 8(z-4) = 0 \Rightarrow 6x + 6y + 8z = 50$; Normal line: $x = 1 + 6t,\ y = 2 + 6t,\ z = 4 + 8t$

7. $\nabla f = (2x+2y)\,\mathbf{i} + (2x-2y)\,\mathbf{j} + 2z\,\mathbf{k} \Rightarrow \nabla f(1,-1,3) = 4\,\mathbf{j} + 6\,\mathbf{k} \Rightarrow$ Tangent plane: $4(y+1) + 6(z-3) = 0$ $\Rightarrow 2y + 3z = 7$; Normal line: $x = 1,\ y = -1 + 4t,\ z = 3 + 6t$

8. $\nabla f = \left(-\pi \sin \pi x + 2xy + ze^{xz}\right)\mathbf{i} + \left(-x^2 + z\right)\mathbf{j} + \left(xe^{xz} + y\right)\mathbf{k} \Rightarrow \nabla f(0,1,2) = 2\,\mathbf{i} + 2\,\mathbf{j} + \mathbf{k} \Rightarrow$ Tangent plane: $2(x-0) + 2(y-1) + 1(z-2) = 0 \Rightarrow 2x + 2y + z - 4 = 0$; Normal line: $x = 2t,\ y = 1 + 2t,\ z = 2 + t$

9. $\nabla f = (2x-y)\,\mathbf{i} - (x+2y)\,\mathbf{j} - \mathbf{k} \Rightarrow \nabla f(1,1,-1) = \mathbf{i} - 3\,\mathbf{j} - \mathbf{k} \Rightarrow$ Tangent plane: $1(x-1) - 3(y-1) - (z+1) = 0 \Rightarrow x - 3y - z = -1$; Normal line: $x = 1 + t,\ y = 1 - 3t,\ z = -1 - t$

10. $\nabla f = \mathbf{i} + \mathbf{j} + \mathbf{k}$ for all points $\Rightarrow$ Tangent plane: $1(x-0) + 1(y-1) + 1(z-0) = 0 \Rightarrow x + y + z - 1 = 0$; Normal line: $x = t,\ y = 1 + t,\ z = t$

11. $\nabla f = (2x - 2y - 1)\,\mathbf{i} + (2y - 2x + 3)\,\mathbf{j} - \mathbf{k} \Rightarrow \nabla f(2,-3,18) = 9\,\mathbf{i} - 7\,\mathbf{j} - \mathbf{k} \Rightarrow$ Tangent plane: $9(x-2) - 7(y+3) - (z-18) = 0 \Rightarrow 9x - 7y - z = 21$; Normal line: $x = 2 + 9t,\ y = -3 - 7t,\ z = 18 - t$

12. $\nabla f = \mathbf{i} + \left(z - 3y^2\right)\mathbf{j} + (y - 4z)\,\mathbf{k} \Rightarrow \nabla f(-4,-2,1) = \mathbf{i} - 11\,\mathbf{j} - 6\,\mathbf{k} \Rightarrow$ Tangent plane: $1(x+4) - 11(y+2) - 6(z-1) = 0 \Rightarrow x - 11y - 6z - 12 = 0$; Normal line: $x = -4 + t,\ y = -2 - 11t,\ z = 1 - 6t$

13. $\nabla f = 2x\,\mathbf{i} + 2y\,\mathbf{j} \Rightarrow \nabla f(1,\sqrt{3},1) = 2\,\mathbf{i} + 2\sqrt{3}\,\mathbf{j}$

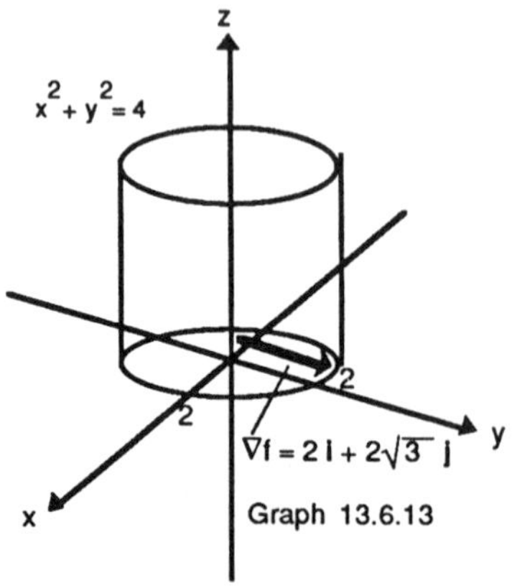

Graph 13.6.13

14. $\nabla f = 2x\,\mathbf{i} + 2z\,\mathbf{k} \Rightarrow \nabla f(1,-1,1) = 2\,\mathbf{i} + 2\,\mathbf{k}$

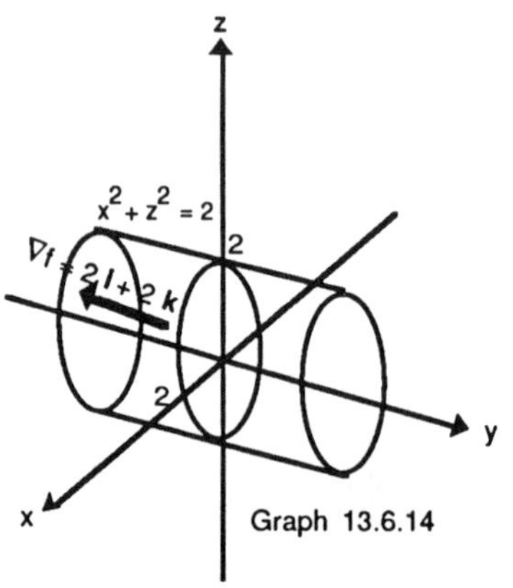

Graph 13.6.14

15. $\nabla f = 2x\,\mathbf{i} + 2y\,\mathbf{j} - \mathbf{k} \Rightarrow \nabla f(1,-1,1) = 2\,\mathbf{i} - 2\,\mathbf{j} - \mathbf{k}$

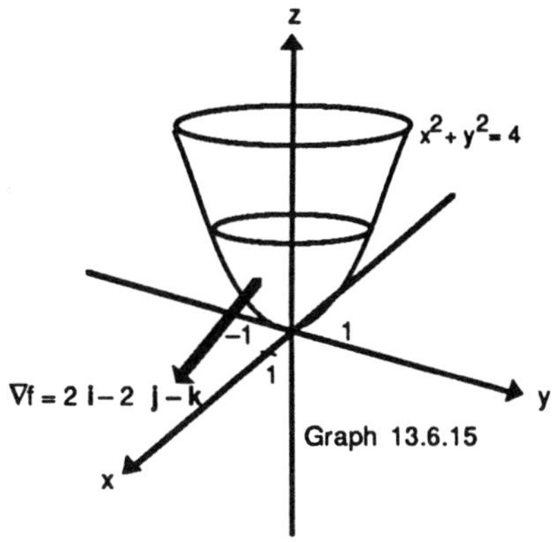

Graph 13.6.15

16. $\nabla f = 2x\,\mathbf{i} + 2y\,\mathbf{j} + \mathbf{k} \Rightarrow \nabla f(-1,-1,2) = -2\,\mathbf{i} - 2\,\mathbf{j} + \mathbf{k}$

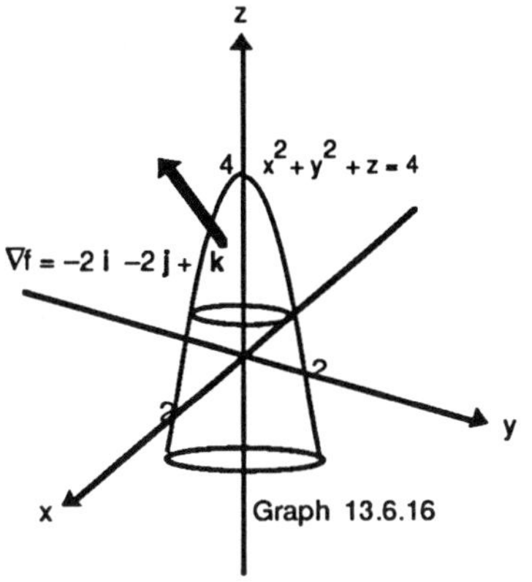

Graph 13.6.16

17. $\nabla f = 2x\,\mathbf{i} + 2y\,\mathbf{j} - 2z\,\mathbf{k} \Rightarrow \nabla f(2,-1,-\sqrt{5}) =$

$4\,\mathbf{i} - 2\,\mathbf{j} + 2\sqrt{5}\,\mathbf{k}$

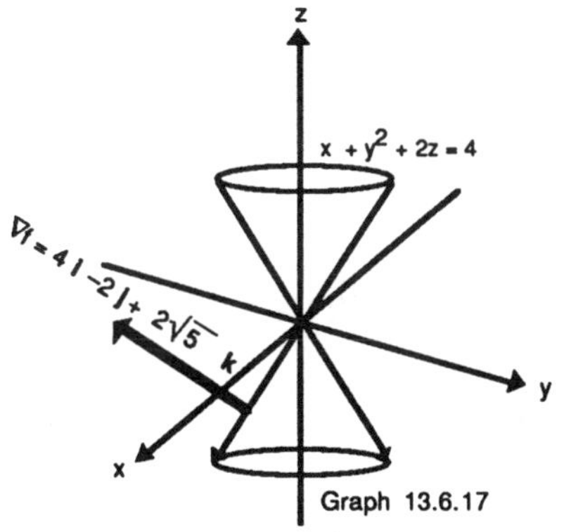

Graph 13.6.17

18. $\nabla f = 8y\,\mathbf{j} + 18z\,\mathbf{k} \Rightarrow \nabla f(0,\frac{3}{2},\sqrt{3}) = 12\,\mathbf{j} + 18\sqrt{3}\,\mathbf{k}$

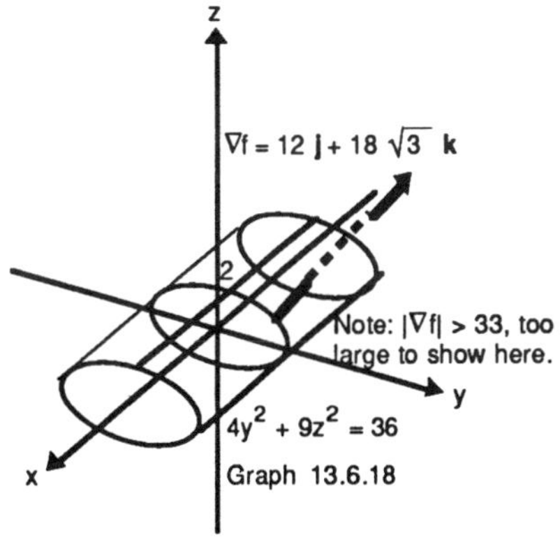

Graph 13.6.18

19. $\nabla f = \mathbf{i} + 2y\,\mathbf{j} + 2\,\mathbf{k} \Rightarrow \nabla f(1,1,1) = \mathbf{i} + 2\,\mathbf{j} + 2\,\mathbf{k}$. $\nabla g = \mathbf{i}$ for all points P. $\mathbf{v} = \nabla f \times \nabla g \Rightarrow$

$$\mathbf{v} = \begin{vmatrix} \mathbf{i} & \mathbf{j} & \mathbf{k} \\ 1 & 2 & 2 \\ 1 & 0 & 0 \end{vmatrix} = 2\,\mathbf{j} - 2\,\mathbf{k} \Rightarrow \text{Tangent line: } x = 1,\ y = 1 + 2t,\ z = 1 - 2t$$

20. $\nabla f = yz\,\mathbf{i} + xz\,\mathbf{j} + xy\,\mathbf{k} \Rightarrow \nabla f(1,1,1) = \mathbf{i} + \mathbf{j} + \mathbf{k}$. $\nabla g = 2x\,\mathbf{i} + 4y\,\mathbf{j} + 6z\,\mathbf{k} \Rightarrow \nabla g(1,1,1) = 2\,\mathbf{i} + 4\,\mathbf{j} + 6\,\mathbf{k}$.

$$\mathbf{v} = \nabla f \times \nabla g \Rightarrow \mathbf{v} = \begin{vmatrix} \mathbf{i} & \mathbf{j} & \mathbf{k} \\ 1 & 1 & 1 \\ 2 & 4 & 6 \end{vmatrix} = 2\,\mathbf{i} - 4\,\mathbf{j} + 2\,\mathbf{k} \Rightarrow \text{Tangent line: } x = 1 + 2t,\ y = 1 - 4t,\ z = 1 + 2t$$

21. $\nabla f = 2x\,\mathbf{i} + 2\,\mathbf{j} + 2\,\mathbf{k} \Rightarrow \nabla f(1,1,\frac{1}{2}) = 2\,\mathbf{i} + 2\,\mathbf{j} + 2\,\mathbf{k}$. $\nabla g = \mathbf{j}$ for all points P. $\mathbf{v} = \nabla f \times \nabla g \Rightarrow$

$$\mathbf{v} = \begin{vmatrix} \mathbf{i} & \mathbf{j} & \mathbf{k} \\ 2 & 2 & 2 \\ 0 & 1 & 0 \end{vmatrix} = -2\,\mathbf{i} + 2\,\mathbf{k} \Rightarrow \text{Tangent line: } x = 1 - 2t,\ y = 1,\ z = \frac{1}{2} + 2t$$

22. $\nabla f = \mathbf{i} + 2y\,\mathbf{j} + \mathbf{k} \Rightarrow \nabla f\left(\frac{1}{2},1,\frac{1}{2}\right) = \mathbf{i} + 2\,\mathbf{j} + \mathbf{k}$. $\nabla g = \mathbf{j}$ for all points P. $\mathbf{v} = \nabla f \times \nabla g \Rightarrow$

$$\mathbf{v} = \begin{vmatrix} \mathbf{i} & \mathbf{j} & \mathbf{k} \\ 1 & 2 & 1 \\ 0 & 1 & 0 \end{vmatrix} = -\mathbf{i} + \mathbf{k} \Rightarrow \text{Tangent line: } x = \frac{1}{2} - t,\ y = 1,\ z = \frac{1}{2} + t$$

23. $\nabla f = \left(3x^2 + 6xy^2 + 4y\right)\mathbf{i} + \left(6x^2y + 3y^2 + 4x\right)\mathbf{j} - 2z\,\mathbf{k} \Rightarrow \nabla f(1,1,3) = 13\,\mathbf{i} + 13\,\mathbf{j} - 6\,\mathbf{k}$.

$\nabla g = 2x\,\mathbf{i} + 2y\,\mathbf{j} + 2z\,\mathbf{k} \Rightarrow \nabla g(1,1,3) = 2\,\mathbf{i} + 2\,\mathbf{j} + 6\,\mathbf{k}$. $\mathbf{v} = \nabla f \times \nabla g \Rightarrow \mathbf{v} = \begin{vmatrix} \mathbf{i} & \mathbf{j} & \mathbf{k} \\ 13 & 13 & -6 \\ 2 & 2 & 6 \end{vmatrix} =$

$90\,\mathbf{i} - 90\,\mathbf{j} \Rightarrow$ Tangent line: $x = 1 + 90t,\ y = 1 - 90t,\ z = 3$

24. $\nabla f = 2x\,\mathbf{i} + 2y\,\mathbf{j} \Rightarrow \nabla f\left(\sqrt{2},\sqrt{2},4\right) = 2\sqrt{2}\,\mathbf{i} + 2\sqrt{2}\,\mathbf{j}$. $\nabla g = 2x\,\mathbf{i} + 2y\,\mathbf{j} - \mathbf{k} \Rightarrow \nabla g\left(\sqrt{2},\sqrt{2},4\right) = 2\sqrt{2}\,\mathbf{i} +$

$2\sqrt{2}\,\mathbf{j} - \mathbf{k}$. $\mathbf{v} = \nabla f \times \nabla g \Rightarrow \mathbf{v} = \begin{vmatrix} \mathbf{i} & \mathbf{j} & \mathbf{k} \\ 2\sqrt{2} & 2\sqrt{2} & 0 \\ 2\sqrt{2} & 2\sqrt{2} & -1 \end{vmatrix} = -2\sqrt{2}\,\mathbf{i} + 2\sqrt{2}\,\mathbf{j} \Rightarrow$ Tangent line: $x = \sqrt{2} - 2\sqrt{2}\,t$,

$y = \sqrt{2} + 2\sqrt{2}\,t,\ z = 4$

25. $\nabla f = 2x\,\mathbf{i} + 2y\,\mathbf{j} \Rightarrow \nabla f(\sqrt{2},\sqrt{2}) = 2\sqrt{2}\,\mathbf{i} + 2\sqrt{2}\,\mathbf{j} \Rightarrow$ Tangent line: $2\sqrt{2}\left(x - \sqrt{2}\right) + 2\sqrt{2}\left(y - \sqrt{2}\right) = 0 \Rightarrow$
$\sqrt{2}\,x + \sqrt{2}\,y = 4$

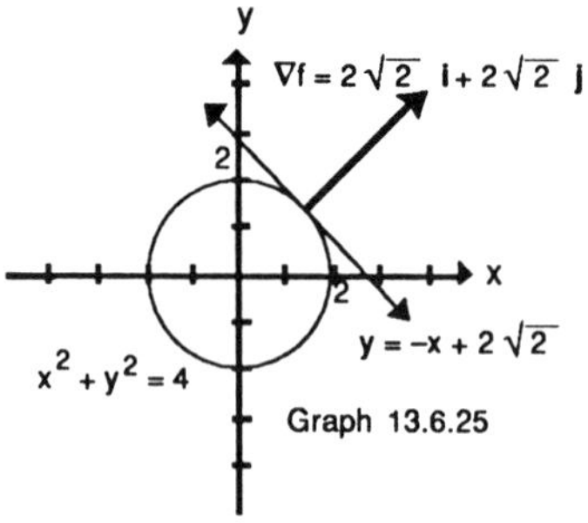

Graph 13.6.25

26. $\nabla f = 2x\,\mathbf{i} + y\,\mathbf{j} \Rightarrow \nabla f(1,2) = 2\,\mathbf{i} + 2\,\mathbf{j} \Rightarrow$ Tangent line: $2(x - 1) + 2(y - 2) = 0 \Rightarrow x + y = 3$

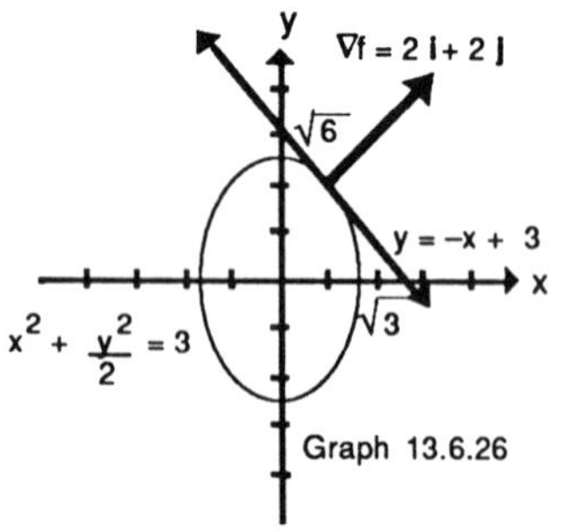

Graph 13.6.26

27. $\nabla f = 8x\,\mathbf{i} + 18y\,\mathbf{j} \Rightarrow \nabla f\left(2, \frac{2\sqrt{5}}{3}\right) = 16\,\mathbf{i} + 12\sqrt{5}\,\mathbf{j} \Rightarrow$ Tangent line: $16(x-2) + 12\sqrt{5}\left(y - \frac{2\sqrt{5}}{3}\right) = 0 \Rightarrow$ $4x + 3\sqrt{5}\,y = 18$.

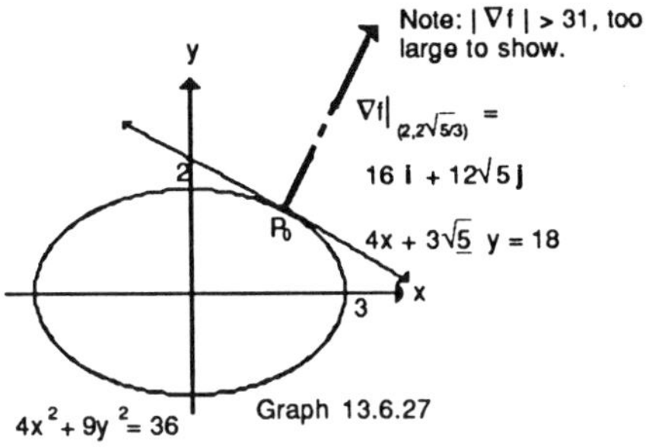

28. $\nabla f = 2x\,\mathbf{i} - \mathbf{j} \Rightarrow \nabla f(\sqrt{2},1) = 2\sqrt{2}\,\mathbf{i} - \mathbf{j} \Rightarrow$ Tangent line: $2\sqrt{2}\left(x - \sqrt{2}\right) - (y-1) = 0 \Rightarrow y = 2\sqrt{2}\,x - 3$

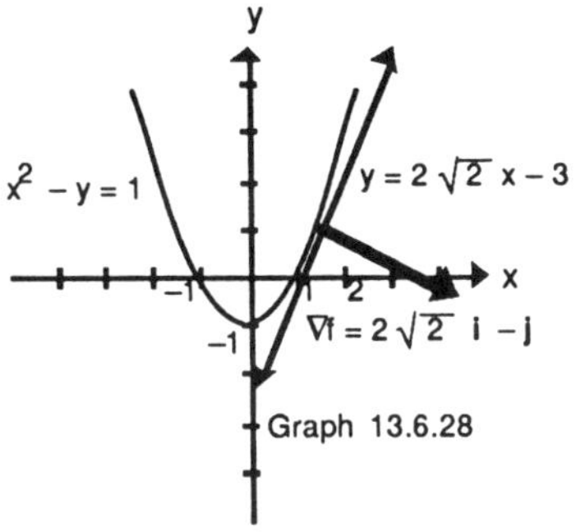

29. $\nabla f = y\,\mathbf{i} + x\,\mathbf{j} \Rightarrow \nabla f(2,-2) = -2\,\mathbf{i} + 2\,\mathbf{j} \Rightarrow$ Tangent line: $-2(x-2) + 2(y+2) = 0 \Rightarrow y = x - 4$

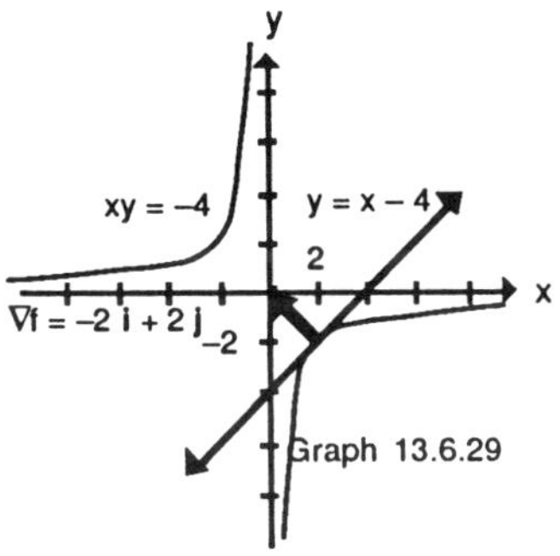

30. $\nabla f = (2x-y)\,\mathbf{i} + (2y-x)\,\mathbf{j} \Rightarrow \nabla f(-1,2) = -4\,\mathbf{i} + 5\,\mathbf{j} \Rightarrow$ Tangent line: $-4(x+1) + 5(y-2) = 0 \Rightarrow$ $-4x + 5y - 14 = 0$

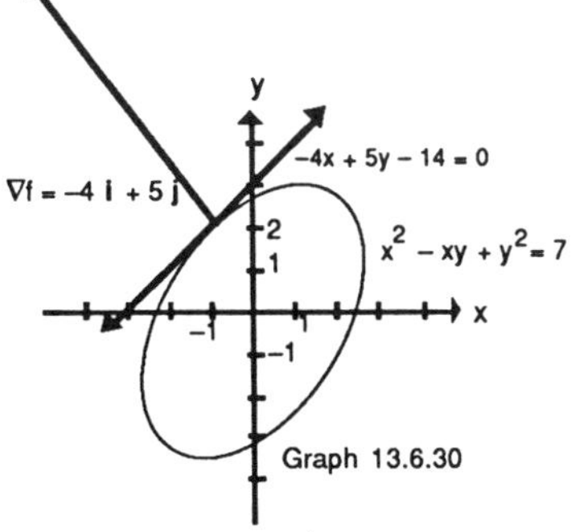

31. $\mathbf{r} = \sqrt{t}\,\mathbf{i} + \sqrt{t}\,\mathbf{j} - \frac{1}{4}(t+3)\,\mathbf{k} \Rightarrow \mathbf{v} = \frac{1}{2}t^{-1/2}\,\mathbf{i} + \frac{1}{2}t^{-1/2}\,\mathbf{j} - \frac{1}{4}\mathbf{k}$. $t = 1 \Rightarrow x = 1,\ y = 1,\ z = -1 \Rightarrow P_0 = (1,1,-1)$.

Also $t = 1 \Rightarrow \mathbf{v}(1) = \frac{1}{2}\mathbf{i} + \frac{1}{2}\mathbf{j} - \frac{1}{4}\mathbf{k}$. $f(x,y,z) = x^2 + y^2 - z - 3 = 0 \Rightarrow \nabla f = 2x\,\mathbf{i} + 2y\,\mathbf{j} - \mathbf{k} \Rightarrow \nabla f(1,1,-1) =$

$2\,\mathbf{i} + 2\,\mathbf{j} - \mathbf{k}$. $\therefore\ \mathbf{v} = \frac{1}{4}(\nabla f) \Rightarrow$ The curve is normal to the surface.

32. $\mathbf{r} = \sqrt{t}\,\mathbf{i} + \sqrt{t}\,\mathbf{j} + (2t - 1)\,\mathbf{k} \Rightarrow \mathbf{v} = \frac{1}{2}t^{-1/2}\,\mathbf{i} + \frac{1}{2}t^{-1/2}\,\mathbf{j} + 2\,\mathbf{k}$. $t = 1 \Rightarrow x = 1, y = 1, z = 1$. Also $t = 1 \Rightarrow$ $\mathbf{v}(1) = \frac{1}{2}\mathbf{i} + \frac{1}{2}\mathbf{j} + 2\,\mathbf{k}$. $f(x,y,z) = x^2 + y^2 - z - 1 = 0 \Rightarrow \nabla f = 2x\,\mathbf{i} + 2y\,\mathbf{j} - \mathbf{k} \Rightarrow \nabla f(1,1,1) = 2\,\mathbf{i} + 2\,\mathbf{j} - \mathbf{k} \Rightarrow$ $\mathbf{v} \cdot \nabla f = \frac{1}{2}(2) + \frac{1}{2}(2) + 2(-1) = 0 \Rightarrow$ The curve is tangent to the surface when $t = 1$.

SECTION 13.7 LINEARIZATION AND DIFFERENTIALS

1. a) $f(0,0) = 1$, $f_x(x,y) = 2x \Rightarrow f_x(0,0) = 0$, $f_y(x,y) = 2y \Rightarrow f_y(0,0) = 0 \Rightarrow L(x,y) = 1 + 0(x - 0) + 0(y - 0) = 1$
   b) $f(1,1) = 3$, $f_x(1,1) = 2$, $f_y(1,1) = 2 \Rightarrow L(x,y) = 3 + 2(x - 1) + 2(y - 1) = 2x + 2y - 1$

2. a) $f(1,1) = 1$, $f_x(x,y) = 3x^2y^4 \Rightarrow f_x(1,1) = 3$, $f_y(x,y) = 4x^3y^3 \Rightarrow f_y(1,1) = 4 \Rightarrow L(x,y) = 1 + 3(x - 1) + 4(y - 1)$ $\Rightarrow 3x + 4y - 6$
   b) $f(0,0) = 0$, $f_x(0,0) = 0$, $f_y(0,0) = 0 \Rightarrow L(x,y) = 0$

3. a) $f(0,0) = 1$, $f_x(x,y) = e^x \cos y \Rightarrow f_x(0,0) = 1$, $f_y(x,y) = -e^x \sin y \Rightarrow f_y(0,0) = 0 \Rightarrow L(x,y) = 1 + 1(x - 0) +$ $0(y - 0) = 1 + x$
   b) $f\left(0,\frac{\pi}{2}\right) = 0$, $f_x\left(0,\frac{\pi}{2}\right) = 0$, $f_y\left(0,\frac{\pi}{2}\right) = -1 \Rightarrow L(x,y) = 0 + 0(x - 0) - 1\left(y - \frac{\pi}{2}\right) = -y + \frac{\pi}{2}$

4. a) $f(0,0) = 4$, $f_x(x,y) = 2(x + y + 2) \Rightarrow f_x(0,0) = 4$, $f_y(x,y) = 2(x + y + 2) \Rightarrow f_y(0,0) = 4 \Rightarrow L(x,y) =$ $4 + 4(x - 0) + 4(y - 0) = 4 + 4x + 4y$
   b) $f(1,2) = 25$, $f_x(1,2) = 10$, $f_y(1,2) = 10 \Rightarrow L(x,y) = 25 + 10(x - 1) + 10(y - 2) = 10x + 10y - 5$

5. a) $f(0,0) = 5$, $f_x(x,y) = 3$ for all (x,y), $f_y(x,y) = -4$ for all $(x,y) \Rightarrow L(x,y) = 5 + 3(x - 0) - 4(y - 0) =$ $5 + 3x - 4y$
   b) $f(1,1) = 4$, $f_x(1,1) = 3$, $f_y(1,1) = -4 \Rightarrow L(x,y) = 4 + 3(x - 1) - 4(y - 1) = 3x - 4y + 5$

6. a) $f(0,0) = 1$, $f_x(x,y) = -e^{2y-x} \Rightarrow f_x(0,0) = -1$, $f_y(x,y) = 2e^{2y-x} \Rightarrow f_y(0,0) = 2 \Rightarrow L(x,y) = 1 - 1(x - 0) +$ $2(y - 0) = 1 - x + 2y$
   b) $f(1,2) = e^3$, $f_x(1,2) = -e^3$, $f_y(1,2) = 2e^3 \Rightarrow L(x,y) = e^3 - e^3(x - 1) + 2e^3(y - 2) = -e^3 x + 2e^3 y - 2e^3$

7. $f(2,1) = 3$, $f_x(x,y) = 2x - 3y \Rightarrow f_x(2,1) = 1$, $f_y(x,y) = -3x \Rightarrow f_y(2,1) = -6 \Rightarrow L(x,y) = 3 + 1(x - 2) - 6(y - 1)$ $= 7 + x - 6y$. $f_{xx}(x,y) = 2$, $f_{yy}(x,y) = 0$, $f_{xy}(x,y) = -3 \Rightarrow M = 3$. $\therefore$ $|E(x,y)| \le \frac{1}{2}(3)(|x - 2| + |y - 1|)^2 \le$ $\frac{3}{2}(0.1 + 0.1)^2 = 0.06$

8. $f(2,2) = 11$, $f_x(x,y) = x + y + 3 \Rightarrow f_x(2,2) = 7$, $f_y(x,y) = x + \frac{1}{2}y - 3 \Rightarrow f_y(2,2) = 0 \Rightarrow L(x,y) = 11 + 7(x - 2) +$ $0(y - 2) = 7x - 3$. $f_{xx}(x,y) = 1$, $f_{yy}(x,y) = \frac{1}{2}$, $f_{xy}(x,y) = 1 \Rightarrow M = 1$. $\therefore$ $|E(x,y)| \le \frac{1}{2}(1)\,(|x - 2| + |y - 2|)^2$ $\le \frac{1}{2}(0.1 + 0.1)^2 = 0.02$

9. $f(0,0) = 1$, $f_x(x,y) = \cos y \Rightarrow f_x(0,0) = 1$, $f_y(x,y) = 1 - x \sin y \Rightarrow f_y(0,0) = 1 \Rightarrow L(x,y) = 1 + 1(x - 0) +$ $1(y - 0) = x + y + 1$. $f_{xx}(x,y) = 0$, $f_{yy}(x,y) = 0$, $f_{xy}(x,y) = -\sin y \Rightarrow M = 1$. $\therefore$ $|E(x,y)| \le \frac{1}{2}(1)(|x| + |y|)^2 \le \frac{1}{2}(0.2 + 0.2)^2 = 0.08$

10. $f(1,1) = 0$, $f_x(x,y) = \frac{1}{x} \Rightarrow f_x(1,1) = 1$, $f_y(x,y) = \frac{1}{y} \Rightarrow f_y(1,1) = 1 \Rightarrow L(x,y) = 0 + 1(x-1) + 1(y-1) = x + y - 2$. $f_{xx}(x,y) = -\frac{1}{x^2}$, $f_{yy}(x,y) = -\frac{1}{y^2}$, $f_{xy}(x,y) = 0$. $|x-1| \le 0.1 \Rightarrow 0.99 \le x \le 1.1 \Rightarrow$ max of $|f_{xx}(x,y)|$ on R is $\frac{1}{(0.99)^2} \le 1.03$. $|y-1| \le 0.1 \Rightarrow 0.99 \le y \le 1.1 \Rightarrow$ max of $|f_{yy}(x,y)|$ on R is $\frac{1}{(0.99)^2} \le 1.03. \Rightarrow M = 1.03$. $\therefore$ $|E(x,y)| \le \frac{1}{2}(1.03)(|x-1| + |y-1|)^2 \le 0.515(0.1 + 0.1)^2 = 0.0206$

11. $f(0,0) = 1$, $f_x(x,y) = e^x \cos y \Rightarrow f_x(0,0) = 1$, $f_y(x,y) = -e^x \sin y \Rightarrow f_y(0,0) = 0 \Rightarrow L(x,y) = 1 + 1(x-0) + 0(y-0) = 1 + x$. $f_{xx}(x,y) = e^x \cos y$, $f_{yy}(x,y) = -e^x \cos y$, $f_{xy}(x,y) = -e^x \sin y$. $|x| \le 0.1 \Rightarrow -0.1 \le x \le 0.1$, $|y| \le 0.1 \Rightarrow -0.1 \le y \le 0.1. \Rightarrow$ max of $|f_{xx}(x,y)|$ on R is $e^{0.1}\cos(0.1) \le 1.11$, max of $|f_{yy}(x,y)|$ on R is $e^{0.1}\cos(0.1) \le 1.11$, max of $|f_{xy}(x,y)|$ on R is $e^{0.1}\sin(0.1) \le 0.002 \Rightarrow M = 1.11$.
$\therefore$ $|E(x,y)| \le \frac{1}{2}(1.11)(|x| + |y|)^2 \le 0.555(0.1 + 0.1)^2 = 0.0222$

12. $f_x(x,y) = 2x(y+1) \Rightarrow f_x(1,0) = 2$, $f_y(x,y) = x^2 \Rightarrow f_y(1,0) = 1 \Rightarrow df = 2\,dx + 1\,dy \Rightarrow df$ is more sensitive to changes in x since $f_x > f_y$.

13. Let the width, w, be the long side. Then $A = lw \Rightarrow dA = A_l\,dl + A_w\,dw \Rightarrow dA = w\,dl + l\,dw$. Since $w > l$, dA is more sensitive to a change in w than l. $\therefore$ pay more attention to the width.

14. a) $f_x(x,y) = 2x(y+1) \Rightarrow f_x(1,0) = 2$, $f_y(x,y) = x^2 \Rightarrow f_y(1,0) = 1$. Then $df = 2\,dx + 1\,dy \Rightarrow df$ is more sensitive to changes in x since $f_x > f_y$.
b) $df = 2\,dx + dy \Rightarrow \frac{df}{dy} = 2\frac{dx}{dy} + 1$. $df = 0 \Rightarrow \frac{df}{dy} = 0 \Rightarrow 2\frac{dx}{dy} + 1 = 0 \Rightarrow \frac{dx}{dy} = -\frac{1}{2}$

15. $T_x(x,y) = e^y + e^{-y}$, $T_y(x,y) = x(e^y - e^{-y}) \Rightarrow dT = T_x(x,y)\,dx + T_y(x,y)\,dy = (e^y + e^{-y})dx + x(e^y - e^{-y})dy \Rightarrow dT|_{(2,\ln 2)} = 2.5\,dx + 3.0\,dy$. If $|dx| \le 0.1$, $|dy| \le 0.02$, then the maximum possible error (estimate) $\le 2.5(0.1) + 3.0(0.02) = 0.31$ in magnitude.

16. $V_r = 2\pi rh$, $V_h = \pi r^2 \Rightarrow dV = V_r\,dr + V_h\,dh \Rightarrow \frac{dV}{V} = \frac{2\pi rh\,dr + \pi r^2\,dh}{\pi r^2 h} = \frac{2}{r}dr + \frac{1}{h}dh$. Now, $\left|\frac{dr}{r} \times 100\right| \le 1$, $\left|\frac{dh}{h} \times 100\right| \le 1$. $\therefore$ $\left|\frac{dV}{V} \times 100\right| \le \left|2\frac{dr}{r} \times 100 + \frac{dh}{h} \times 100\right| \le 2\left|\frac{dr}{r} \times 100\right| + \left|\frac{dh}{h} \times 100\right| \le 2(1) + 1 = 3$ or 3%

17. $V_r = 2\pi rh$, $V_h = \pi r^2 \Rightarrow dV = V_r\,dr + V_h\,dh \Rightarrow dV = 2\pi rh\,dr + \pi r^2\,dh \Rightarrow dV|_{(5,12)} = 120\pi\,dr + 25\pi\,dh$. Since $|dr| \le 0.1$ cm, $|dh| \le 0.1$ cm, $dV \le 120\pi(0.1) + 25\pi(0.1) = 14.5\pi\ \text{cm}^3$. $V(5,12) = 300\pi\ \text{cm}^3 \Rightarrow$ Maximum percentage error $= \pm\frac{14.5\pi}{300\pi} \times 100 = \pm 4.83\%$

18. $V_r = 2\pi rh$, $V_h = \pi r^2 \Rightarrow dV = V_r\,dr + V_h\,dh \Rightarrow dV = 2\pi rh\,dr + \pi r^2\,dh$. Assume $dr = dh \Rightarrow dV = 2\pi rh\,dr + \pi r^2\,dr = (2\pi rh + \pi r^2)\,dr$. For $dV \le 0.1\ \text{m}^3$ when $r = 2$ m, $h = 3$ m, $(2\pi(2)(3) + \pi(2)^2) \le 0.1 \Rightarrow dr \le \frac{0.1}{16\pi} \approx 0.001$ m (rounded down). Thus, the error in measuring r and h should be less than or equal to 0.002 m.

19. $df = f_x(x,y)\,dx + f_y(x,y)\,dy = 3x^2y^4dx + 4x^3y^3dy \Rightarrow df|_{(1,1)} = 3\,dx + 4\,dy$. Let $dx = dy \Rightarrow df = 7\,dx$. $|df| \le 0.1 \Rightarrow 7|dx| \le 0.1 \Rightarrow |dx| \le \frac{0.1}{7} \approx 0.014$. $\therefore$ for the square, let $|x - 1| \le 0.014$, $|y - 1| \le 0.014$

20. a) $\frac{1}{R} = \frac{1}{R_1} + \frac{1}{R_2} \Rightarrow R = \frac{R_1R_2}{R_1 + R_2} \Rightarrow dR = R_{R_1}dR_1 + R_{R_2}dR_2 = \frac{(R_2 + R_1)R_2 - (R_1R_2)}{(R_1 + R_2)^2}dR_1 + \frac{(R_2 + R_1)R_1 - (R_1R_2)}{(R_2 + R_1)^2}dR_2 = \frac{R_2^2}{(R_2 + R_1)^2}dR_1 + \frac{R_1^2}{(R_2 + R_1)^2} = \left(\frac{R}{R_1}\right)^2 dR_1 + \left(\frac{R}{R_2}\right)^2 dR_2$

b) $dR = R^2\left[\left(\frac{1}{R_1^2}\right)dR_1 + \left(\frac{1}{R_2^2}\right)dR_2\right] \Rightarrow dR|_{(100,400)} = R^2\left[\frac{1}{(100)^2}dR_1 + \frac{1}{(400)^2}dR_2\right]$

$\therefore$ R will be more sensitive to a change in R_1 since $\frac{1}{(100)^2} > \frac{1}{(400)^2}$.

21. $dR = \left(\frac{R}{R_1}\right)^2 dR_1 + \left(\frac{R}{R_2}\right)^2 dR_2$ (See Exercise 20 above). R_1 changes from 20 to 20.1 ohms $\Rightarrow dR_1 =$ 0.1 ohms, R_2 changes from 25 to 24.9 ohms $\Rightarrow dR_2 = -0.1$ ohms. $\frac{1}{R} = \frac{1}{R_1} + \frac{1}{R_2} \Rightarrow R = \frac{100}{9}$ ohms.

$dR|_{(20,25)} = \frac{(100/9)^2}{(20)^2}(0.1) + \frac{(100/9)^2}{(25)^2}(-0.1) = 0.011$ ohms $\Rightarrow$ Percentage change $= \frac{dR}{R}\Big|_{(20,25)} \times 100 = \frac{0.011}{100/9} \times 100 \approx 0.099\%$

22. a) $r^2 = x^2 + y^2 \Rightarrow r = \sqrt{x^2 + y^2} \Rightarrow dr = \frac{x}{\sqrt{x^2 + y^2}}dx + \frac{y}{\sqrt{x^2 + y^2}}dy \Rightarrow dr|_{(3,4)} = \frac{3}{\sqrt{3^2 + 4^2}}(\pm 0.01) + \frac{4}{\sqrt{3^2 + 4^2}}(\pm 0.01) = \pm 0.014$. $x = 3$, $y = 4 \Rightarrow r = 5$. $\therefore \left|\frac{dr}{r} \times 100\right| = \left|\frac{\pm 0.014}{5} \times 100\right| = 0.28\%$

$d\theta = \frac{-y/x^2}{\left(\frac{y}{x}\right)^2 + 1}dx + \frac{1/x}{\left(\frac{y}{x}\right)^2 + 1}dy = \frac{-y}{y^2 + x^2}dx + \frac{x}{y^2 + x^2}dy \Rightarrow d\theta|_{(3,4)} = \frac{-4}{25}(\pm 0.01) + \frac{3}{25}(\pm 0.01) = \frac{\mp 0.04}{25} + \frac{\pm 0.03}{25} \Rightarrow$ Maximum change in $d\theta$ occurs when dx and dy have opposite signs ($dx = 0.01$ and $dy = -0.01$ or vice versa) $\therefore d\theta = \frac{\pm 0.07}{25} \approx \pm 0.0028$. $\theta = \tan^{-1}\left(\frac{4}{3}\right) \approx 0.927255218 \Rightarrow$

$\left|\frac{d\theta}{\theta} \times 100\right| = \left|\frac{\pm 0.0028}{0.927255218} \times 100\right| \approx 0.30\%$

b) r is more sensitive to changes in y; θ is more sensitive to changes in x.

23. a) $f(1,1,1) = 3$, $f_x(1,1,1) = y + z|_{(1,1,1)} = 2$, $f_y(1,1,1) = x + z|_{(1,1,1)} = 2$, $f_z(1,1,1) = y + x|_{(1,1,1)} = 2 \Rightarrow$ $L(x,y,z) = 2x + 2y + 2z - 3$

b) $f(1,0,0) = 0$, $f_x(1,0,0) = 0$, $f_y(1,0,0) = 1$, $f_z(1,0,0) = 1 \Rightarrow L(x,y,z) = y + z$

c) $f(0,0,0) = 0$, $f_x(0,0,0) = 0$, $f_y(0,0,0) = 0$, $f_z(0,0,0) = 0 \Rightarrow L(x,y,z) = 0$

24. a) $f(1,1,1) = 3$, $f_x(1,1,1) = 2x|_{(1,1,1)} = 2$, $f_y(1,1,1) = 2y|_{(1,1,1)} = 2$, $f_z(1,1,1) = 2z|_{(1,1,1)} = 2 \Rightarrow$ $L(x,y,z) = 2x + 2y + 2z - 3$

b) $f(0,1,0) = 1$, $f_x(0,1,0) = 0$, $f_y(0,1,0) = 2$, $f_z(0,1,0) = 0 \Rightarrow L(x,y,z) = 2y - 1$

c) $f(1,0,0) = 1$, $f_x(1,0,0) = 2$, $f_y(1,0,0) = 0$, $f_z(1,0,0) = 0 \Rightarrow L(x,y,z) = 2x - 1$

25. a) $f(1,0,0) = 1$, $f_x(1,0,0) = \frac{x}{\sqrt{x^2+y^2+z^2}}\Big|_{(1,0,0)} = 1$, $f_y(1,0,0) = \frac{y}{\sqrt{x^2+y^2+z^2}}\Big|_{(1,0,0)} = 0$, $f_z(1,0,0) = \frac{z}{\sqrt{x^2+y^2+z^2}}\Big|_{(1,0,0)} = 0 \Rightarrow L(x,y,z) = x$

b) $f(1,1,0) = \sqrt{2}$, $f_x(1,1,0) = \frac{1}{\sqrt{2}}$, $f_y(1,1,0) = \frac{1}{\sqrt{2}}$, $f_z(1,1,0) = 0 \Rightarrow L(x,y,z) = \frac{1}{\sqrt{2}}x + \frac{1}{\sqrt{2}}y$

c) $f(1,2,2) = 3$, $f_x(1,2,2) = \frac{1}{3}$, $f_y(1,2,2) = \frac{2}{3}$, $f_z(1,2,2) = \frac{2}{3} \Rightarrow L(x,y,z) = \frac{1}{3}x + \frac{2}{3}y + \frac{2}{3}z$

26. a) $f\left(\frac{\pi}{2},1,1\right) = 1$, $f_x\left(\frac{\pi}{2},1,1\right) = \frac{y\cos xy}{z}\Big|_{(\pi/2,1,1)} = 0$, $f_y\left(\frac{\pi}{2},1,1\right) = \frac{x\cos xy}{z}\Big|_{(\pi/2,1,1)} = 0$, $f_z\left(\frac{\pi}{2},1,1\right) = \frac{-\sin xy}{z^2}\Big|_{(\pi/2,1,1)} = -1 \Rightarrow L(x,y,z)) = 2 - z$

b) $f(2,0,1) = 0$, $f_x(2,0,1) = 0$, $f_y(2,0,1) = 2$, $f_z(2,0,1) = 0 \Rightarrow L(x,y,z) = 2y$

27. a) $f(0,0,0) = 2$, $f_x(0,0,0) = e^x\Big|_{(0,0,0)} = 1$, $f_y(0,0,0) = -\sin(y+z)\Big|_{(0,0,0)} = 0$, $f_z(0,0,0) = -\sin(y+z)\Big|_{(0,0,0)} = 0 \Rightarrow L(x,y,z) = 2 + x$

b) $f\left(0,\frac{\pi}{2},0\right) = 1$, $f_x\left(0,\frac{\pi}{2},0\right) = 1$, $f_y\left(0,\frac{\pi}{2},0\right) = -1$, $f_z\left(0,\frac{\pi}{2},0\right) = -1 \Rightarrow L(x,y,z) = x - y - z + \frac{\pi}{2} + 1$

c) $f\left(0,\frac{\pi}{4},\frac{\pi}{4}\right) = 1$, $f_x\left(0,\frac{\pi}{4},\frac{\pi}{4}\right) = 1$, $f_y\left(0,\frac{\pi}{4},\frac{\pi}{4}\right) = -1$, $f_z\left(0,\frac{\pi}{4},\frac{\pi}{4}\right) = -1 \Rightarrow L(x,y,z) = x - y - z + \frac{\pi}{2} + 1$

28. a) $f(1,0,0) = 0$, $f_x(1,0,0) = \frac{yz}{(xyz)^2+1}\Big|_{(1,0,0)} = 0$, $f_y(1,0,0) = \frac{xz}{(xyz)^2+1}\Big|_{(1,0,0)} = 0$, $f_z(1,0,0) = \frac{xy}{(xyz)^2+1}\Big|_{(1,0,0)} = 0 \Rightarrow L(x,y,z) = 0$

b) $f(1,1,0) = 0$, $f_x(1,1,0) = 0$, $f_y(1,1,0) = 0$, $f_z(1,1,0) = 1 \Rightarrow L(x,y,z) = z$

c) $f(1,1,1) = \frac{\pi}{4}$, $f_x(1,1,1) = 1$, $f_y(1,1,1) = 1$, $f_z(1,1,1) = 1 \Rightarrow L(x,y,z) = x + y + z + \frac{\pi}{4} - 3$

29. $f(a,b,c,d) = \begin{vmatrix} a & b \\ c & d \end{vmatrix} = ad - bc \Rightarrow f_a = d$, $f_b = -c$, $f_c = -b$, $f_d = a \Rightarrow df = d\,da - c\,db - b\,dc + a\,dd$. Since $|a|$ is much greater than $|b|$, $|c|$, and $|d|$, f is most sensitive to a change in d.

30. $p(a,b,c) = abc \Rightarrow p_a = bc$, $p_b = ac$, $p_c = ab \Rightarrow dp = bc\,da + ac\,db + ab\,dc \Rightarrow \frac{dp}{p} = \frac{bc\,da + ac\,db + ab\,dc}{abc} = \frac{da}{a} + \frac{db}{b} + \frac{dc}{c}$. Since $\left|\frac{da}{a} \times 100\right| = 2$, $\left|\frac{db}{b} \times 100\right| = 2$, $\left|\frac{dc}{c} \times 100\right| = 2$, $\left|\frac{dp}{p} \times 100\right| = \left|\frac{da}{a} \times 100 + \frac{db}{b} \times 100 + \frac{dc}{c} \times 100\right| \le \left|\frac{da}{a} \times 100\right| + \left|\frac{db}{b} \times 100\right| + \left|\frac{dc}{c} \times 100\right| = 2 + 2 + 2 = 6$ or 6%

31. $V = lwh \Rightarrow V_l = wh$, $V_w = lh$, $V_h = lw \Rightarrow dV = wh\,dl + lh\,dw + lw\,dh \Rightarrow dV\big|_{(5,3,2)} = 6\,dl + 10\,dw + 15\,dh$

$dl = 1$ in $= \frac{1}{12}$ ft, $dw = 1$ in $= \frac{1}{12}$ ft, $dh = \frac{1}{2}$ in $= \frac{1}{24}$ ft $\Rightarrow dV = 6\left(\frac{1}{12}\right) + 10\left(\frac{1}{12}\right) + 15\left(\frac{1}{24}\right) = \frac{47}{24}$ ft^3

32. $A = \frac{1}{2}ab\sin C \Rightarrow A_a = \frac{1}{2}b\sin C$, $A_b = \frac{1}{2}a\sin C$, $A_C = \frac{1}{2}ab\cos C \Rightarrow dA = \frac{1}{2}b\sin C\,da + \frac{1}{2}a\sin C\,db + \frac{1}{2}ab\cos C\,dC$. $dC = \pm 2° = \pm 0.0349$ radians, $da = \pm 0.5$ ft, $db = \pm 0.5$ ft. At $a = 150$ ft, $b = 200$ ft, $C = 60°$, $dA = \frac{1}{2}(200)\sin 60° (\pm 0.5) + \frac{1}{2}(150)\sin 60° (\pm 0.5) + \frac{1}{2}(200)(150)\cos 60° (\pm 0.0349) = \pm 319.23$ ft^2

33. $u_x = e^y$, $u_y = xe^y + \sin z$, $u_z = y\cos z \Rightarrow du = e^y\,dx + (xe^y + \sin z)dy + (y\cos z)dz \Rightarrow$
$du\big|_{(2,\ln 3,\pi/2)} = 3\,dx + 7\,dy + 0\,dz = 3\,dx + 7\,dy \Rightarrow$ magnitude of the maximum possible error $\le$ $3(0.2) + 7(0.6) = 4.8$

34. $Q_k = \frac{1}{2}(2kM/h)^{-1/2}(2M/h)$, $Q_M = \frac{1}{2}(2kM/h)^{-1/2}(2k/h)$, $Q_h = \frac{1}{2}(2kM/h)^{-1/2}(-2kM/h^2) \Rightarrow$
$dQ = \frac{1}{2}(2kM/h)^{-1/2}(2M/h)\,dk + \frac{1}{2}(2kM/h)^{-1/2}(2k/h)\,dM + \frac{1}{2}(2kM/h)^{-1/2}(-2kM/h^2)\,dh =$
$\frac{1}{2}(2kM/h)^{-1/2}\left[\frac{2M}{h}\,dk + \frac{2k}{h}\,dM - \frac{2kM}{h^2}\,dh\right] \Rightarrow$
$dQ\big|_{(2,20,0.05)} = \frac{1}{2}\left(2(2)(20)/0.05\right)^{-1/2}\left[\frac{2(20)}{0.05}\,dk + \frac{2(2)}{0.05}\,dM - \frac{2(2)(20)}{(0.05)^2}\,dh\right] =$
$(0.0125)(800\,dk + 80\,dM - 32000\,dh)$ $\therefore$ Q is most sensitive to changes in h.

SECTION 13.8 MAXIMA, MINIMA, AND SADDLE POINTS

1. $f_x(x,y) = 2x + y + 3 = 0$ and $f_y(x,y) = x + 2y - 3 = 0 \Rightarrow x = -3,\ y = 3 \Rightarrow$ critical point is $(-3,3)$. $f_{xx}(-3,3) = 2$, $f_{yy}(-3,3) = 2$, $f_{xy}(-3,3) = 1 \Rightarrow f_{xx}f_{yy} - f_{xy}^2 = 3 > 0$ and $f_{xx} > 0 \Rightarrow$ minimum. $f(-3,3) = -5$, absolute minimum.

2. $f_x(x,y) = 2x + 3y - 6 = 0$ and $f_y(x,y) = 3x + 6y + 3 = 0 \Rightarrow x = 15,\ y = -8 \Rightarrow$ critical point is $(15,-8)$. $f_{xx}(15,-8) = 2$, $f_{yy}(15,-8) = 6$, $f_{xy}(15,-8) = 3 \Rightarrow f_{xx}f_{yy} - f_{xy}^2 = 3 > 0$ and $f_{xx} > 0 \Rightarrow$ minimum. $f(15,-8) = -63$, absolute minimum.

3. $f_x(x,y) = 5y - 14x + 3 = 0$ and $f_y(x,y) = 5x - 6 = 0 \Rightarrow x = \frac{6}{5},\ y = \frac{69}{25} \Rightarrow$ critical point is $\left(\frac{6}{5},\frac{69}{25}\right)$. $f_{xx}\left(\frac{6}{5},\frac{69}{25}\right) = -14$, $f_{yy}\left(\frac{6}{5},\frac{69}{25}\right) = 0$, $f_{xy}\left(\frac{6}{5},\frac{69}{25}\right) = 5 \Rightarrow f_{xx}f_{yy} - f_{xy}^2 = -25 < 0 \Rightarrow$ saddle point.

4. $f_x(x,y) = 2y - 10x + 4 = 0$ and $f_y(x,y) = 2x - 4y + 4 = 0 \Rightarrow x = \frac{2}{3},\ y = \frac{4}{3} \Rightarrow$ critical point is $\left(\frac{2}{3},\frac{4}{3}\right)$. $f_{xx}\left(\frac{2}{3},\frac{4}{3}\right) = -10$, $f_{yy}\left(\frac{2}{3},\frac{4}{3}\right) = -4$, $f_{xy}\left(\frac{2}{3},\frac{4}{3}\right) = 2 \Rightarrow f_{xx}f_{yy} - f_{xy}^2 = 36 > 0$and $f_{xx} < 0 \Rightarrow$ maximum (absolute). $f\left(\frac{2}{3},\frac{4}{3}\right) = 0$

5. $f_x(x,y) = 2x + y + 3 = 0$ and $f_y(x,y) = x + 2 = 0 \Rightarrow x = -2,\ y = 1 \Rightarrow$ critical point is $(-2,1)$. $f_{xx}(-2,1) = 2$, $f_{yy}(-2,1) = 0$, $f_{xy}(-2,1) = 1 \Rightarrow f_{xx}f_{yy} - f_{xy}^2 = -1 \Rightarrow$ saddle point.

6. $f_x(x,y) = y - 2 = 0$ and $f_y(x,y) = 2y + x - 2 = 0 \Rightarrow x = -2,\ y = 2 \Rightarrow$ critical point is $(-2,2)$. $f_{xx}(-2,2) = 0$, $f_y(-2,2) = 2$, $f_{xy}(-2,2) = 1 \Rightarrow f_{xx}f_{yy} - f_{xy}^2 = -1 < 0$ saddle point.

7 $f_x(x,y) = 2y - 10x + 4 = 0$ and $f_y(x,y) = 2x - 4y = 0 \Rightarrow x = \frac{4}{9},\ y = \frac{2}{9} \Rightarrow$ critical point is $\left(\frac{4}{9},\frac{2}{9}\right)$. $f_{xx}\left(\frac{4}{9},\frac{2}{9}\right) = -10$, $f_{yy}\left(\frac{4}{9},\frac{2}{9}\right) = -4$, $f_{xy}\left(\frac{4}{9},\frac{2}{9}\right) = 2 \Rightarrow f_{xx}f_{yy} - f_{xy}^2 = 36 > 0$ and $f_{xx} < 0 \Rightarrow$ maximum (absolute). $f\left(\frac{4}{9},\frac{2}{9}\right) = -\frac{252}{81}$

8. $f_x(x,y) = 2y - 2x + 3 = 0$ and $f_y(x,y) = 2x - 4y = 0 \Rightarrow x = 3,\ y = \frac{3}{2} \Rightarrow$ critical point is $\left(3,\frac{3}{2}\right)$.
$f_{xx}\left(3,\frac{3}{2}\right) = -2,\ f_{yy}\left(3,\frac{3}{2}\right) = -4,\ f_{xy}\left(3,\frac{3}{2}\right) = 2 \Rightarrow f_{xx}f_{yy} - f_{xy}^2 = 4 > 0$ and $f_{xx} < 0 \Rightarrow$ maximum (absolute). $f\left(3,\frac{3}{2}\right) = \frac{17}{2}$

9. $f_x(x,y) = 2x - 4y = 0$ and $f_y(x,y) = -4x + 2y + 6 = 0 \Rightarrow x = 2,\ y = 1 \Rightarrow$ critical point is $(2,1)$.
$f_{xx}(2,1) = 2,\ f_{yy}(2,1) = 2,\ f_{xy}(2,1) = -4 \Rightarrow f_{xx}f_{yy} - f_{xy}^2 = -12 \Rightarrow$ saddle point.

10. $f_x(x,y) = 6x + 6y - 2 = 0$ and $f_y(x,y) = 6x + 14y + 4 = 0 \Rightarrow x = \frac{13}{12},\ y = -\frac{3}{4} \Rightarrow$ critical point is $\left(\frac{13}{12},-\frac{3}{4}\right)$.
$f_{xx}\left(\frac{13}{12},-\frac{3}{4}\right) = 6,\ f_{yy}\left(\frac{13}{12},-\frac{3}{4}\right) = 14,\ f_{xy}\left(\frac{13}{12},-\frac{3}{4}\right) = 6 \Rightarrow f_{xx}f_{yy} - f_{xy}^2 = 48 > 0$ and $f_{xx} > 0 \Rightarrow$ minimum (absolute). $f\left(\frac{13}{12},-\frac{3}{4}\right) = -\frac{161}{12}$

11. $f_x(x,y) = 4x + 3y - 5 = 0$ and $f_y(x,y) = 3x + 8y + 2 = 0 \Rightarrow x = 2,\ y = -1 \Rightarrow$ critical point is $(2,-1)$.
$f_{xx}(2,-1) = 4,\ f_{yy}(2,-1) = 8,\ f_{xy}(2,-1) = 3 \Rightarrow f_{xx}f_{yy} - f_{xy}^2 = 29 > 0$ and $f_{xx} > 0 \Rightarrow$
minimum (absolute). $f(2,-1) = -6$.

12. $f_x(x,y) = 8x - 6y - 20 = 0$ and $f_y(x,y) = -6x + 10y + 26 = 0 \Rightarrow x = 1,\ y = -2 \Rightarrow$ critical point is $(1,-2)$.
$f_{xx}(1,-2) = 8,\ f_{yy}(1,-2) = 10,\ f_{xy}(1,-2) = -6 \Rightarrow f_{xx}f_{yy} - f_{xy}^2 = 44 > 0$ and $f_{xx} > 0 \Rightarrow$ minimum (absolute)
$f(1,-2) = -36$.

13. $f_x(x,y) = 2x - 4y + 5 = 0$ and $f_y(x,y) = -4x + 2y - 2 = 0 \Rightarrow x = \frac{1}{6},\ y = \frac{4}{3} \Rightarrow$ critical point is $\left(\frac{1}{6},\frac{4}{3}\right)$.
$f_{xx}\left(\frac{1}{6},\frac{4}{3}\right) = 2,\ f_{yy}\left(\frac{1}{6},\frac{4}{3}\right) = 2,\ f_{xy}\left(\frac{1}{6},\frac{4}{3}\right) = -4 \Rightarrow f_{xx}f_{yy} - f_{xy}^2 = -12 \Rightarrow$ saddle point.

14. $f_x(x,y) = 2x - 2 = 0$ and $f_y(x,y) = -2y + 4 = 0 \Rightarrow x = 1,\ y = 2 \Rightarrow$ critical point is $(1,2)$. $f_{xx}(1,2) = 2$,
$f_{yy}(1,2) = -2,\ f_{xy}(1,2) = 0 \Rightarrow f_{xx}f_{yy} - f_{xy}^2 = -4 \Rightarrow$ saddle point.

15. $f_x(x,y) = 2x - 2y - 2 = 0$ and $f_y(x,y) = -2x + 4y + 2 = 0 \Rightarrow x = 1,\ y = 0 \Rightarrow$ critical point is $(1,0)$.
$f_{xx}(1,0) = 2,\ f_{yy}(1,0) = 4,\ f_{xy}(1,0) = -2 \Rightarrow f_{xx}f_{yy} - f_{xy}^2 = 4 > 0$ and $f_{xx} > 0 \Rightarrow$ minimum (absolute).
$f(1,0) = 0$

16. $f_x(x,y) = 2x + 2y = 0$ and $f_y(x,y) = 2x = 0 \Rightarrow x = 0,\ y = 0 \Rightarrow$ critical point is $(0,0)$. $f_{xx}(0,0) = 2,\ f_{yy}(0,0) = 0$,
$f_{xy}(0,0) = 2 \Rightarrow f_{xx}f_{yy} - f_{xy}^2 = -4 \Rightarrow$ saddle point.

17. $f_x(x,y) = 2 - 4x - 2y = 0$ and $f_y(x,y) = 2 - 2x - 2y = 0 \Rightarrow x = 0,\ y = 1 \Rightarrow$ critical point is $(0,1)$. $f_{xx}(0,1) =$
$-4,\ f_{yy}(0,1) = -2,\ f_{xy}(0,1) = -2 \Rightarrow f_{xx}f_{yy} - f_{xy}^2 = 4 > 0$ and $f_{xx} < 0 \Rightarrow$ maximum (absolute).
$f(0,1) = 4$

18. $f_x(x,y) = 2x + y + 1 = 0$ and $f_y(x,y) = x + 2y - 4 = 0 \Rightarrow x = -2,\ y = 3 \Rightarrow$ critical point is $(-2,3)$. $f_{xx}(-2,3) =$
$2,\ f_{yy}(-2,3) = 2,\ f_{xy}(-2,3) = 1 \Rightarrow f_{xx}f_{yy} - f_{xy}^2 = 3 > 0$ and $f_{xx} > 0 \Rightarrow$ minimum (absolute) .
$f(-2,3) = -2$

19. $f_x(x,y) = 3x^2 - 2y = 0$ and $f_y(x,y) = -3y^2 - 2x = 0 \Rightarrow x = 0,\ y = 0$ or $x = -\frac{2}{3},\ y = \frac{2}{3} \Rightarrow$ critical points are $(0,0)$ and $\left(-\frac{2}{3},\frac{2}{3}\right)$. For $(0,0)$: $f_{xx}(0,0) = 6x\big|_{(0,0)} = 0$, $f_{yy}(0,0) = -6y\big|_{(0,0)} = 0$, $f_{xy}(0,0) = -2 \Rightarrow$ $f_{xx}f_{yy} - f_{xy}^2 = -4 \Rightarrow$ saddle point. For $\left(-\frac{2}{3},\frac{2}{3}\right)$: $f_{xx}\left(-\frac{2}{3},\frac{2}{3}\right) = -4$, $f_{yy}\left(-\frac{2}{3},\frac{2}{3}\right) = -4$, $f_{xy}\left(-\frac{2}{3},\frac{2}{3}\right) = -2$ $\Rightarrow f_{xx}f_{yy} - f_{xy}^2 = 12 > 0$ and $f_{xx} < 0 \Rightarrow$ local maximum. $f\left(-\frac{2}{3},\frac{2}{3}\right) = \frac{170}{27}$

20. $f_x(x,y) = 3x^2 + 6x = 0$ and $f_y(x,y) = 3y^2 - 6y = 0 \Rightarrow x = 0,\ y = 0$ or $x = 0,\ y = 2$ or $x = -2,\ y = 0$ or $x = -2,\ y = 2 \Rightarrow$ critical points are $(0,0)$, $(0,2)$, $(-2,0)$, and $(-2,2)$. For $(0,0)$: $f_{xx}(0,0) = 6x + 6\big|_{(0,0)} = 6$, $f_{yy}(0,0) = 6y - 6\big|_{(0,0)}-6$, $f_{xy}(0,0) = 0 \Rightarrow f_{xx}f_{yy} - f_{xy}^2 = -36 \Rightarrow$ saddle point. For $(0,2)$: $f_{xx}(0,2) = 6$, $f_{yy}(0,2) = 6$, $f_{xy}(0,2) = 0 \Rightarrow f_{xx}f_{yy} - f_{xy}^2 = 36 > 0$ and $f_{xx} > 0 \Rightarrow$ local minimum. $f(0,2) = -12$. For $(-2,0)$: $f_{xx}(-2,0) = -6$, $f_{yy}(-2,0) = -6$, $f_{xy}(-2,0) = 0 \Rightarrow f_{xx}f_{yy} - f_{xy}^2 = 36 > 0$ and $f_{xx} < 0 \Rightarrow$ local maximum. $f(-2,0) = -4$.
For $(-2,2)$: $f_{xx}(-2,2) = -6$, $f_{yy}(-2,2) = 6$, $f_{xy}(-2,2) = 0 \Rightarrow f_{xx}f_{yy} - f_{xy}^2 = -36 \Rightarrow$ saddle point. No absolute extrema.

21. $f_x(x,y) = 12x - 6x^2 + 6y = 0$ and $f_y(x,y) = 6y + 6x = 0 \Rightarrow x = 0,\ y = 0$ or $x = 1,\ y = -1 \Rightarrow$ critical points are $(0,0)$ and $(1,-1)$. For $(0,0)$: $f_{xx}(0,0) = 12 - 12x\big|_{(0,0)} = 12$, $f_{yy}(0,0) = 6$, $f_{xy}(0,0) = 6 \Rightarrow$ $f_{xx}f_{yy} - f_{xy}^2 = 36 > 0$ and $f_{xx} > 0 \Rightarrow$ local minimum. $f(0,0) = 0$. For $(1,-1)$: $f_{xx}(1,-1) = 0$, $f_{yy}(1,-1) = 6$, $f_{xy}(1,-1) = 6 \Rightarrow f_{xx}f_{yy} - f_{xy}^2 = -36 \Rightarrow$ saddle point.

22. $f_x(x,y) = 27x^2 - 4y = 0$ and $f_y(x,y) = y^2 - 4x = 0 \Rightarrow x = 0,\ y = 0$ or $x = \frac{4}{9},\ y = \frac{4}{3} \Rightarrow$ critical points are $(0,0)$ and $\left(\frac{4}{9},\frac{4}{3}\right)$. For $(0,0)$: $f_{xx}(0,0) = 54x\big|_{(0,0)} = 0$, $f_{yy}(0,0) = 2y\big|_{(0,0)} = 0$, $f_{xy}(0,0) = -4 \Rightarrow$ $f_{xx}f_{yy} - f_{xy}^2 = -16 \Rightarrow$ saddle point. For $\left(\frac{4}{9},\frac{4}{3}\right)$: $f_{xx}\left(\frac{4}{9},\frac{4}{3}\right) = 24$, $f_{yy}\left(\frac{4}{9},\frac{4}{3}\right) = \frac{4}{3}$, $f_{xy}\left(\frac{4}{9},\frac{4}{3}\right) = -4 \Rightarrow$ $f_{xx}f_{yy} - f_{xy}^2 = 16 > 0$ and $f_{xx} > 0 \Rightarrow$ local minimum. $f\left(\frac{4}{9},\frac{4}{3}\right) = -\frac{8}{27}$

23. $f_x(x,y) = 3x^2 + 3y = 0$ and $f_y(x,y) = 3x + 3y^2 = 0 \Rightarrow x = 0,\ y = 0$ or $x = -1,\ y = -1 \Rightarrow$ critical points are $(0,0)$ and $(-1,-1)$. For $(0,0)$: $f_{xx}(0,0) = 6x\big|_{(0,0)} = 0$, $f_{yy}(0,0) = 6y\big|_{(0,0)} = 0$, $f_{xy}(0,0) = 3 \Rightarrow$ $f_{xx}f_{yy} - f_{xy}^2 = -9 \Rightarrow$ saddle point. For $(-1,-1)$: $f_{xx}(-1,-1) = -6$. $f_{yy}(-1,-1) = -6$, $f_{xy}(-1,-1) = 3 \Rightarrow$ $f_{xx}f_{yy} - f_{xy}^2 = 27 > 0$ and $f_{xx} < 0 \Rightarrow$ local maximum. $f(-1,-1) = 1$

24. $f_x(x,y) = 4y - 4x^3 = 0$ and $f_y(x,y) = 4x - 4y^3 = 0 \Rightarrow x = 0,\ y = 0$ or $x = 1,\ y = 1 \Rightarrow$ critical points are $(0,0)$ and $(1,1)$. For $(0,0)$: $f_{xx}(0,0) = -12x^2\big|_{(0,0)} = 0$, $f_{yy}(0,0) = -12y^2\big|_{(0,0)} = 0$, $f_{xy}(0,0) = 4 \Rightarrow$ $f_{xx}f_{yy} - f_{xy}^2 = -16 \Rightarrow$ saddle point. For $(1,1)$: $f_{xx}(1,1) = -12$, $f_{yy}(1,1) = -12$, $f_{xy}(1,1) = 4 \Rightarrow$ $f_{xx}f_{yy} - f_{xy}^2 = 128 > 0$ and $f_{xx} < 0 \Rightarrow$ absolute maximum

25.

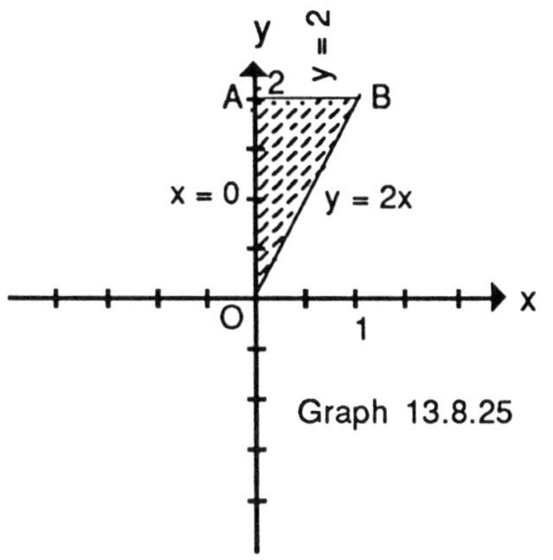

Graph 13.8.25

1. On OA, $f(x,y) = y^2 - 4y + 1 = f(0,y)$ on $0 \le y \le 2$. $y = 0 \Rightarrow f(0,0) = 1$. $y = 2 \Rightarrow f(0,2) = -3$. $f'(0,y) = 2y - 4 = 0 \Rightarrow y = 2 \Rightarrow f(0,2) = -3$.
2. On AB, $f(x,y) = 2x^2 - 4x - 3 = f(x,2)$ on $0 \le x \le 2$. $x = 0 \Rightarrow f(0,2) = -3$. $x = 2 \Rightarrow f(2,2) = -3$. $f'(x,2) = 4x - 4 = 0 \Rightarrow x = 1 \Rightarrow f(1,2) = -5$.
3. On OB, $f(x,y) = 6x^2 - 12x + 1 = f(x,2x)$ on $0 \le x \le 1$. Endpoint values have been found above. $f'(x,2x) = 12x - 12 = 0 \Rightarrow x = 1, y = 2 \Rightarrow (1,2)$, not an interior point of OB.
4. For interior points of the triangular region, $f_x(x,y) = 4x - 4 = 0$ and $f_y(x,y) = 2y - 4 = 0 \Rightarrow x = 1, y = 2 \Rightarrow (1,2)$, not an interior point of the region.

$\therefore$ absolute maximum is 1 at (0,0); absolute minimum is -5 at (1,2)

26.

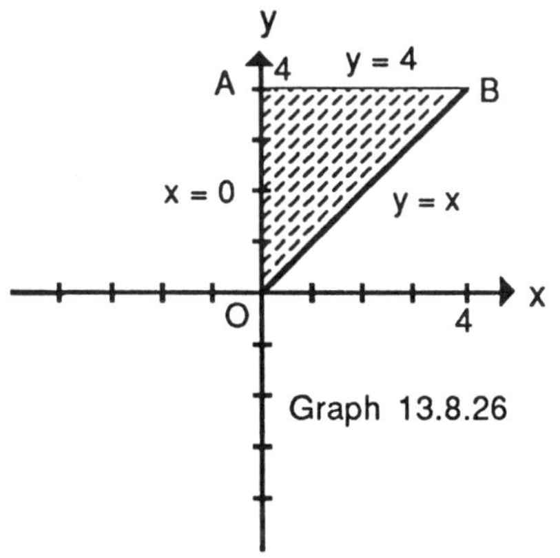

Graph 13.8.26

1. On OA, $D(x,y) = y^2 + 1 = D(0,y)$ on $0 \le y \le 4$. $D'(0,y) = 2y = 0 \Rightarrow y = 0$. $D(0,0) = 1$. $D(0,4) = 17$
2. On AB, $D(x,y) = x^2 - 4x + 17 = D(x,4)$ on $0 \le x \le 4$. $D(4,4) = 17$. $D'(x,4) = 2x - 4 = 0 \Rightarrow x = 2 \Rightarrow (2,4)$ is an interior point of OA. $D(2,4) = 13$
3. On OB, $D(x,y) = x^2 + 1 = D(x,x)$. Endpoint values have been found above. $D'(x,x) = 2x + 1 = 0 \Rightarrow x = -\frac{1}{2}, y = -\frac{1}{2}$, not an interior point of OB.
4. For interior points of the triangular region, $f_x(x,y) = 2x - y = 0$ and $f_y(x,y) = -x + 2y = 0 \Rightarrow x = 0, y = 0$, not an interior point of the region.

$\therefore$ absolute maximum is 17 at (0,4) and (4,4); absolute minimum is 1 at (0,0)

27.

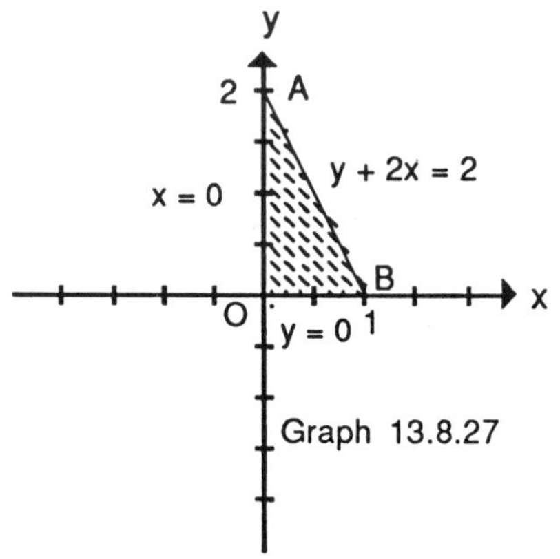

Graph 13.8.27

1. On OA, $f(x,y) = y^2 = f(0,y)$ on $0 \le y \le 2$. $f(0,0) = 0$. $f(0,2) = 4$. $f'(0,y) = 2y = 0 \Rightarrow y = 0, x = 0 \Rightarrow (0,0)$
2. On OB, $f(x,y) = x^2 = f(x,0)$ on $0 \le x \le 1$. $f(1,0) = 1$. $f'(x,0) = 2x = 0 \Rightarrow x = 0, y = 0 \Rightarrow (0,0)$
3. On AB, $f(x,y) = 5x^2 - 8x + 4 = f(x,-2x + 2)$ on $0 \le x \le 1$. $f(0,2) = 4$. $f'(x,-2x + 2) = 10x - 8 = 0 \Rightarrow x = \frac{4}{5}, y = \frac{2}{5}$. $f\left(\frac{4}{5}, \frac{2}{5}\right) = \frac{4}{5}$
4. For interior points of the triangular region, $f_x(x,y) = 2x = 0$ and $f_y(x,y) = 2y = 0 \Rightarrow x = 0, y = 0 \Rightarrow (0,0)$, not an interior point of the region.

$\therefore$ absolute maximum is 4 at (0,2); absolute minimum is 0 at (0,0)

28.

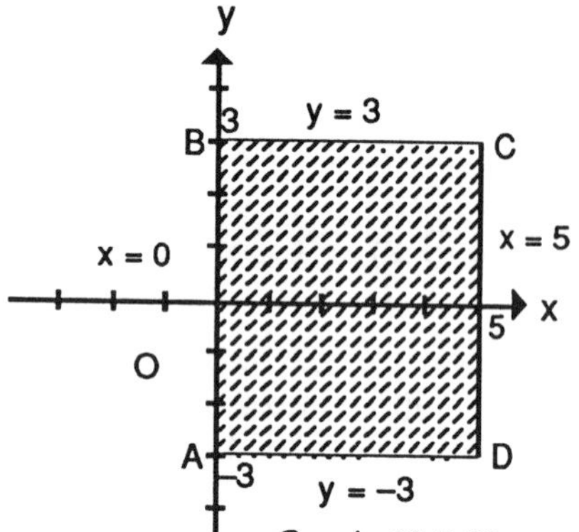

Graph 13.8.28

1. On AB, $T(x,y) = y^2 = T(0,y)$ on $-3 \le y \le 3$. $T(0,-3) = 9$. $T(0,3) = 9$. $T'(0,y) = 2y = 0 \Rightarrow y = 0,\ x = 0$. $T(0,0) = 0$
2. On BC, $T(x,y) = x^2 - 3x + 9 = T(x,3)$ on $0 \le x \le 5$. $T(5,0) = 19$. $T'(x,3) = 2x - 3 = 0 \Rightarrow x = \frac{3}{2},\ y = 3$. $T\left(\frac{3}{2},3\right) = \frac{27}{4}$
3. On CD, $T(x,y) = y^2 + 5y - 5 = T(5,y)$ on $-3 \le y \le 3$. $T(5,-3) = -11$. $T(5,3) = 19$. $T'(5,y) = 2y + 5 = 0 \Rightarrow y = -\frac{5}{2},\ x = 5$ $T\left(5,-\frac{5}{2}\right) = -\frac{45}{4}$
4. On AD, $T(x,y) = x^2 - 9x + 9 = T(x,-3)$ on $0 \le x \le 5$. $T'(x,-3) = 2x - 9 = 0 \Rightarrow x = \frac{9}{2},\ y = -3$. $T\left(\frac{9}{2},-3\right) = -\frac{45}{4}$
5. For interior points of the rectangular region, $T_x(x,y) = 2x + y - 6 = 0$ and $T_y(x,y) = x + 2y = 0 \Rightarrow x = 4$, $y = -2 \Rightarrow (-2,4)$ is an interior critical point. $T(4,-2) = -12$. $\therefore$ absolute maximum is 19 at (5,3); absolute minimum is –12 at (4,–2)

29.

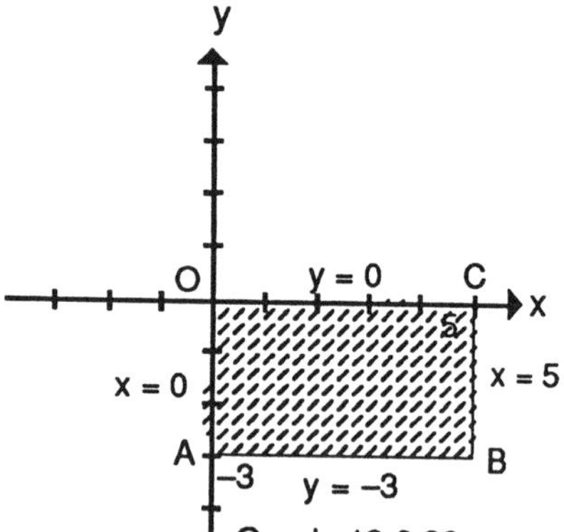

Graph 13.8.29

1. On OC, $T(x,y) = x^2 - 6x + 2 = T(x,0)$ on $0 \le x \le 5$. $T(0,0) = 2$. $T(5,0) = -3$. $T'(x,0) = 2x - 6 = 0 \Rightarrow x = 3,\ y = 0$. $T(3,0) = -7$
2. On CB, $T(x,y) = y^2 + 5y - 3 = T(5,y)$ on $-3 \le y \le 0$. $T(5,-3) = -9$ $T'(5,y) = 2y + 5 = 0 \Rightarrow y = -\frac{5}{2},\ x = 5$. $T\left(5,-\frac{5}{2}\right) = -\frac{37}{4}$
3. On AB, $T(x,y) = x^2 - 9x + 11 = T(x,-3)$ on $0 \le x \le 5$. $T(0,-3) = 11$. $T'(x,-3) = 2x - 9 = 0 \Rightarrow x = \frac{9}{2},\ y = -3$. $T\left(\frac{9}{2},-3\right) = -\frac{37}{4}$
4. On AO, $T(x,y) = y^2 + 2 = T(0,y)$ on $-3 \le y \le 0$. $T'(0,y) = 2y = 0 \Rightarrow y = 0,\ x = 0$. (0,0) not an interior point of AO.
5. For interior points of the rectangular region, $T_x(x,y) = 2x + y - 6 = 0$ and $T_y(x,y) = x + 2y = 0 \Rightarrow x = 4,\ y = -2 \Rightarrow T(4,-2) = -10$.

$\therefore$ absolute maximum is 11 at (0,–3); absolute minimum is –10 at (4,–2).

30.

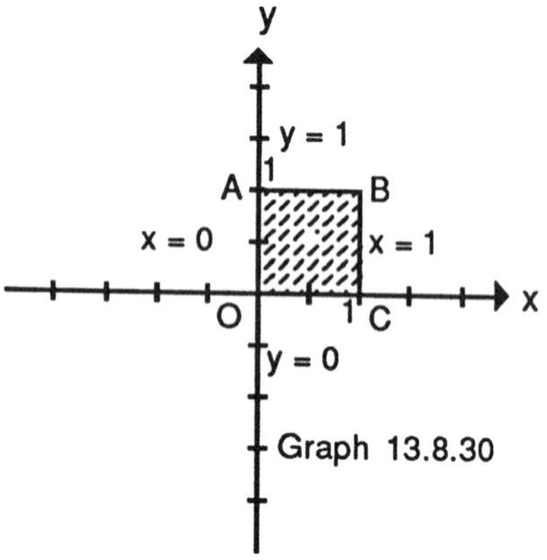

Graph 13.8.30

1. On OA, $f(x,y) = -24y^2 = f(0,y)$ on $0 \le y \le 1$. $f(0,0) = 0$. $f(0,1) = -24$. $f'(0,y) = -48y = 0 \Rightarrow y = 0,\ x = 0$. (0,0) not an interior point of OA.
2. On AB, $f(x,y) = 48x - 32x^3 - 24 = f(x,1)$ on $0 \le x \le 1$. $f(1,1) = -12$. $f'(x,1) = 48 - 96x^2 = 0 \Rightarrow x = \frac{1}{\sqrt{2}},\ y = 1$ or $x = -\frac{1}{\sqrt{2}},\ y = 1$ $\left(-\frac{1}{\sqrt{2}},1\right)$ not in the interior of AB. $f\left(\frac{1}{\sqrt{2}},1\right) = 16\sqrt{2} - 24$.
3. On BC, $f(x,y) = 48y - 32 - 24y^2 = f(1,y)$ on $0 \le y \le 1$. $f(1,0) = -32$. $f'(1,y) = 48 - 48y = 0 \Rightarrow y = 1,\ x = 1$. (1,1) not an interior point of BC.
4. On OC, $f(x,y) = -32x^3 = f(x,0)$ on $0 \le x \le 1$. $f'(x,0) = -96x^2 = 0$ $x = 0,\ y = 0$. (0,0) not an interior point of OC
5. For interior points of the rectangular region, $f_x(x,y) = 48y - 96x^2 = 0$ and $f_y(x,y) = 48x - 48y^2 = 0 \Rightarrow$ $x = 0,\ y = 0$ or $x = \frac{1}{\sqrt[3]{4}},\ y = \frac{1}{\sqrt[3]{2}}$. (0,0) is not an interior point of the region. $f\left(\frac{1}{\sqrt[3]{4}},\frac{1}{\sqrt[3]{2}}\right) = 8 - \frac{12}{\sqrt[3]{2}}$.

$\therefore$ absolute maximum is 0 at (0,0); absolute minimum is –32 at (1,0).

31.

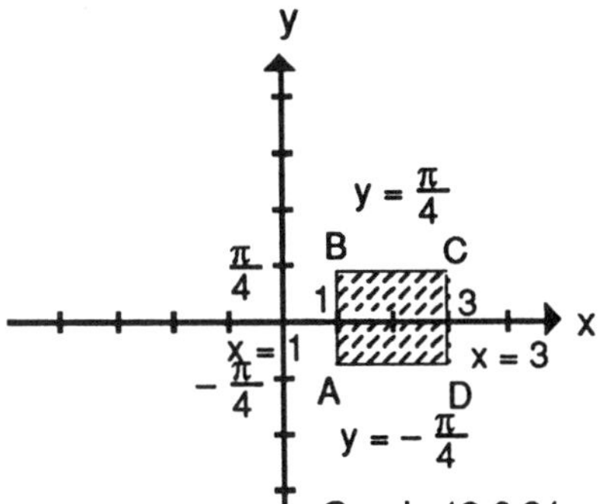

Graph 13.8.31

1. On AB, $f(x,y) = -3\cos y = f(1,y)$ on $-\frac{\pi}{4} \le y \le \frac{\pi}{4}$. $f\left(1,-\frac{\pi}{4}\right) = -\frac{3\sqrt{2}}{2}$. $f\left(1,\frac{\pi}{4}\right) = -\frac{3\sqrt{2}}{2}$. $f'(1,y) = -3\sin y = 0 \Rightarrow y = 0, x = 1$ $f(1,0) = -3$.
2. On CD, $f(x,y) = -3\cos y = f(3,y)$ on $-\frac{\pi}{4} \le y \le \frac{\pi}{4}$. $f\left(3,-\frac{\pi}{4}\right) = -\frac{3\sqrt{2}}{2}$. $f\left(3,\frac{\pi}{4}\right) = -\frac{3\sqrt{2}}{2}$. $f'(3,y) = 3\sin y = 0 \Rightarrow y = 0, x = 3$. $f(3,0) = -3$.
3. On BC, $f(x,y) = \frac{\sqrt{2}}{2}(x^2 - 4x) = f\left(x,\frac{\pi}{4}\right)$ on $1 \le x \le 3$. $f'\left(x,\frac{\pi}{4}\right) = \frac{\sqrt{2}}{2}(2x - 4) = 0 \Rightarrow x = 2, y = \frac{\pi}{4}$. $f\left(2,\frac{\pi}{4}\right) = -2\sqrt{2}$.
4. On AD, $f(x,y) = \frac{\sqrt{2}}{2}(x^2 - 4x) = f\left(x,-\frac{\pi}{4}\right)$ on $1 \le x \le 3$. $f'\left(x,-\frac{\pi}{4}\right) = \frac{\sqrt{2}}{2}(2x - 4) = 0 \Rightarrow x = 2, y = -\frac{\pi}{4}$. $f\left(2,-\frac{\pi}{4}\right) = -2\sqrt{2}$.
5. For interior points of the rectangular region, $f_x(x,y) = (2x - 4)\cos y = 0$ and $f_y(x,y) = -(x^2 - 4x)\sin y = 0 \Rightarrow x = 2, y = 0$. $f(2,0) = -4$.

$\therefore$ absolute maximum is $-\frac{3\sqrt{2}}{2}$ at $\left(3,-\frac{\pi}{4}\right)$, $\left(3,\frac{\pi}{4}\right)$, $\left(1,-\frac{\pi}{4}\right)$, and $\left(1,\frac{\pi}{4}\right)$; absolute minimum is -4 at $(2,0)$.

32.

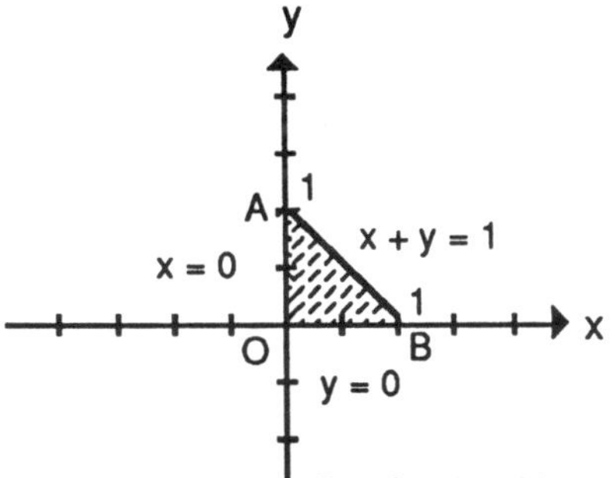

Graph 13.8.32

1. On OA, $f(x,y) = 2y + 1 = f(0,y)$ on $0 \le y \le 1$. $f(0,0) = 1$. $f(0,1) = 3$. $f'(0,y) = 2 \Rightarrow$ no interior critical points.
2. On OB, $f(x,y) = 4x + 1 = f(x,0)$ on $0 \le x \le 1$. $f(1,0) = 5$. $f'(x,0) = 4 \Rightarrow$ no interior critical points.
3. On AB, $f(x,y) = 8x^2 - 6x - 1 = f(x,-x + 1)$ on $0 \le x \le 1$. $f'(x,-x + 1) = 16x - 6 = 0 \Rightarrow x = \frac{3}{8}, y = \frac{5}{8}$. $f\left(\frac{3}{8},\frac{5}{8}\right) = -\frac{17}{8}$.
4. For interior points of the triangular region, $f_x(x,y) = 4 - 8y = 0$ and $f_y(x,y) = -8x + 2 = 0 \Rightarrow y = \frac{1}{2}, x = \frac{1}{4}$. $f\left(\frac{1}{4},\frac{1}{2}\right) = 2$.

$\therefore$ absolute maximum is 5 at $(1,0)$; absolute minimum is $-\frac{17}{8}$ at $\left(\frac{3}{8},\frac{5}{8}\right)$.

33. $T_x(x,y) = 2x - 1 = 0$ and $T_y(x,y) = 4y = 0 \Rightarrow x = \frac{1}{2}, y = 0 \Rightarrow T\left(\frac{1}{2},0\right) = -\frac{1}{4}$. On $x^2 + y^2 = 1$, $T(x,y) = -x^2 - x + 2$ on $-1 \le x \le 1$. $T(-1,0) = 2$, $T(1,0) = 0$. $T'(x,y) = -2x - 1 = 0 \Rightarrow x = -\frac{1}{2}, y = \pm\frac{\sqrt{3}}{2}$. $T\left(-\frac{1}{2},\frac{\sqrt{3}}{2}\right) = \frac{9}{4}$, $T\left(-\frac{1}{2},-\frac{\sqrt{3}}{2}\right) = \frac{9}{4}$. $\therefore$ hottest is $2\frac{1}{4}°$ at $\left(-\frac{1}{2},\frac{\sqrt{3}}{2}\right)$ and $\left(-\frac{1}{2},-\frac{\sqrt{3}}{2}\right)$; coldest is $-\frac{1}{4}°$ at $\left(\frac{1}{2},0\right)$.

34. $f_x(x,y) = y + 2 - \frac{2}{x} = 0$ and $f_y(x,y) = x - \frac{1}{y} = 0 \Rightarrow x = 0$ (does not give a point in the open first quadrant) or $x = \frac{1}{2}$, $y = 2$. $f_{xx}\left(\frac{1}{2},2\right) = \frac{2}{x^2}\Big|_{(1/2,2)} = 8$, $f_{yy}\left(\frac{1}{2},2\right) = \frac{1}{y^2}\Big|_{(1/2,2)} = \frac{1}{4}$, $f_{xy}\left(\frac{1}{2},2\right) = 1 \Rightarrow f_{xx}f_{yy} - f_{xy}^2 = 1 > 0$ and $f_{xx} > 0 \Rightarrow$ minimum.

35. a) $f_x(x,y) = 2x - 4y = 0$ and $f_y(x,y) = 2y - 4x = 0 \Rightarrow x = 0, y = 0$. $f_{xx}(0,0) = 2$, $f_{yy}(0,0) = -4$, $f_{xy}(0,0) = -4$ $\Rightarrow f_{xx}f_{yy} - f_{xy}^2 = -24 \Rightarrow$ saddle point.

b) $f_x(x,y) = 2x - 2 = 0$ and $f_y(x,y) = 2y - 4 = 0 \Rightarrow x = 1, y = 2$. $f_{xx}(1,2) = 2$, $f_{yy}(1,2) = 2$, $f_{xy}(1,2) = 0 \Rightarrow$ $f_{xx}f_{yy} - f_{xy}^2 = 4 > 0$ and $f_{xx} > 0 \Rightarrow$ local minimum at (1,2).

c) $f_x(x,y) = 9x^2 - 9 = 0$ and $f_y(x,y) = 2y + 4 = 0 \Rightarrow x = \pm 1, y = -2$. For (1,–2), $f_{xx}(1,-2) = 18x\big|_{(1,-2)} =$ 18, $f_{yy}(1,-2) = 2$, $f_{xy}(1,-2) = 0 \Rightarrow f_{xx}f_{yy} - f_{xy}^2 = 36 > 0$ and $f_{xx} > 0 \Rightarrow$ local minimum at (1,–2). For (–1,–2), $f_{xx}(-1,-2) = -18$, $f_{yy}(-1,-2) = 2$, $f_{xy}(-1,-2) = 0 \Rightarrow f_{xx}f_{yy} - f_{xy}^2 = -36 \Rightarrow$ saddle point.

36. a) Minimum at (0,0) since $f(x,y) > 0$ for all other (x,y).

b) Maximum of 1 at (0,0) since $f(x,y) < 1$ for all other (x,y).

c) Neither since $f(x,y) < 0$ for $x < 0$ and $f(x,y) > 0$ for $x > 0$.

d) Neither since $f(x,y) < 0$ for $x < 0, y > 0$ and $f(x,y) > 0$ for $x > 0, y > 0$.

e) Neither since $f(x,y) < 0$ for $x < 0, y > 0$ and $f(x,y) > 0$ for $x > 0, y > 0$.

f) Minimum since $f(x,y) > 0$ for all other (x,y).

37. a) $x = 2\cos t$, $y = 2\sin t \Rightarrow f(t) = 4\cos t \sin t \Rightarrow \frac{df}{dt} = y(-2\sin t) + x(2\cos t) = -4\sin^2 t + 4\cos^2 t$. $\frac{df}{dt} = 0 \Rightarrow 4\cos^2 t - 4\sin^2 t = 0 \Rightarrow \cos t = \sin t$ or $\cos t = -\sin t$

i) On the quarter circle $x^2 + y^2 = 4$ in the first quadrant, $\frac{df}{dt} = 0$ at $t = \frac{\pi}{4}$. $f(0) = 0$, $f\left(\frac{\pi}{2}\right) = 0$, $f\left(\frac{\pi}{4}\right) = 2$

$\therefore$ absolute minimum is 0 at $t = 0, \frac{\pi}{2}$; absolute maximum is 2 at $t = \frac{\pi}{4}$.

ii) On the half circle $x^2 + y^2 = 4$, $y \geq 0$, $\frac{df}{dt} = 0 \Rightarrow t = \frac{\pi}{4}, \frac{3\pi}{4}$. $f(0) = 0$, $f\left(\frac{\pi}{4}\right) = 2$, $f\left(\frac{3\pi}{4}\right) = -2$, $f(\pi) = 0$

$\therefore$ absolute minimum is –2 at $t = \frac{3\pi}{4}$; absolute maximum is 2 at $t = \frac{\pi}{4}$.

iii) On the full circle $x^2 + y^2 = 4$, $\frac{df}{dt} = 0 \Rightarrow t = \frac{\pi}{4}, \frac{3\pi}{4}, \frac{5\pi}{4}, \frac{7\pi}{4}$. $f(0) = 0$, $f\left(\frac{\pi}{4}\right) = f\left(\frac{5\pi}{4}\right) = 2$, $f\left(\frac{3\pi}{4}\right) = f\left(\frac{7\pi}{4}\right) = -2$, $f(2\pi) = 0$. $\therefore$ absolute minimum is –2 at $t = \frac{3\pi}{4}, \frac{7\pi}{4}$; absolute maximum is 2 at $t = \frac{\pi}{4}, \frac{5\pi}{4}$.

b) $x = 2\cos t$, $y = 2\sin t \Rightarrow f(t) = 2\cos t + 2\sin t \Rightarrow \frac{df}{dt} = -2\sin t + 2\cos t$. $\frac{df}{dt} = 0 \Rightarrow \cos t = \sin t$

i) On $0 \leq t \leq \frac{\pi}{2}$, $f(0) = 2$, $f\left(\frac{\pi}{2}\right) = 2$. $\frac{df}{dt} = 0 \Rightarrow t = \frac{\pi}{4} \Rightarrow f\left(\frac{\pi}{4}\right) = 2\sqrt{2}$. $\therefore$ absolute minimum is 2 at $t = 0, \frac{\pi}{2}$; absolute maximum is $2\sqrt{2}$ at $t = \frac{\pi}{4}$.

37. (Continued)

ii) On $0 \le t \le \pi$, $f(0) = 2$, $f(\pi) = -2$. $\frac{df}{dt} = 0 \Rightarrow t = \frac{\pi}{4}, \frac{3\pi}{4} \Rightarrow f\left(\frac{\pi}{4}\right) = 2\sqrt{2}$, $f\left(\frac{3\pi}{4}\right) = 0$ $\therefore$ absolute minimum is -2 at $t = \pi$; absolute maximum is $2\sqrt{2}$ at $t = \frac{\pi}{4}$.

iii) On $0 \le t \le 2\pi$, $f(0) = 2$, $f(2\pi) = 2$. $\frac{df}{dt} = 0 \Rightarrow t = \frac{\pi}{4}, \frac{5\pi}{4}$. $f\left(\frac{\pi}{4}\right) = 2\sqrt{2}$, $f\left(\frac{5\pi}{4}\right) = -2\sqrt{2}$. $\therefore$ absolute minimum is $-2\sqrt{2}$ at $t = \frac{5\pi}{4}$; absolute maximum is $2\sqrt{2}$ at $t = \frac{\pi}{4}$.

c) $x = 2 \cos t$, $y = 2 \sin t \Rightarrow f(t) = 8 \cos^2 t + \sin^2 t = 7 \cos^2 t + 1 \Rightarrow \frac{df}{dt} = -8 \cos t \sin t$. $\frac{df}{dt} = 0 \Rightarrow$ $\cos t = 0$ or $\sin t = 0$

i) On $0 \le t \le \frac{\pi}{2}$, $f(0) = 8$, $f\left(\frac{\pi}{2}\right) = 1$. $\frac{df}{dt} = 0 \Rightarrow t = 0, \frac{\pi}{2}$. $\therefore$ absolute minimum is 1 at $t = \frac{\pi}{2}$; absolute maximum is 8 at $t = 0$.

ii) On $0 \le t \le \pi$, $f(0) = 8$, $f(\pi) = 8$. $\frac{df}{dt} = 0 \Rightarrow t = 0, \pi, \frac{\pi}{2}$. $f\left(\frac{\pi}{2}\right) = 1$. $\therefore$ absolute minimum is 1 at $t = \frac{\pi}{2}$; absolute maximum is 8 at $t = 0, \pi$.

iii) On $0 \le t \le 2\pi$, $f(0) = 8$, $f(2\pi) = 8$. $\frac{df}{dt} = 0 \Rightarrow t = 0, \frac{\pi}{2}, \pi, \frac{3\pi}{2}, 2\pi \Rightarrow f\left(\frac{\pi}{2}\right) = 1$, $f(\pi) = 8$, $f\left(\frac{3\pi}{2}\right) = 1$. $\therefore$ absolute minimum is 1 at $t = \frac{\pi}{2}, \frac{3\pi}{2}$; absolute maximum is 8 at $t = 0, \pi, 2\pi$.

38. a) $x = 2t$, $y = t + 1 \Rightarrow f(t) = 2t^2 + 2t \Rightarrow \frac{df}{dt} = 4t + 2$. $\frac{df}{dt} = 0 \Rightarrow t = -\frac{1}{2}$. $f''(t) = 4 \Rightarrow f\left(-\frac{1}{2}\right)$ is a minimum. $f\left(-\frac{1}{2}\right) = -\frac{1}{2}$ is an absolute minimum since $f(t)$ is an upward parabola. No absolute maximum.

b) $x = 2t$, $y = t + 1$ on $-1 \le t \le 0 \Rightarrow t = -\frac{1}{2}$ is a critical number on the interval (see part a) above). $f\left(-\frac{1}{2}\right) = -\frac{1}{2}$. $f(0) = 0$, $f(-1) = 0 \Rightarrow$ absolute minimum is $-\frac{1}{2}$ at $t = -\frac{1}{2}$; absolute maximum is 0 at $t = 0, -1$.

c) $x = 2t$, $y = t + 1$ on $0 \le t \le 1 \Rightarrow$ no critical numbers on the interval (see part a) above). $f(0) = 0$, $f(1) = 4 \Rightarrow$ absolute minimum is 0 at $t = 0$; absolute maximum is 4 at $t = 1$.

39. a) $x = 3 \cos t$, $y = 2 \sin t \Rightarrow f(t) = 9 + 3 \sin^2 t \Rightarrow \frac{df}{dt} = 6 \sin t \cos t$. $\frac{df}{dt} = 0 \Rightarrow \sin t = 0$ or $\cos t = 0$

i) On the quarter ellipse, $\frac{x^2}{9} + \frac{y^2}{4} = 1$, in the first quadrant, $f(0) = 9$, $f\left(\frac{\pi}{2}\right) = 12$. $\frac{df}{dt} = 0 \Rightarrow t = 0, \frac{\pi}{2}$ $\therefore$ absolute maximum is 12 at $t = \frac{\pi}{2}$; absolute minimum is 9 at $t = 0$.

ii) On the half ellipse, $\frac{x^2}{9} + \frac{y^2}{4} = 1$, $y \ge 0$, $f(0) = 9$, $f(\pi) = 9$. $\frac{df}{dt} = 0 \Rightarrow t = 0, \frac{\pi}{2}, \pi$. $f\left(\frac{\pi}{2}\right) = 12$. $\therefore$ absolute minimum is 9 at $t = 0, \pi$; absolute maximum is 12 at $t = \frac{\pi}{2}$.

39. (Continued)

iii) On the full ellipse, $\frac{x^2}{9} + \frac{y^2}{4} = 1$, $f(0) = 9$, $f(2\pi) = 9$. $\frac{df}{dt} = 0 \Rightarrow t = 0, \frac{\pi}{2}, \pi, \frac{3\pi}{2}, 2\pi$. $f\left(\frac{\pi}{2}\right) = 12$, $f(\pi) = 9$, $f\left(\frac{3\pi}{2}\right) = 12$. $\therefore$ absolute maximum is 12 at $t = \frac{\pi}{2}, \frac{3\pi}{2}$; absolute minimum is 9 at $t = 0, \pi, 2\pi$.

b) $x = 3\cos t$, $y = 2\sin t \Rightarrow f(t) = 6\cos t + 6\sin t \Rightarrow \frac{df}{dt} = -6\sin t + 6\cos t$. $\frac{df}{dt} = 0 \Rightarrow \cos t = \sin t$.

i) On the quarter ellipse, $\frac{x^2}{9} + \frac{y^2}{4} = 1$, $f(0) = 6$, $f\left(\frac{\pi}{2}\right) = 6$. $\frac{df}{dt} = 0 \Rightarrow t = \frac{\pi}{4} \Rightarrow f\left(\frac{\pi}{4}\right) = 6\sqrt{2} \Rightarrow$ absolute maximum is $6\sqrt{2}$ at $t = \frac{\pi}{4}$; absolute minimum is 6 at $t = 0, \frac{\pi}{2}$.

ii) On the half ellipse, $\frac{x^2}{9} + \frac{y^2}{4} = 1$, $y \ge 0$, $f(0) = 6$, $f(\pi) = -6$. $\frac{df}{dt} = 0 \Rightarrow t = \frac{\pi}{4} \Rightarrow f\left(\frac{\pi}{4}\right) = 6\sqrt{2} \Rightarrow$ absolute maximum is $6\sqrt{2}$; absolute minimum is -6 at $t = \pi$.

iii) On the full ellipse, $\frac{x^2}{9} + \frac{y^2}{4} = 1$, $f(0) = 6$, $f(2\pi) = 6$. $\frac{df}{dt} = 0 \Rightarrow t = \frac{\pi}{4}, \frac{5\pi}{4} \Rightarrow f\left(\frac{\pi}{4}\right) = 6\sqrt{2}$, $f\left(\frac{5\pi}{4}\right) = -6\sqrt{2} \Rightarrow$ absolute maximum is $6\sqrt{2}$ at $t = \frac{\pi}{4}$; absolute minimum is $-6\sqrt{2}$ at $t = \frac{5\pi}{4}$.

40. $x = \cos t$, $y = \sin t \Rightarrow f(t) = \cos t \sin t \Rightarrow \frac{df}{dt} = -\sin^2 t + \cos^2 t$. On the ellipse $x^2 + 4y^2 = 8$, $0 \le t \le 2\pi$, $\frac{df}{dt} = 0 \Rightarrow -\sin^2 t + \cos^2 t = 0 \Rightarrow \cos t = \sin t$ or $\cos t = -\sin t \Rightarrow t = \frac{\pi}{4}, \frac{5\pi}{4}$ or $t = \frac{3\pi}{4}, \frac{7\pi}{4}$.
$f\left(\frac{\pi}{4}\right) = f\left(\frac{5\pi}{4}\right) = \frac{1}{2}$, $f\left(\frac{3\pi}{4}\right) = f\left(\frac{7\pi}{4}\right) = -\frac{1}{2}$, $f(0) = f(2\pi) = 0 \Rightarrow$ absolute maximum is $\frac{1}{2}$ t at $= \frac{\pi}{4}, \frac{5\pi}{4}$; absolute minimum is $-\frac{1}{2}$ at $t = \frac{3\pi}{4}, \frac{7\pi}{4}$.

41.

k	x_k	y_k	x_k^2	$x_k y_k$
1	−1	2	1	−2
2	0	1	0	0
3	3	−4	9	−12
Σ	2	−1	10	−14

$m = \frac{2(-1) - 3(-14)}{2^2 - 3(10)} \approx -1.5$, $b = \frac{1}{3}(-1 - (-1.5)2) \approx 0.7$

$\therefore$ $y = -1.5x + 0.7$. $y\big|_{x=4} = -5.3$

42.

k	x_k	y_k	x_k^2	$x_k y_k$
1	−2	0	4	0
2	0	2	0	0
3	2	3	4	6
Σ	0	5	8	6

$m = \frac{0(5) - 3(6)}{0^2 - 3(8)} = \frac{3}{4}$, $b = \frac{1}{3}\left(5 - \frac{3}{4}(0)\right) = \frac{5}{3}$

$\therefore$ $y = \frac{3}{4}x + \frac{5}{3}$. $y\big|_{x=4} = \frac{14}{3}$

43.

k	x_k	y_k	x_k^2	$x_k y_k$
1	0	0	0	0
2	1	2	1	2
3	2	3	4	6
Σ	3	5	5	8

$m = \frac{3(5) - 3(8)}{3^2 - 3(5)} = 1.5$, $b = \frac{1}{3}(5 - 1.5(3)) \approx 0.2$

$\therefore$ $y = 1.5x + 0.2$. $y\big|_{x=4} = 6.2$

44.

k	x_k	y_k	x_k^2	$x_k y_k$
1	0	1	0	0
2	2	2	4	4
3	3	2	9	6
Σ	5	5	13	10

$m = \dfrac{5(5) - 3(10)}{5^2 - 3(13)} = \dfrac{5}{14}$, $b = \dfrac{1}{3}\left(5 - \dfrac{5}{14}(5)\right) = \dfrac{15}{14}$

$\therefore\ y = \dfrac{5}{14}x + \dfrac{15}{14}$. $y\big|_{x=4} = \dfrac{35}{14}$

45.

k	x_k	y_k	x_k^2	$x_k y_k$
1	12	5.27	144	63.24
2	18	5.68	324	102.24
3	24	6.25	576	150
4	30	7.21	900	216.3
5	36	8.20	1296	295.2
6	42	8.71	1764	365.82
Σ	162	41.32	5004	1192.8

$m = \dfrac{162(41.32) - 6(1192.8)}{162^2 - 6(5004)} \approx 0.122$,

$b = \dfrac{1}{6}(41.32 - (0.122)(162)) \approx 3.59$

$\therefore\ y = 0.122\,x + 3.59$

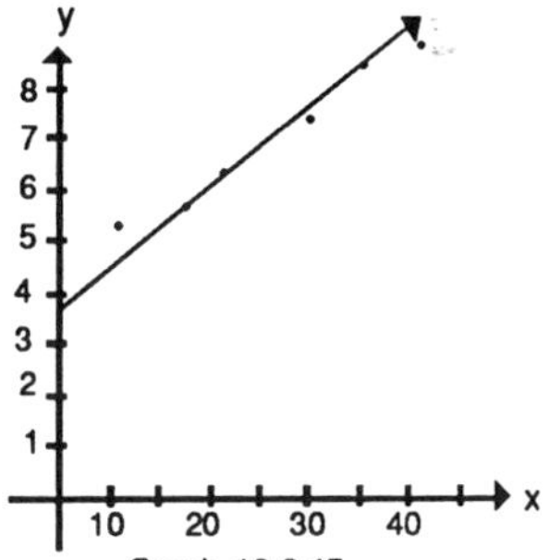

Graph 13.8.45

46.

k	$\left(\frac{1}{D^2}\right)_k$	F_k	$\left(\frac{1}{D^2}\right)_k^2$	$\left(\frac{1}{D^2}\right)_k F_k$
1	0.001	53	0.000001	0.053
2	0.0005	22	0.00000025	0.011
3	0.00025	14	0.000000062	0.0035
4	0.000125	3	0.000000015	0.000375
Σ	0.001875	92	0.000001328	0.067875

$m = \dfrac{(0.001875)(92) - 4(0.06875)}{(0.001875)^2 - 4(0.000001328)} \approx 57000$, $b = \dfrac{1}{4}(92 - 57000(0.001875)) \approx -3.72$.

$\therefore\ F = 57000\,\dfrac{1}{D^2} - 3.72$.

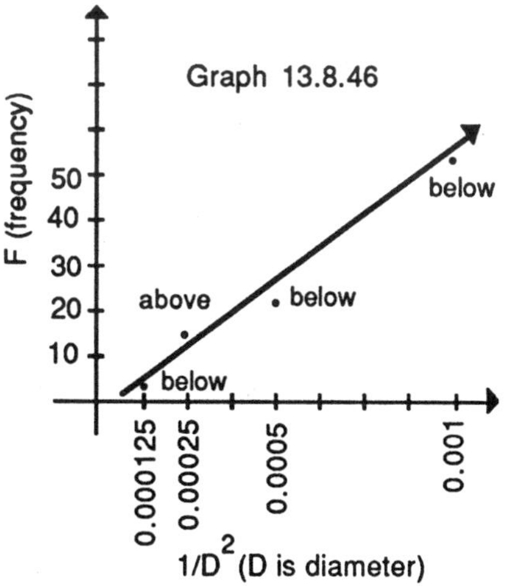

Graph 13.8.46

SECTION 13.9 LAGRANGE MULTIPLIERS

1. $\nabla f = y\,\mathbf{i} + x\,\mathbf{j}$, $\nabla g = 2x\,\mathbf{i} + 4y\,\mathbf{j}$. $\nabla f = \lambda\nabla g \Rightarrow y\,\mathbf{i} + x\,\mathbf{j} = \lambda(2x\,\mathbf{i} + 4y\,\mathbf{j}) \Rightarrow y = 2x\,\lambda$ and $x = 4y\,\lambda \Rightarrow$ $\lambda = \pm\frac{\sqrt{2}}{4}$ or $x = 0$. CASE 1: If $x = 0$, then $y = 0$ but $(0,0)$ not on the ellipse. $\therefore\ x \neq 0$.

   CASE 2: $x \neq 0 \Rightarrow \lambda = \pm\frac{\sqrt{2}}{4} \Rightarrow x = \pm\sqrt{2}\,y \Rightarrow (\pm\sqrt{2}\,y)^2 + 2y^2 = 1 \Rightarrow y = \pm\frac{1}{2}$. $\therefore$ f takes on its extreme values at $\left(\pm\sqrt{2}, \frac{1}{2}\right)$ and $\left(\pm\sqrt{2}, \frac{1}{2}\right) \Rightarrow$ the extreme values of f are $\pm\frac{\sqrt{2}}{2}$.

2. $\nabla f = y\,\mathbf{i} + x\,\mathbf{j}$, $\nabla g = 2x\,\mathbf{i} + 2y\,\mathbf{j}$. $\nabla f = \lambda\,\nabla g \Rightarrow y\,\mathbf{i} + x\,\mathbf{j} = \lambda(2x\,\mathbf{i} + 2y\,\mathbf{j}) \Rightarrow y = 2x\lambda$ and $x = 2y\lambda \Rightarrow x = 0$ or $\lambda = \pm\frac{1}{2}$. CASE 1: $x = 0 \Rightarrow y = 0$. But $(0,0)$ not on $x^2 + y^2 - 10 = 0$. $\therefore\ x \neq 0$. CASE 2: $x \neq 0 \Rightarrow$ $\lambda = \pm\frac{1}{2} \Rightarrow y = 2x\left(\pm\frac{1}{2}\right) = \pm x$. $\therefore\ x^2 + (\pm x)^2 - 10 = 0 \Rightarrow x = \pm\sqrt{5} \Rightarrow y = \pm\sqrt{5}$. $\therefore$ f takes on its extreme values at $(\pm\sqrt{5}, \sqrt{5})$ and $(\pm\sqrt{5}, -\sqrt{5}) \Rightarrow$ the extreme values of f are 5 and -5.

3. $\nabla f = -2x\,\mathbf{i} - 2y\,\mathbf{j}$, $\nabla g = \mathbf{i} + 3\,\mathbf{j}$. $\nabla f = \lambda\,\nabla g \Rightarrow -2x\,\mathbf{i} - 2y\,\mathbf{j} = \lambda(\mathbf{i} + 3\,\mathbf{j}) \Rightarrow x = -\frac{\lambda}{2}$ and $y = -\frac{3\lambda}{2} \Rightarrow \lambda = -2$ $\Rightarrow x = 1$ and $y = 3 \Rightarrow$ f takes on its extreme value at $(1,3) \Rightarrow$ the extreme value of f is 39.

4. Let (x,y) be on the parabola. Then the distance from the point to the line is $d(x,y) = \frac{|x - y + 1|}{\sqrt{2}}$.

   We want to minimize d subject to $y^2 = x \Rightarrow d(y) = \frac{|y^2 - y + 1|}{\sqrt{2}} = \frac{\left|\left(y - \frac{1}{2}\right)^2 + \frac{3}{4}\right|}{\sqrt{2}} = \frac{1}{\sqrt{2}}\left(y^2 - y + 1\right)$.

   $d'(y) = \frac{1}{\sqrt{2}}(2y - 1) \Rightarrow y = \frac{1}{2}$. $d''(y) = 2 \Rightarrow y = \frac{1}{2}$ yields a minimum. $y = \frac{1}{2} \Rightarrow x = \frac{1}{4} \Rightarrow$ the minimum distance is $d\left(\frac{1}{4}, \frac{1}{2}\right) = \frac{\left|\frac{1}{4} - \frac{1}{2} + 1\right|}{\sqrt{2}} = \frac{3\sqrt{2}}{8}$. Note: Lagrange multipliers not used.

5. $\nabla f = 2xy\,\mathbf{i} + x^2\,\mathbf{j}$, $\nabla g = \mathbf{i} + \mathbf{j}$. $\nabla f = \lambda\,\nabla g \Rightarrow 2xy\,\mathbf{i} + x^2\,\mathbf{j} = \lambda(\mathbf{i} + \mathbf{j}) \Rightarrow 2xy = \lambda$ and $x^2 = \lambda \Rightarrow 2xy = x^2 \Rightarrow$ $x = 0, y = 3$ or $x = 2, y = 1$. $\therefore$ f takes on its extreme values at $(0,3)$ and $(2,1)$. $\therefore$ the extreme values of f are $f(0,3) = 0$ and $f(2,1) = 4$.

6. Let $f(x,y) = x^2 + y^2$ be the square of the distance from the origin. The minimize f subject to $g(x,y) = x^2y - 2$. $\nabla f = 2x\,\mathbf{i} + 2y\,\mathbf{j}$, $\nabla g = 2xy\,\mathbf{i} + x^2\,\mathbf{j}$. $\nabla f = \lambda\,\nabla g \Rightarrow 2x = 2xy\,\lambda$ and $2y = x^2\,\lambda \Rightarrow \lambda = \frac{2y}{x^2} \Rightarrow$ $2x = 2xy\left(\frac{2y}{x^2}\right) \Rightarrow x^2 = 2y^2 \Rightarrow 2y^2y - 2 = 0 \Rightarrow y = 1 \Rightarrow x = \pm\sqrt{2}$. $\therefore$ $(\pm\sqrt{2}, 1)$ are the closest points.

7. a) $\nabla f = \mathbf{i} + \mathbf{j}$, $\nabla g = y\,\mathbf{i} + x\,\mathbf{j}$. $\nabla f = \lambda\,\nabla g \Rightarrow \mathbf{i} + \mathbf{j} = \lambda(y\,\mathbf{i} + x\,\mathbf{j}) \Rightarrow 1 = \lambda y$ and $1 = \lambda x \Rightarrow y = \frac{1}{\lambda}$ and $x = \frac{1}{\lambda}$ $\Rightarrow \frac{1}{\lambda^2} = 16 \Rightarrow \lambda = \pm 4$. Use $\lambda = 4$ since $x > 0$, $y > 0$. Then $x = 4$, $y = 4 \Rightarrow$ the minimum value is 8 at $x = 4$, $y = 4$.

   b) $\nabla f = y\,\mathbf{i} + x\,\mathbf{j}$, $\nabla g = \mathbf{i} + \mathbf{j}$. $\nabla f = \lambda\,\nabla g \Rightarrow y\,\mathbf{i} + x\,\mathbf{j} = \lambda(\mathbf{i} + \mathbf{j}) \Rightarrow y = \lambda = x \Rightarrow y = x \Rightarrow y + y = 16 \Rightarrow y = 8$ $\Rightarrow x = 8 \Rightarrow f(8,8) = 64$ is the maximum value.

8. Let $f(x,y) = x^2 + y^2$ be the square of the distance from the origin. Then $\nabla f = 2x\,\mathbf{i} + 2y\,\mathbf{j}$, $\nabla g = (2x + y)\,\mathbf{i} + (2y + x)\,\mathbf{j}$. $\nabla f = \lambda\,\nabla g \Rightarrow 2x = \lambda(2x + y)$ and $2y = \lambda(2y + x) \Rightarrow \frac{2y}{2y + x} = \lambda \Rightarrow 2x = \frac{2y}{2y + x}(2x + y) \Rightarrow y = \pm x$. $y = x \Rightarrow x^2 + x(x) + x^2 - 1 = 0 \Rightarrow x = \pm\frac{1}{\sqrt{3}} \Rightarrow y = \pm\frac{1}{\sqrt{3}}$. $y = -x \Rightarrow x^2 + x(-x) + (-x)^2 - 1 = 0 \Rightarrow x = \pm 1$. $x = 1 \Rightarrow y = -1$, $x = -1 \Rightarrow y = 1$. $f\left(\frac{1}{\sqrt{3}},\frac{1}{\sqrt{3}}\right) = \frac{2}{3} = f\left(-\frac{1}{\sqrt{3}},-\frac{1}{\sqrt{3}}\right)$. $f(1,-1) = 2 = f(-1,1)$. $\therefore$ $(1,-1)$ and $(-1,1)$ are the farthest; $\left(\frac{1}{\sqrt{3}},\frac{1}{\sqrt{3}}\right)$ and $\left(-\frac{1}{\sqrt{3}},-\frac{1}{\sqrt{3}}\right)$ are the closest.

9. $V = \pi r^2 h \Rightarrow 16\pi = \pi r^2 h \Rightarrow 16 = r^2 h \Rightarrow g(r,h) = r^2 h - 16$. $\nabla S = (2\pi h + 4\pi r)\,\mathbf{i} + 2\pi r\,\mathbf{j}$, $\nabla g = 2rh\,\mathbf{i} + r^2\,\mathbf{j}$. $\nabla S = \lambda\,\nabla g \Rightarrow (2\pi rh + 4\pi r)\,\mathbf{i} + 2\pi r\,\mathbf{j} = \lambda(2rh\,\mathbf{i} + r^2\,\mathbf{j}) \Rightarrow 2\pi h + 4\pi r = 2rh\lambda$ and $2\pi r = \lambda r^2 \Rightarrow 0 = \lambda r^2 - 2\pi r \Rightarrow r = 0$ or $\lambda = \frac{2\pi}{r}$. Now $r \neq 0 \Rightarrow \lambda = \frac{2\pi}{r} \Rightarrow 2\pi h + 4\pi r = 2rh\left(\frac{2\pi}{r}\right) \Rightarrow 2r = h \Rightarrow 16 = r^2(2r) \Rightarrow r = 2 \Rightarrow h = 4$. $\therefore$ $r = 2$ cm, $h = 4$ cm give the smallest surface area.

10. $g(x,y) = \frac{x^2}{16} + \frac{y^2}{9} - 1 = 0$. $A = (2x)(2y) = 4xy$. $\nabla A = 4y\,\mathbf{i} + 4x\,\mathbf{j}$, $\nabla g = \frac{x}{8}\,\mathbf{i} + \frac{2y}{9}\,\mathbf{j}$. $\nabla A = \lambda\,\nabla g \Rightarrow 4y\,\mathbf{i} + 4x\,\mathbf{j} = \lambda\left(\frac{x}{8}\,\mathbf{i} + \frac{2y}{9}\,\mathbf{j}\right) \Rightarrow 4y = \frac{x}{8}\lambda$ and $4x = \frac{2y}{9}\lambda \Rightarrow \lambda = \frac{32y}{x} \Rightarrow 4x = \frac{2y}{9}\left(\frac{32y}{x}\right) \Rightarrow y = \pm\frac{3}{4}x \Rightarrow \frac{x^2}{16} + \frac{\left(\frac{\pm 3}{4}x\right)^2}{9} = 1 \Rightarrow x^2 = 8 \Rightarrow x = \pm 2\sqrt{2}$. Use $x = 2\sqrt{2}$ since x represents distance. Then $y = \frac{3}{4}\left(2\sqrt{2}\right) = \frac{3\sqrt{2}}{2}$. $\therefore$ the length is $2x = 4\sqrt{2}$ and the width is $2y = 3\sqrt{2}$.

11. $\nabla T = (8x - 4y)\,\mathbf{i} + (-4x + 2y)\,\mathbf{j}$, $\nabla g = 2x\,\mathbf{i} + 2y\,\mathbf{j}$. $\nabla T = \lambda\,\nabla g \Rightarrow (8x - 4y)\,\mathbf{i} + (-4x + 2y)\,\mathbf{j} = \lambda(2x\,\mathbf{i} + 2y\,\mathbf{j}) \Rightarrow 8x - 4y = 2\lambda x$ and $-4x + 2y = 2\lambda y \Rightarrow y = \frac{-2x}{\lambda - 1}$, $\lambda \neq 1 \Rightarrow 8x - 4\left(\frac{-2x}{\lambda - 1}\right) = 2\lambda x \Rightarrow x = 0$ or $\lambda = 0$ or $\lambda = 5$. $x = 0 \Rightarrow y = 0$. But $(0,0)$ not on $x^2 + y^2 = 25$. $\therefore$ $x \neq 0 \Rightarrow \lambda = 0$ or $\lambda = 5$. $\lambda = 0 \Rightarrow y = 2x \Rightarrow x^2 + (2x)^2 = 25 \Rightarrow x = \pm\sqrt{5} \Rightarrow y = \pm 2\sqrt{5}$. $\lambda = 5 \Rightarrow y = \frac{-2x}{4} = -\frac{1}{2}x \Rightarrow x^2 + \left(-\frac{1}{2}x\right)^2 = 25 \Rightarrow x = \pm 2\sqrt{5}$. $x = 2\sqrt{5} \Rightarrow y = -\sqrt{5}$, $x = -2\sqrt{5} \Rightarrow y = \sqrt{5}$. $T(\sqrt{5},2\sqrt{5}) = 0° = T(-\sqrt{5},-2\sqrt{5})$, the minimum value; $T(2\sqrt{5},-\sqrt{5}) = 125° = T(-2\sqrt{5},\sqrt{5})$, the maximum value.

12. The surface area is given by $S = 4\pi r^2 + 2\pi rh$ subject to the constraint $V(r,h) = \frac{4}{3}\pi r^3 + \pi r^2 h = 8000$. $\nabla S = (8\pi r + 2\pi h)\,\mathbf{i} + 2\pi r\,\mathbf{j}$, $\nabla V = (4\pi r^2 + 2\pi rh)\,\mathbf{i} + \pi r^2\,\mathbf{j}$. $\nabla S = \lambda\,\nabla V \Rightarrow (8\pi r + 2\pi h)\,\mathbf{i} + 2\pi r\,\mathbf{j} = \lambda((4\pi r^2 + 2\pi rh)\,\mathbf{i} + \pi r^2\,\mathbf{j}) \Rightarrow 8\pi r + 2\pi h = \lambda(4\pi r^2 + 2\pi rh)$ and $2\pi r = \lambda\,\pi r^2 \Rightarrow r = 0$ or $2 = r\lambda$. $r \neq 0 \Rightarrow 2 = r\lambda \Rightarrow \lambda = \frac{2}{r} \Rightarrow 4r + h = \frac{2}{r}(2r^2 + rh) \Rightarrow h = 0 \Rightarrow$ The tank is a sphere, there is no cylindrical part.

13. $\nabla f = 2x\,\mathbf{i} + 2y\,\mathbf{j}$, $\nabla g = (2x - 2)\,\mathbf{i} + (2y - 4)\,\mathbf{j}$. $\nabla f = \lambda\,\nabla g \Rightarrow 2x\,\mathbf{i} + 2y\,\mathbf{j} = \lambda((2x - 2)\,\mathbf{i} + (2y - 4)\,\mathbf{j}) \Rightarrow 2x = \lambda(2x - 2)$ and $2y = \lambda(2y - 4) \Rightarrow x = \frac{\lambda}{\lambda - 1}$ and $y = \frac{2\lambda}{\lambda - 1}$, $\lambda \neq 1 \Rightarrow y = 2x \Rightarrow x^2 - 2x + (2x)^2 - 4(2x) = 0 \Rightarrow x = 0, y = 0$ or $x = 2, y = 4$. $\therefore$ $f(0,0) = 0$ is the minimum value, $f(2,4) = 20$ is the maximum value.

14. Let $f(x,y,z) = (x-1)^2 + (y-1)^2 + (z-1)^2$ be the square of the distance from (1,1,1). $\nabla f = 2(x-1)\,\mathbf{i} + 2(y-1)\,\mathbf{j} + 2(z-1)\,\mathbf{k}$, $\nabla g = \mathbf{i} + 2\mathbf{j} + 3\mathbf{k}$. $\nabla f = \lambda \nabla g \Rightarrow 2(x-1)\,\mathbf{i} + 2(y-1)\,\mathbf{j} + 2(z-1)\,\mathbf{k} = \lambda(\mathbf{i} + 2\mathbf{j} + 3\mathbf{k}) \Rightarrow 2(x-1) = \lambda$, $2(y-1) = 2\lambda$, $2(z-1) = 3\lambda \Rightarrow 2(y-1) = 2(2(x-1))$ and $2(z-1) = 3(2(x-1)) \Rightarrow x = \frac{y+1}{2} \Rightarrow z + 2 = 3\left(\frac{y+1}{2}\right) \Rightarrow z = \frac{3y-1}{2} \Rightarrow \frac{y+1}{2} + 2y + 3\left(\frac{3y-1}{2}\right) - 13 = 0 \Rightarrow$ $y = 2 \Rightarrow x = \frac{3}{2}$, $z = \frac{5}{2}$. $\therefore$ the point $\left(\frac{3}{2}, 2, \frac{5}{2}\right)$ is the closest.

15. $\nabla f = \mathbf{i} - 2\mathbf{j} + 5\mathbf{k}$, $\nabla g = 2x\,\mathbf{i} + 2y\,\mathbf{j} + 2z\,\mathbf{k}$. $\nabla f = \lambda \nabla g \Rightarrow \mathbf{i} - 2\mathbf{j} + 5\mathbf{k} = \lambda(2x\,\mathbf{i} + 2y\,\mathbf{j} + 2z\,\mathbf{k}) \Rightarrow 1 = 2x\lambda$, $-2 = 2y\lambda$, and $5 = 2z\lambda \Rightarrow x = \frac{1}{2\lambda}$, $y = -\frac{1}{\lambda} = -2x$, $z = \frac{5}{2\lambda} = 5x \Rightarrow x^2 + (-2x)^2 + (5x)^2 = 30 \Rightarrow x = \pm 1$. $x = 1 \Rightarrow y = -2$, $z = 5$. $x = -1 \Rightarrow y = 2$, $z = -5$. $f(1,-2,5) = 30$, the maximum value; $f(-1,2,-5) = -30$, the minimum value.

16. Let $f(x,y,z) = x^2 + y^2 + z^2$ be the square of the distance from the origin. Then $\nabla f = 2x\,\mathbf{i} + 2y\,\mathbf{j} + 2z\,\mathbf{k}$, $\nabla g = 2x\,\mathbf{i} - 2y\,\mathbf{j} - 2z\,\mathbf{k}$. $\nabla f = \lambda \nabla g \Rightarrow 2x\,\mathbf{i} + 2y\,\mathbf{j} + 2z\,\mathbf{k} = \lambda(2x\,\mathbf{i} - 2y\,\mathbf{j} - 2z\,\mathbf{k}) \Rightarrow 2x = 2x\lambda$, $2y = -2y\lambda$, and $2z = -2z\lambda \Rightarrow x = 0$ or $\lambda = 1$. CASE 1: $\lambda = 1 \Rightarrow 2y = -2y \Rightarrow y = 0$; $2z = -2z \Rightarrow z = 0 \Rightarrow x^2 - 1 = 0$ $\Rightarrow x = \pm 1$. CASE 2: $x = 0 \Rightarrow -y^2 - z^2 = 1$, which has no solution. $\therefore$ the points closest are $(\pm 1,0,0)$

17. Let $f(x,y,z) = x^2 + y^2 + z^2$ be the square of the distance to the origin. Then $\nabla f = 2x\,\mathbf{i} + 2y\,\mathbf{j} + 2z\,\mathbf{k}$, $\nabla g = y\,\mathbf{i} + x\,\mathbf{j} - \mathbf{k}$. $\nabla f = \lambda \nabla g \Rightarrow 2x\,\mathbf{i} + 2y\,\mathbf{j} + 2z\,\mathbf{k} = \lambda(y\,\mathbf{i} + x\,\mathbf{j} - \mathbf{k}) \Rightarrow 2x = \lambda y$, $2y = \lambda x$, and $2z = -\lambda$ $\Rightarrow x = \frac{\lambda y}{2} \Rightarrow 2y = \lambda\left(\frac{\lambda y}{2}\right) \Rightarrow y = 0$ or $\lambda = \pm 2$. $y = 0 \Rightarrow x = 0 \Rightarrow -z + 1 = 0 \Rightarrow z = 1$. $\lambda = 2 \Rightarrow x = y$, $z = -1 \Rightarrow x^2 - (-1) + 1 = 0 \Rightarrow x^2 + 2 = 0$, no solution. $\lambda = -2 \Rightarrow x = -y$, $z = 1 \Rightarrow (-y)y - 1 + 1 = 0 \Rightarrow$ $y = 0$, again. $\therefore$ (0,0,1) is the point on the surface closest to the origin.

18. Let $f(x,y,z) = x^2 + y^2 + z^2$ be the square of the distance to the origin. Then $\nabla f = 2x\,\mathbf{i} + 2y\,\mathbf{j} + 2z\,\mathbf{k}$, $\nabla g = -y\,\mathbf{i} - x\,\mathbf{j} + 2z\,\mathbf{k}$. $\nabla f = \lambda \nabla g \Rightarrow 2x\,\mathbf{i} + 2y\,\mathbf{j} + 2z\,\mathbf{k} = \lambda(-y\,\mathbf{i} - x\,\mathbf{j} + 2z\,\mathbf{k}) \Rightarrow 2x = -y\lambda$, $2y = -x\lambda$, and $2z = 2z\lambda \Rightarrow \lambda = 1$ or $z = 0$. CASE 1: $\lambda = 1 \Rightarrow 2x = -y$ and $2y = -x \Rightarrow y = 0 \Rightarrow x = 0 \Rightarrow z^2 - 4 = 0$ $\Rightarrow z = \pm 2$. CASE 2: $z = 0 \Rightarrow -xy - 4 = 0 \Rightarrow y = -\frac{4}{x}$. Then $2x = \frac{4}{x}\lambda \Rightarrow \lambda = \frac{x^2}{2}$ and $-\frac{8}{x} = -x\lambda \Rightarrow$ $-\frac{8}{x} = -x\left(\frac{x^2}{2}\right) \Rightarrow x^4 = 16 \Rightarrow x = \pm 2$. $x = 2 \Rightarrow y = -4$, $x = -2 \Rightarrow y = 4$. We get four points, (2,–4,0), (–2,4,0), (0,0,2), and (0,0,–2). But the points closest are (0,0,2) and (0,0,–2) since they are 2 units from the origin and the others are $\sqrt{20}$ away.

19. $\nabla f = \mathbf{i} + 2\mathbf{j} + 3\mathbf{k}$, $\nabla g = 2x\,\mathbf{i} + 2y\,\mathbf{j} + 2z\,\mathbf{k}$. $\nabla f = \lambda \nabla g \Rightarrow \mathbf{i} + 2\mathbf{j} + 3\mathbf{k} = \lambda(2x\,\mathbf{i} + 2y\,\mathbf{j} + 2z\,\mathbf{k}) \Rightarrow 1 = 2x\lambda$, $2 = 2y\lambda$, and $3 = 2z\lambda \Rightarrow x = \frac{1}{2\lambda}$, $y = \frac{1}{\lambda}$, and $z = \frac{3}{2\lambda} \Rightarrow y = 2x$ and $z = 3x \Rightarrow x^2 + (2x)^2 + (3x)^2 = 25 \Rightarrow$ $x = \pm\frac{5}{\sqrt{14}}$. $x = \frac{5}{\sqrt{14}} \Rightarrow y = \frac{10}{\sqrt{14}}$, $z = \frac{15}{\sqrt{14}}$. $x = -\frac{5}{\sqrt{14}} \Rightarrow y = -\frac{10}{\sqrt{14}}$, $z = -\frac{15}{\sqrt{14}}$. $f\left(\frac{5}{\sqrt{14}}, \frac{10}{\sqrt{14}}, \frac{15}{\sqrt{14}}\right) =$ $5\sqrt{14}$, the maximum value; $f\left(-\frac{5}{\sqrt{14}}, -\frac{10}{\sqrt{14}}, -\frac{15}{\sqrt{14}}\right) = -5\sqrt{14}$, the minimum value.

20. $\nabla f = 2x\,\mathbf{i} + 2y\,\mathbf{j} + 2z\,\mathbf{k}$, $\nabla g = \mathbf{i} + \mathbf{j} + \mathbf{k}$. $\nabla f = \lambda \nabla g \Rightarrow 2x\,\mathbf{i} + 2y\,\mathbf{j} + 2z\,\mathbf{k} = \lambda(\mathbf{i} + \mathbf{j} + \mathbf{k}) \Rightarrow 2x = \lambda$, $2y = \lambda$, and $2z = \lambda \Rightarrow x = y = z \Rightarrow x + x + x - 9 = 0 \Rightarrow x = 3 \Rightarrow y = 3$, $z = 3$. $\therefore$ Each of the numbers is 3.

21. $\nabla f = yz\,\mathbf{i} + xz\,\mathbf{j} + xy\,\mathbf{k}$, $\nabla g = \mathbf{i} + \mathbf{j} + 2z\,\mathbf{k}$. $\nabla f = \lambda\,\nabla g \Rightarrow yz\,\mathbf{i} + xz\,\mathbf{j} + xy\,\mathbf{k} = \lambda(\mathbf{i} + \mathbf{j} + 2z\,\mathbf{k}) \Rightarrow yz = \lambda$, $xz = \lambda$, and $xy = \lambda \Rightarrow yz = xz \Rightarrow z = 0$ or $y = x$. But $z > 0 \Rightarrow y = x \Rightarrow x^2 = 2z\lambda$ and $xz = \lambda$. Then $x^2 = 2z(xz)$ $\Rightarrow x = 0$ or $x = 2z^2$. But $x > 0 \Rightarrow x = 2z^2 \Rightarrow y = 2z^2 \Rightarrow 2z^2 + 2z^2 + z^2 = 16 \Rightarrow z = \pm\frac{4}{\sqrt{5}}$. Use $z = \frac{4}{\sqrt{5}}$ since $z > 0 \Rightarrow x = \frac{32}{5}$, $y = \frac{32}{5}$. $f\left(\frac{32}{5}, \frac{32}{5}, \frac{4}{\sqrt{5}}\right) = \frac{4096}{25\sqrt{5}}$

22. $\nabla T = 16x\,\mathbf{i} + 4z\,\mathbf{j} + (4y - 16)\,\mathbf{k}$, $\nabla g = 8x\,\mathbf{i} + 2y\,\mathbf{j} + 8z\,\mathbf{k}$. $\nabla T = \lambda\,\nabla g \Rightarrow 16x\,\mathbf{i} + 4z\,\mathbf{j} + (4y - 16)\,\mathbf{k} = \lambda(8x\,\mathbf{i} + 2y\,\mathbf{j} + 8z\,\mathbf{k}) \Rightarrow 16x = 8x\lambda$, $4z = 2y\lambda$, and $4y - 16 = 8z\lambda \Rightarrow \lambda = 2$ or $x = 0$. CASE 1: $\lambda = 2 \Rightarrow 4z = 2y(2) \Rightarrow z = y$. Then $4z - 16 = 16z \Rightarrow z = -\frac{4}{3} \Rightarrow y = -\frac{4}{3}$. Then $4x^2 + \left(-\frac{4}{3}\right)^2 + 4\left(-\frac{4}{3}\right)^2 = 16 \Rightarrow x = \pm\frac{4}{3}$. CASE 2: $x = 0 \Rightarrow \lambda = \frac{2z}{y} \Rightarrow 4y - 16 = 8z\left(\frac{2z}{y}\right) \Rightarrow y^2 - 4y = 4z^2 \Rightarrow 4(0)^2 + y^2 + y^2 - 4y - 16 = 0 \Rightarrow y = 4$ or $y = -2$. $y = 4 \Rightarrow 4z^2 = 4^2 - 4(4) \Rightarrow z = 0$. $y = -2 \Rightarrow 4z^2 = (-2)^2 - 4(-2) \Rightarrow z = \pm\sqrt{3}$ $T\left(\pm\frac{4}{3}, -\frac{4}{3}, -\frac{4}{3}\right) = 642\frac{2}{3}°$; $T(0,4,0) = 600°$; $T(0,-2,\sqrt{3}) = (600 - 24\sqrt{3})°$; $T(0,-2,-\sqrt{3}) = (600 + 24\sqrt{3})° \approx 641.6°$. $\therefore$ $\left(\pm\frac{4}{3}, -\frac{4}{3}, -\frac{4}{3}\right)$ are the hottest places.

23. $\nabla U = (y + 2)\,\mathbf{i} + x\,\mathbf{j}$, $\nabla g = 2\,\mathbf{i} + \mathbf{j}$. $\nabla U = \lambda\,\nabla g \Rightarrow (y + 2)\,\mathbf{i} + x\,\mathbf{j} = \lambda(2\,\mathbf{i} + \mathbf{j}) \Rightarrow y + 2 = 2\lambda$ and $x = \lambda \Rightarrow y + 2 = 2x \Rightarrow y = 2x - 2 \Rightarrow 2x + 2x - 2 = 30 \Rightarrow x = 8 \Rightarrow y = 14$. $\therefore$ $U(8,14) = \$128$, the maximum value of U under the constraint.

24. $\nabla M = (6 + z)\,\mathbf{i} - 2y\,\mathbf{j} + x\,\mathbf{k}$, $\nabla g = 2x\,\mathbf{i} + 2y\,\mathbf{j} + 2z\,\mathbf{k}$. $\nabla M = \lambda\,\nabla g \Rightarrow (6 + z)\,\mathbf{i} - 2y\,\mathbf{j} + x\,\mathbf{k} = \lambda(2x\,\mathbf{i} + 2y\,\mathbf{j} + 2z\,\mathbf{k}) \Rightarrow 6 + z = 2x\lambda$, $-2y = 2y\lambda$, $x = 2z\lambda \Rightarrow \lambda = -1$ or $y = 0$. CASE 1: $\lambda = -1 \Rightarrow 6 + z = -2x$ and $x = -2z \Rightarrow 6 + z = -2(-2z) \Rightarrow z = 2$, $x = -4$. Then $(-4)^2 + y^2 + 2^2 - 36 = 0 \Rightarrow y = \pm 4$. CASE 2: $y = 0$. Then $6 + z = 2x\lambda$ and $x = 2z\lambda \Rightarrow \lambda = \frac{x}{2z} \Rightarrow 6 + z = 2x\left(\frac{x}{2z}\right) \Rightarrow 6z + z^2 = x^2 \Rightarrow 6z + z^2 + 0^2 + z^2 = 36 \Rightarrow z = -6$ or $z = 3$. $z = -6 \Rightarrow x^2 = 0 \Rightarrow x = 0$. $z = 3 \Rightarrow x^2 = 9 \Rightarrow x = \pm 3$. $\therefore$ we have the points $(\pm 3,0,3)$, $(0,0,-6)$, and $(-4,2,\pm 4)$. $M(3,0,3) = 87$; $M(-3,0,3) = 33$; $M(0,0,-6) = 60$; $M(-4,2,4) = 16$; and $M(-4,2,-4) = 48$. $\therefore$ the weakest field will be at $(-4,2,4)$.

25. $\nabla f = \mathbf{i} + \mathbf{j}$, $\nabla g = y\,\mathbf{i} + x\,\mathbf{j}$. $\nabla f = \lambda\,\nabla g \Rightarrow \mathbf{i} + \mathbf{j} = \lambda(y\,\mathbf{i} + x\,\mathbf{j}) \Rightarrow 1 = y\lambda$ and $1 = x\lambda \Rightarrow y = x \Rightarrow y^2 = 16 \Rightarrow y = \pm 4 \Rightarrow x = \pm 4$. But as $x \to \infty$, $y \to 0$ and $f(x,y) \to \infty$; as $x \to -\infty$, $y \to 0$ and $f(x,y) \to -\infty$.

PRACTICE EXERCISES

1.

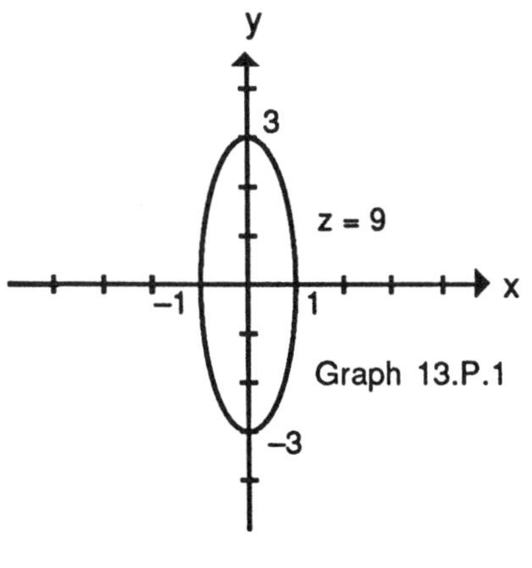

Graph 13.P.1

Domain: All points in the xy–plane
Range: $f(x,y) \geq 0$

Level curves are ellipses with major axis along the y–axis and minor axis along the x–axis.

2.

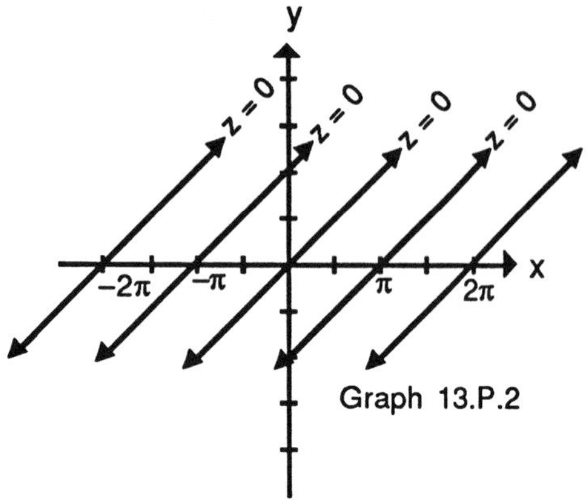

Graph 13.P.2

Domain: All points in the xy–plane
Range: $-1 \le f(x,y) \le 1$

Level curves are lines with m = 1.

3.

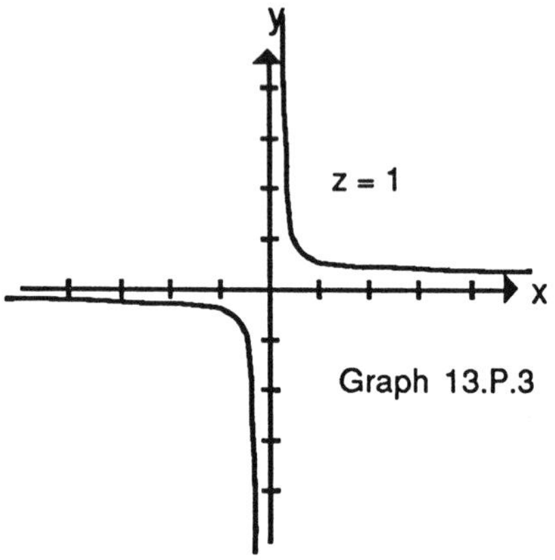

Graph 13.P.3

Domain: All (x,y) such that $x \neq 0$ or $y \neq 0$
Range: $f(x,y) \neq 0$

Level curves are hyperbolas rotated 45° or 135°.

4.

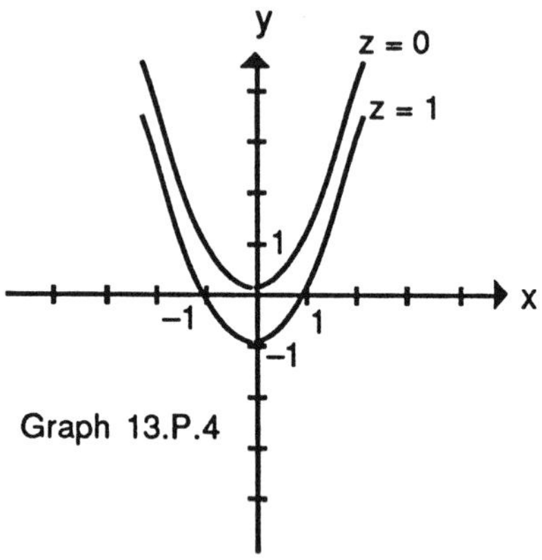

Graph 13.P.4

Domain: All (x,y) so that $x^2 - y \ge 0$
Range: $f(x,y) \ge 0$

Level curves are parabolas with vertex on y–axis.

5.

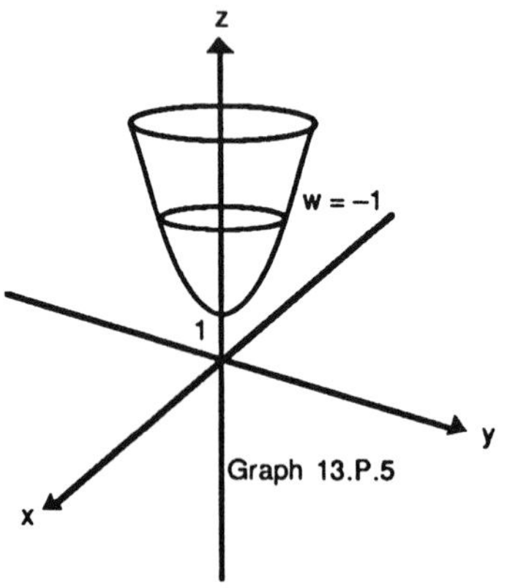

Graph 13.P.5

Domain: All (x,y,z) such (x,y,z ≠ (0,0,0)
Range: All Real Numbers

Level surfaces are paraboloids of revolution with the z–axis as axis.

6.

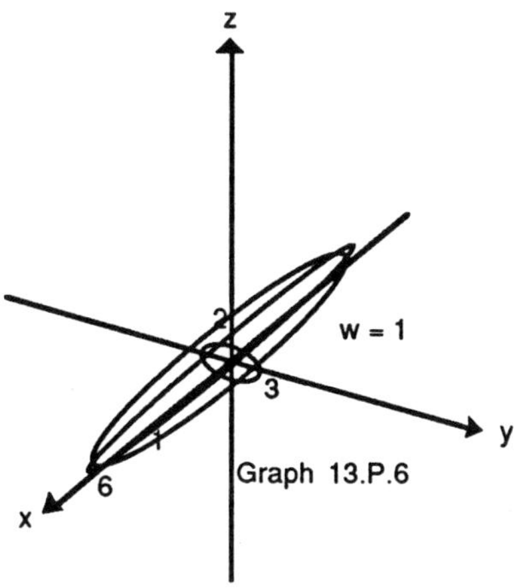

Graph 13.P.6

Domain: All points (x,y,z) in space
Range: $f(x,y,z) \geq 0$

Level surfaces are ellipsoids with center (0,0,0)

7.

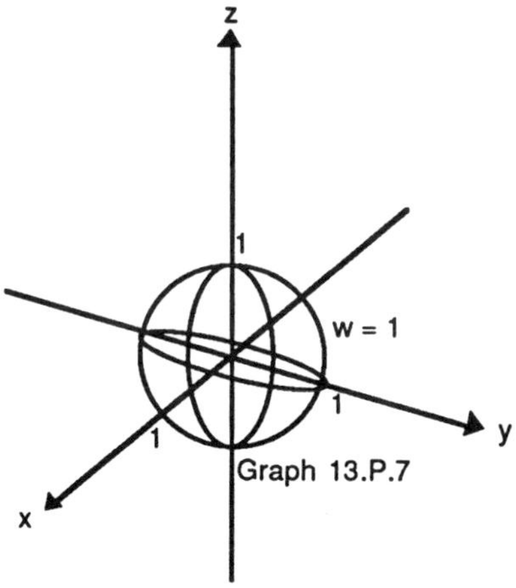

Graph 13.P.7

Domain: All (x,y,z) such that $(x,y,z) \neq (0,0,0)$
Range: $f(x,y,z) > 0$

Level surfaces are spheres with center (0,0,0) and radius $r > 0$.

8.

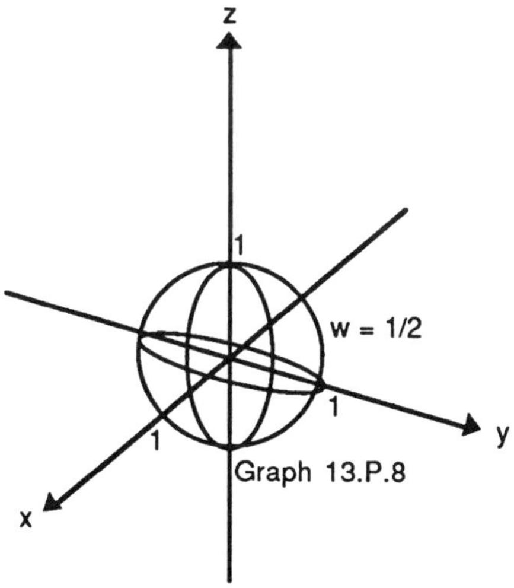

Graph 13.P.8

Domain: All points (x,y,z) in space
Range: $0 < f(x,y,z) \leq 1$

Level surfaces are spheres with center (0,0,0) and radius $r > 0$

9. $\lim_{(x,y)\to(\pi,\ln 2)} e^y \cos x = e^{\ln 2} \cos \pi = -2$

10. $\lim_{(x,y)\to(0,0)} \frac{2+y}{x+\cos y} = \frac{2+0}{0+\cos 0} = 2$

11. $\lim_{(x,y)\to(1,1)} \frac{x^2-y^2}{x-y} = \lim_{(x,y)\to(1,1)} (x+y) = 2$

12. $\lim_{P\to(1,-1,e)} \ln|x+y+z| = \ln|1+(-1)+e| = 1$

13. Let $y = kx^2$, $k \neq 1$. Then $\lim\limits_{\substack{(x,y)\to(0,0)\\ y\neq x^2}} \frac{y}{x^2 - y} = \lim\limits_{(x,kx^2)\to(0,0)} \frac{kx^2}{x^2 - kx^2} = \frac{k}{1 - k^2}$ which gives different limits for different values of k. ∴ the limit doesn't exist.

14. Let $y = kx$. Then $\lim\limits_{\substack{(x,y)\to(0,0)\\ xy\neq 0}} \frac{x^2 + y^2}{xy} = \lim\limits_{(x,kx)\to(0,0)} \frac{x^2 + (kx)^2}{x(kx)} = \frac{1 + k^2}{k}$ which gives different limits for different values of k. ∴ the limit does not exist.

15. $\frac{\partial g}{\partial r} = \cos\theta + \sin\theta$, $\frac{\partial g}{\partial \theta} = -r\sin\theta + r\cos\theta$

16. $f_x(x,y) = \frac{x - y}{x^2 + y^2}$, $f_y(x,y) = \frac{x + y}{x^2 + y^2}$

17. $\frac{\partial f}{\partial R_1} = -\frac{1}{R_1{}^2}$, $\frac{\partial f}{\partial R_2} = -\frac{1}{R_2{}^2}$, $\frac{\partial f}{\partial R_3} = -\frac{1}{R_3{}^2}$

18. $h_x(x,y,z) = 2\pi\cos(2\pi x + y - 3z)$, $h_y(x,y,z) = \cos(2\pi x + y - 3z)$, $h_z(x,y,z) = -3\cos(2\pi x + y - 3z)$

19. $\frac{\partial P}{\partial n} = \frac{RT}{V}$, $\frac{\partial P}{\partial R} = \frac{nT}{V}$, $\frac{\partial P}{\partial T} = \frac{nR}{V}$, $\frac{\partial P}{\partial V} = -\frac{nRT}{V^2}$

20. $f_r(r,l,T,d) = -\frac{1}{2r^2 l}\sqrt{\frac{T}{\pi d}}$, $f_l(r,l,T,d) = -\frac{1}{2rl^2}\sqrt{\frac{T}{\pi d}}$, $f_T(r,l,T,d) = \frac{1}{4\pi rld}\sqrt{\frac{T}{\pi d}}$, $f_d(r,l,T,d) = \frac{-T}{2\pi rld^2}\sqrt{\frac{\pi d}{T}}$

21. $\frac{\partial f}{\partial x} = \frac{1}{y}$, $\frac{\partial f}{\partial y} = 1 - \frac{x}{y^2} \Rightarrow \frac{\partial^2 f}{\partial x^2} = 0$, $\frac{\partial^2 f}{\partial y^2} = \frac{2x}{y^3}$, $\frac{\partial^2 f}{\partial y\partial x} = \frac{\partial^2 f}{\partial x\partial y} = -\frac{1}{y^2}$

22. $f_x(x,y) = e^x + y\cos x$, $f_y(x,y) = \sin x \Rightarrow f_{xx}(x,y) = e^x - y\sin x$, $f_{yy}(x,y) = 0$, $f_{xy}(x,y) = f_{yx}(x,y) = \cos x$

23. $\frac{\partial f}{\partial x} = 1 + y - 15x^2 + \frac{2x}{x^2 + 1}$, $\frac{\partial f}{\partial y} = x \Rightarrow \frac{\partial^2 f}{\partial x^2} = -30x + \frac{2 - 2x^2}{(x^2 + 1)^2}$, $\frac{\partial^2 f}{\partial y^2} = 0$, $\frac{\partial^2 f}{\partial y\partial x} = \frac{\partial^2 f}{\partial x\partial y} = 1$

24. $f_x(x,y) = -3y$, $f_y(x,y) = 2y - 3x - \sin y + 7e^y \Rightarrow f_{xx}(x,y) = 0$, $f_{yy}(x,y) = 2 - \cos y + 7e^y$, $f_{xy}(x,y) = f_{yx}(x,y) = -3$.

25. $\frac{\partial w}{\partial x} = y\cos(xy + \pi)$, $\frac{\partial w}{\partial y} = x\cos(xy + \pi)$, $\frac{dx}{dt} = e^t$, $\frac{dy}{dt} = \frac{1}{t + 1} \Rightarrow \frac{dw}{dt} = y\cos(xy + \pi)\,e^t + x\cos(xy + \pi)\left(\frac{1}{t + 1}\right)$ $= e^t\ln(t + 1)\cos(e^t\ln(t + 1) + \pi) + \frac{e^t}{t + 1}\cos(e^t\ln(t + 1) + \pi) \Rightarrow \left.\frac{dw}{dt}\right|_{t=0} = -1$

26. $\frac{\partial w}{\partial x} = e^y$, $\frac{\partial w}{\partial y} = xe^y + \sin z$, $\frac{\partial w}{\partial z} = y\cos z + \sin z$, $\frac{dx}{dt} = t^{-1/2}$, $\frac{dy}{dt} = 1 + \frac{1}{t}$, $\frac{dz}{dt} = \pi \Rightarrow \frac{dw}{dt} = t^{-1/2}e^y + \left(xe^y + \sin z\right)\left(1 + \frac{1}{t}\right) + (y\cos z + \sin z)\pi$. $t = 1 \Rightarrow x = 2$, $y = 0$, $z = \pi \Rightarrow \left.\frac{dw}{dt}\right|_{t=1} = 5$

27. $\frac{\partial w}{\partial x} = 2\cos(2x - y)$, $\frac{\partial w}{\partial y} = -\cos(2x - y)$, $\frac{\partial x}{\partial r} = 1$, $\frac{\partial x}{\partial s} = \cos s$, $\frac{\partial y}{\partial r} = s$, $\frac{\partial y}{\partial s} = r \Rightarrow \frac{\partial w}{\partial r} = 2\cos(2x - y)(1) + (-\cos(2x - y)(s)) = 2\cos(2r + 2\sin s - rs) - s\cos(2r + 2\sin s - rs) \Rightarrow \left.\frac{\partial w}{\partial r}\right|_{r=\pi,s=0} = 2$.

$\frac{\partial w}{\partial s} = 2\cos(2x - y)(\cos s) + (-\cos(2x - y)(r) = 2\cos(2r + 2\sin s - rs)(\cos s) - r\cos(2r + 2\sin s - rs) \Rightarrow$
$\left.\frac{\partial w}{\partial s}\right|_{r=\pi,s=0} = 2 - \pi$

28. $\frac{\partial w}{\partial u} = \frac{dw}{du}\frac{\partial x}{\partial u} = \left(\frac{x}{1+x^2} - \frac{1}{x^2+1}\right)2e^u \cos v$. $u = v = 0 \Rightarrow x = 2 \Rightarrow \frac{\partial w}{\partial u}\Big|_{u=v=0} = \frac{2}{5}$. $\frac{\partial w}{\partial v} = \frac{dw}{dx}\frac{\partial x}{\partial v} =$ $\left(\frac{x}{1+x^2} - \frac{1}{x^2+1}\right)(-2e^u \sin v)$. $\frac{\partial w}{\partial v}\Big|_{u=v=0} = 0$

29. $F_x = -1 - y\cos xy$, $F_y = -2y - x\cos xy$. $\frac{dy}{dx} = -\frac{F_x}{F_y} = -\frac{-1 - y\cos xy}{-2y - x\cos xy} = \frac{1 + y\cos xy}{-2y - x\cos xy} \Rightarrow$ $\frac{dy}{dx}\Big|_{(x,y)=(0,1)} = -1$

30. $F_x = 2y + e^{x+y}$, $F_y = 2x + e^{x+y} \Rightarrow \frac{dy}{dx} = -\frac{F_x}{F_y} = -\frac{2y + e^{x+y}}{2x + e^{x+y}} \Rightarrow \frac{dy}{dx}\Big|_{(0,\ln 2)} = -(\ln(2) + 1)$

31. $\frac{\partial f}{\partial x} = y + z$, $\frac{\partial f}{\partial y} = x + z$, $\frac{\partial f}{\partial z} = y + x$, $\frac{dx}{dt} = -\sin t$, $\frac{dy}{dt} = \cos t$, $\frac{dz}{dt} = -2\sin 2t \Rightarrow \frac{df}{dt} = -(y+z)\sin t + (x+z)\cos t$ $- 2(y+x)\sin 2t = -(\sin t + \cos 2t)\sin t + (\cos t + \cos 2t)\cos t - 2(\sin t + \cos t)\sin 2t \Rightarrow \frac{df}{dt}\Big|_{t=1} =$ $-(\sin 1 + \cos 2)\sin 1 + (\cos 1 + \cos 2)\cos 1 - 2(\sin 1 + \cos 1)\sin 2$

32. $\frac{\partial w}{\partial x} = \frac{dw}{ds}\frac{\partial s}{\partial x} = 5\frac{dw}{ds}$. $\frac{\partial w}{\partial y} = \frac{dw}{ds}\frac{\partial s}{\partial y} = (1)\frac{dw}{ds} = \frac{dw}{ds}$. $\therefore\ \frac{\partial w}{\partial x} - 5\frac{\partial w}{\partial y} = 5\frac{dw}{ds} - 5\left(\frac{dw}{ds}\right) = 0$

33. $\nabla f = (-\sin x\cos y)\,\mathbf{i} - (\cos x\sin y)\,\mathbf{j} \Rightarrow \nabla f\big|_{(\pi/4,\pi/4)} = -\frac{1}{2}\mathbf{i} - \frac{1}{2}\mathbf{j} \Rightarrow |\nabla f| = \sqrt{\left(-\frac{1}{2}\right)^2 + \left(-\frac{1}{2}\right)^2} = \frac{1}{\sqrt{2}}$ $\mathbf{u} = \frac{\nabla f}{|\nabla f|} = \frac{-\frac{1}{2}\mathbf{i} - \frac{1}{2}\mathbf{j}}{\frac{1}{\sqrt{2}}} = -\frac{\sqrt{2}}{2}\mathbf{i} - \frac{\sqrt{2}}{2}\mathbf{j}$. f increases most rapidly in the direction $\mathbf{u} = -\frac{\sqrt{2}}{2}\mathbf{i} - \frac{\sqrt{2}}{2}\mathbf{j}$; decreases most rapidly in the direction $-\mathbf{u} = \frac{\sqrt{2}}{2}\mathbf{i} + \frac{\sqrt{2}}{2}\mathbf{j}$. $\left(D_{\mathbf{u}}f\right)_{P_0} = \frac{\sqrt{2}}{2}$, $\left(D_{-\mathbf{u}}f\right)_{P_0} = -\frac{\sqrt{2}}{2}$. $\mathbf{u_1} = \frac{\mathbf{A}}{|\mathbf{A}|} = \frac{3\,\mathbf{i} + 4\,\mathbf{j}}{\sqrt{3^2+4^2}} = \frac{3}{5}\mathbf{i} + \frac{4}{5}\mathbf{j}$. $\left(D_{\mathbf{u_1}}f\right)_{P_0} = \nabla f \cdot \mathbf{u_1} = -\frac{7}{10}$.

34. $\nabla f = 2xe^{-2y}\,\mathbf{i} - 2x^2e^{-2y}\,\mathbf{j} \Rightarrow \nabla f\big|_{(1,0)} = 2\,\mathbf{i} - 2\,\mathbf{j} \Rightarrow |\nabla f| = \sqrt{2^2 + (-2)^2} = 2\sqrt{2}$. $\mathbf{u} = \frac{\nabla f}{|\nabla f|} = \frac{2\,\mathbf{i} - 2\,\mathbf{j}}{2\sqrt{2}} =$ $\frac{1}{\sqrt{2}}\mathbf{i} - \frac{1}{\sqrt{2}}\mathbf{j}$. f increases most rapidly in the direction $\mathbf{u} = \frac{1}{\sqrt{2}}\mathbf{i} - \frac{1}{\sqrt{2}}\mathbf{j}$; decreases most rapidly in the direction $-\mathbf{u} = -\frac{1}{\sqrt{2}}\mathbf{i} + \frac{1}{\sqrt{2}}\mathbf{j}$. $\left(D_{\mathbf{u}}f\right)_{P_0} = \nabla f \cdot \mathbf{u} = 2\sqrt{2}$, $\left(D_{-\mathbf{u}}f\right)_{P_0} = -2\sqrt{2}$. $\mathbf{u_1} = \frac{\mathbf{A}}{|\mathbf{A}|} = \frac{\mathbf{i} + \mathbf{j}}{\sqrt{1^2+1^2}}$ $= \frac{1}{\sqrt{2}}\mathbf{i} + \frac{1}{\sqrt{2}}\mathbf{j}$. $\left(D_{\mathbf{u_1}}f\right)_{P_0} = \nabla f \cdot \mathbf{u_1} = 2\left(\frac{1}{\sqrt{2}}\right) + (-2)\left(\frac{1}{\sqrt{2}}\right) = 0$.

35. $\nabla f = \left(\frac{2}{2x+3y+6z}\right)\mathbf{i} + \left(\frac{3}{2x+3y+6z}\right)\mathbf{j} + \left(\frac{6}{2x+3y+6z}\right)\mathbf{k} \Rightarrow \nabla f\big|_{(-1,-1,1)} = 2\,\mathbf{i} + 3\,\mathbf{j} + 6\,\mathbf{k}$. $\mathbf{u} = \frac{\nabla f}{|\nabla f|} = \frac{2\,\mathbf{i} + 3\,\mathbf{j} + 6\,\mathbf{k}}{\sqrt{2^2+3^2+6^2}} = \frac{2}{7}\mathbf{i} + \frac{3}{7}\mathbf{j} + \frac{6}{7}\mathbf{k}$. f increases most rapidly in the direction $\mathbf{u} = \frac{2}{7}\mathbf{i} + \frac{3}{7}\mathbf{j} + \frac{6}{7}\mathbf{k}$; decreases most rapidly in the direction $-\mathbf{u} = -\frac{2}{7}\mathbf{i} - \frac{3}{7}\mathbf{j} - \frac{6}{7}\mathbf{k}$. $\left(D_{\mathbf{u}}f\right)_{P_0} = \nabla f \cdot \mathbf{u} = 7$, $\left(D_{-\mathbf{u}}f\right)_{P_0} = -7$. $\mathbf{u_1} = \frac{\mathbf{A}}{|\mathbf{A}|} = \frac{2}{7}\mathbf{i} + \frac{3}{7}\mathbf{j} + \frac{6}{7}\mathbf{k}$ since $\mathbf{A} = \nabla f$. $\Rightarrow \left(D_{\mathbf{u_1}}f\right)_{P_0} = 7$.

36. $\nabla f = (2x + 3y)\,\mathbf{i} + (3x + 2)\,\mathbf{j} + (1 - 2z)\,\mathbf{k} \Rightarrow \nabla f\big|_{(0,0,0)} = 2\,\mathbf{j} + \mathbf{k}$. $\mathbf{u} = \frac{\nabla f}{|\nabla f|} = \frac{2\,\mathbf{j} + \mathbf{k}}{\sqrt{2^2 + 1^2}} = \frac{2}{\sqrt{5}}\mathbf{j} + \frac{1}{\sqrt{5}}\mathbf{k}$.

f increases most rapidly in the direction $\mathbf{u} = \frac{2}{\sqrt{5}}\mathbf{i} + \frac{1}{\sqrt{5}}\mathbf{k}$; decreases most rapidly in the direction $-\mathbf{u} = -\frac{2}{\sqrt{5}}\mathbf{j} - \frac{1}{\sqrt{5}}\mathbf{k}$. $\left(D_{\mathbf{u}}f\right)_{P_0} = \nabla f \cdot \mathbf{u} = \sqrt{5}$; $\left(D_{-\mathbf{u}}f\right)_{P_0} = -\sqrt{5}$. $\mathbf{u}_1 = \frac{\mathbf{A}}{|\mathbf{A}|} = \frac{\mathbf{i} + \mathbf{j} + \mathbf{k}}{\sqrt{1^2 + 1^2 + 1^2}} = \frac{1}{\sqrt{3}}\mathbf{i} + \frac{1}{\sqrt{3}}\mathbf{j} + \frac{1}{\sqrt{3}}\mathbf{k}$

$\left(D_{\mathbf{u}_1}f\right)_{P_0} = \nabla f \cdot \mathbf{u}_1 = 0\left(\frac{1}{\sqrt{3}}\right) + 2\left(\frac{1}{\sqrt{3}}\right) + (1)\left(\frac{1}{\sqrt{3}}\right) = \sqrt{3}$

37.

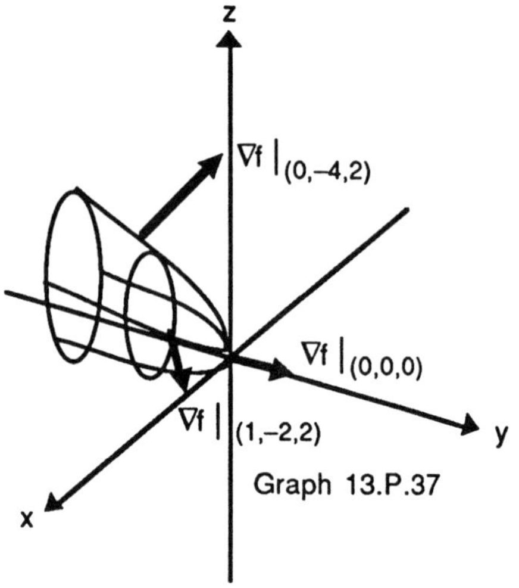

Graph 13.P.37

$\nabla f = 2x\,\mathbf{i} + \mathbf{j} + 2z\,\mathbf{k} \Rightarrow$

$\nabla f\big|_{(1,-2,1)} = 2\,\mathbf{i} + \mathbf{j} + 2\,\mathbf{k}$,

$\nabla f\big|_{(0,0,0)} = \mathbf{j}$,

$\nabla f\big|_{(0,-4,2)} = \mathbf{j} + 4\,\mathbf{k}$

38.

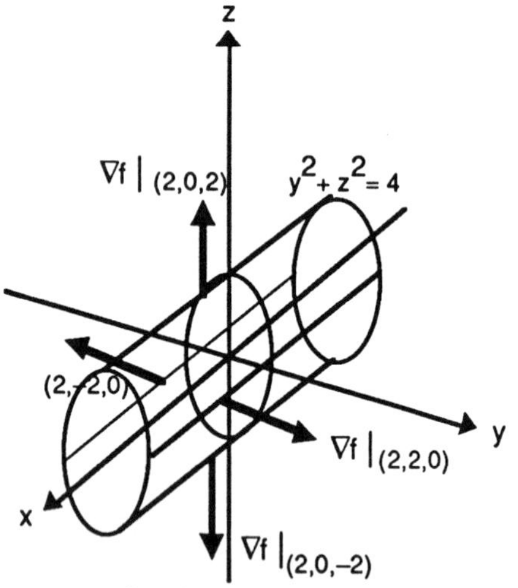

Graph 13.P.38

$\nabla f = 2y\,\mathbf{j} + 2z\,\mathbf{k} \Rightarrow$

$\nabla f\big|_{(2,2,0)} = 4\,\mathbf{j}$,

$\nabla f\big|_{(2,-2,0)} = -4\,\mathbf{j}$,

$\nabla f\big|_{(2,0,2)} = 4\,\mathbf{k}$,

$\nabla f\big|_{(2,0,-2)} = -4\,\mathbf{k}$

39. $\nabla f = 2x\,\mathbf{i} - \mathbf{j} - 5\,\mathbf{k} \Rightarrow \nabla f\big|_{(2,-1,1)} = 4\,\mathbf{i} - \mathbf{j} - 5\,\mathbf{k} \Rightarrow$ Tangent Plane: $4(x - 2) - (y + 1) - 5(z - 1) = 0 \Rightarrow$ $4x - y - 5z = 4$; Normal Line: $x = 2 + 4t$, $y = -1 - t$, $z = 1 - 5t$

40. $\nabla f = 2x\,\mathbf{i} + 2y\,\mathbf{j} + \mathbf{k} \Rightarrow \nabla f\big|_{(1,1,2)} = 2\,\mathbf{i} + 2\,\mathbf{j} + \mathbf{k} \Rightarrow$ Tangent Plane: $2(x - 1) + 2(y - 1) + (z - 2) = 0 \Rightarrow$ $2x + 2y + z - 6 = 0$; Normal Line: $x = 1 + 2t$, $y = 1 + 2t$, $z = 2 + t$

41.

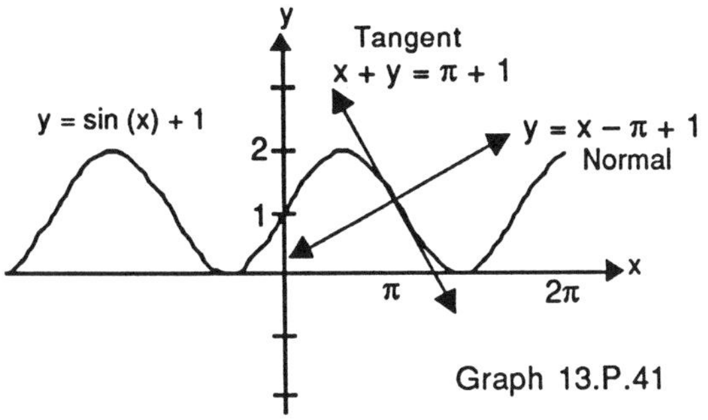

Graph 13.P.41

$\nabla f = (-\cos x)\,\mathbf{i} + \mathbf{j} \Rightarrow \nabla f\big|_{(\pi,1)} = \mathbf{i} + \mathbf{j} \Rightarrow$ Tangent Line: $(x - \pi) + (y - 1) = 0 \Rightarrow x + y = \pi + 1$; Normal Line: $y - 1 = 1(x - \pi) \Rightarrow y = x - \pi + 1$

42.

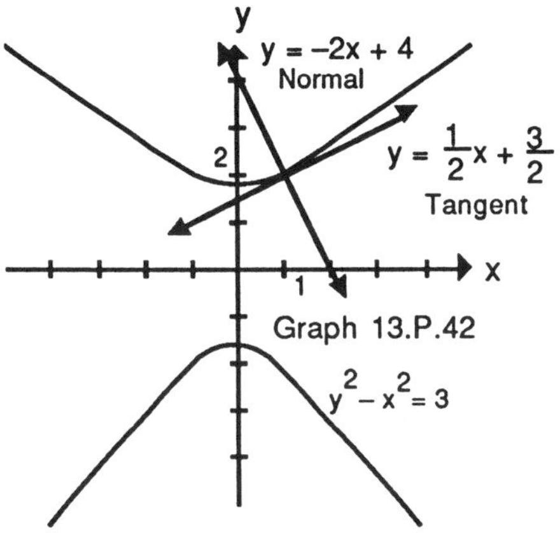

Graph 13.P.42

$\nabla f = -x\,\mathbf{i} + y\,\mathbf{j} \Rightarrow \nabla f\big|_{(1,2)} = -\mathbf{i} + 2\,\mathbf{j} \Rightarrow$ Tangent Line: $-(x - 1) + 2(y - 2) = 0 \Rightarrow y = \frac{1}{2}x + \frac{3}{2}$; Normal Line: $y - 2 = -2(x - 1) \Longrightarrow y = -2x + 4$

43. $f\left(\frac{\pi}{4},\frac{\pi}{4}\right) = \frac{1}{2}$, $f_x\left(\frac{\pi}{4},\frac{\pi}{4}\right) = \cos x \cos y\big|_{(\pi/4,\pi/4)} = \frac{1}{2}$, $f_y\left(\frac{\pi}{4},\frac{\pi}{4}\right) = -\sin x \sin y\big|_{(\pi/4,\pi/4)} = -\frac{1}{2} \Rightarrow L(x,y) = \frac{1}{2} + \frac{1}{2}\left(x - \frac{\pi}{4}\right) - \frac{1}{2}\left(y - \frac{\pi}{4}\right) = \frac{1}{2} + \frac{1}{2}x - \frac{1}{2}y$. $f_{xx}(x,y) = -\sin x \cos y$, $f_{yy}(x,y) = -\sin x \cos y$, $f_{xy}(x,y) = -\cos x \sin y$. $\therefore$ maximum of $|f_{xx}|$, $|f_{yy}|$, and $|f_{xy}|$ is $1 \Rightarrow M = 1 \Rightarrow |E(x,y)| \le \frac{1}{2}(1)\left(\left|x - \frac{\pi}{4}\right| + \left|y - \frac{\pi}{4}\right|\right)^2 \le 0.02$.

44. $f(1,1) = 0$, $f_x(1,1) = y\big|_{(1,1)} = 1$, $f_y(1,1) = x - 6y\big|_{(1,1)} = -5 \Rightarrow L(x,y) = (x - 1) - 5(y - 1) = x - 5y + 4$. $f_{xx}(x,y) = 0$, $f_{yy}(x,y) = -6$, and $f_{xy}(x,y) = 1 \Rightarrow$ maximum of $|f_{xx}|$, $|f_{yy}|$, and $|f_{xy}|$ is $6 \Rightarrow M = 6 \Rightarrow |E(x,y)| \le \frac{1}{2}(6)\left(|x - 1| + |y - 1|\right)^2 \le 0.12$

45. a) $f(1,0,0) = 0$, $f_x(1,0,0) = y - 3z\big|_{(1,0,0)} = 0$, $f_y(1,0,0) = x + 2z\big|_{(1,0,0)} = 1$, $f_z(1,0,0) = 2y - 3x\big|_{(1,0,0)} = -3 \Rightarrow L(x,y,z) = 0(x - 1) + (y - 0) - 3(z - 0) = y - 3z$.

b) $f(1,1,0) = 1$, $f_x(1,1,0) = 1$, $f_y(1,1,0) = 1$, $f_z(1,1,0) = 2 \Rightarrow L(x,y,z) = 1 + (x - 1) + (y - 1) + 2(z - 0) = x + y + 2z - 1$

46. a) $f\left(0,0,\frac{\pi}{4}\right) = 1$, $f_x\left(0,0,\frac{\pi}{4}\right) = -\sqrt{2}\sin x \sin(y + z)\big|_{(0,0,\pi/4)} = 0$, $f_y\left(0,0,\frac{\pi}{4}\right) = \sqrt{2}\cos x \cos(y + z)\big|_{(0,0,\pi/4)}$ 1, $f_z\left(0,0,\frac{\pi}{4}\right) = \sqrt{2}\cos x \cos(y + z)\big|_{(0,0,\pi/4)} = 1 \Rightarrow L(x,y,z) = 1 + 1(y - 0) + 1\left(z - \frac{\pi}{4}\right) = 1 - \frac{\pi}{4} + z$

b) $f\left(\frac{\pi}{4},\frac{\pi}{4},0\right) = \frac{\sqrt{2}}{2}$, $f_x\left(\frac{\pi}{4},\frac{\pi}{4},0\right) = -\frac{\sqrt{2}}{2}$, $f_y\left(\frac{\pi}{4},\frac{\pi}{4},0\right) = \frac{\sqrt{2}}{2}$, $f_z\left(\frac{\pi}{4},\frac{\pi}{4},0\right) = \frac{\sqrt{2}}{2} \Rightarrow L(x,y,z) = \frac{\sqrt{2}}{2} - \frac{\sqrt{2}}{2}\left(x - \frac{\pi}{4}\right) + \frac{\sqrt{2}}{2}\left(y - \frac{\pi}{4}\right) + \frac{\sqrt{2}}{2}(z - 0) = -\frac{\sqrt{2}}{2}x + \frac{\sqrt{2}}{2}y + \frac{\sqrt{2}}{2}z + \frac{\sqrt{2}}{2}$

47. $dV = 2\pi rh\ dr + \pi r^2\ dh \Rightarrow dV\Big|_{(1.5,5280)} = 2\pi(1.5)(5280)\ dr + \pi(1.5)^2\ dh = 15840\pi\ dr + 2.25\pi\ dh$. Be more careful with the diameter since it has a greater effect on dV.

48. $df = (2x - y)dx + (-x + 2y)dy \Rightarrow df\Big|_{(1,2)} = 3\ dy$. $\therefore$ f is more sensitive to changes in y because a change in x does not change f, but a change in y does change f.

49. $dI = \frac{1}{R}dV - \frac{V}{R^2}dR \Rightarrow dI\Big|_{(24,100)} = \frac{1}{100}dV - \frac{24}{100^2}dR \Rightarrow dI\Big|_{dV=-1,dR=-20} = 0.038$. % change in V = $-\frac{1}{24} = -4.17\%$; % change in R = $-\frac{20}{100} = -20\%$. $I = \frac{24}{100} = 0.24 \Rightarrow$ Estimated % change in I = $\frac{dI}{I} \times 100 = \frac{0.038}{0.24} \times 100 = 15.83\%$

50. $A = \pi ab \Rightarrow dA = \pi b\ da + \pi a\ db \Rightarrow dA\Big|_{(10,16);da=\pm 0.05,db=\pm 0.05} = \pi(16)(\pm 0.05) + \pi(10)(\pm 0.05) = \pm 4.0841\ cm^2$. $A = \pi(10)(16) = 160\ \pi\ cm^2$. $\left|\frac{dA}{A} \times 100\right| = \left|\frac{26\pi(\pm 0.05)}{160\pi} \times 100\right| = 0.8125\%$

51. $f_x(x,y) = 2x - y + 2 = 0$ and $f_y(x,y) = -x + 2y + 2 = 0 \Rightarrow x = -2, y = -2 \Rightarrow (-2,-2)$ is the critical point. $f_{xx}(-2,-2) = 2$, $f_{yy}(-2,-2) = 2$, $f_{xy}(-2,-2) = -1 \Rightarrow f_{xx}f_{yy} - f_{xy}^2 = 3 > 0$ and $f_{xx} > 0 \Rightarrow$ Minimum (absolute). $f(-2,-2) = -8$

52. $f_x(x,y) = 10x + 4y + 4 = 0$ and $f_y(x,y) = 4x - 4y - 4 = 0 \Rightarrow x = 0, y = -1 \Rightarrow$ critical point is $(0,-1)$. $f_{xx}(0,-1) = 10$, $f_{yy}(0,-1) = -4$, $f_{xy}(0,-1) = 4 \Rightarrow f_{xx}f_{yy} - f_{xy}^2 = -56 \Rightarrow$ Saddle Point. $f(0,-1) = 2$

53. $f_x(x,y) = 6y - 3x^2 = 0$ and $f_y(x,y) = 6x - 2y = 0 \Rightarrow x = 0, y = 0$ or $x = 6, y = 18 \Rightarrow$ critical points are (0,0) and (6,18). For (0,0): $f_{xx}(0,0) = -6x\Big|_{(0,0)} = 0$, $f_{yy}(0,0) = -2$, $f_{xy}(0,0) = 6 \Rightarrow f_{xx}f_{yy} - f_{xy}^2 = -36 \Rightarrow$ Saddle Point. $f(0,0) = 0$. For (6,18): $f_{xx}(6,18) = -36$, $f_{yy}(6,18) = -2$, $f_{xy}(6,18) = 6 \Rightarrow f_{xx}f_{yy} - f_{xy}^2 = 36 > 0$ and $f_{xx} < 0 \Rightarrow$ maximum (local since $y = 0$ and $x < 0 \Rightarrow f(x,y)$ increases without bound). $f(6,18) = 108$.

54. $f_x(x,y) = 6x^2 + 3y = 0$ and $f_y(x,y) = 3x + 3y^2 = 0 \Rightarrow x = 0, y = 0$ or $x = -\frac{1}{\sqrt[3]{4}}, y = -\frac{1}{\sqrt[3]{2}} \Rightarrow$ critical points are (0,0) and $\left(-\frac{1}{\sqrt[3]{4}}, -\frac{1}{\sqrt[3]{2}}\right)$. For (0,0): $f_{xx}(0,0) = 12x\Big|_{(0,0)} = 0$, $f_{yy}(,0) = 6y\Big|_{(0,0)} = 0$, $f_{xy}(0,0) = 3 \Rightarrow f_{xx}f_{yy} - f_{xy}^2 = -9 \Rightarrow$ Saddle Point. $f(0,0) = 0$. For $\left(-\frac{1}{\sqrt[3]{4}}, -\frac{1}{\sqrt[3]{2}}\right)$: $f_{xx}\left(-\frac{1}{\sqrt[3]{4}}, -\frac{1}{\sqrt[3]{2}}\right) = -\frac{12}{\sqrt[3]{4}}$, $f_{yy}\left(-\frac{1}{\sqrt[3]{4}}, -\frac{1}{\sqrt[3]{2}}\right) = -\frac{6}{\sqrt[3]{2}}$, $f_{xy}\left(-\frac{1}{\sqrt[3]{4}}, -\frac{1}{\sqrt[3]{2}}\right) = 3 \Rightarrow f_{xx}f_{yy} - f_{xy}^2 = 27 > 0$ and $f_{xx} < 0 \Rightarrow$ Maximum (local) $f\left(-\frac{1}{\sqrt[3]{4}}, -\frac{1}{\sqrt[3]{2}}\right) = \frac{1}{2}$

55. $f_x(x,y) = 3x^2 - 3y = 0$ and $f_y(x,y) = 3y^2 - 3x = 0 \Rightarrow x = 0, y = 0$ or $x = 1, y = 1 \Rightarrow$ critical points are (0,0) and (1,1). For (0,0): $f_{xx}(0,0) = 6x\Big|_{(0,0)} = 0$, $f_y(0,0) = 6y\Big|_{(0,0)} = 0$, $f_{xy}(,0) = -3 \Rightarrow f_{xx}f_{yy} - f_{xy}^2 = -9 \Rightarrow$ Saddle Point. $f(0,0) = 15$. For (1,1): $f_{xx}(1,1) = 6$, $f_{yy}(1,1) = 6$, $f_{xy}(1,1) = -3 \Rightarrow f_{xx}f_{yy} - f_{xy}^2 = 27 > 0$ and $f_{xx} > 0 \Rightarrow$ Minimum (local since $y = 0$, $x < 0 \Rightarrow f(x,y)$ decreases without bound). $f(1,1) = 14$.

56. $f_x(x,y) = 3x^2 + 6x = 0$ and $f_y(x,y) = 3y^2 - 6y = 0 \Rightarrow x = 0, y = 0; x = 0, y = 2; x = -2, y = 0;$ or $x = -2, y = 2 \Rightarrow$ critical points are (0,0), (0,2), (−2,0), and (−2,2). For (0,0): $f_{xx}(0,0) = 6x + 6\big|_{(0,0)} = 6$, $f_{yy}(0,0) = 6y - 6\big|_{(0,0)} = -6$, $f_{xy}(0,0) = 0 \Rightarrow f_{xx}f_{yy} - f_{xy}^2 = -36 \Rightarrow$ Saddle Point. $f(0,0) = 0$.
For (0,2): $f_{xx}(0,2) = 6$, $f_{yy}(0,2) = 6$, $f_{xy}(0,2) = 0 \Rightarrow f_{xx}f_{yy} - f_{xy}^2 = 36 > 0$ and $f_{xx} > 0 \Rightarrow$ Minimum (local). $f(0,2) = -4$. For (−2,0): $f_{xx}(-2,0) = -6$, $f_{yy}(-2,0) = -6$, $f_{xy}(-2,0) = 0 \Rightarrow f_{xx}f_{yy} - f_{xy}^2 = 36 > 0$ and $f_{xx} < 0 \Rightarrow$ Maximum (local). $f(-2,0) = 4$. For (−2,2): $f_{xx}(-2,2) = -6$, $f_{yy}(-2,2) = 6$, $f_{xy}(-2,2) = 0 \Rightarrow$ $f_{xx}f_{yy} - f_{xy}^2 = -36 \Rightarrow$ Saddle Point. $f(-2,2) = 0$

57.

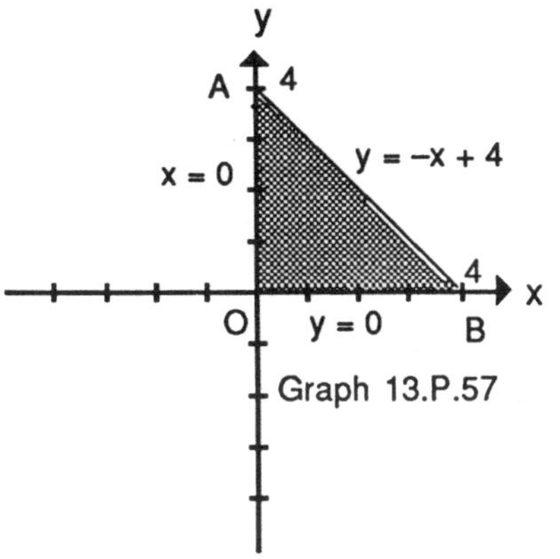

Graph 13.P.57

1. On OA, $f(x,y) = y^2 + 3y = f(0,y)$ for $0 \le y \le 4$. $f(0,0) = 0$, $f(0,4) = 28$. $f'(0,y) = 2y + 3 = 0 \Rightarrow y = -\frac{3}{2}$. But $\left(0,-\frac{3}{2}\right)$ is not in the region.
2. On AB, $f(x,y) = x^2 - 10x + 28 = f(x,-x+4)$ for $0 \le x \le 4$. $f(4,0) = 4$. $f'(x,-x+4) = 2x - 10 = 0 \Rightarrow x = 5, y = -1$. But (5,−1) not in the region.
3. On OB, $f(x,y) = x^2 - 3x = f(x,0)$ for $0 \le x \le 4$. $f'(x,0) = 2x - 3 \Rightarrow x = \frac{3}{2}, y = 0 \Rightarrow \left(\frac{3}{2},0\right)$ is a critical point. $f\left(\frac{3}{2},0\right) = -\frac{9}{4}$
4. For the interior of the triangular region, $f_x(x,y) = 2x + y - 3 = 0$ and $f_y(x,y) = x + 2y + 3 = 0 \Rightarrow x = 3, y = -3$. But (3,−3) is not in the region.

$\therefore$ the absolute maximum is 28 at (0,4); the absolute minimum is $-\frac{9}{4}$ at $\left(\frac{3}{2},0\right)$.

58.

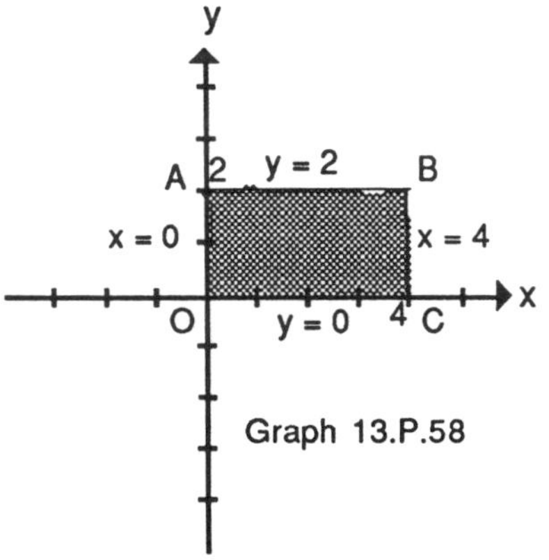

Graph 13.P.58

1. On OA, $f(x,y) = -y^2 + 4y + 1 = f(0,y)$ for $0 \le y \le 2$. $f(0,0) = 1$, $f(0,2) = 5$. $f'(0,y) = -2y + 4 = 0 \Rightarrow y = 2, x = 0 \Rightarrow (0,2)$ which is not in the interior of OA.
2. On AB, $f(x,y) = x^2 - 2x + 5 = f(x,2)$ for $0 \le x \le 4$. $f(4,2) = 13$. $f'(x,2) = 2x - 2 = 0 \Rightarrow x = 1, y = 2 \Rightarrow (1,2)$ is a critical point. $f(1,2) = 4$
3. On BC, $f(x,y) = -y^2 + 4y + 9 = f(4,y)$ for $0 \le y \le 2$. $f(4,0) = 9$. $f'(4,y) = -2y + 4 = 0 \Rightarrow y = 2, x = 4 \Rightarrow (4,2)$ which is not in the interior of BC.
4. On OC, $f(x,y) = x^2 - 2x + 1 = f(x,0)$ for $0 \le x \le 4$. $f'(x,0) = 2x - 2 = 0 \Rightarrow x = 1, y = 0 \Rightarrow (1,0)$ is a critical point. $f(1,0) = 0$.
5. For the interior of the rectangular region, $f_x(x,y) = 2x - 2 = 0$ and $f_y(x,y) = -2y + 4 = 0 \Rightarrow x = 1, y = 2 \Rightarrow (1,2)$ which is not in the interior of the region. $\therefore$ the absolute maximum is 13 at (4,2); the absolute minimum is 0 at (1,0).

59.

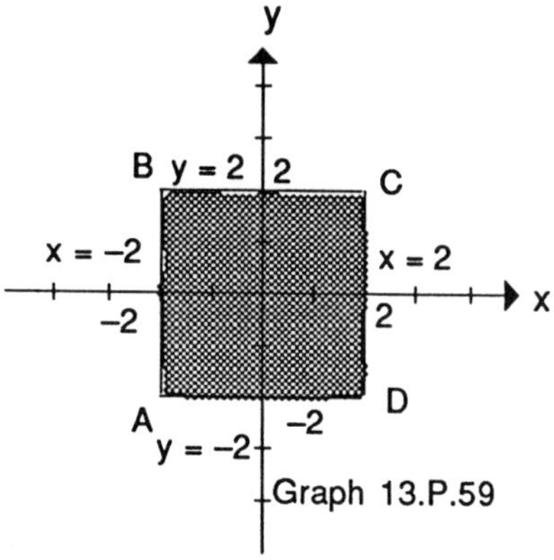

Graph 13.P.59

1. On AB, $f(x,y) = y^2 - y - 4 = f(-2,y)$ for $-2 \le y \le 2$. $f(-2,-2) = 2$, $f(-2,2) = -2$. $f'(-2,y) = 2y - 1 \Rightarrow y = \frac{1}{2}$, $x = -2 \Rightarrow \left(-2,\frac{1}{2}\right)$ is a critical point. $f\left(-2,\frac{1}{2}\right) = -\frac{17}{4}$.
2. On BC, $f(x,y) = -2 = f(x,2)$ for $-2 \le x \le 2$. $f(2,2) = -2$. $f'(x,2) = 0 \Rightarrow$ no critical points in the interior of BC.
3. On CD, $f(x,y) = y^2 - 5y + 4 = f(2,y)$ for $-2 \le y \le 2$. $f(2,-2) = 18$. $f'(2,y) = 2y - 5 = 0 \Rightarrow y = \frac{5}{2}$, $x = 2 \Rightarrow \left(2,\frac{5}{2}\right)$ which is not in the region.
4. On AD, $f(x,y) = 4x + 10 = f(x,-2)$ for $-2 \le x \le 2$. $f'(x,-2) = 4 \Rightarrow$ no critical points in the interior of AD.
5. For the interior of the square, $f_x(x,y) = -y = 2 = 0$ and $f_y(x,y) = 2y - x - 3 = 0 \Rightarrow x = 1$, $y = 2 \Rightarrow (1,2)$ is a critical point. $f(1,2) = -2$

∴ the absolute maximum is 18 at $(2,-2)$; the absolute minimum is $-\frac{17}{4}$ at $\left(-2,\frac{1}{2}\right)$.

60.

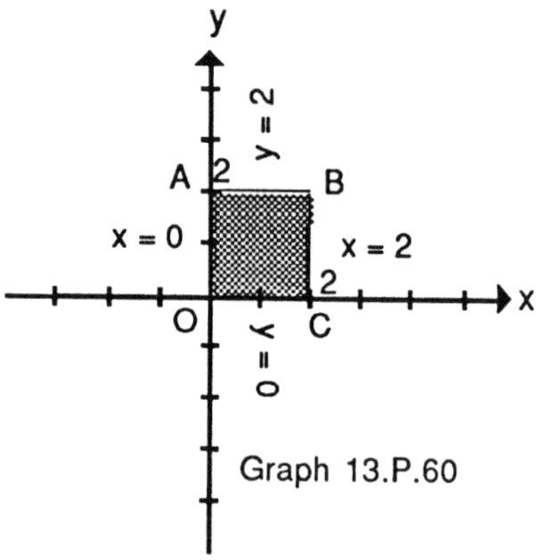

Graph 13.P.60

1. On OA, $f(x,y) = 2y - y^2 = f(0,y)$ for $0 \le y \le 2$. $f(0,0) = 0$, $f(0,2) = 0$. $f'(0,y) = 2 - 2y = 0 \Rightarrow y = 1$, $x = 0 \Rightarrow (0,1)$ is a critical point. $f(0,1) = 1$.
2. On AB, $f(x,y) = 2x - x^2 = f(x,2)$ for $0 \le x \le 2$. $f(2,2) = 0$. $f'(x, 2)$ $2 - 2x = 0 \Rightarrow x = 1$, $y = 2 \Rightarrow (1,2)$ is a critical point. $f(1,2) = 1$
3. On BC, $f(x,y) = 2y - y^2 = f(2,y)$ for $0 \le y \le 2$. $f(2,0) = 0$. $f'(2,y) = 2 - 2y = 0 \Rightarrow y = 1$, $x = 2 \Rightarrow (2,1)$ is a critical point. $f(2,1) = 1$
4. On OC, $f(x,y) = 2x - x^2 = f(x,0)$ for $0 \le x \le 2$. $f'(x,0) = 2 - 2x = 0 \Rightarrow x = 1$, $y = 0 \Rightarrow (1,0)$ is a critical point. $f(1,0) = 1$.
5. For the interior of the rectangular region, $f_x(x,y) = 2 - 2x = 0$ and $f_y(x,y) = 2 - 2y = 0 \Rightarrow x = 1$, $y = 1 \Rightarrow (1,1)$ is a critical point. $f(1,1) = 2$.

∴ the absolute maximum is 2 at $(1,1)$; the absolute minimum is 0 at $(0,0)$, $(0,2)$, $(2,2)$, and $(2,0)$.

61.

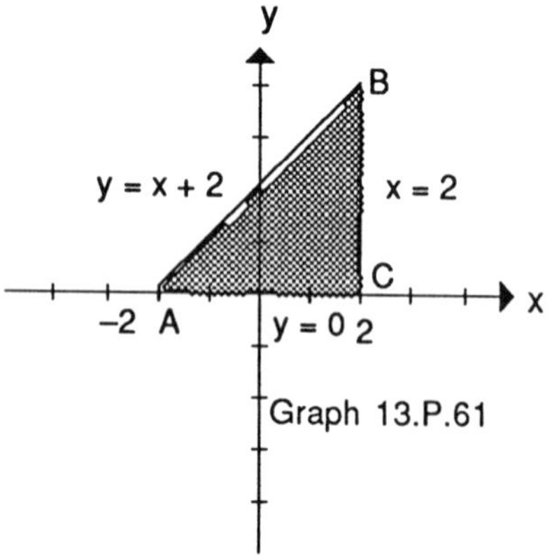

Graph 13.P.61

1. On AB, $f(x,y) = -2x + 4 = f(x,x + 2)$ for $-2 \le x \le 2$. $f(-2,0) = 8$, $f(2,4) = 0$. $f'(x,x + 2) = -2 \Rightarrow$ no critical points in the interior of AB.
2. On BC, $f(x,y) = -y^2 + 4y = f(2,y)$ for $0 \le y \le 4$. $f(2,0) = 0$. $f'(2,y) = -2y + 4 = 0 \Rightarrow y = 2$, $x = 2 \Rightarrow (2,2)$ is a critical point. $f(2,2) = 4$.
3. On AC, $f(x,y) = x^2 - 2x = f(x,0)$ for $-2 \le x \le 2$. $f'(x,0) = 2x - 2 = 0 \Rightarrow x = 1$, $y = 0 \Rightarrow (1,0)$ is a critical point. $f(1,0) = -1$.
4. For the interior of the triangular region, $f_x(x,y) = 2x - 2 = 0$ and $f_y(x,y) = -2y + 4 = 0 \Rightarrow x = 1$, $y = 2 \Rightarrow (1,2)$ is a critical point. $f(1,2) = 3$.

∴ the absolute maximum is 8 at $(-2,0)$; the absolute minimum is -1 at $(1,0)$.

62.

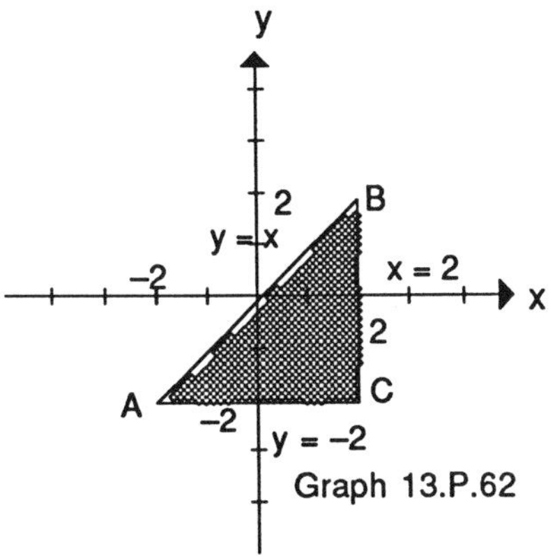

Graph 13.P.62

1. On AB, $f(x,y) = 4x^2 - 2x^4 + 16 = f(x,x)$ for $-2 \le x \le 2$. $f(-2,-2) = 0$, $f(2,2) = 0$. $f'(x,x) = 8x - 8x^3 = 0 \Rightarrow x = 0, y = 0$ or $x = 1, y = 1$ $\Rightarrow (0,0)$ and $(1,1)$ are critical points. $f(0,0) = 16$, $f(1,1) = 18$.
2. On BC, $f(x,y) = 8y - y^4 = f(2,y)$ for $-2 \le y \le 2$. $f(2,-2) = -32$. $f'(2,y) = 8 - 4y^3 = 0 \Rightarrow y = \sqrt[3]{2}$, $x = 2 \Rightarrow (2,\sqrt[3]{2})$ is a critical point. $f(2,\sqrt[3]{2}) = 6\sqrt[3]{2}$.
3. On AC, $f(x,y) = -8x - x^4 = f(x,-2)$ for $-2 \le x \le 2$. $f'(x,-2) = -8 - 4x^3 = 0 \Rightarrow x = \sqrt[3]{-2}$, $y = -2 \Rightarrow (\sqrt[3]{-2},-2)$ is a critical point. $f(\sqrt[3]{-2},-2) = 6\sqrt[3]{2}$.
4. For the interior of the triangular region, $f_x(x,y) = 4y - 4x^3 = 0$ and $f_y(x,y) = 4x - 4y^3 = 0 \Rightarrow$ $x = 0, y = 0$ or $x = 1, y = 1 \Rightarrow (0,0)$ and $(1,1)$, neither of which are interior to the region.

$\therefore$ the absolute maximum is 18 at (1,1); the absolute minimum is –32 at (2,–2).

63. Let $f(x,y) = x^2 + y^2$ be the square of the distance to the origin. $\nabla f = 2x\,\mathbf{i} + 2y\,\mathbf{j}$, $\nabla g = y^2\,\mathbf{i} + 2xy\,\mathbf{j}$. $\nabla f = \lambda \nabla g \Rightarrow 2x\,\mathbf{i} + 2y\,\mathbf{j} = \lambda(y^2\,\mathbf{i} + 2xy\,\mathbf{j}) \Rightarrow 2x = \lambda y^2$ and $2y = 2xy\lambda \Rightarrow 2y = \lambda y^2(y\lambda) \Rightarrow y = 0$ (not on $xy^2 = 54$) or $\lambda^2 y^2 - 2 = 0 \Rightarrow y^2 = \frac{2}{\lambda^2}$. $2y = 2xy\lambda \Rightarrow 1 = x\lambda$ since $y \ne 0 \Rightarrow x = \frac{1}{\lambda}$.
$\therefore \frac{1}{\lambda}\left(\frac{2}{\lambda^2}\right) = 54 \Rightarrow \lambda^3 = \frac{1}{27} \Rightarrow \lambda = \frac{1}{3} \Rightarrow x = 3$, $y^2 = 18 \Rightarrow y = \pm 3\sqrt{2} \Rightarrow$ the points nearest to the origin are $(3,\pm 3\sqrt{2})$.

64. $\nabla f = 3x^2\,\mathbf{i} + 2y\,\mathbf{j}$, $\nabla g = 2x\,\mathbf{i} + 2y\,\mathbf{j}$. $\nabla f = \lambda \nabla g \Rightarrow 3x^2\,\mathbf{i} + 2y\,\mathbf{j} = \lambda(2x\,\mathbf{i} + 2y\,\mathbf{j}) \Rightarrow 3x^2 = 2x\lambda$ and $2y = 2y\lambda$ $\Rightarrow \lambda = 1$ or $y = 0$. CASE 1: $\lambda = 1 \Rightarrow 3x^2 = 2x \Rightarrow x = 0$ or $x = \frac{2}{3}$. $x = 0 \Rightarrow y = \pm 1 \Rightarrow (0,1)$ and $(0,-1)$. $x = \frac{2}{3} \Rightarrow y = \pm\frac{\sqrt{5}}{3} \Rightarrow \left(\frac{2}{3},\frac{\sqrt{5}}{3}\right)$ and $\left(\frac{2}{3},-\frac{\sqrt{5}}{3}\right)$. CASE 2: $y = 0 \Rightarrow x^2 - 1 = 0 \Rightarrow x = \pm 1 \Rightarrow (1,0)$ and $(-1,0)$
$f(0,\pm 1) = 1$, $f\left(\frac{2}{3},\pm\frac{\sqrt{5}}{3}\right) = \frac{23}{27}$, $f(1,0) = 1$, $f(-1,0) = -1$. $\therefore$ the absolute maximum is 1 at $(0,\pm 1)$ and $(1,0)$; the absolute minimum is –1 at $(-1,0)$.

65. $\nabla T = 400yz^2\,\mathbf{i} + 400xz^2\,\mathbf{j} + 800xyz\,\mathbf{k}$, $\nabla g = 2x\,\mathbf{i} + 2y\,\mathbf{j} + 2z\,\mathbf{k}$. $\nabla T = \lambda \nabla g \Rightarrow 400yz^2\,\mathbf{i} + 400xz^2\,\mathbf{j} + 800xyz\,\mathbf{k} = \lambda(2x\,\mathbf{i} + 2y\,\mathbf{j} + 2z\,\mathbf{k}) \Rightarrow 400yz^2 = 2x\lambda$, $400xz^2 = 2y\lambda$, and $800xyz = 2z\lambda$. Solving this system yields the following points: $(0,\pm 1,0)$, $(\pm 1,0,0)$, $\left(\pm\frac{1}{2},\pm\frac{1}{2},\pm\frac{\sqrt{2}}{2}\right)$. $T(0,\pm 1,0) = 0$, $T(\pm 1,0,0) = 0$, $T\left(\pm\frac{1}{2},\pm\frac{1}{2},\pm\frac{\sqrt{2}}{2}\right) = \pm 50$. $\therefore$ 50 is the maximum at $\left(\frac{1}{2},\frac{1}{2},\pm\frac{\sqrt{2}}{2}\right)$ and $\left(-\frac{1}{2},-\frac{1}{2},\pm\frac{\sqrt{2}}{2}\right)$; –50 is the minimum at $\left(\frac{1}{2},-\frac{1}{2},\pm\frac{\sqrt{2}}{2}\right)$ and $\left(-\frac{1}{2},\frac{1}{2},\pm\frac{\sqrt{2}}{2}\right)$.

66. 1. $f(x,y) = x^2 + 3y^2 + 2y$ on $x^2 + y^2 = 1 \Rightarrow \nabla f = 2x\,\mathbf{i} + (6y + 2)\,\mathbf{j}$, $\nabla g = 2x\,\mathbf{i} + 2y\,\mathbf{j}$. $\nabla f = \lambda\,\nabla g \Rightarrow$ $2x\,\mathbf{i} + (6y + 2)\,\mathbf{j} = \lambda(2x\,\mathbf{i} + 2y\,\mathbf{j}) \Rightarrow 2x = 2x\lambda$ and $6y + 2 = 2y\lambda \Rightarrow \lambda = 1$ or $x = 0$. CASE 1: $\lambda = 1 \Rightarrow$ $6y + 2 = 2y \Rightarrow y = -\frac{1}{2}$, $x = \pm\frac{\sqrt{3}}{2} \Rightarrow \left(\pm\frac{\sqrt{3}}{2}, -\frac{1}{2}\right)$. CASE 2: $x = 0 \Rightarrow y^2 = 1 \Rightarrow y = \pm 1 \Rightarrow (0,\pm 1)$. $f\left(\pm\frac{\sqrt{3}}{2}, -\frac{1}{2}\right) = \frac{1}{2}$, $f(0,1) = 5$, $f(0,-1) = 1$. $\therefore$ $\frac{1}{2}$ and 5 are the extreme values on the boundary of the disk.

2. For the interior of the disk, $f_x(x,y) = 2x = 0$ and $f_y(x,y) = 3y + 2 = 0 \Rightarrow x = 0$, $y = -\frac{2}{3}$. $f\left(0,-\frac{2}{3}\right) = 0$.

$\therefore$ the maximum of f on the disk is 5 at (0,1); the absolute minimum of f on the disk is 0 at $\left(0,-\frac{2}{3}\right)$.

CHAPTER 14

MULTIPLE INTEGRALS

14.1 DOUBLE INTEGRALS

1. $\int_0^3 \int_0^2 \left(4 - y^2\right) dy\, dx = \int_0^3 \left[4y - \frac{y^3}{3}\right]_0^2 dx = \frac{16}{3} \int_0^3 dx = 16$

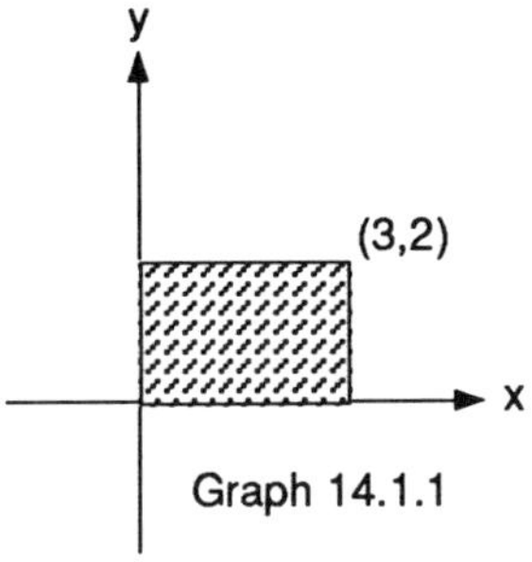

Graph 14.1.1

2. $\int_{-1}^0 \int_{-1}^1 (x + y + 1)\, dx\, dy = \int_{-1}^0 \left[\frac{x^2}{2} + yx + x\right]_{-1}^{1} dy =$

$\int_{-1}^0 2y + 2\, dy = 1$

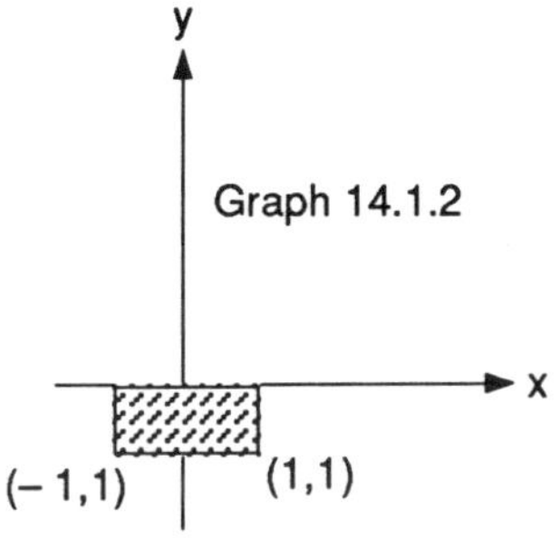

Graph 14.1.2

3. $\int_0^3 \int_{-2}^0 \left(x^2y - 2xy\right) dy\, dx = \int_0^3 \left[\frac{x^2y^2}{2} - xy^2\right]_{-2}^{0} dx =$

$\int_0^3 \left(4x - 2x^2\right) dx = \left[2x^2 - \frac{2x^3}{3}\right]_0^3 = 0$

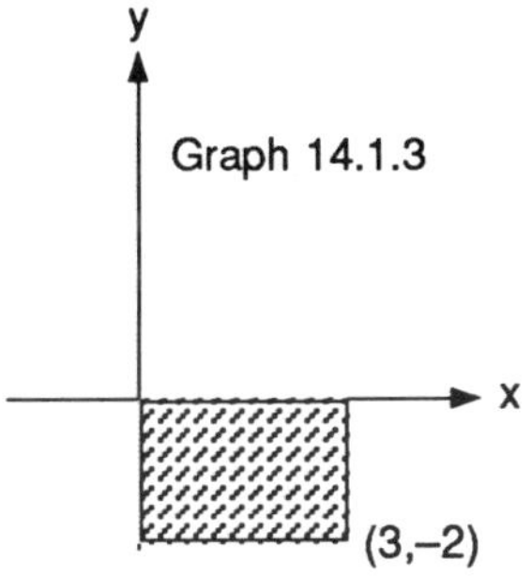

Graph 14.1.3

4. $\int_\pi^{2\pi} \int_0^\pi \sin x + \cos y\, dx\, dy = \int_\pi^{2\pi} \left[- \cos x + (\cos y)x\right]_0^\pi dy =$

$\int_\pi^{2\pi} \pi \cos y + 2\, dy = [\pi \sin y + 2y]_\pi^{2\pi} = 2\pi$

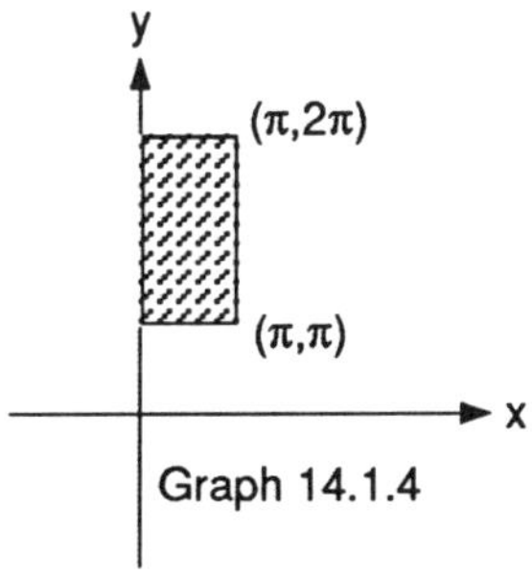

Graph 14.1.4

5. $$\int_0^{\pi}\int_0^{x}(x\sin y)\,dy\,dx = \int_0^{\pi}\left[-x\cos y\right]_0^{x}dx =$$
$$\int_0^{\pi}(x - x\cos x)\,dx = \frac{\pi^2}{2} + 2$$

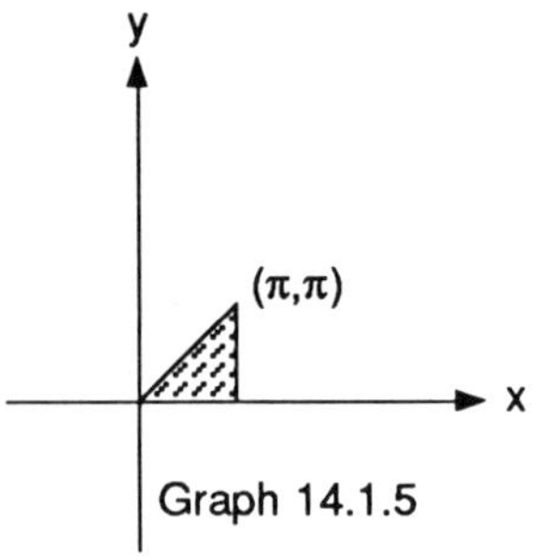

Graph 14.1.5

6. $$\int_1^{\ln 8}\int_0^{\ln y} e^{x+y}\,dx\,dy = \int_1^{\ln 8}\left[e^{x+y}\right]_0^{\ln y}dy =$$
$$\int_1^{\ln 8} y\,e^y - e^y\,dy = \left[(y-1)e^y - e^y\right]_1^{\ln 8} = 8\ln(8) - 16 + e$$

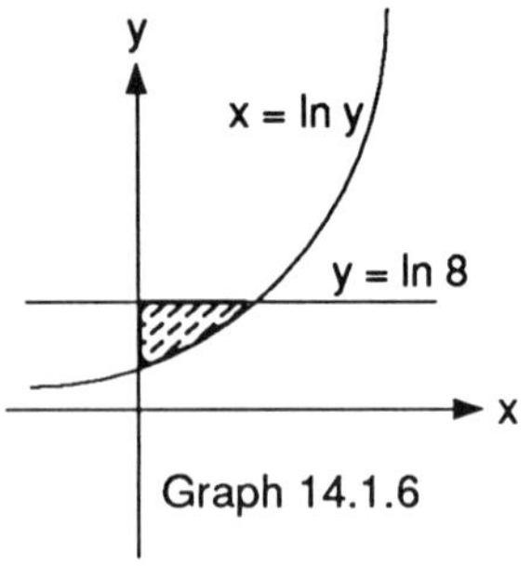

Graph 14.1.6

7. $$\int_0^{\pi}\int_0^{\sin x} y\,dy\,dx = \int_0^{\pi}\left[\frac{y^2}{2}\right]_0^{\sin x}dx =$$
$$\frac{1}{4}\int_0^{\pi}(1 - \cos 2x)\,dx = \frac{\pi}{4}$$

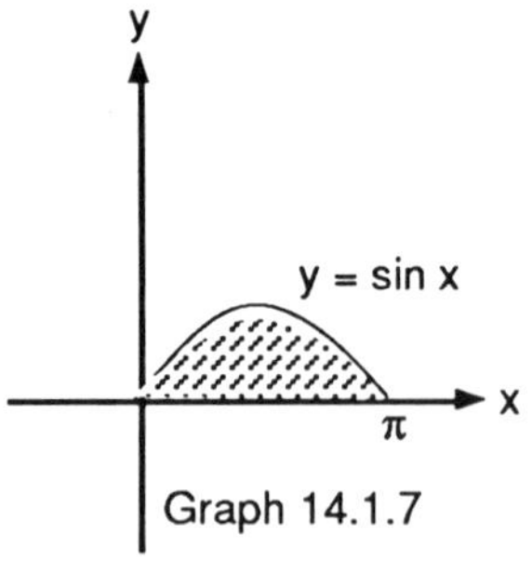

Graph 14.1.7

8. $$\int_1^{2}\int_y^{y^2} dy\,dx = \int_1^{2} y^2 - y\,dy = \left[\frac{y^3}{3} - \frac{y^2}{2}\right]_1^{2} = \frac{5}{6}$$

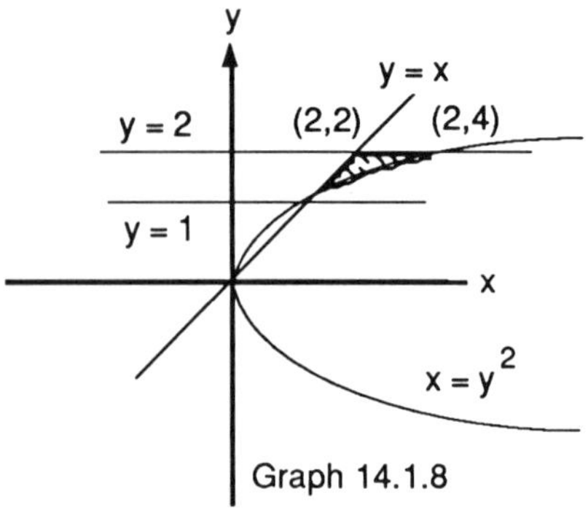

Graph 14.1.8

9. $$\int_1^{2}\int_x^{2x}\frac{x}{y}\,dy\,dx = \int_1^{2}\left[x\ln y\right]_x^{2x}dx = \ln(2)\int_1^{2}x\,dx = \frac{\ln 8}{2}$$

10. $$\int_0^{1}\int_0^{1-y} x^2 + y^2\,dx\,dy = \int_0^{1}\left[\frac{x^3}{3} + y^2x\right]_0^{1-y}dy = \int_0^{1}\frac{(1-y)^3}{3} + y^2 - y^3\,dy = \frac{1}{6}$$

11. $\int_0^1 \int_0^{1-x} y - \sqrt{x}\, dy\, dx = \int_0^1 \left[\frac{y^2}{2} - y\sqrt{x}\right]_0^{1-x} dx = \int_0^1 \frac{1 - 2x + x^2}{2} - \sqrt{x}(1 - x)\, dx = -\frac{1}{10}$

12. $\int_0^1 \int_0^1 x^2 + 3xy\, dx\, dy = \int_0^1 \left[\frac{x^3}{3} + \frac{3x^2y}{2}\right]_0^1 dy = \int_0^1 \frac{1}{3} + \frac{3y}{2}\, dy = \frac{13}{12}$

13. $\int_1^2 \int_1^2 \frac{1}{xy}\, dy\, dx = \int_1^2 \frac{1}{x}(\ln 2 - \ln 1)\, dx = \ln 2 \int_1^2 \frac{1}{x}\, dx = (\ln 2)^2$

14. $\int_0^1 \int_0^{\pi} y \cos xy\, dx\, dy = \int_0^1 [\sin xy]_0^{\pi}\, dy = \int_0^1 \sin \pi y\, dy = \frac{2}{\pi}$

15. $\int_0^4 \int_0^{(4-y)/2} dx\, dy$

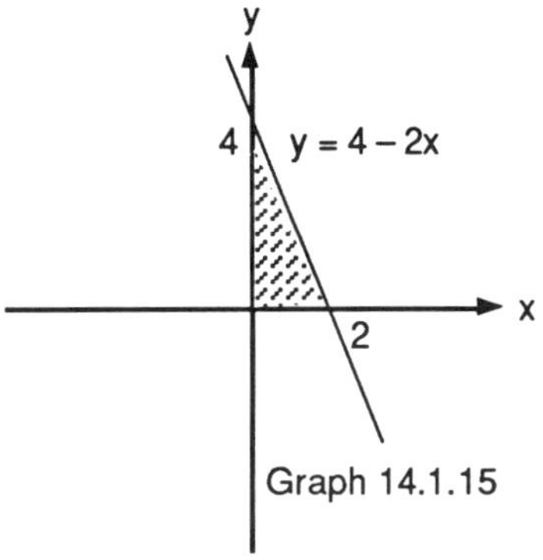

Graph 14.1.15

16. $\int_2^4 \int_0^{(4-y)/2} dx\, dy$

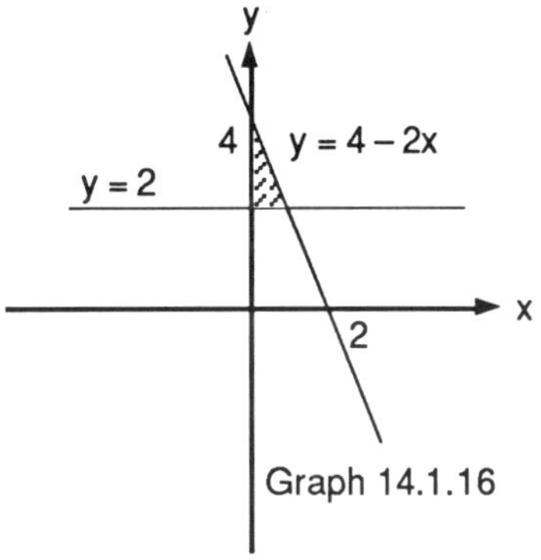

Graph 14.1.16

17. $\int_0^1 \int_{x^2}^{x} dy\, dx$

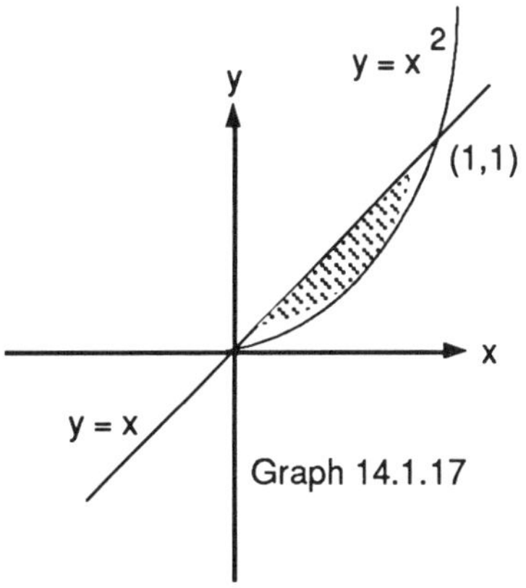

Graph 14.1.17

18. $$\int_0^2 \int_0^{x^2} dy\,dx$$

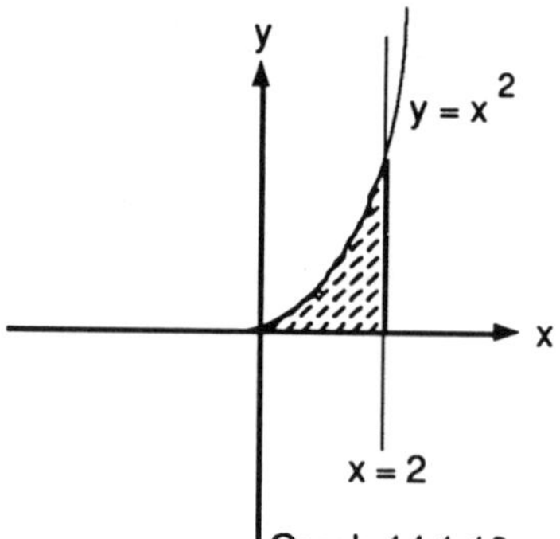

Graph 14.1.18

19. $$\int_1^{\exp(2)} \int_{\ln y}^{2} dx\,dy$$

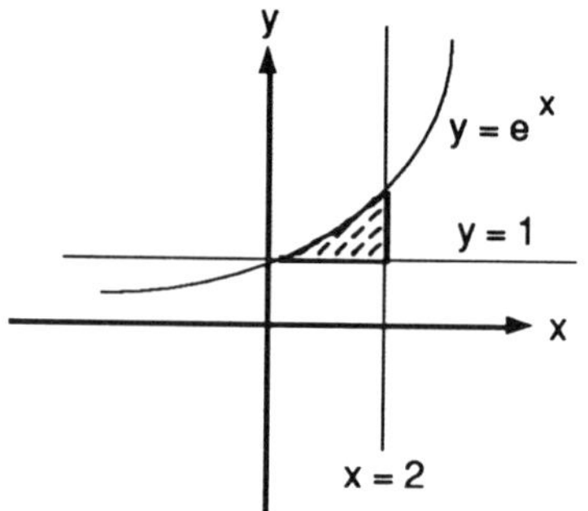

Graph 14.1.19

20. $$\int_0^1 \int_0^{x^2} dy\,dx$$

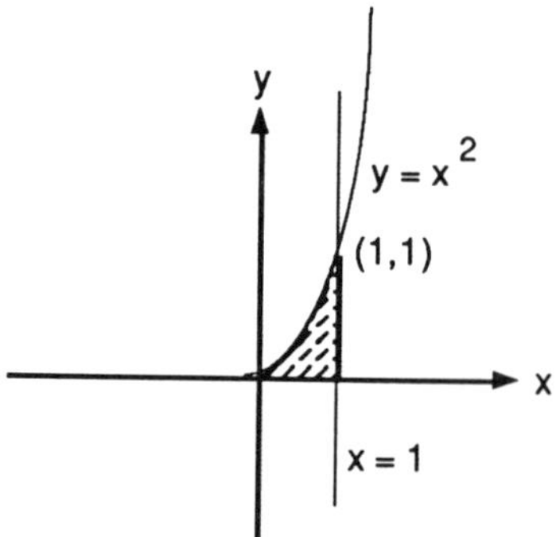

Graph 14.1.20

21. $$\int_0^1 \int_1^{\exp(x)} dy\,dx = \int_1^e \int_{\ln y}^{1} dx\,dy$$

22. $$\int_0^1 \int_{\sqrt{x}}^{1} \cos(x + y)\,dy\,dx = \int_0^1 \int_0^{y^2} \cos(x + y)\,dx\,dy$$

23. $$\int_0^2 \int_0^{x^3} f(x,y)\,dy\,dx = \int_0^8 \int_{\sqrt[3]{y}}^{2} f(x,y)\,dx\,dy$$

24. $\int_0^1 \int_{-\sqrt{y}}^{\sqrt{y}} f(x,y)\,dx\,dy = \int_{-1}^{1} \int_{x^2}^{1} f(x,y)\,dy\,dx$

25. $\int_0^{\pi} \int_x^{\pi} \frac{\sin y}{y}\,dy\,dx = \int_0^{\pi} \int_0^{y} \frac{\sin y}{y}\,dx\,dy = \int_0^{\pi} \sin y\,dy = 2$

26. $\int_0^1 \int_y^1 x^2 e^{xy}\,dx\,dy = \int_0^1 \int_0^x x^2 e^{xy}\,dy\,dx = \int_0^1 x\left[e^{xy}\right]_0^x dx = \int_0^1 x\left(\exp(x^2) - 1\right) dx = \frac{e-2}{2}$

27. $\int_0^2 \int_x^2 2y^2 \sin xy\,dy\,dx = \int_0^2 \int_0^y 2y^2 \sin xy\,dx\,dy = \int_0^2 [-2y \cos xy]_0^y\,dy =$

$\int_0^2 -\cos y^2(2y) + 2y\,dy = 4 - \sin 4$

28. $\int_0^2 \int_0^{4-x^2} \frac{xe^{2y}}{4-y}\,dy\,dx = \int_0^4 \int_0^{\sqrt{4-y}} \frac{xe^{2y}}{4-y}\,dx\,dy = \int_0^4 \left[\frac{x^2 e^{2y}}{2(4-y)}\right]_0^{\sqrt{4-y}} dx = \frac{1}{2}\int_0^4 e^{2y}\,dx = \frac{e^8 - 1}{4}$

29. $V = \int_{-4}^{1} \int_{3x}^{4-x^2} (x+4)\,dy\,dx = \int_{-4}^{1} \Big[(xy + 4y)\Big]_{3x}^{4-x^2} dx = \int_{-4}^{1} \left(-x^3 - 7x^2 - 8x + 16\right) dx = \frac{625}{12}$

30. $V = \int_0^2 \int_0^{\sqrt{4-x^2}} (3-y)\,dy\,dx = \int_0^2 \left[3y - \frac{y^2}{2}\right]_0^{\sqrt{4-x^2}} dx = \int_0^2 3\sqrt{4-x^2} - \left(\frac{4-x^2}{2}\right) dx = \frac{9\pi - 8}{3}$

31. $V = \int_0^2 \int_0^3 4 - y^2\,dx\,dy = \int_0^2 \left[4x - y^2 x\right]_0^3 dy = \int_0^2 \left(12 - 3y^2\right) dy = 16$

32. $V = \int_0^2 \int_0^{4-x^2} 4 - x^2 - y\,dy\,dx = \int_0^2 \left[4y - x^2 y - \frac{y^2}{2}\right]_0^{4-x^2} dx =$

$\int_0^2 8 - 4x^2 + \frac{x^4}{2}\,dx = \frac{128}{15}$

33. $\int_1^3 \int_1^x \frac{1}{xy}\,dy\,dx = 0.603$, use the Calculus Toolkit

34. $\int_0^1 \int_0^1 \exp\left(-\left(x^2 + y^2\right)\right) dy\,dx = 0.557$, use the Calculus Toolkit

35. $\int_0^1 \int_0^1 \tan^{-1} xy\,dy\,dx = 0.233$, use the Calculus Toolkit

36. $\int_{-1}^{1} \int_0^{\sqrt{1-x^2}} 3\sqrt{1 - x^2 - y^2}\,dy\,dx = 3.142$, use the Calculus Toolkit

14.2 AREA, MOMENTS, AND CENTERS OF MASS

1. $\int_0^2 \int_2^{2-x} dy\, dx = \int_0^2 2 - x\, dx = 2$

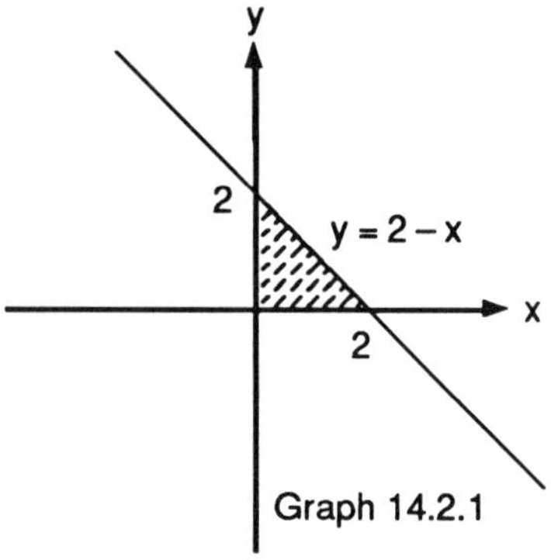

Graph 14.2.1

2. $\int_0^{\ln 2} \int_0^{\exp(x)} dy\, dx = \int_0^{\ln 2} e^x\, dx = 1$

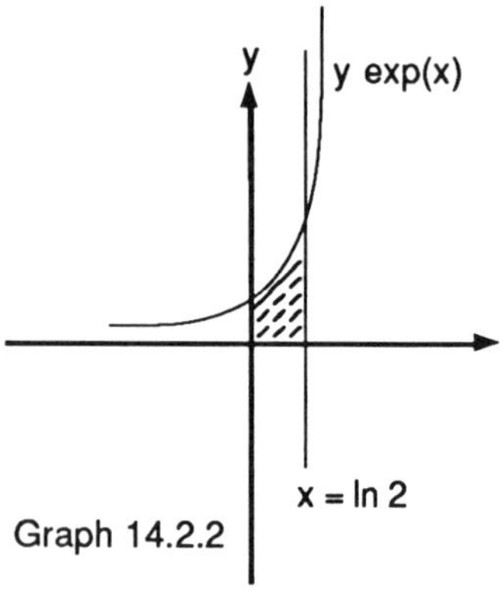

Graph 14.2.2

3. $\int_0^2 \int_{2x}^4 dy\, dx = \int_0^2 4 - 2x\, dx = 4$

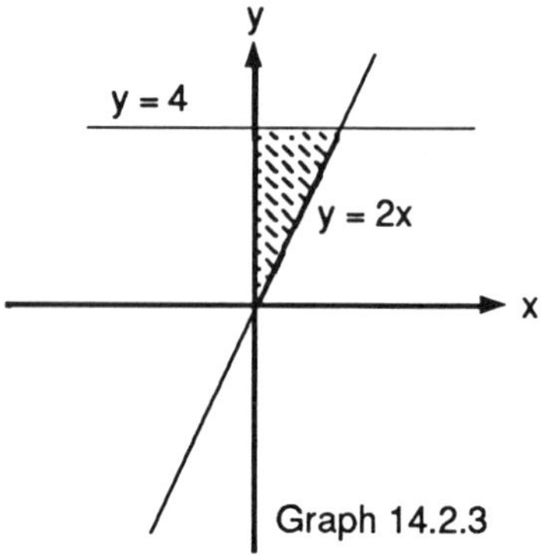

Graph 14.2.3

4. $\int_{-2}^1 \int_{y-2}^{-y^2} dx\, dy = \int_{-2}^1 -y^2 - y + 2\, dy = \frac{9}{2}$

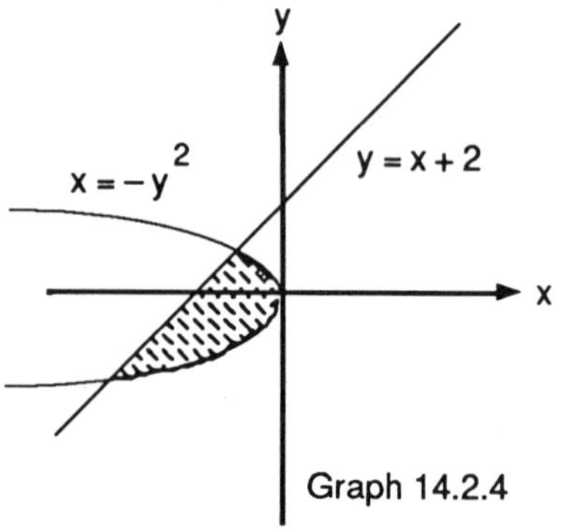

Graph 14.2.4

5. $\displaystyle\int_0^1 \int_{y^2}^{2y-y^2} dx\, dy = \int_0^1 2y - 2y^2\, dy = \frac{1}{3}$

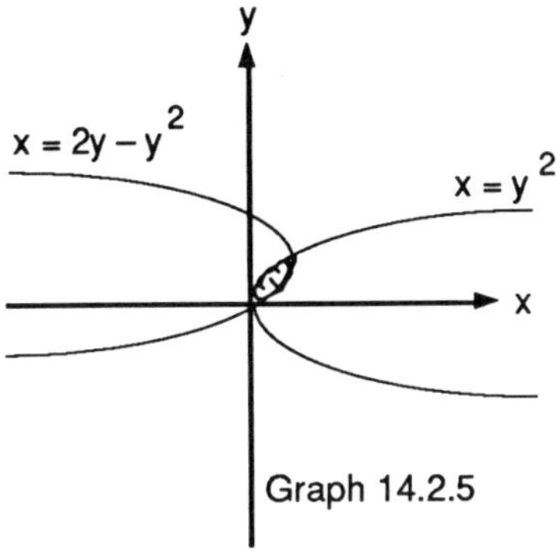

Graph 14.2.5

6. $\displaystyle\int_0^2 \int_{-y}^{y-y^2} dx\, dy = \int_0^2 2y - y^2\, dy = \frac{4}{3}$

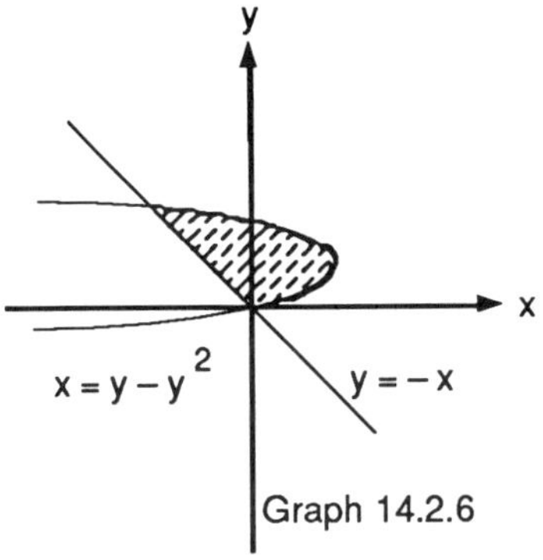

Graph 14.2.6

7. $\displaystyle 2\int_0^1 \int_{-1}^{2\sqrt{1-x^2}} dy\, dx = 2\int_0^1 2\sqrt{1-x^2} + 1\, dx = \pi + 2$

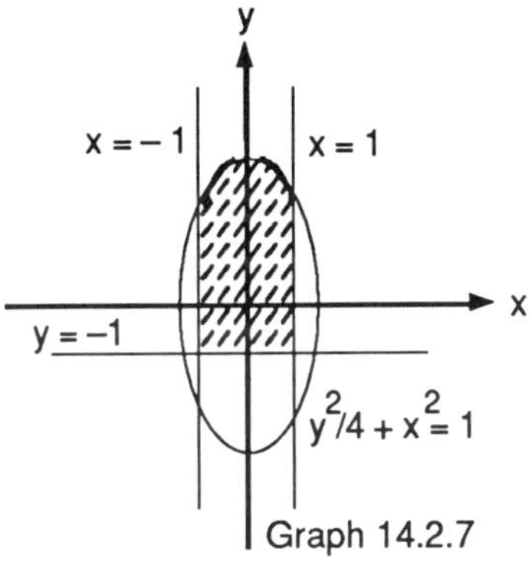

Graph 14.2.7

8. $\displaystyle\int_{-2}^0 \int_{-1}^{x^2} dy\, dx + \int_0^1 \int_{2x-1}^{x^2} dy\, dx =$

$\displaystyle\int_{-2}^0 x^2 + 1\, dx + \int_0^1 x^2 - 2x + 1\, dx = 5$

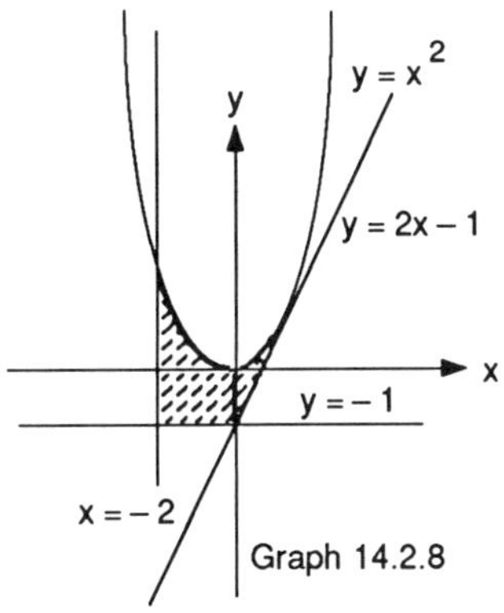

Graph 14.2.8

9. $\displaystyle\int_0^6 \int_{y^2/3}^{2y} dx\, dy = \int_0^6 \left(2y - y^2/3\right) dy = 12$

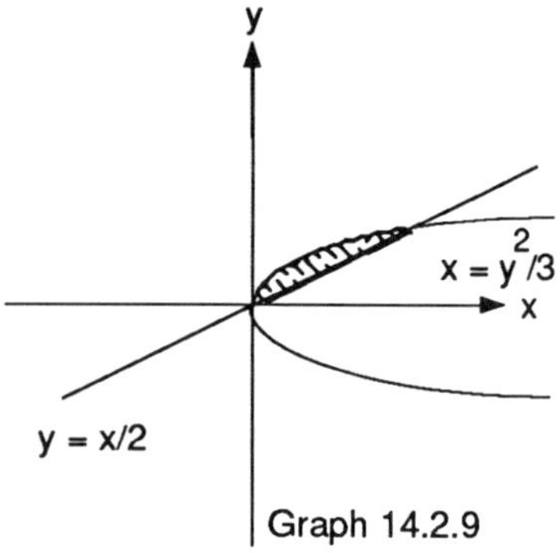

Graph 14.2.9

10. $\displaystyle\int_0^3 \int_{-x}^{2x-x^2} dy\, dx = \int_0^3 3x - x^2\, dx = \frac{9}{2}$

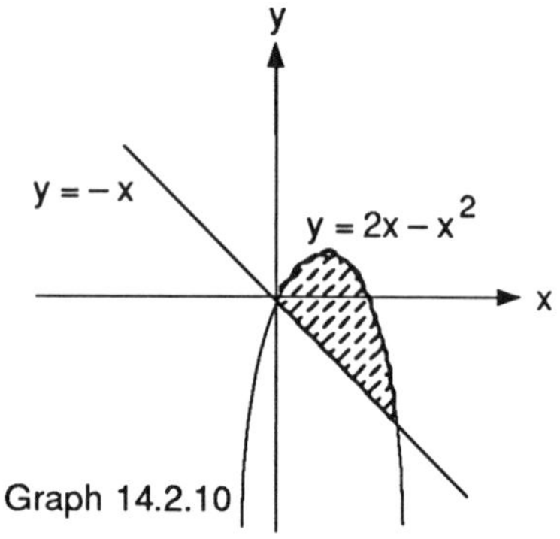

Graph 14.2.10

11. $\displaystyle\int_0^{\pi/4} \int_{\sin x}^{\cos x} dy\, dx = \int_0^{\pi/4} (\cos x - \sin x)\, dx = \sqrt{2} - 1$

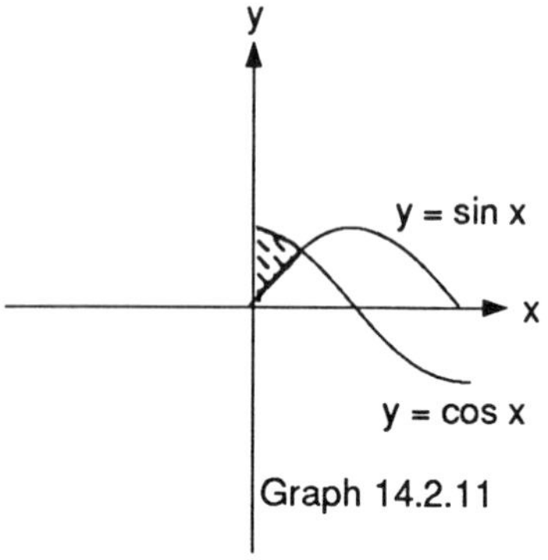

Graph 14.2.11

12. $\displaystyle\int_{-1}^2 \int_{y^2}^{y+2} dx\, dy = \int_{-1}^2 y + 2 - y^2\, dx = 6$

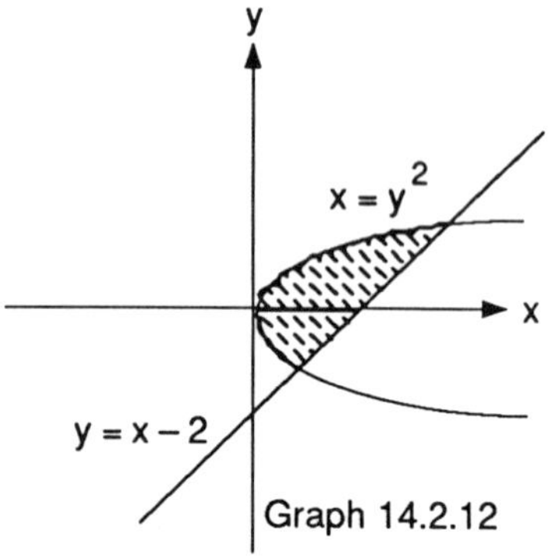

Graph 14.2.12

13. $\displaystyle \int_{-1}^{0} \int_{-2x}^{1-x} dy\,dx + \int_{0}^{2} \int_{-x/2}^{1-x} dy\,dx = \int_{-1}^{0} (1+x)\,dx + \int_{0}^{2} (1-x/2)\,dx = \frac{3}{2}$

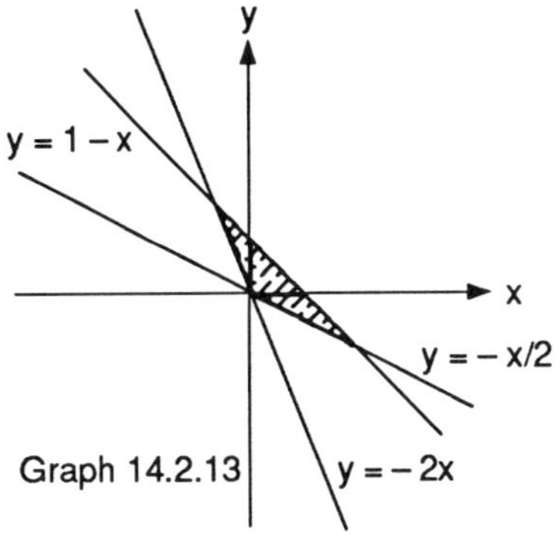

Graph 14.2.13

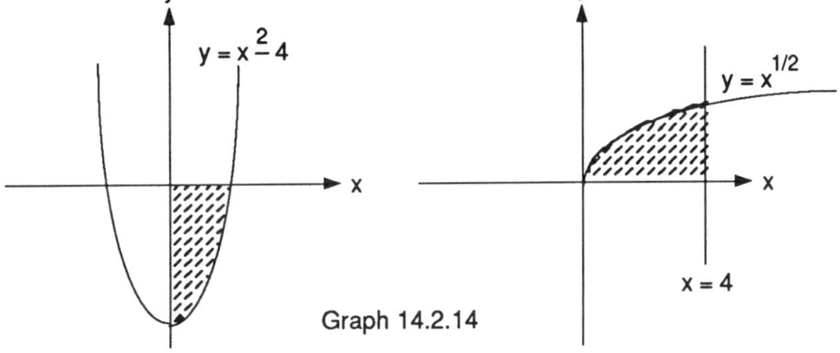

Graph 14.2.14

14. $\displaystyle \int_{0}^{2} \int_{x^2-4}^{0} dy\,dx + \int_{0}^{4} \int_{0}^{\sqrt{x}} dy\,dx = \int_{0}^{2} 4 - x^2\,dx + \int_{0}^{4} x^{1/2}\,dx = \frac{32}{3}$

15. $\displaystyle M = \int_{0}^{1} \int_{x}^{2-x^2} 3\,dy\,dx = 3\int_{0}^{1} 2 - x^2 - x\,dx = \frac{7}{2}$

$\displaystyle M_y = \int_{0}^{1} \int_{x}^{2-x^2} 3x\,dy\,dx = 3\int_{0}^{1} [xy]_{x}^{2-x^2}\,dx = 3\int_{0}^{1} \left(2x - x^3 - x^2\right) dx = \frac{5}{4}$

$\displaystyle M_x = \int_{0}^{1} \int_{x}^{2-x^2} 3y\,dy\,dx = \frac{3}{2}\int_{0}^{1} \left[y^2\right]_{x}^{2-x^2} dx = \frac{3}{2}\int_{0}^{1} 4 - 5x^2 + x^4\,dx = \frac{19}{5} \Rightarrow \overline{x} = \frac{5}{14},\ \overline{y} = \frac{38}{35}$

16. $\displaystyle M = \int_{0}^{3} \int_{0}^{3} dy\,dx = \int_{0}^{3} 3\,dx = 9$

$\displaystyle I_x = \int_{0}^{3} \int_{0}^{3} y^2\,dy\,dx = \int_{0}^{3} \left[\frac{y^3}{3}\right]_{0}^{3} dx = 27\ ,\ R_x = \sqrt{\frac{I_x}{M}} = \sqrt{3}$

$\displaystyle I_y = \int_{0}^{3} \int_{0}^{3} x^2\,dy\,dx = \int_{0}^{3} \left[x^2 y\right]_{0}^{3} dx = \int_{0}^{3} x^2\,dx = 27\ ,\ R_y = \sqrt{\frac{I_y}{M}} = \sqrt{3}$

17. $M = \int_0^2 \int_{y^2/2}^{4-y} dx\, dy = \int_0^2 4 - y - \frac{y^2}{2}\, dy = \frac{14}{3}$

$M_y = \int_0^2 \int_{y^2/2}^{4-y} x\, dx\, dy = \frac{1}{2}\int_0^2 \left[x^2\right]_{y^2/2}^{4-y} dx = \frac{1}{2}\int_0^2 16 - 8y + y^2 - \frac{y^4}{4}\, dy = \frac{128}{15}$

$M_x = \int_0^2 \int_{y^2/2}^{4-y} y\, dx\, dy = \int_0^2 y\left(4 - y - \frac{y^2}{2}\right) dy = \frac{10}{3} \Rightarrow \overline{x} = \frac{64}{35},\ \overline{y} = \frac{5}{7}$

18. $M = \int_0^3 \int_3^{3-x} dy\, dx = \int_0^3 3 - x\, dx = \frac{9}{2}$

$M_y = \int_0^3 \int_0^{3-x} x\, dy\, dx = \int_0^3 [xy]_0^{3-x} dx = \int_0^3 3x - x^2\, dx = \frac{9}{2} \Rightarrow \overline{x} = 1,\ \overline{y} = 1$, by symmetry

19. $M = 2\int_0^1 \int_0^{\sqrt{1-x^2}} dy\, dx = 2\int_0^1 \sqrt{1 - x^2}\, dx = \frac{\pi}{2}$

$M_x = 2\int_0^1 \int_0^{\sqrt{1-x^2}} y\, dy\, dx = \int_0^1 \left[y^2\right]_0^{\sqrt{1-x^2}} dx = \int_0^1 1 - x^2\, dx = \frac{2}{3} \Rightarrow \overline{x} = 0$, by symmetry; $\overline{y} = \frac{4}{3\pi}$

20. $M = \frac{125\delta}{6}$; $M_y = \delta\int_0^5 \int_x^{6x-x^2} x\, dy\, dx = \delta\int_0^5 [xy]_x^{6x-x^2} dx = \delta\int_0^5 5x^2 - x^3\, dx = \frac{625\delta}{12}$

$M_x = \delta\int_0^5 \int_x^{6x-x^2} y\, dy\, dx = \delta\int_0^5 \left[y^2\right]_x^{6x-x^2} dx = \delta\int_0^5 35x^2 - 12x^3 + x^4\, dx = \frac{625\delta}{3} \Rightarrow$

$\overline{x} = \frac{5}{2},\ \overline{y} = 10$

21. $M = \int_0^a \int_0^{\sqrt{a^2-x^2}} dy\, dx = \frac{\pi a^2}{4}$; $M_y = \int_0^a \int_0^{\sqrt{a^2-x^2}} x\, dy\, dx =$

$\int_0^a [xy]_0^{\sqrt{a^2-x^2}} dx = -\frac{1}{2}\int_0^a \sqrt{a^2 - y^2}(-2x)\, dx = \frac{a^3}{3} \Rightarrow \overline{x} = \overline{y} = \frac{4a}{3\pi}$, by symmetry

22. $I_x = \int_{-2}^2 \int_{-\sqrt{4-x^2}}^{\sqrt{4-x^2}} y^2\, dy\, dx = \int_{-2}^2 \left[\frac{y^3}{3}\right]_{-\sqrt{4-x^2}}^{\sqrt{4-x^2}} dx = \frac{2}{3}\int_{-2}^2 \left(4 - x^2\right)^{3/2} dx = 4\pi$;

$I_y = 4\pi$, by symmetry; $I_o = I_x + I_y = 8\pi$

23. $M = \int_0^\pi \int_0^{\sin x} dy\, dx = \int_0^\pi \sin x\, dx = 2$; $M_x = \int_0^\pi \int_0^{\sin x} y\, dy\, dx = \frac{1}{2}\int_0^\pi \left[y^2\right]_0^{\sin x} dx =$

$\frac{1}{4}\int_0^\pi 1 - \cos 2x\, dx = \frac{\pi}{4} \Rightarrow \overline{x} = \frac{\pi}{2},\ \overline{y} = \frac{\pi}{8}$

24. $I_y = \int_{\pi}^{2\pi} \int_0^{(\sin^2 x)/x^2} x^2 \, dy \, dx = \frac{1}{2}\int_{\pi}^{2\pi} (1 - \cos 2x) \, dx = \frac{\pi}{2}$

25. $M = \int_0^2 \int_{-y}^{y-y^2} (x + y) \, dx \, dy = \int_0^2 \left[\frac{x^2}{2} + xy\right]_{-y}^{y-y^2} dy =$

$\int_0^2 \left(\frac{y^4}{2} - 2y^3 + 2y^2\right) dy = \frac{8}{15}$; $I_x = \int_0^2 \int_{-y}^{y-y^2} y^2(x + y) \, dx \, dy = \int_0^2 \left[\frac{x^2y^2}{2} + xy^3\right]_{-y}^{y-y^2} dy =$

$\int_0^2 \left(\frac{y^6}{2} - 2y^5 + 2y^4\right) dy = \frac{64}{105}$; $R_x = \sqrt{\frac{I_x}{M}} = \sqrt{\frac{8}{7}}$

26. $M = \int_{-\sqrt{3}/2}^{\sqrt{3}/2} \int_{4y^2}^{\sqrt{12-4y^2}} 5x \, dx \, dy = \frac{5}{2}\int_{-\sqrt{3}/2}^{\sqrt{3}/2} \left(12 - 4y^2 - 16y^4\right) dy = 23\sqrt{3}$

27. $M = \int_0^1 \int_x^{2-x} (6x + 3y + 3) \, dy \, dx = \int_0^1 - 12x^2 + 12 \, dx = 8$; $M_y = \int_0^1 \int_x^{2-x} x(6x + 3y + 3) \, dy \, dx =$

$\int_0^1 12x - 12x^3 \, dx = 3$; $M_x = \int_0^1 \int_x^{2-x} y(6x + 3y + 3) \, dy \, dx =$

$\int_0^1 14 - 6x - 6x^2 - 2x^3 \, dx = \frac{17}{2} \Rightarrow \overline{x} = \frac{3}{8}, \overline{y} = \frac{17}{16}$

28. $M = \int_0^1 \int_{y^2}^{2y-y^2} y + 1 \, dx \, dy = \int_0^1 2y - 2y^3 \, dy = \frac{1}{2}$; $M_x = \int_0^1 \int_{y^2}^{2y-y^2} y(y + 1) \, dx \, dy =$

$\int_0^1 2y^2 - 2y^4 \, dy = \frac{4}{15}$; $M_y = \int_0^1 \int_{y^2}^{2y-y^2} x(y + 1) \, dx \, dy = \int_0^1 2y^2 - 2y^4 \, dy = \frac{4}{15} \Rightarrow$

$\overline{x} = \frac{8}{15}, \overline{y} = \frac{8}{15}$; $I_x = \int_0^1 \int_{y^2}^{2y-y^2} y^2(y+1) \, dx \, dy = 2\int_0^1 y^3 - y^5 \, dy = \frac{1}{6}$

29. $M = \int_0^1 \int_0^6 (x + y + 1) \, dx \, dy = \int_0^1 (6y + 24) \, dy = 27$

$M_x = \int_0^1 \int_0^6 y(x + y + 1) \, dx \, dy = \int_0^1 y(6y + 24) \, dy = 14$

$M_y = \int_0^1 \int_0^6 x(x + y + 1) \, dx \, dy = \int_0^1 (18y + 90) \, dy = 99 \Rightarrow \overline{x} = \frac{11}{3}, \overline{y} = \frac{14}{27}$

$I_y = \int_0^1 \int_0^6 x^2(x + y + 1) \, dx \, dy = 216\int_0^1 \left(\frac{y}{3} + \frac{11}{6}\right) dy = 432 \Rightarrow R_y = \sqrt{\frac{I_y}{M}} = 4$

30. $M = \int_{-1}^{1} \int_{x^2}^{1} (y + 1)\, dy\, dx = -\int_{-1}^{1} \left(\frac{x^4}{2} + x^2 - \frac{3}{2}\right) dx = \frac{32}{15}$

$M_x = \int_{-1}^{1} \int_{x^2}^{1} y(y + 1)\, dy\, dx = \int_{-1}^{1} \left(\frac{5}{6} - \frac{x^6}{3} - \frac{x^4}{2}\right) dx = \frac{48}{35}$

$M_y = \int_{-1}^{1} \int_{x^2}^{1} x(y + 1)\, dy\, dx = \int_{-1}^{1} \left(\frac{3x}{2} - \frac{x^5}{2} - x^3\right) dx = 0 \Rightarrow \overline{x} = 0,\ \overline{y} = \frac{9}{14}$

$I_y = \int_{-1}^{1} \int_{x^2}^{1} x^2(y + 1)\, dy\, dx = \int_{-1}^{1} \left(\frac{3x^2}{2} - \frac{x^6}{2} - x^4\right) dx = \frac{16}{35} \Rightarrow R_y = \sqrt{\frac{I_y}{M}} = \sqrt{\frac{3}{14}}$

31. $M = \int_{-1}^{1} \int_{0}^{x^2} (7y + 1)\, dy\, dx = \int_{-1}^{1} \left(\frac{7x^4}{2} + x^2\right) dx = \frac{31}{15}$

$M_x = \int_{-1}^{1} \int_{0}^{x^2} y(7y + 1)\, dy\, dx = \int_{-1}^{1} \left(\frac{7x^6}{3} + \frac{x^4}{2}\right) dx = \frac{13}{15}$

$M_y = \int_{-1}^{1} \int_{0}^{x^2} x(7y + 1)\, dy\, dx = \int_{-1}^{1} \left(\frac{7x^5}{2} + x^3\right) dx = 0 \Rightarrow \overline{x} = 0,\ \overline{y} = \frac{13}{31}$

$I_y = \int_{-1}^{1} \int_{0}^{x^2} x^2(7y + 1)\, dy\, dx = \int_{-1}^{1} x^2\left(\frac{7x^4}{2} + x^2\right) dx = \frac{7}{5} \Rightarrow R_y = \sqrt{\frac{I_y}{M}} = \sqrt{\frac{21}{31}}$

32. $M = \int_{0}^{20} \int_{-1}^{1} \left(1 + \frac{x}{20}\right) dy\, dx = \int_{0}^{20} \left(2 + \frac{x}{10}\right) dx = 60$

$M_x = \int_{0}^{20} \int_{-1}^{1} y\left(1 + \frac{x}{20}\right) dy\, dx = \int_{0}^{20} \left[\left(1 + \frac{x}{20}\right)\left(\frac{y^2}{2}\right)\right]_{-1}^{1} dx = 0$

$M_y = \int_{0}^{20} \int_{-1}^{1} x\left(1 + \frac{x}{20}\right) dy\, dx = \int_{0}^{20} x\left(2 + \frac{x}{10}\right) dx = \frac{2000}{3} \Rightarrow \overline{x} = \frac{100}{9},\ \overline{y} = 0$

$I_x = \int_{0}^{20} \int_{-1}^{1} y^2\left(1 + \frac{x}{20}\right) dy\, dx = \frac{2}{3}\int_{0}^{20} \left(1 + \frac{x}{20}\right) dx = 20,\ R_x = \sqrt{\frac{I_x}{M}} = \sqrt{\frac{1}{3}}$

33. $M = \int_0^1 \int_{-y}^{y} (1+y)\,dx\,dy = \int_0^1 \left(2y^2 + 2y\right) dy = \frac{5}{3}$

$M_x = \int_0^1 \int_{-y}^{y} y(1+y)\,dx\,dy = 2\int_0^1 \left(y^3 + y^2\right) dy = \frac{7}{6}$

$M_y = \int_0^1 \int_{-y}^{y} x(1+y)\,dx\,dy = \int_0^1 0\,dy = 0 \Rightarrow \overline{x} = 0, \overline{y} = \frac{7}{10}$

$I_x = \int_0^1 \int_{-y}^{y} y^2(1+y)\,dx\,dy = \int_0^1 y^2\left(2y^2 + 2y\right) dy = \frac{9}{10} \Rightarrow R_x = \sqrt{\frac{I_x}{M}} = \frac{3\sqrt{6}}{10}$

$I_y = \int_0^1 \int_{-y}^{y} x^2(1+y)\,dx\,dy = \frac{1}{3}\int_0^1 y^2\left(2y^2 + 2y\right) dy = \frac{3}{10} \Rightarrow R_y = \sqrt{\frac{I_y}{M}} = \frac{3\sqrt{2}}{10}$

$I_o = I_x + I_y = \frac{6}{5}$ and $R_o = \sqrt{\frac{I_o}{M}} = \frac{3\sqrt{2}}{5}$

34. $M = \int_0^1 \int_{-y}^{y} \left(3x^2 + 1\right) dx\,dy = \int_0^1 \left(2y^3 + 2y\right) dy = \frac{3}{2}$

$M_x = \int_0^1 \int_{-y}^{y} y\left(3x^2 + 1\right) dx\,dy = \int_0^1 y\left(2y^3 + 2y\right) dy = \frac{16}{15}$

$M_y = \int_0^1 \int_{-y}^{y} x\left(3x^2 + 1\right) dx\,dy = 0 \Rightarrow \overline{x} = 0, \overline{y} = \frac{32}{45}$

$I_x = \int_0^1 \int_{-y}^{y} y^2\left(3x^2 + 1\right) dx\,dy = \int_0^1 y^2\left(2y^3 + 2y\right) dy = \frac{5}{6} \Rightarrow R_x = \sqrt{\frac{I_x}{M}} = \frac{\sqrt{5}}{3}$

$I_y = \int_0^1 \int_{-y}^{y} x^2\left(3x^2 + 1\right) dx\,dy = 2\int_0^1 \left(\frac{3}{5}y^5 + \frac{1}{3}y^3\right) dy = \frac{11}{30} \Rightarrow R_y = \sqrt{\frac{I_y}{M}} = \sqrt{\frac{11}{45}}$

$I_o = I_x + I_y = \frac{6}{5}$ and $R_o = \sqrt{\frac{I_o}{M}} = \frac{2}{\sqrt{5}}$

35. $f(a) = I_a = \int_0^4 \int_0^2 (y-a)^2\,dy\,dx = \int_0^4 \left(\frac{(2-a)^3}{3} + \frac{a^3}{3}\right) dx = \frac{4}{3}\left[(2-a)^3 + a^3\right] \Rightarrow f'(a) = 0 \Rightarrow$

$a = 1, f''(a) = 16 > 0 \Rightarrow f(1)$ is minimum

36. $M = \int_0^1 \int_{-1/\sqrt{1-x^2}}^{1/\sqrt{1-x^2}} dy\,dx = \int_0^1 \frac{2}{\sqrt{1-x^2}}\,dx = \pi$

$M_y = \int_0^1 \int_{-1/\sqrt{1-x^2}}^{1/\sqrt{1-x^2}} x\,dy\,dx = \int_0^1 \frac{2x}{\sqrt{1-x^2}}\,dx = 2 \Rightarrow \overline{x} = \frac{2}{\pi}, \overline{y} = 0$ by symmetry

14.3 DOUBLE INTEGRALS IN POLAR FORM

1. $$\int_{-1}^{1}\int_{0}^{\sqrt{1-x^2}} dy\,dx = \int_{0}^{\pi}\int_{0}^{1} r\,dr\,d\theta = \frac{1}{2}\int_{0}^{\pi} d\theta = \frac{\pi}{2}$$

2. $$\int_{-1}^{1}\int_{-\sqrt{1-x^2}}^{\sqrt{1-x^2}} dy\,dx = \int_{0}^{2\pi}\int_{0}^{1} r\,dr\,d\theta = \frac{1}{2}\int_{0}^{2\pi} d\theta = \pi$$

3. $$\int_{0}^{1}\int_{0}^{\sqrt{1-y^2}}\left(x^2+y^2\right) dx\,dy = \int_{0}^{\pi/2}\int_{0}^{1} r^3\,dr\,d\theta = \frac{1}{4}\int_{0}^{\pi/2} d\theta = \frac{\pi}{8}$$

4. $$\int_{-1}^{1}\int_{-\sqrt{1-y^2}}^{\sqrt{1-y^2}}\left(x^2+y^2\right) dx\,dy = \int_{0}^{2\pi}\int_{0}^{1} r^3\,dr\,d\theta = \frac{1}{4}\int_{0}^{2\pi} d\theta = \frac{\pi}{2}$$

5. $$\int_{-a}^{a}\int_{-\sqrt{a^2-x^2}}^{\sqrt{a^2-x^2}} dy\,dx = \int_{0}^{2\pi}\int_{0}^{a} r\,dr\,d\theta = \frac{a^2}{2}\int_{0}^{2\pi} d\theta = \pi a^2$$

6. $$\int_{0}^{2}\int_{0}^{\sqrt{4-y^2}}\left(x^2+y^2\right) dx\,dy = \int_{0}^{\pi/2}\int_{0}^{2} r^3\,dr\,d\theta = 4\int_{0}^{\pi/2} d\theta = 2\pi$$

7. $$\int_{0}^{\pi/4}\int_{0}^{\sqrt{2}} \cos\theta\, r^2\,dr\,d\theta = \frac{2\sqrt{2}}{3}\int_{0}^{\pi/4} \cos\theta\,d\theta = \frac{2}{3}$$

8. $$\int_{0}^{2}\int_{0}^{x} y\,dy\,dx = \int_{0}^{\pi/4}\int_{0}^{2\sec\theta} \sin\theta\, r^2\,dr\,d\theta = \frac{8}{3}\int_{0}^{\pi/4} \tan\theta\sec^2\theta\,d\theta = \frac{4}{3}$$

9. $$\int_{0}^{3}\int_{0}^{\sqrt{3}x} \frac{1}{\sqrt{x^2+y^2}}\,dy\,dx = \int_{0}^{\pi/3}\int_{0}^{3\sec\theta} dr\,d\theta = \int_{9}^{\pi/3} 3\sec\theta\,d\theta = 3\ln\left(2+\sqrt{3}\right)$$

10. $$\int_{0}^{2}\int_{0}^{\sqrt{4-x^2}} \frac{xy}{\sqrt{x^2+y^2}}\,dy\,dx = \int_{0}^{\pi/2}\int_{0}^{2} \sin\theta\cos\theta\, r^2\,dr\,d\theta = \frac{8}{3}\int_{0}^{\pi/2} \sin\theta\cos\theta\,d\theta = \frac{4}{3}$$

11. $$\int_{0}^{1}\int_{0}^{\sqrt{1-x^2}} 5\sqrt{x^2+y^2}\,dy\,dx = \int_{0}^{\pi/2}\int_{0}^{1} 5r^2\,dr\,d\theta = \frac{5}{3}\int_{0}^{\pi/2} d\theta = \frac{5\pi}{6}$$

12. $$\int_{0}^{\infty}\int_{0}^{\infty} \exp\left[-\left(x^2+y^2\right)\right] dx\,dy = \int_{0}^{\pi/2}\int_{0}^{\infty} \exp\left(-r^2\right) r\,dr\,d\theta =$$
$$\int_{0}^{\pi/2} \lim_{t\to\infty}\int_{0}^{t} \exp\left(-r^2\right) r\,dr\,d\theta = \frac{1}{2}\int_{0}^{\pi/2} d\theta = \frac{\pi}{4}$$

13. $$\int_{0}^{\pi/2}\int_{0}^{2\sqrt{2-\sin 2\theta}} r\,dr\,d\theta = 2\int_{0}^{\pi/2}\left(2-\sin 2\theta\right) d\theta = 2(\pi-1)$$

14. $$A = 2\int_{0}^{\pi/2}\int_{1}^{1+\cos\theta} r\,dr\,d\theta = \int_{0}^{\pi/2} 2\cos\theta + \cos^2\theta\,d\theta = \frac{8+\pi}{4}$$

15. $2\int_0^{\pi/6}\int_0^{12\cos 3\theta} r\,dr\,d\theta = 144\int_0^{\pi/6}\cos^2 3\theta\,d\theta = 12\pi$

16. $A = \int_0^{2\pi}\int_0^{4\theta/3} r\,dr\,d\theta = \frac{8}{9}\int_0^{2\pi}\theta^2\,d\theta = \frac{64\pi^3}{27}$

17. $A = \int_0^{\pi/2}\int_0^{1+\sin\theta} r\,dr\,d\theta = \frac{1}{2}\int_0^{\pi/2}\frac{3}{2} + 2\sin\theta - \frac{\cos 2\theta}{2}\,d\theta = \frac{3\pi+8}{8}$

18. $A = 4\int_0^{\pi/2}\int_0^{1-\cos\theta} r\,dr\,d\theta = 2\int_0^{\pi/2}\frac{3}{2} - 2\cos\theta + \frac{\cos 2\theta}{2}\,d\theta = \frac{3\pi-8}{2}$

19. $\int_0^{2\pi}\int_0^{\sqrt{3}/2}\frac{1}{1-r^2}\,r\,dr\,d\theta = \ln(2)\int_0^{2\pi}d\theta = \pi\ln 4$

20. $4\int_0^{\pi/2}\int_1^{e}\frac{\ln\left(r^2\right)}{r}\,dr\,d\theta = 2\int_0^{\pi/2}d\theta = \pi$

21. $M_x = \int_0^{\pi}\int_0^{1-\cos\theta} 3r^2\sin\theta\;dr\,d\theta = \int_0^{\pi}\left(1-\cos\theta\right)^3\sin\theta\,d\theta = 4$

22. $I_x = \int_{-a}^{a}\int_{-\sqrt{a^2-x^2}}^{\sqrt{a^2-x^2}} y^2(k)\left(x^2+y^2\right)dy\,dx = k\int_0^{2\pi}\int_0^{a}\sin^2\theta\,r^5\,dr\,d\theta =$

$\frac{ka^6}{6}\int_0^{2\pi}\frac{1-\cos 2\theta}{2}\,d\theta = \frac{ka^6\pi}{6}$

23. $M = 2\int_0^{\pi}\int_0^{1+\cos\theta} r\,dr\,d\theta = \int_0^{\pi}\left(1+\cos\theta\right)^2 d\theta = \frac{3\pi}{2}$

$M_y = \int_0^{2\pi}\int_0^{1+\cos\theta}\cos\theta\,r^2\,dr\,d\theta = \int_0^{2\pi}\frac{4\cos\theta}{3} + \frac{15}{24} + \cos 2\theta - \sin^2\theta\cos\theta + \frac{\cos 4\theta}{4}\,d\theta =$

$\frac{5\pi}{4} \Rightarrow \overline{x} = \frac{5}{6},\ \overline{y} = 0$ by symmetry

24. $I_o = \int_0^{2\pi}\int_0^{1+\cos\theta} r^3\,dr\,d\theta = \frac{1}{4}\int_0^{2\pi}\left(1+\cos\theta\right)^4 d\theta = \frac{35\pi}{16}$

25. $M = \int_{-\infty}^{0}\int_0^{\exp(x)} dy\,dx = \int_{-\infty}^{0} e^x\,dx = \lim_{t\to-\infty}\int_t^0 e^x\,dx = 1$

$M_y = \int_{-\infty}^{0}\int_0^{\exp(x)} x\,dy\,dx = \int_{-\infty}^{0} x\,e^x\,dx = \lim_{t\to-\infty}\int_t^0 x\,e^x\,dx = -1$

$M_x = \int_{-\infty}^{0}\int_0^{\exp(x)} y\,dy\,dx = \frac{1}{2}\int_{-\infty}^{0} e^{2x}\,dx = \frac{1}{2}\lim_{t\to-\infty}\int_t^0 e^{2x}\,dx = \frac{1}{4} \Rightarrow \overline{x} = -1,\ \overline{y} = \frac{1}{4}$

26. $M_y = \int_{-\infty}^{\infty}\int_0^{\exp(-x^2/2)} x\,dy\,dx = 2\int_0^{\infty}\exp\left(-x^2/2\right)(x)\,dx = 2\lim_{t\to\infty}\int_0^{t}\exp\left(-x^2/2\right)(x)\,dx = 2$

14.4 TRIPLE INTEGRALS IN RECTANGULAR COORDINATES

1. $\int_0^1 \int_0^{1-z} \int_0^2 dx\,dy\,dz = 2\int_0^1 \int_0^{1-z} dy\,dz = 2\int_0^1 1 - z\,dz = 1$

2. $\int_0^1 \int_0^2 \int_0^3 dz\,dy\,dx = \int_0^1 \int_0^2 3\,dy\,dx = \int_0^1 6\,dx = 6,\ \int_0^2 \int_0^1 \int_0^3 dz\,dx\,dy,$

$\int_0^3 \int_0^2 \int_0^1 dx\,dy\,dz,\ \int_0^2 \int_0^3 \int_0^1 dx\,dz\,dy,\ \int_0^3 \int_0^1 \int_0^2 dy\,dx\,dz,\ \int_0^1 \int_0^3 \int_0^2 dy\,dz\,dx$

3. $\int_0^1 \int_0^{2-2x} \int_0^{3-3x-3y/2} dz\,dy\,dx = \int_0^1 \int_0^{2-2x} 3 - 3x - \frac{3}{2}y\,dy\,dx = \int_0^1 3 - 6x + 3x^2\,dx = 1,$

$\int_0^2 \int_0^{x-y/2} \int_0^{3-3x-3y/2} dz\,dx\,dy,\ \int_0^1 \int_0^{3-3x} \int_0^{2-2x-2z/3} dy\,dz\,dx,$

$\int_0^3 \int_0^{1-z/3} \int_0^{2-2x-2z/3} dy\,dx\,dz,\ \int_0^2 \int_0^{3-3y/2} \int_0^{1-y/2-z/3} dx\,dz\,dy,$

$\int_0^3 \int_0^{2-2z/3} \int_0^{1-y/2-z/3} dx\,dy\,dz$

4. $\int_0^2 \int_0^3 \int_0^{\sqrt{4-x^2}} dz\,dy\,dx = \int_0^2 \int_0^3 \sqrt{4-x^2}\,dy\,dx = \int_0^2 3\sqrt{4-x^2}\,dx = 3\pi,$

$\int_0^3 \int_0^2 \int_0^{\sqrt{4-x^2}} dz\,dx\,dy,\ \int_0^2 \int_0^{\sqrt{4-x^2}} \int_0^3 dy\,dz\,dx,$

$\int_0^2 \int_0^{\sqrt{4-x^2}} \int_0^3 dy\,dx\,dz,\ \int_0^2 \int_0^3 \int_0^{\sqrt{4-x^2}} dx\,dy\,dz,$

$\int_0^3 \int_0^2 \int_0^{\sqrt{4-x^2}} dx\,dz\,dy$

5. $\int_0^1 \int_0^1 \int_0^1 x^2 + y^2 + z^2\,dz\,dy\,dx = \int_0^1 \int_0^1 \left(x^2 + y^2 + \frac{1}{3}\right) dy\,dx = \int_0^1 x^2 + \frac{2}{3}\,dx = 1$

6. $\int_0^{\sqrt{2}} \int_0^{3y} \int_{x^2+3y^2}^{8-x^2-y^2} dz\,dx\,dy = \int_0^{\sqrt{2}} \int_0^{3y} 8 - 2x^2 - 4y^2\,dx\,dy = \int_0^{\sqrt{2}} 24y - 18y^3 - 12y^3\,dx = -6$

7. $\int_1^e \int_1^e \int_1^e \frac{1}{xyz}\,dx\,dy\,dz = \int_1^e \int_1^e \frac{\ln x}{yz}\,dy\,dz = \int_1^e \frac{1}{z}\,dz = 1$

8. $\int_0^1 \int_0^{3-3x} \int_0^{3-3x-y} dz\,dy\,dx = \int_0^1 \int_0^{3-3x} 3 - 3x - y\,dy\,dx = \int_0^1 \frac{9}{2} - 9x + \frac{9}{2}x^2\,dx = \frac{3}{2}$

9. $\int_0^1 \int_0^{\pi} \int_0^{\pi} y \sin z\,dx\,dy\,dz = \int_0^1 \int_0^{\pi} \pi y \sin z\,dy\,dz = \frac{\pi^3}{2}\int_0^1 \sin z\,dz = \frac{\pi^3}{2}(1 - \cos 1)$

10. $\int_{-1}^{1}\int_{-1}^{1}\int_{-1}^{1}(x+y+z)\,dy\,dx\,dz = \int_{-1}^{1}\int_{-1}^{1} 2x+2z\,dx\,dz = \int_{-1}^{1} 4z\,dz = 0$

11. $\int_{0}^{3}\int_{0}^{\sqrt{9-x^2}}\int_{0}^{\sqrt{9-x^2}} dz\,dy\,dx = \int_{0}^{3}\int_{0}^{\sqrt{9-x^2}}\sqrt{9-x^2}\,dy\,dx = \int_{0}^{3}\left(9-x^2\right)dx = 18$

12. $\int_{0}^{2}\int_{-\sqrt{4-y^2}}^{\sqrt{4-y^2}}\int_{0}^{2x+y} dz\,dx\,dy = \int_{0}^{2}\int_{-\sqrt{4-y^2}}^{\sqrt{4-y^2}}(2x+y)\,dx\,dy = \int_{0}^{2}\left(4-y^2\right)^{1/2}(2y)\,dy = \frac{16}{3}$

13. $\int_{0}^{1}\int_{0}^{2-x}\int_{0}^{2-x-y} dz\,dy\,dx = \int_{0}^{1}\int_{0}^{2-x} 2-x-y\,dy\,dx = \int_{0}^{1}\frac{x^2}{2}-2x+2\,dx = \frac{7}{6}$

14. $\int_{0}^{1}\int_{0}^{1-x^2}\int_{3}^{4-x^2-y} x\,dz\,dy\,dx = \int_{0}^{1}\int_{0}^{1-x^2} x-x^3-xy\,dy\,dx = \int_{0}^{1}\frac{x}{2}-x^3+\frac{x^5}{2}\,dx = \frac{1}{12}$

15. a) $\int_{-1}^{1}\int_{0}^{1-x^2}\int_{x^2}^{1-z} dy\,dz\,dx$ b) $\int_{0}^{1}\int_{-\sqrt{1-z}}^{\sqrt{1-z}}\int_{x^2}^{1-z} dy\,dx\,dz$

c) $\int_{0}^{1}\int_{0}^{1-z}\int_{-\sqrt{y}}^{\sqrt{y}} dx\,dy\,dz$ d) $\int_{0}^{1}\int_{0}^{1-y}\int_{-\sqrt{y}}^{\sqrt{y}} dx\,dz\,dy$

e) $\int_{0}^{1}\int_{-\sqrt{y}}^{\sqrt{y}}\int_{0}^{1-y} dz\,dx\,dy$

16. a) $\int_{0}^{1}\int_{0}^{1}\int_{-1}^{\sqrt{z}} dy\,dz\,dx$ b) $\int_{0}^{1}\int_{0}^{1}\int_{-1}^{\sqrt{y}} dy\,dx\,dz$

c) $\int_{0}^{1}\int_{0}^{\sqrt{z}}\int_{0}^{1} dx\,dy\,dz$ d) $\int_{-1}^{0}\int_{0}^{y^2}\int_{0}^{1} dx\,dz\,dy$

e) $\int_{-1}^{0}\int_{0}^{1}\int_{0}^{y^2} dz\,dx\,dy$

17. $V = \int_{0}^{1}\int_{-1}^{1}\int_{0}^{y^2} dz\,dy\,dx = \int_{0}^{1}\int_{-1}^{1} y^2\,dy\,dx = \frac{2}{3}\int_{0}^{1} dx = \frac{2}{3}$

18. $V = \int_{0}^{1}\int_{0}^{1-x}\int_{0}^{2-2z} dy\,dz\,dx = \int_{0}^{1}\int_{0}^{1-x} 2-2z\,dz\,dx = \int_{0}^{1} 1-x^2\,dx = \frac{2}{3}$

19. $V = \int_0^4 \int_0^{\sqrt{4-x}} \int_0^{2-y} dz\,dy\,dx = \int_0^4 \int_0^{\sqrt{4-x}} (2-y)\,dy\,dx = \int_0^4 2\sqrt{4-x} - \left(\frac{4-x}{2}\right) dx = \frac{20}{3}$

20. $V = 2\int_0^1 \int_0^{\sqrt{1-x^2}} \int_0^{-y} dz\,dy\,dx = -2\int_0^1 \int_0^{\sqrt{1-x^2}} y\,dy\,dx = -\int_0^1 1 - x^2\,dx = \frac{2}{3}$

21. $V = \int_0^1 \int_0^{2-2x} \int_0^{3-3x-3y/2} dz\,dy\,dx = \int_0^1 \int_0^{2-2x} 3 - 3x - \frac{3}{2}y\,dy\,dx = \int_0^1 3 - 6x + 3x^2\,dx = 1$

22. $V = \int_0^1 \int_0^{1-x} \int_0^{\cos(\pi x/2)} dz\,dy\,dx = \int_0^1 \int_0^{1-x} \left(\cos\left(\frac{\pi x}{2}\right)\right) dy\,dx = \int_0^1 \left(\cos\frac{\pi x}{2}\right)(1-x)\,dx = \frac{4}{\pi^2}$

23. $V = 8\int_0^1 \int_0^{\sqrt{1-x^2}} \int_0^{\sqrt{1-x^2}} dz\,dy\,dx = 8\int_0^1 \int_0^{\sqrt{1-x^2}} \sqrt{1-x^2}\,dy\,dx = 8\int_0^1 1 - x^2\,dx = \frac{16}{3}$

24. $V = \int_0^2 \int_0^{4-x^2} \int_0^{4-x^2-y} dz\,dy\,dx = \int_0^2 \int_0^{4-x^2} 4 - x^2 - y\,dy\,dx = \int_0^2 8 - 4x^2 + \frac{x^4}{2}\,dx = \frac{32}{15}$

25. $V = \int_0^4 \int_0^{\left(\sqrt{16-y^2}\right)/2} \int_0^{4-y} dx\,dz\,dy = \int_0^4 \int_0^{\left(\sqrt{16-y^2}\right)/2} (4-y)\,dz\,dy =$

$\int_0^4 \frac{\sqrt{16-y^2}}{2}(4-y)\,dy = 8\pi + \frac{32}{3}$

26. $V = \int_{-2}^2 \int_{-\sqrt{4-x^2}}^{\sqrt{4-x^2}} \int_0^{3-x} dz\,dy\,dx = \int_{-2}^2 \int_{-\sqrt{4-x^2}}^{\sqrt{4-x^2}} (3-x)\,dy\,dx =$

$\int_{-2}^2 (3-x)\left(2\sqrt{4-x^2}\right) dx = 12\pi$

27. $\text{average} = \frac{1}{8}\int_0^2 \int_0^2 \int_0^2 x^2 + 9\,dz\,dy\,dx = \frac{1}{8}\int_0^2 \int_0^2 \left(2x^2 + 18\right) dy\,dx = \frac{1}{8}\int_0^2 \left(4x^2 + 36\right) dx = \frac{31}{3}$

28. $\text{average} = \frac{1}{2}\int_0^1 \int_0^1 \int_0^2 x + y - z\,dz\,dy\,dx = \frac{1}{2}\int_0^1 \int_0^1 2x + 2y - 2\,dy\,dx = \frac{1}{2}\int_0^1 2x - 1\,dx = 0$

29. $\text{average} = \int_0^1 \int_0^1 \int_0^1 x^2 + y^2 + z^2\,dz\,dy\,dx = \int_0^1 \int_0^1 \left(x^2 + y^2 + \frac{1}{3}\right) dy\,dx = \int_0^1 \left(x^2 + \frac{2}{3}\right) dx = 1$

30. $\text{average} = \frac{1}{8}\int_0^2 \int_0^2 \int_0^2 xyz\,dz\,dy\,dx = \frac{1}{4}\int_0^2 \int_0^2 xy\,dy\,dx = \frac{1}{2}\int_0^2 x\,dx = 1$

14.5 MASSES AND MOMENTS IN THREE DIMENSIONS

1. $I_x = \int_{-c/2}^{c/2}\int_{-b/2}^{b/2}\int_{-a/2}^{a/2}\left(y^2+z^2\right)dx\,dy\,dz = 4a\int_0^{c/2}\int_0^{b/2}\left(y^2+z^2\right)dy\,dz =$
$4a\int_0^{c/2}\left(\frac{b^3}{24}+\frac{z^2b}{2}\right)dx = \frac{abc}{12}\left(b^2+c^2\right) \Rightarrow I_x = \frac{M}{12}\left(b^2+c^2\right);\ R_x = \sqrt{\frac{b^2+c^2}{12}}$,
$R_y = \sqrt{\frac{a^2+c^2}{12}}$, $R_z = \sqrt{\frac{a^2+b^2}{12}}$

2. The plane $z = \frac{4-2y}{3}$ is the top of the wedge. $I_x = \int_{-3}^{3}\int_{-2}^{4}\int_{-4/3}^{(4-2y)/3}\left(y^2+z^2\right)dz\,dy\,dx =$
$\int_{-3}^{3}\int_{-2}^{4}\frac{8y^2}{3}-\frac{2y^3}{3}+\frac{8(2-y)^3}{81}+\frac{64}{81}\,dy\,dx = \int_{-3}^{3}\frac{104}{3}\,dx = 208;$
$I_y = \int_{-3}^{3}\int_{-2}^{4}\int_{-4/3}^{2(2-y)/3}\left(x^2+z^2\right)dz\,dy\,dx = \int_{-3}^{3}\int_{-2}^{4}\frac{8x^2}{3}-\frac{2x^2y}{3}-\frac{8y}{81}+\frac{80}{81}\,dy\,dx =$
$\int_{-3}^{3}12x^2+\frac{432}{81}\,dx = 248;\ I_z = \int_{-3}^{3}\int_{-2}^{4}\int_{-4/3}^{2(2-y)/3}\left(x^2+y^2\right)dz\,dy\,dx =$
$\int_{-3}^{3}\int_{-2}^{4}\left(x^2+y^2\right)\left(\frac{8}{3}-\frac{2y}{3}\right)dy\,dx = 12\int_{-3}^{3}x^2+2\,dx = 360$

3. $I_x = \int_0^a\int_0^b\int_0^c\left(y^2+z^2\right)dz\,dy\,dx = \int_0^a\int_0^b cy^2+\frac{c^3}{3}\,dy\,dx = \int_0^a\frac{cb^3}{3}+\frac{c^3b}{3}\,dx = \frac{abc\left(b^2+c^2\right)}{3} =$
$\frac{M}{3}\left(b^2+c^2\right);\ I_y = \frac{M}{3}\left(a^2+c^2\right)$ and $I_z = \frac{M}{3}\left(a^2+b^2\right)$ by symmetry

4. a) $M = \int_0^1\int_0^{1-x}\int_0^{1-x-y}dz\,dy\,dx = \int_0^1\int_0^{1-x}1-x-y\,dy\,dx = \int_0^1\frac{x^2}{2}-x+\frac{1}{2}\,dx = \frac{1}{6}$
$M_{yz} = \int_0^1\int_0^{1-x}\int_0^{1-x-y}x\,dz\,dy\,dx = \int_0^1\int_0^{1-x}x(1-x-y)\,dy\,dx = \frac{1}{2}\int_0^1 x^3-2x^2+x\,dx = \frac{1}{24} \Rightarrow$
$\overline{x} = \overline{y} = \overline{z} = \frac{1}{4}$ by symmetry; $I_z = \int_0^1\int_0^{1-x}\int_0^{1-x-y}\left(y^2+z^2\right)dz\,dy\,dx =$
$\int_0^1\int_0^{1-x}y^2-xy^2-y^3+\frac{(1-x-y)^3}{3}\,dy\,dx = \frac{1}{6}\int_0^1(1-x)^4\,dx = \frac{1}{30}$, $I_y = I_x = \frac{1}{30}$ by symmetry

b) $R_x = \sqrt{\frac{I_x}{M}} = \sqrt{\frac{1}{5}} \approx 0.4472$, distance from the centroid to the x-axis is $\sqrt{0^2+\frac{1}{16}+\frac{1}{16}} =$
$\sqrt{\frac{1}{8}} \approx 0.3536$

5. $M = 4\int_0^1\int_0^1\int_{4y^2}^4 dz\,dy\,dx = 4\int_0^1\int_0^1 4 - 4y^2\,dy\,dx = 16\int_0^1 \frac{2}{3}dx = \frac{32}{3}$

$M_{xy} = 4\int_0^1\int_0^1\int_{4y^2}^4 z\,dz\,dy\,dx = 2\int_0^1\int_0^1 \left(16 - 16y^4\right) dy\,dx = \frac{128}{5}\int_0^1 dx = \frac{128}{5} \Rightarrow$

$\overline{z} = \frac{12}{5}$, $\overline{x} = \overline{y} = 0$ by symmetry; $I_z = 4\int_0^1\int_0^1\int_{4y^2}^4 \left(x^2 + y^2\right) dz\,dy\,dx =$

$16\int_0^1\int_0^1 x^2 - x^2y^2 + y^2 - y^4\,dy\,dx = 16\int_0^1 \frac{2x^2}{3} + \frac{2}{15}dx = \frac{256}{45}$;

$I_x = 4\int_0^1\int_0^1\int_{4y^2}^4 \left(y^2 + z^2\right) dz\,dy\,dx = 4\int_0^1\int_0^1 \left(4y^2 + \frac{64}{3}\right) - \left(4y^4 + \frac{64y^6}{3}\right) dy\,dx =$

$4\int_0^1 \frac{1976}{105}dx = \frac{7904}{105}$; $I_y = 4\int_0^1\int_0^1\int_{4y^2}^4 \left(x^2 + z^2\right) dz\,dy\,dx =$

$4\int_0^1\int_0^1 \left(4x^2 + \frac{64}{3}\right) - \left(4x^2y^2 + \frac{64y^6}{3}\right) dy\,dx = 4\int_0^1 \frac{8}{3}x^2 + \frac{128}{7}dx = \frac{4832}{63}$

6. a) $M = \int_{-2}^2 \int_{\left(-\sqrt{4-x^2}\right)/2}^{\left(\sqrt{4-x^2}\right)/2} \int_0^{2-x} dz\,dy\,dx = \int_{-2}^2 \int_{\left(-\sqrt{4-x^2}\right)/2}^{\left(\sqrt{4-x^2}\right)/2} (2 - x)\,dy\,dx =$

$\int_{-2}^2 (2 - x)\left(\sqrt{4 - x^2}\right) dx = 4\pi$; $M_{yz} = \int_{-2}^2 \int_{\left(-\sqrt{4-x^2}\right)/2}^{\left(\sqrt{4-x^2}\right)/2} \int_0^{2-x} x\,dz\,dy\,dx =$

$\int_{-2}^2 \int_{\left(-\sqrt{4-x^2}\right)/2}^{\left(\sqrt{4-x^2}\right)/2} x(2 - x)\,dy\,dx = \int_{-2}^2 x(2 - x)\left(\sqrt{4 - x^2}\right) dx = -2\pi$;

$M_{xz} = \int_{-2}^2 \int_{\left(-\sqrt{4-x^2}\right)/2}^{\left(\sqrt{4-x^2}\right)/2} \int_0^{2-x} y\,dz\,dy\,dx = \int_{-2}^2 \int_{\left(-\sqrt{4-x^2}\right)/2}^{\left(\sqrt{4-x^2}\right)/2} y(2 - x)\,dy\,dx =$

$\frac{1}{2}\int_{-2}^2 (2 - x)\left[\frac{4 - x^2}{4} - \frac{4 - x^2}{4}\right] dx = 0 \Rightarrow \overline{x} = -\frac{1}{2}$, $\overline{y} = 0$

b) $M_{xy} = \int_{-2}^2 \int_{\left(-\sqrt{4-x^2}\right)/2}^{\left(\sqrt{4-x^2}\right)/2} \int_0^{2-x} z\,dz\,dy\,dx = \frac{1}{2}\int_{-2}^2 \int_{\left(-\sqrt{4-x^2}\right)/2}^{\left(\sqrt{4-x^2}\right)/2} (2 - x)^2\,dy\,dx =$

$\frac{1}{2}\int_{-2}^2 (2 - x)^2\left(\sqrt{4 - x^2}\right) dx = 5\pi \Rightarrow \overline{z} = \frac{5}{4}$

7. a) $M = 4\int_0^2 \int_0^{\sqrt{4-x^2}} \int_{x^2+y^2}^4 dz\,dy\,dx = 4\int_0^{\pi/2} \int_0^2 \int_{r^2}^4 r\,dz\,dr\,d\theta =$

$4\int_0^{\pi/2} \int_0^2 4r - r^3\,dr\,d\theta = 4\int_0^{\pi/2} 4\,d\theta = 8\pi$; $M_{xy} = \int_0^{2\pi} \int_0^2 \int_{r^2}^4 r\,dz\,dr\,d\theta =$

$\int_0^{2\pi} \int_0^2 \frac{r}{2}\left(16 - r^4\right)\,dr\,d\theta = \frac{32\pi}{3}\int_0^{2\pi} d\theta = \frac{64\pi}{3} \Rightarrow \overline{z} = \frac{8}{3}$, $\overline{x} = \overline{y} = 0$ by symmetry

b) $M = 8\pi$; $4\pi = \int_0^{2\pi} \int_0^{\sqrt{c}} \int_{r^2}^c r\,dz\,dr\,d\theta = \int_0^{2\pi} \int_0^{\sqrt{c}} cr - r^3\,dr\,d\theta = \int_0^{2\pi} \left(\frac{c^2}{4}\right) d\theta = \frac{c^2\pi}{2} \Rightarrow$

$c^2 = 8 \Rightarrow c = 2\sqrt{2}$, since $c > 0$

8. $M = 8$, $M_{xy} = \int_{-1}^1 \int_3^5 \int_{-1}^1 z\,dz\,dy\,dx = \int_{-1}^1 \int_3^5 \left[z^2/2\right]_{-1}^1 dy\,dx = 0$

$M_{yz} = \int_{-1}^1 \int_3^5 \int_{-1}^1 x\,dz\,dy\,dx = 2\int_{-1}^1 \int_3^5 x\,dy\,dx = 4\int_{-1}^1 x^2\,dx = 0$

$M_{xz} = \int_{-1}^1 \int_3^5 \int_{-1}^1 y\,dz\,dy\,dx = 2\int_{-1}^1 \int_3^5 y\,dy\,dx = 16\int_{-1}^1 dx = 32 \Rightarrow \overline{x} = 0,\ \overline{y} = 4,\ \overline{z} = 0$

$I_x = \int_{-1}^1 \int_3^5 \int_{-1}^1 \left(y^2 + z^2\right) dz\,dy\,dx = \int_{-1}^1 \int_3^5 2y^2 + \frac{2}{3}\,dy\,dx = \frac{2}{3}\int_{-1}^1 100\,dx = \frac{400}{3}$

$I_y = \int_{-1}^1 \int_3^5 \int_{-1}^1 \left(x^2 + z^2\right) dz\,dy\,dx = \int_{-1}^1 \int_3^5 2x^2 + \frac{2}{3}\,dy\,dx = \frac{200}{3}\int_{-1}^1 dx = \frac{400}{3}$

$I_z = \int_{-1}^1 \int_3^5 \int_{-1}^1 \left(x^2 + y^2\right) dz\,dy\,dx = 2\int_{-1}^1 \int_3^5 \left(x^2 + y^2\right) dy\,dx = 2\int_{-1}^1 2x^2 + \frac{98}{3}\,dx = \frac{400}{3} \Rightarrow$

$R_x = R_y = R_z = \sqrt{\frac{50}{3}}$; The y–moment will not change, since the y–component of the center of mass is on the y–axis.

9. $I_L = \int_{-2}^2 \int_{-2}^4 \int_{-1}^{1-y/2} \left((y-6)^2 + z^2\right) dz\,dy\,dx = \int_{-2}^2 \int_{-2}^4 \frac{(y-6)^2(4-y)}{2} + \frac{(2-y)^3}{24} + \frac{1}{3}\,dy\,dx =$

$4\int_{-2}^4 \frac{13t^3}{24} + 5t^2 + 16t + \frac{49}{3}\,dt = 1386$, where $t = 2 - y$; $M = 36$, $R_L = \sqrt{\frac{I_L}{M}} = \sqrt{\frac{77}{2}}$

10. The plane $y + 2z = 2$ is the top of the wedge. $I_L = \int_{-2}^2 \int_{-2}^4 \int_{-1}^{(2-y)/2} \left((x-4)^2 + y^2\right) dz\,dy\,dx =$

$\frac{1}{2}\int_{-2}^2 \int_{-2}^4 \left(x^2 - 8x + 16 + y^2\right)(4-y)\,dy\,dx = \int_{-2}^2 9x^2 - 72x - 162\,dx = 696 \Rightarrow R_L = \sqrt{\frac{I_L}{M}} = \sqrt{\frac{29}{6}}$

11. $M = 8$, $I_L = \int_0^4 \int_0^2 \int_0^1 \left(z^2 + (y-2)^2\right) dz\,dy\,dx = \int_0^4 \int_0^2 y^2 - 4y + \frac{13}{3}\,dy\,dx = \frac{10}{3}\int_0^4 dx = \frac{40}{3} \Rightarrow$

$R_L = \sqrt{\frac{I_L}{M}} = \sqrt{\frac{5}{3}}$

12. $M = 8,\ I_L = \int_0^4 \int_0^2 \int_0^1 \left((x-4)^2 + y^2\right) dz\, dy\, dx = \int_0^4 \int_0^2 \left((x-4)^2 + y^2\right) dy\, dx =$

$\int_0^4 \left(2(x-4)^2 + \frac{8}{3}\right) dx = \frac{160}{3} \Rightarrow R_L = \sqrt{\frac{I_L}{M}} = \sqrt{\frac{20}{3}}$

13. $M = \int_0^2 \int_0^{2-x} \int_0^{2-x-y} 2x\, dz\, dy\, dx = \int_0^2 \int_0^{2-x} 4x - 2x^2 - 2xy\, dy\, dx = \int_0^2 x^3 - 4x^2 + 4x\, dx = \frac{4}{3}$

$M_{xy} = \int_0^2 \int_0^{2-x} \int_0^{2-x-y} 2xz\, dz\, dy\, dx = \int_0^2 \int_0^{2-x} x(2-x-y)^2\, dy\, dx = \int_0^2 \frac{x(2-x)^3}{3}\, dx = \frac{8}{15};$

$M_{xz} = \frac{8}{15}$ by symmetry; $M_{yz} = \int_0^2 \int_0^{2-x} \int_0^{2-x-y} 2x^2\, dz\, dy\, dx = \int_0^2 \int_0^{2-x} 2x^2(2-x-y)\, dy\, dx =$

$\int_0^2 \left(2x - x^2\right)^2 dx = \frac{16}{15} \Rightarrow \overline{x} = \frac{4}{5},\ \overline{y} = \overline{z} = \frac{2}{5}$

14. The plane $y + 2z = 2$ is the top of the wedge. $M = \int_{-2}^{2} \int_{-2}^{4} \int_{-1}^{1-y/2} (x+1)\, dz\, dy\, dx =$

$\int_{-2}^{2} \int_{-2}^{4} (x+1)\left(2 - \frac{y}{2}\right) dy\, dx = 36$ by the Calculus Toolkit;

$M_{yz} = \int_{-2}^{2} \int_{-2}^{4} \int_{-1}^{1-y/2} x(x+1)\, dz\, dy\, dx = \int_{-2}^{2} \int_{-2}^{4} x(x+1)\left(2 - \frac{y}{2}\right) dy\, dx = 48$ by the Calculus Toolkit; $M_{xz} = \int_{-2}^{2} \int_{-2}^{4} \int_{-1}^{1-y/2} y(x+1)\, dz\, dy\, dx = \int_{-2}^{2} \int_{-2}^{4} y(x+1)\left(2 - \frac{y}{2}\right) dy\, dx = 0$ by the Calculus Toolkit; $M_{xy} = \int_{-2}^{2} \int_{-2}^{4} \int_{-1}^{1-y/2} z(x+1)\, dz\, dy\, dx = \frac{1}{2}\int_{-2}^{2} \int_{-2}^{4} (x+1)\left(2 - \frac{y}{2}\right) dy\, dx = 0$ by the Calculus Toolkit $\Rightarrow \overline{x} = \frac{4}{3},\ \overline{y} = \overline{z} = 0;\ I_x = \int_{-2}^{2} \int_{-2}^{4} \int_{-1}^{1-y/2} (x+1)\left(y^2 + z^2\right) dz\, dy\, dx =$

$\int_{-2}^{2} \int_{-2}^{4} (x+1)\left(2y^2 + \frac{1}{3} - \frac{y^3}{2} + \frac{1}{3}\left(1 - \frac{y}{2}\right)^3\right) dy\, dx = 90$ by the Calculus Toolkit; $I_y =$

$\int_{-2}^{2} \int_{-2}^{4} \int_{-1}^{1-y/2} (x+1)\left(x^2 + z^2\right) dz\, dy\, dx = \int_{-2}^{2} \int_{-2}^{4} (x+1)\left(2x^2 + \frac{1}{3} - \frac{x^2 y}{2} + \frac{1}{3}\left(1 - \frac{y}{2}\right)^3\right) dy\, dx =$

66 by the Calculus Toolkit; $I_z = \int_{-2}^{2} \int_{-2}^{4} \int_{-1}^{1-y/2} (x+1)\left(x^2 + y^2\right) dz\, dy\, dx =$

$\int_{-2}^{2} \int_{-2}^{4} (x+1)\left(2 - \frac{y}{2}\right)\left(x^2 + y^2\right) dy\, dx = 120$ by the Calculus Toolkit $\Rightarrow R_x = \sqrt{\frac{I_x}{M}} = \sqrt{\frac{5}{2}},$

$R_y = \sqrt{\frac{I_y}{M}} = \sqrt{\frac{11}{6}},\ R_z = \sqrt{\frac{I_z}{M}} = \sqrt{\frac{10}{3}}$

15. $M = \int_0^1 \int_0^1 \int_0^1 (x + y + z + 1)\, dz\, dy\, dx = \int_0^1 \int_0^1 \left(x + y + \frac{3}{2}\right) dy\, dx = \int_0^1 (x + 2)\, dx = \frac{5}{2}$

$M_{xy} = \int_0^1 \int_0^1 \int_0^1 (x + y + z + 1)z\, dz\, dy\, dx = \frac{1}{2}\int_0^1 \int_0^1 \left(x + y + \frac{5}{3}\right) dy\, dx =$

$\frac{1}{2}\int_0^1 \left(x + \frac{13}{6}\right) dx = \frac{4}{3} \Rightarrow M_{xy} = M_{yz} = M_{xz} = \frac{4}{3}$ by symmetry $\therefore \overline{x} = \overline{y} = \overline{z} = \frac{8}{15}$

$I_z = \int_0^1 \int_0^1 \int_0^1 (x + y + z + 1)\left(x^2 + y^2\right) dz\, dy\, dx = \int_0^1 \int_0^1 \left(x + y + \frac{3}{2}\right)\left(x^2 + y^2\right) dy\, dx =$

$\int_0^1 x^3 + 2x^2 + \frac{1}{3}x + \frac{3}{4} dx = \frac{11}{6} \Rightarrow I_x = I_y = I_z = \frac{11}{6}$ by symmetry $\Rightarrow R_x = R_y = R_z = \sqrt{\frac{I_z}{M}} = \sqrt{\frac{11}{15}}$

16. a) $M = \int_0^2 \int_0^{\sqrt{x}} \int_0^{4-x^2} k\, xy\, dz\, dy\, dx = k\int_0^2 \int_0^{\sqrt{x}} xy\left(4 - x^2\right) dy\, dx =$

$\frac{k}{2}\int_0^2 \left(4x - x^3\right) dx = \frac{32k}{15}$

b) $M_{yz} = \int_0^2 \int_0^{\sqrt{x}} \int_0^{4-x^2} k\, x^2y\, dz\, dy\, dx = k\int_0^2 \int_0^{\sqrt{x}} x^2y\left(4 - x^2\right) dy\, dx = (2.666666)k$, by

the Calculus Toolkit $\Rightarrow \overline{x} = 1.25$

c) $M_{xz} = \int_0^2 \int_0^{\sqrt{x}} \int_0^{4-x^2} k\, xy^2\, dz\, dy\, dx = k\int_0^2 \int_0^{\sqrt{x}} xy^2\left(4 - x^2\right) dy\, dx = (1.567267)k$, by

the Calculus Toolkit $\Rightarrow \overline{y} = 0.734656$

d) $M_{xy} = \int_0^2 \int_0^{\sqrt{x}} \int_0^{4-x^2} k\, xyz\, dz\, dy\, dx = k\int_0^2 \int_0^{\sqrt{x}} \left(\frac{xy\left(4 - x^2\right)^2}{2}\right) dy\, dx = (2.438095)k$,

by the Calculus Toolkit $\Rightarrow \overline{z} = 1.142857$

14.6 TRIPLE INTEGRALS IN CYLINDRICAL AND SPHERICAL COORDINATES

1. $\int_0^{2\pi} \int_0^1 \int_r^{\sqrt{2-r^2}} r\, dz\, dr\, d\theta = \int_0^{2\pi} \int_0^1 \left(\left(2 - r^2\right)^{1/2} r - r^2\right) dr\, d\theta = \int_0^{2\pi} \left(\frac{2^{3/2}}{3} - \frac{2}{3}\right) d\theta = \frac{4\pi\left(\sqrt{2} - 1\right)}{3}$

2. $\int_0^{2\pi} \int_0^3 \int_{r^2/3}^{\sqrt{18-r^2}} r\, dz\, dr\, d\theta = \int_0^{2\pi} \int_0^3 \left(\sqrt{18 - r^2} - \frac{r^2}{3}\right) r\, dr\, d\theta =$

$\int_0^{2\pi} \left[-\frac{\left(18 - r^2\right)^{3/2}}{3} - \frac{r^4}{12}\right]_0^3 d\theta = \frac{9\pi\left(8\sqrt{2} - 1\right)}{2}$

3. $\int_0^{2\pi}\int_0^{\theta/2\pi}\int_0^{3+24r^2} r\,dz\,dr\,d\theta = \int_0^{2\pi}\int_0^{\theta/2\pi}\left(3+24r^2\right) r\,dr\,d\theta = \frac{3}{2}\int_0^{2\pi}\frac{\theta^2}{4\pi^2}+\frac{4\theta^4}{16\pi^4}\,d\theta = \frac{17\pi}{5}$

4. $\int_0^{\pi}\int_0^{\theta/\pi}\int_{-\sqrt{4-r^2}}^{3\sqrt{4-r^2}} r\,dz\,dr\,d\theta = 4\int_0^{\pi}\int_0^{\theta/\pi} 4r - r^3\,dr\,d\theta = 4\int_0^{\pi}\frac{2\theta^2}{\pi^2}-\frac{\theta^4}{4\pi^4}\,d\theta = \frac{37\pi}{15}$

5. $\int_0^{2\pi}\int_0^{1}\int_r^{\left(2-r^2\right)^{-1/2}} 3\,r\,dz\,dr\,d\theta = 3\int_0^{2\pi}\int_0^{1}\left(2-r^2\right)^{-1/2}r - r^2\,dr\,d\theta =$

$3\int_0^{2\pi}\left(\sqrt{2}-\frac{4}{3}\right)d\theta = \pi\left(6\sqrt{2}-8\right)$

6. $\int_0^{2\pi}\int_0^{1}\int_{-1/2}^{1/2}\left(r^2\sin^2\theta + z^2\right) r\,dz\,dr\,d\theta = \int_0^{2\pi}\int_0^{1}\left(r^3\sin^2\theta + \frac{r}{12}\right) dr\,d\theta =$

$\int_0^{2\pi}\frac{\sin^2\theta}{4}+\frac{1}{24}\,d\theta = \frac{\pi}{3}$

7. $\int_0^{\pi}\int_0^{\pi}\int_0^{2\sin\phi}\rho^2\sin\phi\,d\rho\,d\phi\,d\theta = \frac{8}{3}\int_0^{\pi}\int_0^{\pi}\sin^4\phi\,\,d\phi\,d\theta = \frac{2}{3}\int_0^{\pi}\frac{3\pi}{2}\,d\theta = \pi^2$

8. $\int_0^{2\pi}\int_0^{\pi/4}\int_0^{2}\rho\cos\phi\,\rho^2\sin\phi\,d\rho\,d\phi\,d\theta = 2\int_0^{2\pi}\int_0^{2\pi}\sin 2\phi\,\,d\phi\,d\theta = \int_0^{2\pi}d\theta = 2\pi$

9. $\int_0^{2\pi}\int_0^{\pi}\int_0^{(1-\cos\phi)/2}\rho^2\sin\phi\,d\rho\,d\phi\,d\theta = \frac{1}{24}\int_0^{2\pi}\int_0^{\pi}\left(1-\cos\phi\right)^3\sin\phi\,\,d\phi\,d\theta =$

$\frac{1}{6}\int_0^{2\pi}d\theta = \frac{\pi}{3}$

10. $\int_0^{3\pi/2}\int_0^{\pi}\int_0^{1} 5\rho^3\sin^3\phi\,d\rho\,d\phi\,d\theta = \frac{5}{4}\int_0^{3\pi/2}\int_0^{\pi}\sin^3\phi\,\,d\phi\,d\theta = \frac{5}{3}\int_0^{3\pi/2}d\theta = \frac{5\pi}{2}$

11. $\int_0^{2\pi}\int_0^{\pi/3}\int_{\sec\phi}^{2} 3\rho^2\sin\phi\,d\rho\,d\phi\,d\theta = \int_0^{2\pi}\int_0^{\pi/3}\left(8-\sec^3\phi\right)\sin\phi\,\,d\phi\,d\theta = \frac{5}{2}\int_0^{2\pi}d\theta = 5\pi$

12. $\int_0^{2\pi}\int_0^{\pi/4}\int_0^{\sec\phi}\rho^3\sin\phi\cos\phi\,d\rho\,d\phi\,d\theta = \frac{1}{4}\int_0^{2\pi}\int_0^{\pi/4}\tan\phi\sec^2\phi\,\,d\phi\,d\theta =$

$\frac{1}{16}\int_0^{2\pi}d\theta = \frac{\pi}{8}$

13. a) $8\int_0^{\pi/2}\int_0^{\pi/2}\int_0^{2}\rho^2\sin\phi\,d\rho\,d\phi\,d\theta$

b) $8\int_0^{\pi/2}\int_0^{2}\int_0^{\sqrt{4-r^2}} r\,dz\,dr\,d\theta$

c) $8\int_0^{2}\int_0^{\sqrt{4-x^2}}\int_0^{\sqrt{4-x^2-y^2}} dz\,dy\,dx$

14. a) $\int_0^{\pi/2}\int_0^{3/\sqrt{2}}\int_r^{\sqrt{9-r^2}} zr\,dz\,dr\,d\theta$

b) $\int_0^{\pi/2}\int_0^{\pi/4}\int_0^{3} \rho^2\sin\phi\,d\rho\,d\phi\,d\theta$

c) $\int_0^{\pi/2}\int_0^{\pi/4}\int_0^{3} \rho^2\sin\phi\,d\rho\,d\phi\,d\theta = 9\int_0^{\pi/2}\int_0^{\pi/4} \sin\phi\,d\phi\,d\theta =$

$-9\int_0^{\pi/2}\left(\frac{1}{\sqrt{2}}-1\right)d\theta = \frac{9\pi\left(2-\sqrt{2}\right)}{4}$

15. $\int_{-\pi/2}^{\pi/2}\int_0^{\cos\theta}\int_0^{3r^2} f(r,\theta,z)\,r\,dz\,dr\,d\theta$

16. $\int_{-\pi/2}^{\pi/2}\int_0^{1}\int_0^{r\cos\theta} r^3\,dz\,dr\,d\theta = \int_{-\pi/2}^{\pi/2}\int_0^{1} r^4\cos\theta\,dr\,d\theta = \frac{1}{5}\int_{-\pi/2}^{\pi/2}\cos\theta\,d\theta = \frac{2}{5}$

17. $v = 4\int_0^{\pi/2}\int_0^{1}\int_0^{r^2} r\,dz\,dr\,d\theta = 4\int_0^{\pi/2}\int_0^{1} r^3\,dr\,d\theta = \int_0^{\pi/4} d\theta = \frac{\pi}{2}$

18. $v = 4\int_0^{\pi/2}\int_0^{1}\int_{x^2+y^2}^{x^2+y^2+1} r\,dz\,dr\,d\theta = 4\int_0^{\pi/2}\int_0^{1} r\,dr\,d\theta = 2\int_0^{\pi/2} d\theta = \pi$

19. $V = 4\int_0^{\pi/2}\int_0^{2}\int_0^{4-r^2} r\,dz\,dr\,d\theta = 4\int_0^{\pi/2}\int_0^{2}\left(4r-r^3\right)dr\,d\theta = 16\int_0^{\pi/2} d\theta = 8\pi$

20. $V = \int_0^{2\pi}\int_0^{2}\int_0^{4-r\sin\theta} r\,dz\,dr\,d\theta = \int_0^{2\pi}\int_0^{2}\left(4-r\sin\theta\right)r\,dr\,d\theta = 8\int_0^{2\pi} 1-\frac{\sin\theta}{3}\,d\theta = 16\pi$

21. $V = 4\int_0^{\pi/2}\int_0^{1}\int_{4r^2}^{5-r^2} r\,dz\,dr\,d\theta = 4\int_0^{\pi/2}\int_0^{1}\left(5-5r^2\right)r\,dr\,d\theta = 5\int_0^{\pi/2} d\theta = \frac{5\pi}{2}$

22. $V = 4\int_0^{\pi/2}\int_1^{3}\int_0^{9-x^2-y^2} r\,dz\,dr\,d\theta = 4\int_0^{\pi/2}\int_1^{3} 9r-r^3\,dr\,d\theta = 64\int_0^{\pi/2} d\theta = 32\pi$

23. $V = 8\int_0^{\pi/2}\int_0^{1}\int_0^{\sqrt{4-r^2}} r\,dz\,dr\,d\theta = 8\int_0^{\pi/2}\int_0^{1}\left(4-r^2\right)^{1/2} r\,dr\,d\theta =$

$-\frac{8}{3}\int_0^{\pi/2}\left(3^{3/2}-8\right)d\theta = \frac{4\pi\left(8-3\sqrt{3}\right)}{3}$

24. $V = 4\int_0^{\pi/2}\int_0^{1}\int_{x^2+y^2}^{\sqrt{2-\left(x^2+y^2\right)}} r\,dz\,dr\,d\theta = 4\int_0^{\pi/2}\int_0^{1}\left(2-r^2\right)^{1/2}(r)-r^3\,dr\,d\theta =$

$4\int_0^{\pi/2}\left(\frac{2\sqrt{2}}{3}-\frac{7}{12}\right)d\theta = \frac{\pi\left(8\sqrt{3}-7\right)}{6}$

25. $\text{average} = \frac{1}{2\pi}\int_0^{2\pi}\int_0^{1}\int_{-1}^{1} r\,dz\,dr\,d\theta = \frac{1}{2\pi}\int_0^{2\pi}\int_0^{1} 2r^2\,dr\,d\theta = \frac{1}{3\pi}\int_0^{2\pi} d\theta = \frac{2}{3}$

26. $\text{average} = \frac{1}{4\pi/3}\int_0^{2\pi}\int_0^1\int_{-\sqrt{1-r^2}}^{\sqrt{1-r^2}} r^2\, dz\, dr\, d\theta = \frac{3}{4\pi}\int_0^{2\pi}\int_0^1 \left(1-r^2\right)^{1/2} 2r^2\, dr\, d\theta = \frac{3}{32}\int_0^{2\pi} d\theta = \frac{3\pi}{16}$

27. $M = 4\int_0^{\pi/2}\int_0^1\int_0^r r\, dz\, dr\, d\theta = 4\int_0^{\pi/2}\int_0^1 r^2\, dr\, d\theta = \frac{4}{3}\int_0^{\pi/2} d\theta = \frac{2\pi}{3}$

$M_{xy} = \int_0^{2\pi}\int_0^1\int_0^r r\, z\, dz\, dr\, d\theta = \frac{1}{2}\int_0^{2\pi}\int_0^1 r^3\, dr\, d\theta = \frac{1}{8}\int_0^{2\pi} d\theta = \frac{\pi}{4} \Rightarrow$

$\overline{z} = \frac{3}{8}$, $\overline{x} = \overline{y} = 0$ by symmetry

28. $M = 4\int_0^{\pi/2}\int_0^2\int_{\sqrt{r}}^2 r\, dz\, dr\, d\theta = 4\int_0^{\pi/2}\int_0^2 \left(2-\sqrt{r}\right) r\, dr\, d\theta = 4\int_0^{\pi/2} 4 - \frac{2^{7/2}}{5}\, d\theta = \frac{8\pi}{5}\left(5-2\sqrt{2}\right)$

$M_{xy} = 4\int_0^{\pi/2}\int_0^2\int_{\sqrt{r}}^2 r\, z\, dz\, dr\, d\theta = 2\int_0^{\pi/2}\int_0^2 4r - r^2\, dr\, d\theta = 2\int_0^{\pi/2} \frac{16}{3}\, d\theta = \frac{16\pi}{3} \Rightarrow$

$\overline{z} = \frac{10}{3\left(5-2\sqrt{2}\right)}$, $\overline{x} = \overline{y} = 0$ by symmetry

29. $M = 12\pi$, $I_z = \int_0^{2\pi}\int_1^2\int_0^4 r^3\, dz\, dr\, d\theta = 4\int_0^{2\pi}\int_1^2 r^3\, dr\, d\theta = 15\int_0^{2\pi} d\theta = 30\pi \Rightarrow$

$R_z = \sqrt{\frac{I_z}{M}} = \sqrt{\frac{5}{2}}$

30. $M = \int_0^{\pi/2}\int_0^2\int_0^{\sqrt{x^2+y^2}} r\, dz\, dr\, d\theta = \int_0^{\pi/2}\int_0^2 r^{3/2}\, dr\, d\theta = \frac{2}{5}\int_0^{\pi/2} 2^{5/2}\, d\theta = \frac{4\sqrt{2}\pi}{5}$

$M_{yz} = \int_0^{\pi/2}\int_0^2\int_0^{\sqrt{x^2+y^2}} r^2\cos\theta\, dz\, dr\, d\theta = \int_0^{\pi/2}\int_0^2 r^{5/2}\cos\theta\, dr\, d\theta = \frac{2^{9/2}}{7}\int_0^{\pi/2} \cos\theta\, d\theta =$

$\frac{16\sqrt{2}}{7} \Rightarrow \overline{x} = \overline{y} = \frac{20}{7\pi}$ by symmetry

$M_{xy} = = \int_0^{\pi/2}\int_0^2\int_0^{\sqrt{x^2+y^2}} r\, z\, dz\, dr\, d\theta = \frac{1}{2}\int_0^{\pi/2}\int_0^2 r^2\, dr\, d\theta = \frac{4}{3}\int_0^{\pi/2} d\theta = \frac{2\pi}{3} \Rightarrow \overline{z} = \frac{5}{6\sqrt{2}}$

31. a) $I_z = \int_0^{2\pi}\int_0^1\int_{-1}^1 r^2\, dz\, dr\, d\theta = 2\int_0^{2\pi}\int_0^1 r^3\, dr\, d\theta = \frac{1}{2}\int_0^{2\pi} d\theta = \pi$

b) $I_x = \int_0^{2\pi}\int_0^1\int_{-1}^1 \left(r^2\sin^2\theta + z^2\right) dz\, dr\, d\theta = \int_0^{2\pi}\int_0^1 \left(2r^2\sin^2\theta + \frac{2}{3}\right) r\, dr\, d\theta =$

$\int_0^{2\pi}\left(\frac{\sin^2\theta}{2} + \frac{1}{3}\right) d\theta = \frac{7\pi}{6}$

32. We orient the cone with its vertex at the origin and note that $\phi = \frac{\pi}{4}$.

$$I_x = \int_0^{2\pi}\int_0^1\int_r^1 \left(r^2\sin^2\theta + z^2\right) r\,dz\,dr\,d\theta = \int_0^{2\pi}\int_0^1 r^3\sin^2\theta + \frac{r}{3} - r^4\sin^2\theta - \frac{r^4}{3}\,dr\,d\theta =$$

$$\int_0^{2\pi} \frac{\sin^2\theta}{20} + \frac{1}{10}\,d\theta = \frac{\pi}{4}$$

33. $$I_z = \int_0^{2\pi}\int_0^a\int_{-\sqrt{a^2-r^2}}^{\sqrt{a^2-r^2}} r^3\,dz\,dr\,d\theta = -\int_0^{2\pi}\int_0^a r^2\left(a^2-r^2\right)^{1/2}(-2r)\,dr\,d\theta = \int_0^{2\pi}\frac{4a^5}{15}\,d\theta = \frac{8\pi a^5}{15}$$

34. $$M = \frac{4\pi}{9},\ M_{yz} = \int_{-\pi/3}^{\pi/3}\int_0^1\int_{-\sqrt{1-r^2}}^{\sqrt{1-r^2}} \left(r^2\cos\theta\right) dz\,dr\,d\theta = 2\int_{-\pi/3}^{\pi/3}\int_0^1 \left(r^2\cos\theta\right)\sqrt{1-r^2}\,dr\,d\theta =$$

$$\frac{\pi}{8}\int_{-\pi/3}^{\pi/3}\cos\theta\,d\theta = \frac{\pi\sqrt{3}}{8} \Rightarrow \overline{x} = \frac{9\sqrt{3}}{32},\ \overline{x} = \overline{y} = 0 \text{ by symmetry}$$

35. $$\int_0^{2\pi}\int_{\pi/3}^{2\pi/3}\frac{a^3}{3}\sin\phi\,d\phi\,d\theta = \frac{a^3}{3}\int_0^{2\pi}d\theta = \frac{2\pi a^3}{3} \Rightarrow V = \frac{4}{3}\pi a^3 - \frac{2\pi a^3}{3} = \frac{2\pi a^3}{3}$$

36. $$V = \int_0^{\pi/6}\int_0^{\pi/2}\int_0^a \rho^2\sin\phi\,d\rho\,d\phi\,d\theta = \frac{a^3}{3}\int_0^{\pi/6}\int_0^{\pi/2}\sin\phi\,d\phi\,d\theta = \frac{a^3}{3}\int_0^{\pi/6}d\theta = \frac{a^3\pi}{18}$$

37. $$V = \int_0^{2\pi}\int_0^{\pi/3}\int_{\sec\phi}^2 \rho^2\sin\phi\,d\rho\,d\phi\,d\theta = \frac{1}{3}\int_0^{2\pi}\int_0^{\pi/3} 8\sin\phi - \tan\phi\sec^2\phi\,d\phi\,d\theta =$$

$$\frac{5}{6}\int_0^{2\pi}d\theta = \frac{5\pi}{3}$$

38. $$V = 4\int_0^{\pi/2}\int_0^{\pi}\int_0^{1-\cos\phi} \rho^2\sin\phi\,d\rho\,d\phi\,d\theta = \frac{4}{3}\int_0^{\pi/2}\int_0^{\pi}\left(1-\cos\phi\right)^3\sin\phi\,d\phi\,d\theta =$$

$$\frac{16}{3}\int_0^{\pi/2}d\theta = \frac{8\pi}{3}$$

39. $$\text{average} = \frac{3}{4\pi}\int_0^{2\pi}\int_0^{\pi}\int_0^1 \rho^3\sin\phi\,d\rho\,d\phi\,d\theta = \frac{3}{16\pi}\int_0^{2\pi}\int_0^{\pi}\sin\phi\,d\phi\,d\theta = \frac{3}{8\pi}\int_0^{2\pi}d\theta = \frac{3}{4}$$

40. $$M = \int_0^{2\pi}\int_{\pi/3}^{\pi/2}\int_0^2 \rho^2\sin\phi\,d\rho\,d\phi\,d\theta = \frac{8}{3}\int_0^{2\pi}\int_{\pi/3}^{\pi/2}\sin\phi\,d\phi\,d\theta = \frac{4}{3}\int_0^{2\pi}d\theta = \frac{8\pi}{3}$$

$$M_{xy} = \int_0^{2\pi}\int_{\pi/3}^{\pi/2}\int_0^2 \rho^3\sin\phi\cos\phi\,d\rho\,d\phi\,d\theta = 2\int_0^{2\pi}\int_{\pi/3}^{\pi/2}\sin 2\phi\,d\phi\,d\theta = \frac{1}{2}\int_0^{2\pi}d\theta = \pi \Rightarrow$$

$$\overline{z} = \frac{3}{8},\ \overline{x} = \overline{y} = 0 \text{ by symmetry}$$

41. $M = \int_0^{2\pi}\int_0^{\pi/4}\int_0^a \rho^2 \sin\phi\, d\rho\, d\phi\, d\theta = \frac{a^3}{3}\int_0^{2\pi}\int_0^{\pi/4} \sin\phi\, d\phi\, d\theta =$

$\frac{a^3}{3}\int_0^{2\pi} \frac{\sqrt{2}-1}{\sqrt{2}}\, d\theta = \frac{\pi a^3\left(2-\sqrt{2}\right)}{3}$

$M_{xy} = \int_0^{2\pi}\int_0^{\pi/4}\int_0^a \rho^3 \sin\phi\cos\phi\, d\rho\, d\phi\, d\theta = \frac{a^3}{4}\int_0^{2\pi}\int_0^{\pi/4} \sin\phi\cos\phi\, d\phi\, d\theta =$

$\frac{a^4}{16}\int_0^{2\pi} d\theta = \frac{\pi a^4}{8} \Rightarrow \overline{z} = \frac{3\left(2+\sqrt{2}\right)a}{16}$, $\overline{x} = \overline{y} = 0$ by symmetry

42. a) $M = \int_0^{2\pi}\int_0^{\pi}\int_0^a \rho^4 \sin\phi\, d\rho\, d\phi\, d\theta = \frac{a^5}{5}\int_0^{2\pi}\int_0^{\pi} \sin\phi\, d\phi\, d\theta = \frac{2a^5}{5}\int_0^{2\pi} d\theta = \frac{4\pi a^5}{5}$

$I_z = \int_0^{2\pi}\int_0^{\pi}\int_0^a \rho^6 \sin^3\phi\, d\rho\, d\phi\, d\theta = \frac{a^7}{7}\int_0^{2\pi}\int_0^{\pi} \left(1-\cos^2\phi\right)\sin\phi\, d\phi\, d\theta = \frac{4a^7}{21}\int_0^{2\pi} d\theta = \frac{8a^7\pi}{21} \Rightarrow$

$R_z = \sqrt{\frac{I_z}{M}} = \sqrt{\frac{10}{21}}\, a$

b) $M = \int_0^{2\pi}\int_0^{\pi}\int_0^a \rho^3 \sin^2\phi\, d\rho\, d\phi\, d\theta = \frac{a^4}{4}\int_0^{2\pi}\int_0^{\pi} \frac{(1-\cos 2\phi)}{2}\, d\phi\, d\theta =$

$\frac{\pi a^4}{8}\int_0^{2\pi} d\theta = \frac{\pi^2 a^4}{4}$; $I_z = \int_0^{2\pi}\int_0^{\pi}\int_0^a \rho^5 \sin^4\phi\, d\rho\, d\phi\, d\theta = \frac{a^6}{6}\int_0^{2\pi}\int_0^{\pi} \sin^4\phi\, d\phi\, d\theta =$

$\frac{a^6}{16}\int_0^{2\pi} \pi\, d\theta = \frac{a^6\pi^2}{8} \Rightarrow R_z = \sqrt{\frac{I_z}{M}} = \frac{a}{\sqrt{2}}$

14.7 SUBSTITUTIONS IN MULTIPLE INTEGRALS

1. $\int_0^4\int_{y/2}^{1+y/2} \frac{2x-y}{2}\, dx\, dy = \int_0^4 \left[\frac{x^2}{2} - \frac{xy}{2}\right]_{y/2}^{1+y/2} dy = \frac{1}{2}\int_0^4 dy = 2$

2. a) $\begin{vmatrix} \cos v & -u\sin v \\ \sin v & u\cos v \end{vmatrix} = u\cos^2 v + u\sin^2 v = u$

b) $\begin{vmatrix} \sin v & u\cos v \\ \cos v & -u\sin v \end{vmatrix} = -u\sin^2 v - u\cos^2 v = -u$

3. a) $x = \frac{u+v}{3}$, $y = \frac{v-2u}{3}$, $J(u,v) = \begin{vmatrix} 1/3 & 1/3 \\ -2/3 & 1/3 \end{vmatrix} = \frac{1}{9} + \frac{2}{9} = \frac{1}{3}$

b) $\int_4^7 \int_{-1}^2 \frac{vu}{3}\, dv\, du = \frac{1}{2}\int_4^7 u\, du = \frac{33}{4}$

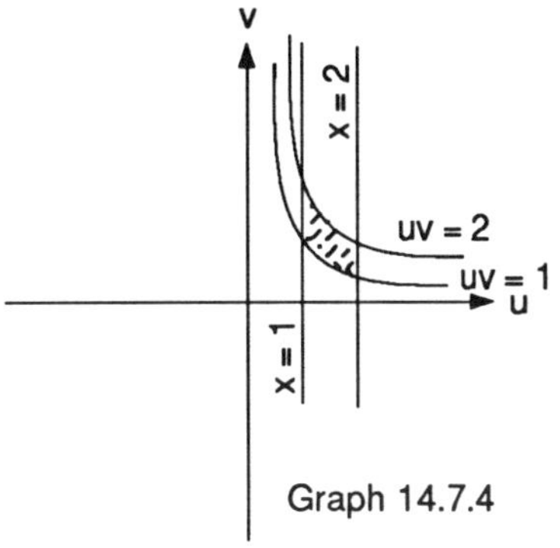

Graph 14.7.4

4. a) $J(u,v) = \begin{vmatrix} 1 & 0 \\ v & u \end{vmatrix} = u$

b) $\int_1^2 \int_{1/u}^{2/u} \left(\frac{uv}{u}\right) u\, dv\, du = \frac{3}{2}\int_a^b \frac{1}{u}\, du = \frac{3 \ln 2}{2}$; $\int_1^2 \int_1^2 \frac{y}{x}\, dy\, dx = \frac{3}{2}\int_1^2 \frac{1}{x}\, dx = \frac{3 \ln 2}{2}$

5. $J(u,v) = \begin{vmatrix} v^{-1} & -uv^{-2} \\ v & u \end{vmatrix} = v^{-1}u + v^{-1}u = \frac{2u}{v}$; $\int_1^1 \int_1^2 (v+u)\left(\frac{2u}{v}\right) dv\, du =$

$\int_1^3 2u + (\ln 4)\, u^2\, du = 8 + \frac{26 \ln 4}{3}$

6. $J(u,v) = \begin{vmatrix} a & 0 \\ 0 & b \end{vmatrix} = ab$; $A = \int_{-a}^{a} \int_{(-b/a)\sqrt{a^2-x^2}}^{(b/a)\sqrt{a^2-x^2}} dy\, dx = \int_{-1}^{1} \int_{-\sqrt{1-u^2}}^{\sqrt{1-u^2}} ab\, du\, dv =$

$2ab\int_{-1}^{1} \sqrt{1-u^2}\, du = ab\pi$

7. $J(r,\theta) = \begin{vmatrix} a\cos\theta & -ar\sin\theta \\ b\sin\theta & br\cos\theta \end{vmatrix} = abr\cos^2\theta + abr\sin^2\theta = abr$

$I_o = \int_{-a}^{a} \int_{(-b/a)\sqrt{a^2-x^2}}^{(b/a)\sqrt{a^2-x^2}} \left(x^2 + y^2\right) dy\, dx = \int_0^{2\pi} \int_0^1 r^3\left(a^2\cos^2\theta + b^2\sin^2\theta\right) abr\, dr\, d\theta =$

$\frac{ab}{4}\int_a^b a^2\cos^2\theta + b^2\sin^2\theta\, d\theta = \frac{\pi ab\left(a^2 + b^2\right)}{4}$

8. $J(u,v) = \begin{vmatrix} 1 & 1/2 \\ 0 & 1 \end{vmatrix} = 1; \int_0^2 \int_{y/2}^{(y+4)/2} y^3(2x-y)\exp\left((2x-y)^2\right) dx\, dy =$

$\int_0^2 \int_0^2 v^3(2u)\exp\left((2u)^2\right) dv\, du = 8\int_0^2 \exp\left(4u^2\right) u\, du = e^{16} - 1$

9. $\begin{vmatrix} \sin\phi\cos\theta & \rho\cos\phi\cos\theta & -\rho\sin\phi\sin\theta \\ \sin\phi\sin\theta & \rho\cos\phi\sin\theta & \rho\sin\phi\cos\theta \\ \cos\phi & -\rho\sin\phi & 0 \end{vmatrix} = \cos\phi \begin{vmatrix} \rho\cos\phi\cos\theta & -\rho\sin\phi\sin\theta \\ \rho\cos\phi\sin\theta & \rho\sin\phi\cos\theta \end{vmatrix} +$

$\rho\sin\phi \begin{vmatrix} \sin\phi\cos\theta & -\rho\sin\phi\sin\theta \\ \sin\phi\sin\theta & \rho\sin\phi\cos\theta \end{vmatrix} = \rho^2\cos\phi\left(\sin\phi\cos\phi\cos^2\theta + \sin\phi\cos\phi\sin^2\theta\right) +$

$\rho^2\sin\phi\left(\sin^2\phi\cos^2\theta + \sin^2\phi\sin^2\theta\right) = \rho^2\sin\phi\cos^2\phi + \rho^2\sin^3\phi =$

$\rho^2\sin\phi\left(\cos^2\phi + \sin^2\phi\right) = \rho^2\sin\phi$

10. $\int_0^3 \int_0^4 \int_{y/2}^{1+y/2} \left(\frac{2x-y}{2} + \frac{z}{3}\right) dx\, dy\, dz = \int_0^3 \int_0^4 \frac{1}{2}(y+1) + \frac{z}{3} - \frac{y}{2} dy\, dz = 4\int_0^3 \frac{1}{2} + \frac{z}{3} dz = 12$

11. $J(u,v,w) = \begin{vmatrix} a & 0 & 0 \\ 0 & b & 0 \\ 0 & 0 & c \end{vmatrix} = abc$. The transformation takes the $\frac{x^2}{a^2} + \frac{y^2}{b^2} + \frac{z^2}{c^2} = 1$ region in the xyz–space into the $u^2 + v^2 + w^2 = 1$ region in the uvw–space which is a unit sphere with $V = \frac{4}{3}\pi$.

$\therefore\ V = \int\int_R\int dx\, dy\, dz = \int\int_G\int abc\, du\, dv\, dw = \frac{4\pi abc}{3}$

12. $J(u,v,w) = \begin{vmatrix} a & 0 & 0 \\ 0 & b & 0 \\ 0 & 0 & c \end{vmatrix} = abc; \int\int\int |xyz|\, dx\, dy\, dz = 8\int_0^1 \int_0^1 \int_0^1 a^2b^2c^2 uvw\, dw\, dv\, du =$

$4a^2b^2c^2 \int_0^1 \int_0^1 uv\, du\, dv = 2a^2b^2c^2 \int_0^1 u\, du = a^2b^2c^2$

13. $J(u,v,w) = \begin{vmatrix} 1 & 0 & 0 \\ -v/u^2 & 1/u & 0 \\ 0 & 0 & 1/3 \end{vmatrix} = \frac{1}{3u}; \int\int_R\int x^2y + 3xyz\, dx\, dy\, dz =$

$\int\int_G\int u^2\left(\frac{v}{u}\right) + 3u\frac{v}{u}\frac{w}{3} J(u,v,w)\, du\, dv\, dw = \frac{1}{3}\int_0^3 \int_0^2 \int_1^2 v + \frac{vw}{u} du\, dv\, dw =$

$\frac{1}{3}\int_0^3 \int_0^2 v + vw\ln 2\, dv\, dw = \frac{1}{3}\int_0^3 2 + (\ln 4)w\, dw = 2 + \ln 8$

PRACTICE EXERCISES

1. $$\int_1^{10}\int_0^{1/y} ye^{xy}\,dy\,dx = \int_1^{10}\left[e^{xy}\right]_0^{1/y} dx = \int_1^{10} (e-1)\,dx = 9e-9$$

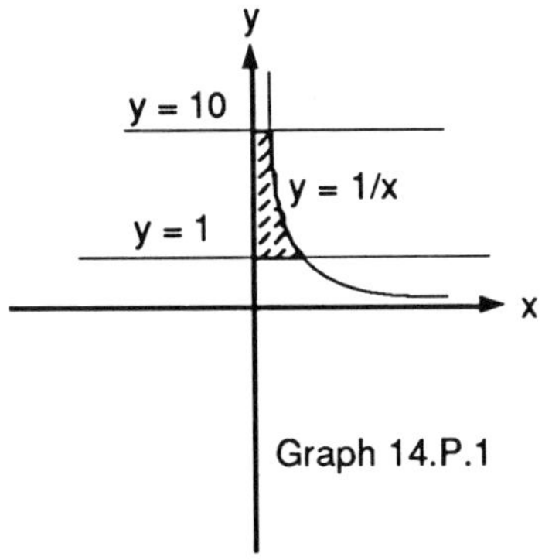

Graph 14.P.1

2. $$\int_0^1\int_0^{x^3} e^{y/x}\,dy\,dx = \int_0^1 x\left[e^{y/x}\right]_0^{x^3} dx = \int_0^1 x\exp\left(x^2\right) - x\,dx = \frac{e-2}{2}$$

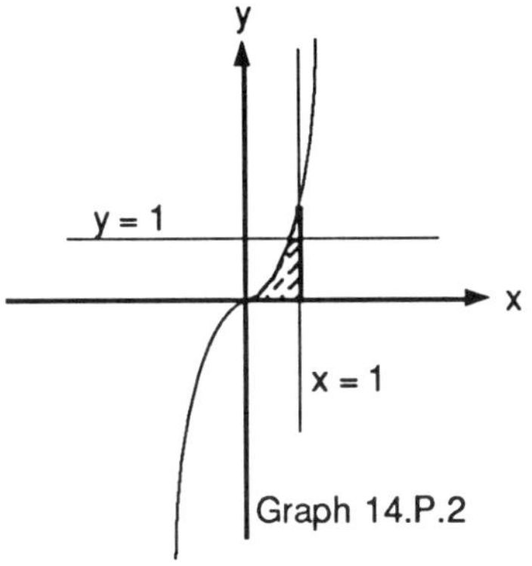

Graph 14.P.2

3. $$\int_0^1\int_y^{\sqrt{y}} f(x,y)\,dx\,dy =$$

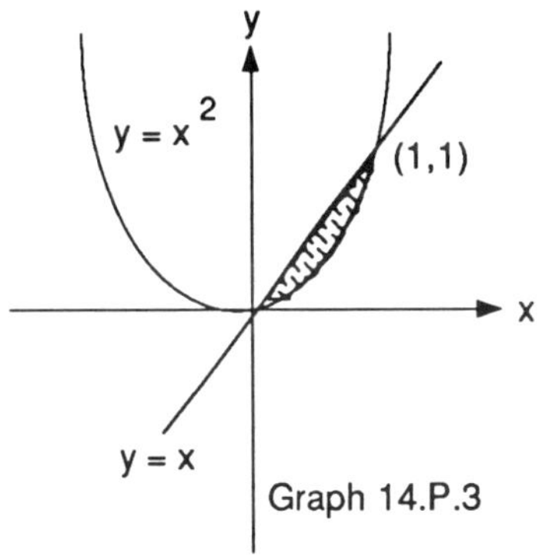

Graph 14.P.3

4. $$\int_0^4\int_{x^2/4}^{x} f(x,y)\,dy\,dx =$$

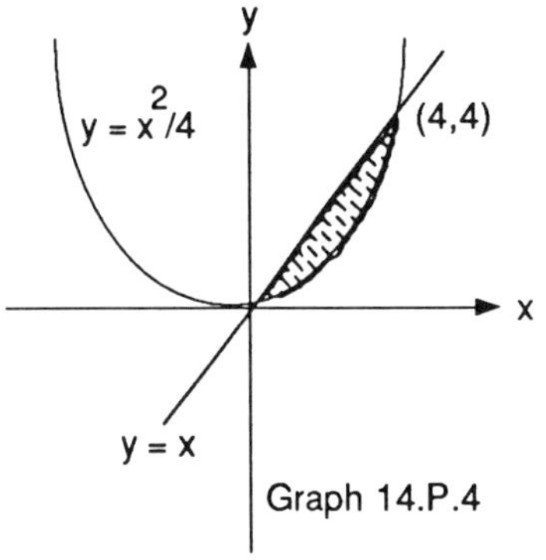

Graph 14.P.4

5. a) $$\int_0^{3/2}\int_{-\sqrt{9-4y^2}}^{\sqrt{9-4y^2}} y\,dx\,dy = \int_0^{3/2}\left[\,yx\,\right]_{-\sqrt{9-4y^2}}^{\sqrt{9-4y^2}}dy =$$

$$\int_0^{3/2} 2y\sqrt{9-4y^2}\,dy = \frac{9}{2}$$

b) $$\int_{-3}^{3}\int_0^{(9-x^2)^{1/2}/2} y\,dy\,dx = \frac{1}{2}\int_{-3}^{3}\left[y^2\right]_0^{(9-x^2)^{1/2}/2}dx =$$

$$\frac{1}{8}\int_{-3}^{3}\left(9-x^2\right)dx = \frac{9}{2}$$

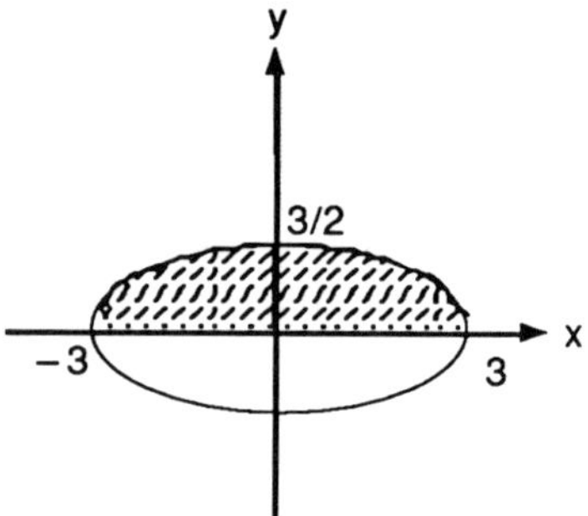

Graph 14.P.5

6. a) $$\int_0^2\int_0^{4-x^2} 2x\,dy\,dx = \int_0^2 [2xy]_0^{4-x^2}dx = \int_0^2 8x - 2x^3\,dx = 8$$

b) $$\int_0^4\int_0^{\sqrt{4-y}} 2x\,dx\,dy = \int_0^4\left[x^2\right]_0^{\sqrt{4-y}}dy = \int_0^4 (4-y)\,dy = 8$$

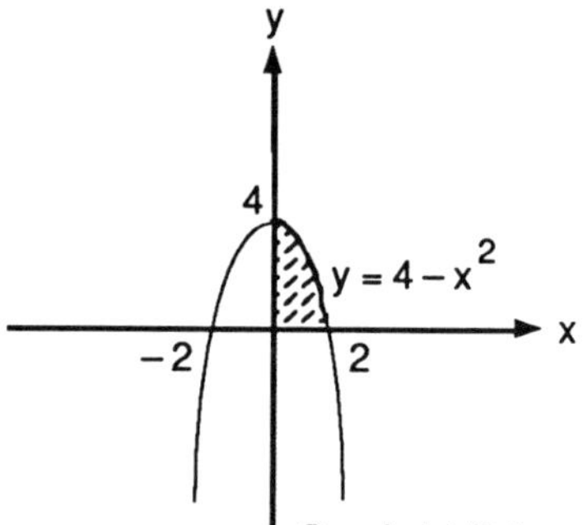

Graph 14.P.6

7. $$\int_0^1\int_{2y}^2 4\cos x^2\,dx\,dy = \int_0^2\int_0^{x/2} 4\cos x^2\,dy\,dx = \int_0^2\left[\left(4\cos x^2\right)y\right]_0^{x/2}dx =$$

$$\int_0^2\left(\cos x^2\right)(2x)\,dx = \sin 4$$

8. $$\int_0^2\int_{y/2}^1 \exp\left(x^2\right)dx\,dy = \int_0^1\int_0^{2x}\exp\left(x^2\right)dy\,dx = \int_0^1\exp\left(x^2\right)(2x)\,dx = e-1$$

9. $$\int_0^8\int_{\sqrt[3]{x}}^2 \frac{1}{y^4+1}\,dy\,dx = \int_0^2\int_0^{y^3}\frac{1}{y^4+1}\,dx\,dy = \frac{1}{4}\int_0^2\frac{4y^3}{y^4+1}\,dy = \frac{\ln 17}{4}$$

10. $$\int_0^1\int_{\sqrt[3]{y}}^1 \frac{2\pi\sin\pi x^2}{x^2}\,dx\,dy = \int_0^1\int_0^{x^3}\left(2\pi\sin\pi x^2\right)/x^2\,dy\,dx = \int_0^1\left(\frac{2\pi\sin\pi x^2}{x^2}\right)x^3\,dx =$$

$$\int_0^1\left(\sin\pi x^2\right)(2\pi x)\,dx = 2$$

11. $$V = \int_0^1\int_x^{2-x} x^2+y^2\,dy\,dx = \int_0^1\left[x^2y+\frac{x^3}{3}\right]_x^{2-x}dx =$$

$$\int_0^1\left(-\frac{8}{3}x^3+4x^2-4x+\frac{8}{3}\right)dx = \frac{4}{3}$$

12. $V = \int_{-3}^{2} \int_{x}^{6-x^2} x^2\, dy\, dx = \int_{-3}^{2} \left[x^2 y\right]_{x}^{6-x^2} dx = \int_{-3}^{2} 6x^2 - x^4 - x^3\, dx = \frac{125}{4}$

13. $A = \int_{-2}^{0} \int_{2x+4}^{4-x^2} dy\, dx = \int_{-2}^{0} -x^2 - 2x\, dx = \frac{4}{3}$

14. $A = \int_{1}^{4} \int_{2-y}^{\sqrt{y}} dx\, dy = \int_{1}^{4} \sqrt{y} - 2 + y\, dy = \frac{13}{6}$

15. average value $= \int_{0}^{1} \int_{0}^{1} xy\, dy\, dx = \int_{0}^{1} \left[\frac{xy^2}{2}\right]_{0}^{1} dx = \int_{0}^{1} \frac{x}{2} dx = \frac{1}{4}$

16. average value $= \frac{1}{\pi/4} \int_{0}^{1} \int_{0}^{\sqrt{1-x^2}} xy\, dy\, dx = \frac{4}{\pi} \int_{0}^{1} \left[\frac{xy^2}{2}\right]_{0}^{\sqrt{1-x^2}} dx = \frac{2}{\pi} \int_{a}^{b} x - x^3\, dx = \frac{1}{2\pi}$

17. $M = \int_{1}^{2} \int_{2/x}^{2} dy\, dx = \int_{1}^{2} 2 - \frac{2}{x} dx = 2 - \ln 4$

$M_y = \int_{1}^{2} \int_{2/x}^{2} x\, dy\, dx = \int_{1}^{2} x\left[2 - \frac{2}{x}\right] dx = 1$

$M_x = \int_{1}^{2} \int_{2/x}^{2} y\, dy\, dx = \int_{1}^{2} 2 - \frac{2}{x^2} dx = 1 \Rightarrow \overline{x} = \frac{1}{2 - \ln 4}, \overline{y} = \frac{1}{2 - \ln 4}$

18. $I_o = \int_{0}^{2} \int_{2x}^{4} \left(x^2 + y^2\right)(3)\, dy\, dx = 3\int_{0}^{2} 4x^2 - \frac{14x^3}{3} + \frac{64}{3} dx = 104$

19. a) $I_o = \int_{-2}^{2} \int_{-1}^{1} \left(x^2 + y^2\right) dy\, dx = \int_{-2}^{2} 2x^2 + \frac{2}{3} dx = \frac{40}{3}$

b) $I_x = \int_{-a}^{a} \int_{-b}^{b} y^2\, dy\, dx = \int_{-a}^{a} \frac{2b^3}{3} dx = \frac{4ab^3}{3}, I_y = \int_{-b}^{b} \int_{-a}^{a} x^2\, dx\, dy =$

$\int_{-b}^{b} \frac{2a^3}{3} dy = \frac{4a^3 b}{3} \Rightarrow I_o = I_x + I_y = \frac{4ab^3}{3} + \frac{4a^3 b}{3} = \frac{4ab\left(b^2 + a^2\right)}{3}$

20. $M = \int_{0}^{3} \int_{0}^{2x/3} dy\, dx = \int_{0}^{3} \frac{2x}{3} dx = 3; I_x = \int_{0}^{3} \int_{0}^{2x/3} y^2\, dy\, dx = \frac{8}{81}\int_{0}^{3} x^3\, dx = 2 \Rightarrow R_x = \sqrt{\frac{2}{3}}$

21. $M = \int_{-1}^{1} \int_{-1}^{1} \left(x^2 + y^2 + \frac{1}{3}\right) dy\, dx = \int_{-1}^{1} 2x^2 + \frac{4}{3} dx = 4$

$M_x = \int_{-1}^{1} \int_{-1}^{1} y\left(x^2 + y^2 + \frac{1}{3}\right) dy\, dx = \int_{-1}^{1} 0\, dx = 0$

$M_y = \int_{-1}^{1} \int_{-1}^{1} x\left(x^2 + y^2 + \frac{1}{3}\right) dy\, dx = \int_{-1}^{1} x\left(2x^2 + \frac{4}{3}\right) dx = 0$

22. $M = \int_0^1 \int_{x^2}^{x} (x+1)\,dy\,dx = \int_0^1 \left(x - x^3\right) dx = \frac{1}{4}$

$M_x = \int_0^1 \int_{x^2}^{x} y(x+1)\,dy\,dx = \frac{1}{2}\int_0^1 \left(x^3 - x^5 + x^2 - x^4\right) dx = \frac{13}{120}$

$M_y = \int_0^1 \int_{x^2}^{x} x(x+1)\,dy\,dx = \int_0^1 x\left(x - x^3\right) dx = \frac{2}{15} \Rightarrow \overline{x} = \frac{8}{15},\ \overline{y} = \frac{13}{30}$

$I_x = \int_0^1 \int_{x^2}^{x} y^2(x+1)\,dy\,dx = \frac{1}{3}\int_0^1 \left(x^4 - x^7 + x^3 - x^6\right) dx = \frac{17}{280} \Rightarrow R_x = \sqrt{\frac{I_x}{M}} = \sqrt{\frac{17}{70}}$

$I_y = \int_0^1 \int_{x^2}^{x} x^2(x+1)\,dy\,dx = \int_0^1 x^2\left(x - x^3\right) dx = \frac{1}{12} \Rightarrow R_y = \sqrt{\frac{I_y}{M}} = \sqrt{\frac{1}{3}}$

23. $\int_{-1}^1 \int_{-\sqrt{1-x^2}}^{\sqrt{1-x^2}} \frac{2}{\left(1 + x^2 + y^2\right)^2}\,dy\,dx = \int_0^{2\pi} \int_0^1 \frac{2r}{\left(1 + r^2\right)^2}\,dr\,d\theta = \frac{1}{2}\int_0^{2\pi} d\theta = \pi$

24. $\int_{-1}^1 \int_{-\sqrt{1-y^2}}^{\sqrt{1-y^2}} \ln\left(x^2 + y^2 + 1\right) dx\,dy = \int_0^{2\pi} \int_0^1 \ln\left(r^2 + 1\right) r\,dr\,d\theta = \frac{1}{2}\int_0^{2\pi} \ln(4) - 1\,d\theta =$
$\left(\ln(4) - 1\right)\pi$

25. $M = 2\int_0^{\pi/2} \int_1^{1+\cos\theta} r\,dr\,d\theta = \int_0^{\pi/2} 2\cos\theta + \frac{1 + \cos 2\theta}{2}\,d\theta = \frac{8+\pi}{4}$

$M_y = \int_{-\pi/2}^{\pi/2} \int_1^{1+\cos\theta} \cos\theta\, r^2\,dr\,d\theta = \int_{-\pi/2}^{\pi/2} \left(\cos^2\theta + \cos^3\theta + \frac{\cos^4\theta}{3}\right) d\theta = \frac{32 + 15\pi}{24} \Rightarrow$
$\overline{x} = \frac{15\pi + 32}{6\pi + 48},\ \overline{y} = 0$ by symmetry

26. $M = \int_{-\pi/3}^{\pi/3} \int_0^3 r\,dr\,d\theta = \frac{9}{2}\int_{-\pi/3}^{\pi/3} d\theta = 3\pi$

$M_y = \int_{-\pi/3}^{\pi/3} \int_0^3 r^2 \cos\theta\,dr\,d\theta = 9\int_{-\pi/3}^{\pi/3} \cos\theta\,d\theta = 9\sqrt{3} \Rightarrow \overline{x} = \frac{3\sqrt{3}}{\pi},\ \overline{y} = 0$ by symmetry

27. $M = \int_0^{\pi/2} \int_1^3 r\,dr\,d\theta = 4\int_0^{\pi/2} d\theta = 2\pi$

$M_y = \int_0^{\pi/2} \int_1^3 r^2 \cos\theta\,dr\,d\theta = \frac{26}{3}\int_0^{\pi/2} \cos\theta\,d\theta = \frac{26}{3} \Rightarrow \overline{x} = \frac{13}{3\pi},\ \overline{y} = \frac{13}{3\pi}$ by symmetry

28. $I^2 = \left(\int_0^\infty \exp\left(-x^2\right) dx\right)\left(\int_0^\infty \exp\left(-y^2\right) dy\right) = \int_0^\infty \int_0^\infty \exp\left[-\left(x^2 + y^2\right)\right] dx\,dy =$
$\int_0^{\pi/2} \int_0^\infty \exp\left(-r^2\right) r\,dr\,d\theta = \int_0^{\pi/2} \left(\lim_{t\to\infty} \left[-\frac{1}{2}\exp\left(-r^2\right)\right]_0^t\right) d\theta = \frac{1}{2}\int_0^{\pi/2} d\theta = \frac{\pi}{4} \Rightarrow$
$I = \int_0^\infty \exp\left(-x^2\right) dx = \frac{\sqrt{\pi}}{2}$

29. $\int_0^{\pi}\int_0^{\pi}\int_0^{\pi}\cos(x+y+z)\,dx\,dy\,dz = \int_0^{\pi}\int_0^{\pi}\sin(z+y+\pi)-\sin(z+y)\,dy\,dz = 0$

$\int_0^{\pi}\left(-\cos(z+2\pi)+\cos(z+\pi)+\cos(z)-\cos(z+\pi)\right)dz =$

30. $\int_0^1\int_0^{1-x^2}\int_{4x^2}^{5-x^2}(x-y+1)\,dz\,dy\,dx = 5\int_0^1\int_0^{1-x^2}(x-y+1)\left(1-x^2\right)dy\,dx =$

$\frac{5}{2}\int_0^1\left(1-x^2\right)^2\left(1+2x+x^2\right)dx = \frac{33}{14}$

31. $V = 2\int_0^{\pi/2}\int_{-\cos y}^{0}\int_0^{-2x}dz\,dx\,dy = -4\int_0^{\pi/2}\int_{-\cos y}^{0}x\,dx\,dy = 2\int_0^{\pi/2}\cos^2 y\,dy = \frac{\pi}{2}$

32. $V = 4\int_0^2\int_0^{\sqrt{4-x^2}}\int_0^{4-x^2}dz\,dy\,dx = 4\int_0^2\int_0^{\sqrt{4-x^2}}4-x^2\,dy\,dx = 4\int_0^2\left(4-x^2\right)^{3/2}dx = 4\pi$

33. a) $4\int_0^{\sqrt{3}}\int_0^{\sqrt{3-x^2}}\int_0^{\sqrt{4-x^2-y^2}}dz\,dy\,dx$

b) $4\int_0^{\pi/2}\int_0^{\sqrt{3}}\int_0^{\sqrt{4-r^2}}r\,dz\,dr\,d\theta$

c) $4\int_0^{\pi/2}\int_0^{\pi/3}\int_{\sec\phi}^{2}\rho^2\sin\phi\,d\rho\,d\phi\,d\theta$

34. a) $I_z = 4\int_0^1\int_0^{\sqrt{1-y^2}}\int_0^{\sqrt{4-x^2-y^2}}\left(x^2+y^2\right)dz\,dy\,dx$

b) $I_z = 4\int_0^{\pi/2}\int_0^1\int_0^{\sqrt{1-r^2}}r^3\,dz\,dr\,d\theta$

c) $I_z = 4\int_0^{\pi/2}\int_0^{\pi/2}\int_0^1\rho^4\sin^3\phi\,d\rho\,d\phi\,d\theta$

35. $V = \int_0^{\pi/2}\int_1^2\int_0^{r^2\sin\theta\cos\theta}r\,dz\,dr\,d\theta = \int_0^{\pi/2}\int_1^2 r^3\sin\theta\cos\theta\,dr\,d\theta =$

$\frac{15}{4}\int_0^{\pi/2}\sin\theta\cos\theta\,d\theta = \frac{15}{8}$

36. $V = 4\int_0^{\pi/2}\int_0^1\int_{r^2}^{\sqrt{2-r^2}}r\,dz\,dr\,d\theta = 4\int_0^{\pi/2}\int_0^1\left(r\sqrt{2-r^2}-r^3\right)dr\,d\theta =$

$\left(\frac{2^{7/2}-7}{3}\right)\int_0^{\pi/2}d\theta = \frac{\pi\left(8\sqrt{2}-7\right)}{6}$

37. $V = 4\int_0^1 \int_0^{\sqrt{1-x^2}} \int_{2x^2+2y^2}^{3-x^2-y^2} dz\,dy\,dx = 4\int_0^{\pi/2} \int_0^1 \int_{2r^2}^{3-r^2} r\,dz\,dr\,d\theta =$

$4\int_0^{\pi/2} \int_0^1 \left(3r - 3r^3\right) dr\,d\theta = 3\int_0^{\pi/2} d\theta = \frac{3\pi}{2}$

38. a) The radius of the sphere is 2; the radius of the hole is 1.

b) $V = 2\int_0^{2\pi} \int_0^{\sqrt{3}} \int_1^{\sqrt{4-z^2}} r\,dr\,dz\,d\theta = \int_0^{2\pi} \int_0^{\sqrt{3}} 3 - z^2\,dz\,d\theta = 2\sqrt{3}\int_0^{2\pi} d\theta = 4\sqrt{3}\pi$

39. a) $M = 4\int_0^{\pi/2} \int_0^1 \int_0^{r^2} z\,r\,dz\,dr\,d\theta = 2\int_0^{\pi/2} \int_0^1 r^5\,dr\,d\theta = \frac{1}{3}\int_0^{\pi/2} d\theta = \frac{\pi}{6}$

$M_{xy} = \int_0^{2\pi} \int_0^1 \int_0^{r^2} z^2\,r\,dz\,dr\,d\theta = \frac{1}{3}\int_0^{2\pi} \int_0^1 r^7\,dr\,d\theta = \frac{1}{24}\int_0^{2\pi} d\theta = \frac{\pi}{12} \Rightarrow \overline{z} = \frac{1}{2},$

$\overline{x} = \overline{y} = 0$ by symmetry; $I_z = \int_0^{2\pi} \int_0^1 \int_0^{r^2} z\,r^3\,dz\,dr\,d\theta = \frac{1}{2}\int_0^{2\pi} \int_0^1 r^7\,dr\,d\theta =$

$\frac{1}{16}\int_0^{2\pi} d\theta = \frac{\pi}{8} \Rightarrow R_z = \sqrt{\frac{I_z}{M}} = \frac{\sqrt{3}}{2}$

b) $M = 4\int_0^{\pi/2} \int_0^1 \int_0^{r^2} r^2\,dz\,dr\,d\theta = 4\int_0^{\pi/2} \int_0^1 r^4\,dr\,d\theta = \frac{4}{5}\int_0^{\pi/2} d\theta = \frac{2\pi}{5}$

$M_{xy} = \int_0^{2\pi} \int_0^1 \int_0^{r^2} z\,r^2\,dz\,dr\,d\theta = \frac{1}{2}\int_0^{2\pi} \int_0^1 r^6\,dr\,d\theta = \frac{1}{14}\int_0^{2\pi} d\theta = \frac{\pi}{7} \Rightarrow \overline{z} = \frac{5}{14}$

$\overline{x} = \overline{y} = 0$ by symmetry; $I_z = \int_0^{2\pi} \int_0^1 \int_0^{r^2} r^4\,dz\,dr\,d\theta = \int_0^{2\pi} \int_0^1 r^6\,dr\,d\theta =$

$\frac{1}{7}\int_0^{2\pi} d\theta = \frac{2\pi}{7} \Rightarrow R_z = \sqrt{\frac{I_z}{M}} = \sqrt{\frac{5}{7}}$

40. a) $M = \int_0^{2\pi} \int_0^1 \int_r^1 z\,r\,dz\,dr\,d\theta = \frac{1}{2}\int_0^{2\pi} \int_0^1 r - r^3\,dr\,d\theta = \frac{1}{8}\int_0^{2\pi} d\theta = \frac{\pi}{4}$

$M_{xy} = \int_0^{2\pi} \int_0^1 \int_r^1 z^2\,r\,dz\,dr\,d\theta = \frac{1}{3}\int_0^{2\pi} \int_0^1 r - r^4\,dr\,d\theta = \frac{1}{10}\int_0^{2\pi} d\theta = \frac{\pi}{5} \Rightarrow \overline{z} = \frac{4}{5},$

$\overline{x} = \overline{y} = 0$ by symmetry; $I_z = \int_0^{2\pi} \int_0^1 \int_r^1 z\,r^3\,dz\,dr\,d\theta = \frac{1}{2}\int_0^{2\pi} \int_0^1 r^3 - r^5\,dr\,d\theta =$

$\frac{1}{24}\int_0^{2\pi} d\theta = \frac{\pi}{12} \Rightarrow R_z = \sqrt{\frac{I_z}{M}} = \sqrt{\frac{1}{3}}$

b) $M = \int_0^{2\pi} \int_0^1 \int_r^1 z^2 \, dz \, dr \, d\theta = \frac{\pi}{5}$, see part (a) for details

$M_{xy} = \int_0^{2\pi} \int_0^1 \int_r^1 z^3 r \, dz \, dr \, d\theta = \frac{1}{4}\int_0^{2\pi} \int_0^1 r - r^5 \, dr \, d\theta = \frac{1}{12}\int_0^{2\pi} d\theta = \frac{\pi}{6} \Rightarrow \overline{z} = \frac{5}{6}$,

$\overline{x} = \overline{y} = 0$ by symmetry; $I_z = \int_0^{2\pi} \int_0^1 \int_r^1 z^2 r^3 \, dz \, dr \, d\theta = \frac{1}{3}\int_0^{2\pi} \int_0^1 r^3 - r^6 \, dr \, d\theta =$

$\frac{1}{28}\int_0^{2\pi} d\theta = \frac{\pi}{14} \Rightarrow R_z = \sqrt{\frac{I_z}{M}} = \sqrt{\frac{5}{14}}$

41. $V = 8\int_0^{\pi/2} \int_0^{\pi/2} \int_0^{2 \sin \phi} \rho^2 \sin \phi \, d\rho \, d\phi \, d\theta = \frac{64}{3}\int_0^{\pi/2} \int_0^{\pi/2} \sin^4\phi \, d\phi \, d\theta = 4\pi\int_0^{\pi/2} d\theta = 2\pi^2$

42. $M = 4\int_0^{\pi/2} \int_0^{\pi/3} \int_0^2 \rho^2 \sin \phi \, d\rho \, d\phi \, d\theta = \frac{32}{3}\int_0^{\pi/2} \int_0^{\pi/3} \sin \phi \, d\phi \, d\theta = \frac{16}{3}\int_0^{\pi/2} d\theta = \frac{8\pi}{3}$

$I_z = \int_0^{2\pi} \int_0^{\pi/3} \int_0^2 \rho^4 \sin^3 \phi \, d\rho \, d\phi \, d\theta = \frac{32}{5}\int_0^{2\pi} \int_0^{\pi/3} \sin^3 \phi \, d\phi \, d\theta =$

$\frac{4}{3}\int_0^{2\pi} d\theta = \frac{8\pi}{3}$

43. $M = \frac{2}{3}\pi a^3$; $M_{xy} = \int_0^{2\pi} \int_0^{\pi/2} \int_0^a \rho^3 \sin \phi \cos \phi \, d\rho \, d\phi \, d\theta = \frac{a^4}{4}\int_0^{2\pi} \int_0^{\pi/2} \sin \phi \cos \phi \, d\phi \, d\theta =$

$\frac{a^4}{8}\int_0^{2\pi} d\theta = \frac{a^4\pi}{4} \Rightarrow \overline{z} = \frac{3a}{8}$, $\overline{x} = \overline{y} = 0$ by symmetry

44. $M = \frac{4}{3}\pi a^3$; $I_z = \int_0^{2\pi} \int_0^{\pi} \int_0^a \rho^4 \sin^3\phi \, d\rho \, d\phi \, d\theta = \frac{a^5}{5}\int_0^{2\pi} \int_0^{\pi} \sin^3\phi \, d\phi \, d\theta = \frac{4a^5}{15}\int_0^{2\pi} d\theta =$

$\frac{8a^5\pi}{15} \Rightarrow R_z = \sqrt{\frac{I_z}{M}} = a\sqrt{\frac{2}{5}}$

CHAPTER 15

VECTOR ANALYSIS

SECTION 15.1 LINE INTEGRALS

1. $\mathbf{r} = t\,\mathbf{i} + (1 - t)\,\mathbf{j} \Rightarrow x = t,\ y = 1 - t \Rightarrow y = 1 - x \Rightarrow$ c

2. $\mathbf{r} = \mathbf{i} + \mathbf{j} + t\,\mathbf{k} \Rightarrow x = 1,\ y = 1,\ z = t \Rightarrow$ e

3. $\mathbf{r} = (2 \cos t)\,\mathbf{i} + (2 \sin t)\,\mathbf{j} \Rightarrow x = 2 \cos t,\ y = 2 \sin t \Rightarrow x^2 + y^2 = 4 \Rightarrow$ g

4. $\mathbf{r} = t\,\mathbf{i} \Rightarrow x = t,\ y = 0,\ z = 0 \Rightarrow$ a

5. $\mathbf{r} = t\,\mathbf{i} + t\,\mathbf{j} + t\,\mathbf{k} \Rightarrow x = t,\ y = t,\ z = t \Rightarrow$ d

6. $\mathbf{r} = t\,\mathbf{j} + (2 - 2t)\,\mathbf{k} \Rightarrow y = t,\ z = 2 - 2t \Rightarrow z = 2 - 2y \Rightarrow$ b

7. $\mathbf{r} = (t^2 - 1)\,\mathbf{j} + 2t\,\mathbf{k} \Rightarrow y = t^2 - 1,\ z = 2t \Rightarrow y = \frac{z^2}{4} - 1 \Rightarrow$ f

8. $\mathbf{r} = (2 \cos t)\,\mathbf{i} + (2 \sin t)\,\mathbf{k} \Rightarrow x = 2 \cos t,\ z = 2 \sin t \Rightarrow x^2 + z^2 = 4 \Rightarrow$ h

9. $\mathbf{r} = t\,\mathbf{i} + (1 - t)\,\mathbf{j} \Rightarrow x = t,\ y = 1 - t,\ z = 0 \Rightarrow f(g(t),h(t),k(t)) = 1.\ \frac{dx}{dt} = 1,\ \frac{dy}{dt} = -1,\ \frac{dz}{dt} = 0 \Rightarrow$

$$\sqrt{\left(\frac{dx}{dt}\right)^2 + \left(\frac{dy}{dt}\right)^2 + \left(\frac{dz}{dt}\right)^2}\ dt = \sqrt{2}\ dt \Rightarrow \int_C f(x,y,z)\ ds = \int_0^1 \sqrt{2}\ dt = \sqrt{2}$$

10. $\mathbf{r} = t\,\mathbf{i} + (1 - t)\,\mathbf{j} + \mathbf{k} \Rightarrow x = t,\ y = 1 - t,\ z = 1 \Rightarrow f(g(t),h(t),k(t)) = 2t - 2.\ \frac{dx}{dt} = 1,\ \frac{dy}{dt} = -1,\ \frac{dz}{dt} = 0 \Rightarrow$

$$\sqrt{\left(\frac{dx}{dt}\right)^2 + \left(\frac{dy}{dt}\right)^2 + \left(\frac{dz}{dt}\right)^2}\ dt = \sqrt{2}\ dt \Rightarrow \int_C f(x,y,z)\ ds = \int_0^1 (2t - 2)\sqrt{2}\ dt = -\sqrt{2}$$

11. $\mathbf{r} = 2t\,\mathbf{i} + t\,\mathbf{j} + (2 - 2t)\,\mathbf{k} \Rightarrow x = 2t,\ y = t,\ z = 2 - 2t \Rightarrow f(g(t),h(t),k(t)) = 2t^2 - t + 2.\ \frac{dx}{dt} = 2,\ \frac{dy}{dt} = 1,\ \frac{dz}{dt} = -2$

$$\Rightarrow \sqrt{\left(\frac{dx}{dt}\right)^2 + \left(\frac{dy}{dt}\right)^2 + \left(\frac{dz}{dt}\right)^2}\ dt = 3\ dt \Rightarrow \int_C f(x,y,z)\ ds = \int_0^1 (2t^2 - t + 2)3\ dt = \frac{13}{2}$$

12. $\mathbf{r} = (4 \cos t)\,\mathbf{i} + (4 \sin t)\,\mathbf{j} + 3t\,\mathbf{k} \Rightarrow x = 4 \cos t,\ y = 4 \sin t,\ z = 3t \Rightarrow f(g(t),h(t),k(t)) = \sqrt{(4 \cos t)^2 + (4 \sin t)^2}$

$= 4.\ \frac{dx}{dt} = -4 \sin t,\ \frac{dy}{dt} = 4 \cos t,\ \frac{dz}{dt} = 3 \Rightarrow \sqrt{\left(\frac{dx}{dt}\right)^2 + \left(\frac{dy}{dt}\right)^2 + \left(\frac{dz}{dt}\right)^2}\ dt = 5\ dt \Rightarrow$

$$\int_C f(x,y,z)\ ds = \int_{-2\pi}^{2\pi} 4(5)\ dt = 80\pi$$

13. $\mathbf{r} = \mathbf{i} + \mathbf{j} + t\,\mathbf{k} \Rightarrow x = 1,\ y = 1,\ z = t \Rightarrow f(g(t),h(t),k(t)) = 3t\sqrt{4+t^2}$. $\frac{dx}{dt} = 0,\ \frac{dy}{dt} = 0,\ \frac{dz}{dt} = 1 \Rightarrow$

$$\sqrt{\left(\frac{dx}{dt}\right)^2 + \left(\frac{dy}{dt}\right)^2 + \left(\frac{dz}{dt}\right)^2}\ dt = 1\ dt = dt \Rightarrow \int_C f(x,y,z)\ ds = \int_{-1}^{1} 3t\sqrt{4+t^2}\ dt = 0$$

14. $\mathbf{r} = t\,\mathbf{i} + t\,\mathbf{j} + t\,\mathbf{k} \Rightarrow x = t,\ y = t,\ z = t \Rightarrow f(g(t),h(t),k(t)) = \frac{\sqrt{3}}{t^2+t^2+t^2} = \frac{\sqrt{3}}{3t^2}$. $\frac{dx}{dt} = 1,\ \frac{dy}{dt} = 1,\ \frac{dz}{dt} = 1 \Rightarrow$

$$\sqrt{\left(\frac{dx}{dt}\right)^2 + \left(\frac{dy}{dt}\right)^2 + \left(\frac{dz}{dt}\right)^2}\ dt = \sqrt{3}\ dt \Rightarrow \int_C f(x,y,z)\ ds = \int_1^2 \left(\frac{\sqrt{3}}{3t^2}\right)\sqrt{3}\ dt = \frac{1}{2}$$

15. C_1: $\mathbf{r} = t\,\mathbf{i} + t^2\,\mathbf{j} \Rightarrow x = t,\ y = t^2,\ z = 0 \Rightarrow f(g(t),h(t),k(t)) = t + \sqrt{t^2} = 2t$ since $0 \le t \le 1$. $\frac{dx}{dt} = 1,\ \frac{dy}{dt} = 2t,$

$\frac{dz}{dt} = 0 \Rightarrow \sqrt{\left(\frac{dx}{dt}\right)^2 + \left(\frac{dy}{dt}\right)^2 + \left(\frac{dz}{dt}\right)^2}\ dt = \sqrt{1+4t^2}\ dt \Rightarrow \int_{C_1} f(x,y,z)\ ds = \int_0^1 2t\sqrt{1+4t^2}\ dt =$

$\frac{1}{6}\left(5^{3/2}\right) - \frac{1}{6} = \frac{5}{6}\sqrt{5} - \frac{1}{6}$. C_2: $\mathbf{r} = \mathbf{i} + \mathbf{j} + t\,\mathbf{k} \Rightarrow x = 1,\ y = 1,\ z = t \Rightarrow f(g(t),h(t),k(t)) = 2 - t^2$. $\frac{dx}{dt} = 0,$

$\frac{dy}{dt} = 0,\ \frac{dz}{dt} = 1 \Rightarrow \sqrt{\left(\frac{dx}{dt}\right)^2 + \left(\frac{dy}{dt}\right)^2 + \left(\frac{dz}{dt}\right)^2}\ dt = 1\ dt = dt \Rightarrow \int_{C_2} f(x,y,z)\ ds = \int_0^1 (2 - t^2)\ dt =$

$\frac{5}{3}$. $\therefore \int_C f(x,y,z)\ ds = \int_{C_1} f(x,y,z)\ ds + \int_{C_2} f(x,y,z)\ ds = \frac{5}{6}\sqrt{5} + \frac{3}{2}$.

16. C_1: $\mathbf{r} = t\,\mathbf{k} \Rightarrow x = 0,\ y = 0,\ z = t \Rightarrow f(g(t),h(t),k(t)) = -t^2$. $\frac{dx}{dt} = 0,\ \frac{dy}{dt} = 0,\ \frac{dz}{dt} = 1 \Rightarrow$

$\sqrt{\left(\frac{dx}{dt}\right)^2 + \left(\frac{dy}{dt}\right)^2 + \left(\frac{dz}{dt}\right)^2}\ dt = 1\ dt \Rightarrow \int_{C_1} f(x,y,z)\ ds = \int_0^1 -t^2\ dt = -\frac{1}{3}$. C_2: $\mathbf{r} = t\,\mathbf{j} + \mathbf{k} \Rightarrow$

$x = 0,\ y = t,\ z = 1 \Rightarrow f(g(t),h(t),k(t)) = \sqrt{t} - 1$. $\frac{dx}{dt} = 0,\ \frac{dy}{dt} = 1,\ \frac{dz}{dt} = 1 \Rightarrow \sqrt{\left(\frac{dx}{dt}\right)^2 + \left(\frac{dy}{dt}\right)^2 + \left(\frac{dz}{dt}\right)^2}\ dt =$

$1\ dt = dt \Rightarrow \int_{C_2} f(x,y,z)\ ds = \int_0^1 (\sqrt{t} - 1)\ dt = -\frac{1}{3}$. C_3: $\mathbf{r} = t\,\mathbf{i} + \mathbf{j} + \mathbf{k} \Rightarrow x = t,\ y = 1,\ z = 1 \Rightarrow$

$f(g(t),h(t),k(t)) = t$. $\frac{dx}{dt} = 1,\ \frac{dy}{dt} = 0,\ \frac{dz}{dt} = 0 \Rightarrow \sqrt{\left(\frac{dx}{dt}\right)^2 + \left(\frac{dy}{dt}\right)^2 + \left(\frac{dz}{dt}\right)^2}\ dt = 1\ dt = dt \Rightarrow$

$\int_{C_3} f(x,y,z)\ ds = \int_0^1 t\ dt = \frac{1}{2}$. $\therefore \int_C f(x,y,z)\ ds = -\frac{1}{3} + \left(-\frac{1}{3}\right) + \frac{1}{2} = -\frac{1}{6}$

17. $\delta(x,y,z) = 2 - z$, $\mathbf{r} = (\cos t)\,\mathbf{j} + (\sin t)\,\mathbf{k}$, $0 \le t \le \pi$, $ds = dt$, $x = 0$, $y = \cos t$, $z = \sin t$, and $M = 2\pi - 2$ are

all given or found in Example 3 in the text, page 946. $I_x = \int_C (y^2 + z^2)\delta\, ds$

$= \int_0^{\pi} (\cos^2 t + \sin^2 t)(2 - \sin t)\, dt = \int_0^{\pi} (2 - \sin t)\, dt = 2\pi - 2 \Rightarrow R_x = \sqrt{\frac{I_x}{M}} = \sqrt{\frac{2\pi - 2}{2\pi - 2}} = 1$

18. $\mathbf{r} = (t^2 - 1)\,\mathbf{j} + 2t\,\mathbf{k} \Rightarrow x = 0,\ y = t^2 - 1,\ z = 2t.\ \frac{dx}{dt} = 0,\ \frac{dy}{dt} = 2t,\ \frac{dz}{dt} = 2 \Rightarrow$

$\sqrt{\left(\frac{dx}{dt}\right)^2 + \left(\frac{dy}{dt}\right)^2 + \left(\frac{dz}{dt}\right)^2}\ dt = 2\sqrt{t^2 + 1}\ dt.\ \therefore\ M = \int_C \delta(x,y,z)\, ds = \int_0^1 \frac{3}{2} t\left(2\sqrt{t^2 + 1}\right) dt =$

$2\sqrt{2} - 1.$

19. Let δ be constant. Let $x = a \cos t$, $y = a \sin t$. Then $\frac{dx}{dt} = -a \sin t,\ \frac{dy}{dt} = a \cos t,\ 0 \le t \le 2\pi,\ \frac{dz}{dt} = 0 \Rightarrow$

$\sqrt{\left(\frac{dx}{dt}\right)^2 + \left(\frac{dy}{dt}\right)^2 + \left(\frac{dz}{dt}\right)^2}\ dt = a\, dt.\ \therefore\ I_z = \int_C (x^2 + y^2)\delta\, ds = \int_0^{2\pi} (a^2 \sin^2 t + a^2 \cos^2 t)a\delta\, dt =$

$\int_0^{2\pi} a^3\delta\, dt = 2\pi a^3\delta.\ M = \int_C \delta(x,y,z)\, ds = \int_0^{2\pi} \delta a\, dt = 2\pi\delta a.\ R_z = \sqrt{\frac{I_z}{M}} = \sqrt{\frac{2\pi a^3\delta}{2\pi a\delta}} = a.$

20. Let $\delta = a$. $\mathbf{r} = t\,\mathbf{j} + (2 - 2t)\,\mathbf{k} \Rightarrow x = 0,\ y = t,\ z = 2 - 2t \Rightarrow \frac{dx}{dt} = 0,\ \frac{dy}{dt} = 1,\ \frac{dz}{dt} = -2 \Rightarrow$

$\sqrt{\left(\frac{dx}{dt}\right)^2 + \left(\frac{dy}{dt}\right)^2 + \left(\frac{dz}{dt}\right)^2}\ dt = \sqrt{5}\ dt.\ M = \int_C \delta(x,y,z)\, ds = \int_0^1 a\sqrt{5}\, dt = a\sqrt{5}.\ I_x =$

$\int_C (y^2 + z^2)\delta\, ds = \int_0^1 (t^2 + (2 - 2t)^2)a\sqrt{5}\, dt = \int_0^1 (5t^2 - 8t + 4)a\sqrt{5}\, dt = \frac{5}{3} a\sqrt{5}.$

$R_x = \sqrt{\frac{I_x}{M}} = \sqrt{\frac{\frac{5}{3}a\sqrt{5}}{a\sqrt{5}}} = \sqrt{\frac{5}{3}}.\ I_y = \int_C (x^2 + z^2)\delta\, ds = \int_0^1 (2 - 2t)^2 a\sqrt{5}\, dt =$

$\int_0^1 (4 - 8t + 4t^2)a\sqrt{5}\, dt = \frac{4}{3} a\sqrt{5}.\ R_y = \sqrt{\frac{I_y}{M}} = \sqrt{\frac{\frac{4}{3}a\sqrt{5}}{a\sqrt{5}}} = \frac{2}{\sqrt{3}}.$

$I_z = \int_C (x^2 + y^2)\delta\, ds = \int_0^1 t^2 a\sqrt{5}\, dt = \frac{1}{3} a\sqrt{5}.\ R_z = \sqrt{\frac{I_z}{M}} = \sqrt{\frac{\frac{1}{3}a\sqrt{5}}{a\sqrt{5}}} = \frac{1}{\sqrt{3}}.$

21. a) $\mathbf{r} = (\cos t)\,\mathbf{i} + (\sin t)\,\mathbf{j} + t\,\mathbf{k} \Rightarrow x = \cos t,\ y = \sin t,\ z = t \Rightarrow \frac{dx}{dt} = -\sin t,\ \frac{dy}{dt} = \cos t,\ \frac{dz}{dt} = 1 \Rightarrow$

$$\sqrt{\left(\frac{dx}{dt}\right)^2 + \left(\frac{dy}{dt}\right)^2 + \left(\frac{dz}{dt}\right)^2}\,dt = \sqrt{2}\,dt.\quad M = \int_C \delta(x,y,z)\,ds = \int_0^{2\pi} \delta\sqrt{2}\,dt = 2\pi\delta\sqrt{2}.$$

$$I_z = \int_C (x^2 + y^2)\delta\,ds = \int_0^{2\pi} (\cos^2 t + \sin^2 t)\delta\sqrt{2}\,dt = \int_0^{2\pi} \delta\sqrt{2}\,dt = 2\pi\delta\sqrt{2}.\quad R_z = \sqrt{\frac{I_z}{M}} =$$

$$\sqrt{\frac{2\pi\delta\sqrt{2}}{2\pi\delta\sqrt{2}}} = 1$$

b) $$M = \int_C \delta(x,y,z)\,ds = \int_0^{4\pi} \delta\sqrt{2}\,dt = 4\pi\delta\sqrt{2}.\quad I_z = \int_C (x^2 + y^2)\delta\,ds = \int_0^{4\pi} \delta\sqrt{2}\,dt = 4\pi\delta\sqrt{2}.$$

$$R_z = \sqrt{\frac{I_z}{M}} = \sqrt{\frac{4\pi\delta\sqrt{2}}{4\pi\delta\sqrt{2}}} = 1$$

22. $\mathbf{r} = (t^2 - 1)\,\mathbf{j} + 2t\,\mathbf{k} \Rightarrow x = 0,\ y = t^2 - 1,\ z = 2t \Rightarrow \frac{dx}{dt} = 0,\ \frac{dy}{dt} = 2t,\ \frac{dz}{dt} = 2$ and $\delta(g(t),h(t),k(t)) = 15\sqrt{t^2 + 1}$

$$\Rightarrow \sqrt{\left(\frac{dx}{dt}\right)^2 + \left(\frac{dy}{dt}\right)^2 + \left(\frac{dz}{dt}\right)^2}\,dt = \sqrt{4t^2 + 4}\,dt = 2\sqrt{t^2 + 1}\,dt.\ \therefore\ \delta\,ds = \left(15\sqrt{t^2 + 1}\right)\left(2\sqrt{t^2 + 1}\right) =$$

$$30\left(t^2 + 1\right).\ M_{yz} = \int_C x\delta\,ds = \int_{-1}^{1} 0\,\delta\,dt = 0.\ M_{xz} = \int_C y\,\delta\,ds = \int_{-1}^{1} (t^2 - 1)\left(30(t^2 + 1)\right)dt$$

$$\int_{-1}^{1} 30(t^4 - 1)\,dt = -48.\ M_{xy} = \int_C z\,\delta\,ds = \int_{-1}^{1} 2t\left(30(t^2 + 1)\right)dt = \int_{-1}^{1} 60\left(t^3 + t\right)dt = 0.$$

$$M = \int_C \delta(x,y,z)\,ds = \int_{-1}^{1} 30\left(t^2 + 1\right)dt = 80.$$

$$\therefore\ \bar{x} = \frac{0}{80} = 0,\ \bar{y} = -\frac{48}{80} = -\frac{3}{5},\ \bar{z} = \frac{0}{80} = 0.\ I_y = \int_C (x^2 + z^2)\delta\,ds = \int_{-1}^{1} 4t^2\left(30(t^2 + 1)\right)dt =$$

$$\int_{-1}^{1} 120\left(t^4 + t^2\right)dt = 128.\ I_z = \int_C (x^2 + y^2)\delta\,ds = \int_{-1}^{1} (t^2 - 1)^2\left(30(t^2 + 1)\right)dt =$$

$$\int_{-1}^{1} 30\left(t^6 - t^4 - t^2 + 1\right)dt = \frac{256}{7}.$$

SECTION 15.2 VECTOR FIELDS, WORK, CIRCULATION, AND FLUX

1. a) $\mathbf{F} = 3t\,\mathbf{i} + 2t\,\mathbf{j} + 4t\,\mathbf{k}$, $\frac{d\mathbf{r}}{dt} = \mathbf{i} + \mathbf{j} + \mathbf{k} \Rightarrow \mathbf{F}\cdot\frac{d\mathbf{r}}{dt} = 9t \Rightarrow W = \int_0^1 9t\,dt = \frac{9}{2}$

b) $\mathbf{F} = 3t^2\,\mathbf{i} + 2t\,\mathbf{j} + 4t^4\,\mathbf{k}$, $\frac{d\mathbf{r}}{dt} = \mathbf{i} + 2t\,\mathbf{j} + 4t^3\,\mathbf{k} \Rightarrow \mathbf{F}\cdot\frac{d\mathbf{r}}{dt} = 7t^2 + 16t^7 \Rightarrow W = \int_0^1 (7t^2 + 16t^7)\,dt = \frac{13}{3}$

c) $\mathbf{F_1} = 3t\,\mathbf{i} + 2t\,\mathbf{j}$, $\frac{d\mathbf{r_1}}{dt} = \mathbf{i} + \mathbf{j} \Rightarrow \mathbf{F_1}\cdot\frac{d\mathbf{r_1}}{dt} = 5t \Rightarrow W_1 = \int_0^1 5t\,dt = \frac{5}{2}$. $\mathbf{F_2} = 3\,\mathbf{i} + 2\,\mathbf{j} + 4t\,\mathbf{k}$, $\frac{d\mathbf{r_2}}{dt} = \mathbf{k} \Rightarrow$

$\mathbf{F_2}\cdot\frac{d\mathbf{r_2}}{dt} = 4t \Rightarrow W_2 = \int_0^1 4t\,dt = 2.$ $\therefore$ $W = W_1 + W_2 = \frac{9}{2}$

2. a) $\mathbf{F} = \left(\frac{1}{t^2+1}\right)\mathbf{j}$, $\frac{d\mathbf{r}}{dt} = \mathbf{i} + \mathbf{j} + \mathbf{k} \Rightarrow \mathbf{F}\cdot\frac{d\mathbf{r}}{dt} = \frac{1}{t^2+1} \Rightarrow W = \int_0^1 \frac{1}{t^2+1}\,dt = \frac{\pi}{4}$

b) $\mathbf{F} = \left(\frac{1}{t^2+1}\right)\mathbf{j}$, $\frac{d\mathbf{r}}{dt} = \mathbf{i} + 2t\,\mathbf{j} + 3t^3\,\mathbf{k} \Rightarrow \mathbf{F}\cdot\frac{d\mathbf{r}}{dt} = \frac{2t}{t^2+1} \Rightarrow W = \int_0^1 \frac{2t}{t^2+1}\,dt = \ln 2$

c) $\mathbf{F_1} = \left(\frac{1}{t^2+1}\right)\mathbf{j}$, $\frac{d\mathbf{r_1}}{dt} = \mathbf{i} + \mathbf{j} \Rightarrow \mathbf{F_1}\cdot\frac{d\mathbf{r_1}}{dt} = \frac{1}{t^2+1}$. $\mathbf{F_2} = \frac{1}{2}\,\mathbf{j}$, $\frac{d\mathbf{r_2}}{dt} = \mathbf{k} \Rightarrow \mathbf{F_2}\cdot\frac{d\mathbf{r_2}}{dt} = 0.$

$\therefore$ $W = \int_0^1 \frac{1}{t^2+1}\,dt = \frac{\pi}{4}$

3. a) $\mathbf{F} = \sqrt{t}\,\mathbf{i} - 2t\,\mathbf{j} + \sqrt{t}\,\mathbf{k}$, $\frac{d\mathbf{r}}{dt} = \mathbf{i} + \mathbf{j} + \mathbf{k} \Rightarrow \mathbf{F}\cdot\frac{d\mathbf{r}}{dt} = 2\sqrt{t} - 2t \Rightarrow W = \int_0^1 (2\sqrt{t} - 2t)\,dt = \frac{1}{3}$

b) $\mathbf{F} = t^2\,\mathbf{i} - 2t\,\mathbf{j} + t\,\mathbf{k}$, $\frac{d\mathbf{r}}{dt} = \mathbf{i} + 2t\,\mathbf{j} + 4t^3\,\mathbf{k} \Rightarrow \mathbf{F}\cdot\frac{d\mathbf{r}}{dt} = 4t^4 - 3t^2 \Rightarrow W = \int_0^1 (4t^4 - 3t^2)\,dt = -\frac{1}{5}$

c) $\mathbf{F_1} = -2t\,\mathbf{j} + \sqrt{t}\,\mathbf{k}$, $\frac{d\mathbf{r_1}}{dt} = \mathbf{i} + \mathbf{j} \Rightarrow \mathbf{F_1}\cdot\frac{d\mathbf{r_1}}{dt} = -2t \Rightarrow W_1 = \int_0^1 -2t\,dt = -1.$ $\mathbf{F_2} = \sqrt{t}\,\mathbf{i} - 2\,\mathbf{j} + \mathbf{k}$,

$\frac{d\mathbf{r_2}}{dt} = \mathbf{k} \Rightarrow \mathbf{F_2}\cdot\frac{d\mathbf{r_2}}{dt} = 1 \Rightarrow W_2 = \int_0^1 dt = 1.$ $\therefore$ $W = W_1 + W_2 = 0$

4. a) $\mathbf{F} = t^2\,\mathbf{i} + t^2\,\mathbf{j} + t^2\,\mathbf{k},\ \frac{d\mathbf{r}}{dt} = \mathbf{i} + \mathbf{j} + \mathbf{k} \Rightarrow \mathbf{F}\cdot\frac{d\mathbf{r}}{dt} = 3t^2 \Rightarrow W = \int_0^1 3t^2\,dt = 1$

b) $\mathbf{F} = t^3\,\mathbf{i} + t^6\,\mathbf{j} + t^5\,\mathbf{k},\ \frac{d\mathbf{r}}{dt} = \mathbf{i} + 2t\,\mathbf{j} + 4t^3\,\mathbf{k} \Rightarrow \mathbf{F}\cdot\frac{d\mathbf{r}}{dt} = t^3 + 2t^7 + 4t^8 \Rightarrow W = \int_0^1 (t^3 + 2t^7 + 4t^8)\,dt = \frac{17}{18}$

c) $\mathbf{F}_1 = t^2\,\mathbf{j},\ \frac{d\mathbf{r}_1}{dt} = \mathbf{i} + \mathbf{j} \Rightarrow \mathbf{F}_1\cdot\frac{d\mathbf{r}_1}{dt} = t^2 \Rightarrow W_1 = \int_0^1 t^2\,dt = \frac{1}{3}$. $\mathbf{F}_2 = \mathbf{i} + t\,\mathbf{j} + t\,\mathbf{k},\ \frac{d\mathbf{r}_2}{dt} = \mathbf{k} \Rightarrow \mathbf{F}_2\cdot\frac{d\mathbf{r}_2}{dt} = t$

$\Rightarrow W_2 = \int_0^1 t\,dt = \frac{1}{2}$. $\therefore\ W = W_1 + W_2 = \frac{5}{6}$

5. a) $\mathbf{F} = (3t^2 - 3t)\,\mathbf{i} + 3t\,\mathbf{j} + \mathbf{k},\ \frac{d\mathbf{r}}{dt} = \mathbf{i} + \mathbf{j} + \mathbf{k} \Rightarrow \mathbf{F}\cdot\frac{d\mathbf{r}}{dt} = 3t^2 + 1 \Rightarrow W = \int_0^1 (3t^2 + 1)\,dt = 2$

b) $\mathbf{F} = (3t^2 - 3t)\,\mathbf{i} + 3t^4\,\mathbf{j} + \mathbf{k},\ \frac{d\mathbf{r}}{dt} = \mathbf{i} + 2t\,\mathbf{j} + 4t^3\,\mathbf{k} \Rightarrow \mathbf{F}\cdot\frac{d\mathbf{r}}{dt} = 6t^5 + 4t^3 + 3t^2 - 3t \Rightarrow$

$W = \int_0^1 \left(6t^5 + 4t^3 + 3t^2 - 3t\right) dt = \frac{3}{2}$

c) $\mathbf{F}_1 = (3t^2 - 3t)\,\mathbf{i} + \mathbf{k},\ \frac{d\mathbf{r}_1}{dt} = \mathbf{i} + \mathbf{j} \Rightarrow \mathbf{F}_1\cdot\frac{d\mathbf{r}_1}{dt} = 3t^2 - 3t \Rightarrow W_1 = \int_0^1 (3t^2 - 3t)\,dt = -\frac{1}{2}$

$\mathbf{F}_2 = 3t\,\mathbf{j} + \mathbf{k},\ \frac{d\mathbf{r}_2}{dt} = \mathbf{k} \Rightarrow \mathbf{F}_2\cdot\frac{d\mathbf{r}_2}{dt} = 1 \Rightarrow W_2 = \int_0^1 dt = 1.\ \therefore\ W = W_1 + W_2 = \frac{1}{2}$

6. a) $\mathbf{F} = 2t\,\mathbf{i} + 2t\,\mathbf{j} + 2t\,\mathbf{k},\ \frac{d\mathbf{r}}{dt} = \mathbf{i} + \mathbf{j} + \mathbf{k} \Rightarrow \mathbf{F}\cdot\frac{d\mathbf{r}}{dt} = 6t \Rightarrow W = \int_0^1 6t\,dt = 3$

b) $\mathbf{F} = (t^2 + t^4)\,\mathbf{i} + (t^4 + t)\,\mathbf{j} + (t^2 + t)\,\mathbf{k},\ \frac{d\mathbf{r}}{dt} = \mathbf{i} + 2t\,\mathbf{j} + 4t^3\,\mathbf{k} \Rightarrow \mathbf{F}\cdot\frac{d\mathbf{r}}{dt} = 6t^5 + 5t^4 + 3t^2 \Rightarrow$

$W = \int_0^1 (6t^5 + 5t^4 + 3t^2)\,dt = 3$

c) $\mathbf{F}_1 = t\,\mathbf{i} + t\,\mathbf{j} + 2t\,\mathbf{k},\ \frac{d\mathbf{r}_1}{dt} = \mathbf{i} + \mathbf{j} \Rightarrow \mathbf{F}_1\cdot\frac{d\mathbf{r}_1}{dt} = 2t \Rightarrow W_1 = \int_0^1 2t\,dt = 1$. $\mathbf{F}_2 = (1 + t)\,\mathbf{i} + (t + 1)\,\mathbf{j} + 2\,\mathbf{k},$

$\frac{d\mathbf{r}_2}{dt} = \mathbf{k} \Rightarrow \mathbf{F}_2\cdot\frac{d\mathbf{r}_2}{dt} = 2 \Rightarrow W_2 = \int_0^1 2\,dt = 2.\ \therefore\ W = W_1 + W_2 = 3$

7. $\mathbf{F} = t^3\,\mathbf{i} + t^2\,\mathbf{j} - t^3\,\mathbf{k}$, $\frac{d\mathbf{r}}{dt} = \mathbf{i} + 2t\,\mathbf{j} + \mathbf{k} \Rightarrow \mathbf{F}\cdot\frac{d\mathbf{r}}{dt} = 2t^3 \Rightarrow W = \int_0^1 2t^3\,dt = \frac{1}{2}$

8. $\mathbf{F} = (2\sin t)\,\mathbf{i} + (3\cos t)\,\mathbf{j} + (\cos t + \sin t)\,\mathbf{k}$, $\frac{d\mathbf{r}}{dt} = (-\sin t)\,\mathbf{i} + (\cos t)\,\mathbf{j} + \frac{1}{6}\,\mathbf{k} \Rightarrow \mathbf{F}\cdot\frac{d\mathbf{r}}{dt} = 5\cos^2 t - 2 + \frac{1}{6}\cos t$

$+ \frac{1}{6}\sin t \Rightarrow W = \int_0^{2\pi} \left(5\cos^2 t - 2 + \frac{1}{6}\cos t + \frac{1}{6}\sin t\right) dt = \pi$

9. $\mathbf{F} = t\,\mathbf{i} + (\sin t)\,\mathbf{j} + (\cos t)\,\mathbf{k}$, $\frac{d\mathbf{r}}{dt} = (\cos t)\,\mathbf{i} - (\sin t)\,\mathbf{j} + \mathbf{k} \Rightarrow \mathbf{F}\cdot\frac{d\mathbf{r}}{dt} = t\cos t - \sin^2 t + \cos t \Rightarrow$

$W = \int_0^{2\pi} (t\cos t - \sin^2 t + \cos t)\,dt = -\pi$

10. $\mathbf{F} = t\,\mathbf{i} + (\cos^2 t)\,\mathbf{j} + (12\sin t)\,\mathbf{k}$, $\frac{d\mathbf{r}}{dt} = (\cos t)\,\mathbf{i} - (\sin t)\,\mathbf{j} + \frac{1}{6}\,\mathbf{k} \Rightarrow \mathbf{F}\cdot\frac{d\mathbf{r}}{dt} = t\cos t - \sin t\cos^2 t + 2\sin t \Rightarrow$

$W = \int_0^{2\pi} \left(t\cos t - \sin t\cos^2 t + 2\sin t\right) dt = 0$

11. $\mathbf{F} = -4t^3\,\mathbf{i} + 8t^2\,\mathbf{j} + 2\,\mathbf{k}$, $\frac{d\mathbf{r}}{dt} = \mathbf{i} + 2t\,\mathbf{j} \Rightarrow \mathbf{F}\cdot\frac{d\mathbf{r}}{dt} = 12t^3 \Rightarrow \text{Flow} = \int_0^2 12t^3\,dt = 48$

12. $\mathbf{F} = 12t^2\,\mathbf{j} + 9t^2\,\mathbf{k}$, $\frac{d\mathbf{r}}{dt} = 3\,\mathbf{j} + 4\,\mathbf{k} \Rightarrow \mathbf{F}\cdot\frac{d\mathbf{r}}{dt} = 72t^2 \Rightarrow \text{Flow} = \int_0^1 72t^2\,dt = 24$

13. $\mathbf{F} = (\cos t - \sin t)\,\mathbf{i} + (\cos t)\,\mathbf{k}$, $\frac{d\mathbf{r}}{dt} = (-\sin t)\,\mathbf{i} + (\cos t)\,\mathbf{k} \Rightarrow \mathbf{F}\cdot\frac{d\mathbf{r}}{dt} = -\sin t\cos t + 1 \Rightarrow$

$\text{Flow} = \int_0^{\pi} (-\sin t\cos t + 1)\,dt = \pi$

14. $\mathbf{F} = (-2\sin t)\,\mathbf{i} - (2\cos t)\,\mathbf{j} + 2\,\mathbf{k}$, $\frac{d\mathbf{r}}{dt} = (2\sin t)\,\mathbf{i} + (2\cos t)\,\mathbf{j} + 2\,\mathbf{k} \Rightarrow \mathbf{F}\cdot\frac{d\mathbf{r}}{dt} = -4\sin^2 t - 4\cos^2 t + 4 = 0$

$\Rightarrow \text{Flow} = 0$

15. a) $\mathbf{F} = (\cos t)\,\mathbf{i} + (\sin t)\,\mathbf{j}$, $\frac{d\mathbf{r}}{dt} = (-\sin t)\,\mathbf{i} + (\cos t)\,\mathbf{j} \Rightarrow \mathbf{F}\cdot\frac{d\mathbf{r}}{dt} = 0 \Rightarrow$ Circulation = 0.

$M = \cos t$, $N = \sin t$, $dx = -\sin t\,dt$, $dy = \cos t\,dt \Rightarrow \text{Flux} = \int_C M\,dy - N\,dx =$

$\int_0^{2\pi} (\cos^2 t + \sin^2 t)\,dt = \int_0^{2\pi} dt = 2\pi$

15. b) $\mathbf{F} = (\cos t)\,\mathbf{i} + (4 \sin t)\,\mathbf{j}$, $\frac{d\mathbf{r}}{dt} = (-\sin t)\,\mathbf{i} + (4 \cos t)\,\mathbf{j} \Rightarrow \mathbf{F}\cdot\frac{d\mathbf{r}}{dt} = 15 \sin t \cos t \Rightarrow$ Circ $=$

$\int_0^{2\pi} 15 \sin t \cos t\, dt = 0$. $M = \cos t$, $N = \sin t$, $dx = -\sin t$, $dy = 4 \cos t \Rightarrow$ Flux $=$

$\int_C M dy - N dx = \int_0^{2\pi} (4 \cos^2 t + 4 \sin^2 t)\, dt = 8\pi$

16. $M_1 = (2a \cos t)$, $N_1 = (3a \sin t)$, $dx = -a \sin t\, dt$, $dy = a \cos t\, dt \Rightarrow$ Flux $= \int_C M dy - N dx =$

$\int_0^{2\pi} (2a^2 \cos^2 t - 3a^2 \sin^2 t)\, dt = -\pi a^2$. $M_2 = (2a \cos t)$, $N_2 = (a \cos t - a \sin t)$, $dx = -a \sin t\, dt$,

$dy = a \cos t\, dt \Rightarrow$ Flux $= \int_C M dy - N dx = \int_0^{2\pi} \left(2a^2 \cos^2 t - (-a^2 \sin t \cos t + a^2 \sin^2 t)\right) dt = \pi a^2$

17. $\mathbf{F_1} = (a \cos t)\,\mathbf{i} + (a \sin t)\,\mathbf{j}$, $\frac{d\mathbf{r}_1}{dt} = (-a \sin t)\,\mathbf{i} + (a \cos t)\,\mathbf{j} \Rightarrow \mathbf{F_1}\cdot\frac{d\mathbf{r}_1}{dt} = 0 \Rightarrow \text{Circ}_1 = 0$. $M_1 = a \cos t$, $N_1 =$

$a \sin t$, $dx = -a \sin t\, dt$, $dy = a \cos t\, dt \Rightarrow \text{Flux}_1 = \int_C M_1 dy - N_1 dx = \int_0^{\pi} (a^2 \cos^2 t + a^2 \sin^2 t)\, dt =$

$\int_0^{\pi} a^2\, dt = a^2\pi$.

$\mathbf{F_2} = t\,\mathbf{i}$, $\frac{d\mathbf{r}_2}{dt} = \mathbf{i} \Rightarrow \mathbf{F_2}\cdot\frac{d\mathbf{r}_2}{dt} = t \Rightarrow \text{Circ}_2 = \int_{-a}^{a} t\, dt = 0$. $M_2 = t$, $N_2 = 0$, $dx = dt$, $dy = 0 \Rightarrow \text{Flux}_2 =$

$\int_C M_2 dy - N_2 dx = \int_{-a}^{a} 0\, dt = 0$. $\therefore$ Circ $= \text{Circ}_1 + \text{Circ}_2 = 0$, Flux $= \text{Flux}_1 + \text{Flux}_2 = a^2\pi$

18. $\mathbf{F_1} = (a^2 \cos^2 t)\,\mathbf{i} + (a^2 \sin^2 t)\,\mathbf{j}$, $\frac{d\mathbf{r}_1}{dt} = (-a \sin t)\,\mathbf{i} + (a \cos t)\,\mathbf{j} \Rightarrow \mathbf{F_1}\cdot\frac{d\mathbf{r}_1}{dt} = -a^3 \sin t \cos^2 t + a^3 \cos t \sin^2 t$

$\Rightarrow \text{Circ}_1 = \int_0^{\pi} (-a^3 \sin t \cos^2 t + a^3 \cos t \sin^2 t)\, dt = -\frac{2a^3}{3}$. $M_1 = a^2 \cos^2 t$, $N_1 = a^2 \sin^2 t$,

$dy = a \cos t$, $dx = -a \sin t\, dt \Rightarrow \text{Flux}_1 = \int_C M_1 dy - N_1 dx = \int_0^{\pi} (a^3 \cos^3 t + a^3 \sin^3 t)\, dt = \frac{4}{3} a^3$.

18. (Continued)

$\mathbf{F_2} = t^2\,\mathbf{i},\ \frac{d\mathbf{r_2}}{dt} = \mathbf{i} \Rightarrow \mathbf{F_2}\cdot\frac{d\mathbf{r_2}}{dt} = t^2 \Rightarrow \text{Circ}_2 = \int_{-a}^{a} t^2\,dt = \frac{2a^3}{3}$. $M_2 = t^2$, $N_2 = 0$, $dy = 0$, $dx = dt \Rightarrow$

$\text{Flux}_2 = \int_C M_2dy - N_2dx = 0$. $\therefore$ $\text{Circ} = \text{Circ}_1 + \text{Circ}_2 = 0$, $\text{Flux} = \text{Flux}_1 + \text{Flux}_2 = \frac{4}{3}a^3$

19. $\mathbf{F_1} = (-a\sin t)\,\mathbf{i} + (a\cos t)\,\mathbf{j},\ \frac{d\mathbf{r_1}}{dt} = (-a\sin t)\,\mathbf{i} + (a\cos t)\,\mathbf{j} \Rightarrow \mathbf{F_1}\cdot\frac{d\mathbf{r_1}}{dt} = a^2\sin^2 t + a^2\cos^2 t = a^2 \Rightarrow$

$\text{Circ}_1 = \int_0^{\pi} a^2\,dt = a^2\pi.$

$\mathbf{F_2} = t\,\mathbf{j},\ \frac{d\mathbf{r_2}}{dt} = \mathbf{i} \Rightarrow \mathbf{F_2}\cdot\frac{d\mathbf{r_2}}{dt} = 0 \Rightarrow \text{Circ}_2 = 0.$ $\therefore$ $\text{Circ} = \text{Circ}_1 + \text{Circ}_2 = a^2\pi$

$M_1 = -a\sin t$, $N_1 = a\cos t$, $dx = -a\sin t$, $dy = a\cos t \Rightarrow \text{Flux}_1 = \int_C M_1dy - N_1dx =$

$\int_0^{\pi} (-a^2\sin t\cos t + a^2\sin t\cos t)\,dt = 0.$ $M_2 = 0$, $N_2 = t$, $dx = dt$, $dy = 0 \Rightarrow \text{Flux}_2 =$

$\int_C M_2dy - N_2dx = \int_{-a}^{a} -t\,dt = 0.$ $\therefore$ $\text{Flux} = \text{Flux}_1 + \text{Flux}_2 = 0$

20. $\mathbf{F_1} = (-a^2\sin^2 t)\,\mathbf{i} + (a^2\cos^2 t)\,\mathbf{j},\ \frac{d\mathbf{r_1}}{dt} = (-a\sin t)\,\mathbf{i} + (a\cos t)\,\mathbf{j} \Rightarrow \mathbf{F_1}\cdot\frac{d\mathbf{r_1}}{dt} = a^3\sin^3 t + a^3\cos^3 t \Rightarrow$

$\text{Circ}_1 = \int_0^{\pi} (a^2\sin^3 t + a^3\cos^3 t)\,dt = \frac{4}{3}a^3$

$\mathbf{F_2} = t^2\,\mathbf{j},\ \frac{d\mathbf{r_2}}{dt} = \mathbf{i} \Rightarrow \mathbf{F_2}\cdot\frac{d\mathbf{r_2}}{dt} = 0 \Rightarrow \text{Circ}_2 = 0.$ $\therefore$ $\text{Circ} = \text{Circ}_1 + \text{Circ}_2 = \frac{4}{3}a^3$

$M_1 = -a^2\sin^2 t$, $N_1 = a^2\cos^2 t$, $dy = a\cos t\,dt$, $dx = -a\sin t\,dt \Rightarrow \text{Flux}_1 = \int_C M_1dy - N_1dx =$

$\int_0^{\pi} (-a^3\cos t\sin^2 t + a^3\sin t\cos^2 t)\,dt = \frac{2}{3}a^3.$ $M_2 = 0$, $N_2 = t^2$, $dy = 0$, $dx = dt \Rightarrow \text{Flux}_2 =$

$\int_C M_2dy - N_2dx = \int_{-a}^{a} -t^2\,dt = -\frac{2}{3}a^3.$ $\therefore$ $\text{Flux} = \text{Flux}_1 + \text{Flux}_2 = 0$

21. $\mathbf{F} = f(t)\,\mathbf{i},\ \frac{d\mathbf{r}}{dt} = \mathbf{i} + f'(t)\,\mathbf{j} \Rightarrow \mathbf{F}\cdot\frac{d\mathbf{r}}{dt} = f(t) \Rightarrow \int_{t=a}^{t=b} \mathbf{F}\cdot d\mathbf{r} = \int_a^b f(t)\,dt$. Since $t > 0$ for $a \le t \le b$, this integral yields the desired area.

SECTION 15.3 GREEN'S THEOREM IN THE PLANE

1. Equation 15: $M = -y = -a \sin t$, $N = x = a \cos t$, $dx = -a \sin t\, dt$, $dy = a \cos t\, dt \Rightarrow \frac{\partial M}{\partial x} = 0, \frac{\partial M}{\partial y} = -1,$ $\frac{\partial N}{\partial x} = 1, \frac{\partial N}{\partial y} = 0 \Rightarrow \oint_C M dy - N dx = \int_0^{2\pi} ((-a \sin\ t)(a \cos t)\, dt - (a \cos t)(-a \sin t)\, dt) =$ $0.$ $\iint_R \left(\frac{\partial M}{\partial x} + \frac{\partial N}{\partial y}\right) dx\, dy = \iint_R 0\, dx\, dy = 0$

Equation 16: $\oint_C M dx + N dy = \int_0^{2\pi} ((-a \sin t)(-a \sin t) + (a \cos t)(a \cos t))\, dt = 2\pi a^2$

$$\iint_R \left(\frac{\partial N}{\partial x} - \frac{\partial M}{\partial y}\right) dx\, dy = \int_{-a}^{a} \int_{-\sqrt{a^2+x^2}}^{\sqrt{a^2+x^2}} 2\, dx\, dy = \int_{-a}^{a} 4\sqrt{a^2 - x^2}\, dx = 2a^2\pi$$

2. $M = y = a \sin t$, $N = 0$, $dx = -a \sin t\, dt$, $dy = a \cos t\, dt \Rightarrow \frac{\partial M}{\partial x} = 0, \frac{\partial M}{\partial y} = 1, \frac{\partial N}{\partial x} = 0, \frac{\partial N}{\partial y} = 0$

Equation 15: $\oint_C M dy - N dx = \int_0^{2\pi} a^2 \sin t \cos t\, dt = 0.$ $\iint_R \left(\frac{\partial M}{\partial x} + \frac{\partial N}{\partial y}\right) dx\, dy =$ $\iint_R 0\, dx\, dy = 0$

Equation 16: $\oint_C M dx + N dy = \int_0^{2\pi} (-a^2 \sin^2 t)\, dt = -a^2\pi.$ $\iint_R \left(\frac{\partial N}{\partial x} - \frac{\partial M}{\partial y}\right) dx\, dy =$ $\iint_R -1\, dx\, dy = -a^2\pi$

3. $M = 2x = 2a \cos t$, $N = -3y = -3a \sin t$, $dx = -a \sin t$, $dy = a \cos t \Rightarrow \frac{\partial M}{\partial x} = 2, \frac{\partial M}{\partial y} = 0, \frac{\partial N}{\partial x} = 0, \frac{\partial N}{\partial y} = -3$

Equation 15: $\oint_C M dy - N dx = \int_0^{2\pi} (2a \cos t(a \cos t) + 3a \sin t(-a \sin t))dt =$ $\int_0^{2\pi} \left(2a^2 \cos^2 t - 3a^2 \sin^2 t\right) dt = -\pi a^2.$ $\iint_R \left(\frac{\partial M}{\partial x} + \frac{\partial N}{\partial y}\right) = \iint_R -1\, dx\, dy =$

3. (Continued)

$$\int_{-a}^{a}\int_{-\sqrt{a^2-x^2}}^{\sqrt{a^2-x^2}} -1 \, dx\, dy = -\pi a^2$$

Equation 16: $\oint_C M dx + N dy = \int_0^{2\pi} (2a \cos t(-a \sin t) + (-3a \sin t)(a \cos t))\, dt =$

$$\int_0^{2\pi} \left(-2a^2 \sin t \cos t - 3a^2 \sin t \cos t\right) dt = 0. \qquad \iint_R 0 \, dx\, dy = 0$$

4. $M = -x^2y = -a^3 \cos^2 t \sin t$, $N = xy^2 = a^3 \cos t \sin^2 t$, $dx = -a \sin t\, dt$, $dy = a \cos t\, dt \Rightarrow \frac{\partial M}{\partial x} =$ $-2a^2 \cos t \sin t$, $\frac{\partial M}{\partial y} = -a^2 \cos^2 t$, $\frac{\partial N}{\partial x} = a^2 \sin^2 t$, $\frac{\partial N}{\partial y} = 2a^2 \cos t \sin t$.

Equation 15: $\oint_C M dy - N dx = \int_0^{2\pi} \left(-a^4 \cos^3 t \sin t + a^4 \cos t \sin^3 t\right) dt = 0$

$$\iint_R \left(\frac{\partial M}{\partial x} + \frac{\partial N}{\partial y}\right) dx\, dy = \iint_R (-2xy + 2xy)\, dx\, dy$$

Equation 16: $\oint_C M dx + N dy = \int_0^{2\pi} \left(a^4 \cos^2 t \sin^2 t + a^4 \cos^2 t \sin^2 t\right) dt = \int_0^{2\pi} (2a^4 \cos^2 t \sin^2 t)\, dt$

$$= \frac{a^4\pi}{2}. \qquad \iint_R \left(\frac{\partial N}{\partial x} - \frac{\partial M}{\partial y}\right) dx\, dy = \iint_R (y^2 + x^2)\, dx\, dy = \int_{-a}^{a}\int_{-\sqrt{a^2-y^2}}^{\sqrt{a^2-y^2}} (y^2 + x^2)\, dx\, dy = \frac{a^4\pi}{2}$$

5. $M = x - y$, $N = y - x \Rightarrow \frac{\partial M}{\partial x} = 1$, $\frac{\partial M}{\partial y} = -1$, $\frac{\partial N}{\partial x} = -1$, $\frac{\partial N}{\partial y} = 1 \Rightarrow$ Flux $= \iint_R 2\, dx\, dy = \int_0^1\int_0^1 2\, dx\, dy$ $= 2$. Circ $= \iint_R (-1 - (-1))\, dx\, dy = 0$

6. $M = x^2 + 4y$, $N = x + y^2 \Rightarrow \frac{\partial M}{\partial x} = 2x$, $\frac{\partial M}{\partial y} = 4$, $\frac{\partial N}{\partial x} = 1$, $\frac{\partial N}{\partial y} = 2y \Rightarrow$ Flux $= \iint_R (2x + 2y)\, dx\, dy =$

6. (Continued)

$$\int_0^1\int_0^1 (2x + 2y)\,dx\,dy = 2. \text{ Circ} = \iint_R (1 - 4)\,dx\,dy = \int_0^1\int_0^1 -3\,dx\,dy = -3$$

7. $M = y^2 - x^2, N = x^2 + y^2 \Rightarrow \frac{\partial M}{\partial x} = -2x, \frac{\partial M}{\partial y} = 2y, \frac{\partial N}{\partial x} = 2x, \frac{\partial N}{\partial y} = 2y \Rightarrow \text{Flux} = \iint_R (-2x + 2y)\,dx\,dy$

$$= \int_0^3\int_0^x (-2x + 2y)dy\,dx = \int_0^3 (-2x^2 + x^2)\,dx = -9. \text{ Circ} = \iint_R (2x - 2y)\,dx\,dy =$$

$$\int_0^3\int_0^x (2x - 2y)\,dy\,dx = \int_0^3 x^2\,dx = 9$$

8. $M = x + y, N = -(x^2 + y^2) \Rightarrow \frac{\partial M}{\partial x} = 1, \frac{\partial M}{\partial y} = 1, \frac{\partial N}{\partial x} = -2x, \frac{\partial N}{\partial y} = -2y \Rightarrow \text{Flux} = \iint_R (1 - 2y)\,dx\,dy =$

$$\int_0^1\int_0^x (1 - 2y)\,dy\,dx = \int_0^1 (x - x^2)\,dx = \frac{1}{6}. \text{ Circ} = \iint_R (-2x - 1)\,dx\,dy = \int_0^1\int_0^x (-2x - 1)\,dy\,dx$$

$$= \int_0^1 (-2x^2 - x)\,dx = -\frac{7}{6}$$

9. $M = xy, N = y^2 \Rightarrow \frac{\partial M}{\partial x} = y, \frac{\partial M}{\partial y} = x, \frac{\partial N}{\partial x} = 0, \frac{\partial N}{\partial y} = 2y \Rightarrow \text{Flux} = \iint_R (y + 2y)\,dy\,dx = \int_0^1\int_{x^2}^x 3y\,dy\,dx$

$$= \int_0^1 \left(\frac{3x^2}{2} - \frac{3x^4}{2}\right)dx = \frac{1}{5}. \text{ Circ} = \iint_R -x\,dy\,dx = \int_0^1\int_{x^2}^x -x\,dy\,dx = \int_0^1 (-x^2 + x^3)\,dx = -\frac{1}{12}$$

10. $M = -\sin y,\ N = x\cos y \Rightarrow \frac{\partial M}{\partial x} = 0,\ \frac{\partial M}{\partial y} = -\cos y,\ \frac{\partial N}{\partial x} = \cos y,\ \frac{\partial N}{\partial y} = -x\sin y \Rightarrow$ Flux =

$$\iint_R (-x\sin y)\,dx\,dy = \int_0^{\pi/2}\int_0^{\pi/2} (-x\sin y)\,dx\,dy = \int_0^{\pi/2} \left(-\frac{\pi^2}{8}\sin y\right)dy = -\frac{\pi^2}{8}$$

$$\text{Circ} = \iint_R (\cos y - (-\cos y))\,dx\,dy = \int_0^{\pi/2}\int_0^{\pi/2} 2\cos y\,dx\,dy = \int_0^{\pi/2} \pi\cos y\,dy = \pi$$

11. $M = y^2,\ N = x^2 \Rightarrow \frac{\partial M}{\partial y} = 2y,\ \frac{\partial N}{\partial x} = 2y \Rightarrow \oint_C y^2\,dx + x^2\,dy = \iint_R (2x - 2y)dy\,dx =$

$$\int_0^1\int_0^{-x+1} (2x - 2y)\,dy\,dx = \int_0^1 (-3x^2 + 4x - 1)\,dx = 0.$$

12. $M = 3y,\ N = 2x \Rightarrow \frac{\partial M}{\partial y} = 3,\ \frac{\partial N}{\partial x} = 2 \Rightarrow \oint_C (3y\,dx + 2x\,dy) = \iint_R (2-3)\,dx\,dy = \int_0^{\pi}\int_0^{\sin x} -1\,dy\,dx$

$= -2$

13. $M = 6y + x,\ N = y + 2x \Rightarrow \frac{\partial M}{\partial y} = 6,\ \frac{\partial N}{\partial x} = 2 \Rightarrow \oint_C (6y + x)dx + (y + 2x)dy = \iint_R (2 - 6)\,dy\,dx =$

-4(Area of the circle) $= -16\pi$

14. $M = 2x + y^2,\ N = 2xy + 3y \Rightarrow \frac{\partial M}{\partial y} = 2y,\ \frac{\partial N}{\partial x} = 2y \Rightarrow \oint_C (2x + y^2)\,dx + (2xy + 3y)\,dy =$

$$\iint_R (2y - 2y)\,dx\,dy = 0$$

15. $M = 2xy^3,\ N = 4x^2y^2 \Rightarrow \frac{\partial M}{\partial y} = 6xy^2,\ \frac{\partial N}{\partial x} = 8xy^2 \Rightarrow \oint_C 2xy^3\,dx + 4x^2y^2\,dy = \iint_R (8xy^2 - 6xy^2)\,dx\,dy$

$$\int_0^1\int_0^{x^3} 2xy^2\,dy\,dx = \int_0^1 \frac{2}{3}x^{10}\,dx = \frac{2}{33}$$

16. $M = 4x - 2y,\ N = 2x - 4y \Rightarrow \frac{\partial M}{\partial y} = -2,\ \frac{\partial N}{\partial x} = 2 \Rightarrow \oint_C (4x - 2y)\,dx + (2x - 4y)\,dy =$

$$\iint_R (2 - (-2))\,dx\,dy = 4 \iint_R dx\,dy = 4(\text{Area of the circle}) = 16\pi$$

17. a) $M = f(x),\ N = g(y) \Rightarrow \frac{\partial M}{\partial y} = 0,\ \frac{\partial N}{\partial x} = 0 \Rightarrow \oint_C f(x)\,dx + g(y)\,dy = \iint_R 0\,dy\,dx = 0$

b) $M = ky,\ N = hx \Rightarrow \frac{\partial M}{\partial y} = k,\ \frac{\partial N}{\partial x} = h \Rightarrow \oint_C ky\,dx + hx\,dy = \iint_R (h - k)\,dx\,dy =$

$(h - k)(\text{Area of the region})$

18. $M = 4x^3y,\ N = x^4 \Rightarrow \frac{\partial M}{\partial y} = 4x^3,\ \frac{\partial N}{\partial x} = 4x^3 \Rightarrow \oint_C 4x^3y\,dx + x^4\,dy = \iint_R (4x^3 - 4x^3)\,dx\,dy = 0$

19. Area $= \frac{1}{2}\oint_C x\,dy - y\,dx$. $M = x = a\cos t,\ N = y = a\sin t \Rightarrow dx = -a\sin t\,dt,\ dy = a\cos t\,dt \Rightarrow$ Area $=$

$$\frac{1}{2}\int_0^{2\pi} \left(a^2\cos^2 t + a^2\sin^2 t\right) dt = \frac{1}{2}\int_0^{2\pi} a^2\,dt = \pi a^2$$

20. Area $= \frac{1}{2}\oint_C x\,dy - y\,dx$. $M = x = a\cos t,\ N = y = b\sin t \Rightarrow dx = -a\sin t\,dt,\ dy = b\cos t\,dt \Rightarrow$

$$\text{Area} = \frac{1}{2}\int_0^{2\pi} (ab\cos^2 t + ab\sin^2 t)\,dt = \frac{1}{2}\int_0^{2\pi} ab\,dt = \pi ab$$

21. Area $= \frac{1}{2}\oint_C x\,dy - y\,dx$. $M = x = \cos^3 t,\ N = y = \sin^3 t \Rightarrow dx = -3\cos^2 t\sin t\,dt,\ dy = 3\sin^2 t\cos t\,dt$

$$\Rightarrow \text{Area} = \frac{1}{2}\int_0^{2\pi} (3\sin^2 t\cos^2 t(\cos^2 t + \sin^2 t))\,dt = \frac{1}{2}\int_0^{2\pi} (3\sin^2 t\cos^2 t)\,dt = \frac{3\pi}{8}$$

22. Area $= \frac{1}{2}\oint_C x\,dy - y\,dx$. $M = x = t^2,\ N = y = \frac{t^3}{3} - t \Rightarrow dx = 2t\,dt,\ dy = (t^2 - 1)\,dt \Rightarrow$

$$\text{Area} = \frac{1}{2}\int_{-\sqrt{3}}^{\sqrt{3}} \left(\frac{1}{3}t^4 + t^2\right) dt = \frac{8}{5}\sqrt{3}$$

SECTION 15.4 SURFACE AREA AND SURFACE INTEGRALS

1. $\mathbf{p} = \mathbf{k}$, $\nabla f = 2x\,\mathbf{i} + 2y\,\mathbf{j} - \mathbf{k} \Rightarrow |\nabla f| = \sqrt{(2x)^2 + (2y)^2 + (-1)^2} = \sqrt{4x^2 + 4y^2 + 1}$. $|\nabla f \cdot \mathbf{p}| = 1 \Rightarrow$

$$S = \iint_R \frac{|\nabla f|}{|\nabla f \cdot \mathbf{p}|}\,dA = \iint_R \sqrt{4x^2 + 4y^2 + 1}\;dx\,dy =$$

$$\iint_R \sqrt{4r^2\cos^2\theta + 4r^2\sin^2\theta + 1}\;r\,dr\,d\theta = \int_0^{2\pi}\int_0^{\sqrt{2}} \sqrt{4r^2 + 1}\;r\,dr\,d\theta = \frac{13}{3}\pi$$

2. $\mathbf{p} = \mathbf{k}$. $\nabla f = 2x\,\mathbf{i} + 2y\,\mathbf{j} - \mathbf{k} \Rightarrow |\nabla f| = \sqrt{4x^2 + 4y^2 + 1}$. $|\nabla f \cdot \mathbf{p}| = 1 \Rightarrow S = \iint_R \frac{|\nabla f|}{|\nabla f \cdot \mathbf{p}|}\,dA =$

$$\iint_R \sqrt{4x^2 + 4y^2 + 1}\;dx\,dy = \iint_R \sqrt{4r^2 + 1}\;r\,dr\,d\theta = \int_0^{2\pi}\int_{\sqrt{2}}^{\sqrt{6}} \sqrt{4r^2 + 1}\;r\,dr\,d\theta =$$

$$\int_0^{2\pi} \frac{98}{12}\,d\theta = \frac{49\pi}{3}$$

3. $\mathbf{p} = \mathbf{k}$. $\nabla f = \mathbf{i} + 2\,\mathbf{j} + 2\,\mathbf{k} \Rightarrow |\nabla f| = 3$. $|\nabla f \cdot \mathbf{p}| = 2 \Rightarrow S = \iint_R \frac{|\nabla f|}{|\nabla f \cdot \mathbf{p}|}\,dA = \iint_R \frac{3}{2}\,dx\,dy =$

$$\int_{-1}^{1}\int_{y^2}^{2-y^2} \frac{3}{2}\,dx\,dy = \int_{-1}^{1} (3 - 3y^2)\,dy = 4$$

4. $\mathbf{p} = \mathbf{k}$. $\nabla f = 2x\,\mathbf{i} - 2\,\mathbf{k} \Rightarrow |\nabla f| = \sqrt{4x^2 + 4} = 2\sqrt{x^2 + 1}$. $|\nabla f \cdot \mathbf{p}| = 2 \Rightarrow S = \iint_R \frac{|\nabla f|}{|\nabla f \cdot \mathbf{p}|}\,dA =$

$$\iint_R \frac{2\sqrt{x^2 + 1}}{2}\,dx\,dy = \int_0^{\sqrt{3}}\int_0^{x} \sqrt{x^2 + 1}\;dy\,dx = \int_0^{\sqrt{3}} x\sqrt{x^2 + 1}\,dx = \frac{7}{3}$$

5. $\mathbf{p} = \mathbf{k}$. $\nabla f = 2x\,\mathbf{i} - 2\,\mathbf{j} - 2\,\mathbf{k} \Rightarrow |\nabla f| = \sqrt{(2x)^2 + (-2)^2 + (-2)^2} = \sqrt{4x^2 + 8}$. $|\nabla f \cdot \mathbf{p}| = 2 \Rightarrow S =$

$$\iint_R \frac{|\nabla f|}{|\nabla f \cdot \mathbf{p}|}\,dA = \iint_R \frac{\sqrt{4x^2+8}}{2}\,dx\,dy = \int_0^2\int_0^{3x} \sqrt{x^2+2}\,dy\,dx = \int_0^2 3x\sqrt{x^2+2}\,dx = 6\sqrt{6} - 2\sqrt{2}$$

6. $\mathbf{p} = \mathbf{k}$. $\nabla f = \left(2x - \frac{2}{x}\right)\mathbf{i} + \sqrt{15}\,\mathbf{j} - \mathbf{k} \Rightarrow |\nabla f| = \sqrt{\left(2x-\frac{2}{x}\right)^2 + \left(\sqrt{15}\right)^2 + (-1)^2} = \sqrt{4x^2 + 8 + \frac{4}{x^2}} = \sqrt{\left(2x+\frac{2}{x}\right)^2} = 2x + \frac{2}{x}$ on $1 \le x \le 2$. $|\nabla f \cdot \mathbf{p}| = 1 \Rightarrow S = \iint_R \frac{|\nabla f|}{|\nabla f \cdot \mathbf{p}|}\,dA = \iint_R (2x + 2x^{-1})\,dx\,dy =$

$$\int_0^1\int_1^2 (2x + 2x^{-1})\,dx\,dy = \int_0^1 (3 + 2\ln 2)\,dy = 3 + 2\ln 2$$

7. $\mathbf{p} = \mathbf{k}$. $\nabla f = 2x\,\mathbf{i} + 2y\,\mathbf{j} + 2z\,\mathbf{k} \Rightarrow |\nabla f| = \sqrt{4x^2+4y^2+4z^2} = \sqrt{8} = 2\sqrt{2}$. $|\nabla f \cdot \mathbf{p}| = 2z \Rightarrow S =$

$$\iint_R \frac{|\nabla f|}{|\nabla f \cdot \mathbf{p}|}\,dA = \iint_R \frac{2\sqrt{2}}{2z}\,dA = \sqrt{2}\iint_R \frac{1}{z}\,dA =$$

$$\sqrt{2}\iint_R \frac{1}{\sqrt{2-(x^2+y^2)}}\,dA = \sqrt{2}\int_0^{2\pi}\int_0^1 \frac{r\,dr\,d\theta}{\sqrt{2-r^2}} = \sqrt{2}\int_0^{2\pi} (-1+\sqrt{2})\,d\theta = 2\pi(2-\sqrt{2})$$

8. $\mathbf{p} = \mathbf{k}$. $\nabla f = c\,\mathbf{i} - \mathbf{k} \Rightarrow |\nabla f| = \sqrt{c^2+1}$. $|\nabla f \cdot \mathbf{p}| = 1 \Rightarrow S = \iint_R \frac{|\nabla f|}{|\nabla f \cdot \mathbf{p}|}\,dA = \iint_R \sqrt{c^2+1}\,dx\,dy =$

$$\int_0^{2\pi}\int_0^1 \sqrt{c^2+1}\,r\,dr\,d\theta = \int_0^{2\pi} \frac{\sqrt{c^2+1}}{2}\,d\theta = \pi\sqrt{c^2+1}$$

9. $\mathbf{p} = \mathbf{k}$. $\nabla f = 2x\,\mathbf{i} + 2z\,\mathbf{j} \Rightarrow |\nabla f| = \sqrt{(2x)^2 + (2z)^2} = 2$. $|\nabla f \cdot \mathbf{p}| = 2z$ for the upper surface, $z \ge 0 \Rightarrow$

$$S = \iint_R \frac{|\nabla f|}{|\nabla f \cdot \mathbf{p}|}\,dA = 2\iint_R \frac{2}{2z}\,dA = 2\iint_R \frac{1}{z}\,dA = 2\iint_R \frac{1}{\sqrt{1-x^2}}\,dy\,dx =$$

$$4\int_{-1/2}^{1/2}\int_0^{1/2} \frac{1}{\sqrt{1-x^2}}\,dy\,dx = 2\int_{-1/2}^{1/2} \frac{1}{\sqrt{1-x^2}}\,dx = \frac{2\pi}{3}$$

10. $\mathbf{p} = \mathbf{j}$. $\nabla f = 2x\,\mathbf{i} + \mathbf{j} + 2z\,\mathbf{k} \Rightarrow |\nabla f| = \sqrt{4x^2+4z^2+1}$. $|\nabla f \cdot \mathbf{p}| = 1 \Rightarrow S = \iint_R \frac{|\nabla f|}{|\nabla f \cdot \mathbf{p}|}\,dA =$

10. (Continued)

$$\iint_R \sqrt{4x^2 + 4z^2 + 1}\, dx\, dz = \int_0^{2\pi}\int_0^1 \sqrt{4r^2 + 1}\, r\, dr\, d\theta = \int_0^{2\pi} \left(\frac{5\sqrt{5} - 1}{12}\right) d\theta = \frac{\pi(5\sqrt{5} - 1)}{6}$$

11. The bottom face of the cube is in the xy–plane $\Rightarrow z = 0 \Rightarrow g(x,y,0) = x + y$ and $f(x,y,z) = z = 0 \Rightarrow$

$\nabla f = \mathbf{k} \Rightarrow |\nabla f| = 1$. $\mathbf{p} = \mathbf{k} \Rightarrow |\nabla f \cdot \mathbf{p}| = 1 \Rightarrow d\sigma = dx\, dy \Rightarrow \iint_{z=0} (x + y)\, dx\, dy = \int_0^1\int_0^1 (x + y)\, dx\, dy =$

1. Because of symmetry, you get 1 over the face of the cube in the xz–plane and 1 over the face of the cube in the yz–plane.
In the top of the cube, $g(x,y,z) = g(x,y,1) = x + y + 1$ and$f(x,y,z) = z = 1 \Rightarrow \nabla f = \mathbf{k} \Rightarrow |\nabla f| = 1$. $\mathbf{p} = \mathbf{k} \Rightarrow$

$|\nabla f \cdot \mathbf{p}| = 1 \Rightarrow d\sigma = dx\, dy \Rightarrow \iint_{z=1} (x + y + 1)\, dx\, dy = \int_0^1\int_0^1 (x + y + 1)\, dx\, dy = 2$. Because of

symmetry, the integral is 2 over each of the other two faces. $\therefore \iint_{\text{cube}} (x + y + z)\, d\sigma = 9$.

12. On the face in the xz–plane, $y = 0 \Rightarrow f(x,y,z) = y = 0$ and $g(x,y,z) = g(x,0,z) = z \Rightarrow \nabla f = \mathbf{j} \Rightarrow |\nabla f| = 1$.

$\mathbf{p} = \mathbf{j} \Rightarrow |\nabla f \cdot \mathbf{p}| = 1 \Rightarrow d\sigma = dx\, dz \Rightarrow \iint_{y=0} (y + z)\, d\sigma = \int_0^1\int_0^2 z\, dx\, dz = 1$

On the face in the xy–plane, $z = 0 \Rightarrow f(x,y,z) = z = 0$ and $g(x,y,z) = g(x,y,0) = y \Rightarrow \nabla f = \mathbf{k} \Rightarrow |\nabla f| = 1$.

$\mathbf{p} = \mathbf{k} \Rightarrow |\nabla f \cdot \mathbf{p}| = 1 \Rightarrow d\sigma = dx\, dy \Rightarrow \iint_{z=0} (y + z)\, d\sigma = \int_0^1\int_0^2 y\, dx\, dy = 1$.

On the triangular face in the plane $x = 2$, $f(x,y,z) = x = 2$ and $g(x,y,z) = g(2,y,z) = y + z \Rightarrow \nabla f = \mathbf{i} \Rightarrow$

$|\nabla f| = 1$. $\mathbf{p} = \mathbf{i} \Rightarrow |\nabla f \cdot \mathbf{p}| = 1 \Rightarrow d\sigma = dz\, dy \Rightarrow \iint_{x=2} (y + z)\, d\sigma = \int_0^1\int_0^{1-y} (y + z)\, dz\, dy = \frac{1}{3}$

On the triangular face in the yz–plane, $x = 0 \Rightarrow f(x,y,z) = x = 0$ and $g(x,y,z) = g(0,y,z) = y + z \Rightarrow \nabla f = \mathbf{i}$

$\Rightarrow |\nabla f| = 1$. $\mathbf{p} = \mathbf{i} \Rightarrow |\nabla f \cdot \mathbf{p}| = 1 \Rightarrow d\sigma = dz\, dy \Rightarrow \iint_{x=0} (y + z)\, d\sigma = \int_0^1\int_0^{1-y} (y + z)\, dz\, dy = \frac{1}{3}$

On the sloped face, $y + z = 1$, $f(x,y,z) = y + z = 1$ and $g(x,y,z) = y + z = 1 \Rightarrow \nabla f = \mathbf{j} + \mathbf{k} \Rightarrow |\nabla f| = \sqrt{2}$.

12. (Continued)

$$\mathbf{p} = \mathbf{k} \Rightarrow |\nabla f \cdot \mathbf{p}| = 1 \Rightarrow d\sigma = \sqrt{2}\, dx\, dy \Rightarrow \iint_{y+z=1} (y + z)\, d\sigma = \int_0^1 \int_0^2 \sqrt{2}\, dx\, dy = 2\sqrt{2}$$

$$\therefore \iint_{\text{wedge}} g(x,y,z)\, d\sigma = \frac{8}{3} + 2\sqrt{2}$$

13. On the faces in the coordinate planes, $g(x,y,z) = 0 \Rightarrow$ the integral over these faces is 0.
On the face, $x = a$, $f(x,y,z) = x = a$ and $g(x,y,z) = g(a,y,z) = ayz \Rightarrow \nabla f = \mathbf{i} \Rightarrow |\nabla f| = 1$. $\mathbf{p} = \mathbf{i} \Rightarrow |\nabla f \cdot \mathbf{p}| = 1$

$$\Rightarrow d\sigma = dy\, dz \Rightarrow \iint_{x=a} xyz\, d\sigma = \int_0^c \int_0^b ayz\, dy\, dz = \frac{ab^2c^2}{4}$$

On the face, $y = b$, $f(x,y,z) = y = b$ and $g(x,y,z) = g(x,b,z) = bxz \Rightarrow \nabla f = \mathbf{j} \Rightarrow |\nabla f| = 1$. $\mathbf{p} = \mathbf{j} \Rightarrow |\nabla f \cdot \mathbf{p}| = 1$

$$\Rightarrow d\sigma = dx\, dz \Rightarrow \iint_{y=b} xyz\, dx\, dz = \int_0^c \int_0^a bxz\, dz\, dx = \frac{a^2bc^2}{4}$$

On the face, $z = c$, $f(x,y,z) = z = c$ and $g(x,y,z) = g(x,y,c) = cxy \Rightarrow \nabla f = \mathbf{k} \Rightarrow |\nabla f| = 1$. $\mathbf{p} = \mathbf{k} \Rightarrow |\nabla f \cdot \mathbf{p}| = 1$

$$\Rightarrow d\sigma = dy\, dx \Rightarrow \iint_{z=c} xyz\, d\sigma = \int_0^b \int_0^a cxy\, dx\, dy = \frac{a^2b^2c}{4}$$

$$\therefore \iint_S g(x,y,z)\, d\sigma = \frac{abc(ab + ac + bc)}{4}$$

14. On the face, $x = a$, $f(x,y,z) = x = a$ and $g(x,y,z) = g(a,y,z) = ayz \Rightarrow \nabla f = \mathbf{i} \Rightarrow |\nabla f| = 1$. $\mathbf{p} = \mathbf{i} \Rightarrow |\nabla f \cdot \mathbf{p}| =$

$1 \Rightarrow d\sigma = dz\, dy \Rightarrow \iint_{x=a} xyz\, d\sigma = \int_{-b}^{b} \int_{-c}^{c} ayz\, dz\, dy = 0$ Because of the symmetry of g on all

the other faces, $\iint_S g(x,y,z)\, d\sigma = 0$

15. $\nabla f = 2\,\mathbf{i} + 2\,\mathbf{j} + \mathbf{k}$ and $g(x,y,z) = x + y + (2 - 2x - 2y) = 2 - x - y \Rightarrow |\nabla f| = 3$. $\mathbf{p} = \mathbf{k} \Rightarrow |\nabla f \cdot \mathbf{p}| = 1 \Rightarrow$

$$d\sigma = 3\, dy\, dx \Rightarrow \iint_S (x + y + z)\, d\sigma = 3 \int_0^1 \int_0^{1-x} (2 - x - y)\, dy\, dx = 2$$

16. $f(x,y,z) = y^2 + 4z = 16 \Rightarrow \nabla f = 2y\,\mathbf{j} + 4\,\mathbf{k} \Rightarrow |\nabla f| = \sqrt{4y^2 + 16} = 2\sqrt{y^2+4}$. $\mathbf{p} = \mathbf{k} \Rightarrow |\nabla f \cdot \mathbf{p}| = 4 \Rightarrow$

$$d\sigma = \frac{2\sqrt{y^2+4}}{4}\,dx\,dy \Rightarrow \iint_R g(x,y,z)\,d\sigma = \int_{-4}^{4}\int_0^1 x\sqrt{y^2+4}\,\frac{2\sqrt{y^2+4}}{4}\,dx\,dy =$$

$$\int_{-4}^{4}\int_0^1 \frac{x(y^2+4)}{2}\,dx\,dy = \frac{56}{3}$$

17. $\nabla G = 2x\,\mathbf{i} + 2y\,\mathbf{j} + 2z\,\mathbf{k} \Rightarrow |\nabla G| = \sqrt{4x^2+4y^2+4z^2} = 2a$. $\mathbf{n} = \dfrac{2x\,\mathbf{i} + 2y\,\mathbf{j} + 2z\,\mathbf{k}}{2\sqrt{x^2+y^2+z^2}} = \dfrac{x\,\mathbf{i} + y\,\mathbf{j} + z\,\mathbf{k}}{a} \Rightarrow$

$\mathbf{F}\cdot\mathbf{n} = \dfrac{z^2}{a}$. $|\nabla G \cdot \mathbf{k}| = 2z \Rightarrow d\sigma = \dfrac{2a}{2z}\,dA = \dfrac{a}{z}\,dA$. $\therefore$ Flux $= \displaystyle\iint_R \frac{z^2}{a}\left(\frac{a}{z}\right)dA = \iint_R z\,dA =$

$$\iint_R \sqrt{a^2-(x^2+y^2)}\,dx\,dy = \int_0^{\pi/2}\int_0^a \sqrt{a^2-r^2}\,r\,dr\,d\theta = \frac{a^3\pi}{6}$$

18. $\nabla G = 2x\,\mathbf{i} + 2y\,\mathbf{j} + 2z\,\mathbf{k} \Rightarrow |\nabla G| = \sqrt{4x^2+4y^2+4z^2} = 2a$. $\mathbf{n} = \dfrac{2x\,\mathbf{i} + 2y\,\mathbf{j} + 2z\,\mathbf{k}}{2\sqrt{x^2+y^2+z^2}} = \dfrac{x\,\mathbf{i} + y\,\mathbf{j} + z\,\mathbf{k}}{a} \Rightarrow$

$\mathbf{F}\cdot\mathbf{n} = \dfrac{-xy}{a} + \dfrac{xy}{a} = 0$. $|\nabla G \cdot \mathbf{k}| = 2z \Rightarrow d\sigma = \dfrac{2a}{2z}\,dA = \dfrac{a}{z}\,dA$. $\therefore$ Flux $= \displaystyle\iint_S \mathbf{F}\cdot\mathbf{n}\,dS =$

$$\iint_S 0\,d\sigma = 0$$

19. $\mathbf{n} = \dfrac{x\,\mathbf{i} + y\,\mathbf{j} + z\,\mathbf{k}}{a}$, $d\sigma = \dfrac{a}{z}\,dA$ (See Exercise 17) and $\mathbf{F}\cdot\mathbf{n} = \dfrac{xy}{a} - \dfrac{xy}{a} + \dfrac{z}{a} = \dfrac{z}{a}$.

$\therefore$ Flux $= \displaystyle\iint_R \frac{z}{a}\left(\frac{a}{z}\right)dA = \iint_R 1\,dA = \frac{\pi a^2}{4}$

20. $\mathbf{n} = \dfrac{x\,\mathbf{i} + y\,\mathbf{j} + z\,\mathbf{k}}{a}$, $d\sigma = \dfrac{a}{z}\,dA$ (See Exercise 17) and $\mathbf{F}\cdot\mathbf{n} = \dfrac{zx^2}{a} + \dfrac{zy^2}{a} + \dfrac{z^3}{a} = z\left(\dfrac{x^2+y^2+z^2}{a}\right) = az \Rightarrow$

Flux $= \displaystyle\iint_R za\left(\frac{a}{z}\right)dx\,dy = \iint_R a^2\,dx\,dy = a^2(\text{Area of R}) = \frac{1}{4}\pi a^4$

21. $\mathbf{n} = \frac{x\,\mathbf{i} + y\,\mathbf{j} + z\,\mathbf{k}}{a}$, $d\sigma = \frac{a}{z}\,dA$ (See Exercise 17) and $\mathbf{F} \cdot \mathbf{n} = \frac{x^2}{a} + \frac{y^2}{a} + \frac{z^2}{a} = a \Rightarrow \text{Flux} = \iint_R a\left(\frac{a}{z}\right) dA$

$$= \iint_R \frac{a^2}{z}\,dA = \iint_R \frac{a^2}{\sqrt{a^2 - (x^2 + y^2)}}\,dA = \int_0^{\pi/2}\int_0^a \frac{a^2}{\sqrt{a^2 - r^2}}\,r\,dr\,d\theta = \frac{a^3\pi}{2}$$

22. $\mathbf{n} = \frac{x\,\mathbf{i} + y\,\mathbf{j} + z\,\mathbf{k}}{a}$, $d\sigma = \frac{a}{z}\,dA$ (See Exercise 17) and $\mathbf{F} \cdot \mathbf{n} = \frac{x^2/a + y^2/a + z^2/a}{\sqrt{x^2 + y^2 + z^2}} = \frac{a^2/a}{a} = 1 \Rightarrow$

$$\text{Flux} = \iint_R \frac{a}{z}\,dx\,dy = \iint_R \frac{a}{\sqrt{a^2 - (x^2 + y^2)}}\,dx\,dy = \int_0^{\pi/2}\int_0^a \frac{a}{\sqrt{a^2 - r^2}}\,r\,dr\,d\theta = \frac{a^2\pi}{2}$$

23. $\nabla G = 2y\,\mathbf{i} + \mathbf{k} \Rightarrow |\nabla G| = \sqrt{4y^2 + 1} \Rightarrow \mathbf{n} = \frac{2y\,\mathbf{i} + \mathbf{k}}{\sqrt{4y^2 + 1}} \Rightarrow \mathbf{F} \cdot \mathbf{n} = \frac{2xy - 3z}{\sqrt{4y^2 + 1}}$. $\mathbf{p} = \mathbf{k} \Rightarrow |\nabla G \cdot \mathbf{k}| = 1 \Rightarrow$

$$d\sigma = \sqrt{4y^2 + 1}\,dA \Rightarrow \text{Flux} = \iint_R \left(\frac{2xy - 3z}{\sqrt{4y^2 + 1}}\right)\sqrt{4y^2 + 1}\,dA = \iint_R (2xy - 3z)\,dA =$$

$$\iint_R (2xy - 3(4 - y^2))\,dA = \int_0^1\int_{-2}^2 (2xy - 12 + 3y^2)\,dy\,dx = -32$$

24. $\nabla G = -2x\,\mathbf{i} - 2y\,\mathbf{j} + \mathbf{k} \Rightarrow |\nabla G| = \sqrt{4x^2 + 4y^2 + 1} = \sqrt{4(x^2 + y^2) + 1}$. $\mathbf{p} = \mathbf{k} \Rightarrow |\nabla G \cdot \mathbf{k}| = 1$

$\mathbf{n} = \frac{2x\,\mathbf{i} + 2y\,\mathbf{j} + \mathbf{k}}{\sqrt{4(x^2 + y^2) + 1}} \Rightarrow \mathbf{F} \cdot \mathbf{n} = \frac{8x^2 + 8y^2 - 2}{\sqrt{4(x^2 + y^2) + 1}}$. $d\sigma = \sqrt{4(x^2 + y^2) + 1}\;dx\,dy \Rightarrow$

$$\text{Flux} = \iint_R \frac{8x^2 + 8y^2 - 2}{\sqrt{4(x^2 + y^2) + 1}}\sqrt{4(x^2 + y^2) + 1}\;dx\,dy = \iint_R (8x^2 + 8y^2 - 2)\,dx\,dy =$$

$$\int_0^{2\pi}\int_0^1 (8r^2 - 2)\,r\,dr\,d\theta = 2\pi$$

25. $\nabla G = -e^x\,\mathbf{i} + \mathbf{j} \Rightarrow |\nabla G| = \sqrt{e^{2x} + 1}$. $\mathbf{p} = \mathbf{i} \Rightarrow |\nabla G \cdot \mathbf{i}| = e^x$. $\mathbf{n} = \frac{e^x\,\mathbf{i} - \mathbf{j}}{\sqrt{e^{2x} + 1}} \Rightarrow \mathbf{F} \cdot \mathbf{n} = \frac{-2e^x - 2y}{\sqrt{e^{2x} + 1}}$.

$$d\sigma = \frac{\sqrt{e^{2x} + 1}}{e^x}\,dA \Rightarrow \text{Flux} = \iint_R \frac{-2e^x - 2y}{\sqrt{e^{2x} + 1}}\left(\frac{\sqrt{e^{2x} + 1}}{e^x}\right) dA = \iint_R -4\,dA = \int_0^1\int_1^2 -4\,dy\,dz$$

$$= -4$$

26. $\nabla G = -\frac{1}{x}\mathbf{i} + \mathbf{j} \Rightarrow |\nabla G| = \sqrt{\frac{1}{x^2} + 1} = \frac{\sqrt{1 + x^2}}{x}$ since $1 \le x \le e$. $\mathbf{p} = \mathbf{j} \Rightarrow |\nabla G \cdot \mathbf{j}| = 1 \Rightarrow \mathbf{n} = \frac{-\frac{1}{x}\mathbf{i} + \mathbf{j}}{\frac{\sqrt{1 + x^2}}{x}} =$

$\frac{-\mathbf{i} + x\,\mathbf{j}}{\sqrt{1 + x^2}} \Rightarrow \mathbf{F} \cdot \mathbf{n} = \frac{2xy}{\sqrt{1 + x^2}}$. $d\sigma = \frac{\sqrt{1 + x^2}}{x}\,dx\,dz \Rightarrow$ Flux $= \iint_R \frac{2xy}{\sqrt{1 + x^2}}\left(\frac{\sqrt{1 + x^2}}{x}\right) dx\,dz =$

$$\int_0^1 \int_1^e 2y\,dx\,dz = \int_1^e \int_0^1 2\ln x\,dz\,dx = 2$$

27. $\nabla F = 2x\,\mathbf{i} + 2y\,\mathbf{j} + 2z\,\mathbf{k} \Rightarrow |\nabla F| = \sqrt{4x^2 + 4y^2 + 4z^2} = 2$. $\mathbf{p} = \mathbf{k} \Rightarrow |\nabla F \cdot \mathbf{k}| = 2z$ since $z \ge 0 \Rightarrow d\sigma =$ $\frac{2}{2z}\,dA = \frac{1}{z}\,dA$. $\therefore$ $M = \iint_S \delta\,d\sigma = \frac{\pi}{2}\delta$. $M_{xy} = \iint_S z\delta\,d\sigma = \delta \iint_S z\left(\frac{1}{z}\right) dA =$

$\delta \int_0^1 \int_0^{\sqrt{1-x^2}} dy\,dx = \frac{\pi}{4}\delta$. $\therefore\ \bar{z} = \frac{\frac{\pi}{4}\delta}{\frac{\pi}{2}\delta} = \frac{1}{2}$. Because of symmetry, $\bar{x} = \bar{y} = \frac{1}{2}$. $\therefore$ Centroid $= \left(\frac{1}{2}, \frac{1}{2}, \frac{1}{2}\right)$

28. $\nabla F = 2y\,\mathbf{i} + 2z\,\mathbf{k} \Rightarrow |\nabla F| = \sqrt{4y^2 + 4z^2} = \sqrt{4(y^2 + z^2)} = 6$. $\mathbf{p} = \mathbf{k} \Rightarrow |\nabla F \cdot \mathbf{k}| = 2z$ since $z \ge 0 \Rightarrow$

$d\sigma = \frac{6}{2z}\,dx\,dy = \frac{3}{z}\,dx\,dy$. $\therefore$ $M = \iint_S 1\,d\sigma = \int_{-3}^{3}\int_0^3 \frac{3}{z}\,dx\,dy = \int_{-3}^{3}\int_0^3 \frac{3}{\sqrt{9 - y^2}}\,dx\,dy = 9\pi.$

$M_{xy} = \iint_S z\,d\sigma = \int_{-3}^{3}\int_0^3 z\left(\frac{3}{z}\right) dx\,dy = 54$. $M_{xz} = \iint_S y\,d\sigma = \int_{-3}^{3}\int_0^3 y\left(\frac{3}{z}\right) dx\,dy =$

$\int_{-3}^{3}\int_0^3 \frac{3y}{\sqrt{9 - y^2}}\,dx\,dy = 0$. $M_{yz} = \iint_S x\,d\sigma = \int_{-3}^{3}\int_0^3 \frac{3x}{\sqrt{9 - y^2}}\,dx\,dy = \frac{27}{2}\pi$

$\therefore\ \bar{x} = \frac{\frac{27}{2}\pi}{9\pi} = \frac{3}{2}$, $\bar{y} = 0$, $\bar{z} = \frac{54}{9\pi} = \frac{6}{\pi}$

29. Because of symmetry, $\bar{x} = \bar{y} = 0$. $M = \iint_S \delta\, d\sigma = \delta \iint_S d\sigma = \delta(\text{Area of S}) = 3\pi\sqrt{2}\,\delta$.

$\nabla F = 2x\,\mathbf{i} + 2y\,\mathbf{j} - 2z\,\mathbf{k} \Rightarrow |\nabla F| = \sqrt{4x^2 + 4y^2 + 4z^2} = 2\sqrt{x^2 + y^2 + z^2}$. $\mathbf{p} = \mathbf{k} \Rightarrow |\nabla F \cdot \mathbf{k}| = 2z \Rightarrow$

$$d\sigma = \frac{2\sqrt{x^2+y^2+z^2}}{2z}\,dA = \frac{\sqrt{x^2+y^2+z^2}}{z}\,dA = \frac{\sqrt{x^2+y^2+(x^2+y^2)}}{z}\,dA = \frac{\sqrt{2}\sqrt{x^2+y^2}}{z}.$$

$$\therefore\ M_{xy} = \delta \iint_S z\left(\frac{\sqrt{2}\sqrt{x^2+y^2}}{z}\right) dA = \delta \iint_S \sqrt{2}\sqrt{x^2+y^2}\,dA = \delta \int_0^{2\pi}\int_1^2 \sqrt{2}\,r^2\,dr\,d\theta =$$

$$\frac{14\pi\sqrt{2}}{3}\delta.\ \ \bar{z} = \frac{\frac{14\pi\sqrt{2}}{3}\delta}{3\pi\sqrt{2}\,\delta} = \frac{14}{9}.\ \therefore\ (\bar{x},\bar{y},\bar{z}) = \left(0,0,\frac{14}{9}\right).\ \ I_z = \iint_S (x^2+y^2)\,\delta\,d\sigma =$$

$$\iint_S (x^2+y^2)\left(\frac{\sqrt{2}\sqrt{x^2+y^2}}{z}\right)\delta\,dA = \delta\sqrt{2}\iint_S (x^2+y^2)\,dA = \delta\sqrt{2}\int_0^{2\pi}\int_1^2 r^3\,dr\,d\theta = \frac{15\pi\sqrt{2}}{2}\delta.$$

$$R_z = \sqrt{I_z/M} = \frac{\sqrt{10}}{2}$$

30. $\nabla f = 8x\,\mathbf{i} + 8y\,\mathbf{j} - 2z\,\mathbf{k} \Rightarrow |\nabla f| = \sqrt{64x^2 + 64y^2 + 4z^2} = 2\sqrt{16x^2 + 16y^2 + z^2} = 2\sqrt{4z^2 + z^2} = 2\sqrt{5}\,z$ since

$z \ge 0$. $\mathbf{p} = \mathbf{k} \Rightarrow |\nabla f \cdot \mathbf{k}| = 2z \Rightarrow d\sigma = \frac{2\sqrt{5}\,z}{2z}\,dA = \sqrt{5}\,dA \Rightarrow I_z = \iint_S (x^2+y^2)\,\delta\,d\sigma =$

$$\delta\sqrt{5}\iint_R (x^2+y^2)\,dx\,dy = \delta\sqrt{5}\int_{-\pi/2}^{\pi/2}\int_0^{2\cos\theta} r^2\,r\,dr\,d\theta = \frac{3\sqrt{5}\,\pi\delta}{2}$$

31. $f_x(x,y) = 2x$, $f_y(x,y) = 2y \Rightarrow \sqrt{f_x^2 + f_y^2 + 1} = \sqrt{4x^2 + 4y^2 + 1} \Rightarrow \text{Area} = \iint_R \sqrt{4x^2+4y^2+1}\,dx\,dy =$

$$\int_0^{2\pi}\int_0^{\sqrt{3}} \sqrt{4r^2+1}\,r\,dr\,d\theta = \frac{\pi}{6}\left(13\sqrt{13} - 1\right)$$

32. $f_x(x,y) = \frac{x}{\sqrt{x^2+y^2}}$, $f_y(x,y) = \frac{y}{\sqrt{x^2+y^2}} \Rightarrow \sqrt{f_x^2 + f_y^2 + 1} = \sqrt{\frac{x^2}{x^2+y^2} + \frac{y^2}{x^2+y^2} + 1} = \sqrt{2} \Rightarrow$

$$\text{Area} = \iint_{R_{xy}} \sqrt{2}\,dx\,dy = \sqrt{2}(\text{Area between the ellipse and the circle}) = \sqrt{2}(6\pi - \pi) = 5\pi\sqrt{2}$$

33. $f_z(y,z) = -2y,\ f_y(y,z) = -2z \Rightarrow \sqrt{f_y^2 + f_z^2 + 1} = \sqrt{4y^2 + 4z^2 + 1} \Rightarrow \text{Area} = \int\int_R \sqrt{4y^2 + 4z^2 + 1}\ dy\,dz$

$= \int_0^{2\pi}\int_0^1 \sqrt{4r^2 + 1}\ r\,dr\,d\theta = \frac{\pi}{6}(5\sqrt{5} - 1)$

34. Over R_{xy}: $z = 2 - \frac{2}{3}x - 2y$ $f_x(x,y) = -\frac{2}{3},\ f_y(x,y) = -2 \Rightarrow \sqrt{f_x^2 + f_y^2 + 1} = \sqrt{\frac{4}{9} + 4 + 1} = \frac{7}{3} \Rightarrow$

$\text{Area} = \int\int_{R_{xy}} \frac{7}{3}\,dA = \frac{7}{3}$(Area of the triangle in the xy–plane) $= \frac{7}{2}$. Over R_{xz}: $y = -\frac{1}{3}x - \frac{1}{2}z \Rightarrow$

$f_x(x,z) = -\frac{1}{3},\ f_z(x,z) = -\frac{1}{2} \Rightarrow \sqrt{f_x^2 + f_z^2 + 1} = \sqrt{\frac{1}{9} + \frac{1}{4} + 1} = \frac{7}{6} \Rightarrow \text{Area} = \int\int_{R_{xz}} \frac{7}{6}\,dA =$

$\frac{7}{6}$(Area of the triangle in the xz–plane) $= \frac{7}{2}$. Over R_{yz}: $x = -3y - \frac{3}{2}z \Rightarrow f_y(y,z) = -3,\ f_z(y,z) = -\frac{3}{2} \Rightarrow$

$\sqrt{f_y^2 + f_z^2 + 1} = \sqrt{9 + \frac{9}{4} + 1} = \frac{7}{2}$

$\Rightarrow \text{Area} = \int\int_{R_{yz}} \frac{7}{2}\,dA = \frac{7}{2}$(Area of the triangle in the yz–plane) $= \frac{7}{2}$

35. $y = \frac{z^2}{2} \Rightarrow f_x(x,z) = 0,\ f_z(x,z) = z \Rightarrow \sqrt{f_x^2 + f_z^2 + 1} = \sqrt{z^2 + 1} \Rightarrow \text{Area} = \int_0^2\int_0^1 \sqrt{z^2 + 1}\ dx\,dz =$

$\sqrt{5} + \frac{1}{2}\ln\left(\sqrt{5} + 2\right)$ (Note: On integrating the second time with respect to z, use the substitution $z = \tan\theta$ which means the integration will go from 0 to $\tan^{-1}2$.)

36. $y = 4 - z \Rightarrow f_x(x,z) = 0,\ f_z(x,z) = -1 \Rightarrow \sqrt{f_x^2 + f_z^2 + 1} = \sqrt{2} \Rightarrow \text{Area} = \int\int_{R_{xz}} \sqrt{2}\,dA =$

$\int_0^2\int_0^{4-z^2} \sqrt{2}\,dx\,dz = \frac{16\sqrt{2}}{3}$

SECTION 15.5 THE DIVERGENCE THEOREM

1. $\frac{\partial}{\partial x}(y-x) = -1, \frac{\partial}{\partial y}(z-y) = -1, \frac{\partial}{\partial z}(y-x) = 0 \Rightarrow \nabla \cdot \mathbf{F} = -2 \Rightarrow \text{Flux} = \int_{-1}^{1}\int_{-1}^{1}\int_{-1}^{1} -2\,dx\,dy\,dz = -16$

2. $\frac{\partial}{\partial x}(x^2) = 2x, \frac{\partial}{\partial y}(y^2) = 2y, \frac{\partial}{\partial x}(z^2) = 2z \Rightarrow \nabla \cdot \mathbf{F} = 2x + 2y + 2z \Rightarrow$

$$\text{Flux} = \int_0^1\int_0^1\int_0^1 (2x + 2y + 2z)\,dx\,dy\,dz = 3.$$

3. $\frac{\partial}{\partial x}(x^2) = 2x, \frac{\partial}{\partial y}(y^2) = 2y, \frac{\partial}{\partial x}(z^2) = 2z \Rightarrow \nabla \cdot \mathbf{F} = 2x + 2y + 2z \Rightarrow$

$$\text{Flux} = \int_{-1}^1\int_{-1}^1\int_{-1}^1 (2x + 2y + 2z)\,dx\,dy\,dz = 0$$

4. $\frac{\partial}{\partial x}(x^2) = 2x, \frac{\partial}{\partial y}(y^2) = 2y, \frac{\partial}{\partial z}(z^2) = 2z \Rightarrow \nabla \cdot \mathbf{F} = 2x + 2y + 2z \Rightarrow$

$$\text{Flux} = \iiint_D (2x + 2y + 2z)\,dx\,dy\,dz = \int_0^1\int_0^{2\pi}\int_0^2 (2r\cos\theta + 2r\sin\theta + 2z)\,r\,dr\,d\theta\,dz = 8\pi$$

5. $\frac{\partial}{\partial x}(y) = 0, \frac{\partial}{\partial y}(xy) = x, \frac{\partial}{\partial z}(-z) = -1 \Rightarrow \nabla \cdot \mathbf{F} = x - 1 \Rightarrow \text{Flux} = \iiint_{\text{solid}} (x-1)\,dz\,dy\,dx =$

$$\int_0^{2\pi}\int_0^2\int_0^{r^2} (r\cos\theta - 1)\,dz\,r\,dr\,d\theta = -8\pi$$

6. $\frac{\partial}{\partial x}(x^2) = 2x, \frac{\partial}{\partial y}(xz) = 0, \frac{\partial}{\partial z}(-3z) = -3 \Rightarrow \nabla \cdot \mathbf{F} = 2x - 3 \Rightarrow \text{Flux} = \iiint_D (2x - 3)\, dV =$

$$\int_0^{2\pi}\int_0^{2\pi}\int_0^{4} (2\,\rho \sin\phi \cos\theta - 3)\,\rho^2 \sin\phi\, d\rho\, d\phi\, d\theta = 0$$

7. $\frac{\partial}{\partial x}(x^2) = 2x, \frac{\partial}{\partial y}(-2xy) = -2x, \frac{\partial}{\partial z}(3xz) = 3x \Rightarrow \text{Flux} = \iiint_D 3x\, dx\, dy\, dz =$

$$\int_0^{\pi/2}\int_0^{\pi/2}\int_0^{2} 3\rho \sin\phi \cos\theta(\rho^2 \sin\phi)\, d\rho\, d\phi\, d\theta = 3\pi$$

8. $\frac{\partial}{\partial x}(6x^2 + 2xy) = 12 + 2y, \frac{\partial}{\partial y}(y + x^2 z) = 2, \frac{\partial}{\partial z}(4x^2y^3) = 0 \Rightarrow \nabla \cdot \mathbf{F} = 14 + 2y \Rightarrow \text{Flux} =$

$$\iiint_D (14 + 2y)\, dV = \int_0^{3}\int_0^{\pi/2}\int_0^{2} (14 + 2r\sin\theta)\, r\, dr\, d\theta\, dz = 42\pi + 16$$

9. $\frac{\partial}{\partial x}(2xz) = 2z, \frac{\partial}{\partial y}(-xy) = -x, \frac{\partial}{\partial z}(-z^2) = -2z \Rightarrow \nabla \cdot \mathbf{F} = -x \Rightarrow \text{Flux} = \iiint_D -x\, dV =$

$$\int_0^{2}\int_0^{\sqrt{16-4x^2}}\int_0^{4-y} -x\, dz\, dy\, dx = -\frac{40}{3}$$

10. $\frac{\partial}{\partial x}(x^3) = 3x^2, \frac{\partial}{\partial y}(y^3) = 3y^2, \frac{\partial}{\partial z}(z^3) = 3z^2 \Rightarrow \nabla \cdot \mathbf{F} = 3x^2 + 3y^2 + 3z^2 = 3a^2$ on the sphere $\Rightarrow$

$$\text{Flux} = \iiint_D 3a^2\, dV = 3a^2\left(\frac{4}{3}\pi a^3\right) = 4\pi a^5$$

11. Let $\rho = \sqrt{x^2 + y^2 + z^2}$. Then $\frac{\partial \rho}{\partial x} = \frac{x}{\rho}, \frac{\partial \rho}{\partial y} = \frac{y}{\rho}, \frac{\partial \rho}{\partial z} = \frac{z}{\rho} \Rightarrow \frac{\partial}{\partial x}(\rho x) = \frac{\partial \rho}{\partial x} x + \rho = \frac{x^2}{\rho} = \rho, \frac{\partial}{\partial y}(\rho y) =$ $\frac{\partial \rho}{\partial y} y + \rho = \frac{y^2}{\rho} + \rho, \frac{\partial}{\partial z}(\rho z) = \frac{\partial \rho}{\partial z} z + \rho = \frac{z^2}{\rho} + \rho \Rightarrow \nabla \cdot \mathbf{F} = \frac{x^2 + y^2 + z^2}{\rho} + 3\rho = 4\rho$ since $\rho = \sqrt{x^2 + y^2 + z^2}$

$$\Rightarrow \text{Flux} = \iiint_D 4\rho \, dV = \iiint_D 4\sqrt{x^2 + y^2 + z^2} \; dx\,dy\,dz =$$

$$\int_0^{2\pi}\int_0^{2\pi}\int_1^{\sqrt{2}} (4\rho)\rho^2 \sin\phi \, d\rho \, d\phi \, d\theta = 0$$

12. Let $\rho = \sqrt{x^2 + y^2 + z^2}$. Then $\frac{\partial \rho}{\partial x} = \frac{x}{\rho}, \frac{\partial \rho}{\partial y} = \frac{y}{\rho}, \frac{\partial \rho}{\partial z} = \frac{z}{\rho} \Rightarrow \frac{\partial}{\partial x}\left(\frac{x}{\rho}\right) = \frac{1}{\rho} - \frac{x}{\rho^2}\frac{\partial \rho}{\partial x} = \frac{1}{\rho} - \frac{x^2}{\rho^3}$. Similarly,

$\frac{\partial}{\partial y}\left(\frac{y}{\rho}\right) = \frac{1}{\rho} - \frac{y^2}{\rho^3}, \frac{\partial}{\partial z}\left(\frac{z}{\rho}\right) = \frac{1}{\rho} - \frac{z^2}{\rho^3} \Rightarrow \nabla \cdot \mathbf{F} = \frac{3}{\rho} - \frac{x^2 + y^2 + z^2}{\rho^3} = \frac{2}{\rho} \Rightarrow \text{Flux} = \iiint_D \frac{2}{\rho} dV =$

$$\int_0^{2\pi}\int_0^{2\pi}\int_1^{4} \left(\frac{2}{\rho}\right)\rho^2 \sin\phi \, d\rho \, d\phi \, d\theta = 0$$

13. $\frac{\partial}{\partial x}(x) = 1, \frac{\partial}{\partial y}(-2y) = -2, \frac{\partial}{\partial z}(z + 3) = 1 \Rightarrow \nabla \cdot \mathbf{F} = 0 \Rightarrow$ Flux = 0 over the solid. In the xy–plane, z = 0, $\mathbf{n} = -\mathbf{k}$, and $\mathbf{F} = x\,\mathbf{i} - 2y\,\mathbf{j} + 3\,\mathbf{k} \Rightarrow \mathbf{F} \cdot \mathbf{n} = -3 \Rightarrow \text{Flux} = \iint_{z=0} -3 \, d\sigma = -3(\text{Area of the square}) =$ -3. In the yz–plane, x = 0, $\mathbf{n} = -\mathbf{i}$, $\mathbf{F} = -2y\,\mathbf{j} + (z + 3)\,\mathbf{k} \Rightarrow \mathbf{F} \cdot \mathbf{n} = 0 \Rightarrow$ Flux = 0.
In the xz–plane, y = 0, $\mathbf{n} = -\mathbf{j}$, $\mathbf{F} = x\,\mathbf{i} + (z + 3)\,\mathbf{k} \Rightarrow \mathbf{F} \cdot \mathbf{n} = 0 \Rightarrow$ Flux = 0. ∴ The total flux = $-3 + 0 + 0 + 1 + (-3) + (\text{Flux of the top}) = 0 \Rightarrow$ Flux of the top = 5

14. $|\mathbf{F} \cdot \mathbf{n}| \le \|\mathbf{F}\|\,\|\mathbf{n}\| \le 1$ since $\|\mathbf{F}\| \le 1$, $\|\mathbf{n}\| = 1$. Then $\iiint_D \nabla \cdot \mathbf{F} \, d\sigma = \iint_S \mathbf{f} \cdot \mathbf{n} \, d\sigma \le$

$$\iint_S |\mathbf{F} \cdot \mathbf{n}| \, d\sigma \le \iint_S 1 \, d\sigma = \text{Area of S}$$

15. a) $\frac{\partial}{\partial x}(x) = 1, \frac{\partial}{\partial y}(y) = 1, \frac{\partial}{\partial z}(z) = 1 \Rightarrow \nabla \cdot \mathbf{F} = 3 \Rightarrow \text{Flux} = \iiint_D 3 \, dV = 3 \iiint_D dV =$
3(Volume of the solid)

b) If **F** is orthogonal to **n** at every point of S, then $\mathbf{F} \cdot \mathbf{n} = 0$ everywhere $\Rightarrow \text{Flux} = \iint_S \mathbf{F} \cdot \mathbf{n} \, d\sigma = 0$.

But the Flux is 3(Volume of the solid) ≠ 0. ∴ **F** is not orthogonal to **n** at every point.

16. a) Flux of ∇f over S = $\iint_S \nabla f \cdot \mathbf{n}\, d\sigma = \iiint_D \nabla \cdot \nabla f\, dV = \iiint_D 0\, dV = 0$ since $\nabla \cdot \nabla f = 0$

b) $\nabla f = \frac{\partial f}{\partial x}\mathbf{i} + \frac{\partial f}{\partial y}\mathbf{j} + \frac{\partial f}{\partial z}\mathbf{k} \Rightarrow f\,\nabla f = f\frac{\partial f}{\partial x}\mathbf{i} + f\frac{\partial f}{\partial y}\mathbf{j} + f\frac{\partial f}{\partial z}\mathbf{k} \Rightarrow \nabla \cdot f\nabla f = \frac{\partial f}{\partial x}\frac{\partial f}{\partial x} + f\frac{\partial^2 f}{\partial x^2} + \frac{\partial f}{\partial y}\frac{\partial f}{\partial y} + f\frac{\partial^2 f}{\partial y^2} + \frac{\partial f}{\partial z}\frac{\partial f}{\partial z} + f\frac{\partial^2 f}{\partial z^2} = \left(\frac{\partial f}{\partial x}\right)^2 + \left(\frac{\partial f}{\partial y}\right)^2 + \left(\frac{\partial f}{\partial z}\right)^2 + f\frac{\partial^2 f}{\partial x^2} + f\frac{\partial^2 f}{\partial y^2} + f\frac{\partial^2 f}{\partial z^2} = |\nabla f|^2 + f(\nabla^2 f) = |\nabla f|^2$ since f is harmonic $\Rightarrow$ $\nabla^2 f = 0$. Then $\iint_S f\nabla f \cdot \mathbf{n}\, d\sigma = \iiint_D (\nabla \cdot f\nabla f)\, dV = \iiint_D |\nabla f|^2\, dV$

SECTION 15.6 STOKES' THEOREM

1. curl $\mathbf{F} = \nabla \times \mathbf{F} = 2\mathbf{k}$, $\mathbf{n} = \mathbf{k} \Rightarrow$ curl $\mathbf{F} \cdot \mathbf{n} = 2 \Rightarrow d\sigma = dx\,dy \Rightarrow \oint_C \mathbf{F} \cdot d\mathbf{r} = \iint_R 2\, dA$ = 2(Area of the ellipse) = 4π

2. curl $\mathbf{F} = \nabla \times \mathbf{F} = \mathbf{k}$, $\mathbf{n} = \mathbf{k} \Rightarrow$ curl $\mathbf{F} \cdot \mathbf{n} = 1 \Rightarrow d\sigma = dx\,dy \Rightarrow \oint_C \mathbf{F} \cdot d\mathbf{r} = \iint_R dx\,dy = 9\pi$

3. curl $\mathbf{F} = \nabla \times \mathbf{F} = -x\,\mathbf{i} - 2x\,\mathbf{j} + (z-1)\,\mathbf{k}$, $\mathbf{n} = \frac{\mathbf{i}+\mathbf{j}+\mathbf{k}}{\sqrt{3}} \Rightarrow$ curl $\mathbf{F} \cdot \mathbf{n} = \frac{1}{\sqrt{3}}(-3x + z - 1) \Rightarrow d\sigma = \frac{\sqrt{3}}{1}\,dA \Rightarrow$

$$\oint_C \mathbf{F} \cdot d\mathbf{r} = \iint_R \frac{1}{\sqrt{3}}(-3x + z - 1)\sqrt{3}\, dA = \int_0^1 \int_0^{1-x} (-3x + z - 1)\, dy\, dx =$$

$$\int_0^1 \int_0^{1-x} (-3x + (1 - x - y) - 1)\, dy\, dx = \int_0^1 \int_0^{1-x} (-4x - y)\, dy\, dx = -\frac{5}{6}$$

4. curl $\mathbf{F} = \nabla \times \mathbf{F} = (2y - 2z)\,\mathbf{i} + (2z - 2x)\,\mathbf{j} + (2x - 2y)\,\mathbf{k}$, $\mathbf{n} = \frac{\mathbf{i}+\mathbf{j}+\mathbf{k}}{\sqrt{3}} \Rightarrow$ curl $\mathbf{F} \cdot \mathbf{n} = \frac{1}{\sqrt{3}}(2y - 2z + 2z - 2x + 2x - 2y) = 0 \Rightarrow \oint_C \mathbf{F} \cdot d\mathbf{r} = \iint_R 0\, d\sigma = 0$

5. curl $\mathbf{F} = \nabla \times \mathbf{F} = (2y - 0)\,\mathbf{i} + (2z - 2x)\,\mathbf{j} + (2x - 2y)\,\mathbf{k} = 2y\,\mathbf{i} + (2z - 2x)\,\mathbf{j} + (2x - 2y)\,\mathbf{k}$, $\mathbf{n} = \mathbf{k} \Rightarrow$ curl $\mathbf{F} \cdot \mathbf{n} = 2x - 2y \Rightarrow d\sigma = dx\,dy \Rightarrow \oint_C \mathbf{F} \cdot d\mathbf{r} = \int_{-1}^1 \int_{-1}^1 (2x - 2y)\, dx\, dy = 0$

6. curl $\mathbf{F} = \nabla \times \mathbf{F} = -3x^2y^2\,\mathbf{k}$, $\mathbf{n} = \mathbf{k} \Rightarrow$ curl $\mathbf{F} \cdot \mathbf{n} = -3x^2y^2 \Rightarrow \oint_C \mathbf{F} \cdot d\mathbf{r} = \iint_R -3x^2y^2\,dA =$

$$\int_0^{2\pi}\int_0^a (-3r^2\cos^2\theta(r^2\sin^2\theta))\,r\,dr\,d\theta = \int_0^{2\pi}\int_0^a (-3r^5\cos^2\theta\sin^2\theta)\,dr\,d\theta = -\frac{\pi a^6}{8}$$

7. $x = 3\cos t$, $y = 2\sin t \Rightarrow \mathbf{F} = (2\sin t)\,\mathbf{i} + (9\cos^2 t)\,\mathbf{j} + (9\cos^2 t + 16\sin^4 t)\sin e^{\sqrt{6\sin t\cos t\,(0)}}$ and $\mathbf{r} = (3\cos t)\,\mathbf{i} + (2\sin t)\,\mathbf{j} \Rightarrow d\mathbf{r} = (-3\sin t)\,dt\,\mathbf{i} + (2\cos t)\,dt\,\mathbf{j} \Rightarrow \mathbf{F} \cdot d\mathbf{r} = -6\sin^2 t\,dt + 18\cos^3 t\,dt \Rightarrow$

$$\iint_S \nabla \times \mathbf{F} \cdot \mathbf{n}\,d\sigma = \int_0^{2\pi} (-6\sin^2 t + 18\cos^3 t)\,dt = -6\pi$$

8. curl $\mathbf{F} = \nabla \times \mathbf{F} = -2x\,\mathbf{j} + 2\,\mathbf{k}$. Flux of $\nabla \times \mathbf{F} = \iint_S \nabla \times \mathbf{F} \cdot \mathbf{n}\,d\sigma = \oint_C \mathbf{F} \cdot d\mathbf{r}$. Let C be $x = a\cos t$, $y = a\sin t$. Then $\mathbf{r} = (a\cos t)\,\mathbf{i} + (a\sin t)\,\mathbf{j} \Rightarrow d\mathbf{r} = (-a\sin t)\,dt\,\mathbf{i} + (a\cos t)\,dt\,\mathbf{j}$. Then $\mathbf{F} \cdot d\mathbf{r} = ay\sin t\,dt + ax\cos t\,dt = a^2\sin^2 t\,dt + a^2\cos^2 t\,dt = a^2 dt$. $\therefore$ Flux of $\nabla \times \mathbf{F} = \oint_C \mathbf{F} \cdot d\mathbf{r} = \int_0^{2\pi} a^2\,dt = 2\pi a^2$

9. Let S_1 and S_2 be oriented surfaces that span C and that induce the same positive direction on C. Then $\iint_{S_1} \nabla \times \mathbf{F} \cdot \mathbf{n}_1\,d\sigma_1 = \int_C \mathbf{F} \cdot d\mathbf{r} = \iint_{S_2} \nabla \times \mathbf{F} \cdot \mathbf{n}_2\,d\sigma_2$

10. a) $\nabla \cdot \mathbf{F} = \left(\frac{\partial}{\partial x}\mathbf{i} + \frac{\partial}{\partial y}\mathbf{j} + \frac{\partial}{\partial z}\mathbf{k}\right) \cdot (x\,\mathbf{i} + y\,\mathbf{j} + z\,\mathbf{k}) = 3$

b) $$\nabla \times \mathbf{F} = \begin{vmatrix} \mathbf{i} & \mathbf{j} & \mathbf{k} \\ \frac{\partial}{\partial x} & \frac{\partial}{\partial y} & \frac{\partial}{\partial z} \\ x & y & z \end{vmatrix} = 0$$

11. a) $\mathbf{F} = M\,\mathbf{i} + N\,\mathbf{j} + P\,\mathbf{k} \Rightarrow$ curl $\mathbf{F} = \left(\frac{\partial P}{\partial y} - \frac{\partial N}{\partial z}\right)\mathbf{i} + \left(\frac{\partial M}{\partial z} - \frac{\partial P}{\partial x}\right)\mathbf{j} + \left(\frac{\partial N}{\partial x} - \frac{\partial M}{\partial y}\right)\mathbf{k}$.

$\nabla \cdot \nabla \times \mathbf{F} = \text{div(curl } \mathbf{F}) = \frac{\partial}{\partial x}\left(\frac{\partial P}{\partial y} - \frac{\partial N}{\partial z}\right) + \frac{\partial}{\partial y}\left(\frac{\partial M}{\partial z} - \frac{\partial P}{\partial x}\right) + \frac{\partial}{\partial z}\left(\frac{\partial N}{\partial x} - \frac{\partial M}{\partial y}\right) = \frac{\partial^2 P}{\partial x \partial y} - \frac{\partial^2 N}{\partial x \partial z} + \frac{\partial^2 M}{\partial y \partial z} - \frac{\partial^2 P}{\partial y \partial x} + \frac{\partial^2 N}{\partial z \partial x} - \frac{\partial^2 M}{\partial z \partial y} = 0$ if the partial derivatives are continuous.

b) $\iint_S \nabla \times \mathbf{F} \cdot \mathbf{n}\,d\sigma = \iiint_D \nabla \cdot \nabla \times \mathbf{F}\,dV = \iiint_D 0\,dV = 0$ if the divergence theorem applies.

12. a) $\mathbf{F} = 2x\,\mathbf{i} + 2y\,\mathbf{j} + 2z\,\mathbf{k} \Rightarrow \text{curl}\,\mathbf{F} = \mathbf{0}$ $\therefore$ $\oint_C \mathbf{F} \cdot d\mathbf{r} = \iint_S \nabla \times \mathbf{F} \cdot \mathbf{n}\, d\sigma = \iint_S 0\, d\sigma = 0$

b) $\mathbf{F} = \nabla(x^2y^2z^3) = 2xy^2z^3\,\mathbf{i} + 2x^2yz^3\,\mathbf{j} + 3x^2y^2z^2\,\mathbf{k} \Rightarrow \text{curl}\,\mathbf{F} = (6x^2yz^2 - 6x^2yz^2)\,\mathbf{i} + (6xy^2z^2 - 6xy^2z^2)\,\mathbf{j} +$ $(4xyz^3 - 4xyz^3)\,\mathbf{k} = \mathbf{0}$. $\therefore$ $\oint_C \mathbf{F} \cdot d\mathbf{r} = \iint_S \nabla \times \mathbf{F} \cdot \mathbf{n}\, d\sigma = \iint_S 0\, d\sigma = 0$

c) $\mathbf{F} = \nabla \times (x\,\mathbf{i} + y\,\mathbf{j} + z\,\mathbf{k}) = \mathbf{0} \Rightarrow \nabla \times \mathbf{F} = \mathbf{0}$. $\therefore$ $\oint_C \mathbf{F} \cdot d\mathbf{r} = \iint_S \nabla \times \mathbf{F} \cdot \mathbf{n}\, d\sigma = \iint_S 0\, d\sigma = 0$

d) $\mathbf{F} = \nabla f \Rightarrow \nabla \times \mathbf{F} = \nabla \times \nabla f = \mathbf{0}$. $\therefore$ $\oint_C \mathbf{F} \cdot d\mathbf{r} = \iint_S \nabla \times \mathbf{F} \cdot \mathbf{n}\, d\sigma = \iint_S 0\, d\sigma = 0$

13. $\mathbf{F} = \nabla f = -\frac{1}{2}(x^2 + y^2 + z^2)^{-3/2}(2x)\,\mathbf{i} - \frac{1}{2}(x^2 + y^2 + z^2)^{-3/2}(2y)\,\mathbf{j} - \frac{1}{2}(x^2 + y^2 + z^2)^{-3/2}(2z)\,\mathbf{k} =$ $-x(x^2 + y^2 + z^2)^{-3/2}\,\mathbf{i} - y(x^2 + y^2 + z^2)^{-3/2}\,\mathbf{j} - z(x^2 + y^2 + z^2)^{-3/2}\,\mathbf{k}$

a) $\mathbf{r} = (a\cos t)\,\mathbf{i} + (a\sin t)\,\mathbf{j},\ 0 \le t \le 2\pi \Rightarrow d\mathbf{r} = (-a\sin t)\,dt\,\mathbf{i} + (a\cos t)\,dt\,\mathbf{j} \Rightarrow$
$\mathbf{F} \cdot d\mathbf{r} = -x(x^2 + y^2 + z^2)^{-3/2}(-a\sin t)\,dt - y(x^2 + y^2 + z^2)^{-3/2}(a\cos t)\,dt =$
$-\frac{a\cos t}{a^3}(-a\sin t)\,dt - \frac{a\sin t}{a^3}(a\cos t)\,dt = 0 \Rightarrow \int_C \mathbf{F} \cdot d\mathbf{r} = 0$

b) $\oint_C \mathbf{F} \cdot d\mathbf{r} = \iint_S \text{curl}\,\mathbf{F} \cdot \mathbf{n}\, d\sigma = \iint_S \text{curl}(\nabla f) \cdot \mathbf{n}\, d\sigma = \iint_S \mathbf{0} \cdot \mathbf{n}\, d\sigma =$ $\iint_S 0\, d\sigma = 0$

14. a) $\text{div}(g\mathbf{F}) = \nabla \cdot g\mathbf{F} = \frac{\partial}{\partial x}(gM) + \frac{\partial}{\partial y}(gN) + \frac{\partial}{\partial z}(gP) = \left(g\frac{\partial M}{\partial x} + M\frac{\partial g}{\partial x}\right) + \left(g\frac{\partial N}{\partial y} + N\frac{\partial g}{\partial y}\right) + \left(g\frac{\partial P}{\partial z} + P\frac{\partial g}{\partial z}\right) =$
$\left(M\frac{\partial g}{\partial x} + N\frac{\partial g}{\partial y} + P\frac{\partial g}{\partial z}\right) + g\left(\frac{\partial M}{\partial x} + \frac{\partial N}{\partial y} + \frac{\partial P}{\partial z}\right) = \mathbf{F} \cdot \nabla g + g(\nabla \cdot \mathbf{F})$

b) $\nabla \times (g\mathbf{F}) = \left(\frac{\partial}{\partial y}(gP) - \frac{\partial}{\partial z}(gN)\right)\mathbf{i} + \left(\frac{\partial}{\partial z}(gM) - \frac{\partial}{\partial x}(gP)\right)\mathbf{j} + \left(\frac{\partial}{\partial x}(gN) - \frac{\partial}{\partial y}(gM)\right)\mathbf{k} =$
$\left(P\frac{\partial g}{\partial y} + g\frac{\partial P}{\partial y} - N\frac{\partial g}{\partial z} - g\frac{\partial N}{\partial z}\right)\mathbf{i} + \left(M\frac{\partial g}{\partial z} + g\frac{\partial M}{\partial z} - P\frac{\partial g}{\partial x} - g\frac{\partial P}{\partial x}\right)\mathbf{j} + \left(N\frac{\partial g}{\partial x} + g\frac{\partial N}{\partial x} - M\frac{\partial g}{\partial y} - g\frac{\partial M}{\partial y}\right)\mathbf{k}$
$= \left(P\frac{\partial g}{\partial y} - N\frac{\partial g}{\partial z}\right)\mathbf{i} + \left(g\frac{\partial P}{\partial y} - g\frac{\partial N}{\partial z}\right)\mathbf{i} + \left(M\frac{\partial g}{\partial z} - P\frac{\partial g}{\partial x}\right)\mathbf{j} + \left(g\frac{\partial M}{\partial z} - g\frac{\partial P}{\partial x}\right)\mathbf{j} + \left(N\frac{\partial g}{\partial x} - M\frac{\partial g}{\partial y}\right)\mathbf{k} +$
$\left(g\frac{\partial N}{\partial x} - g\frac{\partial M}{\partial y}\right)\mathbf{k} = g(\nabla \times \mathbf{F}) + \nabla g \times \mathbf{F}$

c) $$(\mathbf{F_1} \times \mathbf{F_2}) = \begin{vmatrix} \mathbf{i} & \mathbf{j} & \mathbf{k} \\ M_1 & N_1 & P_1 \\ M_2 & N_2 & P_2 \end{vmatrix} = (N_1P_2 - P_1N_2)\,\mathbf{i} - (M_1P_2 - P_1M_2)\,\mathbf{j} + (M_1N_2 - N_1M_2)\,\mathbf{k} \Rightarrow$$

$\nabla \cdot (\mathbf{F_1} \times \mathbf{F_2}) = \nabla \cdot \left((N_1P_2 - P_1N_2)\,\mathbf{i} - (M_1P_2 - P_1M_2)\,\mathbf{j} + (M_1N_2 - N_1M_2)\,\mathbf{k}\right) =$

14. c) (Continued)

$$\frac{\partial}{\partial x}(N_1P_2 - P_1N_2) - \frac{\partial}{\partial y}(M_1P_2 - P_1M_2) + \frac{\partial}{\partial z}(M_1N_2 - N_1M_2) = P_2\frac{\partial N_1}{\partial x} + N_1\frac{\partial P_2}{\partial x} - N_2\frac{\partial P_1}{\partial x} - P_1\frac{\partial N_2}{\partial x}$$
$$- M_1\frac{\partial P_2}{\partial y} - P_2\frac{\partial M_1}{\partial y} + P_1\frac{\partial M_2}{\partial y} + M_2\frac{\partial P_1}{\partial y} + M_1\frac{\partial N_2}{\partial z} + N_2\frac{\partial M_1}{\partial z} - N_1\frac{\partial M_2}{\partial z} - M_2\frac{\partial N_1}{\partial z} =$$
$$M_2\left(\frac{\partial P_1}{\partial y} - \frac{\partial N_1}{\partial z}\right) + N_2\left(\frac{\partial M_1}{\partial z} - \frac{\partial P_1}{\partial x}\right) + P_2\left(\frac{\partial N_1}{\partial x} - \frac{\partial M_1}{\partial y}\right) + M_1\left(\frac{\partial N_2}{\partial z} - \frac{\partial P_2}{\partial y}\right) + N_1\left(\frac{\partial P_2}{\partial x} - \frac{\partial M_2}{\partial z}\right) +$$
$$P_1\left(\frac{\partial M_2}{\partial y} - \frac{\partial N_2}{\partial x}\right) = \mathbf{F_2} \cdot (\nabla \times \mathbf{F_1}) - \mathbf{F_1} \cdot (\nabla \times \mathbf{F_2})$$

SECTION 15.7 PATH INDEPENDENCE, POTENTIAL FUNCTIONS, AND CONSERVATIVE FIELDS

1. $\frac{\partial P}{\partial y} = x = \frac{\partial N}{\partial z}, \frac{\partial M}{\partial z} = y = \frac{\partial P}{\partial x}, \frac{\partial N}{\partial x} = z = \frac{\partial M}{\partial y} \Rightarrow$ Conservative

2. $\frac{\partial P}{\partial y} = x\cos z = \frac{\partial N}{\partial z}, \frac{\partial M}{\partial z} = y\cos z = \frac{\partial P}{\partial x}, \frac{\partial N}{\partial x} = \sin z = \frac{\partial M}{\partial y} \Rightarrow$ Conservative

3. $\frac{\partial P}{\partial y} = -1 \neq \frac{\partial N}{\partial z} \Rightarrow$ Not Conservative

4. $\frac{\partial N}{\partial x} = 1 \neq \frac{\partial M}{\partial y} \Rightarrow$ Not Conservative

5. $\frac{\partial N}{\partial x} = 0 \neq \frac{\partial M}{\partial y} \Rightarrow$ Not Conservative

6. $\frac{\partial P}{\partial y} = 0 = \frac{\partial N}{\partial z}, \frac{\partial M}{\partial z} = 0 = \frac{\partial P}{\partial x}, \frac{\partial N}{\partial x} = -e^x \sin y = \frac{\partial M}{\partial y} \Rightarrow$ Conservative

7. $\frac{\partial f}{\partial x} = 2x \Rightarrow f(x,y,z) = x^2 + g(y,z)$. $\frac{\partial f}{\partial y} = \frac{\partial g}{\partial y} = 3y \Rightarrow g(y,z) = \frac{3y^2}{2} + h(z) \Rightarrow f(x,y,z) = x^2 + \frac{3y^2}{2} + h(z)$. $\frac{\partial f}{\partial z} = h'(z)$ $= 4z. \Rightarrow h(z) = 2z^2 + C \Rightarrow f(x,y,z) = x^2 + \frac{3y^2}{2} + 2z^2 + C$

8. $\frac{\partial f}{\partial x} = y + z \Rightarrow f(x,y,z) = (y+z)x + g(y,z)$. $\frac{\partial f}{\partial y} = x + \frac{\partial g}{\partial y} = x + z \Rightarrow \frac{\partial g}{\partial y} = z \Rightarrow g(y,z) = zy + h(z)$. Then $f(x,y,z) = (y+z)x + zy + h(z)$. $\frac{\partial f}{\partial z} = x + y + h'(z) = x + y \Rightarrow h'(z) = 0 \Rightarrow h(z) = C$. $\therefore$ $f(x,y,z) = (y+z)x + zy + C$

9. $\frac{\partial f}{\partial x} = e^{y+2z} \Rightarrow f(x,y,z) = x\,e^{y+2z} + g(y,z)$. $\frac{\partial f}{\partial y} = x\,e^{y+2z} + \frac{\partial g}{\partial y} = x\,e^{y+2z} \Rightarrow \frac{\partial g}{\partial y} = 0$. Then $f(x,y,z) = x\,e^{y+2z} + h(z)$. $\frac{\partial f}{\partial z} = 2x\,e^{y+2z} + h'(z) = 2x\,e^{y+2z} \Rightarrow h'(z) = 0 \Rightarrow h(z) = C$. $\therefore$ $f(x,y,z) = x\,e^{y+2z} + C$

10. $\frac{\partial f}{\partial x} = y\sin z \Rightarrow f(x,y,z) = xy\sin z + g(y,z)$. $\frac{\partial f}{\partial y} = x\sin z + \frac{\partial g}{\partial y} = x\sin z \Rightarrow \frac{\partial g}{\partial y} = 0 \Rightarrow g(y,z) = h(z)$. Then $f(x,y,z) = xy\sin z + h(z)$. $\frac{\partial f}{\partial z} = xy\cos z + h'(z) = xy\cos z \Rightarrow h'(z) = 0 \Rightarrow h(z) = C$. $\therefore$ $f(x,y,z) = xy\cos z + C$

11. Let $\mathbf{F}(x,y,z) = 2x\,\mathbf{i} + 2y\,\mathbf{j} + 2z\,\mathbf{k} \Rightarrow \frac{\partial P}{\partial y} = 0 = \frac{\partial N}{\partial z}, \frac{\partial M}{\partial z} = 0 = \frac{\partial P}{\partial x}, \frac{\partial N}{\partial x} = 0 = \frac{\partial M}{\partial y} \Rightarrow M\,dx + N\,dy + P\,dz$ is exact. $\frac{\partial f}{\partial x} = 2x \Rightarrow f(x,y,z) = x^2 + g(y,z)$. $\frac{\partial f}{\partial y} = \frac{\partial g}{\partial y} = 2y \Rightarrow g(y,z) = y^2 + h(z) \Rightarrow f(x,y,z) = x^2 + y^2 + h(z)$. $\frac{\partial f}{\partial z} = h'(z) = 2z \Rightarrow h(z) = z^2 + C$. $\therefore$ $f(x,y,z) = x^2 + y^2 + z^2 + C \Rightarrow \int_{(0,0,0)}^{(2,3,-6)} 2x\,dx + 2y\,dy + 2z\,dz = f(2,3,-6) - f(0,0,0) = 49$

12. Let $\mathbf{F}(x,y,z) = yz\,\mathbf{i} + xz\,\mathbf{j} + xy\,\mathbf{k} \Rightarrow \frac{\partial P}{\partial y} = x = \frac{\partial N}{\partial z}, \frac{\partial M}{\partial z} = y = \frac{\partial P}{\partial x}, \frac{\partial N}{\partial x} = z = \frac{\partial M}{\partial y} \Rightarrow M\,dx + N\,dy + P\,dz$ is exact. $\frac{\partial f}{\partial x} = yz \Rightarrow f(x,y,z) = xyz + g(y,z)$. $\frac{\partial f}{\partial y} = xz + \frac{\partial g}{\partial y} = xz \Rightarrow \frac{\partial g}{\partial y} = 0 \Rightarrow g(y,z) = h(z) \Rightarrow f(x,y,z) = xyz + h(z)$. $\frac{\partial f}{\partial z} = xy + h'(z) = xy \Rightarrow h'(z) = 0 \Rightarrow h(z) = C$. $\therefore$ $f(x,y,z) = xyz + C \Rightarrow \int_{(1,1,2)}^{(3,5,0)} yz\,dx + xz\,dy + xy\,dz = f(3,5,0) - f(1,1,2) = -2$

13. Let $\mathbf{F}(x,y,z) = 2xy\,\mathbf{i} + (x^2 - z^2)\,\mathbf{j} - 2yz\,\mathbf{k} \Rightarrow \frac{\partial P}{\partial y} = -2z = \frac{\partial N}{\partial z}, \frac{\partial M}{\partial z} = 0 = \frac{\partial P}{\partial x}, \frac{\partial N}{\partial x} = 2x = \frac{\partial M}{\partial y} \Rightarrow M\,dx + N\,dy + P\,dz$ is exact. $\frac{\partial f}{\partial x} = 2xy \Rightarrow f(x,y,z) = x^2y + g(y,z)$. $\frac{\partial f}{\partial y} = x^2 + \frac{\partial g}{\partial y} = x^2 - z^2 \Rightarrow \frac{\partial g}{\partial y} = -z^2 \Rightarrow g(y,z) = -yz^2 + h(z) \Rightarrow f(x,y,z) = x^2y - yz^2 + h(z)$. $\frac{\partial f}{\partial z} = -2yz + h'(z) = -2yz \Rightarrow h'(z) = 0 \Rightarrow h(z) = C \Rightarrow f(x,y,z) = x^2y - yz^2 + C \Rightarrow \int_{(0,0,0)}^{(1,2,3)} 2xy\,dx + (x^2 - z^2)\,dy - 2yz\,dz = f(1,2,3) - f(0,0,0) = -16$

14. Let $\mathbf{F}(x,y,z) = 3x^2\,\mathbf{i} + \frac{z^2}{y}\,\mathbf{j} + 2z \ln y\,\mathbf{k} \Rightarrow \frac{\partial P}{\partial y} = \frac{2z}{y} = \frac{\partial N}{\partial z}, \frac{\partial M}{\partial z} = 0 = \frac{\partial P}{\partial x}, \frac{\partial N}{\partial x} = 0 = \frac{\partial M}{\partial y} \Rightarrow M\,dx + N\,dy + P\,dz$ is exact. $\frac{\partial f}{\partial x} = 3x^2 \Rightarrow f(x,y,z) = x^3 + g(y,z)$. $\frac{\partial f}{\partial y} = \frac{\partial g}{\partial y} = \frac{z^2}{y} \Rightarrow g(y,z) = z^2 \ln y + h(z) \Rightarrow f(x,y,z) = x^3 + z^2 \ln y + h(z)$. $\frac{\partial f}{\partial z} = 2z \ln y + h'(z) = 2z \ln y \Rightarrow h'(z) = 0 \Rightarrow h(z) = C \Rightarrow f(x,y,z) = x^3 + z^2 \ln y + C \Rightarrow$

$$\int_{(1,1,1)}^{(1,2,3)} 3x^2\,dx + \frac{z^2}{y}\,dy + 2z \ln y\,dz = f(1,2,3) - f(1,1,1) = 9 \ln 2$$

15. Let $\mathbf{F}(x,y,z) = (\sin y \cos x)\,\mathbf{i} + (\cos y \sin x)\,\mathbf{j} + \mathbf{k} \Rightarrow \frac{\partial P}{\partial y} = 0 = \frac{\partial N}{\partial z}, \frac{\partial M}{\partial z} = 0 = \frac{\partial P}{\partial x}, \frac{\partial N}{\partial x} = \cos y \cos x = \frac{\partial M}{\partial y} \Rightarrow$ $M\,dx + N\,dy + P\,dz$ is exact. $\frac{\partial f}{\partial x} = \sin y \cos x \Rightarrow f(x,y,z) = \sin y \sin x + g(y,z)$. $\frac{\partial f}{\partial y} = \cos y \sin x + \frac{\partial g}{\partial y} = \cos y \sin x \Rightarrow \frac{\partial g}{\partial y} = 0 \Rightarrow g(y,z) = h(z) \Rightarrow f(x,y,z) = \sin y \sin x + h(z)$. $\frac{\partial f}{\partial z} = h'(z) = 1 \Rightarrow h(z) = z + C \Rightarrow$ $f(x,y,z) = \sin y \sin x + z + C \Rightarrow \int_{(1,0,0)}^{(0,1,1)} \sin y \cos x\,dx + \cos y \sin x\,dy + dz = f(0,1,1) - f(1,0,0) = 1$

16. Let $\mathbf{F}(x,y,z) = 2x\,\mathbf{i} - y^2\,\mathbf{j} - \frac{4}{1+z^2}\mathbf{k} \Rightarrow \frac{\partial P}{\partial y} = 0 = \frac{\partial N}{\partial z}, \frac{\partial M}{\partial z} = 0 = \frac{\partial P}{\partial x}, \frac{\partial N}{\partial x} = 0 = \frac{\partial M}{\partial y} \Rightarrow M\,dx + N\,dy + P\,dz$ is exact. $\frac{\partial f}{\partial x} = 2x \Rightarrow f(x,y,z) = x^2 + g(y,z)$. $\frac{\partial f}{\partial y} = \frac{\partial g}{\partial y} = -y^2 \Rightarrow g(y,z) = -\frac{y^3}{3} + h(z) \Rightarrow f(x,y,z) = x^2 - y^2 + h(z)$. $\frac{\partial f}{\partial z} = h'(z) = -\frac{4}{1+z^2} \Rightarrow h(z) = -4\tan^{-1} z + C \Rightarrow f(x,y,z) = x^2 - y^2 - 4\tan^{-1} z + C \Rightarrow$

$$\int_{(0,0,0)}^{(3,3,1)} 2x\,dx - y^2\,dy - \frac{4}{1+z^2}dz = f(3,3,1) - f(0,0,0) = -\pi$$

17. Let $\mathbf{F}(x,y,z) = (2\cos y)\,\mathbf{i} + \left(\frac{1}{y} - 2x\sin y\right)\mathbf{j} + \frac{1}{z}\mathbf{k} \Rightarrow \frac{\partial P}{\partial y} = 0 = \frac{\partial N}{\partial z}, \frac{\partial M}{\partial z} = 0 = \frac{\partial P}{\partial x}, \frac{\partial N}{\partial x} = -2\sin y = \frac{\partial M}{\partial y} \Rightarrow$ $M\,dx + N\,dy + P\,dz$ is exact. $\frac{\partial f}{\partial x} = 2\cos y \Rightarrow f(x,y,z) = 2x\cos y + g(y,z)$. $\frac{\partial f}{\partial y} = -2x\sin y + \frac{\partial g}{\partial y} = \frac{1}{y} - 2x\sin y$ $\Rightarrow \frac{\partial g}{\partial y} = \frac{1}{y} \Rightarrow g(y,z) = \ln y + h(z) \Rightarrow f(x,y,z) = 2x\cos y + \ln y + h(z)$. $\frac{\partial f}{\partial z} = h'(z) = \frac{1}{z} \Rightarrow h(z) = \ln z + C \Rightarrow$

$$f(x,y,z) = 2x\cos y + \ln y + \ln z + C \Rightarrow \int_{(0,2,1)}^{(1,\pi/2,2)} 2\cos y\,dx + \left(\frac{1}{y} - 2x\sin y\right)dy + \frac{1}{z}dz =$$

$f(1,\frac{\pi}{2},2) - f(0,2,1) = \ln\frac{\pi}{2}$

18. Let $\mathbf{F}(x,y,z) = (2x\ln y - y^2)\,\mathbf{i} + \left(\frac{x^2}{y} - xz\right)\mathbf{j} - xy\,\mathbf{k}$ (Note: $\mathbf{j}$ component should be as given here; integrand in the text is not exact.) $\Rightarrow \frac{\partial P}{\partial y} = -x = \frac{\partial N}{\partial z}, \frac{\partial M}{\partial z} = -y = \frac{\partial P}{\partial x}, \frac{\partial N}{\partial x} = \frac{2x}{y} - z = \frac{\partial M}{\partial y} \Rightarrow M\,dx + N\,dy + P\,dz$ is exact. $\frac{\partial f}{\partial x} = 2x\ln y - yz \Rightarrow f(x,y,z) = x^2\ln y - xyz + g(y,z)$. $\frac{\partial f}{\partial y} = \frac{x^2}{y} - xz + \frac{\partial g}{\partial y} = \frac{x^2}{y} - xz \Rightarrow g(y,z) = h(z) \Rightarrow$ $f(x,y,z) = x^2\ln y - xyz + h(z)$. $\frac{\partial f}{\partial z} = -xy + h'(z) = -xy \Rightarrow h'(z) = 0 \Rightarrow h(z) = C \Rightarrow f(x,y,z) = x^2\ln y - xyz + C$

$$\Rightarrow \int_{(1,2,1)}^{(2,1,1)} (2x\ln y - yz)\,dx + \left(\frac{x^2}{y} - xz\right)dy - xy\,dz = f(2,1,1) - f(1,2,1) = -\ln 2$$

19. Let $\mathbf{F}(x,y,z) = \frac{1}{y}\mathbf{i} + \left(\frac{1}{z} - \frac{x}{y^2}\right)\mathbf{j} - \frac{y}{z^2}\mathbf{k} \Rightarrow \frac{\partial P}{\partial y} = -\frac{1}{z^2} = \frac{\partial N}{\partial z}, \frac{\partial M}{\partial z} = 0 = \frac{\partial P}{\partial x}, \frac{\partial N}{\partial x} = -\frac{1}{y^2} = \frac{\partial M}{\partial y} \Rightarrow$ $M\,dx + N\,dy + P\,dz$ is exact. $\frac{\partial f}{\partial x} = \frac{1}{y} \Rightarrow f(x,y,z) = \frac{x}{y} + g(y,z)$. $\frac{\partial f}{\partial y} = -\frac{x}{y^2} + \frac{\partial g}{\partial y} = \frac{1}{z} - \frac{x}{y^2} \Rightarrow \frac{\partial g}{\partial y} = \frac{1}{z} \Rightarrow$ $g(y,z) = \frac{y}{z} + h(z) \Rightarrow f(x,y,z) = \frac{x}{y} + \frac{y}{z} + h(z)$. $\frac{\partial f}{\partial z} = -\frac{y}{z^2} + h'(z) = -\frac{y}{z^2} \Rightarrow h'(z) = 0 \Rightarrow h(z) = C \Rightarrow$

$$f(x,y,z) = \frac{x}{y} + \frac{y}{z} + C \Rightarrow \int_{(1,1,1)}^{(2,2,2)} \frac{1}{y}dx + \left(\frac{1}{z} - \frac{x}{y^2}\right)dy - \frac{y}{z^2}dz = f(2,2,2) - f(1,1,1) = 0$$

20. Let $\mathbf{F}(x,y,z) = \frac{2x\,\mathbf{i} + 2y\,\mathbf{j} + 2z\,\mathbf{k}}{x^2+y^2+z^2}$ (and let $\rho^2 = x^2 + y^2 + z^2 \Rightarrow \frac{\partial \rho}{\partial x} = \frac{x}{\rho}, \frac{\partial \rho}{\partial y} = \frac{y}{\rho}, \frac{\partial \rho}{\partial z} = \frac{z}{\rho}$) $\Rightarrow$ $\frac{\partial P}{\partial y} = -\frac{4yz}{\rho^4} = \frac{\partial N}{\partial z}, \frac{\partial M}{\partial z} = -\frac{4xz}{\rho^4} = \frac{\partial P}{\partial x}, \frac{\partial N}{\partial x} = -\frac{4xy}{\rho^4} = \frac{\partial M}{\partial y} \Rightarrow M\,dx + N\,dy + P\,dz$ is exact. $\frac{\partial f}{\partial x} = \frac{2x}{x^2+y^2+z^2} \Rightarrow f(x,y,z) = \ln(x^2+y^2+z^2) + g(y,z)$. $\frac{\partial f}{\partial y} = \frac{2y}{x^2+y^2+z^2} + \frac{\partial g}{\partial y} = \frac{2y}{x^2+y^2+z^2} \Rightarrow \frac{\partial g}{\partial y} = 0$

20. (Continued)

$\Rightarrow g(y,z) = h(z) \Rightarrow f(x,y,z) = \ln(x^2 + y^2 + z^2) + h(z).\ \frac{\partial f}{\partial z} = \frac{2z}{x^2 + y^2 + z^2} + h'(z) = \frac{2z}{x^2 + y^2 + z^2} \Rightarrow h'(z) = 0 \Rightarrow$

$h(z) = C \Rightarrow f(x,y,z) = \ln(x^2 + y^2 + z^2) + C \Rightarrow \int_{(-1,-1,-1)}^{(2,2,2)} \frac{2x\,dx + 2y\,dy + 2z\,dz}{x^2 + y^2 + z^2} = f(2,2,2) - f(-1,-1,-1) = \ln 4$

21. Let $x - 1 = t,\ y - 1 = 2t,\ z - 1 = -2t,\ 0 \le t \le 1 \Rightarrow dx = dt,\ dy = 2\,dt,\ dz = -2\,dt \Rightarrow$

$\int_{(1,1,1)}^{(2,3,-1)} y\,dx + x\,dy + 4\,dz = \int_0^1 (2t + 1)dt + (t + 1)2\,dt + 4(-2\,dt) = \int_0^1 (4t - 5)\,dt = -3$

22. Let $x = 0,\ y = 3t,\ z = 4t,\ 0 \le t \le 1 \Rightarrow dx = 0,\ dy = 3\,dt,\ dz = 4\,dt \Rightarrow \int_{(0,0,0)}^{(0,3,4)} x^2\,dx + yz\,dy + y^2\,dz =$

$\int_0^1 12t^2(3\,dt) + 9t^2(4\,dt) = \int_0^1 72t^2\,dt = 24$

23. $\frac{\partial P}{\partial y} = 0 = \frac{\partial N}{\partial z},\ \frac{\partial M}{\partial z} = 2z = \frac{\partial P}{\partial x},\ \frac{\partial N}{\partial x} = 0 = \frac{\partial M}{\partial y} \Rightarrow M\,dx + N\,dy + P\,dz$ is exact $\Rightarrow$ **F** is conservative $\Rightarrow$ path independence.

24. Let $\mathbf{F}(x,y,z) = \frac{x\,\mathbf{i} + y\,\mathbf{j} + z\,\mathbf{k}}{\left(\sqrt{x^2 + y^2 + z^2}\right)^3} \Rightarrow \frac{\partial P}{\partial y} = -\frac{yz}{\left(\sqrt{x^2 + y^2 + z^2}\right)^3} = \frac{\partial N}{\partial z},\ \frac{\partial M}{\partial z} = -\frac{xz}{\left(\sqrt{x^2 + y^2 + z^2}\right)^3} = \frac{\partial P}{\partial x},$

$\frac{\partial N}{\partial x} = -\frac{xy}{\left(\sqrt{x^2 + y^2 + z^2}\right)^3} = \frac{\partial M}{\partial y}$. $\therefore$ $M\,dx + N\,dy + P\,dz$ is exact and $\mathbf{F} = M\,\mathbf{i} + N\,\mathbf{j} + P\,\mathbf{k}$ is conservative $\Rightarrow$ path independence

25. $\frac{\partial P}{\partial y} = 0,\ \frac{\partial N}{\partial z} = 0,\ \frac{\partial M}{\partial z} = 0,\ \frac{\partial P}{\partial x} = 0,\ \frac{\partial N}{\partial x} = \frac{y^2 - x^2}{(x^2 + y^2)^2},\ \frac{\partial M}{\partial y} = \frac{y^2 - x^2}{(x^2 + y^2)^2} \Rightarrow \text{curl}\,\mathbf{F} = \left(\frac{y^2 - x^2}{(x^2 + y^2)^2} - \frac{y^2 - x^2}{(x^2 + y^2)^2}\right)\mathbf{k}$ $= \mathbf{0}$. $x^2 + y^2 = 1 \Rightarrow \mathbf{r} = (a\cos t)\,\mathbf{i} + (a\sin t)\,\mathbf{j} \Rightarrow d\mathbf{r} = (-a\sin t)\,\mathbf{i} + (a\cos t)\,\mathbf{j} \Rightarrow \mathbf{F} = \frac{-a\sin t}{a^2}\mathbf{i} + \frac{a\cos t}{a^2}\mathbf{j} +$

$z\,\mathbf{k} \Rightarrow \mathbf{F}\cdot d\mathbf{r} = \frac{a^2\sin^2 t}{a^2} + \frac{a^2\cos^2 t}{a^2} = 1 \Rightarrow \int_C \mathbf{F}\cdot d\mathbf{r} = \int_0^{2\pi} 1\,dt = 2\pi$

PRACTICE EXERCISES

1. Path 1: $\mathbf{r} = t\mathbf{i} + t\mathbf{j} + t\mathbf{k} \Rightarrow x = t,\ y = t,\ z = t,\ 0 \le t \le 1 \Rightarrow f(g(t),h(t),h(t)) = 3 - 3t^2$ and $\frac{dx}{dt} = 1,\ \frac{dy}{dt} = 1,$

$$\frac{dz}{dt} = 1 \Rightarrow \sqrt{\left(\frac{dx}{dt}\right)^2 + \left(\frac{dy}{dt}\right)^2 + \left(\frac{dz}{dt}\right)^2}\ dt = \sqrt{3}\ dt \Rightarrow \int_C f(x,y,z)\ ds = \int_0^1 \sqrt{3}\left(3 - 3t^2\right) dt = 2\sqrt{3}$$

Path 2: $\mathbf{r_1} = t\mathbf{i} + t\mathbf{j},\ 0 \le t \le 1 \Rightarrow x = t,\ y = t,\ z = 0 \Rightarrow f(g(t),h(t),h(t)) = 2t - 3t^2 + 3$ and $\frac{dx}{dt} = 1,\ \frac{dy}{dt} = 1,$

$$\frac{dz}{dt} = 0 \Rightarrow \sqrt{\left(\frac{dx}{dt}\right)^2 + \left(\frac{dy}{dt}\right)^2 + \left(\frac{dz}{dt}\right)^2}\ dt = \sqrt{2}\ dt \Rightarrow \int_{C_1} f(x,y,z)\ ds = \int_0^1 \sqrt{2}\left(2t - 3t^2 + 3\right) dt =$$

$3\sqrt{2}$. $\mathbf{r_2} = \mathbf{i} + \mathbf{j} + t\mathbf{k} \Rightarrow x = 1,\ y = 1,\ z = t \Rightarrow f(g(t),h(t),h(t)) = 2 - 2t$ and $\frac{dx}{dt} = 0,\ \frac{dy}{dt} = 0,\ \frac{dz}{dt} = 1 \Rightarrow$

$$\sqrt{\left(\frac{dx}{dt}\right)^2 + \left(\frac{dy}{dt}\right)^2 + \left(\frac{dz}{dt}\right)^2}\ dt = dt \Rightarrow \int_{C_2} f(x,y,z)\ ds = \int_0^1 (2 - 2t)\ dt = 1.\ \therefore \int_C f(x,y,z)\ ds =$$

$$\int_{C_1} f(x,y,z)\ ds + \int_{C_2} f(x,y,z)\ ds = 3\sqrt{2} + 1$$

2. Path 1: $\mathbf{r_1} = t\mathbf{i} \Rightarrow x = 1,\ y = 0,\ z = 0 \Rightarrow f(g(t),h(t),h(t)) = t^2$ and $\frac{dx}{dt} = 1,$

$$\frac{dy}{dt} = 0,\ \frac{dz}{dt} = 0 \Rightarrow \sqrt{\left(\frac{dx}{dt}\right)^2 + \left(\frac{dy}{dt}\right)^2 + \left(\frac{dz}{dt}\right)^2}\ dt = dt \Rightarrow \int_{C_1} f(x,y,z)\ ds = \int_0^1 t^2 dt = \frac{1}{3}$$

$\mathbf{r_2} = t\mathbf{j} \Rightarrow x = 0,\ y = t,\ z = 0 \Rightarrow f(g(t),h(t),h(t)) = t$ and $\frac{dx}{dt} = 0,\ \frac{dy}{dt} = 1,\ \frac{dz}{dt} = 0 \Rightarrow$

$$\sqrt{\left(\frac{dx}{dt}\right)^2 + \left(\frac{dy}{dt}\right)^2 + \left(\frac{dz}{dt}\right)^2}\ dt = dt \Rightarrow \int_{C_2} f(x,y,z)\ ds = \int_0^1 t\ dt = \frac{1}{2}$$

$\mathbf{r_3} = t\mathbf{k} \Rightarrow x = 0,\ y = 0,\ z = t \Rightarrow f(g(t),h(t),h(t)) = -t$ and $\frac{dx}{dt} = 0,\ \frac{dy}{dt} = 0,\ \frac{dz}{dt} = 1 \Rightarrow$

$$\sqrt{\left(\frac{dx}{dt}\right)^2 + \left(\frac{dy}{dt}\right)^2 + \left(\frac{dz}{dt}\right)^2}\ dt = dt \Rightarrow \int_{C_3} f(x,y,z)\ ds = \int_0^1 -t\ dt = -\frac{1}{2}.\ \therefore \int_{\text{Path 1}} f(x,y,z)\ ds =$$

$$\int_{C_1} f(x,y,z)\ ds + \int_{C_2} f(x,y,z)\ ds + \int_{C_3} f(x,y,z)\ ds = \frac{1}{3}$$

Path 2: $\mathbf{r_4} = t\mathbf{i} + t\mathbf{j} \Rightarrow x = t,\ y = t,\ z = 0 \Rightarrow f(g(t),h(t),h(t)) = t^2 + t$ and $\frac{dx}{dt} = 1,\ \frac{dy}{dt} = 1,\ \frac{dz}{dt} = 0 \Rightarrow$

$$\sqrt{\left(\frac{dx}{dt}\right)^2 + \left(\frac{dy}{dt}\right)^2 + \left(\frac{dz}{dt}\right)^2}\ dt = \sqrt{2}\ dt \Rightarrow \int_{C_4} f(x,y,z)\ ds = \int_0^1 \sqrt{2}\left(t^2 + t\right) dt = \frac{5}{6}\sqrt{2}$$

2. (Continued)

$\mathbf{r_3} = t\,\mathbf{k}$ (see above) $\Rightarrow \int_{C_3} f(x,y,z)\,ds = -\frac{1}{2}$. $\therefore \int_{\text{Path 2}} f(x,y,z)\,ds =$

$\int_{C_3} f(x,y,z)\,ds + \int_{C_4} f(x,y,z)\,ds = \frac{5}{6}\sqrt{2} - \frac{1}{2}$

Path 3: $\mathbf{r_5} = \mathbf{r_3} \Rightarrow \int_{C_5} f(x,y,z)\,ds = \int_{C_3} f(x,y,z)\,ds = -\frac{1}{2}$. $\mathbf{r_6} = \mathbf{r_2} \Rightarrow \int_{C_6} f(x,y,z)\,ds =$

$\int_{C_2} f(x,y,z)\,ds = \frac{1}{2}$. $\mathbf{r_7} = \mathbf{r_1} \Rightarrow \int_{C_7} f(x,y,z)\,ds = \int_{C_1} f(x,y,z)\,ds = \frac{1}{3}$. $\therefore \int_{\text{Path 3}} f(x,y,z)\,ds =$

$\int_{C_5} f(x,y,z)\,ds + \int_{C_6} f(x,y,z)\,ds + \int_{C_7} f(x,y,z)\,ds = \int_{C_3} f(x,y,z)\,ds + \int_{C_2} f(x,y,z)\,ds +$

$\int_{C_1} f(x,y,z)\,ds = \frac{1}{3}$

3. $\mathbf{r} = (a\cos t)\,\mathbf{j} + (a\sin t)\,\mathbf{k} \Rightarrow x = 0,\ y = a\cos t,\ z = a\sin t \Rightarrow f(g(t),h(t),h(t)) = \sqrt{a^2\sin^2 t} = a\,|\sin t|$ and

$\frac{dx}{dt} = 0,\ \frac{dy}{dt} = -a\sin t,\ \frac{dz}{dt} = a\cos t \Rightarrow \sqrt{\left(\frac{dx}{dt}\right)^2 + \left(\frac{dy}{dt}\right)^2 + \left(\frac{dz}{dt}\right)^2}\,dt = a\,dt \Rightarrow \int_C f(x,y,z)\,ds =$

$$\int_0^{2\pi} a^2\,|\sin t|\,dt = \int_0^{\pi} a^2\sin t\,dt + \int_{\pi}^{2\pi} -a^2\sin t\,dt = 4a^2$$

4. (See Exercise 14, Section 15.1.) $\mathbf{r} = t\,\mathbf{i} + t\,\mathbf{j} + t\,\mathbf{k},\ 1 \le t \Rightarrow \int_C f(x,y,z)\,ds = \int_1^{\infty} t^{-2}\,dt =$

$$\lim_{k\to\infty} \int_1^{k} t^{-2}\,dt = -1$$

5. a) $\mathbf{r} = \sqrt{2}\,t\,\mathbf{i} + \sqrt{2}\,\mathbf{j} + (4 - t^2)\,\mathbf{k},\ 0 \le t \le 1 \Rightarrow x = \sqrt{2}\,t,\ y = \sqrt{2}\,t,\ z = 4 - t^2 \Rightarrow \frac{dx}{dt} = \sqrt{2},\ \frac{dy}{dt} = \sqrt{2},\ \frac{dz}{dt} = -2t$

$\Rightarrow \sqrt{\left(\frac{dx}{dt}\right)^2 + \left(\frac{dy}{dt}\right)^2 + \left(\frac{dz}{dt}\right)^2}\,dt = \sqrt{4 + t^2}\,dt \Rightarrow M = \int_C \delta(x,y,z)\,ds = \int_0^1 3t\sqrt{4 + 4t^2}\,dt =$

$4\sqrt{2} - 2$

b) $M = \int_C \delta(x,y,z)\,ds = \int_0^1 \sqrt{4 + 4t^2}\,dt = \sqrt{2} + \ln(1 + \sqrt{2})$

6. $\mathbf{r} = t\,\mathbf{i} + 2t\,\mathbf{j} + \frac{2}{3}t^{3/2}\,\mathbf{k},\ 0 \le t \le 2 \Rightarrow x = t,\ y = 2t,\ z = \frac{2}{3}t^{3/2} \Rightarrow \frac{dx}{dt} = 1,\ \frac{dy}{dt} = 2,\ \frac{dz}{dt} = t^{1/2} \Rightarrow$

$\sqrt{\left(\frac{dx}{dt}\right)^2 + \left(\frac{dy}{dt}\right)^2 + \left(\frac{dz}{dt}\right)^2}\,dt = \sqrt{t+5}\,dt \Rightarrow M = \int_C \delta(x,y,z)\,ds = \int_0^2 3\sqrt{5+t}\,\sqrt{t+5}\,dt =$

$\int_0^2 3(t+5)\,dt = 36.$ $M_{yz} = \int_C x\delta\,ds = \int_0^2 3t(t+5)\,dt = 38.$ $M_{xz} = \int_C y\delta\,ds =$

$\int_0^2 6t(t+5)\,dt = 46.$ $M_{xy} = \int_C z\delta\,ds = \int_0^2 2t^{3/2}(t+5)\,dt = \frac{144}{7}\sqrt{2}.$ $\therefore\ \bar{x} = M_{yz}/M = \frac{38}{36} = \frac{19}{18},$

$\bar{y} = M_{xz}/M = \frac{23}{18},\ \bar{z} = M_{xy}/M = \frac{\frac{144}{7}\sqrt{2}}{36} = \frac{4}{7}\sqrt{2}$

7. $\mathbf{r} = t\,\mathbf{i} + \frac{2\sqrt{2}}{3}t^{3/2}\,\mathbf{j} + \frac{t^2}{2}\,\mathbf{k},\ 0 \le t \le 2 \Rightarrow x = t,\ y = \frac{2\sqrt{2}}{3}t^{3/2},\ z = \frac{t^2}{2} \Rightarrow \frac{dx}{dt} = t,\ \frac{dy}{dt} = \sqrt{2}\,t^{1/2},\ \frac{dz}{dt} = t \Rightarrow$

$\sqrt{\left(\frac{dx}{dt}\right)^2 + \left(\frac{dy}{dt}\right)^2 + \left(\frac{dz}{dt}\right)^2}\,dt = \sqrt{(t+1)^2} = |t+1| = t+1$ on the domain given. Then $M_{yz} =$

$\int_C x\delta\,ds = \int_0^2 t\left(\frac{1}{t+1}\right)(t+1)\,dt = \int_0^2 t\,dt = 2.$ $M = \int_C \delta\,ds = \int_0^2 \frac{1}{t+1}(t+1)\,dt = \int_0^2 dt = 2$

$M_{xz} = \int_C y\delta\,ds = \int_0^2 \frac{2\sqrt{2}}{3}t^{3/2}\left(\frac{1}{t+1}\right)(t+1)\,dt = \int_0^2 \frac{2\sqrt{2}}{3}t^{3/2}\,dt = \frac{32}{15}.$ $M_{xy} = \int_C z\delta\,ds =$

$\int_0^2 \frac{t^2}{2}\left(\frac{1}{t+1}\right)(t+1)\,dt = \int_0^2 \frac{t^2}{2}\,dt = \frac{4}{3}.$ $\therefore\ \bar{x} = M_{yz}/M = \frac{2}{2} = 1,\ \bar{y} = M_{xz}/M = \frac{32/15}{2} = \frac{16}{15},\ \bar{z} = M_{xy}/M =$

$\frac{4/3}{2} = \frac{2}{3}.$ $I_x = \int_C (y^2+z^2)\delta\,ds = \int_0^2 \left(\frac{8}{9}t^3 + \frac{t^4}{4}\right)dt = \frac{232}{45}.$ $I_y = \int_C (x^2+z^2)\delta\,ds =$

$\int_0^2 \left(t^2 + \frac{t^4}{4}\right)dt = \frac{64}{15}$ $I_z = \int_C (y^2+x^2)\delta\,ds = \int_0^2 \left(t^2 + \frac{8}{9}t^3\right)dt = \frac{56}{9}.$ $R_x = \sqrt{I_x/M}$

$= \sqrt{\frac{232/45}{2}} = \frac{2}{3}\sqrt{\frac{29}{5}}.$ $R_y = \sqrt{I_y/M} = \sqrt{\frac{64/15}{2}} = 4\sqrt{\frac{2}{15}}.$ $R_z = \sqrt{I_z/M} = \sqrt{\frac{56/9}{2}} = \frac{2}{3}\sqrt{7}$

8. $\mathbf{r} = (e^t \cos t)\,\mathbf{i} + (e^t \sin t)\,\mathbf{j} + e^t\,\mathbf{k},\ 0 \le t \le \ln 2 \Rightarrow x = e^t \cos t,\ y = e^t \sin t,\ z = e^t \Rightarrow \frac{dx}{dt} =$

$(e^t \cos t - e^t \sin t),\ \frac{dy}{dt} = (e^t \sin t + e^t \cos t),\ \frac{dz}{dt} = e^t \Rightarrow \sqrt{\left(\frac{dx}{dt}\right)^2 + \left(\frac{dy}{dt}\right)^2 + \left(\frac{dz}{dt}\right)^2}\ dt =$

$\sqrt{(e^t \cos t - e^t \sin t)^2 + (e^t \sin t + e^t \cos t)^2 + \left(e^t\right)^2}\ dt = \sqrt{3e^{2t}}\ dt = \sqrt{3}\,e^t\,dt.\ \delta\,ds = \sqrt{3}\,e^t\,dt$

Then $M = \int_C \delta\,ds = \int_0^{\ln 2} \sqrt{3}\,e^t\,dt = \sqrt{3}.\ M_{xy} = \int_C z\delta\,ds = \int_0^{\ln 2} \sqrt{3}\,e^t(e^t)\,dt = \int_0^{\ln 2} \sqrt{3}\,e^{2t}\,dt$

$= \frac{3\sqrt{3}}{2} \Rightarrow \bar{z} = M_{xy}/M = \frac{3\sqrt{3}/2}{\sqrt{3}} = \frac{3}{2}.\ I_z = \int_C (x^2 + y^2)\delta\,ds = \int_0^{\ln 2} (e^{2t}\cos^2 t + e^{2t}\sin^2 t)\sqrt{3}\,e^t\,dt =$

$\int_0^{\ln 2} \sqrt{3}\,e^{3t}\,dt = \frac{7\sqrt{3}}{3} \Rightarrow R_z = \sqrt{I_z/M} = \sqrt{\frac{7\sqrt{3}/3}{\sqrt{3}}} = \sqrt{\frac{7}{3}}$

9. a) $x^2 + y^2 = 1 \Rightarrow \mathbf{r} = (\cos t)\,\mathbf{i} + (\sin t)\,\mathbf{j},\ 0 \le t \le \pi \Rightarrow x = \cos t,\ y = \sin t \Rightarrow \mathbf{F} = (\cos t + \sin t)\,\mathbf{i} - \mathbf{j}$ and $\frac{d\mathbf{r}}{dt} = (-\sin t)\,\mathbf{i} + (\cos t)\,\mathbf{j} \Rightarrow \mathbf{F} \cdot \frac{d\mathbf{r}}{dt} = -\sin t \cos t - \sin^2 t - \cos t \Rightarrow$ Flow $=$

$\int_0^{\pi} (-\sin t \cos t - \sin^2 t - \cos t)\,dt = -\frac{1}{2}\pi$

b) $\mathbf{r} = -t\,\mathbf{i},\ -1 \le t \le 1 \Rightarrow x = -t,\ y = 0 \Rightarrow \mathbf{F} = -t\,\mathbf{i} - t^2\,\mathbf{j}$ and $\frac{d\mathbf{r}}{dt} = -\mathbf{i} \Rightarrow \mathbf{F} \cdot \frac{d\mathbf{r}}{dt} = t \Rightarrow$ Flow $= \int_{-1}^{1} t\,dt =$

0

c) $\mathbf{r_1} = (1 - t)\,\mathbf{i} - t\,\mathbf{j},\ 0 \le t \le 1 \Rightarrow \mathbf{F_1} = (1 - 2t)\,\mathbf{i} - (1 - 2t - 2t^2)\,\mathbf{j}$ and $\frac{d\mathbf{r_1}}{dt} = -\mathbf{i} - \mathbf{j} \Rightarrow \mathbf{F_1} \cdot \frac{d\mathbf{r_1}}{dt} = 2t^2 \Rightarrow$

$\text{Flow}_1 = \int_0^1 2t^2\,dt = \frac{2}{3}.\quad \mathbf{r_2} = -t\,\mathbf{i} + (t - 1)\,\mathbf{j},\ 0 \le t \le 1 \Rightarrow \mathbf{F_2} = -\mathbf{i} - (2t^2 - 2t + 1)\,\mathbf{j}$ and $\frac{d\mathbf{r_2}}{dt} = -\mathbf{i} +$

$\mathbf{j} \Rightarrow \mathbf{F_2} \cdot \frac{d\mathbf{r_2}}{dt} = -2t^3 + 4t^2 - 2t + 1 \Rightarrow \text{Flow}_2 = \int_0^1 (-2t^3 + 4t^2 - 2t + 1)\,dt = \frac{1}{3}.\ \therefore\ \text{Flow} = \text{Flow}_1 +$

$\text{Flow}_2 = 1$

10. $\mathbf{r_1} = (\cos t)\,\mathbf{i} + (\sin t)\,\mathbf{j} + t\,\mathbf{k},\ 0 \le t \le \frac{\pi}{2} \Rightarrow \mathbf{F_1} = (2\cos t)\,\mathbf{i} + 2t\,\mathbf{j} + (2\sin t)\,\mathbf{k}$ and $\frac{d\mathbf{r_1}}{dt} = (-\sin t)\,\mathbf{i} + (\cos t)\,\mathbf{j} +$

$\mathbf{k} \Rightarrow \mathbf{F_1} \cdot \frac{d\mathbf{r_1}}{dt} = -2\sin t \cos t + 2t\cos t + 2\sin t \Rightarrow \text{Circ}_1 = \int_0^{\pi/2} (-2\sin t \cos t + 2t\cos t + 2\sin t)\,dt =$

$\pi - 2.\ \mathbf{r_2} = \mathbf{j} + \frac{\pi}{2}(1 - t)\,\mathbf{k},\ 0 \le t \le 1 \Rightarrow \mathbf{F_2} = \pi(1 - t)\,\mathbf{j} + 2\,\mathbf{k}$ and $\frac{d\mathbf{r_2}}{dt} = -\frac{\pi}{2}\,\mathbf{k} \Rightarrow \mathbf{F_2} \cdot \frac{d\mathbf{r_2}}{dt} = -\pi \Rightarrow$

$\text{Circ}_2 = \int_0^1 -\pi\,dt = -\pi.\ \mathbf{r_3} = t\,\mathbf{i} + (1 - t)\,\mathbf{j},\ 0 \le t \le 1 \Rightarrow \mathbf{F_3} = 2t\,\mathbf{i} + 2(1 - t)\,\mathbf{k}$ and $\frac{d\mathbf{r_3}}{dt} = \mathbf{i} - \mathbf{j} \Rightarrow$

10. (Continued)

$$\mathbf{F}_3 \cdot \frac{d\mathbf{r}_3}{dt} = 4t - 2 \Rightarrow \text{Circ}_3 = \int_0^1 (4t - 2)\, dt = 0. \;\therefore\; \text{Circ} = \text{Circ}_1 + \text{Circ}_2 + \text{Circ}_3 = -2$$

11. $M = 2xy + x,\ N = xy - y \Rightarrow \frac{\partial M}{\partial x} = 2y + 1,\ \frac{\partial M}{\partial y} = 2x,\ \frac{\partial N}{\partial x} = y,\ \frac{\partial N}{\partial y} = x - 1 \Rightarrow \text{Flux} =$

$$\iint_R (2y + 1 + x - 1)\, dy\, dx = \int_0^1\int_0^1 (2y + x)\, dy\, dx = \frac{3}{2}.\ \text{Circ} = \iint_R (y - 2x)\, dy\, dx =$$

$$\int_0^1\int_0^1 (y - 2x)\, dy\, dx = -\frac{1}{2}$$

12. $M = y - 6x^2,\ N = x + y^2 \Rightarrow \frac{\partial M}{\partial x} = -12x,\ \frac{\partial M}{\partial y} = 1,\ \frac{\partial N}{\partial x} = 1,\ \frac{\partial N}{\partial y} = 2y \Rightarrow \text{Flux} = \iint_R (-12x + 2y)\, dx\, dy =$

$$\int_0^1\int_y^1 (-12x + 2y)\, dx\, dy = \int_0^1 (4y^2 + 2y - 6)\, dy = -\frac{11}{3}.\ \text{Circ} = \iint_R (1 - 1) dx\, dy = 0$$

13. Let $M = 4x^3y,\ N = x^4 \Rightarrow \frac{\partial M}{\partial y} = 4x^3,\ \frac{\partial N}{\partial x} = 4x^3 \Rightarrow \oint_C 4x^3y\, dx + x^4\, dy = \iint_R (4x^3 - 4x^3)\, dx\, dy = 0$

14. a) Let $M = x,\ N = y \Rightarrow \frac{\partial M}{\partial x} = 1,\ \frac{\partial M}{\partial y} = 0,\ \frac{\partial N}{\partial x} = 0,\ \frac{\partial N}{\partial y} = 1 \Rightarrow \text{Flux} = \iint_R (1 + 1)\, dx\, dy = 2\iint_R dx\, dy$

$= 2$(Area of the region)

b) Let C be a closed curve to which Green's Theorem applies. Let **n** be the unit normal vector to C. Let $\mathbf{F} = x\,\mathbf{i} + y\,\mathbf{j}$. Assume **F** is orthogonal to **n** at every point of C. Then the flux density of **F** at every point of C is 0 since $\mathbf{F} \cdot \mathbf{n} = 0$ at every point of C $\Rightarrow \frac{\partial M}{\partial x} + \frac{\partial N}{\partial y} = 0$ at every point of C. $\therefore$ Flux

$$= \iint_R \left(\frac{\partial M}{\partial x} + \frac{\partial N}{\partial y}\right) dx\, dy = \iint_R 0\, dx\, dy = 0.$$ But part a above states the flux is 2(Area of the region) $\Rightarrow$ the area of the region would be 0 $\Rightarrow$ Contradiction. $\therefore$ **F** cannot be orthogonal to **n** at every point of C.

15. Let $M = 8x \sin y$, $N = -8y \cos x \Rightarrow \frac{\partial M}{\partial y} = 8x \cos y$, $\frac{\partial N}{\partial x} = 8y \sin x \Rightarrow \int_C 8x \sin y \, dx - 8y \cos x \, dy =$

$$\iint_R (8y \sin x - 8x \cos y)\, dy\, dx = \int_0^{\pi/2}\int_0^{\pi/2} (8y \sin x - 8x \cos y)\, dy\, dx = 0$$

16. Let $M = y^2$, $N = x^2 \Rightarrow \frac{\partial M}{\partial y} = 2y$, $\frac{\partial N}{\partial x} = 2x \Rightarrow \int_C y^2\, dx + x^2\, dy = \iint_R (2x - 2y)\, dx\, dy =$

$$\int_{-2}^{2}\int_{-\sqrt{4-x^2}}^{\sqrt{4-x^2}} (2x - 2y)\, dy\, dx = \int_{-2}^{2} 4x\sqrt{4 - x^2}\, dx = 0$$

17. Let $z = 1 - x - y \Rightarrow f_x(x,y) = -1$, $f_y(x,y) = -1 \Rightarrow \sqrt{f_x^{\,2} + f_y^{\,2} + 1} = \sqrt{3} \Rightarrow$ Area $= \iint_R \sqrt{3}\, dx\, dy =$

$\sqrt{3}$(Area of the circlular region in the xy–plane) $= \pi\sqrt{3}$

18. $\nabla f = -3\,\mathbf{i} + 2y\,\mathbf{j} + 2z\,\mathbf{k}$, $\mathbf{p} = \mathbf{i} \Rightarrow |\nabla f| = \sqrt{9 + 4y^2 + 4z^2}$ and $|\nabla f \cdot \mathbf{p}| = 3 \Rightarrow S =$

$$\iint_R \sqrt{9 + 4y^2 + 4z^2}\, dy\, dz = \int_0^{2\pi}\int_0^{\sqrt{3}} \sqrt{9 + 4r^2}\, r\, dr\, d\theta = \int_0^{2\pi} \left(\frac{7}{4}\sqrt{21} - \frac{9}{4}\right) d\theta = \left(\frac{7}{2}\sqrt{21} - \frac{9}{2}\right)\pi$$

19. $\nabla f = 2x\,\mathbf{i} + 2y\,\mathbf{j} + 2z\,\mathbf{k}$, $\mathbf{p} = \mathbf{k} \Rightarrow |\nabla f| = \sqrt{4x^2 + 4y^2 + 4z^2} = 2\sqrt{x^2 + y^2 + z^2} = 2$ and $|\nabla f \cdot \mathbf{p}| = |2z| = 2z$

since $z \ge 0 \Rightarrow S = \iint_R \frac{2}{2z}\, dA = \iint_R \frac{1}{z}\, dA = \iint_R \frac{1}{\sqrt{1 - x^2 - y^2}}\, dx\, dy =$

$$\int_0^{2\pi}\int_0^{1/2} \frac{1}{\sqrt{1 - r^2}}\, r\, dr\, d\theta = \int_0^{2\pi} \left(1 - \frac{\sqrt{3}}{2}\right) d\theta = 2\pi - \pi\sqrt{3}$$

20. $\nabla f = 2x\,\mathbf{i} + 2y\,\mathbf{j} + 2z\,\mathbf{k}$, $\mathbf{p} = \mathbf{k} \Rightarrow |\nabla f| = \sqrt{4x^2 + 4y^2 + 4z^2} = \sqrt{16} = 4$ and $|\nabla f \cdot \mathbf{p}| = 2z$ since $z \ge 0 \Rightarrow$

$$S = \iint_R \frac{4}{2z}\, dA = \iint_R \frac{2}{z}\, dA = 2\int_0^{\pi/2}\int_0^{2\cos\theta} \frac{2}{\sqrt{4 - r^2}}\, r\, dr\, d\theta = 4\pi - 8$$

21. a) $\nabla f = 2y\,\mathbf{j} - \mathbf{k}$, $\mathbf{p} = \mathbf{k} \Rightarrow |\nabla f| = \sqrt{4y^2+1}$ and $|\nabla f \cdot \mathbf{p}| = 1 \Rightarrow d\sigma = \sqrt{4y^2+1}\ dx\,dy \Rightarrow$

$$\iint_S g(x,y,z)\,d\sigma = \iint_S \frac{z}{\sqrt{4y^2+1}}\sqrt{4y^2+1}\,dx\,dy = \iint_S y(y^2-1)\,d\sigma =$$

$$\int_{-1}^{1}\int_{0}^{3} (y^3 - y)\,dx\,dy = 0$$

b) $$\iint_S g(x,y,z)\,d\sigma = \iint_S \frac{z}{\sqrt{4y^2+1}}\sqrt{4y^2+1}\,dx\,dy = \int_{-1}^{1}\int_{0}^{3} (y^2-1)\,dx\,dy = -4$$

22. $\nabla f = 2y\,\mathbf{j} + 2z\,\mathbf{k}$, $\mathbf{p} = \mathbf{k} \Rightarrow |\nabla f| = \sqrt{4y^2+4z^2} = 2\sqrt{y^2+z^2} = 10$ and $|\nabla f \cdot \mathbf{p}| = 2z$ since $z \ge 0 \Rightarrow d\sigma =$

$$\frac{10}{2z}\,dx\,dy = \frac{5}{z}\,dx\,dy \Rightarrow \iint_S g(x,y,z)\,d\sigma = \iint_R x^4y(y^2+z^2)\left(\frac{5}{z}\right)dx\,dy =$$

$$\iint_R x^4y(25)\left(\frac{5}{\sqrt{25-y^2}}\right)dx\,dy = \int_0^4\int_0^1 \frac{125y}{\sqrt{25-y^2}}x^4\,dx\,dy = \int_0^4 \frac{25y}{\sqrt{25-y^2}}\,dy = 50$$

23. On the face, $z = 1$: $G(x,y,z) = G(x,y,1) = z \Rightarrow \nabla G = \mathbf{k} \Rightarrow |\nabla G| = 1$. $\mathbf{n} = \mathbf{k} \Rightarrow \mathbf{F} \cdot \mathbf{n} = 2xz = 2x$ since $z = 1$

$d\sigma = dA \Rightarrow$ Flux $= \iint_R 2x\,dx\,dy = \int_0^1\int_0^1 2x\,dx\,dy = 1$. On the face, $z = 0$: $G(x,y,z) = G(x,y,0) =$

$z \Rightarrow \nabla G = \mathbf{k} \Rightarrow |\nabla G| = 1$. $\mathbf{n} = -\mathbf{k} \Rightarrow \mathbf{F} \cdot \mathbf{n} = -2xz = 0$ since $z = 0 \Rightarrow$ Flux $= \iint_R 0\,dx\,dy = 0$

On the face, $x = 1$: $G(x,y,z) = G(1,y,z) = x \Rightarrow \nabla G = \mathbf{i} \Rightarrow |\nabla G| = 1$. $\mathbf{n} = \mathbf{i} \Rightarrow \mathbf{F} \cdot \mathbf{n} = 2xy = 2y$ since $x = 1$

Flux $= \int_0^1\int_0^1 2y\,dy\,dz = 1$ On the face, $x = 0$: $G(x,y,z) = G(0,y,z) = x \Rightarrow \nabla G = \mathbf{i} \Rightarrow |\nabla G| = 1$. $\mathbf{n} = -\mathbf{i}$

$\Rightarrow \mathbf{F} \cdot \mathbf{n} = -2xy = 0$ since $x = 0 \Rightarrow$ Flux $= 0$. On the face, $y = 1$: $G(x,y,z) = G(x,1,z) = y \Rightarrow \nabla G = \mathbf{j} \Rightarrow |\nabla G|$

$= 1$. $\mathbf{n} = \mathbf{j} \Rightarrow \mathbf{F} \cdot \mathbf{n} = 2yz = 2z$ since $y = 1 \Rightarrow$ Flux $= \int_0^1\int_0^1 2z\,dz\,dx = 1$. On the face, $y = 0$: $G(x,y,z) =$

$G(z,0,z) = y \Rightarrow \nabla G = \mathbf{j} \Rightarrow |\nabla G| = 1$. $\mathbf{n} = -\mathbf{j} \Rightarrow \mathbf{F} \cdot \mathbf{n} = -2yz = 0$ since $y = 0 \Rightarrow$ Flux $= 0$.

$\therefore$ Total Flux $= 3$

24. Across the cap, $x^2 + y^2 + z^2 = 25$: Let $g(x,y,z) = x^2 + y^2 + z^2 - 25 \Rightarrow \nabla g = 2x\,\mathbf{i} + 2y\,\mathbf{j} + 2z\,\mathbf{k} \Rightarrow$ $|\nabla g| = \sqrt{4x^2 + 4y^2 + 4z^2} = 10$. $\mathbf{p} = \mathbf{k} \Rightarrow |\nabla g \cdot \mathbf{p}| = 2z$ since $z \ge 0$. $\mathbf{n} = \frac{2x\,\mathbf{i} + 2y\,\mathbf{j} + 2z\,\mathbf{k}}{10} = \frac{x\,\mathbf{i} + y\,\mathbf{j} + z\,\mathbf{k}}{5}$. $d\sigma = \frac{10}{2z}\,dA = \frac{5}{z}\,dA$. $\mathbf{F} \cdot \mathbf{n} = \frac{x^2z}{5} + \frac{y^2z}{5} + \frac{z}{5}$. $\therefore$ $\text{Flux}_{\text{cap}} = \iint_R \left(\frac{x^2z}{5} + \frac{y^2z}{5} + \frac{z}{5}\right)\left(\frac{5}{z}\right) dA$

$$= \iint_R (x^2 + y^2 + 1)\,dx\,dy = \int_0^{2\pi}\int_0^4 (r^2 + 1)\,r\,dr\,d\theta = \int_0^{2\pi} 72\,d\theta = 144\pi$$

Across the bottom, $z = 3$: $g(x,y,z) = z - 3$, $\mathbf{p} = \mathbf{k} \Rightarrow \nabla g = \mathbf{k} \Rightarrow |\nabla g| = 1$ and $|\nabla g \cdot \mathbf{p}| = 1$. $\mathbf{n} = -\mathbf{k} \Rightarrow$ $\mathbf{F} \cdot \mathbf{n} = -1 \Rightarrow \text{Flux}_{\text{bottom}} = \iint_R -1\,d\sigma = \iint_R -1\,dA = -1(\text{Area of the circular region}) = -16\pi$

$\therefore$ $\text{Flux} = \text{Flux}_{\text{cap}} + \text{Flux}_{\text{bottom}} = 128\pi$

25. Because of symmetry $\bar{x} = \bar{y} = 0$. Let $F(x,y,z) = x^2 + y^2 + z^2 = 25 \Rightarrow \nabla F = 2x\,\mathbf{i} + 2y\,\mathbf{j} + 2z\,\mathbf{k} \Rightarrow$ $|\nabla F| = \sqrt{4x^2 + 4y^2 + 4z^2} = 10$, $\mathbf{p} = \mathbf{k} \Rightarrow |\nabla F \cdot \mathbf{p}| = 2z$ since $z \ge 0 \Rightarrow M = \iint_R \delta(x,y,z)\,d\sigma =$

$\iint_R z\left(\frac{10}{2z}\right) dA = \iint_R 5\,dA = 5(\text{Area of the circular region}) = 80\pi$. $M_{xy} = \iint_R z\delta\,d\sigma =$

$$\iint_R z\,dA = \iint_R 5\sqrt{25 - x^2 - y^2}\,dx\,dy = \int_0^{2\pi}\int_0^4 5\sqrt{25 - r^2}\,r\,dr\,d\theta = \int_0^{2\pi} \frac{490}{3}\,d\theta = \frac{980}{3}\pi$$

$\therefore$ $\bar{z} = \frac{\frac{980}{3}\pi}{80\pi} = \frac{49}{12}$. Thus $(\bar{x},\bar{y},\bar{z}) = \left(0,0,\frac{49}{12}\right)$. $I_z = \iint_R (x^2 + y^2)\delta\,d\sigma = \iint_R 5(x^2 + y^2)\,dx\,dy =$

$$\int_0^{2\pi}\int_0^4 5r^3\,dr\,d\theta = \int_0^{2\pi} 320\,d\theta = 640\pi. \quad R_z = \sqrt{I_z/M} = \sqrt{\frac{640\pi}{80\pi}} = 2\sqrt{2}$$

26. On the face $z = 1$, $g(x,y,z) = z - 1 = 0$, $\mathbf{p} = \mathbf{k} \Rightarrow \nabla g = \mathbf{k} \Rightarrow |\nabla g| = 1$ and $|\nabla g \cdot \mathbf{p}| = 1 \Rightarrow d\sigma = dA \Rightarrow$

$I = \iint_R (x^2 + y^2)\,dA = 2\int_0^{\pi/4}\int_0^{\sec\theta} r^3\,dr\,d\theta = \frac{2}{3}$. On the face $z = 0$, $g(x,y,z) = z = 0 \Rightarrow$

$\nabla g = \mathbf{k}$, $\mathbf{p} = \mathbf{k} \Rightarrow |\nabla g| = 1 \Rightarrow |\nabla g \cdot \mathbf{p}| = 1 \Rightarrow d\sigma = dA \Rightarrow I = \iint_R (x^2 + y^2)\,dA = \frac{2}{3}$.

On the face $y = 0$, $g(x,y,z) = y = 0 \Rightarrow \nabla g = \mathbf{j}$, $\mathbf{p} = \mathbf{j} \Rightarrow |\nabla g| = 1 \Rightarrow |\nabla g \cdot \mathbf{p}| = 1 \Rightarrow d\sigma = dA \Rightarrow$

26. (Continued)

$$I = \iint_R (x^2 + 0^2)\,dA = \int_0^1\int_0^1 x^2\,dx\,dz = \frac{1}{3}.$$ On the face y = 1, $g(x,y,z) = y - 1 = 0 \Rightarrow$

$\nabla g = \mathbf{j}$, $\mathbf{p} = \mathbf{j} \Rightarrow |\nabla g| = 1 \Rightarrow |\nabla g \cdot \mathbf{p}| = 1 \Rightarrow d\sigma = dA \Rightarrow I = \iint_R (x^2 + 1^2)\,dA = \int_0^1\int_0^1 (x^2 + 1)\,dx\,dz$

$= \frac{4}{3}$. On the face x = 1, $g(x,y,z) = x - 1 = 0 \Rightarrow \nabla g = \mathbf{i}$, $\mathbf{p} = \mathbf{i} \Rightarrow |\nabla g| = 1 \Rightarrow |\nabla g \cdot \mathbf{p}| = 1 \Rightarrow d\sigma = dA \Rightarrow$

$$I = \iint_R (1^2 + y^2)\,dA = \int_0^1\int_0^1 (1 + y^2)\,dy\,dz = \frac{4}{3}.$$ On the face x = 0, $g(x,y,z) = x = 0 \Rightarrow$

$\nabla g = \mathbf{i}$, $\mathbf{p} = \mathbf{i} \Rightarrow |\nabla g| = 1 \Rightarrow |\nabla g \cdot \mathbf{p}| = 1 \Rightarrow d\sigma = dA \Rightarrow I = \iint_R (0^2 + y^2)\,dA = \int_0^1\int_0^1 y^2\,dy\,dz = \frac{1}{3}$

$\therefore\ I_z = \frac{14}{3}$

27. $\frac{\partial}{\partial x}(2xy) = 2y$, $\frac{\partial}{\partial y}(2yz) = 2z$, $\frac{\partial}{\partial z}(2xz) = 2x \Rightarrow \nabla \cdot \mathbf{F} = 2x + 2y + 2z \Rightarrow$ Flux $= \iiint_D (2x + 2y + 2z)\,dV$

$$\int_0^1\int_0^1\int_0^1 (2x + 2y + 2z)\,dx\,dy\,dz = \int_0^1\int_0^1 (1 + 2y + 2z)\,dy\,dz = \int_0^1 (2 + 2z)\,dz = 3$$

28. $\frac{\partial}{\partial x}(xz) = z$, $\frac{\partial}{\partial y}(yz) = z$, $\frac{\partial}{\partial z}(1) = 0 \Rightarrow \nabla\cdot\mathbf{F} = 2z \Rightarrow$ Flux $= \iiint_D 2z\,r\,dr\,d\theta\,dz =$

$$\int_0^{2\pi}\int_0^4\int_3^5 2z\,dz\,r\,dr\,d\theta = \int_0^{2\pi}\int_0^4 16r\,dr\,d\theta = \int_0^{2\pi} 128\,d\theta = 256\pi$$

29. $\frac{\partial}{\partial x}(-2x) = -2, \frac{\partial}{\partial y}(-3y) = -3, \frac{\partial}{\partial z}(z) = 1 \Rightarrow \nabla \cdot \mathbf{F} = -4 \Rightarrow \text{Flux} = \iiint_D -4\, dV =$

$-4 \int_0^{2\pi}\int_0^1\int_{r^2}^{\sqrt{2-r^2}} dz\, r\, dr\, d\theta = -4 \int_0^{2\pi}\int_0^1 (r\sqrt{2-r^2} - r^3)\, dr\, d\theta = -4 \int_0^{2\pi} \left(-\frac{7}{12} + \frac{2}{3}\sqrt{2}\right) d\theta =$

$\frac{2}{3}\pi\left(7 - 8\sqrt{2}\right)$

30. $\frac{\partial}{\partial x}(6x + y) = 6, \frac{\partial}{\partial y}(-(x + z)) = 0, \frac{\partial}{\partial z}(4yz) = 4y \Rightarrow \nabla \cdot \mathbf{F} = 6 + 4y \Rightarrow \text{Flux} = \iiint_D (6 + 4y)\, dV =$

$\int_0^1\int_0^{\pi/2}\int_0^1 (6 + 4r\sin\theta)\, r\, dr\, d\theta\, dz = \int_0^1\int_0^{\pi/2} \left(3 + \frac{4}{3}\sin\theta\right) d\theta\, dz = \int_0^1 \left(\frac{3\pi}{2} + \frac{4}{3}\right) dz = \frac{3\pi}{2} + \frac{4}{3}$

31. $\nabla f = 2\,\mathbf{i} + 6\,\mathbf{j} - 3\,\mathbf{k} \Rightarrow \nabla \times \mathbf{F} = -2y\,\mathbf{k}.\ \mathbf{n} = \frac{2\,\mathbf{i} + 6\,\mathbf{j} - 3\,\mathbf{k}}{\sqrt{4 + 36 + 9}} = \frac{2\,\mathbf{i} + 6\,\mathbf{j} - 3\,\mathbf{k}}{7} \Rightarrow \nabla \times \mathbf{F} \cdot \mathbf{n} = \frac{6}{7}y.\ \mathbf{p} = \mathbf{k} \Rightarrow$

$|\nabla f \cdot \mathbf{p}| = 3 \Rightarrow d\sigma = \frac{7}{3}dA \Rightarrow \oint_C \mathbf{F} \cdot d\mathbf{r} = \iint_R \frac{6}{7}y\, d\sigma = \iint_R \frac{6}{7}y\left(\frac{7}{3}dA\right) = \iint_R 2y\, dx\, dy =$

$\int_0^{2\pi}\int_0^1 2r\sin\theta\ r\, dr\, d\theta = \int_0^{2\pi} \frac{2}{3}\sin\theta\, d\theta = 0$

32. curl $\mathbf{F} = (8y - 0)\,\mathbf{i} + (0 - 0)\,\mathbf{j} + (1 - 1)\,\mathbf{k} = 8y\,\mathbf{i}$. $g(x,y,z) = z + y = 0 \Rightarrow \nabla g = \mathbf{i} + \mathbf{j} \Rightarrow |\nabla g| = \sqrt{2} \Rightarrow$

$\mathbf{n} = \frac{\mathbf{i} + \mathbf{j}}{\sqrt{2}} \Rightarrow \nabla \times \mathbf{F} \cdot \mathbf{n} = 0.\ \mathbf{p} = \mathbf{k} \Rightarrow |\nabla g \cdot \mathbf{p}| = \frac{1}{\sqrt{2}} \Rightarrow d\sigma = \frac{\sqrt{2}}{1/\sqrt{2}}dA = 2\, dA \Rightarrow \oint_C \mathbf{F} \cdot d\mathbf{r} =$

$\iint_R \nabla \times \mathbf{F} \cdot \mathbf{n}\, d\sigma = \iint_R 0\, d\sigma = 0$

33. $\frac{\partial P}{\partial y} = 0 = \frac{\partial N}{\partial z}, \frac{\partial M}{\partial z} = 0 = \frac{\partial P}{\partial x}, \frac{\partial N}{\partial x} = 0 = \frac{\partial M}{\partial y} \Rightarrow$ Conservative

34. $\frac{\partial P}{\partial y} = \frac{-zy}{x^2 + y^2 + z^2} = \frac{\partial N}{\partial z}, \frac{\partial M}{\partial z} = \frac{-xz}{x^2 + y^2 + z^2} = \frac{\partial P}{\partial x}, \frac{\partial N}{\partial x} = \frac{-xy}{x^2 + y^2 + z^2} = \frac{\partial M}{\partial y} \Rightarrow$ Conservative

35. $\frac{\partial P}{\partial y} = 0 \neq ye^z = \frac{\partial N}{\partial xz} \Rightarrow$ Not Conservative

36. $\frac{\partial P}{\partial y} = \frac{x}{(x+yz)^2} = \frac{\partial N}{\partial z}$, $\frac{\partial M}{\partial z} = \frac{-y}{(x+yz)^2} = \frac{\partial P}{\partial x}$, $\frac{\partial N}{\partial x} = \frac{-z}{(x+yz)^2} = \frac{\partial M}{\partial y}$ $\Rightarrow$ Conservative

37. $\frac{\partial f}{\partial x} = 2 \Rightarrow f(x,y,z) = 2x + g(y,z)$. $\frac{\partial f}{\partial y} = \frac{\partial g}{\partial y} = 2y + z \Rightarrow g(y,z) = y^2 + zy + h(z) \Rightarrow f(x,y,z) = 2x + y^2 + zy + h(z)$.
$\frac{\partial f}{\partial z} = y + h'(z) = y + 1 \Rightarrow h'(z) = 1 \Rightarrow h(z) = z + C$. $\therefore$ $f(x,y,z) = 2x + y^2 + zy + z + C$

38. $\frac{\partial f}{\partial x} = z\cos xz \Rightarrow f(x,y,z) = \sin xz + g(y,z)$. $\frac{\partial f}{\partial y} = \frac{\partial g}{\partial y} = e^y \Rightarrow g(y,z) = e^y + h(z) \Rightarrow f(x,y,z) = \sin xz + e^y + h(z)$.
$\frac{\partial f}{\partial z} = x\cos xz + h'(z) = x\cos xz \Rightarrow h'(z) = 0 \Rightarrow h(z) = C$. $\therefore$ $f(x,y,z) = \sin xz + e^y + C$

39. $\frac{\partial P}{\partial y} = -\frac{1}{2}(x+y+z)^{-3/2} = \frac{\partial N}{\partial z}$, $\frac{\partial M}{\partial z} = -\frac{1}{2}(x+y+z)^{-3/2} = \frac{\partial P}{\partial x}$, $\frac{\partial N}{\partial x} = -\frac{1}{2}(x+y+z)^{-3/2} = \frac{\partial M}{\partial y}$ $\Rightarrow$ M dx + N dy + P dz is exact. $\frac{\partial f}{\partial x} = \frac{1}{\sqrt{x+y+z}} \Rightarrow f(x,y,z) = 2\sqrt{x+y+z} + g(y,z)$. $\frac{\partial f}{\partial y} = \frac{1}{\sqrt{x+y+z}} + \frac{\partial g}{\partial y} = \frac{1}{\sqrt{x+y+z}}$
$\Rightarrow \frac{\partial g}{\partial y} = 0 \Rightarrow g(y,z) = h(z) \Rightarrow f(x,y,z) = 2\sqrt{x+y+z} + h(z)$. $\frac{\partial f}{\partial z} = \frac{1}{\sqrt{x+y+z}} + h'(z) = \frac{1}{\sqrt{x+y+z}} \Rightarrow h'(z) = 0$
$\Rightarrow h(z) = C \Rightarrow f(x,y,z) = 2\sqrt{x+y+z} + C \Rightarrow$ $\int_{(-1,1,1)}^{(4,-3,0)} \frac{dx + dy + dz}{\sqrt{x+y+z}} = f(4,-3,0) - f(-1,1,1) = 0$

40. $\frac{\partial P}{\partial y} = -\frac{1}{2\sqrt{yz}} = \frac{\partial N}{\partial z}$, $\frac{\partial M}{\partial z} = 0 = \frac{\partial P}{\partial x}$, $\frac{\partial N}{\partial x} = 0 = \frac{\partial M}{\partial y}$ $\Rightarrow$ M dx + N dy + P dz is exact. $\frac{\partial f}{\partial x} = 1 \Rightarrow f(x,y,z) = x +$ $g(y,z)$. $\frac{\partial f}{\partial y} = \frac{\partial g}{\partial y} = -\sqrt{\frac{z}{y}} \Rightarrow g(y,z) = -2\sqrt{yz} + h(z) \Rightarrow f(x,y,z) = x - 2\sqrt{yz} + h(z)$. $\frac{\partial f}{\partial z} = -\sqrt{\frac{y}{z}} + h'(z) =$
$-\sqrt{\frac{y}{z}} \Rightarrow h'(z) = 0 \Rightarrow h(z) = C \Rightarrow f(x,y,z) = x - 2\sqrt{yz} + C$. $\therefore$ $\int_{(1,0,1)}^{(10,3,3)} dx - \sqrt{\frac{z}{y}}\,dy - \sqrt{\frac{y}{z}}\,dz =$
$f(10,3,3) - f(1,0,1) = 3$

41. Over Path 1: $\mathbf{r} = t\,\mathbf{i} + t\,\mathbf{j} + t\,\mathbf{k} \Rightarrow x = t, y = t, z = t$ and $d\mathbf{r} = (\mathbf{i} + \mathbf{j} + \mathbf{k})\,dt \Rightarrow \mathbf{F} = 2t^2\,\mathbf{i} + \mathbf{j} + t^2\,\mathbf{k} \Rightarrow$
$\mathbf{F} \cdot d\mathbf{r} = (3t^2 + 1)\,dt \Rightarrow \text{Work} = \int_0^1 (3t^2 + 1)\,dt = 2$
Over Path 2: $\mathbf{r}_1 = t\,\mathbf{i} + t\,\mathbf{j}$, $0 \le t \le 1 \Rightarrow x = t, y = t, z = 0$ and $d\mathbf{r}_1 = (\mathbf{i} + \mathbf{j})\,dt \Rightarrow \mathbf{F}_1 = 2t^2\,\mathbf{i} + \mathbf{j} + t^2\,\mathbf{k} \Rightarrow$
$\mathbf{F}_1 \cdot d\mathbf{r}_1 = (2t^2 + 1)\,dt \Rightarrow \text{Work}_1 = \int_0^1 (2t^2 + 1)\,dt = \frac{5}{3}$. $\mathbf{r}_2 = \mathbf{i} + \mathbf{j} + t\,\mathbf{k}$, $0 \le t \le 1 \Rightarrow x = 1, y = 1, z = t$ and
$d\mathbf{r}_2 = \mathbf{k}\,dt \Rightarrow \mathbf{F}_2 = 2\,\mathbf{i} + \mathbf{j} + \mathbf{k} \Rightarrow \mathbf{F}_2 \cdot d\mathbf{r}_2 = dt \Rightarrow \text{Work}_2 = \int_0^1 dt = 1$. $\therefore$ $\text{Work} = \text{Work}_1 + \text{Work}_2 = \frac{8}{3}$

42. Over Path 1: $\mathbf{r} = t\,\mathbf{i} + t\,\mathbf{j} + t\,\mathbf{k}$, $0 \le t \le 1 \Rightarrow x = t, y = t, z = t$ and $d\mathbf{r} = (\mathbf{i} + \mathbf{j} + \mathbf{k})\,dt \Rightarrow \mathbf{F} = 2t^2\,\mathbf{i} + t^2\,\mathbf{j} + \mathbf{k}$
$\Rightarrow \mathbf{F} \cdot d\mathbf{r} = (2t^2 + t^2 + 1)\,dt = (3t^2 + 1)\,dt \Rightarrow \text{Work} = \int_0^1 (3t^2 + 1)\,dt = 2$

Over Path 2: Since f is conservative, $\int_{curve} \mathbf{F} \cdot \mathbf{dr} = 0$ around any simple closed curve.

$\int_{curve} \mathbf{F} \cdot \mathbf{dr} = \int_{C_1} \mathbf{F} \cdot \mathbf{dr} + \int_{C_2} \mathbf{F} \cdot \mathbf{dr}$ where C_1 is the path from (0,0,0) to (1,1,0) to (1,1,1) and C_2 is the path from (1,1,1) to (0,0,0). $\int_{C_2} \mathbf{F} \cdot \mathbf{dr} = -2$ (See Path 1 above.) $\therefore$ $\int_{curve} \mathbf{F} \cdot \mathbf{dr} =$

$\int_{C_1} \mathbf{F} \cdot \mathbf{dr} + (-2) = 0 \Rightarrow \int_{C_1} \mathbf{F} \cdot \mathbf{dr} = 2$

43. dx = (–2 sin t + 2 sin 2t) dt, dy = (2 cos t – 2 cos 2t) dt. Area $= \frac{1}{2}\oint_C x\,dy - y\,dx =$

$$\frac{1}{2}\int_0^{2\pi} \big((2\cos t - \cos 2t)(2\cos t - 2\cos 2t) - (2\sin t - \sin 2t)(-2\sin t + 2\sin 2t)\big)\,dt =$$

$$\frac{1}{2}\int_0^{2\pi} (6 - 6\cos t)\,dt = 6\pi$$

44. dx = (–2 sin t – 2 sin 2t) dt, dy = (2 cos t – 2 cos 2t) dt. Area $= \frac{1}{2}\oint_C x\,dy - y\,dx =$

$$\frac{1}{2}\int_0^{2\pi} \big((2\cos t + \cos 2t)(2\cos t - 2\cos 2t) - (2\sin t - \sin 2t)(-2\sin t - 2\sin 2t)\big)\,dt =$$

$$\frac{1}{2}\int_0^{2\pi} (1 - 2\cos^3 t + \cos t + 2\sin^2 t\cos t)\,dt = 2\pi$$

45. dx = cos 2t dt, dy = cos t dt. Area $= \frac{1}{2}\oint_C x\,dy - y\,dx = \frac{1}{2}\int_0^{\pi} \left(\frac{1}{2}\sin 2t\cos t - \sin t\cos 2t\right) dt =$

$$\frac{1}{2}\int_0^{\pi} (-\sin t\cos^2 t + \sin t)\,dt = \frac{2}{3}$$

46. $dx = (-2a \sin t - 2a \cos 2t)\, dt$, $dy = (b \cos t)\, dt$. Area $= \frac{1}{2}\oint_C x\, dy - y\, dx =$

$$\frac{1}{2}\int_0^{2\pi} \Big((2ab \cos^2 t - ab \cos t \sin 2t) - (-2ab \sin^2 t - 2ab \sin t \cos 2t)\Big)\, dt =$$

$$\frac{1}{2}\int_0^{2\pi} (2ab - 5ab \cos^2 t \sin t + 2ab \sin t)\, dt = ab\pi$$

CHAPTER 16

PREVIEW OF DIFFERENTIAL EQUATIONS

SECTION 16.1 SEPARABLE FIRST ORDER EQUATIONS

1. First Order, Non–Linear

2. Third Order, Non–Linear

3. Fourth Order, Linear

4. Second Order, Linear

5. a) $y = x^2 \Rightarrow y' = 2x,\ y'' = 2 \Rightarrow y'' - y' = 2x - 2x = 0$

   b) $y = 1 \Rightarrow y' = 0,\ y'' = 0 \Rightarrow xy'' - y' = 0$

   c) $y = C_1x^2 + C_2 \Rightarrow y' = 2C_1x,\ y'' = 2C_1 \Rightarrow xy'' - y' = x(2C_1) - 2C_1x = 0$

6. a) $y = \frac{x}{2} \Rightarrow y' = \frac{1}{2} \Rightarrow y' + \frac{1}{x}y = \frac{1}{2} + \frac{1}{x}\left(\frac{x}{2}\right) = \frac{1}{2} + \frac{1}{2} = 1$

   b) $y = \frac{1}{x} + \frac{x}{2} \Rightarrow y' = -\frac{1}{x^2} + \frac{1}{2} \Rightarrow y' + \frac{1}{x}y = -\frac{1}{x^2} + \frac{1}{2} + \frac{1}{x}\left(\frac{1}{x} + \frac{x}{2}\right) = -\frac{1}{x^2} + \frac{1}{2} + \frac{1}{x^2} + \frac{1}{2} = 1$

   c) $y = \frac{C}{x} + \frac{x}{2} \Rightarrow y' = -\frac{C}{x^2} + \frac{1}{2} \Rightarrow y' + \frac{1}{x}y = \Rightarrow -\frac{C}{x^2} + \frac{1}{2} + \frac{1}{x}\left(\frac{C}{x} + \frac{x}{2}\right) = -\frac{C}{x^2} + \frac{1}{2} + \frac{C}{x^2} + \frac{1}{2} = 1$

7. a) $y = e^{-x} \Rightarrow y' = -e^{-x} \Rightarrow 2y' + 3y = 2(-e^{-x}) + 3e^{-x} = e^{-x}$

   b) $y = e^{-x} + e^{-3x/2} \Rightarrow y' = -e^{-x} - \frac{3}{2}e^{-3x/2} \Rightarrow 2y' + 3y = 2\left(-e^{-x} - \frac{3}{2}e^{-3x/2}\right) + 3\left(e^{-x} + e^{-3x/2}\right) = e^{-x}$

   c) $y = e^{-x} + Ce^{-3x/2} \Rightarrow y' = -e^{-x} - \frac{3}{2}Ce^{-3x/2} \Rightarrow 2y' + 3y = 2\left(-e^{-x} - \frac{3}{2}Ce^{-3x/2}\right) + 3\left(e^{-x} + Ce^{-3x/2}\right) = e^{-x}$

8. a) $y = 1 \Rightarrow y' = 0,\ y'' = 0 \Rightarrow yy'' = 0$ and $2(y')^2 - 2y' = 2(0)^2 - 2(0) = 0 \Rightarrow yy'' = 2(y')^2 - 2y$

   b) $y = \tan x \Rightarrow y' = \sec^2 x,\ y'' = 2\sec^2 x \tan x \Rightarrow yy'' = \tan x(2\sec^2 x \tan x) = 2\sec^2 x \tan^2 x$ and $2(y')^2 - 2y' = 2(\sec^2 x)^2 - 2\sec^2 x = 2\sec^2 x \tan^2 x$

9. $y'' = -32 \Rightarrow y' = -32x + C_1$. $y'(5) = 0 \Rightarrow -32(5) + C_1 = 0 \Rightarrow C_1 = 160 \Rightarrow y' = -32x + 160 \Rightarrow y = -16x^2 + 160x + C_2$. $y(5) = 400 \Rightarrow -16(5)^2 + 160(5) + C_2 = 400 \Rightarrow C_2 = 0 \Rightarrow y = 160x - 16x^2$

10. $y = 2\left(1 - e^{-3t/2}\right) \Rightarrow \frac{dy}{dt} = 2\left(\frac{3}{2}e^{-3t/2}\right)$ and $y(0) = 2(1 - 1) = 0 \Rightarrow 2\frac{dy}{dt} + 3y = 2\left(2\left(\frac{3}{2}e^{-3t/2}\right)\right) + 3\left(2\left(1 - e^{-3t/2}\right)\right) = 6e^{-3t/2} + 6 - 6e^{-3t/2} = 6$

11. $y = 3\cos 2t - \sin 2t \Rightarrow y(0) = 3\cos 0 - \sin 0 = 3$ and $y' = -6\sin 2t - 2\cos 2t \Rightarrow y'(0) = -6\sin 0 -$

$2\cos 0 = -2$ and $y'' = -12\cos 2t + 4\sin 2t \Rightarrow y'' + 4y = -12\cos 2t + 4\sin 2t + 4(3\cos 2t - \sin 2t) = 0$

12. $y = 2 - \ln(\cos(x-1)) \Rightarrow y(1) = 2 - \ln(\cos(1-1)) = 2$ and $y' = -\frac{1}{\cos(x-1)}(-\sin(x-1)) = \tan(x-1) \Rightarrow$ $y'(1) = \tan(1-1) = 0$. $y'' = \sec^2(x-1) \Rightarrow y'' - (y')^2 = \sec^2(x-1) - (\tan(x-1))^2 = \sec^2(x-1) - \tan^2(x-1) = 1$

13. a) $\frac{dx}{dt} = 1000 + 0.10\,x \Rightarrow dx = (1000 + 0.10\,x)\,dt \Rightarrow \frac{dx}{1000 + 0.10\,x} = dt \Rightarrow \int \frac{dx}{1000 + 0.10\,x} = \int dt$ $\Rightarrow 10\ln(1000 + 0.10\,x) = t + C \Rightarrow 1000 + 0.10\,x = e^{0.10(t+C)} = e^{0.10t + 0.10C} = e^{0.10t}\,e^{0.10C}$ $= C_1 e^{0.10t} \Rightarrow x = 10C_1\,e^{0.10t} - 10000 \Rightarrow x = C_2\,e^{0.10t} - 10000$. $x(0) = 1000 \Rightarrow 1000 = C_2 e^{0.10(0)} - 10000 \Rightarrow C_2 = 11000 \Rightarrow x = 11000\,e^{0.10t} - 10000$

b) $100\,000 = 11000\,e^{0.10t} - 10000 \Rightarrow 10 = e^{0.10t} \Rightarrow \ln 10 = 0.10\,t \Rightarrow t = 10\ln 10 \approx 23.03$

14. $\frac{dT}{dt} = k(T - T_s) \Rightarrow \frac{dT}{T - T_s} = k\,dt \Rightarrow \int \frac{dT}{T - T_s} = \int k\,dt \Rightarrow \ln(T - T_s) = kt + C \Rightarrow T - T_s = C_1\,e^{kt}$

Given $T_s = 20°$ C and the object cools from 100°C to 40°C in 20 minutes, then $T = 100°C \Rightarrow t_0 = 0 \Rightarrow$ $100° - 20° = C_1 e^{k(0)} \Rightarrow C_1 = 80° \Rightarrow T - T_s = 80\,e^{kt}$. $t = 20$ min $\Rightarrow T = 40°C \Rightarrow 40° - 20° = 80\,e^{k(20)}$ $\Rightarrow \frac{1}{4} = e^{20k} \Rightarrow \ln\frac{1}{4} = 20\,k \Rightarrow k = \frac{\ln\frac{1}{4}}{20} \approx -0.069 \Rightarrow T - T_s = 80\,e^{-0.069\,t}$. $T = 60°C \Rightarrow 60° - 20° = 80\,e^{-0.069\,t} \Rightarrow \frac{1}{2} = e^{-0.069\,t} \Rightarrow t = \frac{\ln\frac{1}{2}}{-0.069} \Rightarrow t \approx 10.05$ minutes

15. $(x^2 + y^2)\,dx + xy\,dy = 0 \Rightarrow \left(1 + \frac{y^2}{x^2}\right)dx + \frac{xy}{x^2}\,dy = 0 \Rightarrow \left(1 + \left(\frac{y}{x}\right)^2\right)dx + \frac{y}{x}\,dy = 0 \Rightarrow \Rightarrow \frac{y}{x}\,dy = -\left(1 + \left(\frac{y}{x}\right)^2\right) \Rightarrow \frac{dy}{dx} = \frac{-\left(1 + \left(\frac{y}{x}\right)^2\right)}{y/x} \Rightarrow$ Homogeneous. $v = \frac{y}{x} \Rightarrow F(v) = \frac{-(1+v^2)}{v} \Rightarrow \frac{dx}{x} + \frac{dv}{v + \frac{1+v^2}{v}} = 0 \Rightarrow \frac{dx}{x} + \frac{v\,dv}{1 + 2v^2} = 0 \Rightarrow \ln|x| + \frac{1}{4}\ln(1 + 2v^2) = C \Rightarrow |x|\left(1 + 2v^2\right)^{1/4} = e^C \Rightarrow x^4(1 + 2v^2) = C_1$ where $C_1 = (e^C)^4 \Rightarrow x^4\left(1 + 2\left(\frac{y^2}{x^2}\right)\right) = C_1 \Rightarrow x^4 + 2x^2y^2 = C_1$

16. $(y^2 - xy)\,dx + x^2\,dy = 0 \Rightarrow \left(\frac{y^2}{x^2} - \frac{xy}{x^2}\right)dx + \frac{x^2}{x^2}\,dy = 0 \Rightarrow \left(\left(\frac{y}{x}\right)^2 - \frac{y}{x}\right)dx + dy = 0 \Rightarrow \frac{dy}{dx} = \left(\frac{y}{x} - \left(\frac{y}{x}\right)^2\right) \Rightarrow$ Homogeneous. $v = \frac{y}{x} \Rightarrow F(v) = v - v^2 \Rightarrow \frac{dx}{x} + \frac{dv}{v - (v - v^2)} = 0 \Rightarrow \frac{dx}{x} + \frac{dv}{v^2} = 0 \Rightarrow \ln x - v^{-1} = C \Rightarrow$ $x = C_1 e^{x/y}$

17. $\left(x\,e^{y/x} + y\right)dx - x\,dy = 0 \Rightarrow \left(e^{y/x} + \frac{y}{x}\right)dx - dy = 0 \Rightarrow \frac{dy}{dx} = e^{y/x} + \frac{y}{x} \Rightarrow$ Homogeneous. $v = \frac{y}{x} \Rightarrow$ $F(v) = e^v + v \Rightarrow \frac{dx}{x} + \frac{dv}{v - (e^v + v)} = 0 \Rightarrow \frac{dx}{x} + \frac{dv}{-e^v} = 0 \Rightarrow \ln|x| + e^{-v} = C \Rightarrow \ln|x| + e^{-y/x} = C$

18. $(x - y)\,dx + (x + y)\,dy = 0 \Rightarrow \left(\frac{x}{x} - \frac{y}{x}\right) dx + \left(\frac{x}{x} + \frac{y}{x}\right) dy = 0 \Rightarrow \frac{dy}{dx} = \frac{\left(\frac{y}{x} - 1\right)}{1 + \frac{y}{x}} \Rightarrow$ Homogeneous

$v = \frac{y}{x} \Rightarrow F(v) = \frac{v - 1}{v + 1} \Rightarrow \frac{dx}{x} + \frac{dy}{v - \left(\frac{v - 1}{v + 1}\right)} = 0 \Rightarrow \frac{dx}{x} + \frac{(v + 1)dv}{v^2 + 1} = 0 \Rightarrow \frac{dx}{x} + \frac{v\,dv}{v^2 + 1} + \frac{dv}{v^2 + 1} = 0 \Rightarrow$

$\ln x + \frac{1}{2}\ln(v^2 + 1) + \tan^{-1}(v^2 + 1) = 0$ if $x > 0 \Rightarrow x\sqrt{v^2 + 1} = C_1\, e^{-\tan^{-1}(v^2+1)} \Rightarrow x\sqrt{\left(\frac{y}{x}\right)^2 + 1} =$

$C_1\, e^{-\tan^{-1}((y/x)^2+1)} \Rightarrow \sqrt{y^2 + x^2} = C_1\, e^{-\tan^{-1}((y^2+x^2)/x^2)}$ if $x > 0$.

19. $\frac{dy}{dx} = \frac{y}{x} + \cos\left(\frac{y - x}{x}\right) \Rightarrow \frac{dy}{dx} = \frac{y}{x} + \cos\left(\frac{y}{x} - 1\right) \Rightarrow$ Homogeneous. $v = \frac{y}{x} \Rightarrow F(v) = v + \cos(v - 1) \Rightarrow$

$\frac{dx}{x} + \frac{dv}{v - (v + \cos(v - 1))} = 0 \Rightarrow \frac{dx}{x} - \sec(v - 1)\,dv = 0 \Rightarrow \ln|x| + \ln|\sec(v - 1) + \tan(v - 1)| = C \Rightarrow$

$\frac{x}{\sec(v - 1) + \tan(v - 1)} = \pm e^C \Rightarrow \frac{x}{\sec\left(\frac{y - x}{x}\right) + \tan\left(\frac{y - x}{x}\right)} = C_1 \Rightarrow x = C_1\left(\sec\left(\frac{y - x}{x}\right) + \tan\left(\frac{y - x}{x}\right)\right)$

$y(2) = 2 \Rightarrow 2 = C_1\left(\sec\left(\frac{2 - 2}{2}\right) + \tan\left(\frac{2 - 2}{2}\right)\right) \Rightarrow C_1 = 2 \Rightarrow x = 2\left(\sec\left(\frac{y - x}{x}\right) + \tan\left(\frac{y - x}{x}\right)\right)$

20. $\left(x \sin\frac{y}{x} - y\cos\frac{y}{x}\right) dx + x\cos\frac{y}{x}\,dy = 0 \Rightarrow \left(\sin\frac{y}{x} - \frac{y}{x}\cos\frac{y}{x}\right) dx + \cos\frac{y}{x}\,dy = 0 \Rightarrow$

$\frac{dy}{dx} = \frac{\frac{y}{x}\cos\frac{y}{x} - \sin\frac{y}{x}}{\cos\frac{y}{x}} \Rightarrow$ Homogeneous. $v = \frac{y}{x} \Rightarrow F(v) = \frac{v\cos v - \sin v}{\cos v} = v - \tan v \Rightarrow \frac{dx}{x} + \frac{dv}{v - (v - \tan v)}$

$= 0 \Rightarrow \frac{dx}{x} + \frac{dv}{\tan v} = 0 \Rightarrow \frac{dx}{x} + \frac{\cos v\,dv}{\sin v} = 0 \Rightarrow \ln|x| + \ln|\sin v| = C \Rightarrow x\sin v = \pm e^C \Rightarrow x\sin\frac{y}{x} = C_1$

$y(2) = \pi \Rightarrow 2\sin\frac{\pi}{2} = C_1 \Rightarrow C_1 = 2 \Rightarrow x\sin\frac{y}{x} = 2$

21. $x\,dx + 2y\,dy = 0 \Rightarrow \frac{x^2}{2} + y^2 = C.$ $2y\,dx - x\,dy = 0 \Rightarrow \frac{2\,dx}{x} - \frac{dy}{y} = 0 \Rightarrow 2\ln|x| - \ln|y| = C \Rightarrow \frac{x^2}{|y|} = e^C \Rightarrow$

$\frac{x^2}{y} = \pm e^C = C_1 \Rightarrow x^2 = C_1 y$

22. $3x\,dx + 2y\,dy = 0 \Rightarrow \frac{3x^2}{2} + y^2 = C.$ $2y\,dx - 3x\,dy = 0 \Rightarrow \frac{dx}{3x} = \frac{dy}{2y} \Rightarrow \frac{1}{3}\ln|3x| = \frac{1}{2}\ln|2y| + C \Rightarrow$

$\frac{\sqrt[3]{3x}}{\sqrt{2y}} = \pm e^C = C_1,\ y > 0 \Rightarrow \sqrt[3]{3x} = C_1\sqrt{2y}$

23. $y\,dx + x\,dy = 0 \Rightarrow \frac{dx}{x} + \frac{dy}{y} = 0 \Rightarrow \ln|x| + \ln|y| = C \Rightarrow |xy| = e^C \Rightarrow xy = \pm e^C = C_1 \Rightarrow xy = C_1$, a family of hyperbolas with transverse axis the line $y = x$ or $y = -x$. $x\,dx - y\,dy = 0 \Rightarrow \frac{x^2}{2} - \frac{y^2}{2} = C \Rightarrow x^2 - y^2 = C_1$, a family of hyperbolas with transverse axis the x or y axis.

SECTION 16.2 EXACT DIFFERENTIAL EQUATIONS

1. $\frac{\partial M}{\partial y} = 0, \frac{\partial N}{\partial x} = 0 \Rightarrow$ Exact

2. $\frac{\partial M}{\partial y} = 1, \frac{\partial N}{\partial x} = -1 \Rightarrow$ Not Exact

3. $\frac{\partial M}{\partial y} = -\frac{1}{y^2}, \frac{\partial N}{\partial x} = -\frac{1}{y^2} \Rightarrow$ Exact

4. $\frac{\partial M}{\partial y} = 2y, \frac{\partial N}{\partial x} = 2y \Rightarrow$ Exact

5. $\frac{\partial M}{\partial y} = e^y, \frac{\partial N}{\partial x} = e^y \Rightarrow$ Exact

6. $\frac{\partial M}{\partial y} = \cos xy - yx \sin xy, \frac{\partial N}{\partial x} = \cos xy - xy \sin xy$ $\Rightarrow$ Exact

7. $\frac{\partial M}{\partial y} = 2, \frac{\partial N}{\partial x} = -2 \Rightarrow$ Not Exact

8. $\frac{\partial M}{\partial y} = 2, \frac{\partial N}{\partial x} = -2 \Rightarrow$ Not Exact

9. $\frac{\partial M}{\partial y} = 1 = \frac{\partial N}{\partial x} \Rightarrow$ Exact. $\frac{\partial f}{\partial x} = x + y \Rightarrow f(x,y) = \frac{x^2}{2} + xy + k(y)$. $\frac{\partial f}{\partial y} = x + k'(y) = x + y^2 \Rightarrow k(y) = \frac{y^3}{3} + C \Rightarrow$ $f(x,y) = \frac{x^2}{2} + xy + \frac{y^3}{3} + C = C_1 \Rightarrow 3x^2 + 6xy + 2y^3 = C_2$

10. $\frac{\partial M}{\partial y} = 2x\, e^y = \frac{\partial N}{\partial x} \Rightarrow$ Exact. $\frac{\partial f}{\partial x} = 2x\, e^y + e^x \Rightarrow f(x,y) = x^2e^y + e^x + k(y)$. $\frac{\partial f}{\partial y} = x^2e^y + k'(y) = \left(x^2 + 1\right)e^y \Rightarrow$ $k'(y) = e^y \Rightarrow k(y) = e^y + C \Rightarrow f(x,y) = x^2e^y + e^x + e^y + C = C_1 \Rightarrow x^2e^y + e^x + e^y = C_2$

11. $\frac{\partial M}{\partial y} = 2x + 2y = \frac{\partial N}{\partial x} \Rightarrow$ Exact. $\frac{\partial f}{\partial x} = 2xy + y^2 \Rightarrow f(x,y) = x^2y + xy^2 + k(y)$. $\frac{\partial f}{\partial y} = x^2 + 2xy + k'(y) =$ $x^2 + 2xy - y \Rightarrow k'(y) = -y \Rightarrow k(y) = -\frac{y^2}{2} + C \Rightarrow f(x,y) = x^2y + xy^2 - \frac{y^2}{2} + C = C_1 \Rightarrow 2x^2y + 2xy^2 - y^2 = C_2$

12. $\frac{\partial M}{\partial y} = -1 = \frac{\partial N}{\partial x} \Rightarrow$ Exact. $\frac{\partial f}{\partial x} = x^3 - y \Rightarrow f(x,y) = \frac{x^4}{4} - xy + k(y)$. $\frac{\partial f}{\partial y} = -x + k'(y) = -x \Rightarrow k'(y) = 0 \Rightarrow k(y) = C$ $\Rightarrow f(x,y) = \frac{x^4}{4} - xy + C = C_1 \Rightarrow x^4 - 4xy = C_2$

13. $\frac{\partial M}{\partial y} = \frac{1}{y} + \frac{1}{x} = \frac{\partial N}{\partial x} \Rightarrow$ Exact. $\frac{\partial f}{\partial x} = e^x + \ln y + \frac{y}{x} \Rightarrow f(x,y) = e^x + x \ln y + y \ln x + k(y)$. $\frac{\partial f}{\partial y} = \frac{x}{y} + \ln x + k'(y) =$ $\frac{x}{y} + \ln x + \sin y \Rightarrow k'(y) = \sin y \Rightarrow k(y) = -\cos y + C \Rightarrow f(x,y) = e^x + x \ln y + y \ln x - \cos y + C = C_1 \Rightarrow$ $e^x + x \ln y + y \ln x - \cos y = C_2$

14. $\frac{\partial M}{\partial y} = \cos xy - xy \sin xy = \frac{\partial N}{\partial x} \Rightarrow$ Exact. $\frac{\partial f}{\partial x} = y \cos xy + 1 \Rightarrow f(x,y) = \sin xy + k(y)$. $\frac{\partial f}{\partial y} = x \cos xy + k'(y) =$ $x \cos xy + 2y - 3 \Rightarrow k'(y) = 2y - 3 \Rightarrow k(y) = y^2 - 3y + C \Rightarrow f(x,y) = \sin xy + y^2 - 3y + C = C_1 \Rightarrow$ $\sin xy + y^2 - 3y = C_2$

15. $\rho = \frac{1}{y^2} \Rightarrow \frac{1}{y^2}\left(xy^2 + y\right) dx - \frac{1}{y^2}(x\, dy) = 0 \Rightarrow \left(x + \frac{1}{y}\right) dx - \frac{x}{y^2} dy = 0 \Rightarrow \frac{\partial M}{\partial y} = -\frac{1}{y^2} = \frac{\partial N}{\partial x} \Rightarrow$ Exact. $\frac{\partial f}{\partial x} = x + \frac{1}{y} \Rightarrow f(x,y) = \frac{x^2}{2} + \frac{x}{y} + k(y)$. $\frac{\partial f}{\partial y} = -\frac{x}{y^2} + k'(y) = -\frac{x}{y^2} \Rightarrow k'(y) = 0 \Rightarrow k(y) = C \Rightarrow f(x,y) = \frac{x^2}{2} + \frac{x}{y} + C =$ $C_1 \Rightarrow x^2 + \frac{2x}{y} = C_2$

16. $\rho = \frac{1}{x^3} \Rightarrow \frac{1}{x^3}(x + 2y)\,dx - \frac{1}{x^3}(x\,dy) = 0 \Rightarrow \left(\frac{1}{x^2} + \frac{2y}{x^3}\right)dx - \frac{1}{x^2}\,dy = 0 \Rightarrow \frac{\partial M}{\partial y} = \frac{2}{x^3} = \frac{\partial N}{\partial x} \Rightarrow$ Exact.
$\frac{\partial f}{\partial x} = \frac{1}{x^2} + \frac{2y}{x^3} \Rightarrow f(x,y) = -\frac{1}{x} - \frac{y}{x^2} + k(y).\ \frac{\partial f}{\partial y} = -\frac{1}{x^2} + k'(y) = -\frac{1}{x^2} \Rightarrow k'(y) = 0 \Rightarrow k(y) = C \Rightarrow$
$f(x,y) = -\frac{1}{x} - \frac{y}{x^2} + C = C_1 \Rightarrow -\frac{1}{x} - \frac{y}{x^2} = C_2$

17. a) $\rho = \frac{1}{xy} \Rightarrow \frac{1}{xy}(y\,dx) + \frac{1}{xy}(x\,dy) = 0 \Rightarrow \frac{1}{x}\,dx + \frac{1}{y}\,dy = 0 \Rightarrow \frac{\partial M}{\partial y} = 0 = \frac{\partial N}{\partial x} \Rightarrow$ Exact. $\frac{\partial f}{\partial x} = \frac{1}{x} \Rightarrow f(x,y) =$
$\ln|x| + k(y).\ \frac{\partial f}{\partial y} = k'(y) = \frac{1}{y} \Rightarrow k(y) = \ln|y| + C \Rightarrow f(x,y) = \ln|x| + \ln|y| + C = C_1 \Rightarrow \ln|xy| = C_2 \Rightarrow$
$xy = \pm e^{C_2} \Rightarrow xy = C_3$

b) $\rho = \frac{1}{(xy)^2} \Rightarrow \frac{1}{(xy)^2}(y\,dx) + \frac{1}{(xy)^2}(x\,dy) = 0 \Rightarrow \frac{1}{x^2y}\,dx + \frac{1}{xy^2}\,dy = 0 \Rightarrow \frac{\partial M}{\partial y} = -\frac{1}{x^2y^2} = \frac{\partial N}{\partial x} \Rightarrow$ Exact.
$\frac{\partial f}{\partial x} = \frac{1}{x^2y} \Rightarrow f(x,y) = -\frac{1}{xy} + k(y).\ \frac{\partial f}{\partial y} = \frac{1}{xy^2} + k'(y) = \frac{1}{xy^2} \Rightarrow k'(y) = 0 \Rightarrow k(y) = C \Rightarrow f(x,y) = -\frac{1}{xy} + C = C_1$
$\Rightarrow xy = C_2$

18. a) $\rho = \frac{1}{y^2} \Rightarrow \frac{1}{y^2}(y\,dx) - \frac{1}{y^2}(x\,dy) = 0 \Rightarrow \frac{1}{y}\,dx - \frac{x}{y^2}\,dy = 0 \Rightarrow \frac{\partial M}{\partial y} = -\frac{1}{y^2} = \frac{\partial N}{\partial x} \Rightarrow$ Exact. $\frac{\partial f}{\partial x} = \frac{1}{y} \Rightarrow$
$f(x,y) = \frac{x}{y} + k(y).\ \frac{\partial f}{\partial y} = -\frac{x}{y^2} + k'(y) = -\frac{x}{y^2} \Rightarrow k'(y) = 0 \Rightarrow k(y) = C \Rightarrow f(x,y) = \frac{x}{y} + C = C_1 \Rightarrow \frac{x}{y} = C_2$

b) $\rho = \frac{1}{x^2} \Rightarrow \frac{1}{x^2}(y\,dx) - \frac{1}{x^2}(x\,dy) = 0 \Rightarrow \frac{y}{x^2}\,dx - \frac{1}{x}\,dy = 0 \Rightarrow \frac{\partial M}{\partial y} = \frac{1}{x^2} = \frac{\partial N}{\partial x} \Rightarrow$ Exact.
$\frac{\partial f}{\partial x} = \frac{y}{x^2} \Rightarrow f(x,y) = -\frac{y}{x} + k(y).\ \frac{\partial f}{\partial y} = -\frac{1}{x} + k'(y) = -\frac{1}{x} \Rightarrow k'(y) = 0 \Rightarrow k(y) = C \Rightarrow f(x,y) = -\frac{y}{x} + C = C_1 \Rightarrow$
$-\frac{y}{x} = C_2 \Rightarrow \frac{x}{y} = C_3$

c) $\rho = \frac{1}{xy} \Rightarrow \frac{1}{xy}(y\,dx) - \frac{1}{xy}(x\,dy) = 0 \Rightarrow \frac{1}{x}\,dx - \frac{1}{y}\,dy = 0 \Rightarrow \frac{\partial M}{\partial y} = 0 = \frac{\partial N}{\partial x} \Rightarrow$ Exact.
$\frac{\partial f}{\partial x} = \frac{1}{x} \Rightarrow f(x,y) = \ln|x| + k(y).\ \frac{\partial f}{\partial y} = k'(y) = -\frac{1}{y} \Rightarrow k(y) = -\ln|y| + C \Rightarrow f(x,y) = \ln|x| - \ln|y| + C = C_1 \Rightarrow$
$\ln\left|\frac{x}{y}\right| = C_2 \Rightarrow \frac{x}{y} = \pm e^{C_2} \Rightarrow \frac{x}{y} = C_3$

d) $\rho = \frac{1}{x^2 + y^2} \Rightarrow \frac{1}{x^2 + y^2}(y\,dx) - \frac{1}{x^2 + y^2}(x\,dy) = 0 \Rightarrow \frac{y}{x^2 + y^2}\,dx - \frac{x}{x^2 + y^2}\,dy = 0 \Rightarrow \frac{\partial M}{\partial y} = \frac{x^2 - y^2}{(x^2 + y^2)^2} =$
$\frac{\partial N}{\partial x} \Rightarrow$ Exact. $\frac{\partial f}{\partial x} = \frac{y}{x^2 + y^2} \Rightarrow f(x,y) = \tan^{-1}\frac{x}{y} + k(y).\ \frac{\partial f}{\partial y} = \frac{-x}{x^2 + y^2} + k'(y) = \frac{-x}{x^2 + y^2} \Rightarrow k'(y) = 0 \Rightarrow$
$k(y) = C \Rightarrow f(x,y) = \tan^{-1}\frac{x}{y} + C = C_1 \Rightarrow \tan^{-1}\frac{x}{y} = C_2 \Rightarrow \frac{x}{y} = \tan C_2 \Rightarrow \frac{x}{y} = C_3$

19. $\frac{\partial M}{\partial y} = 2y = \frac{\partial N}{\partial x} \Rightarrow$ Exact. $\frac{\partial f}{\partial x} = x + y^2 \Rightarrow f(x,y) = \frac{x^2}{2} + xy^2 + k(y).\ \frac{\partial f}{\partial y} = 2xy + k'(y) = 2xy + 1 \Rightarrow k'(y) = 1$
$\Rightarrow k(y) = y + C \Rightarrow f(x,y) = \frac{x^2}{2} + xy^2 + y + C = C_1 \Rightarrow x^2 + 2xy^2 + 2y = C_2.\ y(0) = 2 \Rightarrow C_2 = 4 \Rightarrow$
$x^2 + 2xy^2 + 2y = 4$

20. $\frac{\partial M}{\partial y} = e^y = \frac{\partial N}{\partial x} \Rightarrow$ Exact. $\frac{\partial f}{\partial x} = x + e^y \Rightarrow f(x,y) = \frac{x^2}{2} + x\,e^y + k(y).\ \frac{\partial f}{\partial y} = x\,e^y + k'(y) = y + x\,e^y \Rightarrow k'(y) = y \Rightarrow$
$k(y) = \frac{y^2}{2} + C \Rightarrow f(x,y) = \frac{x^2 + y^2}{2} + x\,e^y + C = C_1 \Rightarrow \frac{x^2 + y^2}{2} + x\,e^y = C_2.\ y(2) = 0 \Rightarrow C_2 = 4 \Rightarrow$
$\frac{x^2 + y^2}{2} + x\,e^y = 4 \Rightarrow x^2 + y^2 + 2x\,e^y = 8$

21. $\frac{\partial M}{\partial y} = -1 = \frac{\partial N}{\partial x} \Rightarrow$ Exact. $\frac{\partial f}{\partial x} = \frac{1}{x} - y \Rightarrow f(x,y) = \ln|x| - xy + k(y)$. $\frac{\partial f}{\partial y} = -x + k'(y) = \frac{1}{y} - x \Rightarrow k'(y) = \frac{1}{y} \Rightarrow$ $k(y) = \ln|y| + C \Rightarrow f(x,y) = \ln|x| - xy + \ln|y| + C = C_1 \Rightarrow \ln|xy| = xy + C_2 \Rightarrow xy = \pm e^{xy+C_2} = \pm e^{C_2} e^{xy} \Rightarrow$ $xy = C_3 e^{xy}$. $y(1) = 1 \Rightarrow C_3 = \frac{1}{e} \Rightarrow xy = \frac{1}{e} e^{xy} \Rightarrow xy = e^{xy-1}$

22. $\frac{\partial M}{\partial y} = 1 = \frac{\partial N}{\partial x} \Rightarrow$ Exact. $\frac{\partial f}{\partial x} = 2x + y + 1 \Rightarrow f(x,y) = x^2 + xy + x + k(y)$. $\frac{\partial f}{\partial y} = x + k'(y) = 2y + x + 1 \Rightarrow$ $k'(y) = 2y + 1 \Rightarrow k(y) = y^2 + y + C \Rightarrow f(x,y) = x^2 + xy + y^2 + x + y + C = C_1 \Rightarrow x^2 + y^2 + xy + x + y = C_2$. $y(1) = 5 \Rightarrow C_2 = 37 \Rightarrow x^2 + y^2 + xy + x + y = 37$.

23. $\frac{\partial M}{\partial y} = 2y$, $\frac{\partial N}{\partial x} = ay \Rightarrow a = 2$. $\therefore$ $(3x^2 + y^2)\,dx + 2xy\,dy = 0$ is exact. $\frac{\partial f}{\partial x} = 3x^2 + y^2 \Rightarrow f(x,y) = x^3 + xy^2 +$ $k(y)$. $\frac{\partial f}{\partial y} = 2xy + k'(y) = 2xy \Rightarrow k'(y) = 0 \Rightarrow k(y) = C \Rightarrow f(x,y) = x^3 + xy^2 + C = C_1 \Rightarrow x^3 + xy^2 = C_2$.

24. $\rho = \frac{1}{x^2} \Rightarrow \frac{1}{x^2}\left(x^2 + by^2\right) dx + \frac{1}{x^2}(6xy\,dy) = 0 \Rightarrow \left(1 + \frac{by^2}{x^2}\right) dx + \frac{6y}{x}\,dy = 0 \Rightarrow \frac{\partial M}{\partial y} = \frac{2by}{x^2}, \frac{\partial N}{\partial x} = -\frac{6y}{x^2} \Rightarrow$ $2b = -6$ since ρ makes the equation exact $\Rightarrow b = -3$. $\therefore$ $\left(1 - \frac{3y^2}{x^2}\right) dx + \frac{6y}{x}\,dy = 0$ is exact $\Rightarrow$ $\frac{\partial f}{\partial x} = 1 - \frac{3y^2}{x^2} \Rightarrow f(x,y) = x + \frac{3y^2}{x} + k(y)$. $\frac{\partial f}{\partial y} = \frac{6y}{x} + k'(y) = \frac{6y}{x} \Rightarrow k'(y) = 0 \Rightarrow k(y) = C \Rightarrow f(x,y) = x + \frac{3y^2}{x} + C$ $= C_1 \Rightarrow x + \frac{3y^2}{x} = C_2$

SECTION 16.3 LINEAR FIRST ORDER EQUATIONS

1. $\frac{dy}{dx} + 2y = e^{-x} \Rightarrow P(x) = 2, Q(x) = e^{-x} \Rightarrow \int P(x)\,dx = 2x \Rightarrow \rho(x) = e^{2x} \Rightarrow y = \frac{1}{e^{2x}} \int e^{2x}(e^{-x})\,dx \Rightarrow$ $y = \frac{1}{e^{2x}} \int e^x\,dx = \frac{1}{e^{2x}}\left(e^x + C\right) = e^{-x} + C\,e^{-2x}$

2. $\frac{dy}{dx} - \frac{1}{2}y = \frac{1}{2}e^{x/2} \Rightarrow P(x) = -\frac{1}{2}, Q(x) = \frac{1}{2}e^{x/2} \Rightarrow \int P(x)\,dx = -\frac{1}{2}x \Rightarrow \rho(x) = e^{-x/2} \Rightarrow$ $y = \frac{1}{e^{-x/2}} \int e^{-x/2}\left(\frac{1}{2}e^{x/2}\right) dx = e^{x/2} \int \frac{1}{2}\,dx \Rightarrow y = e^{x/2}\left(\frac{1}{2}x + C\right) = \frac{1}{2}x\,e^{x/2} + C\,e^{x/2}$

3. $\frac{dy}{dx} + \frac{3}{x}y = \frac{\sin x}{x^3} \Rightarrow P(x) = \frac{3}{x}, Q(x) = \frac{\sin x}{x^3} \Rightarrow \int P(x)\,dx = 3\ln|x| = \ln|x|^3 \Rightarrow \rho(x) = e^{\ln|x|^3} = |x|^3 = x^3$ if $x > 0 \Rightarrow y = \frac{1}{x^3} \int x^3\left(\frac{\sin x}{x^3}\right) dx = \frac{1}{x^3} \int \sin x\,dx \Rightarrow y = \frac{1}{x^3}(-\cos x + C) = -\frac{\cos x}{x^3} + \frac{C}{x^3}$

4. $\frac{dy}{dx} + \frac{y}{x} = \cos x \Rightarrow P(x) = \frac{1}{x}, Q(x) = \cos x \Rightarrow \int P(x)\,dx = \ln|x| \Rightarrow \rho(x) = e^{\ln|x|} = x$ if $x > 0 \Rightarrow$ $y = \frac{1}{x} \int x\cos x\,dx = \frac{1}{x}(x\sin x + \cos x + C) \Rightarrow y = \sin x + \frac{\cos x}{x} + \frac{C}{x}$

5. $\frac{dy}{dx} + \frac{4y}{x-1} = \frac{x+1}{(x-1)^3} \Rightarrow P(x) = \frac{4}{x-1}$, $Q(x) = \frac{x+1}{(x-1)^3} \Rightarrow \int P(x)\,dx = 4\ln|x-1| = \ln(x-1)^4 \Rightarrow$
$\rho(x) = e^{\ln(x-1)^4} = (x-1)^4 \Rightarrow y = \frac{1}{(x-1)^4}\int (x-1)^4 \frac{x+1}{(x-1)^3}\,dx = \frac{1}{(x-1)^4}\int (x^2-1)\,dx =$
$\frac{1}{(x-1)^4}\left(\frac{x^3}{3} - x + C\right) \Rightarrow y = \frac{x^3}{3(x-1)^4} - \frac{x}{(x-1)^4} + \frac{C}{(x-1)^4}$

6. $\frac{dy}{dx} + 2y = 2x\,e^{-2x} \Rightarrow P(x) = 2$, $Q(x) = 2x\,e^{-2x} \Rightarrow \int P(x)\,dx = \int 2\,dx = 2x \Rightarrow \rho(x) = e^{2x} \Rightarrow$
$y = \frac{1}{e^{2x}}\int e^{2x}\left(2x\,e^{-2x}\right)dx = \frac{1}{e^{2x}}\int 2x\,dx = e^{-2x}\left(x^2 + C\right) \Rightarrow y = x^2\,e^{-2x} + C\,e^{-2x}$

7. $\frac{dy}{dx} + (\cot x)\,y = \sec x \Rightarrow P(x) = \cot x$, $Q(x) = \sec x \Rightarrow \int P(x)\,dx = \ln|\sin x| \Rightarrow \rho(x) = e^{\ln|\sin x|} = \sin x$ if
$\sin x > 0 \Rightarrow y = \frac{1}{\sin x}\int \sin x(\sec x)\,dx = \frac{1}{\sin x}\int \tan x\,dx = \frac{1}{\sin x}(\ln|\sec x| + C) = \csc x(\ln|\sec x| + C)$

8. $\frac{dy}{dx} + (\tanh x)\,y = e^{-x}\left(\frac{1}{\cosh x}\right) \Rightarrow P(x) = \tanh x$, $Q(x) = e^{-x}\left(\frac{1}{\cosh x}\right) \Rightarrow \int P(x)\,dx = \ln(\cosh x) \Rightarrow$
$\rho(x) = e^{\ln(\cosh x)} = \cosh x \Rightarrow y = \frac{1}{\cosh x}\int \cosh x(e^{-x})\left(\frac{1}{\cosh x}\right)dx = \frac{1}{\cosh x}\int e^{-x}\,dx =$
$\operatorname{sech} x\left(-e^{-x} + C\right) \Rightarrow y = -e^{-x}\operatorname{sech} x + C\operatorname{sech} x$

9. $\frac{dy}{dx} + 2y = x \Rightarrow P(x) = 2$, $Q(x) = x \Rightarrow \int P(x)\,dx = 2x \Rightarrow \rho(x) = e^{2x} \Rightarrow y = \frac{1}{e^{2x}}\int e^{2x}(x\,dx) =$
$\frac{1}{e^{2x}}\left(\frac{1}{2}x\,e^{2x} - \frac{1}{4}e^{2x} + C\right) \Rightarrow y = \frac{1}{2}x - \frac{1}{4} + C\,e^{-2x}$. $y(0) = 1 \Rightarrow -\frac{1}{4} + C = 1 \Rightarrow C = \frac{5}{4} \Rightarrow$
$y = \frac{1}{2}x - \frac{1}{4} + \frac{5}{4}e^{-2x}$

10. $\frac{dy}{dx} + \frac{2y}{x} = x^2 \Rightarrow P(x) = \frac{2}{x}$, $Q(x) = x^2 \Rightarrow \int P(x)\,dx = 2\ln|x| \Rightarrow \rho(x) = e^{\ln x^2} = x^2 \Rightarrow$
$y = \frac{1}{x^2}\int x^2\left(x^2\right)dx = \frac{1}{x^2}\int x^4\,dx = \frac{1}{x^2}\left(\frac{x^5}{5} + C\right) = \frac{x^3}{5} + \frac{C}{x^2}$. $y(2) = 1 \Rightarrow \frac{8}{5} + \frac{C}{4} = 1 \Rightarrow C = \frac{8}{5} \Rightarrow$
$y = \frac{x^3}{5} + \frac{8}{5x^2}$

11. $\frac{dy}{dx} + \frac{y}{x} = \frac{\sin x}{x} \Rightarrow P(x) = \frac{1}{x}$, $Q(x) = \frac{\sin x}{x} \Rightarrow \int P(x)\,dx = \ln|x| \Rightarrow \rho(x) = e^{\ln|x|} = |x| \Rightarrow$
$y = \frac{1}{|x|}\int |x|\frac{\sin x}{x}\,dx = \frac{1}{x}\int x\frac{\sin x}{x}\,dx$ for $x \neq 0 \Rightarrow y = \frac{1}{x}\int \sin x\,dx = \frac{1}{x}(-\cos x + C) \Rightarrow$
$y = -\frac{1}{x}\cos x + \frac{C}{x}$. $y\left(\frac{\pi}{2}\right) = 1 \Rightarrow C = \frac{\pi}{2} \Rightarrow y = -\frac{1}{x}\cos x + \frac{\pi}{2x}$

12. $\frac{dy}{dx} - \frac{2y}{x} = x^2 \sec x \tan x \Rightarrow P(x) = -\frac{2}{x},\ Q(x) = x^2 \sec x \tan x \Rightarrow \int P(x)\,dx = -2\ln|x| \Rightarrow \rho(x) = e^{-2\ln|x|}$
$= x^{-2} \Rightarrow y = \frac{1}{x^{-2}} \int x^{-2}(x^2 \sec x \tan x)\,dx = x^2 \int \sec x \tan x\,dx = x^2(\sec x + C) \Rightarrow$
$y = x^2 \sec x + Cx^2.\ y\left(\frac{\pi}{3}\right) = 2 \Rightarrow C = \frac{18}{\pi^2} - 2 \Rightarrow y = x^2 \sec x + \left(\frac{18}{\pi^2} - 2\right)x^2$

13. $\frac{dy}{dx} - ky = 0 \Rightarrow P(x) = -k,\ Q(x) = 0 \Rightarrow \int P(x)\,dx = -kx \Rightarrow \rho(x) = e^{-kx} \Rightarrow$
$y = \frac{1}{e^{-kx}} \int e^{-kx}(0)\,dx = Ce^{kx}.\ y(0) = y_0 \Rightarrow C = y_0 \Rightarrow y = y_0\,e^{kx}$

14. a) $\frac{dc}{dt} = \frac{G}{100V} - kc \Rightarrow \frac{dc}{dt} + kc = \frac{G}{100V} \Rightarrow P(t) = k,\ Q(t) = \frac{G}{100V} \Rightarrow \int P(t)\,dt = kt \Rightarrow \rho(t) = e^{kt} \Rightarrow$
$c = \frac{1}{e^{kt}} \int e^{kt} \frac{G}{100V}\,dt = \frac{1}{e^{kt}}\left(\frac{1}{k}\left(\frac{G}{100V}\right)e^{kt} + C\right) \Rightarrow c = \frac{1}{k}\left(\frac{G}{100V}\right) + C\,e^{-kt}.\ c(0) = c_0 \Rightarrow$
$C = c_0 - \frac{G}{100Vk} \Rightarrow c = \frac{G}{100Vk} + \left(c_0 - \frac{G}{100Vk}\right) e^{-kt}$

b) $\lim_{t\to\infty} c = \lim_{t\to\infty}\left(\frac{G}{100Vk} + \left(c_0 - \frac{G}{100Vk}\right) e^{-kt}\right) = \frac{G}{100Vk}$

15. $\frac{dp}{dx} + p = 0 \Rightarrow \frac{dp}{dx} = -p \Rightarrow dp = -p\,dx \Rightarrow \frac{1}{p}\,dp = -dx \Rightarrow \ln|p| = -x + C \Rightarrow p = C_1e^{-x} \Rightarrow \frac{dy}{dx} = C_1\,e^{-x} \Rightarrow$
$y = -C_1e^{-x} + C_2 = C_3e^{-x} + C_2$

16. $x\frac{dp}{dx} + 2p = 1 \Rightarrow \frac{dp}{dx} + \frac{2}{x}p = \frac{1}{x} \Rightarrow P(x) = \frac{2}{x},\ Q(x) = \frac{1}{x} \Rightarrow \int P(x)\,dx = 2\ln|x| \Rightarrow \rho(x) = e^{2\ln|x|} = x^2 \Rightarrow$
$p = \frac{1}{x^2} \int x^2\left(\frac{1}{x}\right)dx = \frac{1}{x^2} \int x\,dx = \frac{1}{x^2}\left(\frac{x^2}{2} + C\right) = \frac{1}{2} + \frac{C}{x^2} \Rightarrow \frac{dy}{dx} = \frac{1}{2} + \frac{C}{x^2} \Rightarrow y = \frac{1}{2}x - \frac{C}{x} + C_1$

17. $x\frac{dp}{dx} + p = 0 \Rightarrow \frac{dp}{dx} + \frac{1}{x}p = 0 \Rightarrow P(x) = \frac{1}{x},\ Q(x) = 0 \Rightarrow \int P(x)\,dx = \ln|x| \Rightarrow \rho(x) = e^{\ln|x|} = |x| \Rightarrow$
$p = \frac{1}{|x|} \int |x|\,(0)\,dx = \frac{C}{|x|} = \frac{C}{x}$ if $x > 0 \Rightarrow \frac{dy}{dx} = \frac{C}{x} \Rightarrow y = C\ln x + C_1$

18. $2\frac{dp}{dx} + p^2 = -1 \Rightarrow 2\,dp + p^2\,dx = -\,dx \Rightarrow 2\,dp = -(p^2 + 1)\,dx \Rightarrow \frac{2\,dp}{p^2 + 1} = -\,dx \Rightarrow 2\tan^{-1} p = -x + C \Rightarrow$
$\tan^{-1} p = -\frac{1}{2}(x - C) \Rightarrow p = \tan\left(-\frac{1}{2}(x + C_1)\right) = -\tan\left(\frac{1}{2}(x + C_1)\right) \Rightarrow \frac{dy}{dx} = -\tan\left(\frac{1}{2}(x + C_1)\right) \Rightarrow$
$y = 2\ln\left|\cos\left(\frac{1}{2}(x + C_1)\right)\right| + C_2$

19. $x\frac{dp}{dx} + p = x^2 \Rightarrow \frac{dp}{dx} + \frac{1}{x}p = x \Rightarrow P(x) = \frac{1}{x},\ Q(x) = x \Rightarrow \int P(x)\,dx = \ln x$ if $x > 0 \Rightarrow \rho(x) = e^{\ln x} = x \Rightarrow$
$p = \frac{1}{x} \int x\,(x)\,dx = \frac{1}{x} \int x^2\,dx = \frac{1}{x}\left(\frac{x^3}{3} + C\right) = \frac{x^2}{3} + \frac{C}{x} \Rightarrow \frac{dy}{dx} = \frac{x^2}{3} + \frac{C}{x}.\ y'(1) = 1 \Rightarrow C = \frac{2}{3} \Rightarrow$
$\frac{dy}{dx} = \frac{x^2}{3} + \frac{2}{3x} \Rightarrow y = \frac{x^3}{9} + \frac{2}{3}\ln x + C_1.\ y(1) = 0 \Rightarrow C_1 = -\frac{1}{9} \Rightarrow y = \frac{x^3}{9} + \frac{2}{3}\ln x - \frac{1}{9}$

20. $x\frac{dq}{dx} - 2q = 0 \Rightarrow \frac{dq}{dx} - \frac{2}{x}q = 0 \Rightarrow P(x) = -\frac{2}{x},\ Q(x) = 0 \Rightarrow \int P(x)\,dx = -2\ln|x| \Rightarrow \rho(x) = e^{-2\ln|x|} = x^{-2}$
$\Rightarrow q = \frac{1}{x^{-2}}\int x^{-2}(0)\,dx \Rightarrow q = C\,x^2 \Rightarrow \frac{d^2y}{dx^2} = C\,x^2 \Rightarrow \frac{dy}{dx} = \frac{C\,x^3}{3} + C_1 = C_2\,x^3 + C_1 \Rightarrow y = \frac{C_2\,x^4}{4} + C_1\,x +$
$C_3 \Rightarrow y = C_4\,x^4 + C_1\,x + C_3$

21. Steady State $= \frac{V}{R} \Rightarrow$ we want $i = \frac{1}{2}\left(\frac{V}{R}\right) \Rightarrow \frac{1}{2}\left(\frac{V}{R}\right) = \frac{V}{R}\left(1 - e^{-Rt/L}\right) \Rightarrow \frac{1}{2} = 1 - e^{-Rt/L} \Rightarrow -\frac{1}{2} = -e^{-Rt/L}$
$\Rightarrow \ln\frac{1}{2} = -\frac{Rt}{L} \Rightarrow -\frac{L}{R}\ln\frac{1}{2} = t \Rightarrow t = \frac{L}{R}\ln 2$

22. a) $\frac{di}{dt} + \frac{R}{L}i = 0 \Rightarrow P(t) = \frac{R}{L},\ Q(t) = 0 \Rightarrow \int P(t)\,dt = \frac{R}{L}t \Rightarrow \rho(t) = e^{Rt/L} \Rightarrow i = \frac{1}{e^{Rt/L}}\int e^{Rt/L}(0)\,dt =$
$C\,e^{-Rt/L}$ where C is the initial current, $i_0 \Rightarrow i = i_0\,e^{-Rt/L}$

b) $\frac{1}{2}i_0 = i_0\,e^{-Rt/L} \Rightarrow e^{-Rt/L} = \frac{1}{2} \Rightarrow -\frac{Rt}{L} = \ln\frac{1}{2} = -\ln 2 \Rightarrow t = \frac{L}{R}\ln 2$

c) $t = \frac{L}{R} \Rightarrow i = i_0\,e^{(-Rt/L)(L/R)} = i_0\,e^{-1}$

23. $t = \frac{3L}{R} \Rightarrow i = \frac{V}{R}\left(1 - e^{(-R/L)(3L/R)}\right) = \frac{V}{R}\left(1 - e^{-3}\right) \approx 0.9502\,\frac{V}{R}$ or about 95% of the steady state value.

SECTION 16.4 SECOND ORDER LINEAR HOMOGENEOUS EQUATIONS

1. $r^2 + 2r = 0 \Rightarrow r_1 = 0,\ r_2 = -2 \Rightarrow y = C_1e^{0x} + C_2e^{-2x} = C_1 + C_2e^{-2x}$

2. $r^2 + 5r + 6 = 0 \Rightarrow (r+2)(r+3) = 0 \Rightarrow r_1 = -2,\ r_2 = -3 \Rightarrow y = C_1e^{-2x} + C_2e^{-3x}$

3. $r^2 + 6r + 5 = 0 \Rightarrow (r+5)(r+1) = 0 \Rightarrow r_1 = -5,\ r_2 = -1 \Rightarrow y = C_1e^{-5x} + C_2{}^{-x}$

4. $r^2 - 2r - 3 = 0 \Rightarrow (r-3)(r+1) = 0 \Rightarrow r_1 = 3,\ r_2 = -1 \Rightarrow y = C_1e^{3x} + C_2\,e^{-x}$

5. $r^2 - 4r + 4 = 0 \Rightarrow (r-2)^2 = 0 \Rightarrow r_1 = r_2 = 2 \Rightarrow y = \left(C_1x + C_2\right)e^{2x}$

6. $r^2 + 6r + 9 = 0 \Rightarrow (r+3)^2 = 0 \Rightarrow r_1 = r_2 = -3 \Rightarrow y = \left(C_1x + C_2\right)e^{-3x}$

7. $r^2 - 10r + 25 = 0 \Rightarrow (r-5)^2 = 0 \Rightarrow r_1 = r_2 = 5 \Rightarrow \left(C_1x + C_2\right)e^{5x}$

8. $r^2 - 2\sqrt{2}\,r + 2 = 0 \Rightarrow (r - \sqrt{2}\,)^2 = 0 \Rightarrow r_1 = r_2 = \sqrt{2} \Rightarrow y = \left(C_1x + C_2\right)e^{\sqrt{2}\,x}$

9. $r^2 + r + 1 = 0 \Rightarrow r = \frac{-1 \pm i\sqrt{3}}{2} \Rightarrow y = e^{-x/2}\left(C_1\cos\frac{\sqrt{3}}{2}x + C_2\sin\frac{\sqrt{3}}{2}x\right)$

10. $r^2 - 6r + 10 = 0 \Rightarrow r = 3 \pm i \Rightarrow y = e^{3x}\left(C_1\cos x + C_2\sin x\right)$

11. $r^2 - 2r + 4 = 0 \Rightarrow r = 1 \pm i\sqrt{3} \Rightarrow y = e^{x}\left(C_1\cos\sqrt{3}\,x + C_2\sin\sqrt{3}\,x\right)$

12. $r^2 + 8r + 25 = 0 \Rightarrow r = -4 \pm 3i \Rightarrow y = e^{-4x}\left(C_1 \cos 3x + C_2 \sin 3x\right)$

13. $r^2 - 1 = 0 \Rightarrow r_1 = 1, r_2 = -1 \Rightarrow y = C_1e^x + C_2e^{-x}$. $y(0) = 1 \Rightarrow C_1 + C_2 = 1$. $y' = C_1e^x - C_2e^{-x}$. $y'(0) = -2$ $\Rightarrow -2 = C_1 - C_2$. $\therefore\ C_1 = -\frac{1}{2}, C_2 = \frac{3}{2} \Rightarrow y = -\frac{1}{2}e^x + \frac{3}{2}e^{-x}$

14. $2r^2 - r - 1 = 0 \Rightarrow (2r + 1)(r - 1) = 0 \Rightarrow r_1 = -\frac{1}{2}, r_2 = 1 \Rightarrow y = C_1e^{-x/2} + C_2e^x$. $y(0) = -1 \Rightarrow -1 = C_1 + C_2$ $y' = -\frac{1}{2}C_1 e^{-x/2} + C_2e^x$. $y'(0) = 0 \Rightarrow 0 = -\frac{1}{2}C_1 + C_2$. $\therefore\ C_1 = -\frac{2}{3}, C_2 = -\frac{1}{3} \Rightarrow y = -\frac{2}{3}e^{-x/2} - \frac{1}{3}e^x$

15. $r^2 - 4 = 0 \Rightarrow r_1 = 2, r_2 = -2 \Rightarrow y = C_1e^{2x} + C_2e^{-2x}$. $y(0) = 0 \Rightarrow C_1 + C_2$. $y' = 2C_1e^{2x} - 2C_2e^{-2x}$. $y'(0) = 3 \Rightarrow 3 = 2C_1 - 2C_2$. $\therefore\ C_1 = \frac{3}{4}, C_2 = -\frac{3}{4} \Rightarrow y = \frac{3}{4}e^{2x} - \frac{3}{4}e^{-2x}$

16. $r^2 - 9 = 0 \Rightarrow r_1 = 3, r_2 = -3 \Rightarrow y = C_1e^{3x} + C_2e^{-3x}$. $y(\ln 2) = 1 \Rightarrow 1 = 8C_1 + \frac{1}{8}C_2$. $y' = 3C_1e^{3x} - 3C_2e^{-3x}$. $y'(\ln 2) = -3 \Rightarrow -1 = 8C_1 - \frac{1}{8}C_2$. $\therefore\ C_1 = 0, C_2 = 8 \Rightarrow y = 8e^{-3x}$

17. $r^2 + 2r + 1 = 0 \Rightarrow (r + 1)^2 = 0 \Rightarrow r_1 = r_2 = -1 \Rightarrow y = \left(C_1x + C_2\right)e^{-x}$. $y(0) = 0 \Rightarrow C_2 = 0$. $y' = C_1\left(e^{-x} - x\,e^{-x}\right)$. $y'(0) = 1 \Rightarrow C_1 = 1 \Rightarrow y = x\,e^{-x}$

18. $4r^2 - 2r + 1 = 0 \Rightarrow r = \frac{1 \pm i\sqrt{3}}{4} \Rightarrow y = e^{x/4}\left(C_1 \cos \frac{\sqrt{3}}{4}x + C_2 \sin \frac{\sqrt{3}}{4}x\right)$. $y(0) = 4 \Rightarrow C_1 = 4$. $y' = \frac{1}{4}e^{x/4}\left(C_1 \cos \frac{\sqrt{3}}{4}x + C_2 \sin \frac{\sqrt{3}}{4}x\right) + e^{x/4}\left(-\frac{\sqrt{3}}{4}C_1 \sin \frac{\sqrt{3}}{4}x + \frac{\sqrt{3}}{4}C_2 \cos \frac{\sqrt{3}}{4}x\right)$. $y'(0) = 2 \Rightarrow$ $\frac{1}{4}C_1 + \frac{\sqrt{3}}{4}C_2 = 2 \Rightarrow C_2 = \frac{4}{\sqrt{3}} \Rightarrow y = e^{x/4}\left(4 \cos \frac{\sqrt{3}}{4}x + \frac{4}{\sqrt{3}} \sin \frac{\sqrt{3}}{4}x\right)$

19. $4r^2 + 12r + 9 = 0 \Rightarrow (2r + 3)^2 = 0 \Rightarrow r_1 = r_2 = -\frac{3}{2} \Rightarrow y = \left(C_1x + C_2\right)e^{-3x/2}$. $y(0) = 0 \Rightarrow C_2 = 0 \Rightarrow$ $y = C_1x\,e^{-3x/2} \Rightarrow y' = C_1\left(e^{-3x/2} - \frac{3}{2}x\,e^{-3x/2}\right)$. $y'(0) = -1 \Rightarrow C_1 = -1 \Rightarrow y = -x\,e^{-3x/2}$

20. $y'' = 0 \Rightarrow y' = C$. $y'(0) = 5 \Rightarrow C = 5 \Rightarrow y' = 5 \Rightarrow y = 5x + C_1$. $y(0) = -3 \Rightarrow -3 = C_1 \Rightarrow y = 5x - 3$

21. $r^2 + 4 = 0 \Rightarrow r = \pm 2i \Rightarrow y = C_1 \cos 2x + C_2 \sin 2x$. $y(0) = 0 \Rightarrow C_1 = 0 \Rightarrow y = C_2 \sin 2x$. $y' = 2C_2 \cos 2x$. $y'(0) = 2 \Rightarrow C_2 = 1 \Rightarrow y = \sin 2x$

22. $r^2 + 9 = 0 \Rightarrow r = \pm 3i \Rightarrow y = C_1 \cos 3x + C_2 \sin 3x$. $y(0) = 0 \Rightarrow C_1 = 0 \Rightarrow y = C_2 \sin 3x$. $y' = 3C_2 \cos 3x$ $y'(0) = \sqrt{3} \Rightarrow C_2 = \frac{\sqrt{3}.}{3} \Rightarrow y = \frac{\sqrt{3}}{3} \sin 3x$

23. $r^2 - 2r + 3 = 0 \Rightarrow r = 1 \pm i\sqrt{2} \Rightarrow y = e^x\left(C_1 \cos \sqrt{2}\,x + C_2 \sin \sqrt{2}\,x\right)$. $y(0) = 2 \Rightarrow C_1 = 2 \Rightarrow y =$ $e^x\left(2 \cos \sqrt{2}\,x + C_2 \sin \sqrt{2}\,x\right)$. $y' = e^x\left(2 \cos \sqrt{2}\,x + C_2 \sin \sqrt{2}\,x\right) +$ $e^x\left(-2\sqrt{2} \sin \sqrt{2}\,x + \sqrt{2}\,C_2 \cos \sqrt{2}\,x\right)$. $y'(0) = 1 \Rightarrow 1 = 2 + \sqrt{2}\,C_2 \Rightarrow C_2 = -\frac{1}{\sqrt{2}} \Rightarrow$

23. (Continued)

$y = e^x\left(2\cos\sqrt{2}\,x - \frac{1}{\sqrt{2}}\sin\sqrt{2}\,x\right)$

24. $r^2 - 6r + 10 = 0 \Rightarrow r = 3 \pm i \Rightarrow y = e^{3x}\left(C_1\cos x + C_2\sin x\right)$. $y(0) = 7 \Rightarrow C_1 = 7 \Rightarrow$ $y = e^{3x}\left(7\cos x + C_2\sin x\right)$ $y' = 3e^{3x}\left(7\cos x + C_2\sin x\right) + e^{3x}\left(-7\sin x + C_2\cos x\right)$. $y'(0) = 1 \Rightarrow$ $C_2 = -20 \Rightarrow y = e^{3x}(7\cos x - 20\sin x)$

SECTION 16.5 SECOND ORDER NONHOMOGENEOUS LINEAR EQUATIONS

1. $\frac{d^2y}{dx^2} + \frac{dy}{dx} = 0 \Rightarrow r^2 + r = 0 \Rightarrow r_1 = 0, r_2 = -1 \Rightarrow y_h = C_1 + C_2e^{-x} \Rightarrow u_1 = 1, u_2 = e^{-x} \Rightarrow D = \begin{vmatrix} 1 & e^{-x} \\ 0 & -e^{-x}\end{vmatrix}$ $= -e^{-x} \Rightarrow v'_1 = -\frac{u_2\,F(x)}{D} = x \Rightarrow v_1 = \int x\,dx = \frac{x^2}{2} + C_1$. $v'_2 = \frac{u_1\,F(x)}{D} = -x\,e^x \Rightarrow v_2 = \int x\,e^x\,dx =$ $-x\,e^x + e^x + C_2 \Rightarrow y = \left(\frac{x^2}{2} + C_1\right) + \left(-x\,e^x + e^x + C_2\right)e^{-x} = \frac{x^2}{2} - x + C_3 + C_2\,e^{-x}$

2. $\frac{d^2y}{dx^2} + y = 0 \Rightarrow r^2 + 1 = 0 \Rightarrow r = \pm i \Rightarrow y_h = C_1\cos x + C_2\sin x \Rightarrow u_1 = \cos x, u_2 = \sin x \Rightarrow D =$ $\begin{vmatrix} \cos x & \sin x \\ -\sin x & \cos x\end{vmatrix} = 1 \Rightarrow v'_1 = -\frac{u_2\,F(x)}{D} = -\sin x\tan x \Rightarrow v_1 = \int -\sin x\tan x\,dx = \sin x -$ $\ln(\sec x + \tan x) + C_1$. $v'_2 = \frac{u_1\,F(x)}{D} = \cos x\tan x = \sin x \Rightarrow v_2 = \int \sin x\,dx = -\cos x + C_2 \Rightarrow$ $y = \left(\sin x - \ln(\sec x + \tan x) + C_1\right)\cos x + \left(-\cos x + C_2\right)\sin x = -\cos x\ln(\sec x + \tan x) + C_1\cos x +$ $C_2\sin x$

3. $\frac{d^2y}{dx^2} + y = 0 \Rightarrow r^2 + 1 = 0 \Rightarrow r = \pm i \Rightarrow y_h = C_1\cos x + C_2\sin x \Rightarrow u_1 = \cos x, u_2 = \sin x \Rightarrow D =$ $\begin{vmatrix} \cos x & \sin x \\ -\sin x & \cos x\end{vmatrix} = 1 \Rightarrow v'_1 = -\frac{u_2\,F(x)}{D} = -\sin^2 x \Rightarrow v_1 = \int -\sin^2 x\,dx = -\frac{1}{2}x + \frac{\sin 2x}{4} + C_1$. $v'_2 = \frac{u_1\,F(x)}{D} = \cos x\sin x \Rightarrow v_2 = \int \sin x\cos x\,dx = \frac{\sin^2 x}{2} + C_2 \Rightarrow y = \cos x\left(-\frac{1}{2}x + \frac{\sin 2x}{4} + C_1\right) +$ $\sin x\left(\frac{\sin^2 x}{2} + C_2\right) = -\frac{1}{2}x\cos x + C_1\cos x + C_3\sin x$

4. $\frac{d^2y}{dx^2} + 2\frac{dy}{dx} + y = 0 \Rightarrow r^2 + 2r + 1 = 0 \Rightarrow r_1 = r_2 = -1 \Rightarrow y_h = C_1x\,e^{-x} + C_2\,e^{-x} \Rightarrow u_1 = x\,e^{-x}, u_2 = e^{-x}$ $\Rightarrow D = \begin{vmatrix} x\,e^{-x} & e^{-x} \\ e^{-x} - x\,e^{-x} & -e^{-x}\end{vmatrix} = -e^{-2x} \Rightarrow v'_1 = -\frac{u_2\,F(x)}{D} = e^{2x} \Rightarrow v_1 = \int e^{2x}\,dx = \frac{1}{2}e^{2x} + C_1$. $v'_2 = \frac{u_1\,F(x)}{D} = -x\,e^{2x} \Rightarrow v_2 = \int -x\,e^{2x}\,dx = -\frac{e^{2x}}{4}(2x - 1) + C_2 \Rightarrow y = x\,e^{-x}\left(\frac{1}{2}e^{2x} + C_1\right) -$

4. (Continued)

$$e^{-x}\left(\frac{e^{2x}}{4}(2x-1)+C_2\right)=C_1x\,e^{-x}+\frac{e^x}{4}+C_3e^{-x}$$

5. $\frac{d^2y}{dx^2}+2\frac{dy}{dx}+y=0 \Rightarrow r^2+2r+1=0 \Rightarrow r_1=r_2=-1 \Rightarrow y_h=C_1x\,e^{-x}+C_2\,e^{-x} \Rightarrow u_1=x\,e^{-x},\ u_2=e^{-x}$ $\Rightarrow D=\begin{vmatrix} x\,e^{-x} & e^{-x} \\ e^{-x}-x\,e^{-x} & -e^{-x}\end{vmatrix}=-e^{-2x} \Rightarrow v'_1=-\frac{u_2\,F(x)}{D}=1 \Rightarrow v_1=\int 1\,dx=x+C_1.\ v'_2=\frac{u_1\,F(x)}{D}=$ $-x \Rightarrow v_2=\int -x\,dx=-\frac{x^2}{2}+C_2 \Rightarrow y=\left(x+C_1\right)xe^{-x}+\left(-\frac{x^2}{2}+C_2\right)e^{-x}=C_1x\,e^{-x}+C_2e^{-x}+\frac{1}{2}x^2\,e^{-x}$

6. $\frac{d^2y}{dx^2}-y=0 \Rightarrow r^2-1=0 \Rightarrow r_1=1,\ r_2=-1 \Rightarrow y_h=C_1\,e^x+C_2\,e^{-x} \Rightarrow u_1=e^x,\ u_2=e^{-x} \Rightarrow$ $D=\begin{vmatrix} e^x & e^{-x} \\ e^x & -e^{-x}\end{vmatrix}=-2 \Rightarrow v'_1=-\frac{u_2\,F(x)}{D}=\frac{x\,e^{-x}}{2} \Rightarrow v_1=\int \frac{x\,e^x}{2}dx=\frac{e^{-x}}{2}(-x-1)+C_1.\ v'_2=\frac{u_1\,F(x)}{D}$ $=-\frac{1}{2}x\,e^x \Rightarrow v_2=\int -\frac{1}{2}x\,e^x\,dx=-\frac{1}{2}e^x(x-1)+C_2 \Rightarrow y=e^x\left(\frac{e^{-x}}{2}\right)(-x-1)+C_1e^x+C_2e^{-x}-$ $\frac{1}{2}e^x(x-1)e^{-x}=C_1\,e^x+C_2\,e^{-x}-x$

7. $\frac{d^2y}{dx^2}-y=0 \Rightarrow r^2-1=0 \Rightarrow r_1=1,\ r_2=-1 \Rightarrow y_h=C_1\,e^x+C_2\,e^{-x} \Rightarrow u_1=e^x,\ u_2=e^{-x} \Rightarrow$ $D=\begin{vmatrix} e^x & e^{-x} \\ e^x & -e^{-x}\end{vmatrix}=-2 \Rightarrow v'_1=-\frac{u_2\,F(x)}{D}=\frac{1}{2} \Rightarrow v_1=\frac{1}{2}x+C_1.\ v'_2=\frac{u_1\,F(x)}{D}=-\frac{1}{2}e^{2x} \Rightarrow v_2=-\frac{1}{4}e^{2x}+$ $C_2 \Rightarrow y=\left(\frac{1}{2}x+C_1\right)e^x+\left(-\frac{1}{4}e^{2x}+C_2\right)e^{-x}=\frac{1}{2}x\,e^x+C_3\,e^x+C_2\,e^{-x}$

8. $\frac{d^2y}{dx^2}-y=0 \Rightarrow r^2-1=0 \Rightarrow r_1=1,\ r_2=-1 \Rightarrow y_h=C_1\,e^x+C_2\,e^{-x} \Rightarrow u_1=e^x,\ u_2=e^{-x} \Rightarrow$ $D=\begin{vmatrix} e^x & e^{-x} \\ e^x & -e^{-x}\end{vmatrix}=-2 \Rightarrow v'_1=-\frac{u_2\,F(x)}{D}=\frac{e^{-x}\sin x}{2} \Rightarrow v_1=\int \frac{e^{-x}\sin x}{2}dx=\frac{-e^{-x}(\sin x+\cos x)}{4}+C_1$ $v'_2=\frac{u_1\,F(x)}{D}=-\frac{e^x\sin x}{2} \Rightarrow v_2=\int -\frac{e^x\sin x}{2}dx=\frac{-e^x(\sin x-\cos x)}{4}+C_2 \Rightarrow y=$ $e^x\left(\frac{-e^{-x}(\sin x+\cos x)}{4}+C_1\right)+e^{-x}\left(\frac{-e^x(\sin x-\cos x)}{4}+C_2\right)=C_1\,e^x+C_2\,e^{-x}-\frac{1}{2}\sin x-\frac{1}{2}\cos x$

9. $\frac{d^2y}{dx^2}+4\frac{dy}{dx}+5y=0 \Rightarrow r^2+4r+5=0 \Rightarrow r=-2\pm i \Rightarrow y_h=e^{-2x}\left(C_1\cos x+C_2\sin x\right) \Rightarrow$ $u_1=e^{-2x}\cos x,\ u_2=e^{-2x}\sin x \Rightarrow D=\begin{vmatrix} e^{-2x}\cos x & e^{-2x}\sin x \\ -2e^{-2x}\cos x-e^{-2x}\sin x & -2e^{-2x}\sin x+e^{-2x}\cos x\end{vmatrix}=$ $e^{-4x} \Rightarrow v'_1=-\frac{u_2\,F(x)}{D}=-10\,e^{2x}\sin x \Rightarrow v_1=\int -10\,e^{2x}\sin x\,dx=-2\,e^{2x}(2\sin x-\cos x)+C_1.$ $v'_2=\frac{u_1\,F(x)}{D}=10\,e^{2x}\cos x \Rightarrow v_2=\int 10\,e^{2x}\cos x\,dx=10\left(\frac{e^{2x}}{5}(2\cos x+\sin x)\right)+C_2 \Rightarrow$

9. (Continued)

$y = \left(-2\,e^{2x}(2 \sin x - \cos x) + C_1\right)e^{-2x} \cos x + \left(2\,e^{2x}(2 \cos x + \sin x) + C_2\right)e^{-2x} \sin x =$
$2 + C_1\,e^{-2x} \cos x + C_2\,e^{-2x} \sin x$

10. $\frac{d^2y}{dx^2} - \frac{dy}{dx} = 0 \Rightarrow r^2 - r = 0 \Rightarrow r_1 = 0, r_2 = 1 \Rightarrow y_h = C_1 + C_2\,e^x \Rightarrow u_1 = 1, u_2 = e^x \Rightarrow$

$D = \begin{vmatrix} 1 & e^x \\ 0 & e^x \end{vmatrix} = e^x \Rightarrow v'_1 = -\frac{u_2\,F(x)}{D} = -2^x \Rightarrow v_1 = \int -2^x\,dx = -\frac{2^x}{\ln 2} + C_1.\ v'_2 = \frac{u_1\,F(x)}{D} = \left(\frac{2}{e}\right)^x \Rightarrow$

$v_2 = \int \left(\frac{2}{e}\right)^x dx = \left(\frac{2}{e}\right)^x \left(\frac{1}{\ln 2 - 1}\right) + C_2 \Rightarrow y = \left(-\frac{2^x}{\ln 2} + C_1\right) + e^x\left(\left(\frac{2}{e}\right)^x\left(\frac{1}{\ln 2 - 1}\right) + C_2\right) =$
$-\frac{2^x}{\ln 2} + C_1 + \frac{2^x}{\ln 2 - 1} + C_2\,e^x$

11. $\frac{d^2y}{dx^2} + y = 0 \Rightarrow r^2 + 1 = 0 \Rightarrow r = \pm i \Rightarrow y_h = C_1 \cos x + C_2 \sin x \Rightarrow u_1 = \cos x, u_2 = \sin x \Rightarrow$

$D = \begin{vmatrix} \cos x & \sin x \\ -\sin x & \cos x \end{vmatrix} = 1 \Rightarrow v'_1 = -\frac{u_2\,F(x)}{D} = -\tan x \Rightarrow v_1 = \int -\tan x\,dx = -\ln(\sec x) + C_1.$

$v'_2 = \frac{u_1\,F(x)}{D} = 1 \Rightarrow v_2 = \int dx = x + C_2 \Rightarrow y = \left(-\ln(\sec x) + C_1\right)\cos x + \left(x + C_2\right)\sin x = \cos x \ln(\cos x)$
$+ x \sin x + C_1 \cos x + C_2 \sin x$

12. $\frac{d^2y}{dx^2} - \frac{dy}{dx} = 0 \Rightarrow r^2 - r = 0 \Rightarrow r_1 = 0, r_2 = 1 \Rightarrow y_h = C_1 + C_2\,e^x \Rightarrow u_1 = 1, u_2 = e^x \Rightarrow$

$D = \begin{vmatrix} 1 & e^x \\ 0 & e^x \end{vmatrix} = e^x \Rightarrow v'_1 = -\frac{u_2\,F(x)}{D} = -e^x \cos x \Rightarrow v_1 = \int -e^x \cos x\,dx = -\frac{e^x(\cos x + \sin x)}{2} + C_1.$

$v'_2 = \frac{u_1\,F(x)}{D} = \cos x \Rightarrow v_2 = \int \cos x\,dx = \sin x + C_2 \Rightarrow y = \left(-\frac{e^x(\cos x + \sin x)}{2} + C_1\right) + e^x\left(\sin x + C_2\right)$
$= \frac{e^x \sin x - e^x \cos x}{2} + C_1 + C_2\,e^x$

13. $\frac{d^2y}{dx^2} - 3\frac{dy}{dx} - 10y = 0 \Rightarrow r^2 - 3r - 10 = 0 \Rightarrow r_1 = 5, r_2 = -2 \Rightarrow y_h = C_1\,e^{5x} + C_2\,e^{-2x}.\ y_p = C \Rightarrow$
$\frac{d^2y}{dx^2} = 0, \frac{dy}{dx} = 0 \Rightarrow -10C = -3 \Rightarrow C = \frac{3}{10} \Rightarrow y_p = \frac{3}{10} \Rightarrow y = \frac{3}{10} + C_1\,e^{5x} + C_2\,e^{-2x}$

14. $\frac{d^2y}{dx^2} - 3\frac{dy}{dx} - 10y = 0 \Rightarrow r^2 - 3r - 10 = 0 \Rightarrow r_1 = 5, r_2 = -2 \Rightarrow y_h = C_1\,e^{5x} + C_2\,e^{-2x}.\ y_p = Dx + E \Rightarrow$
$y'_p = D, y''_p = 0 \Rightarrow -3D - 10(Dx + E) = 2x - 3 \Rightarrow D = -\frac{1}{5}, E = \frac{9}{25} \Rightarrow y_p = -\frac{1}{5}x + \frac{9}{25} \Rightarrow$
$y = C_1\,e^{5x} + C_2\,e^{-2x} - \frac{1}{5}x + \frac{9}{25}$

15. $\frac{d^2y}{dx^2} - \frac{dy}{dx} = 0 \Rightarrow r^2 - r = 0 \Rightarrow r_1 = 0, r_2 = 1 \Rightarrow y_h = C_1 + C_2\,e^x.\ y_p = B \cos x + C \sin x \Rightarrow \frac{dy}{dx} = -B \sin x +$
$C \cos x, \frac{d^2y}{dx^2} = -B \cos x - C \sin x \Rightarrow -B \cos x - C \sin x + B \sin x - C \cos x = \sin x \Rightarrow B = \frac{1}{2}, C = -\frac{1}{2} \Rightarrow$
$y_p = \frac{1}{2}\cos x - \frac{1}{2}\sin x \Rightarrow y = \frac{1}{2}\cos x - \frac{1}{2}\sin x + C_1 + C_2\,e^x$

16. $\frac{d^2y}{dx^2} + 2\frac{dy}{dx} + y = 0 \Rightarrow r^2 + 2r + 1 = 0 \Rightarrow r = -1 \Rightarrow y_h = C_1x\,e^{-x} + C_2\,e^{-x}$. $y_p = Dx^2 + Ex + F \Rightarrow$ $y'_p = 2Dx + E$, $y''_p = 2D \Rightarrow 2D + 2(2Dx + E) + Dx^2 + Ex + F = x^2 \Rightarrow D = 1, E = -4, F = 6 \Rightarrow$ $y_p = x^2 - 4x + 6 \Rightarrow y = C_1x\,e^{-x} + C_2\,e^{-x} + x^2 - 4x + 6$

17. $\frac{d^2y}{dx^2} + y = 0 \Rightarrow r^2 + 1 = 0 \Rightarrow r = \pm i \Rightarrow y_h = C_1 \cos x + C_2 \sin x$. $y_p = B\cos 3x + C\sin 3x \Rightarrow y'_p =$ $-3B\sin 3x + 3C\cos 3x \Rightarrow y''_p = -9B\cos 3x - 9C\sin 3x \Rightarrow -9B\cos 3x - 9C\sin 3x + B\cos 3x +$ $C\sin 3x = \cos 3x \Rightarrow B = -\frac{1}{8}, C = 0 \Rightarrow y_p = -\frac{1}{8}\cos 3x \Rightarrow y = -\frac{1}{8}\cos 3x + C_1\cos x + C_2\sin x$

18. $\frac{d^2y}{dx^2} + y = 0 \Rightarrow r^2 + 1 = 0 \Rightarrow r = \pm i \Rightarrow y_h = C_1\cos x + C_2\sin x$. $y_p = A\,e^{2x} \Rightarrow y'_p = 2A\,e^{2x} \Rightarrow y''_p =$ $4A\,e^{2x} \Rightarrow 4A\,e^{2x} + A\,e^{2x} = e^{2x} \Rightarrow A = \frac{1}{5} \Rightarrow y_p = \frac{1}{5}e^{2x} \Rightarrow y = C_1\cos x + C_2\sin x + \frac{1}{5}e^{2x}$

19. $\frac{d^2y}{dx^2} - \frac{dy}{dx} - 2y = 0 \Rightarrow r^2 - r - 2 = 0 \Rightarrow r_1 = 2, r_2 = -1 \Rightarrow y_h = C_1\,e^{2x} + C_2\,e^{-x}$. $y_p = B\cos x + C\sin x \Rightarrow$ $y'_p = -B\sin x + C\cos x \Rightarrow y''_p = -B\cos x - C\sin x \Rightarrow -B\cos x - C\sin x - (-B\sin x + C\cos x) -$ $2(B\cos x + C\sin x) = 20\cos x \Rightarrow B = -6, C = -2 \Rightarrow y_p = -6\cos x - 2\sin x \Rightarrow y = -6\cos x - 2\sin x +$ $C_1\,e^{2x} + C_2\,e^{-x}$

20. $\frac{d^2y}{dx^2} + y = 0 \Rightarrow r^2 + 1 = 0 \Rightarrow r = \pm i \Rightarrow y_h = C_1\cos x + C_2\sin x$. $y_p = D\,x + A\,e^x \Rightarrow y'_p = D + A\,e^x \Rightarrow$ $y''_p = A\,e^x \Rightarrow A\,e^x + D\,x + A\,e^x = 2x + 3\,e^x \Rightarrow A = \frac{3}{2}, D = 2 \Rightarrow y_p = 2x + \frac{3}{2}e^x \Rightarrow y = C_1\cos x + C_2\sin x$ $+ 2x + \frac{3}{2}e^x$

21. $\frac{d^2y}{dx^2} - y = 0 \Rightarrow r^2 - 1 = 0 \Rightarrow r_1 = 1, r_2 = -1 \Rightarrow y_h = C_1\,e^x + C_2\,e^{-x}$. $y_p = Ax\,e^x + Dx^2 + Ex + F \Rightarrow$ $y'_p = A\,e^x + Ax\,e^x + 2Dx + E \Rightarrow y''_p = 2A\,e^x + Ax\,e^x + 2D \Rightarrow 2A\,e^x + Ax\,e^x + 2D -$ $\left(Ax\,e^x + Dx^2 + Ex + F\right) = e^x + x^2 \Rightarrow A = \frac{1}{2}, D = -1, E = 0, F = -2 \Rightarrow y_p = \frac{1}{2}x\,e^x - x^2 - 2 \Rightarrow$ $y = \frac{1}{2}x\,e^x - x^2 - 2 + C_1\,e^x + C_2\,e^{-x}$

22. $\frac{d^2y}{dx^2} + 2\frac{dy}{dx} + y = 0 \Rightarrow r^2 + 2r + 1 = 0 \Rightarrow r = -1 \Rightarrow y_h = C_1x\,e^{-x} + C_2\,e^{-x}$. $y_p = B\cos 2x + C\sin 2x \Rightarrow$ $y'_p = -2B\sin 2x + 2C\cos 2x \Rightarrow y''_p = -4B\cos 2x - 4C\sin 2x \Rightarrow -4B\cos 2x - 4C\sin 2x +$ $2(-2B\sin 2x + 2C\cos 2x) + B\cos 2x + C\sin 2x = 6\sin 2x \Rightarrow B = -\frac{24}{25}, C = -\frac{18}{25} \Rightarrow y_p = -\frac{24}{25}\cos 2x -$ $\frac{18}{25}\sin 2x \Rightarrow y = C_1x\,e^{-x} + C_2\,e^{-x} - \frac{24}{25}\cos 2x - \frac{18}{25}\sin 2x$

23. $\frac{d^2y}{dx^2} - \frac{dy}{dx} - 6y = 0 \Rightarrow r^2 - r - 6 = 0 \Rightarrow r_1 = 3, r_2 = -2 \Rightarrow y_h = C_1\,e^{3x} + C_2\,e^{-2x}$. $y_p = A\,e^{-x} + B\cos x +$ $D\sin x \Rightarrow y'_p = -A\,e^{-x} - B\sin x + C\cos x \Rightarrow y''_p = A\,e^{-x} - B\cos x - C\sin x \Rightarrow A\,e^{-x} - B\cos x -$ $C\sin x - \left(-A\,e^{-x} - B\sin x + C\cos x\right) - 6\left(A\,e^{-x} + B\cos x + C\sin x\right) = e^{-x} - 7\cos x \Rightarrow A = -\frac{1}{4},$

23. (Continued)
$B = \frac{49}{50}$, $C = \frac{7}{50} \Rightarrow y_p = -\frac{1}{4} e^{-x} + \frac{49}{50} \cos x + \frac{7}{50} \sin x \Rightarrow y = -\frac{1}{4} e^{-x} + \frac{49}{50} \cos x + \frac{7}{50} \sin x + C_1 e^{3x} + C_2 e^{-2x}$

24. $\frac{d^2y}{dx^2} + 3\frac{dy}{dx} + 2y = 0 \Rightarrow r^2 + 3r + 2 = 0 \Rightarrow r_1 = -2, r_2 = -1 \Rightarrow y_h = C_1 e^{-2x} + C_2 e^{-x}$. $y_p = Ax e^{-2x} + Bx e^{-x} + C x + D \Rightarrow y'_p = -2Ax e^{-2x} + A e^{-2x} + B e^{-x} - B e^{-x} + C \Rightarrow y''_p = 4Ax e^{-2x} - 4A e^{-2x} - 2B e^{-x} + Bx e^{-x} \Rightarrow 4Ax e^{-2x} - 4A e^{-2x} - 2B e^{-x} + Bx e^{-x} + 3\left(-2Ax e^{-2x} + A e^{-2x} + B e^{-x} - Bx e^{-x} + C\right) + 2\left(Ax e^{-2x} + Bx e^{-x} + Cx + D\right) = e^{-x} + e^{-2x} - x \Rightarrow$
$A = -1, B = 1, C = -\frac{1}{2}, D = \frac{3}{4} \Rightarrow y_p = -x e^{-2x} + x e^{-x} - \frac{1}{2} x + \frac{3}{4} \Rightarrow y = -x e^{-2x} + x e^{-x} - \frac{1}{2} x + \frac{3}{4} + C_1 e^{-2x} + C_2 e^{-x}$

25. $\frac{d^2y}{dx^2} + 5\frac{dy}{dx} = 0 \Rightarrow r^2 + 5r = 0 \Rightarrow r_1 = 0, r_2 = -5 \Rightarrow y_h = C_1 + C_2 e^{-5x}$. $y_p = Dx^3 + Ex^2 + Fx \Rightarrow y'_p = 3Dx^2 + 2Ex + F \Rightarrow y''_p = 6Dx + 2E \Rightarrow 6Dx + 2E + 5\left(3Dx^2 + 2Ex + F\right) = 15x^2 \Rightarrow D = 1, E = -\frac{3}{5}$, $F = \frac{6}{25} \Rightarrow y_p = x^3 - \frac{3}{5} x^2 + \frac{6}{25} x \Rightarrow y = x^3 - \frac{3}{5} x^2 + \frac{6}{25} x + C_1 + C_2 e^{-5x}$

26. $\frac{d^2y}{dx^2} - \frac{dy}{dx} = 0 \Rightarrow r^2 - r = 0 \Rightarrow r_1 = 0, r_2 = 1 \Rightarrow y_h = C_1 + C_2 e^x$. $y_p = Dx^2 + Ex \Rightarrow y'_p = 2Dx + E \Rightarrow y''_p = 2D \Rightarrow 2D - (2Dx + E) = -8x + 3 \Rightarrow D = 4, E = 5 \Rightarrow y_p = 4x^2 + 5x \Rightarrow y = C_1 + C_2\,{}^{x} + 4x^2 + 5x$

27. $\frac{d^2y}{dx^2} - 3\frac{dy}{dx} = 0 \Rightarrow r^2 - 3r = 0 \Rightarrow r_1 = 0, r_2 = 3 \Rightarrow y_h = C_1 + C_2 e^{3x}$. $y_p = Ax e^{3x} + Dx^2 + Ex \Rightarrow y'_p = A e^{3x} + 3Ax e^{3x} + 2Dx + E \Rightarrow y''_p = 6A e^{3x} + 9Ax e^{3x} + 2D \Rightarrow 6A e^{3x} + 9Ax e^{3x} + 2D - 3\left(A e^{3x} + 3Ax e^{3x} + 2Dx + E\right) = e^{3x} - 12x \Rightarrow A = \frac{1}{3}, D = 2, E = \frac{4}{3} \Rightarrow y_p = \frac{1}{3} x e^{3x} + 2x^2 + \frac{4}{3} x \Rightarrow$
$y = \frac{1}{3} x e^{3x} + 2x^2 + \frac{4}{3} x + C_1 + C_2 e^{3x}$

28. $\frac{d^2y}{dx^2} + 7\frac{dy}{dx} = 0 \Rightarrow r^2 + 7r = 0 \Rightarrow r_1 = 0, r_2 = -7 \Rightarrow y_h = C_1 + C_2 e^{-7x}$. $y_p = Dx^3 + Ex^2 + Fx \Rightarrow y'_p = 3Dx^2 + 2Ex + F \Rightarrow y''_p = 6Dx + 2E \Rightarrow 6Dx + 2E + 7\left(3Dx^2 + 2Ex + F\right) = 42x^2 + 5x + 1 \Rightarrow D = 2$, $E = -\frac{1}{2}, F = \frac{2}{7} \Rightarrow y_p = 2x^3 - \frac{1}{2} x^2 + \frac{2}{7} x \Rightarrow y = 2x^3 - \frac{1}{2} x^2 + \frac{2}{7} x + C_1 + C_2 e^{-7x}$

29. $\frac{d^2y}{dx^2} - 5\frac{dy}{dx} = 0 \Rightarrow r^2 - 5r = 0 \Rightarrow r_1 = 0, r_2 = 5 \Rightarrow y_h = C_1 + C_2 e^{5x}$. $y_p = Ax^2 e^{5x} + Bx e^{5x} \Rightarrow y'_p = (2A + 5B)x e^{5x} + 5Ax^2 e^{5x} + B e^{5x} \Rightarrow y''_p = (2A + 10B)e^{5x} + (20A + 25B)x e^{5x} + 25Ax^2 e^{5x} \Rightarrow (2A + 10B)e^{5x} + (20A + 25B)x e^{5x} + 25Ax^2 e^{5x} - 5\left((2A + 5B)e^{5x} + 5Ax^2 e^{5x} + B e^{5x}\right) = x e^{5x} \Rightarrow$
$A = \frac{1}{10}, B = -\frac{1}{25} \Rightarrow y_p = \frac{1}{10} x^2 e^{5x} - \frac{1}{25} x e^{5x} \Rightarrow y = \frac{1}{10} x^2 e^{5x} - \frac{1}{25} x e^{5x} + C_1 + C_2 e^{5x}$

30. $\frac{d^2y}{dx^2} - \frac{dy}{dx} = 0 \Rightarrow r^2 - r = 0 \Rightarrow r_1 = 0,\ r_2 = 1 \Rightarrow y_h = C_1 + C_2\, e^x$. $y_p = A\cos x + B\sin x \Rightarrow y'_p = -A\sin x + B\cos x \Rightarrow y''_p = -A\cos x - B\sin x \Rightarrow -A\cos x - B\sin x - (-A\sin x + B\cos x) = \cos x + \sin x \Rightarrow A = 0,\ B = -1 \Rightarrow y_p = -\sin x \Rightarrow y = C_1 + C_2\, e^x - \sin x$

31. $\frac{d^2y}{dx^2} + y = 0 \Rightarrow r^2 + 1 = 0 \Rightarrow r = \pm i \Rightarrow y_h = C_1\cos x + C_2\sin x$. $y_p = Ax\cos x + Bx\sin x \Rightarrow y'_p = A\cos x - Ax\sin x + B\sin x + Bx\cos x \Rightarrow y''_p = -2A\sin x - Ax\cos x + 2B\cos x - Bx\sin x \Rightarrow -2A\sin x - Ax\cos x + 2B\cos x - Bx\sin x + Ax\cos x + Bx\sin x = 2\cos x + \sin x \Rightarrow A = -\frac{1}{2},\ B = 1 \Rightarrow y_p = -\frac{1}{2}x\cos x + x\sin x \Rightarrow y = -\frac{1}{2}x\cos x + x\sin x + C_1\cos x + C_2\sin x$

32. a) $\frac{d^2y}{dx^2} - 4\frac{dy}{dx} + 4y = 0 \Rightarrow r^2 - 4r + 4 = 0 \Rightarrow r = 2 \Rightarrow y_h = C_1 x\, e^{2x} + C_2\, e^{2x} \Rightarrow u_1 = x\, e^{2x},\ u_2 = e^{2x} \Rightarrow$

$D = \begin{vmatrix} xe^{2x} & e^{2x} \\ e^{2x} + 2xe^{2x} & 2e^{2x} \end{vmatrix} = -e^{4x} \Rightarrow v'_1 = -\frac{u_2\,F(x)}{D} = 2 \Rightarrow v_1 = \int 2\,dx = 2x + C_1.\ v'_2 = \frac{u_1\,F(x)}{D} = -2x \Rightarrow v_2 = \int -2x\,dx = -x^2 + C_2 \Rightarrow y = x\,e^{2x}\left(2x + C_1\right) + e^{2x}\left(-x^2 + C_2\right) = C_1 x\, e^{2x} + C_2\, e^{2x} + x^2\, e^{2x}$

b) $y_h = C_1 x\, e^{2x} + C_2\, e^{2x}$. $y_p = Ax^2\, e^{2x} \Rightarrow y'_p = 2Ax\, e^{2x} + 2Ax^2\, e^{2x} \Rightarrow y''_p = 4Ax^2\, e^{2x} + 8Ax\, e^{2x} + 2A\, e^{2x} \Rightarrow 4Ax^2\, e^{2x} + 8Ax\, e^{2x} + 2A\, e^{2x} - 4\left(2Ax\, e^{2x} + 2Ax^2\, e^{2x}\right) + 4Ax^2\, e^{2x} = 2\, e^{2x} \Rightarrow A = 1 \Rightarrow y_p = x^2\, e^{2x} \Rightarrow y = C_1 x\, e^{2x} + C_2\, e^{2x} + x^2\, e^{2x}$

33. a) $\frac{d^2y}{dx^2} - \frac{dy}{dx} = 0 \Rightarrow r^2 - r = 0 \Rightarrow r_1 = 0,\ r_2 = 1 \Rightarrow y_h = C_1 + C_2\, e^x \Rightarrow u_1 = 1,\ u_2 = e^x \Rightarrow$

$D = \begin{vmatrix} 1 & e^x \\ 0 & e^x \end{vmatrix} = e^x \Rightarrow v'_1 = -\frac{u_2\,F(x)}{D} = -e^x - e^{-x} \Rightarrow v_1 = \int -e^x - e^{-x}\,dx = -e^x + e^{-x} + C_1.$

$v'_2 = \frac{u_1\,F(x)}{D} = 1 + e^{-2x} \Rightarrow v_2 = \int \left(1 + e^{-2x}\right)dx = x - \frac{1}{2}e^{-2x} + C_2 \Rightarrow$

$y = \left(-e^x + e^{-x} + C_1\right) + \left(x - \frac{1}{2}e^{-2x} + C_2\right)e^x = \frac{1}{2}e^{-x} + x\, e^x + C_1 + C_3\, e^x$

b) $y_h = C_1 + C_2\, e^x$. $y_p = Ax\, e^x + B\, e^{-x} \Rightarrow y'_p = A\, e^x + Ax\, e^x - B\, e^{-x} \Rightarrow y''_p = 2A\, e^x + Ax\, e^x + B\, e^{-x} \Rightarrow 2A\, e^x + Ax\, e^x + B\, e^{-x} - \left(A\, e^x + Ax\, e^x - B\, e^{-x}\right) = e^x + e^{-x} \Rightarrow A = 1,\ B = \frac{1}{2} \Rightarrow y_p = x\, e^x + \frac{1}{2}e^{-x} \Rightarrow y = x\, e^x + \frac{1}{2}e^{-x} + C_1 + C_2\, e^x$

34. a) $\frac{d^2y}{dx^2} - 9\frac{dy}{dx} = 0 \Rightarrow r^2 - 9r = 0 \Rightarrow r_1 = 0,\ r_2 = 9 \Rightarrow y_h = C_1 + C_2\, e^{9x} \Rightarrow u_1 = 1,\ u_2 = e^{9x} \Rightarrow$

$D = \begin{vmatrix} 1 & e^{9x} \\ 0 & 9e^{9x} \end{vmatrix} = 9e^{9x} \Rightarrow v'_1 = -\frac{u_2\,F(x)}{D} = -e^{9x} \Rightarrow v_1 = \int -e^{9x}\,dx = -\frac{1}{9}e^{9x} + C_1.$

$v'_2 = \frac{u_1\,F(x)}{D} = 1 \Rightarrow v_2 = \int dx = x + C_2 \Rightarrow y = -\frac{1}{9}e^{9x} + C_1 + e^{9x}\left(x + C_2\right) = C_3\, e^{9x} + C_1 + x\, e^{9x}$

b) $y_h = C_1 + C_2\, e^{9x}$. $y_p = Ax\, e^{9x} \Rightarrow y'_p = A\, e^{9x} + 9Ax\, e^{9x} \Rightarrow y''_p = 18A\, e^{9x} + 81Ax\, e^{9x} \Rightarrow 18A\, e^{9x} + 81Ax\, e^{9x} - 9\left(A\, e^{9x} + 9Ax\, e^{9x}\right) = 9\, e^{9x} \Rightarrow A = 1 \Rightarrow y_p = x\, e^{9x} \Rightarrow y = C_1 + C_2\, e^{9x} + xe^{9x}$

35. a) $\frac{d^2y}{dx^2} - 4\frac{dy}{dx} - 5y = 0 \Rightarrow r^2 - 4r - 5 = 0 \Rightarrow r_1 = 5, r_2 = -1 \Rightarrow y_h = C_1 e^{5x} + C_2 e^{-x} \Rightarrow u_1 = e^{5x}$,

$u_2 = e^{-x} \Rightarrow D = \begin{vmatrix} e^{5x} & e^{-x} \\ 5e^{5x} & -e^{-x} \end{vmatrix} = -6e^{4x} \Rightarrow v'_1 = -\frac{u_2 F(x)}{D} = \frac{1}{6}e^{-4x} + \frac{2}{3}e^{-5x} \Rightarrow$

$v_1 = \int \left(\frac{1}{6}e^{-4x} + \frac{2}{3}e^{-5x}\right) dx = -\frac{1}{24}e^{-4x} - \frac{2}{15}e^{-5x} + C_1.\ v'_2 = \frac{u_1 F(x)}{D} = -\frac{1}{6}e^{2x} - \frac{2}{3}e^{x} \Rightarrow$

$v_2 = \int \left(-\frac{1}{6}e^{2x} - \frac{2}{3}e^{x}\right) dx = -\frac{1}{12}e^{2x} - \frac{2}{3}e^{x} + C_2 \Rightarrow y = \left(-\frac{1}{24}e^{-4x} - \frac{2}{15}e^{-5x} + C_1\right)e^{5x} +$

$\left(-\frac{1}{12}e^{2x} - \frac{2}{3}e^{x} + C_2\right)e^{-x} = -\frac{1}{8}e^{x} - \frac{4}{5} + C_1 e^{5x} + C_2 e^{-x}$

b) $y_h = C_1 e^{5x} + C_2 e^{-x}.\ y_p = A e^{x} + Bx + C \Rightarrow y'_p = A e^{x} + B \Rightarrow y''_p = A e^{x} \Rightarrow A e^{x} - 4(A e^{x} + B) -$

$5(A e^{x} + Bx + C) = e^{x} + 4 \Rightarrow A = -\frac{1}{8}, B = 0, C = -\frac{4}{5} \Rightarrow y_p = -\frac{1}{8}e^{x} - \frac{4}{5} \Rightarrow y = -\frac{1}{8}e^{x} - \frac{4}{5} + C_1 e^{5x} +$

$C_2 e^{-x}$

36. $\frac{d^2y}{dx^2} + y = 0 \Rightarrow r^2 + 1 = 0 \Rightarrow r = \pm i \Rightarrow y_h = C_1 \cos x + C_2 \sin x \Rightarrow u_1 = \cos x, u_2 = \sin x \Rightarrow$

$D = \begin{vmatrix} \cos x & \sin x \\ -\sin x & \cos x \end{vmatrix} = 1 \Rightarrow v'_1 = -\frac{u_2 F(x)}{D} = -1 \Rightarrow v_1 = \int -dx = -x + C_1.\ v'_2 = \frac{u_1 F(x)}{D} = \cot x \Rightarrow$

$v_2 = \int \cot x\, dx = -\ln(\cos x) + C_2 \Rightarrow y = (-x + C_1)\cos x + \sin x(-\ln(\cos x) + C_2) = -x\cos x -$

$\sin x \ln(\cos x)$

37. $\frac{d^2y}{dx^2} + y = 0 \Rightarrow r^2 + 1 = 0 \Rightarrow r = \pm i \Rightarrow y_h = C_1 \cos x + C_2 \sin x \Rightarrow u_1 = \cos x, u_2 = \sin x \Rightarrow$

$D = \begin{vmatrix} \cos x & \sin x \\ -\sin x & \cos x \end{vmatrix} = 1 \Rightarrow v'_1 = -\frac{u_2 F(x)}{D} = -\cos x \Rightarrow v_1 = \int -\cos x\, dx = -\sin x + C_1.$

$v'_2 = \frac{u_1 F(x)}{D} = \csc x - \sin x \Rightarrow v_2 = \int (\csc x - \sin x)\, dx = -\ln|\csc x + \cot x| + \cos x + C_2 \Rightarrow$

$y = (-\sin x + C_1)\cos x + (-\ln|\csc x + \cot x| + \cos x + C_2)\sin x = C_1\cos x + C_2 \sin x -$

$\sin x(\ln|\csc x + \cot x|)$

38. $\frac{d^2y}{dx^2} + 4y = 0 \Rightarrow r^2 + 4 = 0 \Rightarrow r = \pm 2i \Rightarrow y_h = C_1 \cos 2x + C_2 \sin 2x.\ y_p = B\cos x + C\sin x \Rightarrow$

$y'_p = -B\sin x + C\cos x \Rightarrow y''_p = -B\cos x - C\sin x \Rightarrow -B\cos x - C\sin x + 4(B\cos x + C\sin x) = \sin x$

$\Rightarrow B = 0, C = \frac{1}{3} \Rightarrow y_p = \frac{1}{3}\sin x \Rightarrow y = C_1 \cos 2x + C_2 \sin 2x + \frac{1}{3}\sin x$

39. $\frac{d^2y}{dx^2} - 8\frac{dy}{dx} = 0 \Rightarrow r^2 - 8r = 0 \Rightarrow r_1 = 0, r_2 = 8 \Rightarrow y_h = C_1 + C_2 e^{8x}\quad y_p = Ax e^{8x} \Rightarrow y'_p = A e^{8x} + 8Ax e^{8x}$

$\Rightarrow y''_p = 16A e^{8x} + 64Ax e^{8x} \Rightarrow 16A e^{8x} + 64Ax e^{8x} - 8(A e^{8x} + 8Ax e^{8x}) = e^{8x} \Rightarrow A = \frac{1}{8} \Rightarrow$

$y_p = \frac{1}{8}x e^{8x} \Rightarrow y = \frac{1}{8}x e^{8x} + C_1 + C_2 e^{8x}$

40. $\frac{d^2y}{dx^2} + 4\frac{dy}{dx} + 5y = 0 \Rightarrow r^2 + 4r + 5 = 0 \Rightarrow r = -2 \pm i \Rightarrow y_h = e^{-2x}\left(C_1 \cos x + C_2 \sin x\right)$. $y_p = Ax + B \Rightarrow$
$y'_p = A \Rightarrow y''_p = 0 \Rightarrow 4A + 5(Ax + B) = x + 2 \Rightarrow A = \frac{1}{5}, B = \frac{6}{25} \Rightarrow y_p = \frac{1}{5}x + \frac{6}{25} \Rightarrow$
$y = e^{-2x}\left(C_1 \cos x + C_2 \sin x\right) + \frac{1}{5}x + \frac{6}{25}$

41. $\frac{d^2y}{dx^2} - \frac{dy}{dx} = 0 \Rightarrow r^2 - r = 0 \Rightarrow r_1 = 0, r_2 = 1 \Rightarrow y_h = C_1 + C_2 e^x$. $y_p = Dx^4 + Ex^3 + Fx^2 + Gx \Rightarrow$
$y'_p = 4Dx^3 + 3Ex^2 + 2Fx + G \Rightarrow y''_p = 12Dx^2 + 6Ex + 2F \Rightarrow 12Dx^2 + 6Ex + 2F -$
$\left(4Dx^3 + 3Ex^2 + 2Fx + G\right) = x^3 \Rightarrow D = -\frac{1}{4}, E = -1, F = -3, G = -6 \Rightarrow y_p = -\frac{1}{4}x^4 - x^3 - 3x^2 - 6x \Rightarrow$
$y = -\frac{1}{4}x^4 - x^3 - 3x^2 - 6x + C_1 + C_2 e^x$

42. $\frac{d^2y}{dx^2} + 9y = 0 \Rightarrow r^2 + 9 = 0 \Rightarrow r = \pm 3i \Rightarrow y_h = C_1 \cos 3x + C_2 \sin 3x$. $y_p = B \cos x + C \sin x + Dx + E \Rightarrow$
$y'_p = -B \sin x + C \cos x + D \Rightarrow y''_p = -B \cos x - C \sin x \Rightarrow -B \cos x - C \sin x +$
$9(B \cos x + C \sin x + Dx + E) = 9x - \cos x \Rightarrow B = -\frac{1}{8}, C = 0, D = 1, E = 0 \Rightarrow y_p = -\frac{1}{8}\cos x + x \Rightarrow$
$y = C_1 \cos 3x + C_2 \sin 3x - \frac{1}{8}\cos x + x$

43. $\frac{d^2y}{dx^2} + 2\frac{dy}{dx} = 0 \Rightarrow r^2 + 2r = 0 \Rightarrow r_1 = 0, r_2 = -2 \Rightarrow y_h = C_1 + C_2 e^{-2x}$. $y_p = Ax^3 + Bx^2 + Cx + E e^x \Rightarrow$
$y'_p = 3Ax^2 + 2Bx + C + E e^x \Rightarrow y''_p = 6Ax + 2B + E e^x \Rightarrow 6Ax + 2B + E e^x + 2\left(3Ax^2 + 2Bx + C + E e^x\right)$
$= x^2 - e^x \Rightarrow A = \frac{1}{6}, B = -\frac{1}{4}, C = \frac{1}{4}, E = -\frac{1}{3} \Rightarrow y_p = \frac{1}{6}x^3 - \frac{1}{4}x^2 + \frac{1}{4}x - \frac{1}{3}e^x \Rightarrow y = \frac{1}{6}x^3 - \frac{1}{4}x^2 + \frac{1}{4}x -$
$\frac{1}{3}e^x + C_1 + C_2 e^{-2x}$

44. $\frac{d^2y}{dx^2} - 3\frac{dy}{dx} + 2y = 0 \Rightarrow r^2 - 3r + 2 = 0 \Rightarrow r_1 = 1, r_2 = 2 \Rightarrow y_h = C_1 e^x + C_2 e^{2x}$. $y_p = Ax e^x + Bx e^{2x} \Rightarrow$
$y'_p = A e^x + Ax e^x + B e^{2x} + 2Bx e^{2x} \Rightarrow y''_p = 2A e^x + Ax e^x + 4B e^{2x} + 4Bx e^{2x} \Rightarrow 2A e^x + Ax e^x +$
$4B e^{2x} + 4Bx e^{2x} - 3\left(A e^x + Ax e^x + B e^{2x} + 2Bx e^{2x}\right) + 2\left(Ax e^x + Bx e^{2x}\right) = e^x - e^{2x} \Rightarrow A = -1,$
$B = -1 \Rightarrow y_p = -x e^x - x e^{2x} \Rightarrow y = -x e^x - x e^{2x} + C_1 e^x + C_2 e^{2x}$

45. $\frac{d^2y}{dx^2} + y = 0 \Rightarrow r^2 + 1 = 0 \Rightarrow r = \pm i \Rightarrow y_h = C_1 \cos x + C_2 \sin x \Rightarrow u_1 = \cos x, u_2 = \sin x \Rightarrow$
$D = \begin{vmatrix} \cos x & \sin x \\ -\sin x & \cos x \end{vmatrix} = 1 \Rightarrow v'_1 = -\frac{u_2 F(x)}{D} = -\tan^2 x \Rightarrow v_1 = \int -\tan^2 x\, dx = -\tan x + x + C_1.$
$v'_2 = \frac{u_1 F(x)}{D} = \tan x \Rightarrow v_2 = \int \tan x\, dx = \ln|\sec x| + C_2 \Rightarrow y = \left(-\tan x + x + C_1\right)\cos x +$
$\left(\ln|\sec x| + C_2\right)\sin x \Rightarrow y = x \cos x + \sin x \ln(\sec x) + C_1 \cos x + C_3 \sin x$

46. $\frac{dy}{dx} + 4y = 0 \Rightarrow r + 4 = 0 \Rightarrow r = -4 \Rightarrow y_h = C_1 e^{-4x}$. $y_p = Dx + E \Rightarrow y'_p = D \Rightarrow D + 4(Dx + E) = x \Rightarrow D =$
$\frac{1}{4}, E = -\frac{1}{16} \Rightarrow y_p = \frac{1}{4}x - \frac{1}{16} \Rightarrow y = C_1 e^{-4x} + \frac{1}{4}x - \frac{1}{16}$

47. $\frac{dy}{dx} - 3y = 0 \Rightarrow r - 3 = 0 \Rightarrow r = 3 \Rightarrow y_h = C_1\, e^{3x}$. $y_p = A\, e^x \Rightarrow y'_p = A\, e^x \Rightarrow A\, e^x - 3A\, e^x = e^x \Rightarrow A = -\frac{1}{2} \Rightarrow y_p = -\frac{1}{2} e^x \Rightarrow y = C_1\, e^{3x} - \frac{1}{2} e^x$

48. $\frac{dy}{dx} + y = 0 \Rightarrow r + 1 = 0 \Rightarrow r = -1 \Rightarrow y_h = C_1\, e^{-x}$. $y_p = A \cos x + B \sin x \Rightarrow y'_p = -A \sin x + B \cos x \Rightarrow -A \sin x + B \cos x + A \cos x + B \sin x = \sin x \Rightarrow A = -\frac{1}{2},\ B = \frac{1}{2} \Rightarrow y_p = -\frac{1}{2} \cos x + \frac{1}{2} \sin x \Rightarrow y = C_1\, e^{-x} - \frac{1}{2} \cos x + \frac{1}{2} \sin x$

49. $\frac{dy}{dx} - 3y = 0 \Rightarrow r - 3 = 0 \Rightarrow r = 3 \Rightarrow y_h = C_1\, e^{3x}$. $y_p = Ax\, e^{3x} \Rightarrow y'_p = A\, e^{3x} + 3Ax\, e^{3x} \Rightarrow A\, e^{3x} + 3Ax\, e^{3x} - 3Ax\, e^{3x} = 5\, e^{3x} \Rightarrow A = 5 \Rightarrow y_p = 5x\, e^{3x} \Rightarrow y = C_1 e^{3x} + 5x\, e^{3x}$

50. $\frac{d^2y}{dx^2} + y = 0 \Rightarrow r^2 + 1 = 0 \Rightarrow r = \pm i \Rightarrow y_h = C_1 \cos x + C_2 \sin x$. $y_p = A\, e^{2x} \Rightarrow y'_p = 2A\, e^{2x} \Rightarrow y''_p = 4A\, e^{2x} \Rightarrow 4A\, e^{2x} + A\, e^{2x} = e^{2x} \Rightarrow A = \frac{1}{5} \Rightarrow y_p = \frac{1}{5} e^{2x} \Rightarrow y = C_1 \cos x + C_2 \sin x + \frac{1}{5} e^{2x}$. $y(0) = 0 \Rightarrow C_1 \cos 0 + C_2 \sin 0 + \frac{1}{5} e^0 = 0 \Rightarrow C_1 = -\frac{1}{5} \Rightarrow y = -\frac{1}{5} \cos x + C_2 \sin x + \frac{1}{5} e^{2x} \Rightarrow y' = \frac{1}{5} \sin x + C_2 \cos x + \frac{2}{5} e^{2x}$. $y'(0) = \frac{2}{5} \Rightarrow \frac{1}{5} \sin 0 + C_2 \cos 0 + \frac{2}{5} e^0 = \frac{2}{5} \Rightarrow C_2 = 0 \Rightarrow y = -\frac{1}{5} \cos x + \frac{1}{5} e^{2x}$

51. $\frac{d^2y}{dx^2} + y = 0 \Rightarrow r^2 + 1 = 0 \Rightarrow r = \pm i \Rightarrow y_h = C_1 \cos x + C_2 \sin x \Rightarrow u_1 = \cos x,\ u_2 = \sin x \Rightarrow$
$D = \begin{vmatrix} \cos x & \sin x \\ -\sin x & \cos x \end{vmatrix} = 1 \Rightarrow v'_1 = -\frac{u_2 F(x)}{D} = -\sec x \tan x \Rightarrow v_1 = \int -\sec x \tan x\, dx = -\sec x + C_1$. $v'_2 = \frac{u_1 F(x)}{D} = \sec x \Rightarrow v_2 = \int \sec x\, dx = \ln|\sec x + \tan x| + C_2 \Rightarrow y = C_1 \cos x + C_2 \sin x - 1 + \sin x \ln|\sec x + \tan x|$. $y(0) = 1 \Rightarrow C_1 \cos 0 + C_2 \sin 0 - 1 + \sin 0 \ln|\sec 0 + \tan 0| \Rightarrow C_1 = 2 \Rightarrow y = 2 \cos x + C_2 \sin x - 1 + \sin x \ln|\sec x + \tan x| \Rightarrow y' = -2 \sin x + C_2 \cos x + \sec x \sin x + \cos x \ln|\sec x + \tan x|$. $y'(0) = 1 \Rightarrow -2 \sin 0 + C_2 \cos 0 + \sec 0 \sin 0 + \cos 0 \ln|\sec 0 + \tan 0| \Rightarrow C_2 = 1 \Rightarrow y = 2 \cos x + \sin x - 1 + \sin x \ln|\sec x + \tan x|$

52. Step 1: $\frac{1}{y^2}\frac{dy}{dx} + \frac{1}{y} = x^2$ Step 2: Let $u = \frac{1}{y} \Rightarrow \frac{du}{dx} = -\frac{1}{y^2}\frac{dy}{dx} \Rightarrow -\frac{du}{dx} + u = x^2 \Rightarrow \frac{du}{dx} - u = -x^2 \Rightarrow$
$\frac{du}{dx} - u = 0 \Rightarrow r - 1 = 0 \Rightarrow r = 1 \Rightarrow u_h = C_1\, e^x$. $u_p = Ax^2 + Bx + C \Rightarrow u'_p = 2Ax + B \Rightarrow 2Ax + B - (Ax^2 + Bx + C) = -x^2 \Rightarrow A = 1, B = 2, C = 2 \Rightarrow u_p = x^2 + 2x + 2 \Rightarrow u = C_1\, e^x + x^2 + 2x + 2$
Step 3: $u = \frac{1}{y} \Rightarrow y = u^{-1} = \left(C_1\, e^x + x^2 + 2x + 2\right)^{-1}$

53. $y(x) + \int_0^x y(t)\, dt = x \Rightarrow \frac{dy}{dx} + y = 1 \Rightarrow \frac{dy}{dx} = 1 - y \Rightarrow \frac{dy}{1 - y} = dx \Rightarrow -\ln|1 - y| = x + C \Rightarrow 1 - y = C_1\, e^{-x} \Rightarrow y = C_1\, e^{-x} + 1$. $y(0) = 0 \Rightarrow 0 = C_1\, e^0 + 1 \Rightarrow C_1 = -1 \Rightarrow y = -e^{-x} + 1$

SECTION 16.6 OSCILLATION

1. $m\frac{d^2x}{dt^2} + kx = 0 \Rightarrow \frac{d^2x}{dt^2} + \frac{k}{m}x = 0$. Let $\omega = \sqrt{\frac{k}{m}}$. Then $\frac{d^2x}{dt^2} + \omega^2 x = 0 \Rightarrow r^2 + \omega^2 = 0 \Rightarrow r = \pm\,\omega i \Rightarrow$ $x = C_1 \cos \omega t + C_2 \sin \omega t$. $x(0) = x_0 \Rightarrow C_1 \cos 0 + C_2 \sin 0 = x_0 \Rightarrow C_1 = x_0 \Rightarrow x = x_0 \cos \omega t + C_2 \sin \omega t$ $\Rightarrow x' = -x_0\omega \sin \omega t + C_2\omega \cos \omega t$. $x'(0) = v_0 \Rightarrow -x_0\omega \sin 0 + C_2\omega \cos 0 = v_0 \Rightarrow C_2 = \frac{v_0}{\omega} \Rightarrow$ $x = x_0 \cos \omega t + \frac{v_0}{\omega} \sin \omega t$. $x = C \sin(\omega t + \phi)$ where $C = \sqrt{x_0^2 + \left(\frac{v_0}{\omega}\right)^2} = \frac{\sqrt{\omega^2 x_0^2 + v_0^2}}{\omega}$ and $\phi = \tan^{-1}\left(\frac{\omega x_0}{v_0}\right) \Rightarrow x = \frac{\sqrt{\omega^2 x_0^2 + v_0^2}}{\omega} \sin\left(\omega t + \tan^{-1}\left(\frac{\omega x_0}{v_0}\right)\right)$

2. $f(x) = kx \Rightarrow 5 \text{ lbs} = k(0.5 \text{ ft}) \Rightarrow k = 10$ lbs/ft. $x_0 = 0$ ft, $\frac{dx}{dt} = 4$ ft/sec when $t = 0 \Rightarrow \omega = \sqrt{\frac{k}{m}} = \sqrt{\frac{10 \text{ lbs/ft}}{5 \text{ lbs}/32 \text{ ft/sec}^2}} = 8$ sec. $x = C_1 \cos \omega t + C_2 \sin \omega t$. $x_0 = 0 \Rightarrow C_1 = 0 \Rightarrow x = C_2 \sin \omega t \Rightarrow$ $\frac{dx}{dt} = \omega C_2 \cos \omega t \Rightarrow 4 \text{ ft/sec} = (8 \text{ sec})C_2 \Rightarrow C_2 = 0.5 \text{ ft/sec}^2 \Rightarrow x = 0.5 \sin 8t$

3. a) $L\frac{d^2i}{dt^2} + R\frac{di}{dt} + \frac{1}{C}i = \frac{dv}{dt}$. $R = 0$, $\frac{1}{LC} = \omega^2$, v = constant $\Rightarrow L\frac{d^2i}{dt^2} + \frac{1}{C}i = 0 \Rightarrow \frac{d^2i}{dt^2} + \frac{1}{LC}i = 0 \Rightarrow$ $\frac{d^2i}{dt^2} + \omega^2 i = 0 \Rightarrow r = \pm\,\omega i \Rightarrow i = C_1 \cos \omega t + C_2 \sin \omega t$

   b) $L\frac{d^2i}{dt^2} + R\frac{di}{dt} + \frac{1}{C}i = \frac{dv}{dt}$. $R = 0$, $\frac{1}{LC} = \omega^2$, $v = V \sin \alpha t$, $\alpha \ne \omega \Rightarrow \frac{d^2i}{dt^2} + \omega^2 i = \frac{V\alpha}{L} \cos \alpha t \Rightarrow$ $i_h = C_1 \cos \omega t + C_2 \sin \omega t$. $i_p = A \cos \alpha t + B \sin \alpha t \Rightarrow i'_p = -A\alpha \sin \alpha t + B\alpha \cos \alpha t \Rightarrow$ $i''_p = -A\alpha^2 \cos \alpha t - B\alpha^2 \sin \alpha t \Rightarrow -A\alpha^2 \cos \alpha t - B\alpha^2 \sin \alpha t + \omega^2\left(A \cos \alpha t + B \sin \alpha t\right) =$ $\frac{V\alpha}{L} \cos \alpha t \Rightarrow A = \frac{V\alpha}{L\left(\omega^2 - \alpha^2\right)}$, $B = 0 \Rightarrow i_p = \frac{V\alpha}{L\left(\omega^2 - \alpha^2\right)} \cos \alpha t \Rightarrow i = C_1 \cos \omega t + C_2 \sin \omega t +$ $\frac{V\alpha}{L\left(\omega^2 - \alpha^2\right)} \cos \alpha t$

   c) $L\frac{d^2i}{dt^2} + R\frac{di}{dt} + \frac{1}{C}i = \frac{dv}{dt}$ $R = 0$, $\frac{1}{LC} = \omega^2$, $v = V \sin \omega t$, V constant $\Rightarrow L\frac{d^2i}{dt^2} + \frac{1}{C}i = V\omega \cos \omega t \Rightarrow$ $\frac{d^2i}{dt^2} + \omega^2 i = \frac{V\omega}{L} \cos \omega t \Rightarrow i_h = C_1 \cos \omega t + C_2 \sin \omega t$. $i_p = At \cos \omega t + Bt \sin \omega t \Rightarrow i'_p = A \cos \omega t -$ $A\omega t \sin \omega t + B \sin \omega t + B\omega t \cos \omega t \Rightarrow y''_p = -2A\omega \sin \omega t + 2B\omega \cos \omega t - A\omega^2 t \cos \omega t -$ $B\omega^2 t \sin \omega t \Rightarrow -2A\omega \sin \omega t + 2B\omega \cos \omega t - A\omega^2 t \cos \omega t - B\omega^2 t \sin \omega t +$ $\omega^2\left(At \cos \omega t + Bt \sin \omega t\right) = \frac{V\omega}{L} \cos \omega t \Rightarrow A = 0$, $B = \frac{V}{2L} \Rightarrow i_p = \frac{V}{2L} t \sin \omega t \Rightarrow i = C_1 \cos \omega t +$ $C_2 \sin \omega t + \frac{V}{2L} t \sin \omega t$

   d) $L\frac{d^2i}{dt^2} + R\frac{di}{dt} + \frac{1}{C}i = \frac{dv}{dt}$ $R = 50$, $L = 5$, $C = 9 \times 10^{-6}$, v constant $\Rightarrow 5\frac{d^2i}{dt^2} + 50\frac{di}{dt} + \frac{1}{9} \times 10^6\, i = 0 \Rightarrow$ $\frac{d^2i}{dt^2} + 10\frac{di}{dt} + \frac{1}{45} \times 10^6\, i = 0 \Rightarrow r^2 + 10r + \frac{1}{45} \times 10^6 = 0 \Rightarrow r = -5 \pm 5\sqrt{-\frac{7991}{9}} \approx -5 \pm 148.99\, i \Rightarrow$ $i = e^{-5t}\left(C_1 \cos(148.99)t + C_2 \sin(148.99)t\right)$

4. Assume $\theta \approx 0 \Rightarrow \sin\theta \approx \theta \Rightarrow \frac{d^2\theta}{dt^2} + \frac{g}{l}\theta = 0 \Rightarrow r^2 + \frac{g}{l} = 0 \Rightarrow r = \pm i\sqrt{\frac{g}{l}} \Rightarrow \theta = C_1 \cos\sqrt{\frac{g}{l}}\,t +$
$C_2 \sin\sqrt{\frac{g}{l}}\,t.$ $t = 0 \Rightarrow \theta = \theta_0 \Rightarrow \theta_0 = C_1 \Rightarrow \theta = \theta_0 \cos\sqrt{\frac{g}{l}}\,t + C_2 \sin\sqrt{\frac{g}{l}}\,t \Rightarrow \frac{d\theta}{dt} =$
$-\theta_0\sqrt{\frac{g}{l}}\sin\sqrt{\frac{g}{l}}\,t + C_2\sqrt{\frac{g}{l}}\cos\sqrt{\frac{g}{l}}\,t.$ $t = 0 \Rightarrow \frac{d\theta}{dt} = 0 \Rightarrow C_2 = 0 \Rightarrow \theta = \theta_0 \cos\sqrt{\frac{g}{l}}\,t$

5. $\frac{d^2\theta}{dt^2} = -\frac{2k\theta}{mr^2} \Rightarrow \frac{d^2\theta}{dt^2} + \frac{2k}{mr^2}\theta = 0 \Rightarrow r^2 + \frac{2k}{mr^2} = 0 \Rightarrow r = \pm\sqrt{\frac{2k}{mr^2}}\,i \Rightarrow \theta = C_1 \cos\sqrt{\frac{2k}{mr^2}}\,t +$
$C_2 \sin\sqrt{\frac{2k}{mr^2}}\,t.$ $t = 0 \Rightarrow \theta = \theta_0 \Rightarrow C_1 = \theta_0 \Rightarrow \theta = \theta_0 \cos\sqrt{\frac{2k}{mr^2}}\,t + C_2 \sin\sqrt{\frac{2k}{mr^2}}\,t \Rightarrow$
$\frac{d\theta}{dt} = -\theta_0\sqrt{\frac{2k}{mr^2}}\sin\sqrt{\frac{2k}{mr^2}}\,t + C_2\sqrt{\frac{2k}{mr^2}}\cos\sqrt{\frac{2k}{mr^2}}\,t.$ $\frac{d\theta}{dt} = v_0$ at $t = 0 \Rightarrow C_2\sqrt{\frac{2k}{mr^2}} = v_0 \Rightarrow$
$C_2 = v_0\sqrt{\frac{mr^2}{2k}} \Rightarrow \theta = \theta_0 \cos\sqrt{\frac{2k}{mr^2}}\,t + v_0\sqrt{\frac{mr^2}{2k}}\sin\sqrt{\frac{2k}{mr^2}}\,t$

6. $\frac{100}{g}\frac{d^2x}{dt^2} = -16\pi x - c\frac{dx}{dt} \Rightarrow \frac{d^2x}{dt^2} + \frac{gc}{100}\frac{dx}{dt} = -\frac{16\pi g}{100}x \Rightarrow r^2 + \frac{gc}{100}r + \frac{16\pi g}{100} = 0 \Rightarrow 100r^2 + gcr + 16\pi g = 0$
$\Rightarrow r = \frac{-4c \pm 4i\sqrt{200\pi - c^2}}{25} \Rightarrow \omega = \frac{4\sqrt{200\pi - c^2}}{25} \Rightarrow \text{Period} = \frac{2\pi}{\omega} = \frac{2\pi}{\frac{4\sqrt{200\pi - c^2}}{25}} = \frac{25\pi}{2\sqrt{200\pi - c^2}}$

$\therefore\ 1.6 = \frac{25\pi}{2\sqrt{200\pi - c^2}} \Rightarrow c = \sqrt{200\pi - \frac{6.25\pi^2}{10.24}} \approx 5.09$

7. a) $f(t) = A\sin\alpha t,\ \alpha \neq \sqrt{\frac{k}{m}} \Rightarrow \frac{d^2x}{dt^2} + \frac{k}{m}x = \frac{k}{m}(A\sin\alpha t) \Rightarrow r^2 + \frac{k}{m} = 0 \Rightarrow r = \pm i\sqrt{\frac{k}{m}} \Rightarrow$
$x_h = C_1\cos\sqrt{\frac{k}{m}}\,t + C_2\sin\sqrt{\frac{k}{m}}\,t.$ $x_p = B\cos\alpha t + C\sin\alpha t \Rightarrow x'_p = -B\alpha\sin\alpha t + C\alpha\cos\alpha t \Rightarrow$
$x''_p = -B\alpha^2\cos\alpha t - C\alpha^2\sin\alpha t \Rightarrow -B\alpha^2\cos\alpha t - C\alpha^2\sin\alpha t + \frac{k}{m}(B\cos\alpha t + C\sin\alpha t) =$
$\frac{k}{m}(A\sin\alpha t) \Rightarrow B = 0,\ C = \frac{Ak}{k - m\alpha^2} \Rightarrow x_p = \frac{Ak}{k - m\alpha^2}\sin\alpha t \Rightarrow x = \frac{Ak}{k - m\alpha^2}\sin\alpha t + C_1\cos\sqrt{\frac{k}{m}}\,t$
$+ C_2\sin\sqrt{\frac{k}{m}}\,t.$ $x(0) = x_0 \Rightarrow C_1 = x_0 \Rightarrow x = x_0\cos\sqrt{\frac{k}{m}}\,t + C_2\sin\sqrt{\frac{k}{m}}\,t + \frac{Ak}{k - m\alpha^2}\sin\alpha t \Rightarrow$
$\frac{dx}{dt} = -x_0\sqrt{\frac{k}{m}}\sin\sqrt{\frac{k}{m}}\,t + C_2\sqrt{\frac{k}{m}}\cos\sqrt{\frac{k}{m}}\,t + \frac{Ak\alpha}{k - m\alpha^2}\cos\alpha t.$ $x'(0) = 0 \Rightarrow C_2\sqrt{\frac{k}{m}} +$
$\frac{Ak\alpha}{k - m\alpha^2} = 0 \Rightarrow C_2 = -\frac{Ak\alpha}{k - m\alpha^2}\sqrt{\frac{m}{k}} \Rightarrow x = x_0\cos\sqrt{\frac{k}{m}}\,t - \frac{A\alpha\sqrt{mk}}{k - m\alpha^2}\sin\sqrt{\frac{k}{m}}\,t + \frac{Ak}{k - m\alpha^2}\sin\alpha t$

b) $f(t) = A\sin\alpha t,\ \alpha = \sqrt{\frac{k}{m}} \Rightarrow \frac{d^2x}{dt^2} + \frac{k}{m}x = \frac{k}{m}(A\sin\alpha t) \Rightarrow r^2 + \frac{k}{m} = 0 \Rightarrow r = \pm\sqrt{\frac{k}{m}}\,i \Rightarrow$
$x_h = C_1\cos\sqrt{\frac{k}{m}}\,t + C_2\sin\sqrt{\frac{k}{m}}\,t.$ $x_p = Bt\cos\sqrt{\frac{k}{m}}\,t + Ct\sin\sqrt{\frac{k}{m}}\,t \Rightarrow x'_p = B\cos\sqrt{\frac{k}{m}}\,t -$
$B\sqrt{\frac{k}{m}}\,t\sin\sqrt{\frac{k}{m}}\,t + C\sin\sqrt{\frac{k}{m}}\,t + C\sqrt{\frac{k}{m}}\,t\cos\sqrt{\frac{k}{m}}\,t \Rightarrow x''_p = -2B\sqrt{\frac{k}{m}}\sin\sqrt{\frac{k}{m}}\,t +$
$2C\sqrt{\frac{k}{m}}\cos\sqrt{\frac{k}{m}}\,t - B\left(\frac{k}{m}\right)t\cos\sqrt{\frac{k}{m}}\,t - C\left(\frac{k}{m}\right)t\sin\sqrt{\frac{k}{m}}\,t \Rightarrow -2B\sqrt{\frac{k}{m}}\sin\sqrt{\frac{k}{m}}\,t +$

7. b) (Continued)

$$2C\sqrt{\frac{k}{m}}\cos\sqrt{\frac{k}{m}}\,t - B\left(\frac{k}{m}\right)t\cos\sqrt{\frac{k}{m}}\,t - C\left(\frac{k}{m}\right)t\sin\sqrt{\frac{k}{m}}\,t + \frac{k}{m}\left(Bt\cos\sqrt{\frac{k}{m}}\,t + Ct\sin\sqrt{\frac{k}{m}}\,t\right)$$

$= \frac{kA}{m}\sin\alpha t \Rightarrow B = -\frac{A\alpha}{2}, C = 0 \Rightarrow x_p = -\frac{A\alpha}{2}t\cos\alpha t \Rightarrow x = C_1\cos\sqrt{\frac{k}{m}}\,t + C_2\sin\sqrt{\frac{k}{m}}\,t - \frac{A\alpha}{2}t\cos\alpha t$. $x(0) = x_0 \Rightarrow C_1 = x_0 \Rightarrow x = x_0\cos\sqrt{\frac{k}{m}}\,t + C_2\sin\sqrt{\frac{k}{m}}\,t - \frac{A\alpha}{2}t\cos\alpha t \Rightarrow$ $\frac{dx}{dt} = -x_0\alpha\sin\alpha t + C_2\alpha\cos\alpha t - \frac{A\alpha}{2}\cos\alpha t - \frac{A\alpha^2}{2}t\sin\alpha t$. $x'(0) = 0 \Rightarrow C_2\alpha - \frac{A\alpha}{2} = 0 \Rightarrow$ $C_2 = \frac{A}{2} \Rightarrow x = x_0\cos\alpha t + \frac{A}{2}\sin\alpha t - \frac{A\alpha}{2}t\cos\alpha t$

SECTION 16.7 NUMERICAL METHODS

1.

	x_n	y_n
x_0	0	1
x_1	1/5	1.2
x_2	2/5	1.44
x_3	3/5	1.728
x_4	4/5	2.0736
x_5	1	2.48832

Exact Value: $y' = y \Rightarrow \frac{dy}{dx} = y \Rightarrow \frac{dy}{y} = dx \Rightarrow \ln|y| = x + C \Rightarrow e^{\ln|y|} = e^{x+C} \Rightarrow |y| = e^C e^x \Rightarrow y = C_1 e^x$. $y(0) = 1 \Rightarrow C_1 = 1 \Rightarrow y = e^x \Rightarrow y(1) = e^1 = 2.718281828...$

2. $y' = y,\ y(0) = 1,\ h = \frac{1}{n} \Rightarrow x_{n+1} = x_n + h = x_n + \frac{1}{n},\ y_{n+1} = y_n + hy_n = y_n + \frac{1}{n}y_n = \left(1 + \frac{1}{n}\right)y_n$

$x_0 = 0 \Rightarrow y_0 = 1;\ x_1 = 0 + \frac{1}{n} = \frac{1}{n} \Rightarrow y_1 = \left(1 + \frac{1}{n}\right)1 = \left(1 + \frac{1}{n}\right);\ x_2 = \frac{1}{n} + \frac{1}{n} = \frac{2}{n} \Rightarrow y_2 = \left(1 + \frac{1}{n}\right)\left(1 + \frac{1}{n}\right) = \left(1 + \frac{1}{n}\right)^2;\ x_3 = \frac{3}{n} \Rightarrow y_3 = \left(1 + \frac{1}{n}\right)^3$; etc.

$\therefore$ let $n = 1 \Rightarrow y_1 = \left(1 + \frac{1}{n}\right)$. Assume $y_k = \left(1 + \frac{1}{n}\right)^k$ for some $k \ge 1$. Then $y_{k+1} = \left(1 + \frac{1}{n}\right)y_k = \left(1 + \frac{1}{n}\right)\left(1 + \frac{1}{n}\right)^k = \left(1 + \frac{1}{n}\right)^{k+1}$. $\therefore$ by mathematical induction, $y_n = \left(1 + \frac{1}{n}\right)^n$ for $n \ge 1$.

$y(1) = \lim_{n\to\infty} y_n = \lim_{n\to\infty}\left(1 + \frac{1}{n}\right)^n = e$

3.

	x_n	y_n
x_0	0	1
x_1	1/5	1.22
x_2	2/5	1.4884
x_3	3/5	1.815848
x_4	4/5	2.2153346
x_5	1	2.702708163

4.

	x_n		k_i		y_n
x_0	0.2	k_1	0.2		
		k_2	0.22		
		k_3	0.222		
		k_4	0.2444	y_1	1.2214
x_2	0.4	k_1	0.24428		
		k_2	0.268708		
		k_3	0.2711508		
		k_4	0.2985101	y_2	1.4918179
x_3	0.6	k_1	0.2983635		
		k_2	0.3281999		
		k_3	0.3311835		
		k_4	0.3646002	y_3	1.8221063
x_4	0.8	k_1	0.3644212		
		k_2	0.4008633		
		k_3	0.4045075		
		k_4	0.4453227	y_4	2.2255205
x_5	1	k_1	0.4451041		
		k_2	0.4896145		
		k_3	0.4940655		
		k_4	0.5439172	y_5	2.7182507

5. $y' = x^2 + y^2 \Rightarrow x_{n+1} = x_n + h \Rightarrow y_{n+1} = y_n + h\left(x_n^{\,2} + y_n^{\,2}\right)$. $Y' = Y^2 \Rightarrow x_{n+1} = x_n + h \Rightarrow Y_{n+1} = Y_n + h(Y_n^{\,2})$. $y_0 = h(x_0^{\,2} + y_0^{\,2}) = Y_0 + h(Y_0)$ since $x_0 = 0 \Rightarrow y_1 = Y_1$. $y_1 + h\left(x_1^{\,2} + y_1^{\,2}\right) > Y_1 + h(Y_1^{\,2})$ since $x_1 > 0 \Rightarrow y_2 > Y_2$. And from here on up to $x = 1$, $y_{n+1} > Y_{n+1}$.
$Y' = Y^2 \Rightarrow \frac{dY}{dx} = Y^2 \Rightarrow \frac{1}{Y^2}\,dY = dx \Rightarrow -\frac{1}{Y} = x + C \Rightarrow \frac{1}{Y} = -x - C \Rightarrow Y = -\frac{1}{x + C}$. $y(0) = 1 \Rightarrow$
$-\frac{1}{C} = 1 \Rightarrow C = -1 \Rightarrow Y = -\frac{1}{x - 1} \Rightarrow y \rightarrow \infty$ as $x \rightarrow 1^-$. Since $y_{n+1} > Y_{n+1}$, $y_{n+1} \rightarrow \infty$ as $x \rightarrow 1^-$

6. a) $\frac{dy}{dx} = 1 + y^2 \Rightarrow \frac{1}{1 + y^2}\,dy = dx \Rightarrow \tan^{-1} y = x + C \Rightarrow y = \tan(x + C)$. $y(0) = 0 \Rightarrow C = 0 \Rightarrow$
$y = \tan x$

b) Let $y = -\frac{u'}{u} \Rightarrow \frac{dy}{dx} = \frac{-uu'' + (u')^2}{u^2}$. Then $\frac{-uu'' + (u')^2}{u^2} = 1 + \left(-\frac{u'}{u}\right)^2 \Rightarrow u'' + u = 0 \Rightarrow r^2 + 1 = 0 \Rightarrow$
$r = \pm i \Rightarrow u = C_1 \cos x + C_2 \sin x \Rightarrow u' = -C_1 \sin x + C_2 \cos x$. Now $y(0) = 0 \Rightarrow -\frac{u'}{u} = 0$ when $x = 0$
$\Rightarrow u'(0) = 0$. Then $0 = -C_1 \sin 0 + C_2 \cos 0 \Rightarrow C_2 = 0$. Then $u = C_1 \cos x$ and $u' = -C_1 \sin x \Rightarrow$
$y = -\frac{u'}{u} = -\frac{-C_1 \sin x}{C_1 \cos x} = \tan x$

Note: For Exercises 7–12, the Calculus Tookit was used.

7. $y = 0.571428572$

8. $y = -0.175611085$

9. $y = 0.810263855$

10. $y = 0.810317873$

11. a) $y = 0.841470985$ b) $y = 0.841470983$

12. a) $y = 0.367879689$ b) $y = 0.367879688$

PRACTICE EXERCISES

1. $e^{y-2}\,dx - e^{x+2y}\,dy = 0 \Rightarrow e^{-x}\,dx - \frac{e^{2y}}{e^{y-2}}\,dy = 0 \Rightarrow e^{-x}\,dx - e^{y+2}\,dy = 0 \Rightarrow -e^{-x} - e^{y+2} = C.$ $y(0) = -2$ $\Rightarrow -e^{0} - e^{-2+2} = C \Rightarrow C = -2 \Rightarrow e^{-x} + e^{y+2} = 2$

2. $y \ln y\,dx + (1 + x^2)\,dy = 0 \Rightarrow \frac{dx}{1 + x^2} + \frac{dy}{y \ln y} = 0 \Rightarrow \tan^{-1} x + \ln(\ln y) = C.$ $y(0) = e \Rightarrow \tan^{-1} 0 + \ln(\ln e) =$ $C \Rightarrow C = 0 \Rightarrow \tan^{-1} x + \ln(\ln y) = 0$

3. $\frac{dy}{dx} = \frac{x^2 + y^2}{2xy} \Rightarrow \frac{dy}{dx} = \frac{1 + \left(\frac{y}{x}\right)^2}{2\left(\frac{y}{x}\right)} \Rightarrow$ Homogeneous. $v = \frac{y}{x} \Rightarrow F(v) = \frac{1 + v^2}{2v} \Rightarrow \frac{dx}{x} + \frac{dv}{v - \left(\frac{1 + v^2}{2v}\right)} = 0 \Rightarrow$ $\frac{dx}{x} + \frac{2v\,dv}{v^2 - 1} = 0 \Rightarrow \ln|x| + \ln|v^2 - 1| = C \Rightarrow x(v^2 - 1) = C_1 \Rightarrow x\left(\left(\frac{y}{x}\right)^2 - 1\right) = C_1 \Rightarrow \frac{y^2}{x} - x = C_1.$ $y(5) = 0$ $\Rightarrow 0 - 5 = C_1 \Rightarrow C_1 = -5 \Rightarrow \frac{y^2}{x} - x = -5$

4. $\frac{dy}{dx} = \frac{y(1 + \ln y - \ln x)}{x(\ln y - \ln x)} \Rightarrow \frac{dy}{dx} = \frac{y}{x}\left(\frac{1 + \ln \frac{y}{x}}{\ln \frac{y}{x}}\right) \Rightarrow$ Homogeneous. $v = \frac{y}{x} \Rightarrow F(v) = v\left(\frac{1 + \ln v}{\ln v}\right) \Rightarrow$ $\frac{dx}{x} + \frac{dv}{v - v\left(\frac{1 + \ln v}{\ln v}\right)} = 0 \Rightarrow \frac{dx}{x} + \frac{\ln v\,dv}{-v} = 0 \Rightarrow \ln|x| - (\ln|v|)^2 + C = 0 \Rightarrow \ln|x| - \left(\ln\left|\frac{y}{x}\right|\right)^2 + C = 0 \Rightarrow$ $\ln(x) - \left(\ln\left(\frac{y}{x}\right)\right)^2 + C = 0$ if $x,y > 0$. $y(1) = 1 \Rightarrow \ln 1 - (\ln 1)^2 + C = 0 \Rightarrow C = 0 \Rightarrow \ln x - \left(\ln\left(\frac{y}{x}\right)\right)^2 = 0$

5. $(x^2 + y)\,dx + \left(e^y + x\right)dy = 0 \Rightarrow \frac{\partial M}{\partial y} = 1 = \frac{\partial N}{\partial x} \Rightarrow$ Exact. $\frac{\partial f}{\partial x} = x^2 + y \Rightarrow f(x,y) = \frac{x^3}{3} + xy + k(y) \Rightarrow$ $\frac{\partial f}{\partial y} = x + k'(y) = e^y + x \Rightarrow k'(y) = e^y \Rightarrow k(y) = e^y + C \Rightarrow f(x,y) = \frac{x^3}{3} + xy + e^y + C \Rightarrow \frac{x^3}{3} + xy + e^y = C_1.$ $y(3) = 0 \Rightarrow 9 + 1 = C_1 \Rightarrow C_1 = 10 \Rightarrow \frac{x^3}{3} + xy + e^y = 10$

6. $(e^x + \ln y)\,dx + \left(\frac{x + y}{y}\right)dy = 0 \Rightarrow \frac{\partial M}{\partial y} = \frac{1}{y} = \frac{\partial N}{\partial x} \Rightarrow$ Exact. $\frac{\partial f}{\partial x} = e^x + \ln y \Rightarrow f(x,y) = e^x + x \ln y + k(y) \Rightarrow$ $\frac{\partial f}{\partial y} = \frac{x}{y} + k'(y) = \frac{x + y}{y} \Rightarrow k'(y) = 1 \Rightarrow k(y) = y + C \Rightarrow f(x,y) = e^x + x \ln y + y + C \Rightarrow e^x + x \ln y + y + C = 0.$ $y(\ln 2) = 1 \Rightarrow 2 + 1 + C = 0 \Rightarrow C = -3 \Rightarrow e^x + x \ln y + y - 3 = 0$

7. $(x+1)\frac{dy}{dx} + 2y = x \Rightarrow P(x) = 2,\ Q(x) = x \Rightarrow \int P(x)\,dx = \int 2\,dx = 2x \Rightarrow \rho(x) = e^{2x} \Rightarrow$
$y = \frac{1}{e^{2x}} \int e^{2x}(x\,dx) = \frac{1}{e^{2x}}\left(\frac{1}{2}x\,e^{2x} - \frac{1}{4}e^{2x} + C\right) \Rightarrow y = \frac{1}{2}x - \frac{1}{4} + C\,e^{-2x}.\ y(0) = 1 \Rightarrow 1 = -\frac{1}{4} + C \Rightarrow$
$C = \frac{5}{4} \Rightarrow y = \frac{1}{2}x - \frac{1}{4} + \frac{5}{4}e^{-2x}$

8. $x\frac{dy}{dx} + 2y = x^2 + 1 \Rightarrow \frac{dy}{dx} + \frac{2}{x}y = x + \frac{1}{x} \Rightarrow P(x) = \frac{2}{x},\ Q(x) = x + \frac{1}{x} \Rightarrow \int P(x)\,dx = \int \frac{2}{x}\,dx = 2\ln|x| =$
$\ln x^2 \Rightarrow \rho(x) = e^{\ln x^2} = x^2 \Rightarrow y = \frac{1}{x^2}\int x^2\left(x + \frac{1}{x}\right)dx = \frac{1}{x^2}\left(\frac{x^4}{4} + \frac{x^2}{2} + C\right) = \frac{x^2}{4} + \frac{1}{2} + \frac{C}{x^2}.\ y(1) = 1 \Rightarrow$
$1 = \frac{1}{4} + \frac{1}{2} + C \Rightarrow C = \frac{1}{4} \Rightarrow y = \frac{x^2}{4} + \frac{1}{2} + \frac{1}{4x^2}$

9. $\frac{d^2y}{dx^2} - \left(\frac{dy}{dx}\right)^2 = 1.$ Let $p = \frac{dy}{dx} \Rightarrow \frac{dp}{dx} - p^2 = 1 \Rightarrow \frac{dp}{1+p^2} = dx \Rightarrow \tan^{-1} p = x + C \Rightarrow p = \tan(x + C) \Rightarrow$
$\frac{dy}{dx} = \tan(x + C) \Rightarrow y = \ln|\sec(x + C)| + C_1.\ y'\left(\frac{\pi}{3}\right) = \sqrt{3} \Rightarrow \sqrt{3} = \tan\left(\frac{\pi}{3} + C\right) \Rightarrow \tan^{-1}\sqrt{3} = \frac{\pi}{3} + C \Rightarrow$
$\frac{\pi}{3} = \frac{\pi}{3} + C \Rightarrow C = 0 \Rightarrow y = \ln|\sec x| + C_1.\ y\left(\frac{\pi}{3}\right) = 0 \Rightarrow 0 = \ln\left|\sec\frac{\pi}{3}\right| + C_1 \Rightarrow C_1 = -\ln 2 \Rightarrow$
$y = \ln|\sec x| - \ln 2$

10. $x^2\frac{d^2y}{dx^2} + x\frac{dy}{dx} = 1.$ Let $p = \frac{dy}{dx} \Rightarrow x^2\frac{dp}{dx} + xp = 1 \Rightarrow \frac{dp}{dx} + \frac{1}{x}p = \frac{1}{x^2} \Rightarrow P(x) = \frac{1}{x},\ Q(x) = \frac{1}{x^2} \Rightarrow \int P(x)\,dx$
$= \ln|x| \Rightarrow \rho(x) = e^{\ln|x|} = |x| \Rightarrow p = \frac{1}{|x|}\int |x|\left(\frac{1}{x^2}\right)dx = \frac{1}{x}\int x\left(\frac{1}{x^2}\right)dx = \frac{1}{x}\int \frac{1}{x}\,dx = \frac{1}{x}(\ln x + C)$ if $x > 0$
$\Rightarrow p = \frac{\ln x}{x} + \frac{C}{x} \Rightarrow \frac{dy}{dx} = \frac{\ln x}{x} + \frac{C}{x} \Rightarrow y = (\ln x)^2 + C\ln x + C_1.\ y'(1) = 1 \Rightarrow 1 = \ln 1 + C \Rightarrow C = 1 \Rightarrow$
$y = (\ln x)^2 + \ln x + C_1.\ y(1) = 1 \Rightarrow 1 = (\ln 1)^2 + \ln 1 + C_1 \Rightarrow C_1 = 1 \Rightarrow y = (\ln x)^2 + \ln x + 1$

11. $\frac{d^2y}{dx^2} - 4\frac{dy}{dx} + 3y = 0 \Rightarrow r^2 - 4r + 3 = 0 \Rightarrow r_1 = 3,\ r_2 = 1 \Rightarrow y = C_1 e^{3x} + C_2 e^{x} \Rightarrow y' = 3C_1 e^{3x} + C_2 e^{x}.$
$y'(0) = -2 \Rightarrow -2 = 3C_1 + C_2.\ y(0) = 2 \Rightarrow 2 = C_1 + C_2 \Rightarrow C_1 = -2,\ C_2 = 4 \Rightarrow y = -2e^{3x} + 4e^{x}$

12. $\frac{d^2y}{dx^2} + 5\frac{dy}{dx} + 6y = 0 \Rightarrow r^2 + 5r + 6 = 0 \Rightarrow r_1 = -2,\ r_2 = -3 \Rightarrow y = C_1 e^{-2x} + C_2 e^{-3x} \Rightarrow y' = -2C_1 e^{-2x} -$
$3C_2{}^{-3x}.\ y'(0) = -2 \Rightarrow -2 = -2C_1 - 3C_2.\ y(0) = \frac{5}{6} \Rightarrow \frac{5}{6} = C_1 + C_2 \Rightarrow C_1 = \frac{1}{2},\ C_2 = \frac{1}{3} \Rightarrow y = \frac{1}{2}e^{-2x} +$
$\frac{1}{3}e^{-3x}$

13. $\frac{d^2y}{dx^2} + 4\frac{dy}{dx} + 4y = 0 \Rightarrow r_1 = r_2 = -2 \Rightarrow y = \left(C_1x + C_2\right)e^{-2x}.\ y(0) = 0 \Rightarrow C_2 = 0 \Rightarrow y = C_1x\,e^{-2x} \Rightarrow$
$y' = C_1\left(e^{-2x} - 2x\,e^{-2x}\right).\ y'(0) = 7 \Rightarrow C_1 = 7 \Rightarrow y = 7x\,e^{-2x}$

14. $\frac{d^2y}{dx^2} - 8\frac{dy}{dx} + 16y = 0 \Rightarrow r^2 - 8r + 16 = 0 \Rightarrow r_1 = r_2 = 4 \Rightarrow y = \left(C_1x + C_2\right)e^{4x}.\ y(0) = 4 \Rightarrow C_2 = 4 \Rightarrow$
$y = \left(C_1x + 4\right)e^{4x} \Rightarrow y' = C_1e^{4x} + 4\left(C_1x + 4\right)e^{4x}.\ y'(0) = -4 \Rightarrow -4 = C_1 + 4(0 + 4) \Rightarrow C_1 = -20 \Rightarrow$
$y = (-20x + 4)e^{4x}$

15. $\frac{d^2y}{dx^2} + 2\frac{dy}{dx} + 2y = 0 \Rightarrow r^2 + 2r + 2 = 0 \Rightarrow r = -1 \pm i \Rightarrow y = e^{-x}\left(C_1 \cos x + C_2 \sin x\right)$. $y(0) = 1 \Rightarrow$ $e^0\left(C_1 \cos 0 + C_2 \sin 0\right) = 1 \Rightarrow C_1 = 1 \Rightarrow y = e^{-x}\left(\cos x + C_2 \sin x\right) \Rightarrow y' = -e^{-x}\left(\cos x + C_2 \sin x\right) +$ $e^{-x}\left(-\sin x + C_2 \cos x\right)$. $y'(0) = 0 \Rightarrow 0 = -\left(\cos 0 + C_2 \sin 0\right) + \left(-\sin 0 + C_2 \cos 0\right) \Rightarrow C_2 = 1 \Rightarrow$ $y = e^{-x}(\cos x + \sin x)$

16. $\frac{d^2y}{dx^2} - 2\frac{dy}{dx} - 4y = 0 \Rightarrow r^2 - 2r - 4 = 0 \Rightarrow r = 1 \pm \sqrt{5} \Rightarrow y = C_1 e^{(1+\sqrt{5})x} + C_2 e^{(1-\sqrt{5})x}$. $y(0) = 1 \Rightarrow$ $1 = C_1 + C_2$. $y' = \left(1 + \sqrt{5}\right)C_1 e^{(1+\sqrt{5})x} + \left(1 - \sqrt{5}\right)C_2 e^{(1-\sqrt{5})x}$. $y'(0) = -1 \Rightarrow -1 = \left(1 + \sqrt{5}\right)C_1 +$ $\left(1 - \sqrt{5}\right)C_2 \Rightarrow C_1 = \frac{-2 + \sqrt{5}}{2\sqrt{5}}$, $C_2 = \frac{2 + \sqrt{5}}{2\sqrt{5}} \Rightarrow y = \left(\frac{-2 + \sqrt{5}}{2\sqrt{5}}\right)e^{(1+\sqrt{5})x} + \left(\frac{2 + \sqrt{5}}{2\sqrt{5}}\right)e^{(1-\sqrt{5})x}$

17. $\frac{d^2y}{dx^2} + 2\frac{dy}{dx} = 4x \Rightarrow r^2 + 2r = 0 \Rightarrow r_1 = 0, r_2 = -2 \Rightarrow y_h = C_1 + C_2 e^{-2x}$. $y_p = Ax^2 + Bx \Rightarrow y'_p = 2Ax + B$. $y''_p = 2A \Rightarrow 2A + 2(2Ax + B) = 4x \Rightarrow A = 1, B = -1 \Rightarrow y_p = x^2 - x \Rightarrow y = C_1 + C_2 e^{-2x} + x^2 - x$. $y(0) = 1 \Rightarrow C_1 + C_2 = 1$. $y' = -2C_2 e^{-2x} + 2x - 1$. $y'(0) = -3 \Rightarrow -2C_2 - 1 = -3 \Rightarrow C_1 = 0, C_2 = 1 \Rightarrow$ $y = e^{-2x} + x^2 - x$

18. $\frac{d^2y}{dx^2} + y = \csc x \Rightarrow r^2 + 1 = 0 \Rightarrow r = \pm i \Rightarrow y_h = C_1 \cos x + C_2 \sin x \Rightarrow u_1 = \cos x, u_2 = \sin x \Rightarrow$

$$D = \begin{vmatrix} \cos x & \sin x \\ -\sin x & \cos x \end{vmatrix} = 1 \Rightarrow v'_1 = -\frac{u_2 F(x)}{D} = -\sin x \csc x = -1 \Rightarrow v_1 = \int -dx = -x + C_1.$$

$v'_2 = \frac{u_1 F(x)}{D} = \cos x \csc x = \cot x \Rightarrow v_2 = \int \cot x\, dx = \ln(\sin x) + C_2$ if $\sin x > 0 \Rightarrow y = \cos x\left(-x + C_1\right)$ $+ \sin x\left(\ln(\sin x) + C_2\right) = -x \cos x + C_1 \cos x + \sin x \ln(\sin x) + C_2 \sin x$. $y\left(\frac{\pi}{2}\right) = 1 \Rightarrow -\frac{\pi}{2}\cos\frac{\pi}{2} +$ $C_1 \cos\frac{\pi}{2} + \sin\frac{\pi}{2}\ln\left(\sin\frac{\pi}{2}\right) + C_2 \sin\frac{\pi}{2} = 1 \Rightarrow C_2 = 1 \Rightarrow y = -x \cos x + C_1 \cos x + \sin x \ln(\sin x) + \sin x$ $\Rightarrow y' = \cos x + x \sin x - C_1 \sin x + \cos x \ln(\sin x)$. $y'\left(\frac{\pi}{2}\right) = \frac{\pi}{2} \Rightarrow \cos\frac{\pi}{2} + \frac{\pi}{2}\sin\frac{\pi}{2} - C_1 \sin\frac{\pi}{2} +$ $\cos\frac{\pi}{2}\ln\left(\sin\frac{\pi}{2}\right) = \frac{\pi}{2} \Rightarrow C_1 = 0 \Rightarrow y = -x \cos x + \sin x \ln(\sin x) + \sin x$

19. $\frac{d^2y}{dx^2} - \frac{dy}{dx} - 2y = 3e^{2x} \Rightarrow r^2 - r - 2 = 0 \Rightarrow r_1 = 2, r_2 = -1 \Rightarrow y_h = C_1 e^{2x} + C_2 e^{-x}$. $y_p = Ax e^{2x} \Rightarrow$ $y'_p = A e^{2x} + 2Ax e^{2x} \Rightarrow y''_p = 4A e^{2x} + 4Ax e^{2x} \Rightarrow 4A e^{2x} + 4Ax e^{2x} - \left(A e^{2x} + 2Ax e^{2x}\right) -$ $2\left(Ax e^{2x}\right) = 3e^{2x} \Rightarrow A = 1 \Rightarrow y_p = x e^{2x} \Rightarrow y = x e^{2x} + C_1 e^{2x} + C_2 e^{-x}$. $y(0) = -2 \Rightarrow C_1 + C_2 = -2$. $y' = 2C_1 e^{2x} - C_2 e^{-x} + e^{2x} + 2x e^{2x}$. $y'(0) = 0 \Rightarrow 2C_1 - C_2 + 1 = 0 \Rightarrow C_1 = -1, C_2 = -1 \Rightarrow$ $y = -e^{2x} - e^{-x} + x e^{2x}$

20. $\frac{d^2y}{dx^2} - 2\frac{dy}{dx} + 5y = 4e^{-x} \Rightarrow r^2 - 2r + 5 = 0 \Rightarrow r = 1 \pm 2i \Rightarrow y_h = e^x\left(C_1 \cos 2x + C_2 \sin 2x\right)$. $y_p = A e^{-x}$ $\Rightarrow y'_p = -A e^{-x} \Rightarrow y''_p = A e^{-x} \Rightarrow A e^{-x} - 2\left(-A e^{-x}\right) + 5A e^{-x} = 4e^{-x} \Rightarrow A = \frac{1}{2} \Rightarrow y_p = \frac{1}{2}e^{-x} \Rightarrow$

20. (Continued)

$y = e^x \left(C_1 \cos 2x + C_2 \sin 2x\right) + \frac{1}{2} e^{-x}$. $y(0) = 1 \Rightarrow C_1 + \frac{1}{2} = 1 \Rightarrow C_1 = \frac{1}{2} \Rightarrow$
$y = e^x \left(\frac{1}{2} \cos 2x + C_2 \sin 2x\right) + \frac{1}{2} e^{-x} \Rightarrow y' = e^x \left(\frac{1}{2} \cos 2x + C_2 \sin 2x\right) + e^x \left(- \sin 2x + 2C_2 \cos 2x\right) -$
$\frac{1}{2} e^{-x}$. $y'(0) = -\frac{1}{2} \Rightarrow \frac{1}{2} + 2C_2 - \frac{1}{2} = -\frac{1}{2} \Rightarrow C_2 = -\frac{1}{4} \Rightarrow y = e^x \left(\frac{1}{2} \cos 2x - \frac{1}{4} \sin 2x\right) + \frac{1}{2} e^{-x}$

APPENDIX A.7 DETERMINANTS AND CRAMER'S RULE

1. $\begin{vmatrix} 2 & 3 & 1 \\ 4 & 5 & 2 \\ 1 & 2 & 3 \end{vmatrix} \begin{matrix} 2 & 3 \\ 4 & 5 \\ 1 & 2 \end{matrix} = 30 + 6 + 8 - 5 - 8 - 36 = -5$

2. $\begin{vmatrix} 2 & -1 & -2 \\ -1 & 2 & 1 \\ 3 & 0 & -3 \end{vmatrix} = 0$ (Third column is a multiple of the first column.)

3. $\begin{vmatrix} 1 & 2 & 3 & 4 \\ 0 & 1 & 2 & 3 \\ 0 & 0 & 2 & 1 \\ 0 & 0 & 3 & 2 \end{vmatrix} = 1\begin{vmatrix} 1 & 2 & 3 \\ 0 & 2 & 1 \\ 0 & 3 & 2 \end{vmatrix} = 1\begin{vmatrix} 2 & 1 \\ 3 & 2 \end{vmatrix} = 1$

4. $\begin{vmatrix} 1 & -1 & 2 & 3 \\ 2 & 1 & 2 & 6 \\ 1 & 0 & 2 & 3 \\ -2 & 2 & 0 & -5 \end{vmatrix} = \begin{vmatrix} 1 & -1 & 2 & 3 \\ 3 & 0 & 4 & 9 \\ 1 & 0 & 2 & 3 \\ 0 & 0 & 4 & 1 \end{vmatrix} = -(-1)\begin{vmatrix} 3 & 4 & 9 \\ 1 & 2 & 3 \\ 0 & 4 & 1 \end{vmatrix} = \begin{vmatrix} 0 & -2 & 0 \\ 1 & 2 & 3 \\ 0 & 4 & 1 \end{vmatrix} = -1\begin{vmatrix} -2 & 0 \\ 4 & 1 \end{vmatrix} =$

$-1(-2) = 2$

5. a) $\begin{vmatrix} 2 & -1 & 2 \\ 1 & 0 & 3 \\ 0 & 2 & 1 \end{vmatrix} = \begin{vmatrix} 2 & -5 & 2 \\ 1 & -6 & 3 \\ 0 & 0 & 1 \end{vmatrix} = 1\begin{vmatrix} 2 & -5 \\ 1 & -6 \end{vmatrix} = -7$

b) $\begin{vmatrix} 2 & -1 & 2 \\ 1 & 0 & 3 \\ 0 & 2 & 1 \end{vmatrix} = \begin{vmatrix} 2 & -1 & 2 \\ 1 & 0 & 3 \\ 4 & 0 & 5 \end{vmatrix} = -(-1)\begin{vmatrix} 1 & 3 \\ 4 & 5 \end{vmatrix} = -7$

6. a) $\begin{vmatrix} 1 & 0 & -1 \\ 0 & 2 & -2 \\ 2 & 0 & 1 \end{vmatrix} = \begin{vmatrix} 3 & 0 & -1 \\ 4 & 2 & -2 \\ 0 & 0 & 1 \end{vmatrix} = 1\begin{vmatrix} 3 & 0 \\ 4 & 2 \end{vmatrix} = 6$

b) $\begin{vmatrix} 1 & 0 & -1 \\ 0 & 2 & -2 \\ 2 & 0 & 1 \end{vmatrix} = 2\begin{vmatrix} 1 & -1 \\ 2 & 1 \end{vmatrix} = 2(3) = 6$

7. a) $\begin{vmatrix} 1 & 1 & 0 & 0 \\ 0 & 0 & -2 & 1 \\ 0 & -1 & 0 & 7 \\ 3 & 0 & 2 & 1 \end{vmatrix} = \begin{vmatrix} 1 & 1 & 0 & 7 \\ 0 & 0 & -2 & 1 \\ 0 & -1 & 0 & 0 \\ 3 & 0 & 2 & 1 \end{vmatrix} = -(-1)\begin{vmatrix} 1 & 0 & 7 \\ 0 & -2 & 1 \\ 3 & 2 & 1 \end{vmatrix} = \begin{vmatrix} 1 & 0 & 7 \\ 0 & -2 & 1 \\ 0 & 2 & -20 \end{vmatrix} =$

$1\begin{vmatrix} -2 & 1 \\ 2 & -20 \end{vmatrix} = 38$

b) $\begin{vmatrix} 1 & 1 & 0 & 0 \\ 0 & 0 & -2 & 1 \\ 0 & -1 & 0 & 7 \\ 3 & 0 & 2 & 1 \end{vmatrix} = \begin{vmatrix} 1 & 1 & 0 & 0 \\ 0 & 0 & -2 & 1 \\ 1 & 0 & 0 & 7 \\ 3 & 0 & 2 & 1 \end{vmatrix} = -1\begin{vmatrix} 0 & -2 & 1 \\ 1 & 0 & 7 \\ 3 & 2 & 1 \end{vmatrix} = -1\begin{vmatrix} 0 & -2 & 1 \\ 1 & 0 & 7 \\ 0 & 2 & -20 \end{vmatrix} =$

$-1(-1)\begin{vmatrix} -2 & 1 \\ 2 & -20 \end{vmatrix} = 38$

8. a) $\begin{vmatrix} 0 & 1 & 0 & 0 \\ 0 & 1 & 1 & 0 \\ 1 & 1 & 1 & 1 \\ 1 & 1 & 0 & 0 \end{vmatrix} = \begin{vmatrix} 0 & 1 & 0 & 0 \\ 0 & 1 & 1 & 0 \\ 1 & 0 & 0 & 0 \\ 1 & 0 & -1 & -1 \end{vmatrix} = 1\begin{vmatrix} 1 & 0 & 0 \\ 1 & 1 & 0 \\ 0 & -1 & -1 \end{vmatrix} = -1\begin{vmatrix} 1 & 0 \\ 1 & 1 \end{vmatrix} = -1$

b) $\begin{vmatrix} 0 & 1 & 0 & 0 \\ 0 & 1 & 1 & 0 \\ 1 & 1 & 1 & 1 \\ 1 & 1 & 0 & 0 \end{vmatrix} = \begin{vmatrix} 0 & 1 & 0 & 0 \\ 0 & 0 & 1 & 0 \\ 1 & 0 & 1 & 1 \\ 1 & 0 & 0 & 0 \end{vmatrix} = -1\begin{vmatrix} 0 & 1 & 0 \\ 1 & 1 & 1 \\ 1 & 0 & 0 \end{vmatrix} = -1(-1)\begin{vmatrix} 1 & 1 \\ 1 & 0 \end{vmatrix} = -1$

9. $D = \begin{vmatrix} 1 & 8 \\ 3 & -1 \end{vmatrix} = -25.$ $x = \dfrac{\begin{vmatrix} 4 & 8 \\ -13 & -1 \end{vmatrix}}{-25} = \dfrac{100}{-25} = -4,$ $y = \dfrac{\begin{vmatrix} 1 & 4 \\ 3 & -13 \end{vmatrix}}{-25} = \dfrac{-25}{-25} = 1$

10. $D = \begin{vmatrix} 2 & 3 \\ 3 & -1 \end{vmatrix} = -11.$ $x = \dfrac{\begin{vmatrix} 5 & 3 \\ 2 & -1 \end{vmatrix}}{-11} = \dfrac{-11}{-11} = 1,$ $y = \dfrac{\begin{vmatrix} 2 & 5 \\ 3 & 2 \end{vmatrix}}{-11} = \dfrac{-19}{-11} = \dfrac{19}{11}$

11. $D = \begin{vmatrix} 4 & -3 \\ 3 & -2 \end{vmatrix} = 1.$ $x = \dfrac{\begin{vmatrix} 6 & -3 \\ 5 & -2 \end{vmatrix}}{1} = 3,$ $y = \dfrac{\begin{vmatrix} 4 & 6 \\ 3 & 5 \end{vmatrix}}{1} = 2$

12. $D = \begin{vmatrix} 1 & 1 & 1 \\ 2 & -1 & 1 \\ 1 & 2 & -1 \end{vmatrix} = \begin{vmatrix} 3 & 0 & 2 \\ 2 & -1 & 1 \\ 5 & 0 & 1 \end{vmatrix} = -1\begin{vmatrix} 3 & 2 \\ 5 & 1 \end{vmatrix} = 7.$ $x = \dfrac{\begin{vmatrix} 2 & 1 & 1 \\ 0 & -1 & 1 \\ 4 & 2 & -1 \end{vmatrix}}{7} = \dfrac{\begin{vmatrix} 2 & 1 & 1 \\ 0 & -1 & 1 \\ 0 & 0 & -3 \end{vmatrix}}{7} =$

$\dfrac{2\begin{vmatrix} -1 & 1 \\ 0 & -3 \end{vmatrix}}{7} = \dfrac{6}{7},$ $y = \dfrac{\begin{vmatrix} 1 & 2 & 1 \\ 2 & 0 & 1 \\ 1 & 4 & -1 \end{vmatrix}}{7} = \dfrac{\begin{vmatrix} 1 & 2 & 1 \\ 2 & 0 & 1 \\ -1 & 0 & -3 \end{vmatrix}}{7} = \dfrac{-2\begin{vmatrix} 2 & 1 \\ -1 & -3 \end{vmatrix}}{7} = \dfrac{10}{7},$ $z = \dfrac{\begin{vmatrix} 1 & 1 & 2 \\ 2 & -1 & 0 \\ 1 & 2 & 4 \end{vmatrix}}{7}$

$= \dfrac{\begin{vmatrix} 1 & 1 & 2 \\ 2 & -1 & 0 \\ -1 & 0 & 0 \end{vmatrix}}{7} = \dfrac{2\begin{vmatrix} 2 & -1 \\ -1 & 0 \end{vmatrix}}{7} = -\dfrac{2}{7}$

13. $D = \begin{vmatrix} 2 & 1 & -1 \\ 1 & -1 & 1 \\ 2 & 2 & 1 \end{vmatrix} = \begin{vmatrix} 2 & 1 & -1 \\ 3 & 0 & 0 \\ 4 & 3 & 0 \end{vmatrix} = -1\begin{vmatrix} 3 & 0 \\ 4 & 3 \end{vmatrix} = -9.$ $x = \dfrac{\begin{vmatrix} 2 & 1 & -1 \\ 7 & -1 & 1 \\ 4 & 2 & 1 \end{vmatrix}}{-9} = \dfrac{\begin{vmatrix} 2 & 1 & -1 \\ 9 & 0 & 0 \\ 6 & 3 & 0 \end{vmatrix}}{-9} =$

$\dfrac{-1\begin{vmatrix} 9 & 0 \\ 6 & 3 \end{vmatrix}}{-9} = 3,$ $y = \dfrac{\begin{vmatrix} 2 & 2 & -1 \\ 1 & 7 & 1 \\ 2 & 4 & 1 \end{vmatrix}}{-9} = \dfrac{\begin{vmatrix} 2 & 2 & -1 \\ 3 & 9 & 0 \\ 4 & 6 & 0 \end{vmatrix}}{-9} = \dfrac{-1\begin{vmatrix} 3 & 9 \\ 4 & 6 \end{vmatrix}}{-9} = -2,$ $z = \dfrac{\begin{vmatrix} 2 & 1 & 2 \\ 1 & -1 & 7 \\ 2 & 2 & 4 \end{vmatrix}}{-9} =$

$\dfrac{\begin{vmatrix} 3 & 0 & 9 \\ 1 & -1 & 7 \\ 4 & 0 & 18 \end{vmatrix}}{-9} = \dfrac{-1\begin{vmatrix} 3 & 9 \\ 4 & 18 \end{vmatrix}}{-9} = 2$

14. $D = \begin{vmatrix} 2 & -4 & 0 \\ 1 & 1 & 1 \\ 0 & 5 & 7 \end{vmatrix} = \begin{vmatrix} 0 & -6 & -2 \\ 1 & 1 & 1 \\ 0 & 5 & 7 \end{vmatrix} = -1\begin{vmatrix} -6 & -2 \\ 5 & 7 \end{vmatrix} = 32.$ $x = \dfrac{\begin{vmatrix} 6 & -4 & 0 \\ 1 & 1 & 1 \\ 10 & 5 & 7 \end{vmatrix}}{32} =$

$\dfrac{\begin{vmatrix} 6 & -4 & 0 \\ 1 & 1 & 1 \\ 3 & -2 & 0 \end{vmatrix}}{32} = \dfrac{-1\begin{vmatrix} 6 & -4 \\ 3 & -2 \end{vmatrix}}{32} = 0,$ $y = \dfrac{\begin{vmatrix} 2 & 6 & 0 \\ 1 & 1 & 1 \\ 0 & 10 & 7 \end{vmatrix}}{32} - \dfrac{\begin{vmatrix} 0 & 4 & -2 \\ 1 & 1 & 1 \\ 0 & 10 & 7 \end{vmatrix}}{32} = \dfrac{-1\begin{vmatrix} 4 & -2 \\ 10 & 7 \end{vmatrix}}{32} = -\dfrac{3}{2},$

14. (Continued)

$$z = \frac{\begin{vmatrix} 2 & -4 & 6 \\ 1 & 1 & 1 \\ 0 & 5 & 10 \end{vmatrix}}{32} = \frac{\begin{vmatrix} 0 & -6 & 4 \\ 1 & 1 & 1 \\ 0 & 5 & 10 \end{vmatrix}}{32} = \frac{-1\begin{vmatrix} -6 & 4 \\ 5 & 10 \end{vmatrix}}{32} = \frac{5}{2}$$

15. $D = \begin{vmatrix} 1 & 0 & -1 \\ 0 & 2 & -2 \\ 2 & 0 & 1 \end{vmatrix} = 2\begin{vmatrix} 1 & -1 \\ 2 & 1 \end{vmatrix} = 6.$ $x = \frac{\begin{vmatrix} 3 & 0 & -1 \\ 2 & 2 & -2 \\ 3 & 0 & 1 \end{vmatrix}}{6} = \frac{2\begin{vmatrix} 3 & -1 \\ 3 & 1 \end{vmatrix}}{6} = 2,$ $y = \frac{\begin{vmatrix} 1 & 3 & -1 \\ 0 & 2 & -2 \\ 2 & 3 & 1 \end{vmatrix}}{6} =$

$$\frac{\begin{vmatrix} 1 & 3 & -1 \\ 0 & 2 & -2 \\ 0 & -3 & 3 \end{vmatrix}}{6} = \frac{1\begin{vmatrix} 2 & -2 \\ -3 & 3 \end{vmatrix}}{6} = 0, \quad z = \frac{\begin{vmatrix} 1 & 0 & 3 \\ 0 & 2 & 2 \\ 2 & 0 & 3 \end{vmatrix}}{6} = \frac{2\begin{vmatrix} 1 & 3 \\ 2 & 3 \end{vmatrix}}{6} = -1$$

16. $D = \begin{vmatrix} 1 & 1 & -1 & 1 \\ 1 & -1 & 1 & 1 \\ 1 & 1 & 1 & -1 \\ 1 & 0 & 1 & 1 \end{vmatrix} = \begin{vmatrix} 2 & 0 & 0 & 2 \\ 1 & -1 & 1 & 1 \\ 2 & 0 & 2 & 0 \\ 1 & 0 & 1 & 1 \end{vmatrix} = -1\begin{vmatrix} 2 & 0 & 2 \\ 2 & 2 & 0 \\ 1 & 1 & 1 \end{vmatrix} = -1\begin{vmatrix} 2 & 0 & 0 \\ 2 & 2 & -2 \\ 1 & 1 & 0 \end{vmatrix} =$

$$-1(2)\begin{vmatrix} 2 & -2 \\ 1 & 0 \end{vmatrix} = -4. \quad x_1 = \frac{\begin{vmatrix} 2 & 1 & -1 & 1 \\ -1 & -1 & 1 & 1 \\ 2 & 1 & 1 & -1 \\ -1 & 0 & 1 & 1 \end{vmatrix}}{-4} = \frac{\begin{vmatrix} 1 & 0 & 0 & 2 \\ -1 & -1 & 1 & 1 \\ 1 & 0 & 2 & 0 \\ -1 & 0 & 1 & 1 \end{vmatrix}}{-4} = \frac{-1\begin{vmatrix} 1 & 0 & 2 \\ 1 & 2 & 0 \\ -1 & 1 & 1 \end{vmatrix}}{-4} =$$

$$\frac{-1\begin{vmatrix} 1 & 0 & 0 \\ 1 & 2 & -2 \\ -1 & 1 & 3 \end{vmatrix}}{-4} = \frac{-1(1)\begin{vmatrix} 2 & -2 \\ 1 & 3 \end{vmatrix}}{-4} = 2, \quad x_2 = \frac{\begin{vmatrix} 1 & 2 & -1 & 1 \\ 1 & -1 & 1 & 1 \\ 1 & 2 & 1 & -1 \\ 1 & -1 & 1 & 1 \end{vmatrix}}{-4} = \frac{\begin{vmatrix} 1 & 2 & -1 & 1 \\ 2 & 1 & 0 & 2 \\ 2 & 4 & 0 & 0 \\ 2 & 1 & 0 & 2 \end{vmatrix}}{-4} =$$

$$\frac{-1\begin{vmatrix} 2 & 1 & 2 \\ 2 & 4 & 0 \\ 2 & 1 & 2 \end{vmatrix}}{-4} = 0, \quad x_3 = \frac{\begin{vmatrix} 1 & 1 & 2 & 1 \\ 1 & -1 & -1 & 1 \\ 1 & 1 & 2 & -1 \\ 1 & 0 & -1 & 1 \end{vmatrix}}{-4} = \frac{\begin{vmatrix} 2 & 0 & 1 & 2 \\ 1 & -1 & -1 & 1 \\ 2 & 0 & 1 & 0 \\ 1 & 0 & -1 & 1 \end{vmatrix}}{-4} = \frac{-1\begin{vmatrix} 2 & 1 & 2 \\ 2 & 1 & 0 \\ 1 & -1 & 1 \end{vmatrix}}{-4} =$$

16. (Continued)

$$\frac{-1\begin{vmatrix} 3 & 0 & 3 \\ 3 & 0 & 1 \\ 1 & -1 & 1 \end{vmatrix}}{-4} = \frac{-1(1)\begin{vmatrix} 3 & 3 \\ 3 & 1 \end{vmatrix}}{-4} = -\frac{3}{2},\ x_4 = \frac{\begin{vmatrix} 1 & 1 & -1 & 2 \\ 1 & -1 & 1 & -1 \\ 1 & 1 & 1 & 2 \\ 1 & 0 & 1 & -1 \end{vmatrix}}{-4} = \frac{\begin{vmatrix} 2 & 0 & 0 & 1 \\ 1 & -1 & 1 & -1 \\ 2 & 0 & 2 & 1 \\ 1 & 0 & 1 & -1 \end{vmatrix}}{-4} =$$

$$\frac{-1\begin{vmatrix} 2 & 0 & 1 \\ 2 & 2 & 1 \\ 1 & 1 & -1 \end{vmatrix}}{-4} = \frac{-1\begin{vmatrix} 2 & 0 & 1 \\ 0 & 0 & 3 \\ 1 & 1 & -1 \end{vmatrix}}{-4} = \frac{-1(-1)\begin{vmatrix} 2 & 1 \\ 0 & 3 \end{vmatrix}}{-4} = -\frac{3}{2}$$

17. $D = \begin{vmatrix} 2 & h \\ 1 & 3 \end{vmatrix} = 6 - h = 0 \Rightarrow h = 6.$ x: $\begin{vmatrix} 8 & h \\ k & 3 \end{vmatrix} = 24 - hk = 24 - 6k = 0 \Rightarrow k = 4$

a) When $h = 6$, $k = 4$, there are infinitely many solutions.

b) When $h = 6$, $k \neq 4$, there are no solutions.

18. $\begin{vmatrix} x & x & 1 \\ 2 & 0 & 5 \\ 6 & 7 & 1 \end{vmatrix}\begin{matrix} x & x \\ 2 & 0 \\ 6 & 7 \end{matrix} = 0 + 30x + 14 - 0 - 35x - 2x = 0 \Rightarrow -7x + 14 = 0 \Rightarrow x = 2$

19. $au + bv + cw = 0 \Rightarrow v = \frac{-au - cw}{b}$, $au' + bv' + cw' = 0 \Rightarrow v' = \frac{-au' - cw'}{b}$, $au'' + bv'' + cw'' = 0 \Rightarrow$

$$v'' = \frac{-au'' - cw''}{b} \Rightarrow \begin{vmatrix} u & v & w \\ u' & v' & w' \\ u'' & v'' & w'' \end{vmatrix} = \begin{vmatrix} u & \frac{-au - cw}{b} & w \\ u' & \frac{-au' - cw'}{b} & w' \\ u'' & \frac{-au'' - cw''}{b} & w'' \end{vmatrix} = \frac{1}{b}\begin{vmatrix} u & -au - cw & w \\ u' & -au' - cw' & w' \\ u'' & -au'' - cw'' & w'' \end{vmatrix} =$$

$\frac{1}{b}(u(-au' - cw')w'' + u''(-au - cw)w' + w(-au'' - cw'')u' - u''(-au' - cw')w - u'(-au - cw)w'' - u(-au'' - cw'')w') = \frac{1}{b}(0) = 0$

20. a) $\frac{ax + b}{(x - r_1)(x - r_2)} = \frac{C}{x - r_1} + \frac{D}{x - r_2} \Rightarrow \frac{ax + b}{(x - r_1)(x - r_2)} = \frac{C(x - r_2)}{(x - r_1)(x - r_2)} + \frac{D(x - r_1)}{(x - r_1)(x - r_2)} =$

$\frac{(C + D)x - Cr_2 - Dr_1}{(x - r_1)(x - r_2)} \Rightarrow C + D = a, -Cr_2 - Dr_1 = b$

b) $\begin{vmatrix} 1 & 1 \\ r_2 & r_1 \end{vmatrix} = r_1 - r_2 \neq 0$ when $r_1 \neq r_2$

APPENDIX A.8 LAGRANGE MULTIPLIERS WITH TWO CONSTRAINTS

1. Let $f(x,y,z) = x^2 + y^2 + z^2$. Maximize f subject to $g_1(x,y,z) = y + 2z - 12 = 0$ and $g_2(x,y,z) = x + y - 6 = 0$
   $\nabla f = 2x\,\mathbf{i} + 2y\,\mathbf{j} + 2z\,\mathbf{k}$, $\nabla g_1 = \mathbf{j} + 2\,\mathbf{k}$, $\nabla g_2 = \mathbf{i} + \mathbf{j}$. Then $2x\,\mathbf{i} + 2y\,\mathbf{j} + 2z\,\mathbf{k} = \lambda(\mathbf{j} + 2\,\mathbf{k}) + \mu(\mathbf{i} + \mathbf{j}) \Rightarrow$
   $2x\,\mathbf{i} + 2y\,\mathbf{j} + 2z\,\mathbf{k} = \mu\,\mathbf{i} + (\lambda + \mu)\,\mathbf{j} + 2\lambda\,\mathbf{k} \Rightarrow 2x = \mu,\ 2y = \lambda + \mu,\ 2z = 2\lambda \Rightarrow x = \frac{\mu}{2},\ z = \lambda \Rightarrow$
   $2x = 2y - z \Rightarrow x = \frac{2y - z}{2}$. Then $y + 2z - 12 = 0$ and $\frac{2y - z}{2} + y - 6 = 0 \Rightarrow 4y - z - 12 = 0 \Rightarrow z = 4 \Rightarrow$
   $y = 4 \Rightarrow x = 2 \Rightarrow (2,4,4)$ is the point closest.

2. Let $g_1(x,y,z) = 2x - y = 0$ and $g_2(x,y,z) = y + z = 0 \Rightarrow \nabla g_1 = 2\,\mathbf{i} - \mathbf{j}$, $\nabla g_2 = \mathbf{j} + \mathbf{k}$. $\nabla f = 2x\,\mathbf{i} + 2\,\mathbf{j} - 2z\,\mathbf{k} \Rightarrow$
   $2x\,\mathbf{i} + 2\,\mathbf{j} - 2z\,\mathbf{k} = \lambda(2\,\mathbf{i} - \mathbf{j}) + \mu(\mathbf{j} + \mathbf{k}) \Rightarrow 2x\,\mathbf{i} + 2\,\mathbf{j} - 2z\,\mathbf{k} = 2\lambda\,\mathbf{i} + (\mu - \lambda)\,\mathbf{j} + \mu\,\mathbf{k} \Rightarrow 2x = 2\lambda,\ 2 = \mu - \lambda,$
   $-2z = \mu \Rightarrow x = \lambda$. Then $2 = -2z - x \Rightarrow x = -2z - 2 \Rightarrow 2(-2z - 2) - y = 0 \Rightarrow -4z - 4 - y = 0$. This
   equation along with $y + z = 0 \Rightarrow z = -\frac{4}{3},\ y = \frac{4}{3}$. Then $x = \frac{2}{3} \Rightarrow \left(\frac{2}{3}, \frac{4}{3}, -\frac{4}{3}\right)$ is the point that gives the
   maximum value $\Rightarrow$ Maximum value of f is $\frac{4}{3}$.

3. Let $g_1(x,y,z) = z - 1 = 0$ and $g_2(x,y,z) = x^2 + y^2 + z^2 - 10 = 0 \Rightarrow \nabla g_1 = \mathbf{k}$, $\nabla g_2 = 2x\,\mathbf{i} + 2y\,\mathbf{j} + 2z\,\mathbf{k}$.
   $\nabla f = 2xyz\,\mathbf{i} + x^2z\,\mathbf{j} + x^2y\,\mathbf{k} \Rightarrow 2xyz\,\mathbf{i} + x^2z\,\mathbf{j} + x^2y\,\mathbf{k} = \lambda(\mathbf{k}) + \mu(2x\,\mathbf{i} + 2y\,\mathbf{j} + 2z\,\mathbf{k}) \Rightarrow 2xyz = 2x\mu,\ x^2z =$
   $2y\mu,\ x^2y = 2z + \lambda \Rightarrow xyz = x\mu \Rightarrow x = 0$ or $yz = \mu \Rightarrow \mu = y$ since $z = 1$. $x = 0 \Rightarrow 2y\mu = 0$ and $2z + \lambda = 0$
   $\Rightarrow z = 1 \Rightarrow \lambda = -2 \Rightarrow y^2 - 9 = 0 \Rightarrow y = \pm 3 \Rightarrow (0,\pm 3,1)$. $\mu = y \Rightarrow x^2z = 2y^2 \Rightarrow x^2 = 2y^2$ since $z = 1$
   $\Rightarrow 2y^2 + y^2 + 1 - 10 = 0 \Rightarrow 3y^2 - 9 = 0 \Rightarrow y = \pm\sqrt{3} \Rightarrow \mu = \pm\sqrt{3} \Rightarrow x^2 = 2\left(\pm\sqrt{3}\right)\left(\pm\sqrt{3}\right) = 6 \Rightarrow x =$
   $\pm\sqrt{6} \Rightarrow (\pm\sqrt{6}, \pm\sqrt{3}, 1)$. $f(0,\pm 3,1) = 1$. $f(\pm\sqrt{6},\pm\sqrt{3},1) = 6\left(\pm\sqrt{3}\right) + 1 = 1 \pm 6\sqrt{3} \Rightarrow$ Maximum of f
   is $1 + 6\sqrt{3}$ at $(\pm\sqrt{6},\sqrt{3},1)$; minimum of f is $1 - 6\sqrt{3}$ at $(\pm\sqrt{6},-\sqrt{3},1)$

4. a) Let $g_1(x,y,z) = x + y + z - 40 = 0$ and $g_2(x,y,z) = x + y - z = 0 \Rightarrow \nabla g_1 = \mathbf{i} + \mathbf{j} + \mathbf{k}$, $\nabla g_2 = \mathbf{i} + \mathbf{j} - \mathbf{k}$.
   $\nabla w = yz\,\mathbf{i} + xz\,\mathbf{j} + xy\,\mathbf{k} \Rightarrow yz\,\mathbf{i} + xz\,\mathbf{j} + xy\,\mathbf{k} = \lambda(\mathbf{i} + \mathbf{j} + \mathbf{k}) + \mu(\mathbf{i} + \mathbf{j} - \mathbf{k}) \Rightarrow yz = \lambda + \mu,\ xz = \lambda + \mu,\ xy =$
   $\lambda - \mu \Rightarrow yz = xz \Rightarrow z = 0$ or $y = x$. $z = 0 \Rightarrow x + y - 40,\ x + y = 0 \Rightarrow$ No solution. $x = y \Rightarrow$
   $2x + z - 40 = 0,\ 2x - z = 0 \Rightarrow x = 10,\ y = 10 \Rightarrow 20 - z = 0 \Rightarrow z = 20 \Rightarrow (10,10,20) \Rightarrow$
   $w = (10)(10)(20) = 2000$

   b) $\mathbf{N} = \begin{vmatrix} \mathbf{i} & \mathbf{j} & \mathbf{k} \\ 1 & 1 & 1 \\ 1 & 1 & -1 \end{vmatrix} = -2\,\mathbf{i} + 2\,\mathbf{j}$. $\therefore$ the line is $x = -2t + 10,\ y = 2t + 10,\ z = 20$. Since $z = 20$,

   $w = xyz = (-2t + 10)(2t + 10)20 = (-4t^2 + 100)20$ which will have its maximum when $t = 0 \Rightarrow$
   $x = 10,\ y = 10,\ z = 20$

5. Let $g_1(x,y,z) = y - x = 0 \Rightarrow x = y$. Let $g_2(x,y,z) = x^2 + y^2 + z^2 - 4 = 0 \Rightarrow \nabla g_2 = 2x\,\mathbf{i} + 2y\,\mathbf{j} + 2z\,\mathbf{k}$. $\nabla g_1 =$
   $-\mathbf{i} + \mathbf{j}$; $\nabla f = y\,\mathbf{i} + x\,\mathbf{j} + 2z\,\mathbf{k} \Rightarrow y\,\mathbf{i} + x\,\mathbf{j} + 2z\,\mathbf{k} = \lambda(-\mathbf{i} + \mathbf{j}) + \mu(2x\,\mathbf{i} + 2y\,\mathbf{j} + 2z\,\mathbf{k}) \Rightarrow y = -\lambda + 2x\mu,\ x =$
   $\lambda + 2y\mu,\ 2z = 2z\mu \Rightarrow z = z\mu \Rightarrow z = 0$ or $\mu = 1$. $z = 0 \Rightarrow x^2 + y^2 - 4 = 0 \Rightarrow 2x^2 - 4 = 0$ (since $x = y$) $\Rightarrow$
   $x^2 = 2 \Rightarrow x = \pm\sqrt{2} \Rightarrow y = \pm\sqrt{2} \Rightarrow (\pm\sqrt{2},\pm\sqrt{2},0)$. $\mu = 1 \Rightarrow y = -\lambda + 2x$ and $x = \lambda + 2y \Rightarrow x + y =$
   $2(x + y) \Rightarrow 2x = 2(2x)$ since $x = y \Rightarrow x = 0 \Rightarrow y = 0 \Rightarrow z^2 - 4 = 0 \Rightarrow z = \pm 2 \Rightarrow (0,0,\pm 2)$.
   $f(0,0,\pm 2) = 4$, $f(\pm\sqrt{2},\pm\sqrt{2}, 0) = 2 \Rightarrow$ Maximum value of f is 4 at $(0,0,\pm 2)$; minimum value of f is 2 at
   $(\pm\sqrt{2},\pm\sqrt{2},0)$